Implantable Sensors and Systems

Guang-Zhong Yang
Editor

Implantable Sensors and Systems

From Theory to Practice

Springer

Editor
Guang-Zhong Yang
The Hamlyn Centre
Imperial College London
London
UK

ISBN 978-3-030-09909-1 ISBN 978-3-319-69748-2 (eBook)
https://doi.org/10.1007/978-3-319-69748-2

Softcover re-print of the Hardcover 1st edition 2018

Printed on acid-free paper

This Springer imprint is published by Springer Nature
The registered company is Springer International Publishing AG
The registered company address is: Gewerbestrasse 11, 6330 Cham, Switzerland

To my mother, who fought a courageous battle against cancer.

Preface

From passive devices to current generation instruments integrated with sensing, actuation, and drug delivery, surgical implants have undergone a major transformation in the last two decades due to advances in micro-nanofabrication, new materials, microelectronics, wireless transmission, and real-time data analytics. This book is a sequel to the book on *Body Sensor Networks*, first published in 2006 and followed by its second edition in 2013.

By following the principle of "*using the body as the medium, inspiration, and a source of energy to provide continuous sensing, monitoring, and intervention*", our first book was mainly focussed on the development of wearable sensors for providing continuous monitoring of physical, physiological, and biochemical parameters in any environment without activity or behavioral restrictions. With the rapid advances and clinical uptake of low-power, miniaturized, smart implants, it is timely to review and assess the unique challenges imposed and outline future development pathways and new opportunities.

Of particular importance is the use of these devices in surgical applications. For both acute and chronic scenarios, implantable sensing is increasingly used for transient or long-term monitoring of *in vivo* physiological, bioelectrical, biochemical, and metabolic changes. Underpinned by increasingly small, smart, and energy efficient designs, they have become an integral part of surgical prostheses or implants, supporting optimized, context-aware sensing, feedback, or stimulation with due consideration of system-level impact.

Key topics covered in this book include electrochemical sensor designs; electrical and physical sensors; sensor embodiment and flexible electronics; ultralow-power Application-Specific Integrated Circuits (ASICs); optical sensors; power harvesting; and data exchange links, as well as wireless data paths and security. It provides an in-depth explanation of both the theoretical and practical considerations of developing novel implantable sensors and can serve as a comprehensive reference for those working at the forefront of implantable sensing or as an introductory reference for those about to enter this exciting field of research and development. I hope this book will act as a valuable resource to a wide spectrum of readers interested in, or inspired by, this intrinsically multidisciplinary topic.

There has been tremendous effort from all contributors to this book, and I would like to express my sincere thanks to my team at the Hamlyn Centre. Without their enthusiasm, support, and dedication in meeting the tight publishing schedule, this book would not have become possible. I would also like to thank the editorial staff of Springer, the publisher of this volume. In particular, I am grateful to Helen Desmond and her colleagues in helping with all the editorial matters.

This work would not have been possible without the research support from all the funding bodies that supported our work, particularly the UK Engineering and Physical Sciences Research Council (EPSRC). Their ongoing research funding has allowed us to establish and promote this exciting field of research.

Finally, I would like to dedicate this book to my mother, who fought a courageous battle against cancer. Sadly, advances in medicine have not yet reached the stage of eradicating this dreadful disease, but the development of sensing technologies and smart implants could contribute to early detection as well as a targeted and personalized treatment of this disease; prolonging life whilst ensuring improved quality of life for those surviving.

London, UK
August 2017

Guang-Zhong Yang, CBE, FREng

Contents

Editor and Contributors

About the Editor

Prof. Guang-Zhong Yang, CBE, FREng is director and co-founder of the Hamlyn Centre for Robotic Surgery. The Hamlyn Centre (http://www.imperial.ac.uk/hamlyn-centre/) has been established for developing safe, effective, and accessible technologies that can reshape the future of health care for both developing and developed countries. Focusing on technological innovation but with a strong emphasis on clinical translation and direct patient benefit with a global impact, the center is at the forefront of research in imaging, sensing, and robotics for addressing global health challenges associated with demographic, environment, social, and economic changes. The center plays an active role in international collaboration and outreach activities, as well as in the training of surgeons and engineers in sensing, imaging, and robotic technologies, thereby facilitating a fully integrated clinical approach.

Professor Yang's main research interests are in medical imaging, sensing, and robotics. He is a pioneer in wearable and implantable body sensor networks and is internationally recognized for developing novel sensing solutions for providing personalized monitoring platforms that are pervasive, intelligent, context-aware, and "invisible" to the patient. In imaging, he is credited for a number of novel MR phase contrast velocity imaging and computational modeling techniques that have transformed in vivo blood flow quantification and visualization. In robotics, his concept on perceptual docking represents a paradigm shift of learning and knowledge acquisition of motor and perceptual/cognitive behavior for robotics.

Professor Yang has received numerous awards for his work in imaging, sensing, and robotics. These include the ISMRM I. I. Rabi Award, Medical Futures Innovation Award, the Bluetooth Innovation World Cup, and the IEEE EMBS Technical Achievement Award. He is a fellow of the Royal Academy of Engineering, fellow of IEEE, IET, AIMBE, IAMBE, MICCAI, City, and Guilds, and a recipient of the Royal Society Research

Merit Award and listed in The Times Eureka "Top 100" in British Science. He is the founding editor of Science Robotics (http://robotics.sciencemag.org/)—a journal of the Science family dedicated to the latest advances in robotics and how it enables or underpins new scientific discoveries. He was awarded a CBE in Her Majesty the Queen's 2017 New Year Honour for his contribution to biomedical engineering.

Contributors

S. Anastasova The Hamlyn Centre, Imperial College London, London, UK

C. M. Chen The Hamlyn Centre, Imperial College London, London, UK

B. Gil The Hamlyn Centre, Imperial College London, London, UK

H. Ip The Hamlyn Centre, Imperial College London, London, UK

P. Kassanos The Hamlyn Centre, Imperial College London, London, UK

A. J. Thompson The Hamlyn Centre, Imperial College London, London, UK

Guang-Zhong Yang The Hamlyn Centre, Imperial College London, London, UK

Chapter 1
Introduction

Guang-Zhong Yang

1.1 Why Implantable Sensing?

The development of surgical implants has a long history. Over 4500 years ago, the ancient Egyptians were already using gold ligature wires to stabilize teeth. Archeological studies have revealed the skills of Mayas in sophisticated dentistry and the use of shells for dental implants. Modern dental implants thrived from the 1960s as the choices of materials, structural design, surface finish, and coating become more versatile. Today, the new implants developed allow not only seamless structural support but also tissue integration, with sophisticated functional coatings to combat infection and promote bone regeneration [1]. In orthopedics, since the pioneering work by Theophilus Gluck over a century ago using ivory as an implant affixed with plaster and rosin, up to 90% of the current generation implants can survive more than 10 years [2] and new material science and regenerative medicine are further propelling the continuous innovations of the field, leading to increasingly robust, long-lasting joint replacements. Similar advances can be found in cardiothoracic, neurological, and gastrointestinal implants, supporting the restoration of function and improvement of quality of life after surgery.

To provide structural and functional support or restoration yet being fully integrated with the local environment is the primary goal of implantable devices. Increasingly, new implants are moving from passive devices to active instrument, being able to sense, adapt, and actuate. This is also driven by the need to provide continuous sensing directly at the target location in vivo, such that targeted interventional or therapeutic measures can be deployed at the onset of adverse events. One clinical example is the management of surgical site infection, which occurs in 2–5% of surgical patients. Surgical site infections, catheter-related sepsis, wound dehiscence and gastrointestinal anastomotic leakage are recognised complications

G.-Z. Yang (✉)
The Hamlyn Centre, Imperial College London, London, UK
e-mail: g.z.yang@imperial.ac.uk

G.-Z. Yang (ed.), *Implantable Sensors and Systems*
https://doi.org/10.1007/978-3-319-69748-2_1

following surgical interventions or invasive monitoring of critically ill surgical patients. The severity of these complications can be mild, necessitating antibiotic therapy and close observation in hospital, but can also be serious, with life-threatening sepsis requiring repeated surgical and radiological interventions. Clinically, surgical site infections include superficial/deep incisional or organ/space infections and affect 2–5% of surgical patients, giving an attributable mortality of 0.9% [3]. For example, deep incisional surgical site infection is common following small bowel (33.9%) and large bowel surgery (26.6%), extending the post-operative length-of-stay to over three weeks [3]. Early detection of surgical site infections, catheter-related sepsis, wound dehiscence and anastomotic failure are critical to patient management to ensure prompt instigation of appropriate therapy and avoid the associated mortality.

The need for more accurate and clinically relevant physiological recordings specific to certain conditions in the vicinity of where a physiological and clinically relevant event takes place and providing real-time closed loop intervention or therapy has led to the development of implantable sensors. Implantable sensing has thus far moved far beyond simple biophysical and bioelectrical sensing. Specific biomarkers include, for example, oxygen, glucose, pH, lactate, uncoupling of cations (Ca^{2+}, Mg^{2+}, K^+/Na^+), and inflammatory cytokines, as well as reactive oxygen and nitrogen species produced by the host immune and white cell response to infection, and innate markers such as lysozyme or lactoferrin. The purpose of this book is to provide detailed technical background to, as well as in-depth analysis of, the current state of the art in implantable sensing. Key topics covered include: electrochemical sensor designs, electrical and physical sensors, sensor embodiment and flexible electronics, ultra-low-power Application Specific Integrated Circuits (ASICs), optical sensors, power harvesting, and data exchange links, as well as wireless data paths and security.

1.2 Electrochemical Sensor Designs for Biomedical Implants

In Chap. 2, we describe in detail electrochemical methods, including potentiometry, amperometry, and voltammetry commonly used for in vivo sensing. This is due to their simplicity in design and ease of fabrication, as well as low-power operation. The chapter includes the basic principles of sensor operation and fabrication, with due consideration when these sensors are embodied as an implantable system. Potentiometric sensors represent an important family of sensors typically used for the detection of ions. Monitoring of various ion concentrations can provide valuable insights into the state of tissue and the underlying biochemical processes. Different ion selective electrode designs are discussed, providing theoretical consideration and practical insights for these sensors used in implants.

The operation of amperometric sensors, on the other hand, is based on applying a potential between two electrodes and measurement of the resulting current, typically via a third electrode. The development of amperometric sensors typically focuses on different electrode materials, enzyme immobilisation techniques, mediators and coenzymes, sensor stability and lifetime, miniaturisation and implementation. The most popular amperometric sensors are sensors for the detection of glucose and other popular target analytes, such as oxygen and lactate. Figure 1.1 provides several examples of these types of sensors and a fully integrated telemetrically powered implantable sensor for continuous blood glucose monitoring.

For biosensing, affinity-based measurement techniques are also widely used. With this approach, a binding molecule [e.g. antibodies or other proteins, aptamers, ss-DNA (single stranded deoxyribonucleic acid) and RNA (ribonucleic acid)] are used as part of the transduction process to enhance selectivity and specificity. These sensors are mainly used for single-use disposable sensors, and thus are difficult to

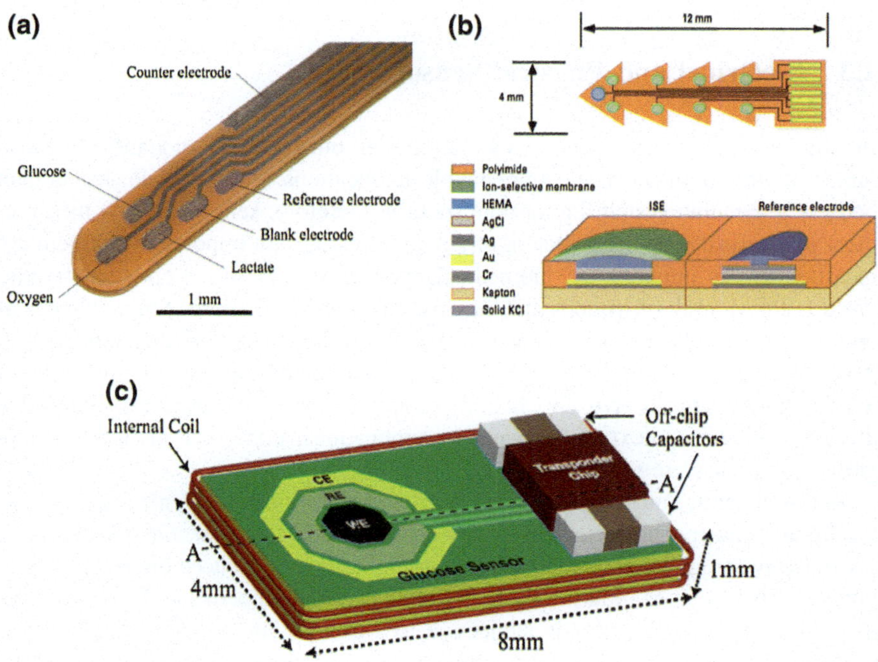

Fig. 1.1 **a** A multi-layer amperometric structure for the detection of glucose and lactate composed of a polyimide substrate, deposited Pt electrodes, insulation from spin-coated epoxy and a laminated dry-film resist. Reprinted with permission from [4], © IEEE 2014. **b** Illustration of a flexible platform for pH and K^+ measurements with a common reference electrode for in vivo measurements. Reprinted with permission from [5], © American Chemical Society 2001. **c** A telemetrically powered and interrogated implantable continuous blood glucose monitoring system with a three electrode amperometric glucose sensor and ASIC. Reprinted with permission from [6], © IEEE 2009

be used for implants that require continuous sensing. The difficulties are associated with the in situ regeneration of antigen binding sites once a detection event has taken place; i.e., the antigen-bio-recognition element binding is very specific with strong bond formation and is essentially irreversible. Breaking that bond to free up the detection site and reuse the biosensor (without damaging the binding sites) is extremely challenging. However, for the sake of completeness we discuss in this chapter a variety of methods that can be used in affinity-based sensing. Affinity based-sensing can be performed via quartz crystal microbalance (QCM), surface acoustic wave (SAW), surface plasmon resonance (SPR), electric impedance spectroscopy (EIS), voltammetry or amperometry, micro-electro-mechanical systems (MEMS)-based cantilever sensors read out electronically (resistor bridge or capacitive sensing) or optically, and atomic force microscopy (AFM). Developing complex microfluidics for affinity-based sensors could potentially allow the use of these in implantable applications. Affinity-based sensing can be used for detecting a plethora of clinically relevant targets, such as bacteria and cancer bio-markers; thus their potential use in implantable applications should not be overlooked.

1.3 Electrical and Physical Sensors

In addition to electrochemical sensors, a range of other sensing modalities are also important for biomedical and implantable applications. For example, the human body is a dynamic system in constant motion; therefore, sensors for the measurement of physical properties such as strain and pressure are important. Temperature measurement is important in biomedical applications. Tissue temperature is elevated in the vicinity of wounds and may indicate infection. Temperature changes of only 2 °C can induce protein denaturation and many sensors are affected by temperature changes. Compensating for this variation requires the measurement of temperature in situ, in vivo. In Chap. 3, various electrical and physical sensors are discussed in the context of implantable applications, either for monolithic integration or as discrete sensor solutions.

In this chapter, much of the focus is on electrical bioimpedance measurement, as the frequency-dependent electrical properties of tissues are essential for assessing various physiological parameters. For example, ischemia induces biochemical and physiological changes in tissues, which influence tissue impedance. In general, the magnitude of bioimpedance is subjective to variations in the extracellular environment, while the phase is mostly influenced by variations of the membrane of the cell and its integrity. Edema and tissue necrosis, for example, have been shown to be associated with a reduction of tissue electrical impedance. Metabolic processes that take place during ischemia produce ions and ion shifts, which can alter the high-frequency intracellular resistance. In this chapter, fundamental theories and frequency-dependent electrical properties of tissue are discussed. Electrical bioimpedance is a non-invasive, rapid, low-cost technique suitable for long-term continuous monitoring. It is well suited for low-power miniaturized wearable and implantable applications. Different bioimpedance sensors, as well as modeling and optimization schemes, are discussed. Furthermore, this chapter provides practical considerations of bioimpedance instrumentation to facilitate clinical applications.

A very important class of sensors is those assessing tissue or organ mechanics. Within this class, we include strain and stress, pressure, velocity, gyroscopes, and acceleration. Figure 1.2 illustrates an example pressure monitoring platform and an on-chip pressure sensor integrated with signal processing electronics. Micro-electro-mechanical systems (MEMS) and simple thin film planar and flexible electronics technologies play an important role in the development of such sensors. In this section the chapter focuses on stress/strain and pressure sensors. A plethora of MEMS sensors, including accelerometers, gyroscopes, and magnetometers are available and they are increasingly used with implantable sensors to provide context-aware sensing such that the mechanical properties of the target tissue is known. Advances in micro/nano-fabrication have allowed the fabrication of MEMS-based sensing devices on the same substrate of the ASICs, allowing high miniaturization, sensitivity and low-power operation, and low cost.

Other topics covered by Chap. 3 include electrode design and different materials, including stainless steel, tungsten, platinum, platinum-iridium alloys, iridium oxide,

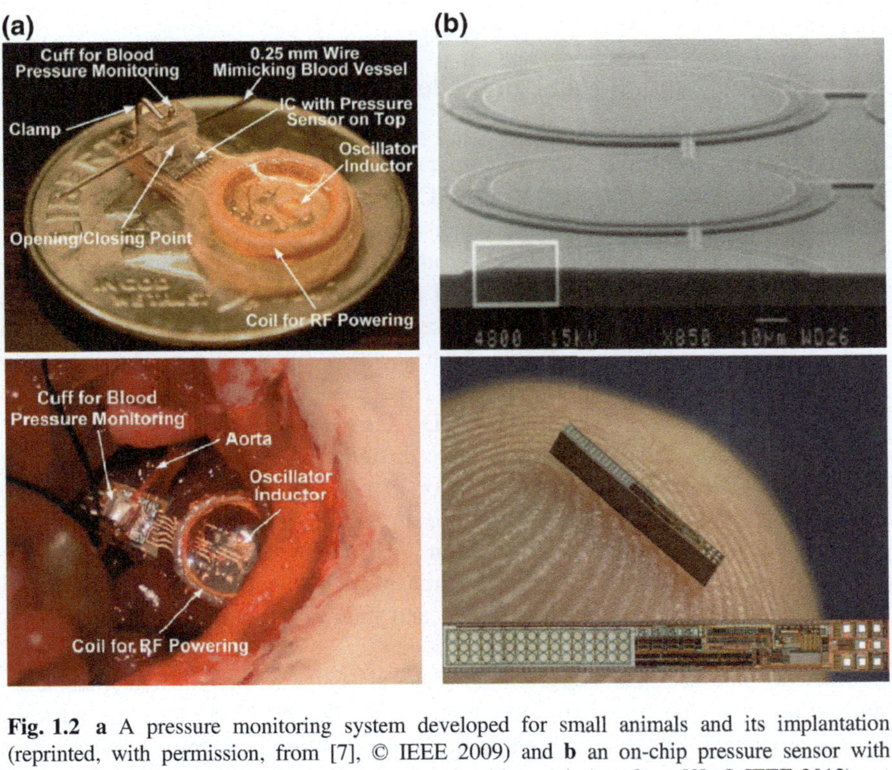

Fig. 1.2 **a** A pressure monitoring system developed for small animals and its implantation (reprinted, with permission, from [7], © IEEE 2009) and **b** an on-chip pressure sensor with integrated signal processing electronics (reprinted, with permission, from [8], © IEEE 2012)

titanium nitride, and PEDOT. Advanced materials such as biocompatible carbon nanotube (CNT) fiber microelectrodes can be used for neural stimulation and recording. The CNT fibers are used without additional surface modification and high charge density, offering low impedance and flexible microelectrodes for stable interface manipulation of the activity of neural ensembles. Carbon-based materials present a very high surface area, resulting in large charge capacity. CNT-based fibers offer long-term stability and biocompatibility, and for long-term implantation liquid crystal polymer (LCP) could be utilized instead of polyimide as an electrode substrate due to its low water absorption rate.

Real-time sensing in vivo is complicated by many factors, including sensor fouling, drift, and motion artefacts. An important and natural step forward from the classical systems of employing a single sensing element is to employ a variety of sensors for the detection of the same target marker with complementary information. This is important, as many metabolites are interdependent and patient dependent, while allowing spatial and temporal relationships to be established. The use of sensor array and multi-sensor fusion therefore provides an effective way of mitigating errors and individual sensor failures. In this chapter, we also define a number of widely used metrics for assessing sensor performance. These include sensitivity, limit of detection, stability, linear range, selectivity and specificity, and biocompatibility.

1.4 Sensor Embodiment and Flexible Electronics

For reliable in vivo deployment of surgical implants, appropriate device embodiment and packaging are essential. In Chap. 4, key issues related to sensor embodiment and flexible electronics are discussed. Over the years, microfabrication techniques have advanced significantly, and many of the processes used for the fabrication of semiconductor devices and integrated circuits can be used. Silicon wafers are often used as a substrate because of their wide availability, low cost, and surface flatness, but other materials, such as glass, quartz and plastics, are also common. Conductors and insulators can be deposited on the wafer via methods such as physical vapor deposition (PVD), e.g., sputtering, or chemical vapor deposition (CVD). Spin coating can be used to deposit on the wafer initially thin films of poly-dimethylsiloxane (PDMS) or PI prior to other material deposition. Photolithography is then used to introduce on the top deposited thin film masking patterns. These methods have been extensively used for fabricating micro-engineering surfaces, sensors and transistors, as well as analog, digital, and mixed-signal integrated circuits, MEMS, and micro-opto-electro-mechanical systems (MOEMS).

Recently, the use of unconventional materials (such as organic materials, conductive polymers, and graphene among others) for flexible electronics has attracted extensive interest. The development of flexible and shape-conforming electronics is particularly relevant for implantable sensors, because if the device is to be integrated into a soft, biological structure, then an important feature would be the stretchability of the material. Thus far, different materials have been investigated, including CNT for flexible and stretchable electronics and CNT networks as a channel semiconductor in flexible thin-film transistors and as stretchable conductors. In Chap. 4, example platforms and the current state of the art are discussed and Fig. 1.3 shows several examples of flexible electronics and a high-resolution multiplexed electrode array developed for implantable applications.

In addition to fabrication protocols and the selection of substrate and device materials, another consideration for enhancing the tensile strength and thus the flexibility and stretchability of devices is through effective geometrical design. In this chapter, practical considerations on interconnect design, geometry, and optimisation are discussed. As inorganic semiconductors are brittle, fracturing with strains as low as 1%, a major challenge is to enhance via optimized mechanical design, particularly the stretchability of these materials, which is crucial for device development. A good architecture ensures interconnects deform upon stretching whilst maintaining minimal deformation for active devices, and thus the stability of sensing output.

Of course, the technical hurdles in the development of implantable medical devices extend beyond those in the design of interconnects, sensors, readout electronics, and powering. When all the components are put in place, they need to

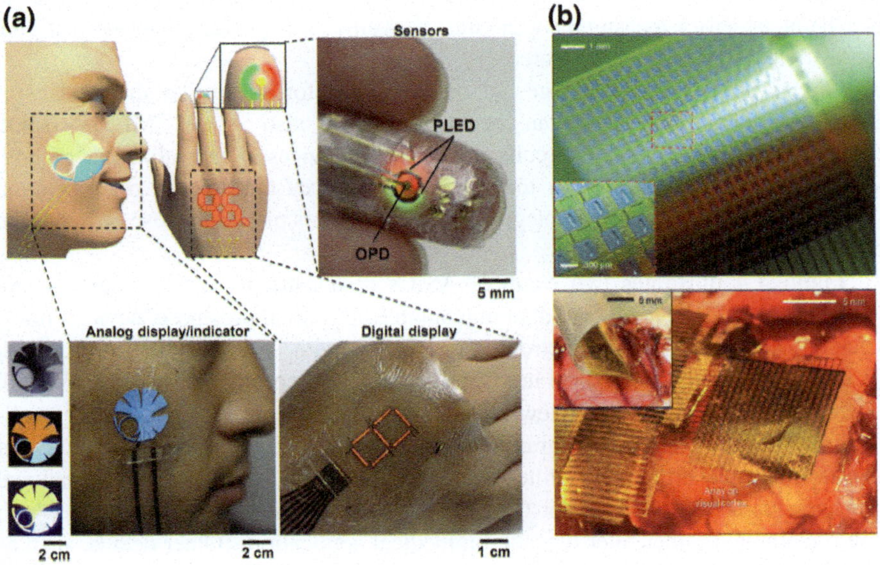

Fig. 1.3 **a** Example flexible electronics developed for optoelectronic skin system showing the organic optical sensor attached to a finger, a color-changing indicator and flexible digital display. From Science Advances [9], with permission and licensed under BY-NC (http://creativecommons.org/licenses/by-nc/4.0/). **b** A flexible, high-resolution multiplexed electrode array, applied to the visual cortex of a feline model. Reprinted by permission from Macmillan Publishers Ltd on behalf of Cancer Research UK: Nature Neuroscience [10], © 2011

be appropriately packaged prior to implantation. In this chapter, issues related to biocompatibility and hermeticity of the encapsulation of the final device are also discussed. More specifically, consideration must be given to appropriate packaging materials. They must be non-inflammatory, non-toxic, non-allergenic, non-carcinogenic and non-thrombogenic when located within the bloodstream. We will discuss polymer encapsulation and packaging as an example, and we have also presented detailed bio-compatible standards for medical devices to guide the development of clinically applicable implantable devices.

1.5 Ultra-Low-Power ASICs

For implantable sensing and closed loop control, the development of ASIC plays a pivotal role, as it allows a much reduced footprint and low-power operation. Furthermore, commercial systems-on-chip (SoC) encompassing digital microcontrollers, radio, and analogue-digital converters, as well as re-configurable amplifier circuits, are now widely available. In Chap. 5, we begin with a survey of the latest development of ASICs and related integrated systems from literature. This is followed by an overview of technological trends in integrated circuit and sensor fabrication. It addresses ultra-low-power ASIC circuits appropriate for implantable sensors and sensor front-ends. It also covers bioimpedance and neural and electrochemical sensor measurement circuits, as well as low-power analogue-to-digital converter design and architectures.

One of the major advantages of using ASICs for sensing is the density of electronic integration that can be achieved. It also allows MEMS-CMOS co-fabrication of transistor circuits and MEMS devices. The realization of both MEMS sensors and ASIC on the same chip reduces the amount of integration required and improves signal integrity, with the sensors placed in close proximity to the readout circuits.

Figure 1.4 illustrates two example ASICs demonstrating a 52 channel neural probe on all CMOS process and an implantable epi-retinal prosthesis device. In this chapter, analog front-end electronics interfacing with sensors will be discussed. A common task for biomedical implant involves the recording of voltages. We will therefore begin with an overview of common recording amplifiers used for applications such as electrocardiography, electromyography, electroneurography, and bioimpedance. Furthermore, filtering techniques such as linearized operational transconductance amplifiers for G_m-C filter applications will be discussed, which are preferable to other filter types in biomedical applications. Other details, such as offset compensation and leakage in switches in deep sub-micron technologies, will also be addressed. Linking to previous chapters, realization of sensing methodologies such as bioimpedance, voltammetry and amperometry and voltage-to-frequency converters will be presented. Furthermore, a brief overview of stimulators and capacitive sensor interfacing circuits is provided.

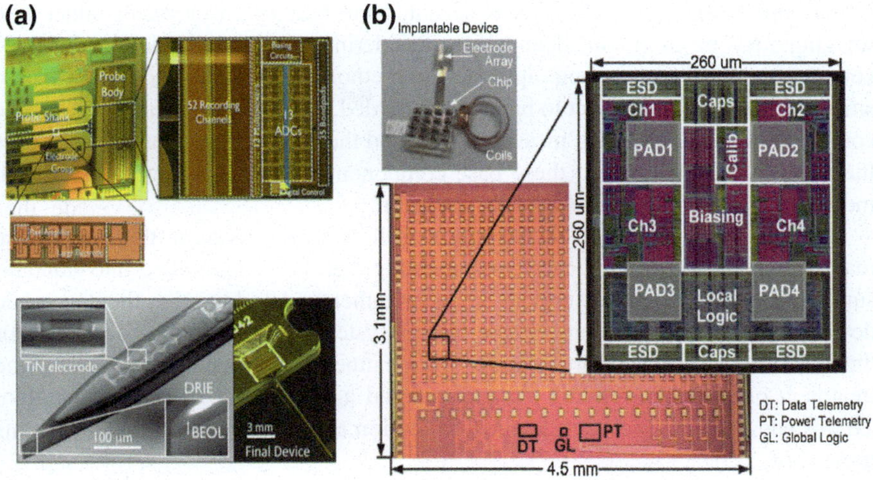

Fig. 1.4 **a** A 52 channel neural probe on all CMOS process (reprinted, with permission, from [11], © IEEE 2014). **b** A die micrograph and implantable epi-retinal prosthesis device (reprinted, with permission, from [12], © IEEE 2013)

1.6 Implantable Optical Sensors

The use of optical sensors has many advantages, as it avoids direct tissue contact and many of the issues associated with sensor fouling and drift can be better managed. Currently, many optical sensors used in vivo are in the form of tethered devices due to power and miniaturization issues. However, recent advances in low-power optoelectronics have made fully integrated optical implants possible. This chapter starts from endoscopy and capsule endoscopy to look at the current state of the art in optical sensing and imaging for ingestible and implantable applications. The chapter focuses on the development of optical instruments that can provide in situ diagnoses at the point of care, with an emphasis on current progress in miniature devices that may function as implants in the future.

Clinically, both tethered and fully implantable optical systems will have a significant clinical impact, and the aim of this chapter is to provide an overview of optical technologies with clinical applications that are currently in use or under development. For optical biopsy, both endomicroscopy and OCT have been used in a number of different clinical applications—ranging from gastrointestinal endoscopy for the detection of premalignant tissue changes to intravascular imaging for the diagnosis of atherosclerosis—and their use is continually increasing. Other imaging and spectroscopic techniques are also emerging and a number of capsule- and integrated chip-based devices have also been reported. These include laser speckle imaging (LSI), diffuse reflectance spectroscopy, fluorescence spectroscopy (including time resolved fluorescence measurements), and Raman spectroscopy.

To develop an implantable optical sensor, point-based measurements rather than imaging may be more practical when considering power utilization and device complexity. For spectroscopic approaches, methods related to spectral decomposition are discussed. They act to break the recorded data into a small number of base components that describe all measurements, and the discriminant analysis then uses the relative contributions of these base components to make a diagnosis for each measurement site based solely on its spectrum.

It is worth noting that all of the optical techniques discussed in this chapter are readily compatible with optical fibers, allowing them to be embodied into standard surgical appliances such as surgical drains or catheters. In addition to the diagnostic devices described above, implantable optical systems have also been deployed for therapeutic purposes, in particular for the treatment of vision loss, including the treatment of cataracts, retinitis pigmentosa, and age-related macular degeneration. To this end, Chap. 6 provides detailed discussion about the current progress on this topic (Fig. 1.5).

In this chapter, we also address the current state of the art in optogenetics. Optogenetics involves the genetic modification of nerve cells, most often neurons in the brain, in order to render them sensitive to light. Subsequently, the sensitized neurons can be addressed optically, allowing control of particular functions or

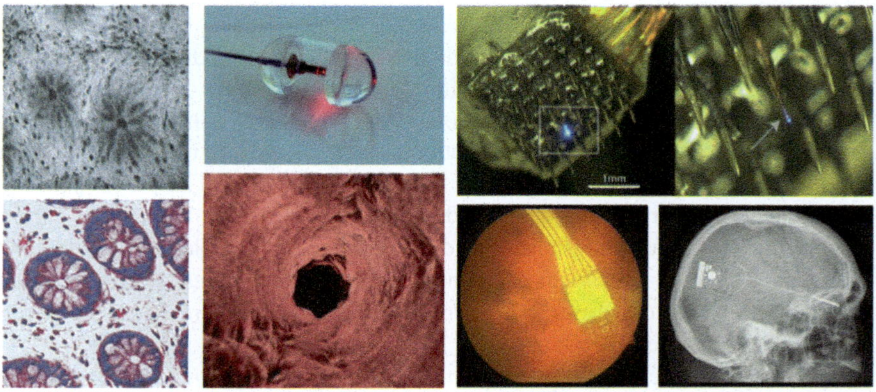

Fig. 1.5 Example optical sensing platforms used either as tethered or fully implantable devices. 1st column (left): confocal endomicroscopy (upper image) of normal GI tissue reveals similar features to those observed in histology (lower image). Reproduced from reference [13], © 2007, with permission from Elsevier. 2nd column: image of a tethered OCT capsule for upper GI imaging along with a 3D rendered view of a Barrett's segment of esophagus collected with this device. Reproduced with permission from Macmillan Publishers Ltd: Nature Medicine [14], © 2013. Upper right: an electrode measurement array for use in optogenetic experiments, where one electrode has been replaced with an optical fiber to allow optical stimulation. Reproduced with permission from reference [15], © IOP Publishing 2012. All rights reserved. Lower right: photograph of a retinal prosthesis implanted below the fovea in a human subject, alongside an X-ray showing the location of the implant and the subcutaneous coil for wireless power delivery. Reproduced from reference [16] under the terms of the Creative Commons Attribution license CC BY 4.0 (https://creativecommons.org/licenses/by/4.0/), © 2013

events, similar to the use of electrical activation of neurons used in the visual prostheses. In recent years, optogenetics has been widely applied in the field of neuroscience, where many in vitro cell studies and in vivo animal experiments have been conducted to elucidate and provide further insights to signaling pathways in the brain. They have shown great promise, not only in the study of neurological function, but also in the investigation of the causes and effects of a multitude of brain disorders. The growing use of optogenetic techniques has triggered much discussion about the potential development of neural implants based on optogenetics that would allow treatment of a wide range of ailments, including brain injuries, depression and even drug/alcohol dependency. In addition to the general topic of optogentics covered, this chapter also briefly discusses optogenetic retinal prostheses for restoring vision to the blind and we also discuss the possible future development of more ambitious fully implantable systems for use in the brain.

1.7 Power Harvesting and Data Exchange Links

Power management and effective wireless links are essential for data exchange of fully implantable devices. Low power consumption is of paramount importance in implantables to assure long-term function of the sensor and patient safety. In practice, the operational longevity of the implantable sensors is typically dictated by the time between battery replacements. As the dimensions of implantable devices decrease, so does energy consumption. The development of batteryless solutions has clear advantages in terms of both safety, long-term operation, and miniaturization. In Chap. 7, we will describe the current and future approaches to wireless powering and data exchange for implantable sensors. Strategies devoted to reducing power demand include low-power transducing techniques (capacitive and inductive), integrated electronics and smart compression algorithms on measurement and data transmission.

Wireless powering via inductive means is now ubiquitous with wirelessly chargeable smart devices currently widely available in the consumer market. Standards such as Qi and WPC are widely adopted in commercially available integrated circuits, hence making wireless charging/powering accessible to the vast majority of device designers. Before inductive power transfer became popular amongst wearable gadgets and IoT devices, it had been one of the most successfully deployed wireless powering modalities used in transcutaneous implants, such as cochlea prosthetics. The latest research on inductive powering aims to improve efficiency from low to moderate power consumption, facilitating a simultaneous high-speed data link.

In this chapter, the main sources of power harvesting for implantable devices are discussed. These include electrochemical, photovoltaic, thermoelectric, mechanical, inductive, RF, and ultrasound. Mechanical harvesters can use body and tissue motions to convert to an electrical equivalent as long as the operation of the harvester does not interfere with the well-being of the patient. The low frequency contents of the movements limit the output power obtained to the μW range, although a stable output can be achieved if using a cardiac pacemaker that permanently produces variations in the entire cardiovascular system that can be converted to electrical energy.

Whilst RF transmission in the open air is well studied, its propagation inside the human body is different, mainly due to the differences in the dielectric permittivity of body tissues recorded in the far-field regime. Antenna design is thus one of the most critical aspects during the projection of any RF-operating implantable devices, not only due to dimension constraints and/or biocompatibility of the casing, but also the effective radiative power within safety limits. Power harvesting from external RF sources is an alternative to inductive powering. In fact, the inaccurate geometrical alignment between implantable sensor and external RF devices no longer limits the efficiency in power transference, making it more orientationally immune to larger operational distances. These topics and many others involved in RF powering and data exchange are addressed in this chapter (Fig. 1.6).

In this chapter, we also discuss the recent surge of interest in ultrasound for power harvesting purposes, especially for delivering energy to implanted devices due to its long transmission depths in tissues and electromagnetic immunity. Ultrasound uses the piezoelectricity property found in some materials to convert

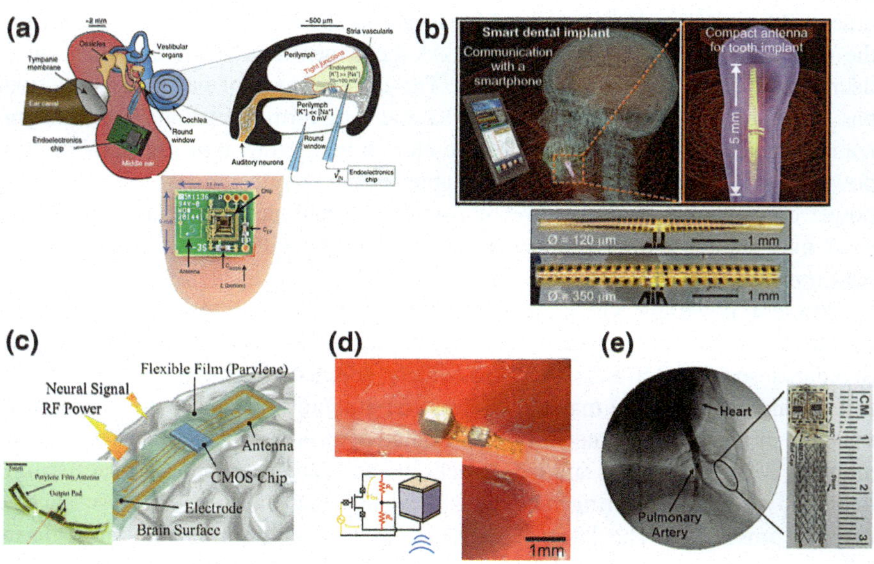

Fig. 1.6 Power harvest and data exchange. **a** Cochlear implant that harvests power from the electrochemical gradient located on the inner ear [17], © Nature Publishing Group 2012. **b** Compact helical antenna for the wireless tooth implant [18], © Nature Publishing Group 2015. **c** RF-powered implantable neural interface with thin film flexible antenna [19], © IEEE 2015. **d** Neural dust: an ultrasonic, low-power solution for brain–machine interfaces, deployed on the sciatic nerve of a rat [20]. Reprinted with acknowledgement of the authors. **e** Implantable cardiovascular pressure sensing system integrated with a medical stent [21], © IEEE 2010

acoustic waves into electrical energy. As such, piezoelectric transducers replace the coils employed in the inductive link. For the external device, the transducer converts electrical power into a pressure wave that is transmitted transcutaneously and collected by the implant transducer, where the inverse process takes place and energy is transformed back to electrical. The power transfer efficiency is higher than the inductive and RF links and it does not suffer from signal interference arising from devices at close range, either inside or outside the human body.

1.8 Wireless Data Paths and Security

With increasing use of implantable devices, particularly those with closed-loop control and wireless powering and reconfiguration capabilities, reliability and security are major issues to consider. Regulatory bodies such as the FDA in the United States and EMA in Europe have already started issuing guidelines to manufacturers to take effective measures regarding cybersecurity. There are many fundamental issues to address and wireless data paths and security represent the central focus of Chap. 8. It identifies some of the security questions and provides a detailed technical overview of the latest developments in lightweight security systems suitable for implants.

The chapter begins with an analysis of the relationship between the interacting parties of an implantable device, followed by a qualitative overview of the strategies commonly taken to provide a secure implant system. Implantable devices are highly specialized and designed with the smallest amount of resources indispensable to their operation in order to minimize the physical space and save battery life. Memory and computation capabilities are required to perform any security-oriented operation inside the implant. Long computation cycles can easily drain batteries, and the same is true for operations involving access to the memory.

Figure 1.7 illustrates key considerations for wireless data paths and security, showing how implantable and wearable devices inside a personal healthcare system may be vulnerable to remote attacks, as well as example body-coupled communication and security schemes. In practice, the security mechanisms available to implants are diverse and originally derived from common defense models deployed in the much broader area of sensor networking. However, the field of implantable technology is unique, and reliable solutions to the security problem must be derived from analysis of the patient's behavior towards the surrounding environment and how this interaction can help promote (or degrade) security. For implantable devices, this is particularly relevant when the environment changes from home to the emergency room. In a changing world, where technologies evolve at unprecedented rates, the behavior of the patient can change, and the same is valid for the attacker towards what might be truly desirable to protect or hack, respectively.

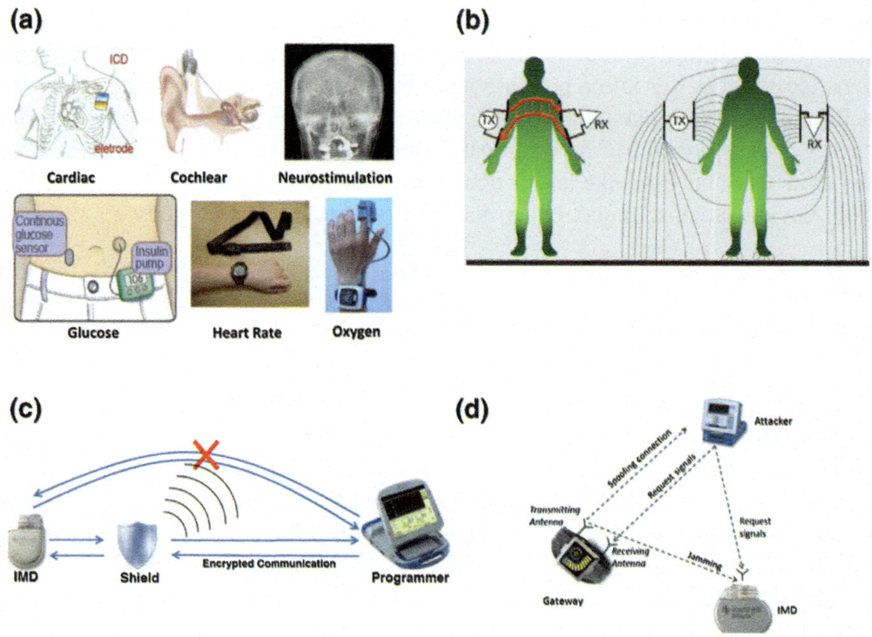

Fig. 1.7 Wireless data paths and security. **a** Implantable and wearable devices inside a personal healthcare system vulnerable to remote attacks. **b** Body-coupled communication for transmission of signals through the body as opposed to on-air communication [22]. **c** Shield security scheme that jams all IMD transmissions and unwanted commands [22], © IEEE 2014. **d** BodyDoubler security scheme that establishes a spoofing connection with the attacker while jamming the signal at the IMD antenna [23], © IEEE 2014

In this chapter, different security mechanisms are discussed. Particular attention is directed to schemes that are suitable for implant deployment. We focus in this chapter on 'lightweight' methods that require minimum hardware logic, implementable on low-power microcontrollers. These should be well suited to remotely powered implants that do not rely on batteries or other long-term energy storage elements and are designed to consume microwatts of power or even less. We also highlight in this chapter some of the cryptanalysis techniques that are relevant for the evaluation of lightweight, implant-ready security schemes. The cryptographic mechanisms ultimately constitute the last refuge for securing patient's data when all spatial and physical constrains have been broken by attackers. The cryptanalysis described in this chapter helps to understand some of the pitfalls of the lightweight algorithms and future research directions. Whereas established cryptography protocols are difficult to be mapped onto the limited implant hardware, these algorithms have survived the test of time. In this context, research into more advanced techniques for implant security therefore presents both challenges and new opportunities.

1.9 Conclusions

With increasing incidences of cancer, cardiovascular disease, and chronic symptoms such as diabetes, COPD, and Alzheimer's disease, the demand for both wearable and implantable sensors will be on the rise. This is further propelled by an increasing demand for continuous monitoring for postoperative care and rehabilitation, as more attention is paid to quantitative improvement of functional recovery and quality of life. A myriad of implantable sensors are already used in clinical practice, ranging from implantable pH sensors for monitoring gastro-oesophageal reflux, implantable arterial or pulmonary pressure sensors, neurostimulators for electrocorticographic activity monitoring and stopping seizure, to sensor-enabled prostheses. Improved sensor design plays a key part in the long-term stability and bio-integration of the devices, and issues related to biofouling, sensor drift, and calibration are important. Biocompatibility, integratability, and resorbability are therefore active research topics in the field of medical implants. There has also been significant attention to the use of biodegradable materials for the development of transient implants to avoid the problem of long-term effect on the body. Increasingly sophisticated implantable devices are creating ever more demanding requirements for their continuous, long-lasting, and reliable energy supply. In addition to the power harvesting schemes discussed in this book, there are also other schemes being explored in the research community. These include the use of biocatalytic fuel cells consisting of immobilised microorganisms or enzymes acting as catalysts, with glucose as a fuel to produce electricity. Such enzymatic micro-batteries offer the prospect of dramatically reduced sensor size and can potentially be also used for microbots in vivo.

In this book, we have highlighted the vulnerabilities of existing implantable medical devices. The security of implantable devices is in its infancy. The difficulties stem from the scarcity of power and computational resources, the conflicting goals of availability (e.g. in emergency) and security, and the large range of contexts in which the devices need to operate. Simultaneously protecting the device and the confidentiality and integrity of the data, as well as the privacy of the user, are important research areas and the topics covered in this book represent only the tip of an iceberg—for wider deployment of implantable sensors that are smart, context-aware, and adaptive to different environments, the issue of security is of paramount importance.

Ultimately, implantable systems need to be *reconfigurable* to cater for different sensing environment, usage context, and noise, and *recalibratable* to combat sensor drift, fouling, and motion autonomously. They need to be built with *redundancy* to ensure reliable long-term operation in vivo and be *resilient* to the uninviting and often hostile environment within the human body as well as the external security risks. With the development of new materials, they can be made to be *resorbable* when deployed as a transient implant and *regeneratable* through effective biointegration when used as a long-term device in vivo. In this regard, the topics covered in this book including sensor designs, sensor embodiment and flexible electronics, ultra-low-power ASICs, power harvesting, and wireless data paths and security are essential ingredients for reaching such goals.

References

1. C.M. Abraham, A brief historical perspective on dental implants, their surface coatings and treatments. Open Dent. J. **8**(Suppl 1–M2), 50–55 (2014)
2. A. Barg, M.D. Wimmer, M. Wiewiorski, D.C. Wirtz, G.I. Pagenstert, V. Valderrabano, Total ankle replacement—indications, implant designs, and results. Dtsch Arztebl Int **112**, 177–184 (2015)
3. R. Coello, A. Charlett, J. Wilson, V. Ward, A. Pearson, P. Borriello, adverse impact of surgical site infections in english hospitals. J. Hosp. Infect. **60**(2), 93–103 (2005)
4. Weltin, B. Enderle, J. Kieninger, G.A. Urban, Multiparametric, flexible microsensor platform for metabolic monitoring. IEEE Sens. J. **14**(10), 3345–3351 (2014)
5. E. Lindner, R. Buck, Microfabricated potentiometric electrodes and their in vivo applications. Anal. Chem. **72**(9), 336A–345A (2000)
6. M.M. Ahmadi, G.A. Jullien, A wireless-implantable microsystem for continuous blood glucose monitoring. IEEE Trans. Biomed. Circuits Syst. **3**(3), 169–180 (2009)
7. P. Cong, N. Chaimanonart, W.H. Ko, D.J. Young, A wireless and batteryless 10-bit implantable blood pressure sensing microsystem with adaptive RF Powering for real-time laboratory mice monitoring. IEEE J. Solid State Circuits **44**(12), 3631–3644 (2009)
8. N.J. Cleven et al., A novel fully implantable wireless sensor system for monitoring hypertension patients. IEEE Trans. Biomed. Eng. **59**(11), 3124–3130 (2012)
9. T. Yokota, P. Zalar, M. Kaltenbrunner et al., Ultraflexible organic photonic skin. Sci. Adv. **2** (4), e1501856 (2016)
10. J. Viventi, D.H. Kim, L. Vigeland et al., Flexible, foldable, actively multiplexed, high-density electrode array for mapping brain activity in vivo. Nat. Neurosci. **14**(12), 1599–1605 (2011)
11. C.M. Lopez, A. Andrei, S. Mitra et al., An implantable 455-active-electrode 52-channel CMOS neural probe. IEEE J. Solid State Circuits **49**(1), 248–261 (2014)
12. M. Monge, M. Raj, M.H. Nazari, H.C. Chang, Y. Zhao, J.D. Weiland, M.S. Humayun, Y.C. Tai, A. Emami, A fully intraocular high-density self-calibrating epiretinal prosthesis. IEEE Trans. Biomed. Circuits Syst. **7**(6), 747–760 (2013)
13. R. Kiesslich, M. Goetz, K. Lammersdorf, C. Schneider, J. Burg, M. Stolte, M. Vieth, B. Nafe, P.R. Galle, M.F. Neurath, Chromoscopy-guided endomicroscopy increases the diagnostic yield of intraepithelial neoplasia in ulcerative colitis. Gastroenterology **132**(3), 874–882 (2007)
14. M.J. Gora, J.S. Sauk, R.W. Carruth, K.A. Gallagher, M.J. Suter, N.S. Nishioka, L.E. Kava, M. Rosenberg, B.E. Bouma, G.J. Tearney, Tethered capsule endomicroscopy enables less invasive imaging of gastrointestinal tract microstructure. Nat. Med. **19**(2), 238–240 (2013)
15. J. Wang, F. Wagner, D.A. Borton, J.Y. Zhang, I. Ozden, R.D. Burwell, A.V. Nurmikko, R. van Wagenen, I. Diester, and K. Deisseroth, Integrated device for combined optical neuromodulation and electrical recording for chronic in vivo applications. J. Neural Eng. **9**(1), 016001 (2012). https://doi.org/10.1088/1741-2560/9/1/016001
16. K. Stingl, K.U. Bartz-Schmidt, D. Besch, A. Braun, A. Bruckmann, F. Gekeler, U. Greppmaier, S. Hipp, G. Hortdorfer, C. Kernstock, A. Koitschev, A. Kusnyerik, H. Sachs, A. Schatz, K.T. Stingl, T. Peters, B. Wilhelm, E. Zrenner, Artificial vision with wirelessly powered subretinal electronic implant alpha-ims. Proc. Royal Soc. B **280**(1757), 20130077 (2013)

17. P.P. Mercier, A.C. Lysaght, S. Bandyopadhyay, A.P. Chandrakasan, K.M. Stankovic, Energy extraction from the biologic battery in the inner ear. Nat. Biotechnol. **30**(12), 1240–1243 (2012)
18. D.D. Karnaushenko, D. Karnaushenko, D. Makarov, O.G. Schmidt, Compact helical antenna for smart implant applications. Nat. Publ. Group Asia Mater. **7**, e188 (2015)
19. K. Okabe, I. Akita, S. Yamagiwa, T. Kawano, M. Ishida, A thin film flexible antenna with CMOS rectifier chip for RF-powered Implantable neural interfaces, in *International Conference on Solid-State Sensors, Actuators and Microsystems (TRANSDUCERS)* (2015)
20. D. Seo, J.M. Carmena, J.M. Rabaey, E. Alon, M.M. Maharbiz, Neural dust: an ultrasonic, low power solution for chronic brain–machine interfaces. arXiv:1307.2196 [q-bio.NC] (2013)
21. E.Y. Chow, A.L. Chlebowski, S. Chakraborty, W.J. Chappell, P.P. Irazoqui, Fully wireless implantable cardiovascular pressure monitor integrated with a medical stent. IEEE Trans. Biomed. Eng. **57**(6), 1487–1496 (2010)
22. M. Zhang, A. Raghunathan, N.K. Jha, Trustworthiness of medical devices and body area networks. Proc. IEEE **102**(8), 1174–1188 (2014)
23. G. Zheng, G. Fang, M.A. Orgun, R. Shankaran, A non-key based security scheme supporting emergency treatment of wireless implants, in *IEEE ICC Symposium on Communication and Information Systems Security* (2014)

Chapter 2
Electrochemical Sensor Designs for Biomedical Implants

S. Anastasova, P. Kassanos and Guang-Zhong Yang

List of Acronyms

AIROF	Anodic iridium oxide film
Al_2O_3	Aluminium oxide
ATP	Adenosine triphosphate
CMOS	Complementary metal-oxide-semiconductor
CNS	Central nervous system
DOS	Bis(2-ethylhexyl)sebacate
ESD	Electrostatic discharge
FEM	Finite element analysis
FET	Field-effect transistor
FG-MOS	Floating-gate MOS
GA	Glutaraldehyde
GERD	Gastroesophageal reflux disease
GI tract	Gastro intestinal tract
IrO_x	Iridium oxide
ISE	Ion-selective electrode
ISFET	Ion-sensitive field-effect transistor
LPCVD	Low pressure chemical vapour deposition
L-PEI	Linear polyethylenimine polymer
LPF	Low-pass filter
L-PPI	Linear polypropyleneimine polymer
MMO	Metal-metal-oxide
MNOS	Metal-nitride-oxide-silicon
MOSFET	Metal oxide semiconductor field effect transistor
MWCNT	Multi-walled carbon nanotubes
NMOS	N-type metal-oxide-semiconductor
OCP	Open circuit potential

S. Anastasova (✉) · P. Kassanos · G.-Z. Yang
The Hamlyn Centre, Imperial College London, London, UK
e-mail: s.anastasova-ivanova@imperial.ac.uk

G.-Z. Yang (ed.), *Implantable Sensors and Systems*
https://doi.org/10.1007/978-3-319-69748-2_2

o-NPOE	2-nitrophenyloctyl ether
p(HEMA)	Poly(2-hydroxyethyl methacrylate)
PECVD	Low temperature chemical vapour deposition
PEDOT	Poly(3,4-ethylenedioxythiophene) polymer
PEG	Poly(ethylene glycol)
PMOS	P-type metal-oxide-semiconductor
PVC	Polyvinylchloride
RE	Reference electrode
ReFET	Reference FET
Si_3N_4	Silicon nitride
SIROF	Sputtered Ir(Ox) films
SNP	Single-nucleotide polymorphism
SsDNA	Single-Stranded Deoxyribonucleic acid
Ta_2O_5	Tantalum oxide
UV exposure	Ultraviolet exposure
WE	Working electrode

2.1 Introduction

The need to record directly the sensing target of interest in the vicinity of where a physiological and clinically relevant event takes place, rather than indirectly or through surrogate measures, has led to the need for implantable monitoring devices. In addition to ensuring the sensitivity and specificity of sensor responses, issues related to sensor fouling, drift, biocompatibility, and hermeticity of the packaging are important considerations. This chapter examines the current state of the art of sensing techniques, focusing on electrochemical methods (potentiometry, amperometry, and voltammetry), due to their simplicity in design and fabrication [1], as well as low-power operation. The basic principles of sensor operation and fabrication, with due consideration of these sensors when embodied as an implantable system, will be discussed.

2.2 Potentiometric Ionic Sensing

Potentiometric sensors represent an important family of sensors typically used for the detection of ions. Monitoring of various ion concentrations can provide valuable insights into the state of tissue and the underlying biochemical processes. Here we will discuss different potential ionic targets and the basic theory of potentiometric sensing, as well as clinical examples.

2.2.1 Ionic Targets

The most common application of ion sensors is their use in the clinical analysis for example in blood. Electrolytes such as Na^+, K^+, Cl^-, Ca^{2+}, Mg^{2+} and pH are routinely measured using ion sensors in clinical analyzers (HTA) and point-of-care (POC) discrete instruments [2]. Magnesium (Mg^{2+}) is the eleventh most common element, the fourth most common cation in the human body, and the second most common intracellular cation [3]. It is essential to all cells, as more than 300 enzymes need magnesium ions for their catalytic action (including all enzymes synthesizing or utilizing ATP, or those that synthesize DNA or RNA). Intracellular Mg^{2+} correlates with intracellular potassium (K^+), which is also essential for the function of all living cells. Figure 2.1 illustrates the intracellular and extracellular ions, their concentrations and ionic shifts taking place.

Another common element in the human body is K^+. It is the main intracellular cation and is essential in neuronal function and in the osmotic balance between cells and the extracellular fluid. K^+ is the predominant ion inside a cell (typically 140 mM [4]), being one to several orders of magnitude higher than Na^+, Ca^{2+}, and Cl^-. Thus, intracellular K^+ essentially determines cytoplasm volume. The extracellular space, however, contains low levels of K^+ [5, 6]. Cellular loss of K^+, leading to its extracellular accumulation, is one of the most prominent ionic shifts in early ischemia [7]. In general, cellular loss of K^+ takes place in three distinct phases: (1) an initial rapid increase of extracellular K^+ is followed by (2) a plateau and (3) a secondary increase. The latter is only partially reversible and hence marks the beginning of irreversible ischemic damage [7]. Thus, extracellular K^+ increases due to the loss of the intracellular K^+. According to [8], dysfunction of the sodium–potassium pumps due to ATP reduction and activation of potassium channels where the intracellular compartments are filled with water are key factors for the increase of K^+ in extracellular space during ischemia. According to [9], in cellular cultures, potassium leaks out of dying cells, leading to elevated extracellular potassium ranging between 1 and 150 mM K^+.

On the other hand, Na^+ is the main extracellular cation. The concentration difference between K^+ and Na^+ leads to a potential difference, known as the

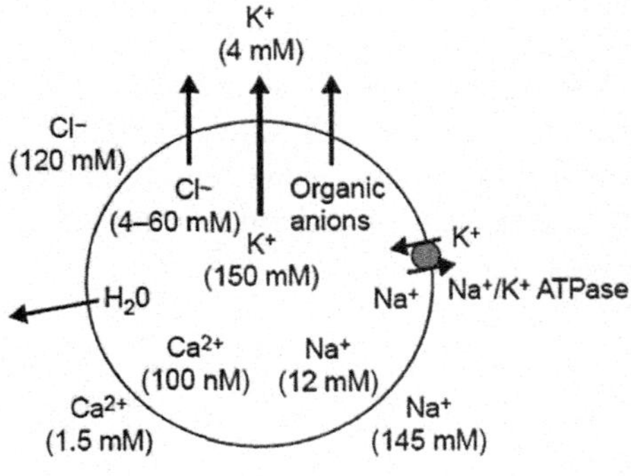

Fig. 2.1 Intracellular and extracellular ions and shifts. Reprinted from [4], © 2001 with permission from Elsevier

membrane potential. Na^+ regulates blood volume, pressure, osmotic equilibrium, and pH. "*Cellular and whole body* Na:K *ratios are crucial to the maintenance of normal blood pressure*" [3]. The Na:K ratio may become too high via high Na^+ or low K^+, or indirectly by Mg^{2+} deficiency, leading to hypertension.

Extracellular Ca^{2+} concentration increases at sites of infection from its typical baseline, which ranges between 0.9 and 1.2 mM [10]. In addition, calcifications at sites of chronic inflammation or ischemic necrosis have led to the conclusion that extracellular Ca^{2+} concentrations also increase in sterile inflammation in addition to infectious inflammation [10]. According to [10], increased extracellular Ca^{2+} concentration can lead to the amplification of the inflammatory response.

Monitoring of tissue pH can be used as a means of assessing the ischemic state of tissues, especially since its value follows cell metabolism, and it can thus be considered as a direct method for monitoring ischemia [11–14]. A lack of oxygen delivery to the tissue, due to ischemia, can lead to a decrease of intracellular pH. This is due to glycolytic lactate acid generation and ATP depletion processes [8]. The tissue acid–base balance is maintained by the release of energy by ATP (which releases protons) and the re-synthesis of ATP (which consumes protons). When oxygen delivery to the tissue is inadequate and fails to synthesize the amount of ATP to address the metabolic needs of the tissue, the rate of proton release exceeds the rate of proton consumption. As a result, the pH decreases proportionally. A significant blood loss induced by trauma, for example, can lead to a reduced perfusion of tissue. This reduced supply of blood to tissue has as a consequence the increase of CO_2 and lactic acid, leading inevitably to a pH decrease. Measurement of tissue pH provides localized assessment of metabolic and blood flow abnormalities. If the body pH is significantly reduced below 7.4, acidosis takes place. Many chemical reactions taking place in the human body, especially those related to proteins, are pH dependent. If the pH decreases below 6.8 or increases above 7.9, cell death occurs. As discussed in [15], once the tissue becomes acidotic, pH cannot offer any more information regarding the progression of ischemic injury. According to [8], it is the end of the active metabolic processes (glycolytic lactate acid generation and the ATP depletion) that gives rise to the decrease in pH that leads to a plateau in the measured pH value. Measurements of pH are regularly used to monitor tissue hypoxia [8], while low pH values have also been found in tumor tissue. Nevertheless, it is affected by other parameters, such as temperature, and its application is further restricted as it provides information only for the area in contact with the pH sensor. From the above, it is evident that human health is closely related to pH homeostasis, and its fluctuations, in addition to indicating ischemia, are also related to atherosclerotic plaque development, inflammation, and tumor growth [16]. Ionic sensors are therefore an important class of sensors for providing valuable information regarding the state of tissues and patients.

2.2.2 *Basic Theory of Ion Sensing*

The first significant commercial contribution in the field of potentiometric sensing was in 1932, when Arnold Beckman developed the pH glass-based liquid electrode [17]. pH provides a measure of the amount of hydrogen ions (H^+) and hydroxide ions (OH^-) in an aqueous solution. A solution is defined as being neutral when the hydrogen ion concentration is equal to the hydroxide ion concentration. When this is the case, the pH is equal to 7. This is the case for water at 25 °C, where the concentrations of hydrogen and hydroxide ions are both 1×10^{-7} mol l^{-1}. If the H^+ ion concentration increases, pH decreases and when the concentration of OH^- ions increases pH increases. A low pH value (between 0 and 7) describes an acidic solution, while a high pH describes an alkaline solution (between 7 and 14). pH is given by the decimal logarithm of the reciprocal of the hydrogen ion activity, a_H+, in a solution.

$$pH = -\log_{10}(a_{H^+}) \tag{2.1}$$

A standard pH potentiometric system consists of a reference electrode (RE) and a working electrode (WE). The former generates a constant potential irrespective of the ionic composition of the solution, while the latter generates a voltage which is a function of the hydrogen activity of the solution. Thus, a pH sensor can be considered as a voltage source, i.e. a battery. The series resistance of the source depends on the electrodes composition and configuration. The lower the pH, the larger the potential difference measured between the two electrodes and vice versa.

When equilibrium has been reached, the expressions for the ion-selective membrane phase (usually organic) and electrochemical potential in the aqueous phase are equal, so the phase-boundary potential (E_{PB}) can be expressed as:

$$E_{PB} = \Delta_{\varphi} = -\frac{\mu^0_{org} - \mu^0_{aq}}{z_I F} + \frac{RT}{z_I F}\ln\frac{\alpha_{I,(aq)}}{\alpha_{I,(org)}} \tag{2.2}$$

and when $\mu^0_{org} = \mu^0_{aq}$

$$E_{PB} = \frac{RT}{z_I F}\ln\frac{k_I \alpha_{I,(aq)}}{\alpha_{I,(org)}}, \tag{2.3}$$

where μ is the chemical potential (μ^0 under standard conditions), k_I desctribes the phase transfer energy, z_I is the charge, and α_I is the activity of ion I, ϕ is the electrical potential, T is the temperature, F = 96,485 C mol^{-1} is the Faraday constant, R = 8.314 J $(K\ mol)^{-1}$ is the gas constant and k_I is the phase transfer energy. The Nernst equation is expressed as:

$$E_M = E_0 + \frac{RT}{z_I F}\ln(\alpha_{I,(aq)}), \tag{2.4}$$

E_0 includes all the constant contributions, where the phase boundary potential is proportional to the logarithmic value of the activity ideally with a slope of 59.154 mV per decade for monovalent ions and for divalent ions 29.6 mV per decade at 25 °C. The modified Nernst equation is usually called the Nicolskii–Eisenman equation, and it is given by

$$E = E_{\mathrm{I}}^{0} + \frac{RT}{z_{\mathrm{I}}F}\ln\left(\alpha_{\mathrm{I,(IJ)}} + K_{\mathrm{IJ}}^{\mathrm{pot}}\alpha_{\mathrm{J,(IJ)}}^{\frac{z_{\mathrm{I}}}{z_{\mathrm{J}}}}\right), \tag{2.5}$$

where, $\alpha_{\mathrm{I,(IJ)}}$ and $\alpha_{\mathrm{J,(IJ)}}$ are the activities of the primary (I) and interfering ions (J), $K_{\mathrm{IJ}}^{\mathrm{pot}}$ represents the selectivity coefficient. The activity of $\alpha_{\mathrm{I,(I)}}$ is related to the mixed ion activity $\alpha_{\mathrm{I,(IJ)}}$ according to

$$\alpha_{\mathrm{I,(I)}} = \alpha_{\mathrm{I,(IJ)}} + K_{\mathrm{IJ}}^{\mathrm{pot}}\alpha_{\mathrm{J,(IJ)}}^{\frac{z_{\mathrm{I}}}{z_{\mathrm{J}}}} \tag{2.6}$$

The smaller the coefficient, the better the selectivity of the sensor is. When $K_{\mathrm{IJ}}^{\mathrm{pot}}$ is very small, the part on the right-hand side of the equation approaches the primary ion activity $\alpha_{\mathrm{I,(I)}}$ without interfering ions, and therefore interference is negligible. If there is more than one interfering ion in the solution, the sum of the selectivity coefficients $\left(\sum_{\mathrm{J}} K_{\mathrm{IJ}}^{\mathrm{pot}}\alpha_{\mathrm{J}}\right)$ must be used. In mixed solutions, the potential response of an electrode obtained according to the Nicolskii–Eisenman equation will be different depending on the charges of primary and interfering ions [18].

If we consider pH sensors according to the Nernst equation, a pH sensor generates an ideal 59.154 mV pH^{-1} slope at 25 °C

$$E = E^{0} - 2.303\frac{RT}{F}pH = E^{o} - 0.05916 \cdot pH, \tag{2.7}$$

where E is the voltage of the hydrogen electrode and pH is the hydrogen ion concentration being measured. E^0 is the standard electrode potential. In metal–metal-oxide (MMO) electrodes, E^0 is a constant that lumps together the standard potentials, the metal oxide solubility product, and the water ionization product [19]. Generally speaking, it is a constant potential difference that is independent of the sample composition, but depends on the temperature and the type of the reference electrode [20]. According to [21], in the case of a standard hydrogen electrode (SHE) E^0 has a value of 926 mV and in case of using a Ag/AgCl RE has a value of 577 mV [22]. The intercept of the mV versus pH curve of the sensors response with the y-axis for pH of zero indicates the value of E^0. Variations in E^0 can in the case of IrO_x electrodes, for example, be "*due to variations in the stoichiometry of oxide compounds and the difference in oxidation states of iridium oxides*" [22]. Additional factors related to the mechanical and chemical properties of the films as well as the hydrodynamics related to the redox processes taking place will influence the response of the sensors and lead to deviations between identical sensors [22]. The slope of the measured response is calculated by

$$s = -\frac{V_1 - V_2}{pH_1 - pH_2}, \tag{2.8}$$

where V_1 is the potential difference measured with a solution with pH_1 and V_2 is the potential difference measured with a solution of pH_2. The intercept for a temperature (T) in kelvin is found by:

$$V'' = \frac{V_1 - S \cdot pH_1}{T}, \tag{2.9}$$

where V'' is the potential difference measured in the solution being characterized.

According to the Nernst equation, at $T = 25$ °C, where the sensitivity of the sensor is equal to ±59.14 mV pH^{-1}, for pH between 0 and 14, the voltage measurement can range between ±414 mV, and at $T = 80$ °C, where the sensitivity of the probe is ±70 mV pH^{-1}, between ±490 mV. A typical pH measurement system produces at pH = 7 a zero voltage, has an accuracy of 0.05 pH between $T = 20$–25 °C, a resolution of 0.01 pH (=>0.1 mV) and a reaction time less than 1 s for 95% of the final value [23]. The pH sensitivity of −59 mV pH^{-1} at room temperature is well known as a Nernstian slope. However, depending on the surface and the formed microstructures and oxidation states of the ion-sensitive films and membranes, and thus the fabrication process, this slope may vary. Therefore, in [21] a sensitivity of −49.7 mV pH^{-1} was demonstrated, while, as will be discussed in following sections, certain films such as IrO_x produce super-Nernstian responses.

All interfaces in the galvanic cell must be dominated by fast, reversible, and well-established faradaic charge transfer [2]. A practical potentiometric measurement of the potential difference across a galvanic cell requires an open circuit potential (OCP) measurement. Potentiometric measurements are performed with a zero faradaic current. Thus, high input impedance is mandatory for the measurement. This is in contrast with the amperometric, conductimetric, and impedimetric type of sensors.

The definition of the response time of a pH sensor is given as the time needed for the potential change induced by a pH unit change to reach 90% of its final equilibrium value. Issues regarding stability and repeatability in the measurements are related to potential fluctuation (ΔV), potential deviation (δV) and potential drift (V') and hysteresis (dV) as discussed in [22]. These are illustrated in Fig. 2.2. ΔV is a small non-random voltage fluctuation caused by electronic noise or interference and motion artifacts such as liquid motion. This can be in the range of ±0.3 mV and ±1 mV. δV is defined as the difference in measured potential responses between different tests when using the same electrode in the same solution. Reported values are less than 5 mV [22]. For recorded potentials varying between −0.07 and 0.46 V for pH between 1.5 and 12.1, the proposed ideal resolution is 0.02 pH mV^{-1}. Nevertheless, due to a $\delta V = 5$ mV, the minimum pH sensing resolution is 0.1 unit of pH. The difference between the peak recorded potential and the 90% value of the saturated recorded potential is defined as the V'. This can range between 3 and 10 mV, with the potential stabilizing within 5 s. Hysteresis, is defined as the dV obtained when comparing measurements obtained from solutions with the same value of pH within the same experiment. Reported values range from 0.3 mV and can be as large as 23.7 mV [22]. In iridium-based pH sensors, dV can be minimized by creating high-quality IrO_x films in terms of thickness, amorphousness and porosity [22]. The theoretical temperature dependence of potentiometric pH sensors is 0.3, 0.8, 1.3, and 2 mV °C^{-1} for pH of 2, 4, 7, and 10, respectively. Drift is a change in baseline potential over time, while the loss of sensitivity has a decreasing slope [20].

2.2.3 *Ion Selective Electrodes*

There are many different ways to develop an ion sensor. The classical pH electrode is the glass-based electrode sensor, which, however, has a number of drawbacks when considering its use for biomedical applications. These include slow response times, instability in fluoride, hydrofluoric acid, and silane solutions, the need for recalibration, and their susceptibility to membrane fouling, leading to loss of precision and accuracy. In physiological measurements, rapid pH measurements are of interest, as is the steady state value. Because of the very high impedance of these electrodes (multiple GΩs), these sensors require a high impedance meter [24], complicating the recording electronics. Due to their mechanical fragility, large size, and lack of deformability [22], they are unsuitable for in vivo and implantable applications.

Ion Selective Electrodes (ISEs) were first described by Cremer [25]. The discovery of antibiotics and selective binding cations gave a push to the field of ISEs. In the last two decades, solid state pH sensors have been developed in an effort to tackle the issues associated with glass-based electrodes [26]. The ionic markers discussed earlier can be monitored using solid contact ISE sensors instead of glass-based ones. Similarly to standard glass-based sensors, these are transducers,

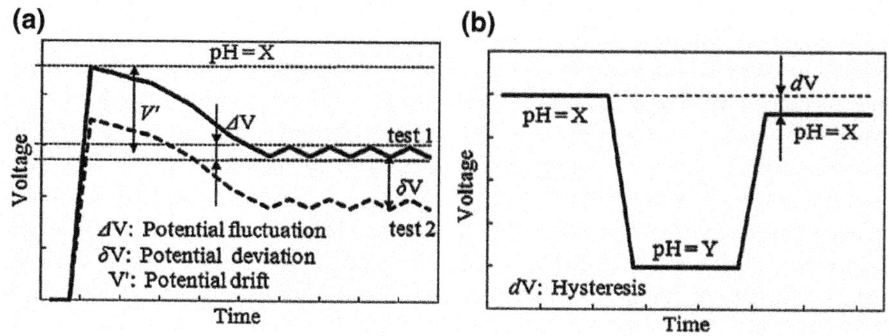

Fig. 2.2 **a** Potential fluctuation (δV), deviation (ΔV) and drift (V') are characteristics defining stability and repeatability, while **b** hysteresis (dV) a defines the reversibility of the sensor. Reprinted from [22], © 2011, with permission from Elsevier

which produce a change in their equilibrium d.c. electrical potential, which is related to the activity of a specific ion in the solution. The potential is measured as the difference between an RE of constant potential and a working electrode, the surface of which is modified by an ion-specific membrane, since this process has to be selective. ISEs are very simple, robust sensors, which can be easily miniaturized. In addition, they are very sensitive, reaching sensing limits as low as parts-per-billion (ppb) [20]. They are based on the ion transfer processes taking place between the solution and the ion-sensitive membrane. The interaction of the analyte with the membrane leads to a non-uniform charge distribution, which gives rise to a change in electrode potential. As discussed in [27], the response of potentiometric sensors, i.e. the recorded signal, does not depend on the size of the active area of the sensor. Thus, due to their size independence, the response of these sensors will not change as their dimensions are reduced to micro and nano scales. It is important to note that when multiple ISEs are used there is a need for only one reference electrode. Nevertheless, each ISE will have its own baseline and slope [20].

Noble metals with stable oxides and reasonable conductance are one possible implementation of the ion-selective membrane, forming what are known as metal/metal-oxide and metal-oxide/metal-oxide electrodes [28]. They have an intrinsic mechanical stability and they can be easily miniaturized using semiconductor fabrication technology, making them CMOS compatible. Metal oxides absorb hydrogen atoms at the surface sites, changing the valency of the oxygen atom. This leads to the formation of an OCP difference between the two electrodes, the magnitude of which is proportional to the solutions pH [24]. Iridium oxide is a well-documented material [28–33], which can be formed hydrated or unhydrated [28, 30], with the hydrated form exhibiting super-Nernstian responses.

An alternative to MMO electrodes is solid-contact conductive polymer electrodes. These have been extensively reviewed in [34, 35]. Functional polymers (such as conductive polymers like polyaniline and polypyrrole) can be designed to selectively swell and shrink so that, depending on the analyte's concentration. They can change their mass and elasticity. Polymer pH sensors react to these changes due to the protonation and deprotonation of nitrogen atoms in the polymer film. In conductive polymers, protons added to the material make them more conductive. Linear polyethylenimine (L-PEI) and linear polypropylenimine (L-PPI) polymers are particularly suitable for biosensing applications due to their strong bonding to electrode surfaces [24].

For the detection of K^+, Na^+ and ammonium (NH_4^+), valinomycin, monensin and nonactin are respectively used [17, 36–38]. In comparison to glass electrodes, in ionophore-based ISEs the chemical structure of the ionophore can be tailored such that a very high selectivity is achieved [39, 40]. Today there are ISEs for more than 60 different ions. ISEs can be used for the measurement of polyions, such as heparin and protamine, inorganic ions [41, 42], and also neutral species such as CO_2, O_2, SO_2 [43–45], water, ammonia, organic amines, alcohol and non-ionic surfactants [41]. Advances in the host–guest chemistry for ISEs and the understanding of their theoretical response mechanisms led to the development of the first clinical analyser and its commercialization in 1972. Nowadays ISEs are almost irreplaceable in clinical analysis for detection of ions of clinical relevance, such as K^+, Na^+, Ca^{2+}, Mg^{2+}, and Cl^-. ISEs determine the concentration of free ions in comparison with other methods, such as atomic absorption spectrophotometry (AAS) or inductively coupled plasma mass spectrometry (ICP-MS) techniques used to determine total ion concentration.

Lipophilicity of the ISE's membrane is defined as $p = \log k$, with k being the distribution coefficient of the different species between water and normal octanol [46]. Ionophores and ion-exchangers of low lipophilicity ($p < 6$), e.g. unsubstituted tetrakis(phenyl)borates, can leach from the ISE membranes relatively fast, so that the bulk resistance of the membranes increases 2–3 times within a few days, even at room temperature. Most of the used ionophores (charged or neutral), as well as ion exchangers, are highly lipophilic, with $p = 8$.

No matter what technique is used for the functionalization of the electrodes, deposited ion-sensitive membranes should have good adhesion to the sensor substrate. An issue with ISEs is the membrane lifetime. Membrane failure is mainly caused by partial detachment of the sensing membrane from the conductive surface or by the loss of the ionophore, plasticizer or carrier from the polymeric membrane film used due to leaching of some of the components into the surround solution. Leaching of the ionophore from its polymer matrix can influence the lifetime of a selective membrane. Insufficient lipophilicity of the membrane components is a dominant reason for the limited lifetime of ISEs; the more lipophilic the ionophore, the longer the lifetime of an ISE. Several approaches to how to improve it are discussed in [47]. For example, in the application of ISEs having a minimum lifetime of one month in clinical analysis, it requires a lipophilicity ($\log P_c$) in regular 200 μm thick membranes of 11.3 and greater. It is possible to increase the lipophilicity with attachment of longer alkyl chains to the ionophore's body. The attachment of an ionophore to a plasticizer-free polymer matrix has also been reported with minimal or no loss of the ionophore's functionality [48–50]. The lower detection limit is dictated by the presence of the complexed primary ions in the membrane, where the upper limit is determined from the fact that the counterions must not enter from the sample side. When simultaneous extraction of a primary ion and its counterion (coextraction) occurs, the response slope is changed [51].

Ion exchangers allow for a proper electrode response and also have another important role. Membranes containing only ionic sites in a polymer matrix will still exhibit some selectivity based on the lipophilicity of ions due to their ion exchange properties. The more lipophilic ions have a lower hydration energy and partition more easily into the organic membrane phase from the aqueous solution [52, 53]. The physical and chemical properties, such as the polarity, and lipophilicity can have a significant effect on the ion-selective electrode selectivity and response time [54, 55]. Apolar and polar plasticizers can be distinguished based on the dielectric constant. Apolar plasticizers, such as bis(2-ethylhexyl)sebacate (DOS), promote the formation of ion pairs within the membrane [56].

The formation of ion pairs can significantly influence the slope of the response. If a divalent ion M^{2+} forms ion pairs with a monovalent ion A^{-}, the phase transfer equilibria are dictated by MA^{+} as the predominant species, and the slope, characteristic for monovalent ion, can be obtained [57]. A polar membrane containing 2-nitrophenyloctyl ether (o-NPOE) shows an increase in selectivity for alkali ions due to decreasing coordinating ability [58], which is the reason for improvement of the detection limit [59]. There are several problems associated with using plasticized PVC. Some disadvantages are related to the possible leaching of the plasticizer from the membrane into the sample. While it shortens the lifetime of an electrode, it can significantly affect in vivo measurements due to inflammation and toxicity to the surrounding tissue [60]. In the case of microelectrodes, the specific resistance of the membrane can be increased, which can affect the electrode response [61]. The adhesion towards the matrix is important because the formation of an aqueous layer between the ion-selective solid-based membrane and metallic conductor can significantly impair the response [62]. The development of polymers that are self-plasticizing, such as polyurethanes [63], polysiloxanes [64], and polyacrylates [65], is a significant and active area of research. Other plasticizer-free polymers, prepared by one-step polymerization, are methacrylate and methacrilate–acrilate copolymers [66].

Nowadays ISEs are miniaturized, robust, and solid state and use integrated solid state REs. The main sensors developed for use in clinical environments are H^{+}, Na^{+}, K^{+}, Ca^{2+}, and Cl^{-}; also used are Li^{+}, Mg^{2+}, NH^{4+}, trace metal ion detection, and organic ion detection. Improved selectivity and sensitivity are achieved with new affinity ligands [67].

Another important family of potentiometric sensors are ion-sensitive field effect transistors (ISFETs). As discussed in [68], the line separating ISFETs (and thus also ChemFETs) from ISEs with integrated electronics is very thin. In an ISE, there is a conductor between the selective sensing membrane and the transistor gate (similarly to an extended gate ISFET), and an operational amplifier-based voltage follower is used instead of a single transistor. Active ISEs demonstrate improved thermal stability and reduced photo-induced junction currents. ISE membrane deposition is performed following the CMOS fabrication process. Active ISEs demonstrate an improved response time to chemical changes in contrast to passive ones. CMOS ISEs have been reported with a lifetime in excess of 120 days, with response time within 5% of the ideal and a 10 μM detection limit.

2.2.3.1 Iridium Oxide MMO pH Sensors

In vivo pH sensing is an important target for implantable systems. In general, pH sensing can be achieved by using iridium oxide metal–metal-oxide (MMO) sensors or polymer-based pH sensors. The ideal characteristics of these MMOs are discussed in [19]. According to [16, 19, 69], the advantages of IrO_x films over other pH-sensitive oxides include a wide response range, fast response time, high pH sensitivity (a super Nernstian response, i.e. greater than 59 mV pH^{-1}), minimal potential drift, low sensitivity to redox pair interference (thus high chemical selectivity), and finally high durability. IrO_x has also demonstrated an outstanding stability over wide pH ranges and in different solutions. They are also characterized by a low impedance and low temperature dependence [33]. Thus far, there are many methods for the fabrication of IrO_x functionalized electrodes for pH sensing. Sputtered IrO_x (SIROF) is costly and the fabrication protocol is hard to optimize. Thermal oxidation requires very high temperatures (500–800 °C) and the film may crack if it is incompatible with a CMOS process. Anodic deposition based on electrolysis is an economic solution; however, process parameters can easily affect the deposition efficiency. Electroplating allows the formation of hydrous (electrochemical growth in aqueous solutions) IrO_x on target planar microelectrodes selectively, using relatively inexpensive compounds of iridium. Hydrous IrO_x leads to higher sensitivities [16]. According to [69], the above make IrO_x "*one of the best choices as sensitive material for* pH *electrodes*". Its use has also been discussed within the context of muscle and nerve electrical stimulation. Another advantage of (electrochemically generated, AIROF, not thermally) iridium-based pH sensing is the compatibility of the material with CMOS processes.

The authors in [70] discuss in detail how different applied potentials for the oxidation and formation of the IrO_x film change the oxidation state of the film. At a constant pH the change is reflected in the OCP. A higher OCP corresponds to an oxide with higher valence and leads to higher sensitivity. Thus, the OCP at pH 0 increased from 490 to 825 mV and the slope from 62 mV pH^{-1} to 73.6 mV pH^{-1}. In this way, the OCP recorded can be fine-tuned, e.g., for a specific power supply, so that the sensor response can be compatible with the output swing of the pH recording amplifier. In addition, the thicker the oxide the slower the response of the sensor to a pH change (350 vs. 40 ms), while in terms of drift the sensors are characterized as electrochemically stable, with a drift less than 300 $\mu V\ h^{-1}$ [70].

The characteristics of these devices and the variety of different fabrication methods of IrO_x pH sensors are discussed and summarized in [71]. As discussed in this paper, the potential drift, which causes errors in the measurement of pH, depends on the oxidation state and hydration of the oxide film and the preparation method used for the film. The fabricated sensor exhibited OCP changes between 700 and −100 mV for pH 0 to pH 14 and thus a slope of 59 mV pH^{-1}, with no hysteresis and no changes in sensitivity, but a small 5 mV drift within 100 days. During a two-day continuous experiment at a pH of 6.6 the sensor demonstrated a relatively constant potential of 321.2 mV with a deviation of less than ±0.2 mV. Part of this could be due to temperature variations. The response time of the sensor to a pH change was considered to be less than 0.2 s. In [72] a drift of ±0.03 pH h^{-1} was reported for AEIROF and responses of −71 mV pH^{-1} and −61 mV pH^{-1} varying from 700 to 120 mV between pH of 3 and 10.

Sol-gel is a simpler and economical approach. Sol-gel based IrO_x has low temperature dependence, low interference and low voltage drifts. Potential differences from 500 to −100 mV between pH 0 and 12 and a 58.5 mV pH^{-1} sensitivity is demonstrated. However, both AIROF and SIROF films are more sensitive due to their higher porosity. In [22, 73], response times from 2 to 60 s were reported for flexible iridium oxide fabricated with a sol-gel process. The response time is mainly affected by the porosity of the sensing membrane. The sol-gel process results in a lower porosity (reduced response time). Lower porosity, however, means lower sensitivity. The measured potential differences ranged from about 500 to −200 mV and sensitivities from 57 mV pH^{-1} to 63.4 mV pH^{-1}.

In [74], the authors again used an IrO_x pH electrode and a Ag/AgCl reference with sensitivities of 69–71 mV pH^{-1} and OCP varying between 500 mV and 0 V. In order to protect the electrode from corrosion by physiological electrolytes a biocompatible ionophoric coating was used (Nafion). In [75], implanted pH sensors were used to assess in vivo tissue trauma in the brain. Following characterization, the sensors were dipped in Nafion. AIROF electrodes were also coated with Nafion in [19] in order to stabilize the response of the sensors and to protect them from chloride and protein adsorption. Nafion coatings were also used in [76] and once again improved the stability and selectivity of the sensor. IrO_x was once again used with a 80–90 mV pH^{-1} sensitivity and an E^0 from 800 to 400 mV. Nafion was also used in [77] in AIROF electrodes. According to [78], most metal oxides, such as IrO_x, are electronic conductors and thus they will respond to solution redox species. This will lead to large measurement errors. Many of these interferences can be eliminated or attenuated by coating the electrode with Nafion. There is, however, a tradeoff. The thicker the Nafion coating is, the greater the attenuation and thus elimination of the effect of interference, but the slower the response time of the electrode will be to a change in pH. This equilibration time was found to be a function of pH and it can reach a maximum value of 2 min [78].

The cytotoxicity of IrO_x has been examined and has been found not to be toxic, with cells adhering and surviving on Ir surfaces. Table 2.1 compares and summarizes the published literature in terms of MMO pH sensor characteristics and fabrication methodology. Different methods are used, such as electrochemical growth (AIROF), electrodeposition, sputtered coating (SIROF), thermal, and printing methods.

2.2.3.2 Polymer-Based pH Sensing

Polyvinylchloride (PVC) is the most used polymer for solid state ISEs. Immobilization of the PVS membrane is via adsorption through Van der Waals interactions; thus adhesion to the surface is poor. In this case, stability and reproducibility could be a problem [91]. Leaching of components from the membrane will lead to a short lifetime and a loss of functionality. Leaching from covalently linked benzo-18-crown-6 with multi-walled carbon nanotubes (MWCNT) was studied in [92]. It is known that PVC is not haemocompatible [93]. Possible rejection and passivation of the sensor can be observed during blood fouling [94]. Thick PVC layers can hinder ion diffusion through the membrane. The thermal stability of these is low; thus one must consider the application they are intended to be used for [90].

Table 2.1 Comparison of some of the published pH sensors and their characteristics

Paper	Slope mV/pH unit	V-range mV	pH range	Material	Dimensions	Response time	Drift	Substrate
[29]	59.8	600–250	2–10	IrO_x	n/a	n/a	2	n/a
[30]	72–77	900 → −100	0–14	IrO_x	0.5 mm diam.	3 min	1 mV min^{-1}	n/a
[31]	59	300 → −200	2–11	IrO_x	1 cm^2	n/a	n/a	Ti
[32]	62–74	750–350	2.5–8.7	IrO_x	3 mm diam.	n/a	n/a	Ir
[33]	81–75.5	782 → −200	0–12	IrO_x	1 mm diam.	1 min	n/a	Ir
[28]	70–92	600, 1246 → 0	0–11	IrO_x	3 mm diam.	n/a	n/a	Ti, Au, Pt
[19]	59.5–67.8	428.5	6.5–8	IrO_x	0.125, 0.25 mm	2 s	n/a	Pt, Ir
[79]	n/a	n/a	n/a	Poly-Hema + PVC	0.25 mm	n/a	n/a	Polyimide Au → Ag → AgCl
[80]	58.5	75 → −100	6.5–8.5	Poly-Hema + PVC	1, 0.5, 0.25 mm	n/a	0.12 mV h^{-1}	Polyimide Au → Ag → AgCl
[81]	n/a	n/a	n/a	Poly-Hema + ETH 5294, o-NPOE	0.5, 0.25 mm	n/a	0.25 mV h^{-1}	Polyimide Cr → Au → Ag → AgCl
[77]	63.5	550 → −100 340–200	2–10 6–8	IrO_x + Nafion	1 mm	5 s	0.2 mV h^{-1}	Polyimide → Ti → Pt
[82]	n/a	n/a	7.4–6.4	IrO_x	n/a	n/a	n/a	Polyimide
[21]	49.7	500 → −50	2–12	IrO_x	2 mm × 2 mm	n/a	5 mV	Polyimide + Cr + Au
[20]	n/a	n/a	n/a	PVC-NPOE-THF	n/a	n/a	n/a	n/a
[83]	59.8	300 → −150	2–9	PVC-NPOE-THF	1 mm	n/a	10 s	Au-PEDOT
[73]	57–61	500 → −200	2–12	IrO_x	1 mm × 1 mm	30–50 s	n/a	Polyimide + Cr + Au
[22]	51	500 → −100	2–12	IrO_x	2 mm × 2 mm	2 s	3–10 mV, 5 s	Polyimide + Cr + Au
[84]	n/a	n/a	n/a	Pyrrole-$LiClO_4$ + PVC	300 µm × 300 µm	1 s	250 µV d^{-1}	Polymer (Upilex) − Ti − Pt

(continued)

Table 2.1 (continued)

Paper	Slope mV/pH unit	V-range mV	pH range	Material	Dimensions	Response time	Drift	Substrate
[85]	51	n/a	n/a	IrO_x	n/a	n/a	n/a	n/a
[41]	70	450–0	3.5–9.5	IrO_x	400 μm × 400 μm	n/a	n/a	PDMS – Cr – Au
[69]	77.6	450–0	4–10	IrO_x + Nafion	20 × 40, 20 × 20, 5 × 10 μm	150–200 s	2–3 mV	Glass – Ti – Pt
[70]	69	400 → −200	2–11	IrO_x	1 mm × 1 mm	0.35 s	0.3 mV h^{-1}	Si/SiO_2 – Ti – Ag
[71]	58.92	700 → −100	0–14	IrO_x	0.25 mm diam.	<0.2 s	<0.1 mV d^{-1}	None
[72]	71	700–100	2.5–10.5	IrO_x	100 μm diam.	n/a	0.03 pH h^{-1}	Glass – SiN – Ti – Pt
[74]	71	500–100	4–10	IrO_2 + Nafion	5 μm diam.	n/a	n/a	Glass or Kapton – Ti – Pt
[86]	~70	~500 (pH 0)	6–8.5	IrO_x	500 μm × 2 μm	n/a	n/a	Glass – Ti – Au
[87]	73–81	400 → −200	4–11	IrO_x	75 μm diam.	<2 s	3 mV min^{-1} = 0.04 pH min^{-1}	Au
[75]	85.9	791.5	n/a	IrO_2 + Nafion	10 μm	n/a	0.6 mV h^{-1}	Si
[76]	n/a	n/a	n/a	n/a	n/a	n/a	n/a	n/a
[88]	70	n/a	n/a	IrO_x	10 μm × 200 μm, 20 μm × 400 μm	6–15 s	2–3 mV $month^{-1}$	Glass – Ti – Pt
[89]	70	n/a	4–10	IrO_x	20 μm × 60 μm, 20 μm × 400 μm	5–10 s	3–5 mV $month^{-1}$	Glass – Ti – Pt
[78]	51–60	222–730 (for E^0), 600 → −500	2–12	IrO_x + Nafion		0.3–2 min	n/a	Au
[90]	54, 53, 42	−20 → −110, 180–30, 140–10	6.5–8.25, 5–8, 5–8	PVC, PEDOT, PEG + PEDOT		200, 40, 10 s	n/a	Au

Modification of the ISE surface is discussed in [95]. Poly(ethylene glycol) (PEG) and phosphorylcholine [96, 97] are used as hydrophilic materials so as to improve the sensor biocompatibility. Phosphorylcholine mimics the zwitterion nature of phospholipid moieties which is observed on the cell membrane surface [98]. PEG binds water molecules and creates a hydrophilic barrier which prevents protein adsorption.

It is evident from the above discussions that the field of polymeric pH sensors is an active research field. In [83], a Nernstian slope of 59.9 mV pH^{-1} was achieved between pH 2 and 9 with a measured potential difference between 200 and −150 mV. In [20], a conductive polymer pH sensor had a response time of $t_{90\%} = 45$ s and slopes between 59.8 and 62 mV pH^{-1}. In [84], a pH sensor was developed using a metal Pt electrode electropolymerized by a pyrrole conductive polymer and a drop-cast PVC neutral carrier membrane. This sensor exhibited a Nernstian response, a response time less than 1 s, an impedance of 10 MΩ (the measurements are performed without sophisticated shielding and integrated amplifiers), and a potential drift of less than 250 μV d^{-1}, corresponding to a pH drift of 5/1000 pH units day^{-1}.

2.2.3.3 In Vivo pH Sensing

There is significant interest in the use of pH sensing for tissue ischemia, as pH can be used as a marker for ischemia. In [13], tissue pH due to ischemia changes from about 7 to 6.5 within an hour of induced ischemia. In [99], within 45 min of induced ischemia, the measured pH changed by 0.8 (from 7.4 to 6.6) and K^+ by 30 mM. Measurements were performed using commercially available ion-selective electrodes. In [8], pH changed from 7.4 to 6.4 within 40 min of induced ischemia. In a number of papers (e.g. [8, 15, 99]), the measured tissue pH initially decreased rapidly on the onset of ischemia and it subsequently stabilized to a constant value. This is presumably due to the tissue becoming acidic, marking the end of all metabolic processes giving rise to the pH decrease. This was argued to be a disadvantage of pH measurement when compared to tissue bio-impedance. Intracellular pH falls by 0.5–1 pH units during cerebral ischemia and the extracellular space becomes more acidic within 20 s of ischemia. According to [100–102], one of the main problems with tissue pH measurements is that they are not sufficiently stable during long-term measurements. Drifts, which have been described as unacceptable, greater than 0.48 pH units per 100 h were reported in [101]. According to [101, 103], the deposition of proteins on the surface of glass pH sensors during tissue pH measurements leads to measurement errors, producing a rise in measured pH. This protein deposition potentially creates an additional junction potential, which in turn leads to this pH increase. Other sources of drift are temperature variations, patient movement, and RE junction potential variation. A second pH sensor was used to simultaneously measure the pH of healthy muscle tissue (pH between 7.38 and 7.44) in order to compare this with ischemic muscle tissue pH. The pH measured from the healthy site remained somewhat constant throughout the experiment, between pH 7 and 7.2. The pH of the compromised tissue was initially at a lower pH value of 6.9. At the moment of occlusion, the pH started to decrease. Within one hour of occlusion, the pH dropped to 6.5. Within minutes from the release of occlusion, the pH increased to 6.9. In [101], 120 h of pH measurements from a patient with good post-operative recovery remained essentially constant and averaged a pH of 7.4, which is within healthy levels. Results obtained from a patient with an ischemic episode indicated a pH decrease 5 h following surgery. During this period, pH decreased by 0.4 pH units. Paper [104] reports pH changes of pH 0.73 between pH 7.08 to pH 6.35 within three hours (i.e. −0.0041 pH min^{-1}) of tourniquet on canine hind limbs, a drop of pH 0.66 within an hour (i.e. −0.011 pH min^{-1}) via arterial occlusion, a drop of pH 0.27 within an hour (i.e. −0.0045 pH min^{-1}) via venous occlusion and a drop of pH 0.55 within an hour (i.e. −0.0092 pH min^{-1}) via total occlusion of rat anterior thigh flaps. Other reported values are a drop of pH 0.80 in tissue pH within 240 min (i.e. 0.0033 pH min^{-1}) in an amputated rat hind limb and a decrease of pH 0.35 from pH 7.4 to pH 7.05 (i.e. −0.0058 pH min^{-1}) following occlusion of the femoral artery of a rabbit hind limb. An average decline of −0.00677 pH min^{-1} was also reported using a similar setup. An average per minute change of muscle tissue pH ranging between pH −0.33 to pH −0.11 is possible. In [105], the pH probe used exhibited a drift of less than 0.01 pH units over a six-hour period.

Of interest is the intestinal luminal pH of the normal healthy gastrointestinal (GI) tract. The colonic mucosa is covered by a surface of mucus with a thickness of 100–800 μm [106]. As discussed in [107], the colonic epithelium maintains a constant pH up to 840 μm above the cell lining. Failure of the anastomosis in the GI tract following colorectal resection, for example, leads to anastomotic leakage, which is considered one of the most challenging complications in GI surgery. Intramucosal pH at the anastomosis was found to be an independent factor for leakage development [108]. In patients with an anastomotic intramucosal pH which was less than 7.28 units within the first 24 h after surgery a 22 times higher risk of leakage is reported [108]. Patients with anastomotic leak had a mean pH of 7.05, while patients without a leak had a pH of 7.32 in the first 24 h [108]. In [105], the threshold value of intramural pH for ischemia and thus failure in the anastomosis was found to be less than pH 7. Compared to laser Doppler velocimetry, which correctly predicted the outcome of the anastomosis by 70%, tissue pH measurements succeeded in 93% of the cases [105]. It is, however, important to note that normal intestinal mucosal pH levels are not constant throughout the length of the intestines. In [107] the authors assessed the pH in different areas of the colon leading to what they termed as "*the closest approximation of* pH *values of superficial colonic epithelium*". The measurements follow luminal colonic measurements (such as those obtained with pH sensing capsules), which nevertheless are more acidic in all colonic segments. It was found that the rectum (pH 7.15 ± 0.44) and the right colon (pH 7.05 ± 0.32) were relatively acidic, while the transverse (pH 7.42 ± 0.51), left (pH 7.46 ± 0.60), and sigmoid (pH 7.38 ± 0.59) colon were more alkaline. According to [107], the colonic mucosa tends to be more alkaline relative to the luminal colonic pH. In [15], pH in the intestine significantly dropped after ischemia, from 7.6 to 6.8 within 180 min of induced ischemia. Arterial pH followed a similar trend (7.44–6.95). According to [108], a pH less than 7.28 in the first 24 h following surgery is associated with a 22 times higher risk of anastomotic leakage.

In addition, continuous pH measurements provide an important parameter in detection of gastroesophageal reflux disease (GERD) [109] and also drug activity, which depend on saliva pH [110]. In [85], a system for GERD employing both impedance and pH measurements was demonstrated. The pH sensor was based on [22], which provided a slope of 51 mV pH^{-1}, a voltage ranged between 0.5, and −0.1 V for pH values between 1 and 12. A second example of an ingestible capsule incorporating both impedance and pH for GERD is presented in [111]. The pH sensor had a sensitivity of 33.3 mV pH^{-1} and recorded voltages of 350 mV for pH 7 and 250 mV for pH 4. The combination of multichannel impedance with pH monitoring helps to detect nonacid reflux.

2.2.4 K^+ Sensors

As discussed earlier, K^+ is an important ion in biomedical applications. The cellular K^+ level is a clinically important parameter for assessing cellular function. In ischemic myocardium for example, K^+ loss is observed [112]. A connection between the cellular K^+ loss and the balance in the charge during lactate transport has been hypothesized [113]. Passive movement of K^+ in response to cellular loading of Na^+ may enhance the loss of cellular potassium ions [114]. Cellular K^+ loss in the ischemic heart is also related to the pH level [115].

An enhanced selectivity to K^+ was achieved using a solid state-based internal electrolyte, such as polypyrrole stabilized with cobalt organic counterion and using calyx[4]arene ester [116]. In [117], micromolar detection was achieved where the silicon atom is part of the ether ring. In [91], the use of valinomycin for highly selective measurements of K^+ is discussed, where different changes in the internal layer can give an extended sensitivity limit. Another internal solid contact electrode is based on the use of poly(3-octylthiophene) [118]. Using a lipophilic ion conductor avoids the drift related to the development of a water layer between the ionophore layer and the substrate. Graphene has also been proposed as the solid contact material [119]. An example of early multiparametric chemical tissue sensing is shown in Fig. 2.3.

In [120], a K^+ solid contact microelectrode is described. The sensor is based on a neutral carrier and demonstrated operation within a pH range of 3–11, detection limit of 1.8×10^{-6} M, and working range between 6×10^{-6} to 1×10^{-1} M with a Nernstian slope of 52 mV $decade^{-1}$. Valinomycin-based electrodes for online detection of intravascular and myocardial K^+ ions was described in [91]. In [82], K^+ sensitive electrodes were presented with a slope of 51 mV per decade (potential differences between 0 and 250 mV) for concentrations between 6 μM and 100 mM, a life time exceeding 40 days and a response time of $t_{95\%} = 14.2$ s.

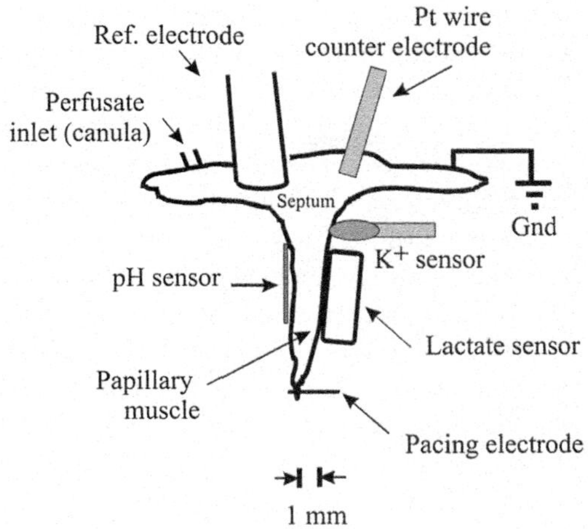

Fig. 2.3 Isolated rabbit papillary muscle, which was arterially perfused, with lactate, pH, K^+ sensors and with both the reference and counter electrodes positioned at the septum. Reprinted from [82], © 2002, with permission from Elsevier

2.2.5 *Na^+ Sensors*

Na^+ selective electrodes were traditionally made using variant glass electrodes. However, nowadays one uses PVC-based membranes loaded with ionophores like monensin, an antibiotic, or its derivatives. Urine can be tested for Na^+ in the case that a nonpolar pre-extraction is done, because the lipophilic PVC membrane can extract nonpolar species that interfere with the response towards the primary ion, which leads to a drift and loss of selectivity [91].

2.2.6 *In Vivo Ionic Sensing Using ISEs*

In the last few decades, extensive progress has been made through the development of portable blood-gas and electrolyte devices. However, these remain unsuitable for continuous in vivo measurements and thus unable to provide real-time diagnostic information for rapid therapeutic decisions. Current ambulatory pH recordings require the passing of the sensor transanally or transnasally using a catheter. These are uncomfortable and conspicuous, thus affecting the daily life of patients. The Medtronic Bravo pH monitoring system is a solution which is attached to the mucosal wall to measure its pH and transmits data wirelessly to an external receiving device [121, 122]. Its dimensions are 6 mm × 5.5 mm × 25 mm and has an antimony pH sensor and a RE. The entire system is contained within epoxy.

A number of flexible implantable ion-sensitive electrode arrays mainly based on polymeric membranes have also been fabricated for myocardial measurements. In [79], flexible arrays of eight ion-sensitive 250 μm diameter electrodes were presented. The electrodes were functionalized such that they are sensitive to pH (55 mV pH^{-1}), K^+ (58.1 mV/K^+) and Ca^{2+} (28.9 mV/Ca^{2+}). The sensors demonstrated a reduced lifetime when used in serum and whole blood measurements. According to the authors, interferences in living tissue are more complex due to the inflammatory response. Protein adsorption and other effects in tissues can reduce the sensitivity of the sensors. In vivo experiments were performed in rabbits.

In [80], in vivo experiments were performed in porcine models. pH sensors with 500 μm diameter electrodes demonstrated linear responses between pH 4 and 12, with a slope of 58.5 mV pH^{-1}. The reported lifetime of the electrodes was about 3 months while the reported drift was equal to 0.12 mV h^{-1}. The authors reported a myocardial pH decrease during induced ischemia from pH 7.4 to 7.25 and from 7.4 to 7.05 (at a second site closer to the ischemic region). In [81], the same authors reported results from a similar electrode array for pH and K^+ sensing (Fig. 2.4b, c). A slope of 58.9 mV pH^{-1} with a drift of 0.25 mV h^{-1} and 58.3 mV/K^+ with a drift of 0.08 mV h^{-1} were reported and the electrodes exhibited equivalent resistances of less than 10 MΩ. The platform was used to detect acute ischemia in several sites of the heart. During an 18 min experiment in porcine hearts, pH changed from about 7.3 to 6.25 and K^+ from 3 to 12 mM. Once again, electrodes closer to the ischemic region demonstrated a greater change in tissue pH. The same group reported on the use of metal–metal-oxide (MMO) anodic electrodeposited iridium oxide film (AIROF) pH electrodes which exhibited a super-Nernstian response of −63.5 mV pH^{-1}, linear between pH of 2–10, a lifetime of one month, a fast response and an accuracy of 0.02 pH units [77] (Fig. 2.4a). The measured potentials varied between about 300 mV for pH 2 to −100 mV for pH 10. During ischemia, a significant increase of CO_2 takes place. This can diffuse through the polymeric membrane to the inner hydrogel layer of polymeric pH sensors leading to pH measurement errors. This is potentially eliminated by MMO pH sensors. Generally, during ischemia, pH may decline from 7.2 to 6.3 and even as low as 5.7 [77, 82]. Simultaneously, extracellular K^+ may vary between 3.5 and 9 mM [77, 82]. In [8, 99, 125–128] the authors built upon previous work by incorporating in their system impedance measurements in addition to pH and K^+. Nevertheless, the ionic and impedance sensors were on two different substrates. The ionic sensors were realized with ISFET-based structures (Fig. 2.4e).

Rogers' group [16, 129] also uses IrO_x, for pH measurements. They reported super-Nernstian sensitivities of 68.9 mV pH^{-1} in [129] (Fig. 2.5) and 69.9 mV pH^{-1} in [16] (Fig. 2.4f) and a temperature dependence of −1.6 mV $°C^{-1}$, corresponding to a change of pH of 0.02 for a 1 °C change in temperature. Thus temperature changes within typical physiological temperature variations will not affect pH measurements significantly. Measurements on a rabbit heart gave a baseline pH of 7.4 and 7.32 in [129] and [16] respectively. During induced ischemia, this dropped to 6.22 in [129] and to 5.6 in [16] within 30 min. A significant advance in the field is presented in Fig. 2.5 [129], where an array employing a variety of sensing methodologies from electrical, chemical, and optical was developed for cardiac monitoring.

An endoscopic sensor array (Fig. 2.5d) for use with a laparoscopic teleoperated robot gastroendoscope was presented in [90, 124, 130] for the detection of ischemia in the gastrointestinal system. The electrodes used were commercially available

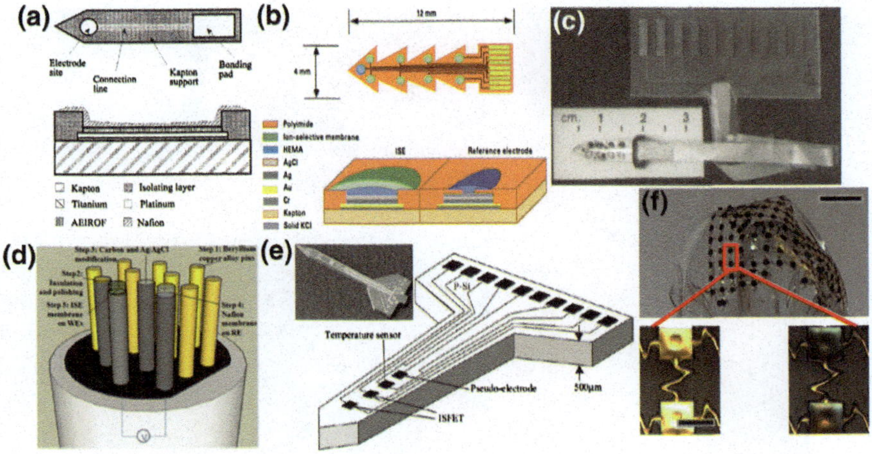

Fig. 2.4 **a** A flexible IrO_x pH sensor for measurements at the myocardium. Reprinted with permission from [77], © American Chemical Society 1998. **b** Top: illustration of a flexible platform for pH and K^+ measurements with a common reference electrode for in vivo measurements. Bottom: Sensor cross-section. Reprinted with permission from [123], © American Chemical Society 2001 and **c** its realization, with connecting wires. Reprinted with permission from [81], © American Chemical Society 1995. The sensors of (**a**–**c**) are some examples of the first generation of flexible sensors for in vivo measurements developed by R. Buck's group in the late 1990s for heart ischemia monitoring. **d** Ionic sensor array-enabled endoscopic probe developed for measurements in the upper GI tract. Reprinted from [124], © 2014, with permission from Elsevier. **e** A pH and K^+ ISFET and temperature sensor array probe for ischemia monitoring from the early 2000s. Reprinted from [125], © 2001, with permission from Elsevier. **f** A flexible and stretchable array of IrO_x pH sensors developed by J. Rogers' group for heart ischemia monitoring. Reprinted with permission from [16], © Wiley 2013

3 mm long, 600 μm diameter beryllium copper alloy pins. For the RE the pin was covered by a carbon ink and subsequently by Ag/AgCl and finally by Nafion. The authors studied the interference of K^+ and Na^+ and the adhesion of the sensing films to the substrate at low pH. In in vivo applications, membrane stability and leaching of membrane constituents are important issues which must be addressed. A response time of about 18 s and a maximum sensitivity of 60 mV pH^{-1} were reported in [130], while sensitivities of 13.07 mV pH^{-1} and 17.8 mV/K^+ were reported in [124] and 43.23 mV pH^{-1}, 52.62 mV pH^{-1} and 53.88 mV pH^{-1} in [90]. Polymeric sensing films were used in all of these ISEs using PVC, PEDOT, and PEDOT-PEG [90] membranes and the authors report that the eletropolymerized PEDOT membranes exhibit a significantly smaller impedance (about 100 times less, 1.3×10^6 Ω vs. 1.4×10^4 Ω). PEDOT is strongly attached on Au surfaces used as a substrate through the dative binding of the sulfur groups in the PEDOT molecule [90], making the membrane very stable. In another approach the authors used as a crosslinker poly (ethylene glycol) diglycidilether (PEG) mixed with PEDOT.

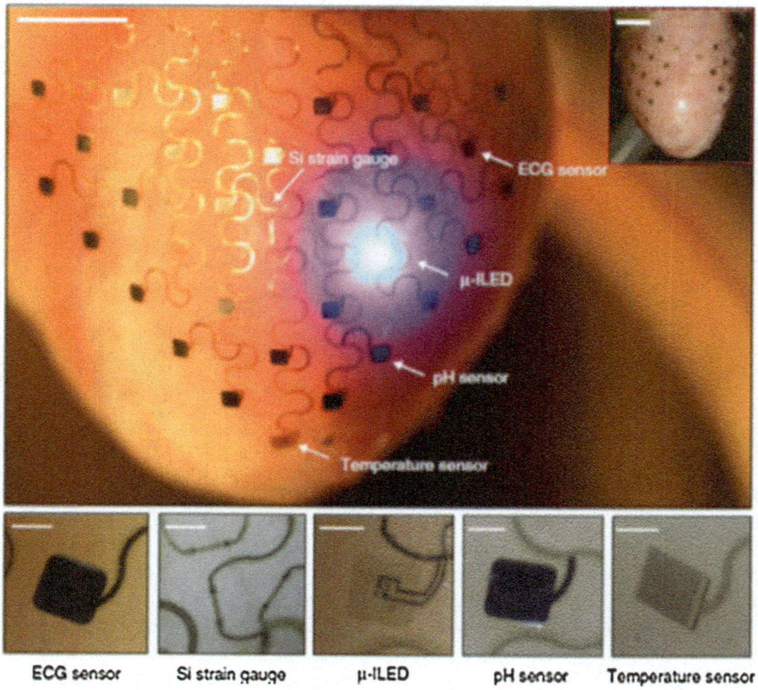

Fig. 2.5 Multisensor array for monitoring of the epicardium for ischemia and other applications from Rogers' group. Reprinted with permission from [129]. © Wiley 2015

2.2.7 ISFETs

An extension to potentiometric sensors, such as those discussed in previous sections, is the ISFET. This was first proposed in the 1970s by Piet Bergveld as a MOSFET without a gate metallization [131], as shown in Fig. 2.6. The exposed oxide is then placed in direct contact with the solution/sample under test. Charge accumulates at the exposed oxide insulator of the ISFET via the specific adsorption of hydrogen ions available in the solution [132]. This build-up of charge is a function of ionic concentration (pH or others depending on the sensing membrane) and leads to the formation of an electrical double layer capacitance at the oxide/solution interface due to charge separation and polarization. Because of this, a potential difference is established across the interface and thus at the transistor's gate oxide. The ISFET equivalent circuit will hence include a number of different capacitances and voltage sources.

CMOS transistors can be operated in two modes. In strong inversion the device is operated above its threshold voltage, and its operation is characterized by the drift of electrons across an electric field established in the transistor channel.

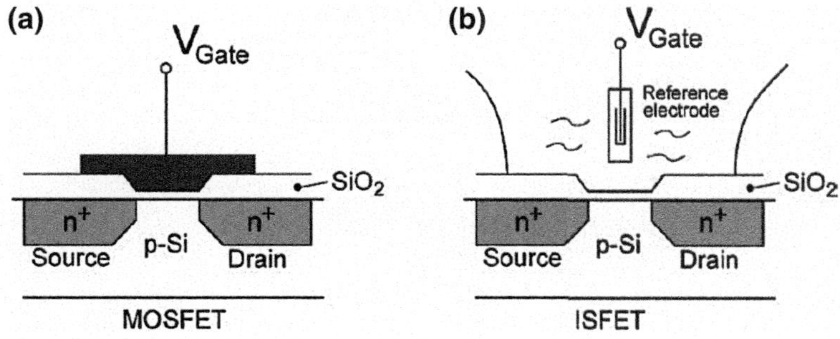

Fig. 2.6 **a** Simplified schematic of a MOSFET. **b** The ISFET structure. Reprinted from [131], © 2003, with permission from Elsevier

Its current-voltage characteristic when the device is operated in saturation is defined by the classic square law equation:

$$I_D = \frac{1}{2}\mu_{n,p} C_{ox} \frac{W}{L} (V_{GS} - V_T)^2 (1 + \lambda V_{DS}). \tag{2.10}$$

If it is operated in the strong inversion triode region, then it is characterized by

$$I_D = \mu_{n,p} C_{ox} \frac{W}{L} \left[(V_{GS} - V_T) V_{DS} - \frac{1}{2} V_{DS}^2 \right]. \tag{2.11}$$

In weak inversion, the operation of the device is characterized by the diffusion of electrons across an electric field established in the transistor channel due to the very low gate voltage (e.g. $V_{GS} < V_T$). The drain current in this regime is typically in the nano-ampere scale or lower and the transconductance to current ratio is maximized and a constant for a given current, leading to a maximum intrinsic voltage gain for the device. Due to the low currents, this regime is unsuitable for high-frequency application, but for potentiometric sensing, which involves the recording of d.c. voltages, it is ideal, leading to ultra-low-power instrumentation. The low terminal voltage required for operation in weak inversion and the gate capacitance being minimized in this regime leads to a minimum input noise density for a given drain current and a minimum input offset voltage due to low voltage mismatch [133]. However, the output noise current is maximized and so is the current mismatch, which is mainly due to V_T mismatch between identical devices. The current-voltage characteristic follows an exponential relationship similarly to bipolar transistors, thus allowing translinear and log domain processing circuit techniques to be employed. If the transistor source is tied to the bulk, the current-voltage characteristic of a MOS operated in weak inversion is approximated by [134]

$$I_D = I_o e^{\left(\frac{V_{GS}-V_T}{nU_t}\right)}\left(1 - e^{\left(\frac{-V_{DS}}{U_t}\right)}\right), \tag{2.12}$$

and if $V_{DS} > 4U_t$, the transistor is in saturation and the last term is ignored since $e^{-4} = 0.018$, leading to

$$I_D = I_0 e^{\left(\frac{V_{GS}-V_T}{nU_t}\right)}, \tag{2.13}$$

where I_0 is also known as the specific current (I_s or I_{spec}), defined as the current when $V_{GS} = V_T$. This is equal to

$$I_0 = I_{spec} = 2n\mu_{n,p}C_{ox}U_T^2\frac{W}{L} = I'_{spec}\frac{W}{L}. \tag{2.14}$$

Parameter I'_{spec} is the specific current of a square device and is technology dependent, and n is known as the slope or substrate factor or gate coupling coefficient, and is typically considered to be constant. It represents the loss of "*coupling efficiency between gate and channel caused by the substrate or body, which act as a back gate*" [134]. The slope factor in weak inversion corresponds to the capacitive division between the gate voltage and the silicon surface potential resulting from the gate oxide (C_{ox}), the depletion below the channel (C_d) and the interface state (C_{int}) capacitances, with the effect of the latter being negligible; thus, this is expressed as [134]

$$n = \frac{C_{ox} + C_d + C_{int}}{C_{ox}} \approx 1 + \frac{C_d}{C_{ox}}. \tag{2.15}$$

Its value can vary between 1 and 1.5 depending on the technology used. Parameter C_{ox} is the oxide capacitance per unit area, which is transistor (as a process may have transistors with different gate oxide thicknesses) and process-dependent; thus the gate capacitance is defined by

$$C'_{ox} = \frac{\varepsilon_{SiO_2}\varepsilon_0}{t_{ox}}WL = C_{ox}WL, \tag{2.16}$$

where ε_{SiO_2} is the SiO_2 permittivity, ε_0 is the permittivity of free space and t_{ox} is the thickness of the gate oxide SiO_2 layer; $\mu_{n,p}$ is the carrier mobility (electrons for NMOS and holes for PMOS, $\mu_n > \mu_p$), W and L are the width and length of the transistor, respectively, V_{GS} is the transistor gate-to-source voltage, V_{DS} the drain to source voltage and $V_{Tn,p}$ is the transistor's threshold voltage. The subscript n, p indicates that these parameters will be different for NMOS and PMOS devices. C_{ox} and $\mu_{n,p}$ are typically provided by the CMOS foundry for a specific process as is $V_{Tn,p}$ for a regular MOSFET. The product $\mu_{n,p}C_{ox}$ is known as the transistor gain factor and is typically given as $K_{n,p}$ or as $\beta_{n,p}$. All other variables are design parameters. If V_{DS} is kept constant by circuit design techniques, the only device input variable is V_{GS}. It has thus been debated whether the sensitivity of the device to ionic changes should be defined as an additional variable changing V_{GS} or V_T [131]. The measurement is performed against a RE, such that a two electrode potentiometric cell is formed. Typically, a non-polarizable RE is used, such as Ag/AgCl, in order to form a stable potential difference at its interface with the solution in contact. If the RE biasing the otherwise floating solution, is defined as a remote transistor gate connected to the gate oxide via the intermediate ionic solution, then the ionic sensitivity of the device can be described as a change in V_T, changing as a function of the concentration of a target ion [131]. The perception that the device threshold modulation leads to detectable changes in the I-V characteristic of the device has been the dominant method to describe the device operation. The RE voltage is capacitively coupled thought the electrolyte to the ion sensitive passivation surface. The ion-dependent charge separation of this interface modulates the threshold voltage of the device. A MOSFET's V_T (V_{T_MOS}) is defined by [131]

$$V_{T_{MOS}} = \frac{\Phi_M - \Phi_{Si}}{q} - \frac{Q_{Tot}}{C_{ox}} + 2\varphi_f. \tag{2.17}$$

The first term defines the difference between the gate metal (Φ_M) and silicon (Φ_{Si}) work functions, with q being the electron charge. The second term is due to the total semiconductor charge density (Q_{Tot}), which is equal to the sum of the accumulated charge in the oxide insulator (Q_{ox}), the trapped charge in the oxide-silicon interface (Q_{SS}) and the depletion charge in the silicon bulk (Q_B), i.e. $Q_{Tot} = Q_{ox} + Q_{SS} + Q_B$. The final term determines the onset of inversion depending on the silicon doping level [131, 135]. In the ISFET structure, however, the V_T expression is defined by [133]

$$\begin{aligned} V_{T_{ISFET}} &= V_{ref} - \Psi + \chi_{Sol} - \frac{\Phi_{Si}}{q} - \frac{Q_{Tot}}{C_{ox}} + 2\varphi_f \\ &= \left(V_{ref} - \Psi + \chi_{Sol} - \frac{\Phi_M}{q}\right) + V_{T_{MOS}} \\ &= V_{Chem} + V_{T_{MOS}}, \end{aligned} \tag{2.18}$$

where V_{ref} is the constant potential formed at the RE/solution interface, Ψ is the chemical potential across the electrolyte/passivation interface, which is a function of pH and χ_{Sol} is the surface dipole potential of the electrolyte in contact with the sensing passivation layer, a constant. Thus, Ψ is the only pH dependent term modeled by the combination of the site binding theory and the Gouy–Chapman–Stern double-layer model. According to the former, the insulator surface becomes charged once an equilibrium is reached between the insulator surface binding sites and the bulk solution hydrogen ions. The charge on both sides of the interface is matched and on the solution side it spreads across the double layer. According to the double layer model, there is a tightly packed compact layer close to the interface. This is known as the Helmholtz layer, the potential across which rises linearly.

This is followed by a diffuse outer layer of charge which follows a Boltzmann distribution. This is known as the Gouy–Chapman layer, the potential across which rises exponentially. This eventually is replaced by an even distribution of charge in the bulk of the electrolyte. The double layer can thus be modeled by a linear constant capacitance C_{Helm} and a nonlinear capacitance $C_{\text{Gouy-Chap}}$, as shown in the ISFET macro-model of Fig. 2.7. Following the above, it can be shown [131, 136] that Ψ is given by

$$\Psi = 2.303\alpha U_{\text{t}}\left(pH_{\text{pzc}} - pH\right), \tag{2.19}$$

where pH_{pzc} is the pH value with which the oxide surface is electrically neutral, U_{t} is the thermal voltage ($U_{\text{t}} = kT/q \approx 25.85$ mV at room temperature), while α is related to the intrinsic surface proton buffer capacity and the double layer capacity and can take a value between 0 and 1 [136]. Parameter α is a dimensionless sensitivity parameter describing the reduction of sensitivity by relating the ISFET sensitivity (S_{ISFET}) to the ideal Nernstian response ($S_{\text{Nernst}} = 2.303U_{\text{t}} \approx 59.2$ mV pH^{-1} at 25 °C).

$$0 \leq \alpha = \frac{S_{\text{ISFET}}}{S_{\text{Nernst}}} \leq 1. \tag{2.20}$$

For $\alpha = 1$, the device has the ideal Nernstian sensitivity, while for $\alpha = 0$, the device is insensitive to pH changes. Using the above, V_{Chem} can be simplified by introducing the term γ, which groups all the pH-independent terms [133, 136], leading to

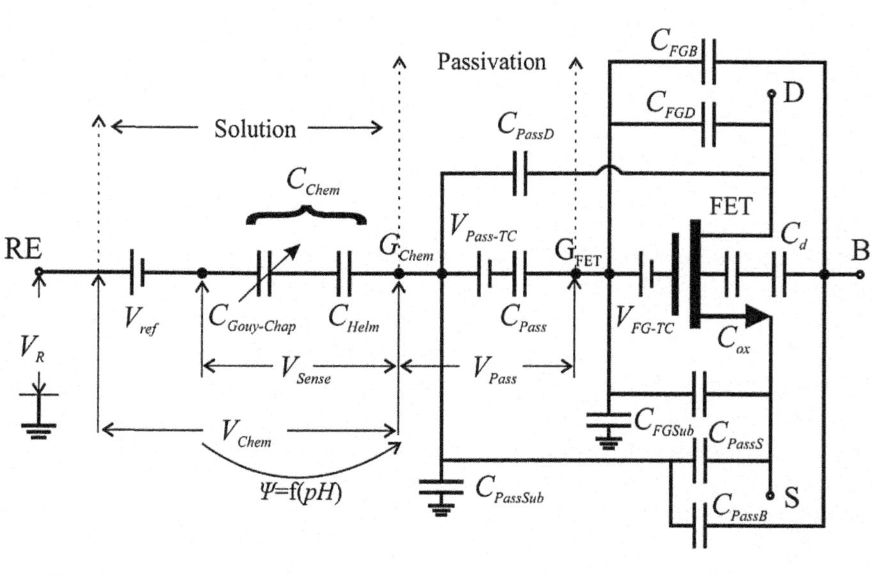

Fig. 2.7 ISFET macro-model based on [133, 135, 137, 138, 140]

$$V_{Chem} = \gamma + 2.303\alpha U_t pH, \quad (2.21)$$

According to which, if α is considered a constant, V_T and pH are linearly related. The temperature dependence of this should not be overlooked.

As mentioned previously, the sensitivity of the device to ionic changes can also be defined as an additional variable changing V_{GS} instead of a change in V_T [131, 135]. This can be performed by considering the ISFET as a floating-gate MOS (FG-MOS) device with a single input capacitively coupled through the passivation and defined by the passivation interface chemistry to the floating gate and considering V_T as a process-dependent constant; thus, in this scenario $V_T = V_{T_{MOS}}$. An ISFET macro-model is shown in Fig. 2.7. There is a potential applied to the RE (V_R). The RE/solution interface is characterized by a double layer model. Depending on the electrode material used (polarizable, e.g. Pt, or non-polarizable, e.g. Ag/AgCl), different elements of the electrode's equivalent circuit model may dominate. A polarizable electrode is characterized by charge storage, demonstrating a high resistance to charge transfer through the interface; thus the model is described by the double layer capacitance. In a non-polarizable electrode, charge transfer dominates, in which the interface is characterized by a small resistance. In either case, a potential drop across the interface will be developed, and in the case of a non-polarizable electrode, an (ideally) constant potential difference with the solution (V_{ref}) is formed across the interface, unaffected by changes in solution composition, temperature or other factors. At the interface between the solution and the ISFET sensing membrane, there is a second double layer capacitive interface, elements of which, however, are pH dependent, leading to a pH dependent potential formed at this interface (V_{Sense}, which we termed earlier as Ψ), i.e. across the interfacial capacitances C_{Helm} and $C_{Gouy\text{-}Chap}$. The series combination of these two capacitances (C_{Chem}) is estimated to be equal to 0.14 pF μm^{-2} [133] and 1 pF μm^{-2} [137]. Subsequently, there is the passivation capacitance (C_{Pass}) coupling the latter potential to the floating gate. There are trapped charges both in the passivation ($V_{Pass\text{-}TC}$) and in the floating gate structure ($V_{FG\text{-}TC}$) due to the fabrication process. Additional parasitic capacitances exist between the transistor's electrical gate and its other nodes (body, C_{FGB}, substrate, C_{FGSub}, drain, C_{FGD}, and source, C_{FGS}) due to the overlap of conductive areas between the gate and these nodes and the operating region of the device [138], as well as between the chemical gate passivation and these same nodes (C_{PassB}, $C_{PassSub}$, C_{PassD}, C_{PassS}), which are all process and device geometry dependent. Finally, there is also C_{ox} and the bias dependent depletion capacitance (C_d) of the device (which is a constant in the deep weak inversion region of operation) [133, 137]. Taking into account the above, it leads to the equivalent circuit of Fig. 2.7. All these capacitances collectively influence the sensor performance nonlinearly [137]. The passivation capacitance couples the chemical dependent interface potential to the floating gate and influences the chemical response of the sensor. This leads to a capacitive division of the chemical dependent voltage developed at the interface and explains the reduced transconductance of ISFETs when compared to a MOSFET with the same dimensions [133, 137]. This capacitance depends on the top metal layer geometry which defines the chemical sensing area of the device, i.e., the chemical gate. It can be modeled, to a first level approximation, as a series of capacitances due to the two dielectrics (e.g. SiO_2 and Si_3N_4), using (2.22), however, the fringing field at the edges will essentially modify the geometric factor of the equation and can be more accurately modeled using finite element analysis (FEM) [137]. This fringing field effect is modeled by the δ power factor (2.22) and it was experimentally found to be in agreement with FEM simulations which indicated that it is equal to 0.7. Approximate values for C_{Pass} given in [138] are 100 aF for 0.35 and 0.18 μm and 200 aF for 0.13 and 0.09 μm CMOS processes. The value of C_{Pass} can be in the region of 22.65 μF m^{-2} for a standard 0.35 μm process according to [139, 140], giving for a $W_S = L_S = 2$ μm a capacitance of 90.6 aF. Thus, since capacitance is defined by $C_x = \varepsilon_0 \varepsilon_x W_s L_S / t_x$, where ε_x the permittivity of the x dielectric layer, t is the thickness of the x dielectric layer, with x being either SiO_2 or Si_3N_4, and W_S and L_S are the width and length of the top metal layer defining the chemical gate [133, 137, 138]. C_{Pass} is given by

$$C_{\mathrm{Pass}} = \frac{C_{\mathrm{SiO_2}} C_{\mathrm{Si_3N_4}}}{C_{\mathrm{SiO_2}} + C_{\mathrm{Si_3N_4}}} = \frac{\varepsilon_{\mathrm{SiO_2}} \varepsilon_{\mathrm{Si_3N_4}}}{\varepsilon_{\mathrm{SiO_2}} t_{\mathrm{Si_3N_4}} + \varepsilon_{\mathrm{Si_3N_4}} t_{\mathrm{SiO_2}}} \varepsilon_0 (W_S L_S)^{\delta}. \tag{2.22}$$

C_{d} is given by [139]

$$C_{\mathrm{d}} = (n-1) C_{\mathrm{ox}}. \tag{2.23}$$

From the above discussion and the equivalent model, we can calculate the voltage at the interface as [139]

$$V_{\mathrm{Pass}} = V_{\mathrm{R}} - \underbrace{(V_{\mathrm{ref}} + V_{\mathrm{Sense}})}_{V_{\mathrm{Chem}}} = V_{\mathrm{R}} - V_{\mathrm{Chem}}. \tag{2.24}$$

As mentioned previously, there are trapped charges in the passivation and in the floating gate [137]. If these are modeled as a single voltage source ($V_{\mathrm{TC-total}}$) prior to the passivation capacitance, then from [133] we have

$$V_{\mathrm{Pass}} = V_{\mathrm{R}} - \underbrace{(V_{\mathrm{ref}} + V_{\mathrm{Sense}})}_{V_{\mathrm{Chem}}} - V_{\mathrm{TC-total}} = V_{\mathrm{R}} - V_{\mathrm{Chem}} - V_{\mathrm{TC-total}}. \tag{2.25}$$

From the simple equivalent model, the potential formed at the floating gate (V_{FG}) can be deduced from the following, if the total capacitance (C_{total}) seen between the nodes is defined as $C_{\mathrm{total-1}}$:

$$C_{\mathrm{total-1}} = C_{\mathrm{Pass}} + C_{\mathrm{ox,d}} = C_{\mathrm{Pass}} + \frac{C_{\mathrm{ox}} C_{\mathrm{d}}}{C_{\mathrm{ox}} + C_{\mathrm{d}}} \tag{2.26}$$

$$V_{FG} = V_{Pass}\frac{C_{Pass}}{C_{total-1}} = V_{Pass}\frac{C_{Pass}}{C_{Pass}+C_{ox,d}} = V_{Pass}\frac{1}{1+\frac{C_{ox,d}}{C_{Pass}}} = V_{Pass}\frac{1}{1+\zeta}. \quad (2.27)$$

For more discussions regarding the term ζ, refer to [139, 140]. The model can become increasingly complex if the intrinsic parasitics of the MOSFET are considered. In this case C_{total} is equal to $C_{total\text{-}2}$ below

$$C_{total-2} = C_{Pass} + C_{ox,d} + C_{FGD} + C_{FGS} + C_{FGB} + C_{FGSub}, \quad (2.28)$$

In which case V_{FG} becomes [135]

$$V_{FG} = \frac{V_{Pass}C_{Pass} + V_D C_{FGD} + V_S C_{FGS} + V_B C_{FGB} + V_{Sub} C_{FGSub}}{C_{total-2}}. \quad (2.29)$$

Typically, the last two terms in the numerator are ignored, as they may be grounded, while C_{FGSub} may be one and the same with C_{FGB} if the device is not on a separate well. An additional capacitive term, C_P, was introduced in [137] to account for the increase in floating gate capacitance due to the larger top metal layer defining the chemical gate of the device. The ratio of C_{Pass} to C_{total} (the total capacitance seen by the floating gate) scales down the effect of V_{Chem}, and thus the device's pH sensitivity [135]. Maximizing this ratio by increasing the chemical gate area formed by the top metal layer ($W_S L_S$) and making it much larger than the transistor's electrical gate area (WL) improves the pH sensitivity. In this case, the ISFET characteristics will approach those of the underlying MOSFET, since V_{FG} will become equal to V_{Pass} [133]. This is one of the reasons for the sub-Nernstian pH sensitivity of unmodified CMOS process-compatible ISFETs [135]. This capacitive division due to passivation thus leads to a reduction of the device transconductance when referred to the remote gate (i.e. the RE). Increasing C_{Pass} will increase it and will also decrease V_T [140]. In addition, the subthreshold slope will also be affected, shown to be inversely proportional to the dimensions of the chemical gate and thus C_{Pass} [133, 137]. In [137], this was found to be larger in ISFETs than in MOSFETs with the same electrical dimensions. Thus, a smaller $W_S L_S$ will lead to a larger linear voltage operating range, requiring larger voltages to generate the same drain current [133]. The factor n is modified due to the additional capacitances present and thus according to [133] it becomes

$$n' = 1 + \frac{C_d}{C_{eff}} = 1 + C_d\left(\frac{1}{C_{ox}} + \frac{1}{C_{Pass}} + \frac{1}{C_{Gouy-Chap}} + \frac{1}{C_{Helm}}\right). \quad (2.30)$$

The first two terms of C_{eff} are several orders of magnitude smaller than the other two associated with the chemically sensitive interface, and thus they dominate.

The noise of an ISFET is considered to be the same as that of a MOSFET in terms of the intrinsic device properties. Chemical noise, which is attributed to electrode degradation over time and surface chemical noise, will add up to this, as reported in [137, 138]. This additional noise contribution was defined as K/*f*, where *f* is the frequency and K was found experimentally to be equal to 1 nV^2 Hz. The ISFET noise was found thus to be an order of magnitude larger than that of a MOSFET. Increasing the chemical gate dimensions was suggested as a means of reducing the random offset associated with trapped charge in [137]. The interested reader may refer to [137, 138] for further analysis, discussion, and equations, while the effect of the capacitances between passivation and all transistor nodes to V_{FG} are also taken into account in [137]. When silicon nitride is expose in an aqueous solution, a thin hydrated surface layer is formed as hydrogen ions diffuse into the material. This changes the effective layer of the passivation and thus also the potential drop across it. Changes of this over time lead to drift in the recorded signal. In [137], it was found that drift had a negligible dimensional relationship to physical dimensions, ranging between 1 and 1.4 μV s^{-1}, three orders of magnitude larger than that of fabricated MOSFETs of the same dimensions, while it exhibited a relaxed exponential characteristic.

2.2.7.1 CMOS-Compatible ISFETs

In [141], ISFET structures where developed using an unmodified standard commercial CMOS process. The polysilicon gate was not removed and it formed a floating gate electrode together with the two metal layers of the process on top of the gate as in Fig. 2.8. Above the top metal layer, the insulating pH sensitive oxynitride layer was located [131, 141]. In this way, the top layer nitride passivation was connected to the gate oxide through the stack of the process's metal layers. The top metal layer essentially defines the chemical sensing area of the device. Thus two gates may be defined; one is the electrical gate, defined by the transistor polysilicon gate and the other is the chemical sensing gate defined by the top metal layer [135]. This separation allows the two to be dimensionally different [135]. The top passivation layer, which is typically silicon nitride on top of silicon oxynitride in commercially available CMOS processes, is used to protect the silicon chip from the environment [132]. It is deposited using low-temperature (300–400 °C) chemical vapor deposition (PECVD) [142]. Non-CMOS-compatible ISFETs typically use low-pressure chemical vapor deposition (LPCVD) to deposit the silicon nitride pH sensing film. This process allows a higher deposition rate and leads to dense films with superior purity and uniformity and few pinholes. However, it is incompatible with film deposition on metals as a result of the high temperatures required (700–800 °C). Si_3N_4 films deposited with PECVD, on the other hand, lead to films with lower oxynitride content, i.e. low density and consequently fewer hydrogen ion binding sites, and thus a sub-Nernstian pH sensitivity [133], while they also contain pinholes. Silicon nitride provides a linear response with pH with a sensitivity typically within 45–56 mV pH^{-1}, with the sensitivity as a function of the oxide's stoichiometry [142]. These led to a linear sub-Nernstian surface potential/pH slope (less than 59 mV pH^{-1}) [143]. This structure is essentially an ISE with FET detection via an electrically floating transistor polysilicon gate [141]. Another advantage of keeping the polysilicon gate is reduced light sensitivity [141]. A structure where the chemically sensitive layer is deposited on a metal, which is located at a distance from the MOS structure and is connected to the transistor gate through a metal line, is known as an extended gate ISFET [141]. However, if the length of this wire is reduced and is connected vertically to the transistor gate the compact device of Fig. 2.8 is obtained. The use of the passivation opening mask typically used in standard CMOS processes to define bond-pad openings in the top passivation layer was proposed in [144]. This was used as a means of minimizing the use of a full stack of metals and inter-metal insulators in a technology. In this way, the ISFET consisted of the gate oxide, the polysilicon gate, the first aluminum metal layer and the first inter-metal silicon dioxide dielectric, and thus the second and third metal and their intermediate inter-metal dielectric as well as the top layer insulation were removed without additional fabrication steps. This led to ISFETs with similar chemical sensitivity, improved effective transconductance (approximately three times larger), and reduced threshold voltage (approximately a third smaller). Classic CMOS compatible ISFETs exhibit a low transconductance due to the capacitive voltage divider formed by the stack of capacitances between the gate and the sensing membrane. In [144], this was reduced by reducing the sensing membrane by 5–11%. Nevertheless, the device noise was increased by two orders of magnitude, but still within acceptable levels for chemical sensing application according to [144], while the drift was dramatically increased, making the device inappropriate for absolute concentration measurements and limiting its usefulness for detecting the occurrence of an event (such as ssDNA binding).

Another issue with CMOS-compatible ISFETs is related to the low-temperature PECVD process used in CMOS for the deposition of the passivation sensing layers. While the quality of the deposited passivation may be adequate for typical CMOS device and circuit performance, it may potentially fall short in chemical sensing applications. This is due to the low density and the porosity of the resulting sensing film. Because of these characteristics, the passivation layer may over time "absorb

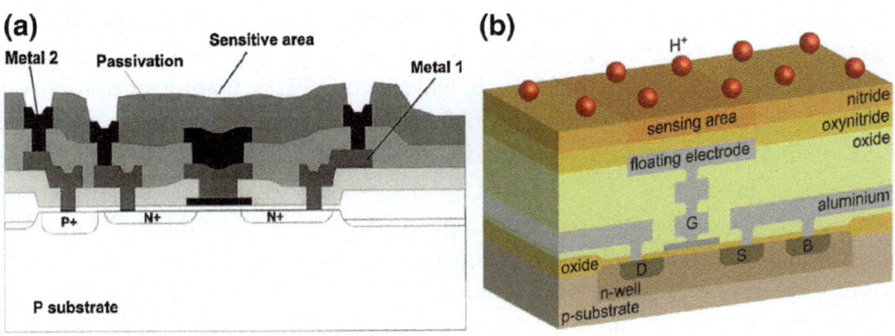

Fig. 2.8 **a** Illustration of the cross-section of the first reported linear n-type ISFET using an unmodified CMOS process. Reprinted from [141], © 1999, with permission from Elsevier and **b** a p-type example. Reprinted from [142]. © IEEE 2008

and become saturated with analytes or other substances in the solution, which may in turn cause an undesirable time-varying drift in the chemFETs" V_T [145, 146]. Consequently, this may lead to a loss of accuracy in the measurement and monitoring of the target analyte.

2.2.7.2 ReFETs and Differential Measurements

An ISFET insensitive to any ions is known as a reference FET, i.e. a ReFET, and such a device offers important advantages. Identical ISFET and ReFET pairs can allow differential measurements to be performed, leading to common mode errors, such as drift and temperature effects, to be eliminated while also relaxing the RE specifications. Thus pseudo-reference electrodes, i.e. polarizable electrodes, made out of noble materials such as Au or Pt, can be used and the non-constant interface potential formed at their interface with the solution/sample under test will also be a common mode signal that can be removed by the instrumentation. One such circuit is shown in Fig. 2.10a. A ReFET must be as similar as possible to the ISFET it is being used with; it should thus have the same electrical and chemical gate dimensions and it should be laid out similarly, such that common mode interferences influence the devices similarly. The challenge is in making it insensitive to ions. Many methods have been proposed. One option is to apply a membrane on top of the chemical gate such that the device's pH sensitivity is reduced ideally to zero. Hydrophobic membranes with very low buffer capacity, such that $\alpha = 0$, have been proposed [131]. Examples of such membranes are parylene and Teflon. Another approach is to add a diffusion barrier, such that the ISFET illustrates a delayed response [131]. According to [132], a polymer membrane must be used which prevents hydrogen ions from reaching the pH-sensitive membrane. However, this should not block any other ions to "preserve the electrical characteristics of the underlying ISFET" [132], such that there is no potential drop across this additional membrane. In [132] a PVC-based membrane was used which has a very low sensitivity to H^+, Na^+, and K^+. The ISFET and the device to be used as a ReFET were matched through UV exposure; however, a V_T mismatch between the two was established once the PVC membrane was deposited. Another approach used to perform differential measurements is to have two ISFETs with different sensitivities. In [147] for example, the first ISFET had a Ta_2O_5/SiO_2 gate with a sensitivity of 58–59 mV pH^{-1} and the second one used a $SiO_xN_y/Si_3N_4/SiO_2$ gate with a sensitivity of 18–20 mV pH^{-1}. Each of the two devices was used as one of the input transistors in the differential pair of an operational amplifier configured as a source follower and their outputs were subsequently subtracted by a differential amplifier resulting in an overall sensitivity of 40–43 mV pH^{-1}. Another approach is to use the same ISFETs, with the same sensitivity and sensing films, but to use circuit techniques to decrease the sensitivity of one of the two devices. This was implemented in [148], where the authors used different gains for two ISFET amplifiers by controlling the width of the ISFET pMOS load. The two topologies resulted in sensitivities of 50 and 30 mV pH^{-1}. Similarly to [147], a differential amplifier subtracted the signals from the two ISFET amplifiers and amplified the resulting signal leading to an overall sensitivity of 40–45 mV pH^{-1}. These techniques are not ideal; they may relax the specifications of the RE, but since the devices are not identical and their sensitivities different [147] or because their response is scaled [148], the effect of common mode interferences will not be completely eliminated.

2.2.7.3 Sensing Oxides

The use of different insulators and other layers allows sensitivities to different target analytes, similarly to standard potentiometric sensors. The first ISFET gate material used was silicon dioxide [131]. This was followed by double layer insulator structures which resulted in better pH sensitivities and stability. The sandwich of silicon dioxide with silicon nitride (Si_3N_4) as the top insulator is an example [131]. Other top layer insulators used include tantalum oxide (Ta_2O_5) and aluminum oxide (Al_2O_3) [132]; these are examples of oxide insulators used for pH sensitive ISFETs. This standard ISFET, however, is not CMOS process compatible. Early examples of ISFET systems with recoding electronics on the same substrate required additional fabrication steps; e.g. in [147] four extra masks were required for the ISFET fabrication. In CMOS processes a polysilicon electrode is required in the gate region to determine the self-aligned source and drain regions of the transistors [141]. Since this is not in agreement with the original ISFET device structure, e.g. in [147], the polysilicon gate and the underlying gate oxide were removed by wet etching and a thinner gate oxide was regrown in the gate region of the device followed by the deposition of different gate dielectrics to form ISFETs with different pH sensitivities.

2.2.7.4 Trapped Charge

One of the main advantages of the ISFET is high miniaturization not only of the sensor (allowing measurements of nano and picoliter volumes [139]), but also of the complete system, since sensor and recording (typically analog) front-end electronics and back-end digital circuits can coexist on the same silicon chip. Exploiting CMOS technology economies of scale with ISFETs in unmodified CMOS processes can lead to mass-produced and low-cost miniature systems for point-of-care and implantable systems. ISFETs are, however, associated with a number of disadvantages. One parameter which is typically ignored according to [131] is the potential electrostatic damage of the gate oxide due to high external electric fields [131]. Electrostatic discharge (ESD) may cause charge to be trapped in the floating gate. This will lead to large shifts in the ISFET's V_T (residual charges may remain on different structures comprising the ISFET, e.g. oxides) and may even permanently damage (physically and chemically) the device. Steps in the CMOS fabrication process such as plasma metal etching, wafer cleaning, dicing, packaging/wire bonding, and handling may negatively affect the fabricated devices [146]. The etching process, for example, used to define the metals of the floating gate may induce the accumulation of charge trapped in the metal layers that constitute the floating gate. This is known as the antenna effect [145, 146]. If there are no paths for these to be removed the charge may be injected to the oxide. Thus, charge can be trapped in the passivation and/or on the floating gate. Charge in the gate, being floating, has no path to ground and remains there similarly to EEPROM [132]. Trapped charge at the typically dual passivation layer stack has mechanisms similar to the one used in metal-nitride-oxide-silicon (MNOS) transistors used for non-volatile memory [133]. Fixed charge in the passivation is due to the intrinsic dangling bonds within the silicon nitride sensing films and buried sites or surface defects due to extrinsic dangling bonds [137]. The trapped charge, and thus the V_T shift, is indeterminable, leading to random offsets and a large V_T, which may lead to an inability of the instrumentation to record the sensor output signal within its recording range. For example the p-type ISFET of [132] was found to have a V_T of −5 V, when the proposed system was designed for 3 V supply operation, while in [142] the V_T of an array of ISFETs was found to range from −4 to −1 V. In addition, geometrically identical ISFETs comprising arrays on the same die will demonstrate different V_T. Thus, ISFETs comprising large sensor arrays will also behave differently among them. An ESD event may occur also due to the solution under test floating, and thus it is recommended to ground the solution via the RE prior to ISFET contact, or to insert the RE and ISFET within the solution simultaneously [131].

Fabrication-based solutions to trapped charge

A thinner gate oxide can allow an adequate amount of charge accumulated on the floating gate to move to the substrate [145]. Such a device can be used either as an ISFET or as an additional device connected to the gate of an ISFET. In order to pattern metal structures using plasma etching, a photoresist is deposited on the metal. An additional dielectric, such as an oxide between metal and photoresist, can be used in order to reduce the charge trapped during the plasma etching process. This dielectric may be left behind and form part of the sensing membrane of the final device. Alternatively, the etching rate may be modified by using e.g. lower power densities and thus potentially lead to reduction in the trapped charge. Instead of plasma etching, wet chemistry or ion-beam milling may be used [145, 146], leading to no charge being trapped during the metal patterning step since these techniques do not rely on an electric field. However, these require alterations to the CMOS process.

Modifications in the wafer post-fabrication handling and packaging steps may also be employed in order to reduce trapped charge [145]. Examples include using anti-static dicing tape to hold wafers in place and replacing conventional deionized water for cooling the dicing saws, which is of high resistivity, with lower resistivity ones [146].

Post-processing solutions to trapped charge

Post-processing techniques include gas annealing using a heated hydrogen and nitrogen gas mixture following wafer dicing into individual chips. This neutralizes trapped charges. Another approach is using UV radiation. For this to be effective, a hole or window, i.e. an aperture in the top metal layer within the vicinity of the floating gate is required, such that the radiation reaches the gate oxide and other layers [142]. An EPROM eraser was used, and following three hours of UV exposure the V_T was increased from −5 to −1 V in [132]. In [142] the UV wavelength used was 253.7 nm with an exposure level of 4 mW cm^{-2}. Appropriate biasing of the device during UV illumination facilitates the reduction of the trapped charge. Applying a positive voltage to the bulk with respect to the substrate (V_{BS} = 5 V), and thus with the floating gate, during UV exposure, allowed further increase of V_T in [132] to about 3 V. This voltage changes the conduction and valence band level in the bulk of the device. UV light excites the charges trapped at the gate such that they overcome the oxide energy barrier and discharge the gate to reach equilibrium with those in the bulk, permanently changing the ISFET's V_T [132]. The effect of the device biasing during UV illumination to V_T modification demonstrated that it is in fact the removal of trapped charge in the floating gate rather than the fixed charge in the passivation layer that mainly affects the V_T. UV exposure of 17 h led to a minimum 15 mV V_T mismatch between identical ISFETs in [132]. In [142] during UV exposure a constant 20 μA current was supplied to the source of every ISFET and the drain and bulk voltage was set to ground. After 10 h the V_T of all the ISFETs converge to about −1 V with an offset less than 10 mV.

Design-based solutions to trapped charge

A number of solutions have been proposed, which focus on the use of on-chip ESD protection techniques. Such solutions may involve the use of protection diodes (Zener or avalanche diode structures) connected to electrode (metal) structures partially or entirely encompassing (rings or other geometries) the floating gate. The diodes can allow conduction above a certain predetermined breakdown voltage [131]. The metal structure should be in close proximity to the floating gate; this may be connected to ground, to a power supply directly, or through a passive device (resistor or capacitor) or diodes and a rail clamp circuit, which may be internal or external to the chip. Thus, a path of least resistance to a large ESD current is formed which will be preferred against the path towards the floating gate [149]. Alternatively a diode may be connected directly between the floating gate and the substrate [146]. Another solution to ESD was presented in the early paper of [150]. An ISFET structure was presented, which is similar to those proposed in unmodified CMOS processes, as there is a polysilicon gate electrode, which is in turn in contact with a silicon nitride pH-sensitive top passivation layer. The polysilicon gate is connected to a MOSFET switch which allows the connection of the ISFET gate to a potential, in which case the device can be operated as a regular MOSFET. However, the off resistance of the switch is finite and there will be a leakage current. This leakage current needs to be minimized. If the switch is located in the immediate vicinity of the gate and connected to the diffusion of the switch's source, then a pn-junction diode exists between gate and substrate. This diode provides additional ESD protection as its breakdown voltage (35–45 V) can be substantially smaller than the gate oxide breakdown voltage (100 V) [150]. According to [141], if an ISFET gate is connected to a diode, switch [150, 151] or other structures, then it is not completely floating. The solution proposed in [149] allows the gate to remain completely floating.

Floating gate circuit design solutions to trapped charge

Nevertheless, the ISFET V_T can be programmed without the need for device post-processing techniques. The use of well-known techniques developed for the programmability of FG-MOS devices and systems can be used. FG-MOS transistors can have multiple capacitively coupled input nodes resulting in a floating gate potential determined by their weighted sum. The effective weight of each input is proportional to its coupling capacitance. The operating point of the device can hence be controlled by applying a voltage to one such control gate. In [152] a second capacitively coupled input was used for the floating gate of the ISFET, the capacitance of which was made large such that it may be programmed by a voltage within the rails of the proposed system. Thus, in an array of ISFETs, different programming voltages may be used in order to match their V_{TS}, while known drift signals may be cancelled out. The proposed device allowed a 2.16 V decrease in V_T for an increase of 500 mV in the programming voltage [152]. This, however, can lead to a reduction in the strength of the chemical signal, thus reducing the chemical sensitivity of the device [153]. Hot-electron injection [154] and electron tunneling [153] are techniques that have been proposed as alternatives enabling V_T programmability while the device is in a pH-buffered saline solution. Hot-electron injection can take place in a p-type ISFET when a large V_{SD} across the device channel is applied, such that a large lateral electric field is established. If such a current injection is induced, charge storage takes place on the gate, counteracting the trapped charge. In [154], the solution RE was grounded and a V_{SD} ranging from 500 mV to 8 V was applied for 1 min. In this way the V_T was increased and shifted by up to 2.8 V. This method cannot decrease the V_T by removing electrons; electron tunneling, however, allows this. This was demonstrated in [153, 155] where the use of additional diodes and voltages capacitively coupled to the gate allowed electrons to be removed from the floating ISFET gate. These capacitively coupled inputs are similar to the one used in FG-MOS structures and in [152]. During normal device operation they are grounded and all other device inputs are used as usual. During threshold programmability the RE, drain and source voltages are grounded and the programming input nodes are set to the desired voltage for 80 s in [155] and 5 min in [153]. Tunneling voltages between 6.9 and 8.5 V allowed a V_T shift of −3.45 V in [155].

ISFET-based biomedical devices

A popular recent application of ISFETs is in large arrays for DNA detection. In [156], a system with 1.2 million sensing wells was demonstrated which allowed the sequencing of three bacterial and one human genomes. In [157] the authors presented a system that not only performed DNA detection but also real-time DNA amplification using PCR and isothermal amplification. DNA amplification was monitored and performed using on-chip temperature sensors and resistive heating. Differential measurements were performed by using a regular ISFET identical with the rest but on a separate well, such that the ISFET can record baseline signals and reference measurements can be performed to eliminate common mode interferences. The device was used to genotype and discriminate unique single-nucleotide polymorphism (SNP) from human saliva. A recent system for DNA detection which combined a CMOS image sensor with the ISFET array was presented in [158]. By deposition of ion-selective membranes on the insulator sensing membrane, structures sensitive to ions such as Na^+, K^+ and Ca^{2+} can be obtained. These devices are known as ChemFETs. More relevant to implantable applications is the work presented in [125, 159–161]. Flexible organic ISFETs are presented in [159] for the detection of pH and K^+ changes induced by myocardial ischemia in the extracellular matrix of the myocardium. Once again, valinomycin was used for the K^+ sensor. A wireless diagnostic capsule for ISFET-based pH monitoring in the GI tract was presented in [161] for Gastroesophageal Reflux Disease (GERD) and inflammatory bowel diseases such as Crohn's and ulcerative colitis. The system had two ISFETs and a ReFET and could perform single-ended and differential measurements with the use of multiplexers. The system also incorporated an 8-bit ADC and a microcontroller. The ISFET demonstrated a 48 mV pH^{-1} sensitivity. Another pH capsule was presented in [160], where the ISFET demonstrated a 44.95 mV pH^{-1} sensitivity. Surrounding an ISFET by a thin film of an electrolyte solution and by enclosing this in a gas-permeable membrane allows the development of gas-sensitive structures known as GasFETs. The immobilization of enzymes or enzyme-containing layers directly on the inorganic gate insulator of an ISFET leads to a device known as the EnFET, originally proposed in 1976 [162]. EnFETs for the detection of penicillin, glucose, sucrose, urea, and creatinine, among others, have been developed over the years [162]. If the enzymatic reaction leads to the production of hydrogen or other ions, these can then be subsequently detected by the underlying ISFET of ChemFET. Major problems associated with EnFETs are related to enzyme immobilization (covalent attachment and polymer entrapment) and associated nonlinearities, with stability and reproducibility of the results being major issues [162]. Recently an EnFET for ATP detection was presented in [163]. Similarly to the EnFET, ISFETs can be used for label-free DNA hybridization detection and immunological sensing, e.g. through antibodies [162].

2.2.8 Reference Electrodes

It is evident from previous discussions that in electroanalytical measurements the RE plays a very important role. Such an electrode should be insensitive to changes in the ionic concentration, be reversible (thus a low charge transfer resistance is required), and demonstrate a stable constant reference potential and a reproducible response [27, 164, 165]. A well-defined RE is necessary in all potentiometric measurements. As discussed in [27], there is always a potential at any interface; however, if this potential is not dominated by a reaction with high exchange current density it cannot be stable. The traditional glass-based Ag/AgCl RE addresses the above. However, careful control of its working conditions is required. This is because it requires a liquid reservoir with a positive hydrostatic pressure of the inner electrolyte. The inner solution is moving in one direction through a liquid junction so as to keep a high and constant concentration of the electrolyte at the solution interface.

Thus, one of the main limitations of potentiometric sensors lies in the fundamental need of this family of electrochemical sensors for such an RE. This is particularly problematic in miniaturized biocompatible applications. In such applications, the use of conventional glass-based REs is not possible due to their large size, fragility and use of the liquid filling solution. Such conventional REs cannot be miniaturized and are unpractical. The RE has to have similar dimensions to the ion sensors used in order to be practical.

Emerging technologies, such as wearable and embedded electrochemical sensors, point-of-care, and telehealth have driven the development of suitable REs. Several approaches are suggested: using solid polymers [83, 166], solid KCl [167], micro-fabricated liquid junctions, agarose gels with KCl and heterogeneous membranes separating a KCl reservoir. Solid-state technologies [168, 169] have played an important role in RE technology. Several solutions have been proposed, such as solid-state junctions based on carbon-fiber nanocomposite membranes incorporating poly(methyl methacrylate) and carbon graphene stacked nanofibers [170], membranes prepared by aromatic polyurethane [171], metal alloys and composite materials [172]. A solid-state RE made with vinyl ester resin doped with KCl has been reported in [173]. Combination of PVC membrane and ionic liquid is reported in [174, 175], while other groups [176] used a screen-printed insulation paste. In all these approaches there is always a range in which the performance of the sensor is interfered with by lipophilic species or even Ag ions in solution. Different types of polymers, such as polyacrilate microspheres and polypyrrole microcapsules, were used to improve the stability of the potential [177, 178]. Single-wall carbon nanotubes (CNTs) are used [179, 180] as potentiometric transducers so that the Ag/AgCl/Cl^- system is encapsulated in a polyacrylate membrane [181, 182]. Recently, a new reference system was developed in [183], where polyvinyl acetate with KCl was used.

The development of an RE based on polyvinyl butyral (PVB) membranes is reported in [184]. The membrane was made by dispersing Ag, AgCl, and NaCl in a

PVB solution with methanol. The observed stability of the signal is related to the formation of nanopores on the surface. PVB has high polarity, making it highly resistant to fouling, which can be caused by non-polar substances [185]. Polar substances and ionic salts can be incorporated in PVB, which can increase the ionic conductivity of the polymer [186]. In another example, the use of REs prepared using a poly(vinyl chloride) membrane on top of conductive poly(3, 4-ethylenedioxythiophene) was reported in [177]. In [177] the authors used poly (n-butyl acrylate). Polyacrilates are promising materials showing a lower detection limit and lower ionic activity compared with PVC, and they can be produced by photopolymerization.

In [187], four chemical designs are compared as REs. In one of the approaches the membranes were non-responsive to ions, but the system wasn't evaluated. A different concept involves the doping of the membrane with an inert electrolyte, such as ions with medium lipophilicity, which can partioning in the sample and become the potential determining ion. In this case the membrane responds only to the membrane ions and does not respond to ions in the sample. Kakiuchi et al. [188] used Ag/AgCl electrodes and IL as a salt bridge. The authors reported that the saturation of the ionic liquid with AgCl is crucial for proper functioning of the RE. In [175] the authors reported a liquid-junction-free polymer membrane, where an intermediate layer is used together with a PVC-IL membrane on top as reference. Another type of reference electrode was proposed using ionic liquid [175, 188] where gelled and non-gelled ionic liquid is used so as to form a liquid junction reference electrode.

Solid-contact RE based on a PEDOT layer electrochemically deposited on a carbon layer and an ionogel membrane consisting of tris(penta–fluoroethyl) trifluorophosphate $[HMIM]^{+}[FAP]^{-}$ in poly(butyl-co-decyl–methacrylate) is discussed in [189]. Other reported ionic liquids are tributyl (2-methoxyethyl) phosphonium bis(penthafluoroethanesulfonyl)amide [190] and bis(tritrifluoromethane sulfonyl) amides with 1-alkyl-3-methyl-imidazolium, phosphonium or ammonium cations [175]. Another type of RE was proposed using poly(n-butyl acrylate)-based membrane on top of glassy carbon electrode with polypyrrole as a transducing layer [177]. REs based on a polymer, an inorganic salt, and an Ag/AgCl element showed excellent stability, long lifetime, and insensitivity towards the ions in the sample. The basic principles of REs are discussed in [191] and shown in Fig. 2.9.

In [78], an Ag/AgCl RE was coated with a polymer containing chloride ions, which was encapsulated by a Nafion outer layer. Nafion blocks the chloride anion diffusion to the test solution maintaining a constant chloride concentration on the AgCl electrode surface. An Ag/AgCl RE with a Nafion coating to increase its stability was reported in [124, 130].

Despite all the aforementioned effort in the development of a miniature RE, many issues related to the long-term behavior of these have not been solved yet. The electrochemical kinetics need to be investigated further to clarify the processes related with the mobility and exchange between the solids and dissolved ions. Figure 2.9 summarizes and illustrates the various methodologies followed for RE development.

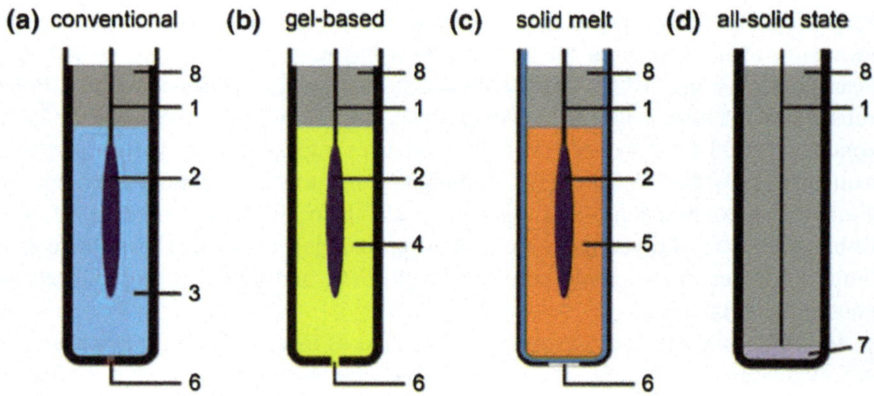

Fig. 2.9 Principle of RE (Ag/Ag/Cl reference electrode), 1: metal lead (Ag), 2: metal salt (AgCl), 3: aqueous solution of KCl, 4: hydrogel-containing KCl solution, 5: solid melt of KCl, 6: junction containing diaphragm, porous ceramic or opening, 7: solid state reference element, 8: insulating encapsulation. Reprinted from [191] with permission of Springer. © 2008

2.2.9 *Read-Out Circuits and Instrumentation*

In the preceding sections, the technology regarding the potentiometric sensing element was discussed. It is obvious that these are not standalone devices, but they rather require additional circuitry in order to record their output signal. The following is a very brief discussion of the challenges in recording from potentiometric sensors and some examples of recording electronics used in the literature using commercially available discreet electronic components.

A practical potentiometric measurement of the potential difference across a galvanic cell requires an OCP measurement. Typical glass electrode pH probes have a very high output impedance ($R_0 > 1$ GΩ), which is in series with the pH voltage source. This is an important parameter, as the input impedance of the circuits interfacing with ion-selective sensors must have an input impedance which is at least three orders of magnitude higher than the electrode resistance [20], sinking minimal current (I_{in}) to prevent loading. This is because potentiometric cells are galvanostatic cells and no current should flow through the ISE or RE used in the cell. If not, an unwanted potential drop will be created leading to a measurement error: e.g. for a glass electrode with $R_0 = 1$ GΩ and $I_{in} = 200$ fA, a voltage offset error of 0.2 mV is generated which is translated into an error of 0.0037 pH [23]. Thus, recording from such electrodes can be achieved with the use of high input impedance ($>10^{14}$ Ω) and hence low input bias currents (<100 fA) amplifiers. Input currents greater than a pA will disturb the electrochemical potential established. An instrumentation amplifier typically satisfies this condition. Polymer membrane ISEs have a membrane resistance of the order of 1–10 MΩ. Thus, in such a case, the input impedance of the circuit used to interface with the electrode must be larger than 10^{10} Ω. The signal of interest is at DC. A low-pass filter (LPF) at the output of the recording amplifier reduces the harmonic content of the signal. This is especially important with human body measurements, as physiological signals in this case will act as interference and thus need to be cancelled out, as do motion artifacts.

In [100], the voltage of the pH electrode was first buffered by a high input impedance, low input bias current operation amplifier (op-amp), in order to transform the high output impedance signal to a low impedance signal. Subsequently, the signal was amplified differentially with respect to the RE potential by a gain of 14.7 with an instrumentation amplifier. The IA was used to reduce common-mode interference. An active Butterworth LPF with a cutoff frequency of 1 Hz was used before digitization.

The polymer-based pH sensor demonstrated in [20] was sensitive between pH 2 and pH 9.5. This leads to a change of measured potential of from 400 to −20 mV and from 770 to 320 mV for the two different types of electrodes used. Hence, the signal conditioning circuitry needs a dynamic range of 450 mV. The circuit must be able to adjust the baseline and thus offset the signal such that the lowest potential difference is about 0 V and also to amplify the signal by 10 V/V for a 5 V supply. Because negative potential differences could be generated, a split ±5 V supply was used. In order to offset the signal so that it is within the supply of the subsequent stages (microcontroller), which had a 0–5 V supply, the offset pin of the IA was used to apply a d.c. voltage offset generated through a trimmer.

In [21], the pH electrode is connected to a relaxation oscillator to convert the electrochemical potential into a frequency varying signal. This leads to a varying modulating frequency from 22 to 15.9 kHz for pH levels of 2–12. This system was inductively powered. As discussed in [27], drift can be minimized by decreasing the distance between the sensor and the amplifier, as the parasitics will be reduced. Measurements using a porcine heart were collected in [77] using an amplification of 10 V/V and an LPF with a cutoff of 1 Hz. For a resolution of 0.1 pH units a voltage resolution of 5.9 mV is required for the typical Nernstian response of 59 mV pH^{-1}.

In practice, the readout circuits commonly used for ISFETs are different. An ISFET is an impedance transformer, so the high impedance to low impedance node transformation is by default dealt with. As discussed in previous sections, one approach to examining the pH sensitivity of an ISFET is to consider the pH needed to alter the V_T of the device. In order to properly assess this it is evident from the transistor equations that the device needs to be biased at a constant I_D and V_{DS}, such that as V_T varies, the RE to source potential (V_{RS}) changes by an equal amount to compensate, since I_D is constant. V_R is constant, thus V_S will change and consequently also V_D will change, tracking V_S to maintain a constant V_{DS}. This is achieved by the circuit of Fig. 2.10a [132]. The current source on the top forces a constant current through the ISFET and the bottom one forces the same current to flow through resistor R_{DS}. The constant voltage across this resistor is forced across the ISFET via the two operation amplifier buffers. This circuit has been used with both with an ISFET and a ReFET and the difference between the two was obtained using an instrumentation amplifier (IA). In [132] the ISFET was operated in the saturation region, while in [192] it operated in the linear region. A variation of this uses the instrumentation amplifier concept as in [131]. Both these circuits can be implemented using discrete component or on-chip solutions. Another classic ISFET circuit is the source-drain follower of [141] (Fig. 2.10b). More advanced circuit techniques implemented in CMOS are out of the scope of this chapter and will not be discussed. The interested reader can refer to the referenced ISFET literature for recent circuit implementations.

2.3 Amperometric Sensors

The development of amperometric sensors focuses on different electrode materials, enzyme immobilization techniques, mediators and coenzymes, sensor stability and lifetime, miniaturization, and implementation. The operation of amperometric sensors is based on applying a potential between two electrodes and measuring the resulting current [193], typically via a third electrode. The working electrode (typically denoted as W or WE), is the electrode that is being characterized or that on which specific processes are characterized. The purpose of the second electrode is to balance the charge added or removed by redox processes taking place at the WE and also to maintain a constant interfacial potential. However, it is challenging to perform both these tasks; thus these are separated between two electrodes, with the task of maintaining a contact interfacial potential being carried out by the reference electrode (R or RE) and separated by charge transfer kinetics, which are dealt with by the third electrode, known as the counter electrode (C or CE) [194, 195]. The latter has typically a much larger area compared to the WE, such that the effect of its interfacial impedance is minimal. No current should pass through the RE if it is to act as a true reference electrode. The most popular amperometric

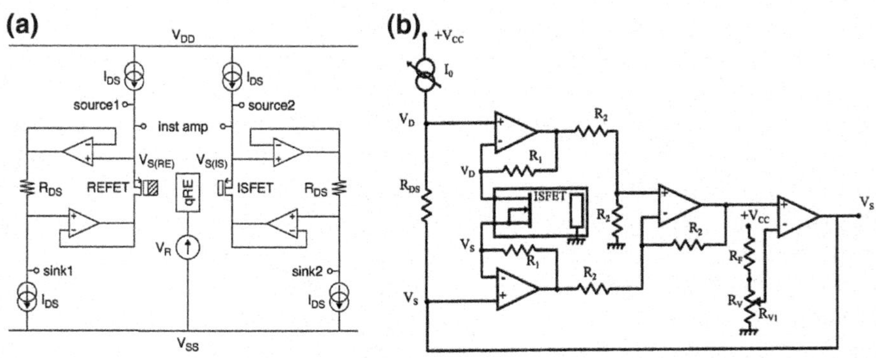

Fig. 2.10 **a** The differential ISFET-REFET topology. Reprinted with permission from [132]. © IEEE 2004 and **b** the ISFET source-drain follower circuit. Reprinted from [141], © 1999, with permission from Elsevier

sensors are sensors for the detection of glucose. Such sensors are also the most popular and successful electrochemical sensors in general. As such, the following discussion of amperometric sensors will begin with glucose before expanding to other popular target analytes such as oxygen and lactate.

2.3.1 Glucose

Often glucose concentration is determined using the immobilization of glucose oxidase in a membrane. The current between the electrodes is measured as a function of the oxygen reduction concentration as it diffuses through the bioactive membrane to the cathode as in (2.31) and (2.32).

$$\text{Pt anode}: H_2O_2 \rightarrow O_2 + 2H^+ + 2e^- \tag{2.31}$$

$$\text{Ag cathode}: 2AgCl + 2e^- \rightarrow 2Ag^0 + 2Cl^- \tag{2.32}$$

One approach is to measure the production of H_2O_2 by applying a potential of +0.65 V versus a *Pt* electrode. A major drawback is the dependence of the biosensor from the dissolved oxygen concentration. A possible method to overcome this issue is to use mediators playing the role of transfering electrons directly to the electrode. These mediators are important to react quickly with the reduced form of the enzyme and to be sufficiently soluble in both the oxidized and reduced form, such that they diffuse rapidly between the electrode surface and the active site of the enzyme. One very important aspect of using mediators is associated with their solubility; it must be non-toxic if there is a possibility of losing the mediator in the biosensor microenvironment. Some of the requirements are related to the reduced form of the mediator, which should not readily react with oxygen; also the generation of the oxidized mediator at the electrode surface should be low and independent of pH. Ferrocene is a commonly used mediator. A schematic figure of the different techniques in the development of amperometric biosensors is shown in Fig. 2.11 [196].

Currently, electrodes are developed so that the electrons are directly removed from the reduced enzymes without the need for mediators. The current (i) is related to the rate of reaction (v_A) by:

$$i = nFAv_A, \tag{2.33}$$

where n represents the number of electrons transferred, A is the electrode area, and F is the Faraday constant. External membranes are used to achieve a diffusional controlled rate of reaction. The electrical current produced is proportional to the analyte concentration and independent of the electrochemical kinetics.

Roche Diagnostics [197] advanced the realization of redox enzyme-based systems for glucose and lactate. They developed stable, thick films with good adhesion

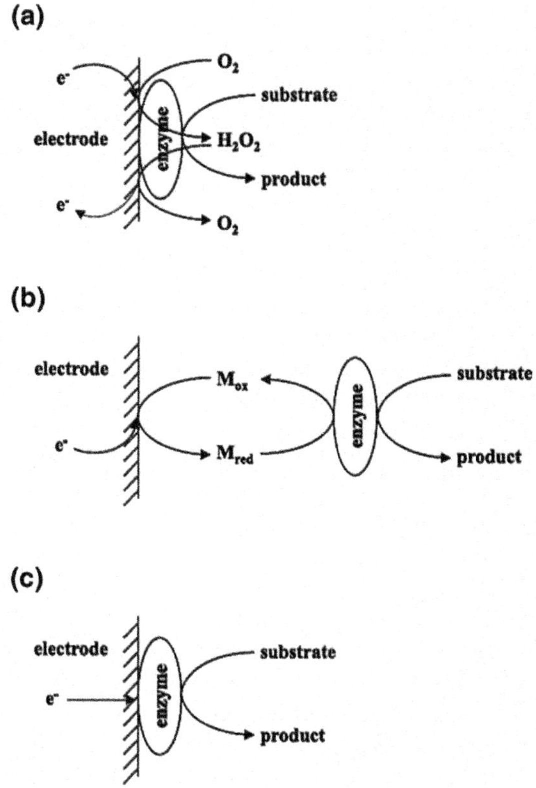

Fig. 2.11 Schematic representation of different biosensor generations. Reprinted with permission from [196]. © 2003 from JBCS

over screen-printed working electrodes using the oxidase enzyme. A biocompatible membrane consisting of a PVC copolymer with vinyl acetate/vinyl maleine was used. Minimization of the electrochemical interferences through measurements at reduced polarized voltages was achieved using an MnO_2-carbon ink composite electrode, where oxygen from the enzyme reaction aids in replenishing the substrate such that the linear working range is extended.

In [198], the authors used SiO_2 nanoparticles (NPs) to control the transmembrane potential by using a combination of an enzyme, an electron mediator, and a selectivity membrane barrier. Permeability is one of the important features for sensor performance. In a modeling study [199] it was found that a greater barrier is achieved with layer-by-layer deposition related to a bigger response. The reason is the reduced permeability to H_2O_2 and its accumulation in the sensing layer.

It was observed that using single-wall CNTs and glucose oxidase entrapped within a polypyrrole layer considerably extends the upper linear range, with detection limits down to the micromolar level [200]. In another case, lowering the polarization voltage was achieved using carboxylated graphene, enabling self-assembly (enzyme prosthetic group with direct electron exchange) of the glucose oxidase.

The pyrroloquinoline quinone (PQQ) cofactor is also used as an electron carrier to a mediator or directly to the polarized electrode, where O_2 dependence is not observed. Ruthenium hexamine and osmium complexes are used to construct commercial glucose strips [201]. Direct electron transfer is also achieved by using nanoporous carbon electrodes. The reported PPQ-glucose dehydrogenase enzyme is not as selective as glucose oxidase, but it was modified genetically. FAD-glucose dehydrogenase is modified to give better glucose sensitivity [202].

A nonenzymatic glucose sensor with wide linear range, low detection limit and good selectivity, based on in situ synthesis of Au NPs on reduced graphene oxide (*rGO*) was reported in [203]. In [204], polymer-enzyme-metallic NP composite films are prepared for glucose and galactose. In [205] perovskite-type oxide $La_{0.88}Sr_{0.12}MnO_3$ (LSMO) nanofibers made by electrospinning and calcination, demonstrated wide linear range, high sensitivity, and a rapid response time.

A xerogel-based amperometric sensor was developed in [206] (Fig. 2.12), which demonstrated a linear range of detection (>24–28 mM glucose), fast response time, and discrimination against interferences, such as acetaminophen, ascorbic acid (AA), sodium nitrite, oxalic acid, and uric acid.

A remotely powered telemetric implantable sensor chip (Fig. 2.13) for subcutaneous applications with integrated electronics was presented in [207]. The potentiostat developed generated an output current which in turn was converted into an output frequency. The signal was used together with a load modulator circuit to transmit the data. The sensor was comprised of a Au counter electrode and an Au working electrode with an enzyme-immobilized layer on top and an Ag RE. The sensors had a glucose detection range of 0–40 mM.

2.3.2 Commercially Available Implantable Glucose Sensors

Monitoring glucose levels of diabetic patients is an enormous market, which is constantly increasing. As a result, the implantable sensors research field as well as the market is driven by the development of glucose monitoring systems. It comes thus as no surprise that the first FDA-approved and CE-marked commercially available implantable sensing products are for glucose monitoring. Such an approach is particularly important for Type I diabetes patients, who depend on daily insulin injections. These patients require rigorous monitoring of their glucose levels (ideally about every 5 min), as large fluctuations of blood glucose levels take place due to the insulin therapy. The Medtronic Enlite [208], the Dexcom S5 [209], and the Abbott Freestyle Libre [210] (Fig. 2.14) are examples of such commercially available implantable glucose sensing systems. The sensors use glucose oxidase catalyzed oxidation of glucose and are implanted transcutaneously so they do not measure blood glucose directly. Instead they measure the interstitial fluid of the subcutaneous tissue. Nevertheless, prick finger stick tests are still necessary, as they are required to calibrate the implanted sensor. The Enlite recommends 3–4 prick tests a day and the Dexcom at least 2 prick tests a day, while the Libre claims that it

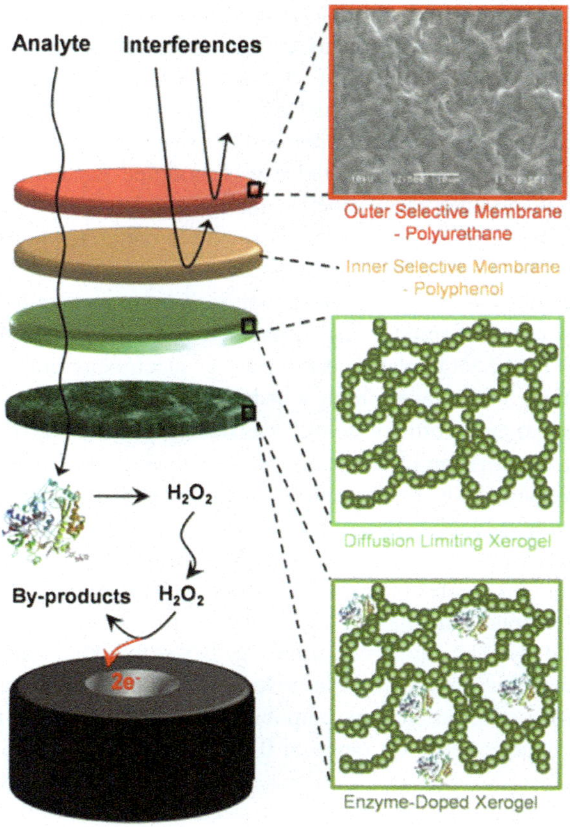

Fig. 2.12 Schematic of a layered xerogel-based amperometric sensor consisting of enzyme doped, diffusion limiting xerogel layers, semipermeable eletropolymerized polyphenol and a polyurethane outer membrane. Reprinted with permission from [206]. © American Chemical Society 2015

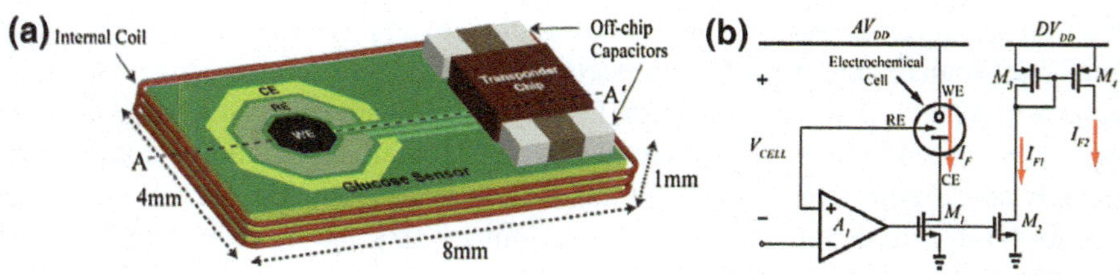

Fig. 2.13 **a** The telemetrically powered and interrogated implantable continuous blood glucose monitoring system with a three-electrode amperometric glucose sensor and ASIC and the front-end potentiostat electronics implemented as in **b** using a novel current mirror-based topology. Reprinted with permission from [207]. © IEEE 2009

does not need any. Nevertheless, according to the product's website: prick tests are required "during times of rapidly changing glucose levels when interstitial fluid glucose levels may not accurately reflect blood glucose levels, or if hypoglycemia or impending hypoglycemia is reported or the symptoms do not match the system readings" [210]. The Enlite sensor can be implanted for up to 6 days, the Dexcomm for up to 7 days and the Libre for up to 14 days. These sensors can be linked to mobile phone and tablet applications and also have their own devices with which they can communicate if needed to provide the patient with warning and glucose level trends and recommendations. Such implantable systems can be linked with insulin pump systems to essentially create an artificial pancreas, which (depending on the implant sensor readings) can deliver the necessary amount of insulin immediately when needed. One example of such a system is the Medtronic MiniMed 530G pump, which works together with the Medtronic Enlite sensor. The system provides patients with a 30 min warning as to whether their glucose levels are going up or down. Information is updated every 5 min to provide a real-time and realistic image of the trends and the insulin delivery is controlled according to the sensor readings and preset settings.

2.3.3 Lactate

Lactate is another important analyte. Lactic acid production can be increased in the case of haemorrhagic shock or pulmonary embolism [211] related to anaerobic metabolism. Increased levels of lactate can be observed in pathological conditions such as respiratory failure [211], liver disease [212], endotoxin [213] or cardiogenic [214] shocks, and tissue hypoxia [215]. Lactate levels can increase and lead to acidosis in the event of insufficient oxygen supply during childbirth. Hypoxic/acidotic distress in fetuses during labour [216] has been detected from fetal scalp blood measurements of lactate and pH. The possibility of developing metastasis in cancer and tumour cells is associated with high levels of lactate [217]. Acidification around the cancer cells and tumours is related to lactate levels because of changes in the ratio of lactate production to lactate consumption [218]. Lactate is related to brain metabolism and plays an important role in brain tissue, [219] and its

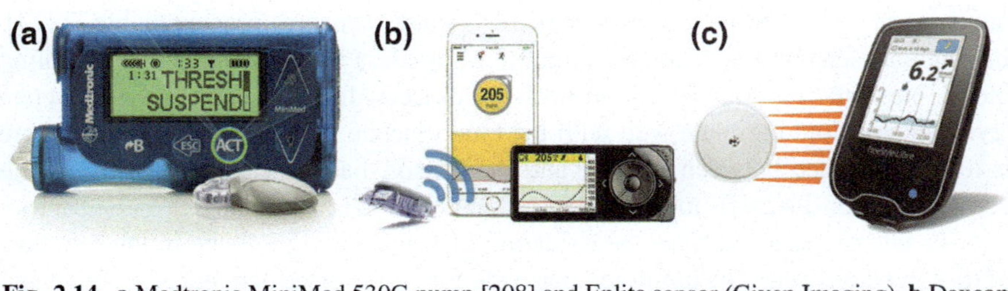

Fig. 2.14 **a** Medtronic MiniMed 530G pump [208] and Enlite sensor (Given Imaging), **b** Dexcom S5 CGM [209] (Given Imaging), and **c** the Abbott Freestyle Libre [210] (Given Imaging). FreeStyle and related brand marks are the trademarks of Abbott Diabetes Care in various jurisdictions

measurement would provide important information regarding stroke and head trauma [220, 221]. It was demonstrated in [222] that the extracellular lactate level is related to sleep: it is increases while awake and decreases during sleep. Commonly used enzymes for lactate sensors are lactate oxidase and lactate dehydrogenase. It is very important to have a high transport rate of oxygen versus substrate so as to maximize the linear operational range. A study [223] showed that oxygen concentration in the enzyme layer is minimally changed when compared with the bulk solution. The operating voltage was reduced when using a cobalt phthalocyanine as an electron mediator [224] on modified screen-printed electrodes. Stabilization of the enzyme can be achieved by crosslinking. Enhanced enzyme stability when mucin is employed in the crosslinking has been reported [225]. Immobilized lactate oxidase on a 3D macroporous Au construct showed that electrode architecture and surface area are important [226].

A possible way to increase the sensitivity of enzymatic biosensors is to increase the surface area using nanomaterials such Au NPs [227], Pt NPs [228, 229], CNTs [230, 231], NPs of metal oxides such as molybdenum oxide [232] or niobium oxide [233] and NPs of semiconducting materials like zinc oxide [234]. An alternative is to use a bioenzyme system, where for example horseradish peroxidase is combined with lactate oxidase [235]. For long-term use, enzyme stability is critical. There are different ways to achieve this, such as crosslinking [236, 237], entrapment in conductive or nonconductive polymers [236, 238], entrapment in sol-gel [239, 240], and immobilization with another biological component, such as bovine serum albumin and covalent attachment of the enzyme to the electrode [241]. Leakage can be prevented by using a combination of a sol-gel matrix and covalent attachment of the enzyme [242], where the carboxylic group of lactate dehydrogenase is covalently linked to the amino group in the organoalkoxysilane precursor. Higher sensitivity and stability are reported based on molecular interactions with chitosan [243]. A number of different approaches have been investigated to reduce the interfering species, one of which involves the use of a mediator, such as phenazine methosulfate [244]. Alternatively, protection and reduction of interference can be achieved by the use of permselective membranes [245, 246] or multilayer polymers [247, 248]. The membranes will serve as a barrier and also as an antifouling component [249]. Such approaches also aid in increasing the linear response of the sensor. Nafion is one such commonly used polymer [223, 250–252].

Thanks to improvements in sensitivity, portability and low detection limits a number of commercial electrochemical lactate sensors are available [223]. Implantable sensors sampling interstitial fluids are less invasive than those sampling blood. Nevertheless, this is still an invasive process. It has been found that lactate levels in tear fluid correlate well with the blood lactate level, which, provided there is a safe interface between the eye and the sensor, has the potential of providing completely non-invasive measurements [223].

L-lactate oxidase catalyses the oxidation of L-lactate to pyruvate in the presence of dissolved oxygen. The hydrogen peroxide formed is electrochemically active and the produced current is proportional to the L-lactate concentration as in

$$\text{L-Lactate} + O_2 \xrightarrow{\text{L-lactate oxidase}} \text{Pyruvate} + H_2O_2 \tag{2.34}$$

$$H_2O_2 \rightarrow 2H^+ + 2e^- \tag{2.35}$$

The electrons are transferred in the enzyme structure through a FAD cofactor. Because of the oxygen participation in the reaction and the related difficulties of additional methods to detect the oxygen level, another approach was investigated. This involved the use of L-lactate dehydrogenase instead of L-lactate oxidase. This type of enzyme needs a coenzyme (NADH or NADPH):

$$\text{L-Lactate} + NAD^+ \xrightarrow{\text{L-lactate dehydrogenase}} \text{pyruvate} + NADH + H^+ \tag{2.36}$$

The coenzyme acts as a mediator and transports the electrons between the electrode surface and the enzyme. However, the use of coenzymes in vivo is related to challenges associated with high oxidation potential and an additional need for optimization [233].

2.3.4 Oxygen

Physiological oxygenation is fundamental to tissue metabolism, providing important information regarding the oxygen supply and local tissue oxygen uptake. Oxygen partial pressure is the standard measure of the oxygenation state. The common pathway of oxygen supply is from the arterial circulation via cellular and extracellular capillaries through diffusive transport in the interstitial tissue space to individual cells. Hb saturation measurement using pulse oximetry has proved a valuable, non-invasive measurement of the oxygenation state, but it is focused on the arterial compartment.

The Clark oxygen electrode is well known, where reduction of oxygen is detected on Pt surface as in (2.37) and (2.38). The applied potential is −0.65 V versus Ag/AgCl RE and the detected current is proportional to the oxygen concentration. Both electrodes are separated from the solution by an oxygen-permeable plastic membrane such as polytetrafluoroethylene or Teflon.

$$\text{Ag anode}: \ 4Ag^0 + 4Cl^- \rightarrow 4AgCl + 4e^- \tag{2.37}$$

$$\text{Pt cathode}: \ O_2 + 4H^+ + 4e^- \rightarrow 2H_2O \tag{2.38}$$

The rate of the electrochemical reaction depends on the diffusion rate of oxygen from the bulk of the solution, which is related to the concentration gradient and bulk oxygen concentration. Oxygen electrodes measure the rate of the process, that is far from equilibrium (in contrast to ISEs which measure close to equilibrium conditions), because they are more sensitive to temperature, as discussed in [193].

The Clark oxygen electrode has addressed the major challenges of biofouling, selectivity and sample convection through the use of a hydrophobic gas permeable membrane with an internal electrolyte film for stable electrochemistry. However, miniaturization of this type of electrode is challenging. This is mainly due to the problem of bonding and integrating the hydrophobic gas membrane with the body of the device, where the internal electrolyte can be removed during the sterilization process. A microfabricated version containing three electrodes, a glass substrate with a fluorinated ethylene propylene oxygen-permeable membrane, and a PDMS (polydimethylsiloxane) tank used for storing solution has been described [253]. Miniaturization using a glass capillary platform has been achieved [254]; however, due to the fragility of the glass used these are not suitable for in vivo use. Other polarographic electrodes, e.g. measuring current that is generated when pair a cathode and anode pair are immersed in an electrolyte solution [255], have been described, where oxygen diffuses through a polyethylene wall into the electrolyte solution, but relative long response times are observed here. A commercial variant of the Clark electrode has been produced for brain monitoring [256]. The sensor is based on an O_2 permeable polypropylene tube, rather than a planar membrane, to house the inner electrolyte solution.

Various experimental physiological studies have been conducted using Clark electrode variants. Muscle tissue has been monitored to assess treatment in septic shock, as well as human head and neck solid cancer tissue. The Eppendorf electrode consists of a central Au and Pt wire encased in glass [257] with the overall sensing electrode constructed within a protective metal needle of 300 μm o.d. Another Licox 0.8 mm tip diameter electrode has been used to monitor pO_2 of central retinal vein occlusion during vitrectomy [258, 259].

A non-membrane alternative for protecting the electrochemical tip surface is to incorporate it into a recess. Recessed polarographic cathode Au microelectrodes have been reported in [260]. The construct used was a metal alloy within a glass micropipette; the device is quite complicated and in vivo use would be difficult because of likely damage during percutaneous insertion. Additional problems with the etched needle microelectrode design include hydration and weakening of the glass insulation over time and membrane disruption during insertion [261].

To limit the diffusion of oxygen and control the oxygen consumption, a p(HEMA)-based hydrogel was used. The constructed sensor was used in combination with other sensors for detection of analytes in tissue brain. In [262], a sensor for oxygen detection in the brain using carbon paste electrodes was presented. Such oxygen sensors have the advantage of being less prone to surface poisoning, thus requiring (in contrast to metal-based sensors) no additional protective membranes. The sensors were fabricated by using Teflon-covered silver wires and carbon paste prepared with graphite powder and silicone oil. Electrochemical reduction of O_2 took place at a potential of −650 mV and the sensor demonstrated a sensitivity of $-1.49\ \text{nA}\ \mu\text{M}^{-1}$. In vivo experiments in rat brains for monitoring tissue O_2 in the brain's extracellular fluid, where hypoxia and hyperoxia were artificially induced, validated the sensor operation. The sensor performance demonstrated no sensitivity to pH, temperature, or flow and an in vivo stability over 12 weeks.

2.3.5 *Oxygen Dependence*

The majority of amperometric sensors depend on oxidase enzymes which catalyze analyte oxidation to form hydrogen peroxide, which is in turn amperometrically recorded at the working electrode. This transduction process in most oxidase enzymes depends on the local oxygen concentration [1]. In implantable applications this is an important sensor failure parameter, as the dissolved body oxygen is significantly lower than that of target analytes. Exogenous environmental conditions (e.g. anesthesia), endogenous physiological responses (e.g. due to physical exercise) and local inflammatory response during implantation, reduce oxygen levels further. The latter is a significant source of oxygen exhaustion [263]. This will be an issue when ischemic events are studied, as by definition tissue oxygen levels are low. These will affect the accuracy of the sensor, as oxygen inadequacy will saturate the sensor at high concentrations of the analyte [1]. The concentrations of glucose and lactate in the brain and in the interstitial fluid of the transcutaneous tissue are high and in the mM range. According to [264], tissue oxygen concentration is an order of magnitude lower than glucose.

Many techniques have been proposed which aim at decreasing the dependence of the transduction process on the local oxygen concentration and can be separated into three generations of sensors Fig. 2.11. Second-generation sensors decrease oxygen dependence by means of synthetic redox mediators to compete with oxygen concentration. In third-generation sensors, the redox cofactor of the enzyme is bound directly to the working electrode to re-reduce or re-oxidize the enzyme. According to [1], the second and third generations have never been tested in vivo due to toxicity concerns. In addition, the need for additional outer membranes to reduce the risk of biofouling and leaching makes first-generation sensors more appealing. This family of sensors uses such membranes in order to reduce the flux of the target analyte to the sensor enzyme layer by, for example, 100, while at the same time the flux of oxygen is effectively unaffected, leading to an enzyme layer rich in oxygen. An additional advantage of this is that the sensor's response time is no longer dependent on the kinetics of the enzymatic reaction, which is temperature and concentration dependent, but rather by mass transfer, i.e. the influx of the target analyte into the sensor's enzyme layer [1, 263]. While these analyte flux-limiting membranes address the issue with oxygen, they lead to decreased sensor sensitivity and an increased response time [1, 263].

2.3.6 Multi-parametric Amperometric Systems

Guiseppi-Elli et al. have presented a number of amperometric systems for implantable applications [265–268]. In [265] the authors present a 2 × 4 × 0.5 mm chip with two Pt working electrodes (Fig. 2.15a). Each working electrode is covered by a passivation layer, which has a circular opening to form Pt microdiscs. This minimizes the diffusional limitation associated with enzyme kinetics [265]. Each working electrode is complemented by a large Pt counter electrode while a single Ag/AgCl electrode works as a common RE to both electrode sensing sets. A 1 μm thick layer was deposited on the working electrode, comprised of "tetraethylene glycol (TEGDA) cross-linked poly(2-hydroxyethyl methacrylate) that also contained a derivatized polypyrrole component and a biomimetic methacrylate component with pendant phosphorylcholine groups" [265], providing interference screening and in vivo biocompatibility, and used to immobilize glucose oxidase and lactate oxidase to create glucose and lactate sensors. The glucose sensor had a linear range of 0.1–13 mM and the lactate sensor up to 90 mM. The lactate linear range sensing response was 30-fold greater than that obtained from a sensor without the microdisc passivation openings. The sensors retained 80% of their initial performance after 5 days of continuous measurements in vitro. A similar topology was presented in [266, 267], with the relevant potentiostat electronics (Fig. 2.15b) in [266] and its use in lactate and glucose sensing in [267]. The same electrode array was used in [268, 269] together with a commercially available wireless potentiostat from Pinnacle Technology, Inc. [270] designed for tethered implantable in vivo central nervous system (CNS) applications in rats (Fig. 2.15c). The authors focused in [269] on trauma-induced hemorrhage applications and the system was validated in vivo with a hemorrhage model, where tissue lactate rose more rapidly. Immobilized on each working electrode were glucose oxidase or lactate oxidase within a biorecognition layer with a thickness of 1.0–5.0 μm. This was a 3 mol% tetraethyleneglycol diacrylate cross-linked p(HEMA-co-PEGMA-co-HMMA-co-SPA)-p(Py-co-PyBA) electroconductive hydrogel. The device was subsequently coated with a bioactive hydrogel layer containing phosphoryl choline and polyethylene glycol pendant moieties [p(HEMA-co-PEGMA-co-HMMA-co-MPC)] to increase the biocompatibility of the implanted sensor platform. The sensors demonstrated linear ranges of 0.1–13 mM concentration range for glucose and 1–7 mM for lactate with response times of 50 and 35–30 s respectively. Extensive in vitro experimentation demonstrated the biocompatibility of the implanted sensors, while in vivo experiments of up to two weeks demonstrated only a thin band of encapsulation and reduced inflammation of poly(2-hydroxyethyl methacrylate) (p(HEMA)) hydrogels containing phosphorylchlorine in contrast to unmodified p(HEMA).

A multisite microelectrode array for in vivo measurements was presented in [252] (Fig. 2.16a) where the authors demonstrated its use for monitoring rapid endogenous changes in brain tissue. The polyimide platform of Fig. 2.16b was used

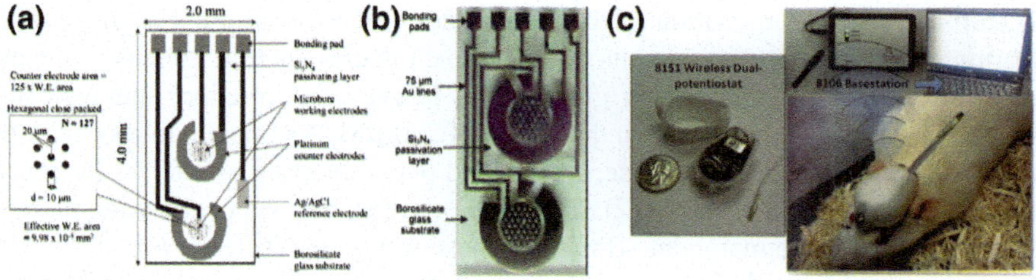

Fig. 2.15 **a** The three-electrode amperometric sensing device with two sets of Pt CE and micropore WEs and a common Ag/AgCl RE. Reprinted with permission from [265]. © IEEE 2005. **b** Amperometric sensors. Reprinted with permission from [266]. © Springer 2008. Each amperometric sensing system has its own RE, which are, however, shorted together. Similarly to (a), these are microdisc WEs; in this case there are 37 microdisc openings in the passivation 50 μm in diameter and with a pitch of 100 μm. **c** Lactate and glucose measurements in the trapezius muscle of a rat via sensors similar to those of (**b**) and with a head-mounted wireless dual channel potentiostat. Reprinted with permission from [268]. © Springer 2010

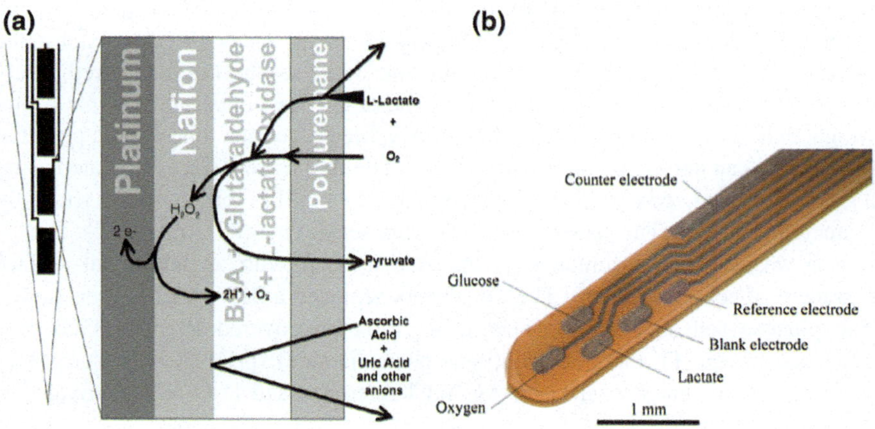

Fig. 2.16 **a** Ceramic-based multisite microelectrode with 50 μm × 150 μm recording sites. The various sensor layers are shown on the right. Reprinted from [252], © 2005, with permission from Elsevier. **b** Cross-sectional view of the multi-layer structure, which is composed of polyimidesubstrate, deposited Pt electrodes, insulation from spin-coated epoxy and a laminated dry-film resist. All electrodes are modified with hydrogel membranes. Reprinted with permission from [271]. © IEEE 2014

for detection of glucose, lactate, glutamate, and O_2 in soft tissues, such as muscles or abdominal tissues and also brain tissue [271].

Amperometric sensors are also applicable to experimental neuroscience applications, where for example researchers are interested in the CNS [262, 272, 273] and where microdialysis systems lack the temporal resolution necessary for the rapid changes in brain metabolism [272]. The brain requires up to 25% of total glucose consumption, being the main energy substrate for glial cells and neurons, with lactate and pyruvate following at a smaller extent in the adult brain [272].

Disturbances in the regulation of the brain energy metabolism are associated with learning and memory impairments and neuropathologies such as epilepsy, meningitis, and ischemia [272]. Brain glucose and lactate concentrations are typically between 1–2 mM and pyruvate in the region of 200 μM. An implantable system for amperometric detection of glucose, lactate, and pyruvate using needle Pt electrodes (Fig. 2.17) was presented in [272]. Electrodes were coated via cyclic voltammetry with a permselective membrane (PmPD). These were then coated manually with a hydrogel of either GOx, LOx, or POx, cross-linked with glutaraldehyde (GA) and bovine serum albumin (BSA). An electrode with just the PmPD layer was also included to record the background (BG). All sensors demonstrated limits of detection (LOD) lower than 5 μM. The lactate sensor used, was loaded by a 0.4 U μL^{-1} of LOx and had an additional external layer of polyacrylonitrile (PAN) and demonstrated a LOD of 0.86 μM, a linear range of 1.13 mM and a sensitivity of 241.14 nA mM^{-1}. The system was tested in vivo with rats to monitor the levels of these markers in their brains during induced hypo- and hyper-glycaemia, with all sensors retaining high sensitivity and selectivity after implantation. The lactate sensors were the least affected by the in vivo environment due to the additional PAN layer. Similarly in [273], carbon and Pt-based sensors were used for the amperometric detection of AA, O_2 and glucose in the striatum of untethered, freely moving rats using a biotelemetric device. One of the most important low molecular weight antioxidants in the brain is vitamin AA. Electrodes at a mild anodic potential (+120 mV) can detect AA amperometrically. The AA sensors were fabricated using a graphite-loaded epoxy which was patterned with a drill to form a needle with a tip diameter of 25 μm. The sensor demonstrated a sensitivity of 7.3 pA μM^{-1}. The O_2 sensor was fabricated similarly to the AA sensors, with an additional surface treatment of cellulose nitrate. The O_2 sensors achieved a 213 pA μM^{-1} sensitivity. The glucose sensor was fabricated using a Teflon-covered Pt wire. GOx was deposited on the Pt and poly(ortho-phenylenediamine) (p-OPD) was electrosynthesized on the sensor against an Ag/AgCl reference and H_2O_2 electro-oxidation

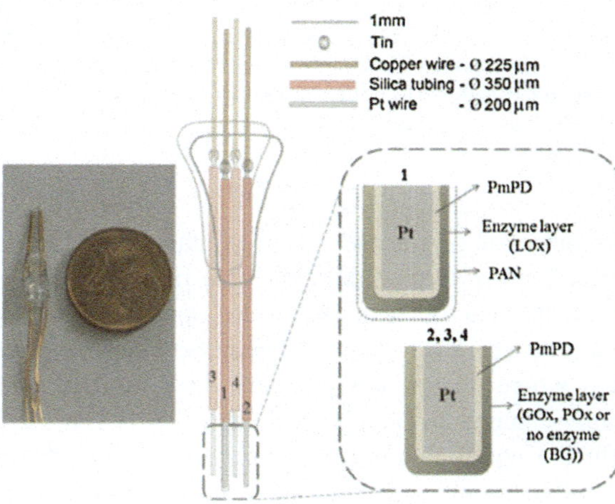

Fig. 2.17 The multisensory amperometric platform developed. Reprinted with permission from [272], © Elsevier 2015

was subsequently carried out, finalizing the glucose sensor fabrication. The sensor demonstrated a linear region with a slope of 15.2 nA mM^{-1} at concentrations between 0 and 2 mM.

2.3.7 Voltammetry

Voltammetry is an analytical method that, similarly to amperometry, measures the current through an electrochemical cell under an applied potential. The difference is that while in amperometry the applied potential is constant, in voltammetry the applied potential changes linearly over time. This is known as linear sweep voltammetry (LSV). The linear rate of change of the applied potential can vary from a few μV s^{-1} to mV s^{-1} and it is known as the scan rate. The sensitivity of the measurement can be improved by increasing the scan rate, as more redox reactions will take place per second on the WE. For example, the peak current may be proportional to the square root of the scan rate. As in amperometry, a three-electrode system is typically used. The concentration of the material undergoing the redox reaction and the faradaic current are according to

$$i_f = -nFAD\frac{\partial C(x,t)}{\partial x}\bigg|_{x=0}, \tag{2.39}$$

where A is the area of the electrode involved in the reaction, D is the diffusion coefficient, and $C(x,t)$ is a function of concentration of the analyte at a distance from the electrode at time t. During the measurements, redox reactions must occur at both the WE and CE. In most cases, the WE is a polarizable electrode (i.e. from a noble metal, such as Au or Pt, or carbon-based materials, such as glassy carbon and graphite), which must also be electrochemically inert, unless it is the electrode itself that is to be characterized. The voltage drop across the double layer provides the energy for the electron transfer. Considering a generic redox reaction, then:

$$\mathrm{O} + ne^- \Leftrightarrow \mathrm{R}, \tag{2.40}$$

where O is the oxidized form, R is the reduced form of the analyte, and n is the number of electrons (e^-) exchanged in the reaction. When equilibrium is achieved, the potential difference, E_M, between the solution and the electrode depends on electrode material and the concentration of the analyte. The potential is the Nernst potential and given by the Nernst equation:

$$E_M = E^0 + \frac{RT}{nF}\ln\left(\frac{C_O}{C_R}\right) \tag{2.41}$$

where E^0 is the standard potential of the reaction, R is the molar gas constant (8.3145 J $Kmol^{-1}$), T is temperature, n is the number of electrons exchanged in the reaction, F is Faraday's constant (9.6485×10^4 C mol^{-1}), C_O is the concentration of the oxidized form, and C_R is the concentration of the reduced form [274]. In more concentrated solutions the concentrations C_R and C_O must be replaced by chemical activities. For solid phase reactants at $T = 298$ K, the Nernst equation becomes:

$$E_M = E^0 + \frac{59.2\text{mV}}{n} \log_{10}\left(\frac{C_O}{C_R}\right) \tag{2.42}$$

For a normal hydrogen electrode, E^0 is defined to be 0 V. When the cell is away from equilibrium:

$$\frac{C_O}{C_R} = \exp\left(\frac{(E^0 - E)nF}{RT}\right) \tag{2.43}$$

Different parameters control the rate of the reaction, such as the applied potential between the working and counter electrodes, the mass transfer to the surface of the electrode and of the product, adsorption of ions on the electrode surface, and the electron kinetics of the redox reaction. The Cottrell equation is used for quantiative analysis and to measure the electrode area, where the I_{pa} is related to the concentration of the measured redox molecule [275]

$$I_{pa} = nFAC_jD_j^{1/2}(\pi t)^{-1/2} \tag{2.44}$$

where n is the number of electrons interchanged on the redox reaction, F is the Faraday constant (C/mol), A is the area of the working electrode (cm^2), C is the concentration (mol/cm^3), D is the diffusion coefficient of the redox species (cm^2/s), and t is time (s).

There are many variations of this electroanalytical technique that fall under the term "voltammetry". Different voltammetric techniques arise, depending on how the applied potential changes. Thus, in addition to LSV, techniques include staircase voltammetry, square wave voltammetry (SWV), differential pulse voltammetry, cyclic voltammetry (CV), and fast-scan cyclic voltammetry (FSCV). Staircase voltammetry is a variation of LSV, where, as the name suggests, the applied voltage is increased in steps. Measurements are recorded at the end of each step. This allows enough time for the signal to settle, in order to minimize contributions from capacitive charging (i.e. non-faradaic currents). A variation of these methods is SWV, which is a type of differential pulse voltammetry, where the pulse width is 50% of the cycle time of the square wave used. Basically, a square wave stimulus is superimposed on a linear staircase waveform. Similarly to staircase voltammetry, non-faradaic current contributions are minimized. Measurements are performed at the end of the forward and reverse half cycles of the stimulus. The two recorded currents are similar to those obtained from CV, as will be explained subsequently. However, in differential pulse voltammetry, the difference of these currents is plotted instead. Peaks and troughs are associated with redox reactions taking place at specific potentials and the magnitude of the current is proportional to the concentration of the redox active species.

CV is an extension of LSV where the voltage is immediately swept back at the same rate instead of sweeping the voltage in one direction, as in Fig. 2.18a. A typical cyclic voltammogram is shown in Fig. 2.18b. CV has several advantages; the reverse sweep gives additional information regarding the reactions taking place within the electrochemical cell. These include whether the reaction is reversible (in which case the reverse sweep converts the material back to its initial form), information about the transfer kinetics of the reaction, and the intermediate products of the reaction [274, 276, 277]. As shown in Fig. 2.18b, troughs and peaks in the recorded current response of a typical CV voltammogram correspond to the potential at which oxidation or reduction of species takes place, respectively. The peak currents increase with the increasing concentration of the target analyte and with increasing scan rate if the concentration is kept constant. FSCV is used as a term to described measurements undertaken with scan rates in the mV s^{-1} region. Such high scan rates are used to perform rapid measurements for high temporal resolution. As such, FSCV is useful for in vivo measurements for the detection of neurotransmitters, metabolites and hormones. This is because at these scan rates, FSCV can perform voltammetric measurements at ~100 ms which is on the same timescale as that of neurotransmission, thus enabling real-time measurements of neurochemical dynamics. Since a cyclic voltammogram is obtained, FSCV offers greater specificity, as the chemical signatures of different species at different potentials can be identified. In such applications, carbon fiber microelectrodes are typically used and since each species of interest will have unique redox potentials, a high specificity is achieved. Nevertheless, the applied potential must not exceed 1.23 V, where water electrolysis takes place. On the same time, the current flowing though the cell, which is a function of the applied potential and the electrode tissue interface impedances, among others, must also remain low to avoid cell lysis and depolarization.

From the above, we can see that voltammetry provides us with important information regarding the kinetic and thermodynamic parameters of electron-transfer events. In more detail, CV is typically used for the characterization of an electrochemical system and provides information regarding electron

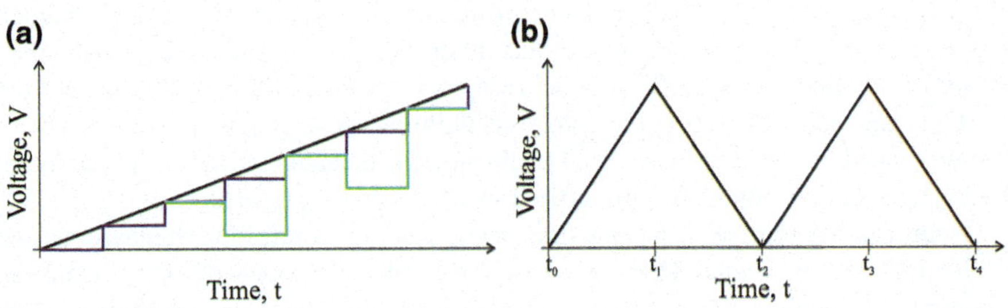

Fig. 2.18 **a** Black: LSV, blue: staircase voltammetry, green: SWV stimulus waveforms. **b** CV stimulus waveform

stoichiometry, diffusion coefficients, reversibility, and the redox potential of the process being examined. Identifying the latter allows one to properly perform amperometric measurements. The two techniques, in terms of signals applied and measured, are similar in that the current is measured while a voltage is applied, but in amperometry the applied voltage is a constant d.c. potential, while in cyclic voltammetry it is a triangular waveform. Thus, the same front-end electronics can be used in both applications.

An important implantable application in which CV and amperometry are used is in neurotransmitter sensing, where they can be combined with traditional electrical action potential recordings for neuroprosthetics applications and basic neuroscience research.

2.3.8 Neurotransmitter Sensing

The majority of the communication in the nervous system is via electrical impulses. Many systems have been developed for the recording of these for various implantable and benchtop/basic research applications [278]. An example is shown in Fig. 2.19a, b. Communication between neurons is performed at sites known as synapses. These allow a neuron to transfer information to another neuron. At a synapse the plasma membranes of the two neurons are in close proximity, and various molecular machinery link the two membranes together for the purposes of the signaling process. The neuron transferring the signal is known as the presynaptic neuron (the transmitter in electronic engineering terms) and the target neuron is known as the postsynaptic cell (i.e. the receiver). This is illustrated in Fig. 2.19c. There are two types of synapses, electrical and chemical. In the former, cell membranes are connected via gap junctions. These are conductive channels capable of passing electrical current. In chemical synapses, however, there are no conductive pathways, and communication is established via the release or uptake of chemicals known as neurotransmitters. Neurotransmitters play the role of chemical messengers which send the signal between neurons and other cells. The activation of voltage-gated calcium channels allows the conversion of electrical activity in the presynaptic neuron into the release of excitatory or inhibitory neurotransmitters into the synapse. Thus, electrical and chemical neural activities are tightly coupled together [279]. The released neurotransmitter diffuses towards the postsynaptic neuron and binds to specific protein receptors in the plasma membrane of the postsynaptic cell. This causes excitatory or inhibitory postsynaptic potentials. These are integrated by the neuron and lead to physiological changes which will determine whether an action potential will be fired by the postsynaptic neuron.

Neurotransmitters play a very important role in controlling behavioral and physiological conditions. They are involved in the processes of sleep, learning, appetite, and memory [280]. Detection and monitoring of neurotransmitters, such as nitric oxide, glutamate, gamma-aminobutyric acid(γ-aminobutyric acid) (GABA), acetylcholine, norepinephrine, and dopamine, is extremely important for research in

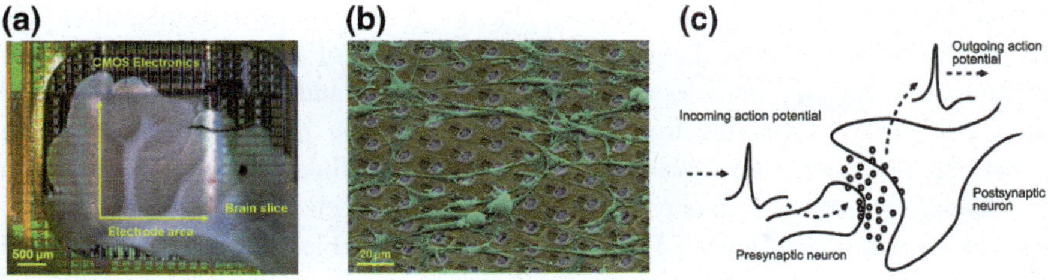

Fig. 2.19 Examples of multichannel CMOS ASICs for the measurement of action potentials from electrogenic cells of **a** brain slices and **b** neuronal cell cultures in vitro and directly on the ASIC via on-chip electrode arrays and recording electronics. Reprinted with permission from [278]. © IEEE 2011. **c** Neurochemical signaling between two neurons at a synapse. Reprinted with permission from [279]. © IEEE 2009

the neurochemistry of the nervous system as it can reveal how information is transmitted and processed. Recording these at different areas of the brain would help us understand further the neurophysiology of sensory-motor systems and encoding in the auditory nerve and visual cortex [281]. This would be also important since various neuropathologies, such as Parkinson's disease, epilepsy, schizophrenia, substance addiction, and stroke are associated with neurotransmitters, while it may lead to advances also in neural prostheses (closed-loop systems), artificial limbs, brain-computer interfaces (BCI), and tissue engineering [282].

Electroactive monoamines such as serotonin, norepinephrine, epinephrine, and dopamine are of interest. Because these neurotransmitters are redox active, they can be measured by electrochemical methods at their oxidation or reduction potential. In the brain other electroactive neurochemicals are present, which can interfere with the detection of neurotransmitters. Common inteferants include AA and uric acid. The extracellular concentration of neurotransmitters is low and their concentrations can change rapidly, which is why their in vivo measurement is challenging [283]. Optical and chromatographic techniques, positron emission tomography (PET), and single photon emission computerized tomography (SPECT) have been used for the detection and analysis of neurotransmitters. These are, however, large, high-power, complicated, and expensive techniques, most capable of detecting the reaction products of neurotransmitters, with low quantitative, temporal, and spatial resolution [272, 282].

Depending on the target analyte, highly sensitive front-end sensor recording electronics (potentiostats) with current measuring capabilities ranging from fA to mA may be required; thus the recording electronics need to have a wide dynamic range. At the same time, the sensors themselves must demonstrate fast response times and low hysteresis. As mentioned previously, in a typical amperometric/voltammetric cell there are three electrodes. No current should flow through the RE, to prevent an ohmic drop forming across the solution which would influence the potential between reference and working electrodes. However, in most neurotransmitter detection applications, the currents can be very low, leading to negligible ohmic drops across the solution. Hence, reference and counter electrode terminals can be short-circuited and two electrodes can thus be used instead of three [281, 282]. Thakor et al. have developed a number of CMOS-based integrated potentiostats for implantable neurochemical sensing and neuroprosthetics [279, 281, 282, 284]. A two-electrode system was similarly used in [285]. The telemetrically operated system was used to record extracellular levels of dopamine in the caudate-putamen of an anesthetized rat, elicited via electrical stimulation of the medial forebrain bundle. The system was capable of performing both chrono-amperometric measurements and fast-scan cyclic voltammetry (FSCV). CMOS on-chip recordings of transmitter release were presented in [286], via post-processing of the chips such that aluminum electrodes are converted into Ti/Au electrodes. A 96 channel potentiostat with on-chip Au electrodes and microfluidics (Fig. 2.20a–c) incorporated on the CMOS die was demonstrated in [287]. Flat 2D electrodes (Fig. 2.20a) were developed on chip for neural culture measurements, while 3D electrode spikes (Fig. 2.20b) were developed for brain slice measurements. An amperometric CMOS ASIC was presented in [288] together with a microfabricated multi-electrode platform (Fig. 2.20d–f) for neurotransmitter sensing. The Au working electrodes of the array (Fig. 2.20d) were functionalized with additional layers for the detection of dopamine and glutamate (Fig. 2.20e, f). Since dopamine is electroactive, a MWCNT-loaded Nafion layer is sufficient for its detection at the appropriate redox potential. AA and uric acid have similar redox potentials and this layer increases the specificity to dopamine. Glutamate is not electroactive; thus, an enzymatic approach was implemented with the use of glutamate dehydrogenase (GLDH) enzyme that produces electroactive nicotinamide adenine dinucleotide (NADH). A composite film was developed with MWCNT, chitosan (CHIT), meldola's blue (MDB), and GLDH. An automatic spotting machine was used to dropcast these membranes onto the Au electrodes.

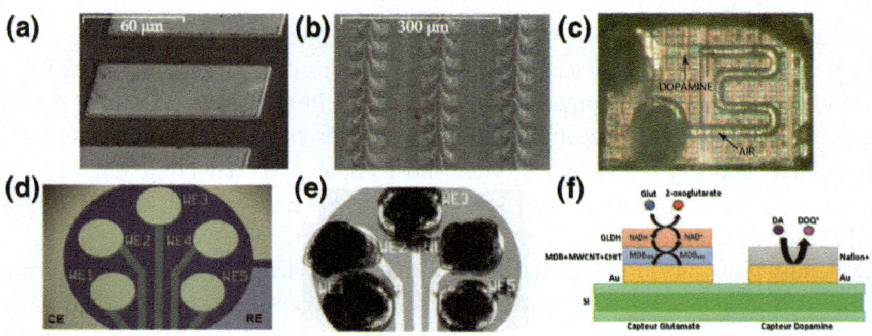

Fig. 2.20 **a** On-chip 2D and **b** 3D Au electrodes for neural cell and brain slices, respectively, **c** neurotransmitter sensing and microfluidic chamber on the CMOS die. Reprinted with permission from [287]. © IEEE 2013. **d** The sensors with five Au WEs, common Pt RE and Au CE. **e** The Au electrodes are covered with CNT-based films. Electrodes 1, 3, and 5 with MWCNT-loaded Nafion and electrode 2 and 4 with MDB-CHIT-MWCNT-GLDH. **f** Schematic sideviews of the glutamate and dopamine sensors. Reprinted with permission from [288]. © IEEE 2016

Some of the most used electrochemical sensors for neurotransmitter analysis are carbon-based, because it has been shown that the surface oxides facilitate the electrostatic interactions [289] and since carbon biocompatibility nanomaterials with enhanced electrochemical properties [290]. The nanostructured surface is related with larger specific surface area, increased adsorption capabilities, and enhanced electro catalytic activity. For neurotransmitter detection the use of the CNT-based electrodes is related to increased sensitivity and selectivity rapid electron transfer and reduced electrode fouling [291]. New carbon-hybrid materials like CNTs or carbon nanofibers directly on tetrahedral amorphous carbon have also been proposed. These demonstrated high sensitivity and selectivity in the presence of AA [292]. The authors discussed the possibility of non-enzymatic detection of glutamate at physiological pH.

Similarly to glutamate, lactate is also an electrochemically inactive molecule. Specific enzymes to generate redox active molecules can be used to determine it. Glutamate measurements with a high sensitivity of 1.95 nA mm^{-2} μM^{-1} were demonstrated in [271], with a linear range of up to 100 μM. It was discussed that using glutaraldehyde for cross-linking results in a thinner membrane and that a higher concentration of enzyme could be immobilized. This increased the sensitivity, but reduced the linear working range and long-term stability. The sensor was used in measurements of extracellular concentration in the brain over 3 days. Acetylcholine (ACh) is another important neurotransmitter. Choline (Ch) is a neurotransmitter at nicotinic receptor sites [293]. Detection of Ch is done by using detectable peroxidase through the use of choline oxidase [294, 295]. Determination of ACh is connected with conversion of Ch, which leads to peroxide in two steps.

Determination of ACh needs to be corrected for the endogenous Ch, which means that two sensors are needed, one for total ACh + Ch and one for Ch. Some sensors are reported to measure neurotransmitters released in individual tissues, such as the hippocampus [296, 297]. Multielectrode arrays of planar CNTs on indium tin oxide microelectrodes have also been proposed for long-term recording at nM concentrations of dopamine and changes in the extracellular chemical microenvironment in mouse striatal brain slices [297]. Some issues are related to non-specific adsorption and reduced tissue reaction and also interference from electroactive molecules in the physiological environment. There is thus a need for the design and development of novel approaches and materials for selective membranes.

Implantable neural signal recording for neuroprosthetic applications and diagnostics is a large research area that has drawn considerable attention in recent years. Such devices are used in both the CNS and the peripheral nervous system (PNS). At the core of these technologies lie the electrodes necessary for recording action potentials and for electrical stimulation. Pt and Au, conducting polymers (e.g. PEDOT), CNTs, and graphene are some of the materials currently being used for the electrodes. A discussion on the technology of these is out of the scope of this chapter; the reader is referred to [298] on recent advances on flexible neural interfaces using organic and inorganic materials.

2.3.9 Read-Out Circuits for Amperometric and Voltammetric Sensors

As mentioned earlier, three-electrode systems are typically used in amperometric applications. For steady state amperometry a stable voltage source is required that can respond rapidly to current loading of many orders of magnitude. The circuit needs to ensure that the cell potential is kept constant with respect to the RE, and feedback is used to achieve this irrespective of any changes in the cell impedance. In other words, the cell potential, i.e. the potential difference between WE and RE, needs to remain constant [195], and this is done by controlling the cell current (I_{cell}) [299]. The basic setup of the driving circuit of a potentiostat is shown in Fig. 2.21a. The equivalent circuit of the three electrode electrochemical cell is shown in (b) [299]. R_{CE}, R_{WE}, and R_{RE} are the interfacial charge transfer resistances at the CE/solution, WE/solution, and RE/solution interfaces, respectively. C_{CE} and C_{WE} are the double layer capacitances at those interfaces. R_{S1} is the solution resistance between CE and RE and R_{S2} is the solution resistance between RE and WE. As can be seen, a voltage signal is applied to the CE and the circuit also needs to supply the required current (I_{cell}) for the electrochemical reaction being studied. This is so that the CE can provide sufficient current to compensate for the redox reaction currents at the WE. The circuits of Fig. 2.21 thus compensate for the potential drops across the impedances along the CE-RE path [195]. Placing the RE close to the WE minimizes potential errors due to the voltage drop across R_{S2}, which is the only uncompensated impedance [195]. If the CE is large enough compared to the WE, then the effects of its interfacial impedance on the circuit operation and redox processes taking place there are minimized and can be ignored. As mentioned earlier, no current should flow through the RE, to avoid RE polarization, which would influence its potential, which should be constant. To achieve this, the input impedance of the amplifier in Fig. 2.21a should be very large and its input current should be minimal. Depending on the application, a number of other design parameters may become important. These include output voltage swing, input referred offset and noise, gain, bandwidth, and slew rate. The circuit of Fig. 2.21c is a common potentiostat configuration. A_2, A_3, and A_4 are unity gain buffers used to isolate the electrochemical cell from the rest of the circuit, and R_1 and R_2 are summing resistors, with a low resistance in the range of 15 kΩ to minimize thermal noise [195].

In the circuits of Fig. 2.21, there is a current flow (I_{cell}) from the CE to the WE, which needs to be recorded. This is typically done by converting this into a voltage, which is linearly proportional to I_{cell}, by a transimpedance amplifier (TIA), implemented in Fig. 2.21c by A_5, R_3, and C. This comes with the benefit of holding the working electrode at virtual ground, measuring with reference to ground and measuring small currents by switching between different values of feedback resistors. The output voltage of the TIA ($V_{out\text{-}TIA}$) is given by

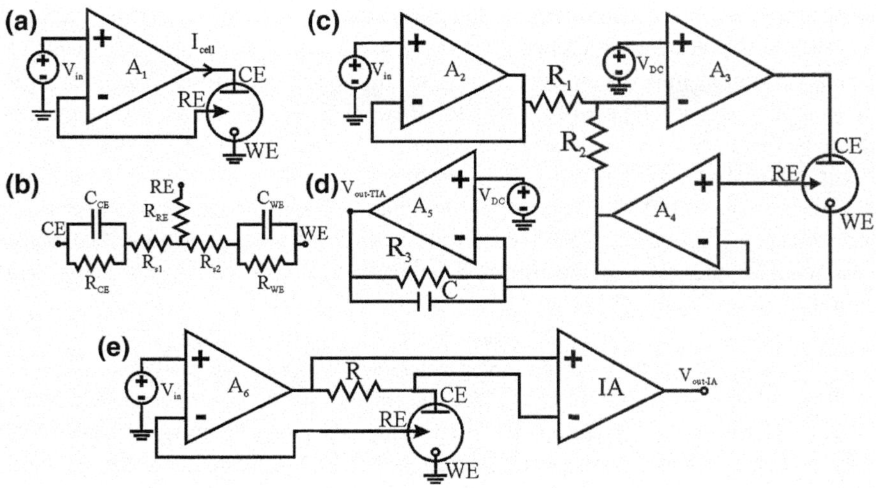

Fig. 2.21 **a** The basic grounded WE topology, **b** The equivalent circuit of a three-electrode electrochemical cell [300], **c** The same topology with buffers for isolating the reference electrode and a TIA for recording of the current through the cell. **d** Recording the current through a current-sense resistor with an instrumentation amplifier (IA)

$$V_{\text{out}-\text{TIA}} = -I_{\text{cell}} \cdot R_3 \tag{2.45}$$

In order for A_5 to provide a good virtual ground, the input-referred offset voltage and noise current need to be minimized. The latter is particularly important as the electrode dimensions are reduced. This is because electrode impedance increases and the current being recorded may be very low (fA levels). In these cases active shielding is necessary and current follower circuits are advantageous, as leakage currents to ground are minimized. In addition, since I_{cell} must all flow through R_3, the input bias current of the amplifier should be minimized and the input impedance maximized. The input impedance of a TIA is low at low frequencies, thus behaving inductively. At higher frequencies it increases and is in series with the impedance of the electrochemical cell, which, as shown in Fig. 2.21b, is mainly capacitive. This may lead to instabilities in the feedback circuit [299]. The feedback capacitor, C, is used to stabilize the amplifier by reducing the bandwidth. The input impedance issue on both A_4 and A_5 is particularly important when measuring very low currents, which will require large amplifications via a large R_3. As discussed in [299], in single-supply systems the non-inverting terminal of TIAs must be biased at an appropriate level to allow all voltages to lie within the power supply range.

The current can also be recorded by adding a resistor between the driving circuit output and the CE or between WE and ground and measuring the potential drop across the resistor. The former is preferable, as the direct connection of the WE to ground ensures that its potential is constant. Also, because this is a true ground connection, as opposed to that in the TIA circuits, this topology is insensitive to noise and EMI. An instrumentation amplifier (IA) is used as in Fig. 2.21d to measure the potential difference between the two terminals of the resistor. In addition to input impedance, input bias current, and offsets, which are once again important, resistor tolerance and resistor thermal noise must also be considered. Because there are no active components in the feedback path, this topology is more stable, while both voltage and current are measured with respect to ground [299]. A disadvantage of this topology, is that *R* is part of the feedback loop, which limits the voltage compliance across the cell [195, 299]. When connecting the resistor between WE and ground, the recorded voltage needs to be fed back to the drive circuit to ensure that the WE potential is constant as discussed in [299].

It should also be noted that apart from the circuits described above, which consider only the scenario of having a grounded WE, other topologies also exist, where the CE can be grounded as discussed briefly in [299] and in more detail in [301]. According to these papers, when shielding and screening of the WE to reduce the effect of electromagnetic interference (EMI) is challenging, the grounded CE approach can achieve a better performance.

2.4 A Note on Affinity-Based Biosensing

Affinity-based measurement techniques, where a binding molecule [e.g. antibodies or other proteins, aptamers, ss-DNA (single stranded deoxyribonucleic acid), or RNA (ribonucleic acid)] is used as part of the transduction process to enhance selectivity and specificity, are mainly single use disposable types of sensors. A discussion on these is thus excluded from this chapter, as are sensors requiring agents and labels for detection of target analytes (thus only reagentless and label-free techniques are discussed). This is due to the inherent limitation of these devices in their implementation as implantable sensors. These hurdles are associated with the in situ regeneration of antigen binding sites once a detection event has taken place; i.e. the antigen–bio-recognition element binding is very specific with strong bond formation and is essentially irreversible [263]. Breaking that bond to free up the detection site and reuse the biosensor (without damaging the binding sites) is extremely challenging. Applying high temperatures, chemicals, and electromagnetic fields are some options examined in the literature. Nevertheless, it is obvious that these are not suitable for implantable applications. In addition, using (redox) agents and labels for detection is also not possible in an implantable in vivo environment. For the sake of completeness we mention a variety of methods that can be used in affinity-based sensing. Affinity based-sensing can be performed via quartz crystal microbalance (QCM), surface acoustic wave (SAW), surface plasmon resonance (SPR), impedance spectroscopy (EIS), voltammetry or amperometry, micro-electro-mechanical systems (MEMS) based cantilever sensors read out electronically (resistor bridge or capacitive sensing), or optically and atomic force microscopy (AFM). Developing complex microfluidics for affinity-based sensors could potentially allow the use of these in implantable applications. Affinity-based sensing can be used for detecting a plethora of clinically relevant targets, such as bacteria (e.g. for infection diagnostics) and cancer biomarkers (e.g. certain proteins [302]); thus their potential use in implantable applications should not be overseen.

2.5 Conclusion

Electrochemical sensors are an important family of sensing techniques invaluable for the detection of a wide range of physiologically critical markers. Their advantages include their ability to be highly miniaturized, high dynamic range and sensitivity, high specificity through biorecognition elements and membranes, and their low-power operation. These make them vital for biomedical applications and especially for implantable systems. Potentiometric sensors are two-electrode galvanostatic systems used for the detection of vital ions such as pH, Na^+, Ca^{2+}, Mg^{2+}, NH_3 and K^+. Polymeric membranes are typically used on metal electrodes to create miniaturizable electrodes known as ISEs, which circumvent the disadvantages of traditional glass-based sensors such as fragility, slow response, and large size. MMO pH sensors are an alternative with IrO_x being a common MMO used, demonstrating super-Nernstian responses (sensitivities above the Nernstian 59 mV pH^{-1}). The galvanostatic operational mode of potentiometric sensors requires high input impedance electronics to minimize the current through the electrochemical cell. ISFETs are an important family of potentiometric chemical sensors that readily provide this functionality, as an ISFET is by default an impedance converter. ISFET technology allows high-density miniaturized chemical arrays to be developed using common VLSI technologies for mass-produced, low-cost chemical analysis systems. Charge trapped during fabrication hinders the performance of ISFETs and their matching; however, various techniques can be used during and following fabrication together with circuit design techniques to minimize the effects of trapped charge. Voltammetric techniques are essential in the characterization and development of sensors and materials, especially since it allows the identification of the redox potential of target species and a voltammetric sensor's response at that potential. These are typically three-electrode systems, and in contrast to potentiometric sensors, there is a current flow through the electrochemical cell that is being recorded. CV is the most common voltammetric technique. FSCV is a high-speed variant which, when used in combination with carbon-based electrodes, is a powerful technique for the detection of electroactive species and in particular various neurotransmitters. Such sensors can be used to monitor the nervous system chemically and can also be used in combination with action potential electrical recordings. Knowledge of the redox potential allows simple chrono-amperometric measurements at a constant potential. A wide range of amperometric sensors have been developed for the detection of important analytes, such as lactate, O_2, and glucose, the latter type being the most successful, with a wide range of commercially available systems.

Electrochemical sensors are constantly evolving. In recent years, the introduction of various nanomaterials, such as graphene, CNTs, and NPs has led to new developments. These together with flexible, stretchable, and transient/resorbable materials are paving the way for further developments in the field, with a particular focus on implantable medical devices.

References

1. S. Vaddiraju, I. Tomazos, D.J. Burgess, F.C. Jain, F. Papadimitrakopoulos, Emerging synergy between nanotechnology and implantable biosensors: a review. Biosens. Bioelectron. **25**(7), 1553–1565 (2010)
2. A. Lewenstam, Routines and challenges in clinical application of electrochemical ion-sensors. Electroanalysis **26**(6), 1171–1181 (2014)
3. K. Ueshima, Magnesium and ischemic heart disease: a review of epidemiological, experimental, and clinical evidences. Magnes. Res. **18**(4), 275–284 (2005)
4. S.P. Yu, L.M.T. Canzoniero, D.W. Choi, Ion homeostasis and apoptosis. Curr. Opin. Cell Biol. **13**(4), 405–411 (2001)
5. S.P. Yu, Regulation and critical role of potassium homeostasis in apoptosis. Prog. Neurobiol. **70**(4), 363–386 (2003)
6. J. Flores, D.R. DiBona, N. Frega, D.A. Leaf, Cell volume regulation and ischemic tissue damage. J. Membr. Biol. **10**(1), 331–343 (1972)
7. W.E. Cascio, G.X. Yan, A.G. Kléber, Early changes in extracellular potassium in ischemic rabbit myocardium. The role of extracellular carbon dioxide accumulation and diffusion. Circ. Res. **70**(2), 409–422 (1992)
8. A. Sola et al., Multiparametric monitoring of ischemia-reperfusion in rat kidney: effect of ischemic preconditioning. Transplantation **75**(6), 744–749 (2003)
9. O.T. Guenat, S. Generelli, N.F. de Rooij, M. Koudelka-Hep, F. Berthiaume, M.L. Yarmush, Development of an array of ion-selective microelectrodes aimed for the monitoring of extracellular ionic activities. Anal. Chem. **78**(21), 7453–7460 (2006)
10. M. Rossol et al., Extracellular Ca^{2+} is a danger signal activating the NLRP3 inflammasome through G protein-coupled calcium sensing receptors. Nat. Commun. **3**, 1329 (2012)
11. S. Kun, B. Ristic, R.A. Peura, R.M. Dunn, Algorithm for tissue ischemia estimation based on electrical impedance spectroscopy. IEEE Trans. Biomed. Eng. **50**(12), 1352–1359 (2003)
12. J. Wtorek, L. Jozefiak, A. Polinski, J. Siebert, An averaging two-electrode probe for monitoring changes in myocardial conductivity evoked by ischemia. IEEE Trans. Biomed. Eng. **49**(3), 240–246 (2002)
13. B. Ristic, S. Kun, R.A. Peura, Muscle tissue ischemia monitoring using impedance spectroscopy: quantitative results of animal studies, in *Proceedings of the* 19*th Annual International Conference of the IEEE Engineering in Medicine and Biology Society,* vol. 5 (1997), pp. 2108–2111, 5
14. S. Kun, R.A. Peura, Tissue ischemia detection using impedance spectroscopy, in *Proceedings of the* 16*th Annual International Conference of the IEEE Engineering in Medicine and Biology Society,* 1994. *Engineering Advances: New Opportunities for Biomedical Engineers*, vol. 2 (1994), pp. 868–869
15. C.A. González, C. Villanueva, S. Othman, R. Narváez, E. Sacristán, Impedance spectroscopy for monitoring ischemic injury in the intestinal mucosa. Physiol. Meas. **24** (2), 277 (2003)
16. H.-J. Chung et al., Stretchable, multiplexed pH sensors with demonstrations on rabbit and human hearts undergoing Ischemia. Adv. Healthc. Mater. **3**(1), 59–68 (2014)

17. M.S. Frant, Historical perspective. History of the early commercialization of ion-selective electrodes. Analyst **119**(11), 2293–2301 (1994)
18. E. Bakker, R. Meruva, E. Pretsch, M. Meyerhoff, Selectivity of polymer membrane-based ion-selective electrodes: self-consistent model describing the potentiometric response in mixed ion solutions of different charge. Anal. Chem. **66**(19), 3021–3030 (1994)
19. D. O'Hare, K.H. Parker, C.P. Winlove, Metal–metal oxide pH sensors for physiological application. Med. Eng. Phys. **28**(10), 982–988 (2006)
20. C. Fay et al., Wireless ion-selective electrode autonomous sensing system. IEEE Sens. J. **11** (10), 2374–2382 (2011)
21. W.-D. Huang, S. Deb, Y.-S. Seo, S. Rao, M. Chiao, J.C. Chiao, A passive radio-frequency pH-sensing tag for wireless food-quality monitoring. IEEE Sens. J. **12**(3), 487–495 (2012)
22. W.-D. Huang, H. Cao, S. Deb, M. Chiao, J.C. Chiao, A flexible pH sensor based on the iridium oxide sensing film. Sens. Actuators Phys. **169**(1), 1–11 (2011)
23. CN–0326: isolated low power pH monitor with temperature compensation. Analog Devices, Inc., 2013
24. O. Korostynska, K. Arshak, E. Gill, A. Arshak, Review paper: materials and techniques for in vivo pH monitoring. IEEE Sens. J. **8**(1), 20–28 (2008)
25. M. Cremer, Origin of electromotor properties of tissues, and instructional contribution for polyphasic electrolyte chains. Z. Für Biol. **47**, 562–608 (1906)
26. J. Ruzicka, The seventies—golden age for ion selective electrodes. J. Chem. Educ. **74**(2), 167 (1997)
27. J. Janata, Potentiometric microsensors. Chem. Rev. **90**(5), 691–703 (1990)
28. P. Steegstra, E. Ahlberg, Influence of oxidation state on the pH dependence of hydrous iridium oxide films. Electrochim. Acta **76**, 26–33 (2012)
29. A. Fog, R.P. Buck, Electronic semiconducting oxides as pH sensors. Sens. Actuators **5**(2), 137–146 (1984)
30. L.D. Burke, J.K. Mulcahy, D.P. Whelan, Preparation of an oxidized iridium electrode and the variation of its potential with pH. J. Electroanal. Chem. Interfacial Electrochem. **163**(1), 117–128 (1984)
31. K. Kinoshita, M.J. Madou, Electrochemical measurements on Pt, Ir, and Ti oxides as pH probes. J. Electrochem. Soc. **131**(5), 1089–1094 (1984)
32. E. Kinoshita, F. Ingman, G. Edwall, S. Thulin, S. Głąb, Polycrystalline and monocrystalline antimony, iridium and palladium as electrode material for pH-sensing electrodes. Talanta **33** (2), 125–134 (1986)
33. M.L. Hitchman, S. Ramanathan, Evaluation of iridium oxide electrodes formed by potential cycling as pH probes. Analyst **113**(1), 35–39 (1988)
34. J. Bobacka, Conducting polymer-based solid-state ion-selective electrodes. Electroanalysis **18**(1), 7–18 (2006)
35. J. Bobacka, A. Ivaska, A. Lewenstam, Potentiometric ion sensors. Chem. Rev. **108**(2), 329–351 (2008)
36. Z. Štefanac, W. Simon, Ion specific electrochemical behavior of macrotetrolides in membranes. Microchem. J. **12**(1), 125–132 (1967)
37. Z. Stefanac, W. Simon, Highly selective cation electrode systems based on in-vitro behavior of macrotetrolides in membranes. Chimica **20** (1966)
38. L.A.R. Pioda, H.A. Wachter, R.E. Dohner, W. Simon, Complexes of nonactin and monactin with sodium, potassium and ammonium ions. Helv. Chim. Acta **50**, 1373–1375 (1967)
39. A.C. Ion, E. Bakker, E. Pretsch, Potentiometric Cd^{2+}-selective electrode with a detection limit in the low ppt range. Anal. Chim. Acta **440**(2), 71–79 (2001)
40. A. Ceresa, E. Bakker, B. Hattendorf, D. Günther, E. Pretsch, Potentiometric polymeric membrane electrodes for measurement of environmental samples at trace levels: new requirements for selectivities and measuring protocols, and comparison with ICPMS. Anal. Chem. **73**(2), 343–351 (2001)
41. E. Bakker, M.E. Meyerhoff, Ionophore-based membrane electrodes: new analytical concepts and non-classical response mechanisms. Anal. Chim. Acta **416**(2), 121–137 (2000)

42. A. Shvarev, E. Bakker, Reversible electrochemical detection of nonelectroactive polyions. J. Am. Chem. Soc. **125**(37), 11192–11193 (2003)
43. M.E. Collison, G.V. Aebli, J. Petty, M.E. Meyerhoff, Potentiometric combination ion-carbon dioxide sensors for in vitro and in vivo blood measurements. Anal. Chem. **61**(21), 2365–2372 (1989)
44. D.L. Simpson, R.K. Kobos, Potentiometric microbiological assay of gentamicin, streptomycin, and neomycin with a carbon dioxide gas-sensing electrode. Anal. Chem. **55**(12), 1974–1977 (1983)
45. A.F. Bradley, Determination of blood-gases utilizing specially designed electrodes for PCO_2, PO_2, PO_2 and pH. Biomed. Sci. Instrum. **3**, 181–188 (1966)
46. R. Zahradník, P. Hobza, Z. Slanina, Calculations of Henry constants and partition coefficients using quantum chemical approach, in *Quantitative Structure-Activity Relationships*, ed. by M. Tichý (Birkhäuser Basel, 1976), p. 217–230
47. I.A. Pechenkina, K.N. Mikhelson, Materials for the ionophore-based membranes for ion-selective electrodes: problems and achievements (review paper). Russ. J. Electrochem. **51**(2), 93–102 (2015)
48. A. Radu, Y. Qin, S. Peper, A. Ceresa, E. Bakker, Improving the low detection limit of polymer-based ion selective electrodes with a plasticizer-free polymer containing a covalently immobilized Ca^{2+}-selective ionophore. Abstr. Pap. Am. Chem. Soc. **226**, U105–U105 (2003)
49. L.Y. Heng, E.A.H. Hall, One-step synthesis of K^{+}-selective methacrylic-acrylic copolymers containing grafted ionophore and requiring no plasticizer. Electroanalysis **12**(3), 178–186 (2000)
50. E. Malinowska, L. Gawart, P. Parzuchowski, G. Rokicki, Z. Brzózka, Novel approach of immobilization of calix[4]arene type ionophore in 'self-plasticized' polymeric membrane. Anal. Chim. Acta **421**(1), 93–101 (2000)
51. E. Bakker, P. Bühlmann, E. Pretsch, Carrier-based ion-selective electrodes and bulk optodes. 1. General characteristics. Chem. Rev. **97**(8), 3083–3132 (1997)
52. W.E. Morf, *The Principles of Ion-Selective Electrodes and of Membrane Transport* (Elsevier, New York, 2012)
53. E. Bakker, M. Willer, M. Lerchi, K. Seiler, E. Pretsch, Determination of complex formation constants of neutral cation-selective ionophores in solvent polymeric membranes. Anal. Chem. **66**(4), 516–521 (1994)
54. P. Anker, E. Wieland, D. Ammann, R.E. Dohner, R. Asper, W. Simon, Neutral carrier based ion-selective electrode for the determination of total calcium in blood serum. Anal. Chem. **53** (13), 1970–1974 (1981)
55. D. Ammann et al., Preparation of neutral ionophores for Alkali and Alkaline earth metal cations and their application in ion selective membrane electrodes. Helv. Chim. Acta **58**(6), 1535–1548 (1975)
56. Y. Qin, E. Bakker, Evaluation of the separate equilibrium processes that dictate the upper detection limit of neutral ionophore-based potentiometric sensors. Anal. Chem. **74**(13), 3134–3141 (2002)
57. E. Lindner, K. Toth, E. Pungor, Lead-selective neutral carrier based liquid membrane electrode. Anal. Chem. **56**(7), 1127–1131 (1984)
58. E. Bakker, A. Xu, E. Pretsch, Optimum composition of neutral carrier based pH electrodes. Anal. Chim. Acta **295**(3), 253–262 (1994)
59. M. Telting-Diaz, E. Bakker, Effect of lipophilic ion-exchanger leaching on the detection limit of carrier-based ion-selective electrodes. Anal. Chem. **73**(22), 5582–5589 (2001)
60. E. Lindner et al., Ion-selective membranes with low plasticizer content: electroanalytical characterization and biocompatibility studies. J. Biomed. Mater. Res. **28**(5), 591–601 (1994)
61. R. Lenigk, H. Zhu, T.-C. Lo, R. Renneberg, Recessed microelectrode array for a micro flow-through system allowing on-line multianalyte determination in vivo. Fresenius J. Anal. Chem. **364**(1–2), 66–71 (1999)

62. R.E. Gyurcsányi, N. Rangisetty, S. Clifton, B.D. Pendley, E. Lindner, Microfabricated ISEs: critical comparison of inherently conducting polymer and hydrogel based inner contacts. Talanta **63**(1), 89–99 (2004)
63. S.Y. Yun et al., Potentiometric properties of ion-selective electrode membranes based on segmented polyether urethane matrices. Anal. Chem. **69**(5), 868–873 (1997)
64. D.N. Reinhoudt et al., Development of durable K^+-selective chemically modified field effect transistors with functionalized polysiloxane membranes. Anal. Chem. **66**(21), 3618–3623 (1994)
65. G.J. Moody, B. Saad, J.D.R. Thomas, Glass transition temperatures of poly(vinyl chloride) and polyacrylate materials and calcium ion-selective electrode properties. Analyst **112**(8), 1143–1147 (1987)
66. L.Y. Heng, E.A.H. Hall, Methacrylic–acrylic polymers in ion-selective membranes: achieving the right polymer recipe. Anal. Chim. Acta **403**(1–2), 77–89 (2000)
67. E. Bakker, P. Bühlmann, E. Pretsch, Polymer membrane ion-selective electrodes-what are the limits? Electroanalysis **11**(13), 915–933 (1999)
68. S. Joo, R.B. Brown, Chemical sensors with integrated electronics. Chem. Rev. **108**(2), 638–651 (2008)
69. I.A. Ges, B.L. Ivanov, D.K. Schaffer, E.A. Lima, A.A. Werdich, F.J. Baudenbacher, Thin-film IrOx pH microelectrode for microfluidic-based microsystems. Biosens. Bioelectron. **21**(2), 248–256 (2005)
70. W. Olthuis, M.A.M. Robben, P. Bergveld, M. Bos, W.E. van der Linden, pH sensor properties of electrochemically grown iridium oxide. Sens. Actuators B Chem. **2**(4), 247–256 (1990)
71. S. Yao, M. Wang, M. Madou, A pH electrode based on melt-oxidized iridium oxide. J. Electrochem. Soc. **148**(4), H29–H36 (2001)
72. J. Kieninger, A. Marx, F. Spies, A. Weltin, G.A. Urban, G. Jobst, pH micro sensor with micro-fluidic liquid-junction reference electrode on-chip for cell culture applications. IEEE Sens. **2009**, 2009–2012
73. C.M. Nguyen et al., Sol-Gel iridium oxide-based pH sensor array on flexible polyimide substrate. IEEE Sens. J. **13**(10), 3857–3864 (2013)
74. M. Kubon et al., A microsensor system to probe physiological environments and tissue response. IEEE Sens. **2010**, 2607–2611 (2010)
75. M.D. Johnson, O.E. Kao, D.R. Kipke, Spatiotemporal pH dynamics following insertion of neural microelectrode arrays. J. Neurosci. Methods **160**(2), 276–287 (2007)
76. X. Yue et al., A real-time multi-channel monitoring system for stem cell culture process. IEEE Trans. Biomed. Circuits Syst. **2**(2), 66–77 (2008)
77. S.A.M. Marzouk, S. Ufer, R.P. Buck, T.A. Johnson, L.A. Dunlap, W.E. Cascio, Electrodeposited iridium oxide pH electrode for measurement of extracellular myocardial acidosis during acute ischemia. Anal. Chem. **70**(23), 5054–5061 (1998)
78. P.J. Kinlen, J.E. Heider, D.E. Hubbard, A solid-state pH sensor based on a Nafion-coated iridium oxide indicator electrode and a polymer-based silver chloride reference electrode. Sens. Actuators B Chem. **22**(1), 13–25 (1994)
79. E. Lindner et al., In vivo and in vitro testing of microelectronically fabricated planar sensors designed for applications in cardiology. Fresenius J. Anal. Chem. **346**(6–9), 584–588 (1993)
80. V.V. Cosofret, E. Lindner, T.A. Johnson, M.R. Neuman, Planar micro sensors for in vivo myocardial pH measurements. Talanta **41**(6), 931–938 (1994)
81. V.V. Cosofret, M. Erdosy, T.A. Johnson, R.P. Buck, R.B. Ash, M.R. Neuman, Microfabricated sensor arrays sensitive to pH and K^+ for ionic distribution measurements in the beating heart. Anal. Chem. **67**(10), 1647–1653 (1995)
82. S.A.M. Marzouk, R.P. Buck, L.A. Dunlap, T.A. Johnson, W.E. Cascio, Measurement of extracellular pH, K^+, and lactate in ischemic heart. Anal. Biochem. **308**(1), 52–60 (2002)
83. S. Anastasova-Ivanova et al., Development of miniature all-solid-state potentiometric sensing system. Sens. Actuators B Chem. **146**(1), 199–205 (2010)

84. G. Urban et al., Miniaturized multi-enzyme biosensors integrated with pH sensors on flexible polymer carriers for in vivo applications. Biosens. Bioelectron. **7**(10), 733–739 (1992)
85. H. Cao et al., An implantable, batteryless, and wireless capsule with integrated impedance and pH sensors for gastroesophageal reflux monitoring. IEEE Trans. Biomed. Eng. **59**(11), 3131–3139 (2012)
86. E. Bitziou, D. O'Hare, B.A. Patel, Spatial changes in acid secretion from isolated stomach tissue using a pH-histamine sensing microarray. Analyst **135**(3), 482–487 (2010)
87. E. Bitziou, D. O'Hare, B.A. Patel, Simultaneous detection of pH changes and histamine release from oxyntic glands in isolated stomach. Anal. Chem. **80**(22), 8733–8740 (2008)
88. I.A. Ges, B.L. Ivanov, A.A. Werdich, F.J. Baudenbacher, Differential pH measurements of metabolic cellular activity in nl culture volumes using microfabricated iridium oxide electrodes. Biosens. Bioelectron. **22**(7), 1303–1310 (2007)
89. I.A. Ges, I.A. Dzhura, F.J. Baudenbacher, On-chip acidification rate measurements from single cardiac cells confined in sub-nanoliter volumes. Biomed. Microdevices **10**(3), 347–354 (2008)
90. M. Mir, R. Lugo, I.B. Tahirbegi, J. Samitier, Miniaturizable ion-selective arrays based on highly stable polymer membranes for biomedical applications. Sensors **14**(7), 11844–11854 (2014)
91. W. Qin, T. Zwickl, E. Pretsch, Improved detection limits and unbiased selectivity coefficients obtained by using ion-exchange resins in the inner reference solution of ion-selective polymeric membrane electrodes. Anal. Chem. **72**(14), 3236–3240 (2000)
92. E.J. Parra, P. Blondeau, G.A. Crespo, F.X. Rius, An effective nanostructured assembly for ion-selective electrodes. An ionophore covalently linked to carbon nanotubes for Pb^{2+} determination. Chem. Commun. **47**(8), 2438–2440 (2011)
93. M.A. Simon, R.P. Kusy, Plasticizer-level study of poly(vinyl chloride) ion-selective membranes. J. Biomed. Mater. Res. **30**(3), 313–320 (1996)
94. V.G. Gavalas, M.J. Berrocal, L.G. Bachas, Enhancing the blood compatibility of ion-selective electrodes. Anal. Bioanal. Chem. **384**(1), 65–72 (2005)
95. M. Pawlak, E. Bakker, Chemical modification of polymer ion-selective membrane electrode surfaces. Electroanalysis **26**(6), 1121–1131 (2014)
96. S. Yajima, Y. Sonoyama, K. Suzuki, K. Kimura, Ion-sensor property and blood compatibility of neutral-carrier-type poly(vinyl chloride) membranes coated by phosphorylcholine polymers. Anal. Chim. Acta **463**(1), 31–37 (2002)
97. M.J. Berrocal, R.D. Johnson, I.H.A. Badr, M. Liu, D. Gao, L.G. Bachas, Improving the blood compatibility of ion-selective electrodes by employing poly(MPC-co-BMA), a copolymer containing phosphorylcholine, as a membrane coating. Anal. Chem. **74**(15), 3644–3648 (2002)
98. J.A. Hayward, D. Chapman, Biomembrane surfaces as models for polymer design: the potential for haemocompatibility. Biomaterials **5**(3), 135–142 (1984)
99. A. Ivorra et al., Minimally invasive silicon probe for electrical impedance measurements in small animals. Biosens. Bioelectron. **19**(4), 391–399 (2003)
100. G.P. Gumbrell, R.A. Peura, S. Kun, R.M. Dunn, Development of a pH based tissue ischemia monitor: hardware realization, in *Proceedings of the* 1996 *IEEE Twenty-Second Annual Northeast Bioengineering Conference*, 54–55 (1996)
101. G.P. Gumbrell, R.A. Peura, S. Kun, R.M. Dunn, Development of a minimally invasive microvascular ischemia monitor: drift reduction results, in *Proceedings of the* 19*th Annual International Conference of the IEEE Engineering in Medicine and Biology Society*, vol. 1, 25–27 (1997)
102. G.P. Gumbrell, R.A. Peura, S. Kun, R.M. Dunn, Development of a pH based tissue ischemia monitor: preliminary clinical results, in *Proceedings of the* 18*th Annual International Conference of the IEEE Engineering in Medicine and Biology Society,* 1996. *Bridging Disciplines for Biomedicine*, vol. 1, 40–41 (1996)

103. A.H. Auerbach, B.R. Soller, R.A. Peura, Sources of error in measuring tissue pH with microsensors, in *Proceedings of the* 20*th Annual Northeast Bioengineering Conference,* 108–109 (1994)
104. J. Songer, Tissue ischemia monitoring using impedance spectroscopy. MSc. Thesis, Worcester Polytechnic Institute, 2001
105. A. Senagore, D.J.W. Milsom, R.K. Walshaw, R. Dunstan, W.P. Mazier, I.H. Chaudry, Intramural pH: a quantitative measurement for predicting colorectal anastomotic healing. Dis. Colon Rectum **33**(3), 175–179 (1990)
106. S.G. Nugent, D. Kumar, D.S. Rampton, D.F. Evans, Intestinal luminal pH in inflammatory bowel disease: possible determinants and implications for therapy with aminosalicylates and other drugs. Gut **48**(4), 571–577 (2001)
107. D.C.J. McDougall, R. Wong, P. Scudera, M. Lesser, J.J. DeCosse, Colonic mucosal pH in humans. Dig. Dis. Sci. **38**(3), 542–545 (1993)
108. M. Millan, E. García-Granero, B. Flor, S. García-Botello, S. Lledo, Early prediction of anastomotic leak in colorectal cancer surgery by intramucosal pH. Dis. Colon Rectum **49**(5), 595–601 (2006)
109. C.A. Eckley, H.O. Costa, Comparative study of salivary pH and volume in adults with chronic laryngopharyngitis by gastroesophageal reflux disease before and after treatment. Rev. Bras. Otorrinolaringol. **72**(1), 55–60 (2006)
110. S. Ghimenti et al., Measurement of warfarin in the oral fluid of patients undergoing anticoagulant oral therapy. PLoS ONE **6**(12), e28182 (2011)
111. J.L. Gonzalez-Guillaumin, D.C. Sadowski, K.V.I.S. Kaler, M.P. Mintchev, Ingestible capsule for impedance and pH monitoring in the esophagus. IEEE Trans. Biomed. Eng. **54** (12), 2231–2236 (2007)
112. G.X. Yan, K.A. Yamada, A.G. Kléber, J. McHowat, P.B. Corr, Dissociation between cellular K^+ loss, reduction in repolarization time, and tissue ATP levels during myocardial hypoxia and ischemia. Circ. Res. **72**(3), 560–570 (1993)
113. J.N. Weiss, S.T. Lamp, K.I. Shine, Cellular K^+ loss and anion efflux during myocardial ischemia and metabolic inhibition. Am. J. Physiol. Heart Circ. Physiol. **256**(4), H1165–H1175 (1989)
114. B.F. Palmer, Regulation of potassium homeostasis. Clin. J. Am. Soc. Nephrol. **10**(6), 1050–1060 (2015)
115. A.G. Kléber, C.B. Riegger, M.J. Janse, Extracellular K^+ and H^+ shifts in early ischemia: mechanisms and relation to changes in impulse propagation. J. Mol. Cell. Cardiol. **19**(Suppl 5), 35–44 (1987)
116. I.A. Marques de Oliveira et al., Sodium ion sensitive microelectrode based on a p-tert-butylcalix[4]arene ethyl ester. Sens. Actuators B Chem. **130**(1), 295–299 (2008)
117. S. Chandra, H. Lang, A new sodium ion selective electrode based on a novel silacrown ether. Sens. Actuators B Chem. **114**(2), 849–854 (2006)
118. K.Y. Chumbimuni-Torres, N. Rubinova, A. Radu, L.T. Kubota, E. Bakker, Solid contact potentiometric sensors for trace level measurements. Anal. Chem. **78**(4), 1318–1322 (2006)
119. F. Li et al., All-solid-state potassium-selective electrode using graphene as the solid contact. Analyst **137**(3), 618–623 (2012)
120. N. Zine et al., Potassium-ion selective solid contact microelectrode based on a novel 1,3-(di-4-oxabutanol)-calix[4]arene-crown–5 neutral carrier. Electrochim. Acta **51**(24), 5075–5079 (2006)
121. J.E. Pandolfino, J.E. Richter, T. Ours, J.M. Guardino, J. Chapman, P.J. Kahrilas, Ambulatory esophageal pH monitoring using a wireless system. Am. J. Gastroenterol. **98** (4), 740–749 (2003)

122. G. Karamanolis et al., Bravo 48-hour wireless pH monitoring in patients with non-cardiac chest pain. Objective gastroesophageal reflux disease parameters predict the responses to proton pump inhibitors. J. Neurogastroenterol. Motil. **18**(2), 169–173 (2012)
123. E. Lindner, R. Buck, Microfabricated potentiometric electrodes and their in vivo applications. Anal. Chem. **72**(9), 336 A–345 A (2000)
124. I.B. Tahirbegi, M. Mir, S. Schostek, M. Schurr, J. Samitier, In vivo ischemia monitoring array for endoscopic surgery. Biosens. Bioelectron. **61**, 124–130 (2014)
125. A. Errachid, A. Ivorra, J. Aguiló, R. Villa, N. Zine, J. Bausells, New technology for multi-sensor silicon needles for biomedical applications. Sens. Actuators B Chem. **78**(1–3), 279–284 (2001)
126. R. Gómez et al., A SiC microdevice for the minimally invasive monitoring of ischemia in living tissues. Biomed. Microdevices **8**(1), 43–49 (2006)
127. M. Genescà et al., Electrical bioimpedance measurement during hypothermic rat kidney preservation for assessing ischemic injury. Biosens. Bioelectron. **20**(9), 1866–1871 (2005)
128. M. Tijero et al., SU–8 microprobe with microelectrodes for monitoring electrical impedance in living tissues. Biosens. Bioelectron. **24**(8), 2410–2416 (2009)
129. L. Xu et al., 3D multifunctional integumentary membranes for spatiotemporal cardiac measurements and stimulation across the entire epicardium. Nat. Commun. **5**, 3329 (2014)
130. I.B. Tahirbegi, M. Mir, J. Samitier, Real-time monitoring of ischemia inside stomach. Biosens. Bioelectron. **40**(1), 323–328 (2013)
131. P. Bergveld, Thirty years of ISFETOLOGY: what happened in the past 30 years and what may happen in the next 30 years. Sens. Actuators B Chem. **88**(1), 1–20 (2003)
132. P.A. Hammond, D. Ali, D.R.S. Cumming, Design of a single-chip pH sensor using a conventional 0.6 μm CMOS process. IEEE Sens. J. **4**(6), 706–712 (2004)
133. P. Georgiou, C. Toumazou, ISFET characteristics in CMOS and their application to weak inversion operation. Sens. Actuators B Chem. **143**(1), 211–217 (2009)
134. D. Binkley, *Tradeoffs and optimization in analog CMOS design* (Wiley, Chichester, West Sussex, UK, 2008)
135. A. Al-Ahdal, C. Toumazou, ISFET-based chemical switch. IEEE Sens. J. **12**(5), 1140–1146 (2012)
136. L.M. Shepherd, C. Toumazou, A biochemical translinear principle with weak inversion ISFETs. IEEE Trans. Circuits Syst. Regul. Pap. **52**(12), 2614–2619 (2005)
137. Y. Liu, P. Georgiou, T. Prodromakis, T.G. Constandinou, C. Toumazou, An extended CMOS ISFET model incorporating the physical design geometry and the effects on performance and offset variation. IEEE Trans. Electron Devices **58**(12), 4414–4422 (2011)
138. N. Miscourides, P. Georgiou, Impact of technology scaling on ISFET performance for genetic sequencing. IEEE Sens. J. **15**(4), 2219–2226 (2015)
139. M. Sohbati, Y. Liu, P. Georgiou, C. Toumazou, An ISFET design methodology incorporating CMOS passivation, *IEEE Biomedical Circuits and SystemsConference (BioCAS),* 65–68 (2012)
140. M. Sohbati, C. Toumazou, Dimension and shape effects on the ISFET performance. IEEE Sens. J. **15**(3), 1670–1679 (2015)
141. J. Bausells, J. Carrabina, A. Errachid, A. Merlos, Ion-sensitive field-effect transistors fabricated in a commercial CMOS technology. Sens. Actuators B Chem. **57**(1–3), 56–62 (1999)
142. M.J. Milgrew, D.R.S. Cumming, Matching the transconductance characteristics of CMOS ISFET arrays by removing trapped charge. IEEE Trans. Electron Devices **55**(4), 1074–1079 (2008)
143. M.J. Milgrew, M.O. Riehle, D.R.S. Cumming, A large transistor-based sensor array chip for direct extracellular imaging. Sens. Actuators B Chem. **111–112**, 347–353 (2005)

144. T. Prodromakis, Y. Liu, T. Constandinou, P. Georgiou, C. Toumazou, Exploiting CMOS technology to enhance the performance of ISFET sensors. IEEE Electron Device Lett. **31**(9), 1053–1055 (2010)
145. J.M. Rothberg, W. Hinz, K.L. Johnson, J. Bustillo, Methods and apparatus for measuring analytes using large scale fet arrays. CA2672315 A1, 26 Jun 2008
146. M. Milgrew, J. Bustillo, T. Rearick, Chemically-sensitive field effect transistor based pixel array with protection diodes. US20130001653 A1, 03 Jan 2013
147. H.-S. Wong, M.H. White, A CMOS-integrated `ISFET-operational amplifier' chemical sensor employing differential sensing. IEEE Trans. Electron Devices **36**(3), 479–487 (1989)
148. V.P. Chodavarapu, A.H. Titus, A.N. Cartwright, Differential read-out architecture for CMOS ISFET microsystems. Electron. Lett. **41**(12), 698–699 (2005)
149. D. Garner, H. Bai, Electrostatic discharge protection. US20130188288 A1, 25 July 2013
150. R. Smith, R.J. Huber, J. Janata, Electrostatically protected ion sensitive field effect transistors. Sens. Actuators **5**(2), 127–136 (1984)
151. Y. Hu, P. Georgiou, A robust ISFET pH-measuring front-end for chemical reaction monitoring. IEEE Trans. Biomed. Circuits Syst. **8**(2), 177–185 (2014)
152. P. Georgiou, C. Toumazou, CMOS-based programmable gate ISFET. Electron. Lett. **44**(22), 1289–1290 (2008)
153. A. Al-Ahdal, P. Georgiou, C. Toumazou, ISFET's threshold voltage control using bidirectional electron tunnelling, in *2012 IEEE Biomedical Circuits and Systems Conference (BioCAS)*, 172–175 (2012)
154. P. Georgiou, C. Toumazou, ISFET threshold voltage programming in CMOS using hot-electron injection. Electron. Lett. **45**(22), 1112–1113 (2009)
155. A.G. Al-Ahdal, C. Toumazou, ISFET threshold voltage programming in CMOS using electron tunnelling. Electron. Lett. **47**(25), 1398–1399 (2011)
156. J.M. Rothberg et al., An integrated semiconductor device enabling non-optical genome sequencing. Nature **475**(7356), 348–352 (2011)
157. C. Toumazou et al., Simultaneous DNA amplification and detection using a pH-sensing semiconductor system. Nat. Methods **10**(7), 641–646 (2013)
158. X. Huang, H. Yu, X. Liu, Y. Jiang, M. Yan, D. Wu, A dual-mode large-arrayed CMOS ISFET sensor for accurate and high-throughput pH sensing in biomedical diagnosis. IEEE Trans. Biomed. Eng. **62**(9), 2224–2233 (2015)
159. P. Rai, S. Jung, T. Ji, V.K. Varadan, Drain current centric modality: instrumentation and evaluation of ISFET for monitoring myocardial ischemia like variations in pH and potassium ion concentration. IEEE Sens. J. **9**(12), 1987–1995 (2009)
160. F. Xu, G. Yan, Z. Wang, P. Jiang, Continuous accurate pH measurements of human GI tract using a digital pH-ISFET sensor inside a wireless capsule. Measurement **64**, 49–56 (2015)
161. P.A. Hammond, D. Ali, D.R.S. Cumming, A system-on-chip digital pH meter for use in a wireless diagnostic capsule. IEEE Trans. Biomed. Eng. **52**(4), 687–694 (2005)
162. C.-S. Lee, S.K. Kim, M. Kim, Ion-sensitive field-effect transistor for biological sensing. Sensors **9**(9), 7111–7131 (2009)
163. S. Migita, K. Ozasa, T. Tanaka, T. Haruyama, Enzyme-based field-effect transistor for adenosine triphosphate (ATP) sensing. Anal. Sci. **23**(1), 45–48 (2007)
164. A. Bratov, N. Abramova, A. Ipatov, Recent trends in potentiometric sensor arrays—a review. Anal. Chim. Acta **678**(2), 149–159 (2010)
165. J. Janata, *Principles of Chemical Sensors* (Springer Science & Business Media, 2010)
166. T.-W. Huang, J.-C. Chou, T.-P. Sun, S.-K. Hsiung, Fabrication of a screen-printing reference electrode for potentiometric measurement. Sens. Lett. **6**(6), 860–863 (2008)
167. W. Vonau, W. Oelßner, U. Guth, J. Henze, An all-solid-state reference electrode. Sens. Actuators B Chem. **144**(2), 368–373 (2010)
168. J. Sutter, E. Lindner, R.E. Gyurcsányi, E. Pretsch, A polypyrrole-based solid-contact Pb^{2+}-selective PVC-membrane electrode with a nanomolar detection limit. Anal. Bioanal. Chem. **380**(1), 7–14 (2004)

169. A. Cadogan, Z. Gao, A. Lewenstam, A. Ivaska, D. Diamond, All-solid-state sodium-selective electrode based on a calixarene ionophore in a poly(vinyl chloride) membrane with a polypyrrole solid contact. Anal. Chem. **64**(21), 2496–2501 (1992)
170. G.D. O'Neil, R. Buiculescu, S.P. Kounaves, N.A. Chaniotakis, Carbon-nanofiber-based nanocomposite membrane as a highly stable solid-state junction for reference electrodes. Anal. Chem. **83**(14), 5749–5753 (2011)
171. H.J. Yoon et al., Solid-state ion sensors with a liquid junction-free polymer membrane-based reference electrode for blood analysis. Sens. Actuators B Chem. **64**(1–3), 8–14 (2000)
172. G. Valdés-Ramírez, G.A. Álvarez-Romero, C.A. Galán-Vidal, P.R. Hernández-Rodríguez, M.T. Ramírez-Silva, Composites: a novel alternative to construct solid state Ag/AgCl reference electrodes. Sens. Actuators B Chem. **110**(2), 264–270 (2005)
173. D. Rehm, E. McEnroe, D. Diamond, An all solid-state reference electrode based on a potassium chloride doped vinyl ester resin. Anal. Proc. Anal. Commun. **32**(8), 319–322 (1995)
174. R. Mamińska, A. Dybko, W. Wróblewski, All-solid-state miniaturised planar reference electrodes based on ionic liquids. Sens. Actuators B Chem. **115**(1), 552–557 (2006)
175. D. Cicmil et al., Ionic liquid-based, liquid-junction-free reference electrode. Electroanalysis **23**(8), 1881–1890 (2011)
176. Ł. Tymecki, E. Zwierkowska, R. Koncki, Screen-printed reference electrodes for potentiometric measurements. Anal. Chim. Acta **526**(1), 3–11 (2004)
177. A. Kisiel, A. Michalska, K. Maksymiuk, E.A.H. Hall, All-solid-state reference electrodes with poly(n-butyl acrylate) based membranes. Electroanalysis **20**(3), 318–323 (2008)
178. A. Kisiel, M. Donten, J. Mieczkowski, F.X. Rius-Ruiz, K. Maksymiuk, A. Michalska, Polyacrylate microspheres composite for all-solid-state reference electrodes. Analyst **135**(9), 2420–2425 (2010)
179. G.A. Crespo, S. Macho, F.X. Rius, Ion-selective electrodes using carbon nanotubes as ion-to-electron transducers. Anal. Chem. **80**(4), 1316–1322 (2008)
180. G.A. Crespo, S. Macho, J. Bobacka, F.X. Rius, Transduction mechanism of carbon nanotubes in solid-contact ion-selective electrodes. Anal. Chem. **81**(2), 676–681 (2009)
181. F.X. Rius-Ruiz, D. Bejarano-Nosas, P. Blondeau, J. Riu, F.X. Rius, Disposable planar reference electrode based on carbon nanotubes and polyacrylate membrane. Anal. Chem. **83** (14), 5783–5788 (2011)
182. F.X. Rius-Ruiz, A. Kisiel, A. Michalska, K. Maksymiuk, J. Riu, F.X. Rius, Solid-state reference electrodes based on carbon nanotubes and polyacrylate membranes. Anal. Bioanal. Chem. **399**(10), 3613–3622 (2011)
183. Z. Mousavi, K. Granholm, T. Sokalski, A. Lewenstam, An analytical quality solid-state composite reference electrode. The Analyst **138**(18), 5216–5220 (2013)
184. T. Guinovart, G.A. Crespo, F.X. Rius, F.J. Andrade, A reference electrode based on polyvinyl butyral (PVB) polymer for decentralized chemical measurements. Anal. Chim. Acta **821**, 72–80 (2014)
185. A.K. Ghosh, V. Ramachandhran, M.S. Hanra, B.M. Misra, Studies on fouling and gel polarisation aspects of polyvinyl butyral blended cellulose acetate ultrafiltration membrane by resistance model approach. Indian J. Chem. Technol. **7**(2), 55–60 (2000)
186. Y. Saito, M. Okano, K. Kubota, T. Sakai, J. Fujioka, T. Kawakami, Evaluation of interactive effects on the ionic conduction properties of polymer gel electrolytes. J. Phys. Chem. B **116** (33), 10089–10097 (2012)
187. E. Bakker, Hydrophobic membranes as liquid junction-free reference electrodes. Electroanalysis **11**(10–11), 788–792 (1999)
188. T. Kakiuchi, T. Yoshimatsu, N. Nishi, New class of Ag/AgCl electrodes based on hydrophobic ionic liquid saturated with AgCl. Anal. Chem. **79**(18), 7187–7191 (2007)
189. C. Zuliani, G. Matzeu, D. Diamond, A liquid-junction-free reference electrode based on a PEDOT solid-contact and ionogel capping membrane. Talanta **125**, 58–64 (2014)

190. M. Shibata, H. Sakaida, T. Kakiuchi, Determination of the activity of hydrogen ions in dilute sulfuric acids by use of an ionic liquid salt bridge sandwiched by two hydrogen electrodes. Anal. Chem. **83**(1), 164–168 (2011)
191. U. Guth, F. Gerlach, M. Decker, W. Oelßner, W. Vonau, Solid-state reference electrodes for potentiometric sensors. J. Solid State Electrochem. **13**(1), 27–39 (2008)
192. B. Palán, F.V. Santos, J.M. Karam, B. Courtois, M. Husák, New ISFET sensor interface circuit for biomedical applications. Sens. Actuators B Chem. **57**(1–3), 63–68 (1999)
193. M.F. Chaplin, C. Bucke, *Enzyme Technology* (CUP Archive, 1990)
194. G.N. Meloni, Building a microcontroller based potentiostat: a inexpensive and versatile platform for teaching electrochemistry and instrumentation. J. Chem. Educ. **93**(7), 1320–1322 (2016)
195. M.D.M. Dryden, A.R. Wheeler, DStat: a versatile, open-source potentiostat for electroanalysis and integration. PLoS ONE **10**(10), e0140349 (2015)
196. R.S. Freire, C.A. Pessoa, L.D. Mello, L.T. Kubota, Direct electron transfer: an approach for electrochemical biosensors with higher selectivity and sensitivity. J. Braz. Chem. Soc. **14**(2), 230–243 (2003)
197. B.P. Schaffar, Thick film biosensors for metabolites in undiluted whole blood and plasma samples. Anal. Bioanal. Chem. **372**(2), 254–260 (2002)
198. Z. Chen, C. Fang, H. Wang, J. He, Disposable glucose test strip for whole blood with integrated sensing/diffusion-limiting layer. Electrochim. Acta **55**(2), 544–550 (2009)
199. R.A. Croce, S. Vaddiraju, F. Papadimitrakopoulos, F.C. Jain, Theoretical analysis of the performance of glucose sensors with layer-by-layer assembled outer membranes. Sensors **12** (10), 13402–13416 (2012)
200. F. Valentini, L. Galache Fernàndez, E. Tamburri, G. Palleschi, Single Walled Carbon Nanotubes/polypyrrole-GOx composite films to modify gold microelectrodes for glucose biosensors: Study of the extended linearity. Biosens. Bioelectron. **43**, 75–78 (2013)
201. A. Heller, B. Feldman, Electrochemical glucose sensors and their applications in diabetes management. Chem. Rev. **108**(7), 2482–2505 (2008)
202. Z. Zhou, L. Xu, S. Wu, B. Su, A novel biosensor array with a wheel-like pattern for glucose, lactate and choline based on electrochemiluminescence imaging. Analyst **139**(19), 4934–4939 (2014)
203. Y. Liu, Y. Dong, C.X. Guo, Z. Cui, L. Zheng, C.M. Li, Protein-directed in situ synthesis of gold nanoparticles on reduced graphene oxide modified electrode for nonenzymatic glucose sensing. Electroanalysis **24**(12), 2348–2353 (2012)
204. Y. Fu et al., One-pot preparation of polymer–enzyme–metallic nanoparticle composite films for high-performance biosensing of glucose and galactose. Adv. Funct. Mater. **19**(11), 1784–1791 (2009)
205. D. Xu, L. Luo, Y. Ding, P. Xu, Sensitive electrochemical detection of glucose based on electrospun $La_{0.88}Sr_{0.12}MnO_3$ nanofibers modified electrode. Anal. Biochem. **489**, 38–43 (2015)
206. N.G. Poulos, J.R. Hall, M.C. Leopold, functional layer-by-layer design of xerogel-based first-generation amperometric glucose biosensors. Langmuir **31**(4), 1547–1555 (2015)
207. M.M. Ahmadi, G.A. Jullien, A wireless-implantable microsystem for continuous blood glucose monitoring. IEEE Trans. Biomed. Circuits Syst. **3**(3), 169–180 (2009)
208. Medtronic MiniMed 530G. [Online]. Available: http://www.medtronicdiabetes.com/products/minimed–530g-diabetes-system-with-enlite. Accessed: 28 Jan 2016
209. Dexcom G5TM Mobile Continuous Glucose Monitoring (CGM) System. [Online]. Available: http://www.dexcom.com/g5-mobile-cgm. Accessed: 28 Jan 2016
210. FreeStyle Libre| FreeStyle Blood Glucose Meters. [Online]. Available: https://abbottdiabetescare.co.uk/our-products/freestyle-libre. Accessed: 26 Jan 2016
211. D. De Backer, J. Creteur, H. Zhang, M. Norrenberg, J.-L. Vincent, Lactate production by the lungs in acute lung injury. Am. J. Respir. Crit. Care Med. **156**(4), 1099–1104 (1997)
212. J.A. Kruse, S.A.J. Zaidi, R.W. Carlson, Significance of blood lactate levels in critically III patients with liver disease. Am. J. Med. **83**(1), 77–82 (1987)

213. S. Mm, M. Pn, Adenine nucleotide and lactate metabolism in the lung in endotoxin shock. Circ. Shock **8**(6), 657–666 (1980)
214. J. Karlsson, J.T. Willerson, S.J. Leshin, C.B. Mullins, J.H. Mitchell, Skeletal muscle metabolites in patients with cardiogenic shock or severe congestive heart failure. Scand. J. Clin. Lab. Invest. **35**(1), 73–79 (1975)
215. R. Rimachi, F. Bruzzi de Carvahlo, C. Orellano-Jimenez, F. Cotton, J.L. Vincent, D. De Backer, Lactate/pyruvate ratio as a marker of tissue hypoxia in circulatory and septic shock. Anaesth. Intensive Care **40**(3), 427–432 (2012)
216. B. Carbonne, K. Pons, E. Maisonneuve, Foetal scalp blood sampling during labour for pH and lactate measurements. Best Pract. Res. Clin. Obstet. Gynaecol. **30**, 62–67 (2016)
217. R.K.D. Suveera Dhup, Multiple biological activities of lactic acid in cancer: influences on tumor growth, angiogenesis and metastasis. Curr. Pharm. Des. **18**(10), 1319–1330 (2012)
218. K.M. Kennedy, M.W. Dewhirst, Tumor metabolism of lactate: the influence and therapeutic potential for MCT and CD147 regulation. Future Oncol. **6**(1), 127–148 (2009)
219. R.M.A. Bhatia et al., Application of rapid-sampling, online microdialysis to the monitoring of brain metabolism during aneurysm surgery. Neurosurgery **58**(4) (2006)
220. J.C. Goodman, A.B. Valadka, S.P. Gopinath, M. Uzura, C.S.M. Robertson, Extracellular lactate and glucose alterations in the brain after head injury measured by microdialysis. [Miscellaneous Article]. Crit. Care Med. **27**(9), 1965–1973 (1999)
221. E.L. Cureton, R.O. Kwan, K.C. Dozier, J. Sadjadi, J.D. Pal, G.P. Victorino, A different view of lactate in trauma patients: protecting the injured brain. J. Surg. Res. **159**(1), 468–473 (2010)
222. E. Naylor et al., Lactate as a biomarker for sleep. Sleep **35**(9), 1209–1222 (2012)
223. L. Rassaei, W. Olthuis, S. Tsujimura, E.J.R. Sudhölter, A. van den Berg, Lactate biosensors: current status and outlook. Anal. Bioanal. Chem. **406**(1), 123–137 (2013)
224. G.F. Manbeck, E. Fujita, A review of iron and cobalt porphyrins, phthalocyanines and related complexes for electrochemical and photochemical reduction of carbon dioxide. J. Porphyr. Phthalocyanines **19**(01–03), 45–64 (2015)
225. M.R. Romero, F. Ahumada, F. Garay, A.M. Baruzzi, Amperometric biosensor for direct blood lactate detection. Anal. Chem. **82**(13), 5568–5572 (2010)
226. S.A. Bhakta, E. Evans, T.E. Benavidez, C.D. Garcia, Protein adsorption onto nanomaterials for the development of biosensors and analytical devices: a review. Anal. Chim. Acta **872**, 7–25 (2015)
227. M. Gamero, F. Pariente, E. Lorenzo, C. Alonso, Nanostructured rough gold electrodes for the development of lactate oxidase-based biosensors. Biosens. Bioelectron. **25**(9), 2038–2044 (2010)
228. X.-R. He et al., Amperometric L-lactate biosensor based on sol-gel film and multi-walled carbon nanotubes/platinum nanoparticles enhancement. Chin. J. Anal. Chem. Chin. Version **38**(1), 57–61 (2010)
229. Y. Yu, Y. Yang, H. Gu, T. Zhou, G. Shi, Size-tunable Pt nanoparticles assembled on functionalized ordered mesoporous carbon for the simultaneous and on-line detection of glucose and L-lactate in brain microdialysate. Biosens. Bioelectron. **41**, 511–518 (2013)
230. J.M. Goran, J.L. Lyon, K.J. Stevenson, Amperometric detection of l-Lactate using nitrogen-doped carbon nanotubes modified with lactate oxidase. Anal. Chem. **83**(21), 8123–8129 (2011)
231. L. Agüí, M. Eguílaz, C. Peña-Farfal, P. Yáñez-Sedeño, J.M. Pingarrón, Lactate dehydrogenase biosensor based on an hybrid carbon nanotube-conducting polymer modified electrode. Electroanalysis **21**(3–5), 386–391 (2009)
232. I. Shakir, M. Shahid, H.W. Yang, S. Cherevko, C.-H. Chung, D.J. Kang, α-MoO3 nanowire-based amperometric biosensor for l-lactate detection. J. Solid State Electrochem. **16**(6), 2197–2201 (2012)
233. A.C. Pereira, A. Kisner, C.R.T. Tarley, L.T. Kubota, Development of a carbon paste electrode for lactate detection based on Meldola's blue adsorbed on silica gel modified with niobium oxide and lactate oxidase. Electroanalysis **23**(6), 1470–1477 (2011)

234. Y.T. Wang et al., A novel l-lactate sensor based on enzyme electrode modified with ZnO nanoparticles and multiwall carbon nanotubes. J. Electroanal. Chem. **661**(1), 8–12 (2011)
235. U. Spohn, D. Narasaiah, L. Gorton, The influence of the carbon paste composition on the performance of an amperometric bienzyme sensor for L-lactate. Electroanalysis **8**(6), 507–514 (1996)
236. M.R. Romero, F. Garay, A.M. Baruzzi, Design and optimization of a lactate amperometric biosensor based on lactate oxidase cross-linked with polymeric matrixes. Sens. Actuators B Chem. **131**(2), 590–595 (2008)
237. M.M. Rahman, M.J.A. Shiddiky, M.A. Rahman, Y.-B. Shim, A lactate biosensor based on lactate dehydrogenase/nictotinamide adenine dinucleotide (oxidized form) immobilized on a conducting polymer/multiwall carbon nanotube composite film. Anal. Biochem. **384**(1), 159–165 (2009)
238. E. Al-Jawadi, S. Pöller, R. Haddad, W. Schuhmann, NADH oxidation using modified electrodes based on lactate and glucose dehydrogenase entrapped between an electrocatalyst film and redox catalyst-modified polymers. Microchim. Acta **177**(3–4), 405–410 (2012)
239. E.I. Yashina et al., Sol-gel immobilization of lactate oxidase from organic solvent: toward the advanced lactate biosensor. Anal. Chem. **82**(5), 1601–1604 (2010)
240. M. Tsuchiya, H. Matsuhisa, Y. Hasebe, Selective amperometric response to hydrogen peroxide at a protein-incorporated sol-gel hybrid film-modified platinum electrode. Bunseki Kagaku **61**(5), 425–428 (2012)
241. F. Palmisano, G.E.D. Benedetto, C.G. Zambonin, Lactate amperometric biosensor based on an electrosynthesizedbilayer film with covalently immobilized enzyme. Analyst **122**(4), 365–369 (1997)
242. C.-L. Lin, C.-L. Shih, L.-K. Chau, Amperometric l-Lactate sensor based on sol-gel processing of an enzyme-linked silicon alkoxide. Anal. Chem. **79**(10), 3757–3763 (2007)
243. E. Szymańska, K. Winnicka, Stability of chitosan—a challenge for pharmaceutical and biomedical applications. Mar. Drugs **13**(4), 1819–1846 (2015)
244. R. Garjonyte, V. Melvydas, A. Malinauskas, Mediated amperometric biosensors for lactic acid based on carbon paste electrodes modified with baker's yeast *Saccharomyces cerevisiae*. Bioelectrochemistry **68**(2), 191–196 (2006)
245. M. Piano, S. Serban, R. Pittson, G.A. Drago, J.P. Hart, Amperometric lactate biosensor for flow injection analysis based on a screen-printed carbon electrode containing Meldola's Blue-Reinecke salt, coated with lactate dehydrogenase and NAD^+. Talanta **82**(1), 34–37 (2010)
246. N. Hamdi, J. Wang, H.G. Monbouquette, Polymer films as permselective coatings for H_2O_2-sensing electrodes. J. Electroanal. Chem. **581**(2), 258–264 (2005)
247. K. Bridge, F. Davis, S. Collyer, S.P.J. Higson, Flexible ultrathin PolyDVB/EVB composite membranes for the optimization of a whole blood glucose sensor. Electroanalysis **19**(4), 487–495 (2007)
248. K. Bridge, F. Davis, S.D. Collyer, S.P.J. Higson, Flexible ultrathin PolyDVB/EVB composite membranes for the optimization of a lactate sensor. Electroanalysis **19**(5), 567–574 (2007)
249. S. Cosnier, Biosensors based on electropolymerized films: new trends. Anal. Bioanal. Chem. **377**(3), 507–520 (2003)
250. C. Qin et al., Amperometric enzyme electrodes of glucose and lactate based on poly (diallyldimethylammonium)-alginate-metal ion-enzyme biocomposites. Anal. Chim. Acta **720**, 49–56 (2012)
251. A. Radoi, D. Moscone, G. Palleschi, Sensing the lactic acid in probiotic yogurts using an L-Lactate biosensor coupled with a microdialysis fiber inserted in a flow analysis system. Anal. Lett. **43**(7–8), 1301–1309 (2010)
252. J.J. Burmeister, M. Palmer, G.A. Gerhardt, l-lactate measures in brain tissue with ceramic-based multisite microelectrodes. Biosens. Bioelectron. **20**(9), 1772–1779 (2005)
253. J. Park, J. Chang, M. Choi, J.J. Pak, D.-Y. Lee, Y.K. Pak, Microfabirated clark-type sensor for measuring dissolved oxygen. IEEE Sens. 1412–1415 (2007)

254. K.K. Tremper, T.W. Rutter, J.A. Wahr, Monitoring oxygenation. Curr. Anaesth. Crit. Care **4** (4), 213–222 (1993)
255. C. Cody, D.J. Buggy, B. Marsh, D.C. Moriarity, Subcutaneous tissue oxygen tension after coronary revascularisation with and without cardiopulmonary bypass. Anaesthesia **59**(3), 237–242 (2004)
256. I. Bromley, Transcutaneous monitoring—understanding the principles. Infant **4**(3), 95–98 (2008)
257. L.S. Mortensen et al., Identifying hypoxia in human tumors: a correlation study between 18F-FMISO PET and the Eppendorf oxygen-sensitive electrode. Acta Oncol. **49**(7), 934–940 (2010)
258. T.H. Williamson, J. Grewal, B. Gupta, B. Mokete, M. Lim, C.H. Fry, Measurement of PO2 during vitrectomy for central retinal vein occlusion, a pilot study. Graefes Arch. Clin. Exp. Ophthalmol. **247**(8), 1019–1023 (2009)
259. Y.-H. Park, Y.-B. Shui, D.C. Beebe, Comparison of two probe designs for determining intraocular oxygen distribution. Br. J. Ophthalmol. p. bjo.2010.186064 (2010)
260. L. Toma-Daşu, A. Waites, A. Daşu, J. Denekamp, Theoretical simulation of oxygen tension measurement in tissues using a microelectrode: I. The response function of the electrode. Physiol. Meas. **22**(4), 713–725 (2001)
261. R.A. Linsenmeier, C.M. Yancey, Improved fabrication of double-barreled recessed cathode O_2 microelectrodes. J. Appl. Physiol. Bethesda Md 1985, **63**(6), 2554–2557 (1987)
262. F.B. Bolger et al., Characterisation of carbon paste electrodes for real-time amperometric monitoring of brain tissue oxygen. J. Neurosci. Methods **195**(2), 135–142 (2011)
263. G.S. Wilson, R. Gifford, Biosensors for real-time in vivo measurements. Biosens. Bioelectron. **20**(12), 2388–2403 (2005)
264. G.S. Wilson, M. Ammam, In vivo biosensors. FEBS J. **274**(21), 5452–5461 (2007)
265. A. Guiseppi-Elie, S. Brahim, G. Slaughter, K.R. Ward, Design of a subcutaneous implantable biochip for monitoring of glucose and lactate. IEEE Sens. J. **5**(3), 345–355 (2005)
266. A.R.A. Rahman, G. Justin, A. Guiseppi-Elie, Towards an implantable biochip for glucose and lactate monitoring using microdisc electrode arrays (MDEAs). Biomed. Microdevices **11**(1), 75–85 (2008)
267. A.R.A. Rahman, G. Justin, A. Guiseppi-Wilson, A. Guiseppi-Elie, Fabrication and packaging of a dual sensing electrochemical biotransducer for glucose and lactate useful in intramuscular physiologic status monitoring. IEEE Sens. J. **9**(12), 1856–1863 (2009)
268. A. Guiseppi-Elie, An implantable biochip to influence patient outcomes following trauma-induced hemorrhage. Anal. Bioanal. Chem. **399**(1), 403–419 (2010)
269. C.N. Kotanen, A. Guiseppi-Elie, Characterization of a wireless potentiostat for integration with a novel implantable biotransducer. IEEE Sens. J. **14**(3), 768–776 (2014)
270. Pinnacle Technology, Inc. [Online]. Available: http://www.pinnaclet.com/. Accessed: 28 Jan 2016
271. A. Weltin, B. Enderle, J. Kieninger, G.A. Urban, Multiparametric, flexible microsensor platform for metabolic monitoring. IEEE Sens. J. **14**(10), 3345–3351 (2014)
272. C.A. Cordeiro, M.G. de Vries, W. Ngabi, P.E. Oomen, T.I.F.H. Cremers, B.H.C. Westerink, In vivo continuous and simultaneous monitoring of brain energy substrates with a multiplex amperometric enzyme-based biosensor device. Biosens. Bioelectron. **67**, 677–686 (2015)
273. G. Calia et al., Biotelemetric monitoring of brain neurochemistry in conscious rats using microsensors and biosensors. Sensors **9**(4), 2511–2523 (2009)
274. A.J. Bard, L.R. Faulkner, *Electrochemical Methods: Fundamentals and Applications*, 2nd edn. (Wiley, USA)
275. P.T. Kissinger, W.R. Heineman, Cyclic voltammetry. J. Chem. Educ. **60**(9), 702 (1983)
276. P. Kissinger, W.R. Heineman, *Laboratory Techniques in Electroanalytical Chemistry, Second Edition, Revised and Expanded* (CRC Press, New York, 1996)
277. A.J. Bard, M. Stratmann, F. Scholz, C.J. Pickett, *Encyclopedia of Electrochemistry*, 7A, Inorganic Chemistry

278. A. Hierlemann, U. Frey, S. Hafizovic, F. Heer, Growing cells atop microelectronic chips: interfacing electrogenic cells in vitro with CMOS-based microelectrode arrays. Proc. IEEE **99**(2), 252–284 (2011)
279. M. Mollazadeh, K. Murari, G. Cauwenberghs, N. Thakor, Wireless micropower instrumentation for multimodal acquisition of electrical and chemical neural activity. IEEE Trans. Biomed. Circuits Syst. **3**(6), 388–397 (2009)
280. M. Sprenger, *Learning and Memory: The Brain in Action*. ASCD (1999)
281. R. Genov, M. Stanacevic, M. Naware, G. Cauwenberghs, N.V. Thakor, 16-channel integrated potentiostat for distributed neurochemical sensing. IEEE Trans. Circuits Syst. Regul. Pap. **53**(11), 2371–2376 (2006)
282. K. Murari, M. Stanacevic, G. Cauwenberghs, N.V. Thakor, Integrated potentiostat for neurotransmitter sensing. IEEE Eng. Med. Biol. Mag. **24**(6), 23–29 (2005)
283. J. Lerma, A.S. Herranz, O. Herreras, V. Abraira, R.M. del Rio, In vivo determination of extracellular concentration of amino acids in the rat hippocampus. A method based on brain dialysis and computerized analysis. Brain Res. **384**(1), 145–155 (1986)
284. M. Stanacevic, K. Murari, A. Rege, G. Cauwenberghs, N.V. Thakor, VLSI potentiostat array with oversampling gain modulation for wide-range neurotransmitter sensing. IEEE Trans. Biomed. Circuits Syst. **1**(1), 63–72 (2007)
285. M. Roham et al., A wireless IC for wide-range neurochemical monitoring using amperometry and fast-scan cyclic voltammetry. IEEE Trans. Biomed. Circuits Syst. **2**(1), 3–9 (2008)
286. S. Ayers, K. Berberian, K.D. Gillis, M. Lindau, B.A. Minch, Post-CMOS fabrication of working electrodes for on-chip recordings of transmitter release. IEEE Trans. Biomed. Circuits Syst. **4**(2), 86–92 (2010)
287. M.H. Nazari, H. Mazhab-Jafari, L. Leng, A. Guenther, R. Genov, CMOS neurotransmitter microarray: 96-channel integrated potentiostat with on-die microsensors. IEEE Trans. Biomed. Circuits Syst. **7**(3), 338–348 (2013)
288. G. Massicotte, S. Carrara, G. Di Micheli, M. Sawan, A CMOS amperometric system for multi-neurotransmitter detection. IEEE Trans. Biomed. Circuits Syst. 99, 1–1 (2016)
289. C. Yang, M.E. Denno, P. Pyakurel, B.J. Venton, Recent trends in carbon nanomaterial-based electrochemical sensors for biomolecules: a review. Anal. Chim. Acta **887**, 17–37 (2015)
290. S. Sainio et al., Integrated carbon nanostructures for detection of neurotransmitters. Mol. Neurobiol. **52**(2), 859–866 (2015)
291. N. Xiao, B.J. Venton, Rapid, sensitive detection of neurotransmitters at microelectrodes modified with self-assembled SWCNT forests. Anal. Chem. **84**(18), 7816–7822 (2012)
292. S.B. Hočevar, J. Wang, R.P. Deo, M. Musameh, B. Ogorevc, Carbon nanotube modified microelectrode for enhanced voltammetric detection of dopamine in the presence of ascorbate. Electroanalysis **17**(5–6), 417–422 (2005)
293. K.M. Mitchell, Acetylcholine and choline amperometric enzyme sensors characterized in vitro and in vivo. Anal. Chem. **76**(4), 1098–1106 (2004)
294. M.G. Garguilo, A.C. Michael, Enzyme-modified electrodes for peroxide, choline, and acetylcholine. TrAC Trends Anal. Chem. **14**(4), 164–169 (1995)
295. J. Cui, N.V. Kulagina, A.C. Michael, Pharmacological evidence for the selectivity of in vivo signals obtained with enzyme-based electrochemical sensors. J. Neurosci. Methods **104**(2), 183–189 (2001)
296. I. Suzuki, M. Fukuda, K. Shirakawa, H. Jiko, M. Gotoh, Carbon nanotube multi-electrode array chips for noninvasive real-time measurement of dopamine, action potentials, and postsynaptic potentials. Biosens. Bioelectron. **49**, 270–275 (2013)
297. M. Ganesana, J.S. Erlichman, S. Andreescu, Real-time monitoring of superoxide accumulation and antioxidant activity in a brain slice model using an electrochemical cytochrome c biosensor. Free Radic. Biol. Med. **53**(12), 2240–2249 (2012)
298. P. Fattahi, G. Yang, G. Kim, M.R. Abidian, A review of organic and inorganic biomaterials for neural interfaces. Adv. Mater. **26**(12), 1846–1885 (2014)

299. M.M. Ahmadi, G.A. Jullien, Current-mirror-based potentiostats for three-electrode amperometric electrochemical sensors. IEEE Trans. Circuits Syst. Regul. Pap. **56**(7), 1339–1348 (2009)
300. M.M. Ahmadi, G.A. Jullien, A very low power CMOS potentiostat for bioimplantable applications, in *Fifth International Workshop on System-on-Chip for Real-Time Applications (IWSOC'05)* (2005), pp. 184–189
301. L. Busoni, M. Carlà, L. Lanzi, A comparison between potentiostatic circuits with grounded work or auxiliary electrode. Rev. Sci. Instrum. **73**(4), 1921–1923 (2002)
302. P. Kassanos, R.K. Iles, R.H. Bayford, A. Demosthenous, Towards the development of an electrochemical biosensor for hCGβ detection. Physiol. Meas. **29**(6), S241 (2008)

Chapter 3
Electrical and Physical Sensors for Biomedical Implants

P. Kassanos, S. Anastasova and Guang-Zhong Yang

List of Acronyms

AA	Ascorbic acid
AC	Alternating current
AM	Amplitude modulation
AMR	Anisotropic magnetoresistance
ASIC	Application specific integrated circuit
ATP	Adenosine triphosphate
BGA	Ball grid array
BJT	Bipolar junction transistor
BPF	Bandpass filter
CF	Crest factor
CIN	Cervical intraepithelial neoplasia
CM	Conformal mapping
CNT	Carbon nanotube
CMOS	Complementary metal-oxide semiconductor
CMRR	Common-mode rejection ratio
CTAT	Complementary to absolute temperature
CVD	Cardiovascular disease
DAC	Digital-to-analog converter
DBS	Deep brain stimulation
DC	Direct current
DDS	Direct digital synthesis
DIBS	Discrete interval binary sequence
EIS	Electrical impedance spectroscopy
EIT	Electrical impedance tomography
EVD	External ventricular drain
FEM	Finite element method
FET	Field-effect transistor

P. Kassanos (✉) · S. Anastasova · G.-Z. Yang
The Hamlyn Centre, Imperial College London, London, UK
e-mail: p.kassanos@imperial.ac.uk

G.-Z. Yang (ed.), *Implantable Sensors and Systems*
https://doi.org/10.1007/978-3-319-69748-2_3

FFT	Fast Fourier transform
FOG	Freezing of gait
FPGA	Field programmable gate arrays
GF	Gauge factor
GI	Gastro-intestinal
GMR	Giant magnetoresistance
GOD	Glucose oxidase
HEX	Hexokinase
HPF	High-pass filter
I	In-phase
ICP	Intracranial pressure
IOP	Intraocular pressure
ISE	Ion selective electrode
ISFET	Ion-sensitive field-effect transistor
LOD	Limit of detection
LPF	Low-pass filter
MEG	Magneto-encephalography
MEMS	Micro-electro-mechanical systems
MLBS	Maximum length binary sequence
MOSFET	Metal-oxide-semiconductor field-effect transistor
MWCNT	Multi-walled carbon nanotube
MP	Magnitude/phase
NMRR	Normal mode rejection ratio
NTC	Negative temperature coefficient
OTA	Operational transconductance amplifier
PAC	Patient auxiliary currents
PCB	Printed circuit board
PDMS	Polydimethylsiloxane
PEN	Poly (ethylene naphthalate)
PET	Polyethylene terephthalate
PI	Polyimide
PSA	Prostate specific antigen
PTAT	Proportional to absolute temperature
Q	Quadrature
RE	Reference electrode
RF	Radio frequency
RMS	Root mean square
RRF	Resonance response frequency
RTD	Resistance temperature detector
SAW	Surface acoustic wave
SD	Synchronous demodulation
SEM	Scanning electron microscope
SMRR	Series mode rejection ratio

SNR	Signal to noise ratio
SOI	Silicon-on-insulator
SQUID	Superconducting quantum interference devices
SO	Sphincter of Oddi
SOM	Sphincter of Oddi manometry
SRR	Split ring resonator
SS	Synchronous sampling
ssDNA	Single stranded deoxyribonucleic acid
SSI	Surgical site infection
SWCNT	Single-walled carbon nanotube
TCR	Temperature coefficient of resistance
TRUS	Transrectal ultrasound
UA	Uric acid
VCCS	Voltage controlled current source
VOR	Vestibulo-ocular reflex
ZOH	Zero-order hold

3.1 Introduction

In addition to the electrochemical sensors discussed in Chap. 2, a range of other sensing modalities are also important for biomedical and implantable applications. The frequency-dependent electrical properties of tissues are essential for assessing various physiological parameters. This, for example, can be quantified via electrical bioimpedance measurements, which can be combined and corroborated with electrochemical sensors. The human body is a dynamic system in constant motion; therefore, sensors for the measurement of physical properties such as strain and pressure are also important. Sensors for these applications rely on the measurement of resistance, capacitance, and sometimes inductance, and these will also be discussed in this chapter for completeness. Temperature is an important health marker for various applications, and consequently the current state of the art in temperature sensors is also discussed, in terms of both monolithic integration and discrete sensor solutions. Monitoring of the electrical response of the nervous system and the delivery of stimuli represent an important family of applications for neuroscience research and neuroprosthetic devices. These will be addressed in this chapter, along with various application scenarios. Other aspects to be discussed include sensor metrics, such as sensitivity, limit of detection, stability, linear range, selectivity, and specificity.

3.2 Electrical Bioimpedance

In this section, the use of electrical impedance spectroscopy (EIS) for the interrogation of tissues is examined. Electrical bioimpedance is a non-invasive, low-cost technique appropriate for long-term continuous monitoring. It can be miniaturized and can be low-power, making it suitable for wearable and implantable applications. The basic theory of impedance measurements, the state of the art in impedance sensors for tissue characterization, and the necessary instrumentation are discussed in detail.

3.2.1 *Bioimpedance Fundamentals*

3.2.1.1 What Is Impedance and How Can It Be Measured

According to Ohm's law, the constant current (direct current, DC) through a material is directly proportional to the voltage (*V*) and inversely proportional to electrical resistance (*R*). This can be extended to when frequency-dependent signals (alternating current, AC) are used. If these are a function of frequency (*f*), then the resistance is known as an impedance (*Z*) and is found by the complex ratio of *V* to the current (*I*), again using Ohm's law

$$Z(f) = \frac{1}{Y(f)} = \frac{V_0 \sin(2\pi ft)}{I_0 \sin(2\pi ft + \varphi)} = R + jX = |Z| e^{j \arg(Z)} = |Z| e^{j \arctan\left(\frac{X}{R}\right)}, \quad (3.1)$$

where V_0 and I_0 are the peak voltage and current respectively, *t* is time, φ is the phase shift between the voltage and current, *Y* is the admittance, *R* is the real part of the impedance (Re(Z)), which is the resistance and *X* is the imaginary part of the impedance (Im(Z)), known as the reactance, $|Z|$ is the magnitude of the impedance, arg(*Z*) gives φ, and the angular frequency is $\omega = 2\pi f$. As a result, impedance can be described either by considering the magnitude and phase or equivalently the real and imaginary parts of the impedance. This is illustrated in Fig. 3.1 [1].

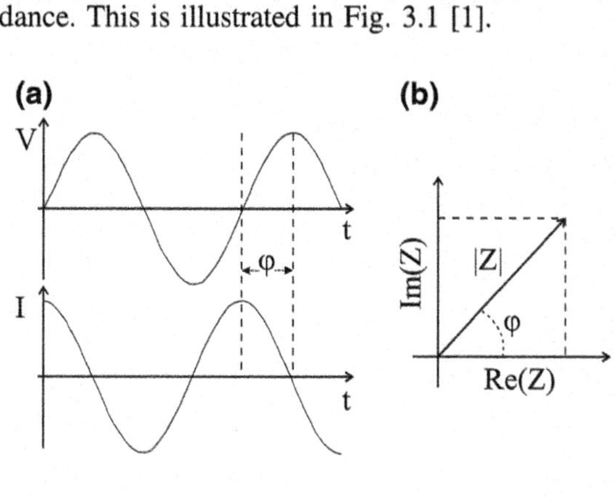

Fig. 3.1 Voltage and current time plots with a phase delay and a graphical representation of the complex impedance plane

If X is zero, then the impedance is frequency-independent and purely resistive and is given by

$$R = Z_R = \rho \frac{l}{A} = \frac{1}{\sigma} \frac{l}{A} = \frac{1}{G}, \tag{3.2}$$

where ρ is the materials resistivity and σ is its inverse, known as conductivity, and they define the ability of a material to pass current; l is the length of the conductor, A is its cross-sectional area and G is the inverse of R, known as the conductance. Materials, however, can also be described by their ability to not pass current (i.e. a very small σ). A dielectric is a material with no free charges or charged species, i.e. an insulator, that can store energy capacitively without dissipating it and the ability to do so is defined by its dielectric permittivity, ε, which is the product of ε_0, the permittivity of free space (8.854×10^{-12} F m^{-1}), and the relative permittivity (or dielectric constant) of the material, ε_r. If a potential V is applied across such material via two parallel plate electrodes, a simple parallel plate capacitor is formed, with a charge q stored. If the material is purely capacitive, $R = 0$ and the voltage across it lags the current through it by a phase of −90°. Then

$$I(t) = C \frac{\mathrm{d}V(t)}{\mathrm{d}t}, \tag{3.3}$$

$$X = Z_C = \frac{1}{\mathrm{j}\omega C} \tag{3.4}$$

$$C = \frac{q}{V} = \varepsilon_0 \varepsilon_r \frac{A}{d}, \tag{3.5}$$

where C is the capacitance and d is the distance between the two parallel plate electrodes. If the device/material is purely inductive, the voltage across it leads the current through it by 90°; X is zero and

$$V(t) = L \frac{\mathrm{d}I(t)}{\mathrm{d}t}, \tag{3.6}$$

$$X = Z_L = \mathrm{j}\omega L. \tag{3.7}$$

The (self) inductance (L) of a simple coil with a length, l, greater than its diameter, wound without a central magnetic core (i.e. air core), is defined as the ratio of the magnetic flux, Φ, through the coil for a given current through each turn. In such a coil the magnetic flux density, B, is constant and for a cross-sectional area of A, these are related by the following

$$B = \frac{\mu_o n I(t)}{l} \tag{3.8}$$

$$\Phi = B \cdot A \tag{3.9}$$

$$L = \frac{n\Phi}{I(t)} = \frac{\mu_0 N^2 A}{l}, \tag{3.10}$$

where n is the number of turns and μ_0 is the vacuum permeability. In reality there is no such thing as a perfect resistor, capacitor, or inductor, but rather one of these characteristics will dominate.

Due to this frequency dependency, it is thus necessary in many applications to measure Z at different frequencies to properly characterize a material and to extract enough information to obtain a sample's conductivity and permittivity and to obtain an equivalent electrical circuit for modeling its electrical properties. This is particularly important when composite or unknown materials are examined or when various physical phenomena (temperature, humidity, redox processes, etc.) and their effect on material properties are studied, as they may influence these differently at different excitation frequencies. Depending on the application, the impedance can be measured at a single frequency at a time or, if rapidly changing phenomena are investigated, at multiple frequencies simultaneously [2]. The need to measure impedance at various frequencies gives rise to the term impedance spectroscopy.

Hence, in order to calculate the impedance at a particular frequency, we need to know the voltage and the current. Consequently, there are two ways in which this can be performed. One can either apply a voltage and measure the resulting current flowing through the electrodes and sample under evaluation (AC potentiostatic measurement) or inject a current and measure the resulting voltage (AC galvanostatic measurement). For safety reasons, when measuring the impedance of living tissues the galvanostatic approach is preferred, as the current flowing though the tissue must be accurately controlled and limited. The amount of current allowed to be injected in living tissues is a function of the signal frequency and this is dictated by certain guidelines. The IEC601 and the equivalent BS5724 standards describe the AC currents applied to tissue via electrodes as patient auxiliary currents (PACs), and these are given as 100 μA from 0.1 Hz to 1 kHz; 100f μA from 1 to 100 kHz, where f is the frequency in kHz, and 10 mA from 100 kHz and above [3]. Thus, in order to measure the impedance of a sample, a circuit injecting to the sample a constant amplitude AC current throughout a specific bandwidth of interest (governed by the application and the physical phenomena under investigation) and a circuit measuring a potential difference across it must be used. In addition, a set of electrodes are necessary to interface the electronics with the sample under test. All these will be examined is subsequent sections.

3.2.1.2 Frequency-Dependent Electrical Properties of Tissue

Living tissue is an electrolytic conductor, as both intracellular and extracellular liquids are ionic. These ions are charged (positive, cations; negative, anions) and will thus migrate in an electric field. Hence, in an electrolyte, the charge carriers are ions and not electrons or holes. In contrast to electronic conduction, electrolytic conduction involves the transport of substance (i.e. charged particles) and not charges. However, some materials may demonstrate both electrolytic and electronic conduction. Different ions will demonstrate different mobilities (positive or negative or ions of different species), thus each carrying different portions of the total current. All free ions will contribute to the sample's conductivity and according to their concentration, charge, and mobility [4], while temperature will also influence conduction.

Extracellular conductance is primarily defined, due to their abundance, by Na^+ and Cl^-. Proteins in body liquids can also be considered as a 'colloidal electrolyte solute' [4] and free proteins in plasma are negatively charged, also contributing to conductance. In reality, as mentioned previously, every material can be characterized by both σ and ε, and thus all can be characterized by a frequency-dependent complex conductivity or equivalently by a complex permittivity. If a material is considered as predominantly a conductor with capacitive properties, then a complex conductivity is considered. This will also contain parameters accounting for the displacement currents. If the material is considered as an insulator with losses, then a complex permittivity is considered, which will contain parameters accounting for the material's conductance. These are defined by the following equations [4]:

$$\sigma(\omega) = \sigma' + \mathrm{j}\sigma'' \tag{3.11}$$

and

$$\varepsilon(\omega) = \varepsilon' - \mathrm{j}\varepsilon'' = \left(\varepsilon_r' - j\varepsilon_r''\right)\varepsilon_0, \tag{3.12}$$

where

$$\sigma'' = \omega\varepsilon' \tag{3.13}$$

and

$$\varepsilon'' = \frac{\sigma'}{\omega}. \tag{3.14}$$

As discussed in [4], saline solution, for example, with a conductivity of 1 S m^{-1} and a dielectric constant of 80, is a conductor for frequencies below 250 MHz and a dielectric above 250 MHz, and even metals can become dielectric at high frequencies. Of interest in this chapter are the electrical properties of biomaterials. According to [4], muscle tissue is mainly conductive with some capacitive properties and the stratum corneum is more of a dielectric with some conductive properties.

Impedance spectroscopy studies of biological tissues typically result in impedance spectra demonstrating a high dielectric constant at low frequencies (10^6 or 10^7 at frequencies below 100 Hz [5]), which is reduced typically in distinct steps with increasing frequency, leading to distinctly identifiable regions. The transition from one level to the other is known as dispersion. These are known as the α, β, and γ-dispersion regions, each related to various tissue properties. The median value between these two levels occurs at what is known as the characteristic frequency. The α-dispersion region (mHz–kHz) is associated with active cell membrane effects, ionic diffusion, and gated membrane channels. In the β-dispersion region (0.1–100 MHz) spectra are related to relaxations caused mainly by the polarization of cell membranes [5]. A δ-dispersion region has also been identified and related to protein molecules, protein-bound water, and intracellular organelle (e.g. mitochondria) membranes [6]. The γ-dispersion (0.1–100 GHz) is related mainly to the abundant tissue water and to other dipolar mechanisms in polar media (e.g. proteins and salts). The exact value of the α and β-dispersion region characteristic frequencies may vary dramatically from sample to sample and from tissue to tissue. However, the γ-dispersion is relatively constant [7]. The different dispersion regions are shown in Fig. 3.2.

Tissue is a highly inhomogeneous and anisotropic medium. Cell membranes are mainly insulating with a high membrane capacitance (C_{Membr}) and small conductance (R_{Membr}). Because of that, at low frequencies, this acts as a barrier to the flow of current and thus most of the current interrogates the extracellular space (R_{Extra}). As the frequency is increased, and thus the membrane impedance to current flow is decreased, progressively more current will flow through the cell membrane and into the mainly conductive intracellular space (R_{Intra}). This is illustrated in Fig. 3.3 in a

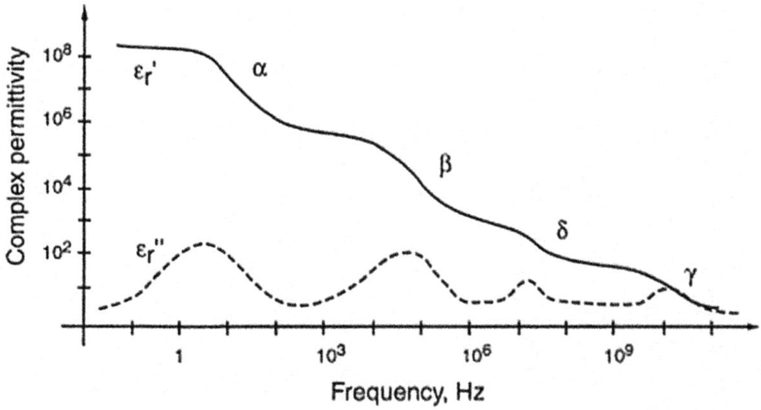

Fig. 3.2 Tissue dispersion regions with tissue DC conductance subtracted from the ε_r'' values [6]. Republished with permission of Taylor and Francis Group LLC Books, from [6]; permission conveyed through Copyright Clearance Center, Inc.

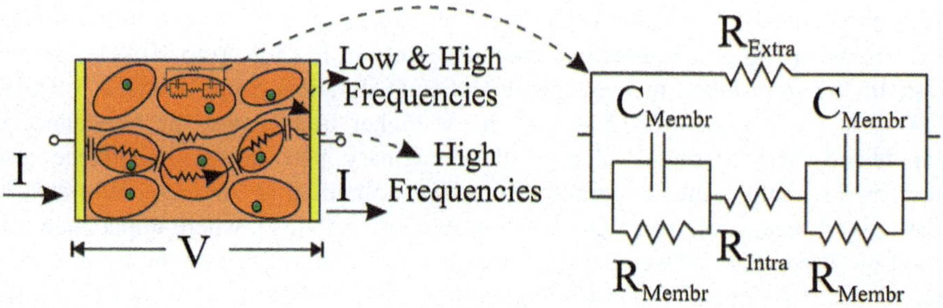

Fig. 3.3 Low- and high-frequency current paths within the tissue and an equivalent circuit of the cell and extracellular space impedance

similar manner as in [8]. Thus, information regarding cell membranes and the intracellular space will also be obtained. Larger scale morphologies, such as membranes compartmentalizing organs (e.g. the lungs, heart, stomach, and brain), will also contribute to similar processes [4].

Tissue impedance correlates with functional abnormalities and ionic and metabolic changes in tissues and organs [9]. Consequently, it has been applied successfully to a wide range of biomedical applications, including 2D and 3D imaging [electrical impedance tomography (EIT)] [10], respiration and cardiac output monitoring, tissue hydration monitoring and assessment, body composition analysis (hemodialysis applications and athletes) [4], and cancerous tissue detection [11–13]. Of particular importance is tissue ischemia monitoring.

3.2.1.3 Bioimpedance Monitoring of Ischemia

Electrical bioimpedance has been used in several studies in the literature for the detection of ischemia in tissues. Ischemia induces biochemical and physiological changes in tissues, which influence tissue impedance. The majority of published studies have focused on the myocardium; nevertheless, it has also been applied to the liver, kidney [14], and the gastrointestinal track via catheters [15–24]. In [25], an increase in the real part of the measured impedance was recorded at the low-frequency end of the spectrum examined (1 kHz), at which the frequency characteristics of the extracellular space dominate the measured impedance. At 10 kHz, the phase provided a similar response to the real part. In [9, 26] arterial occlusion significantly increased the measured impedance, reaching a subsequent plateau (11–20% in [9]). In [27], magnitude impedance measurements at 1 kHz were capable of differentiating between different tissue states. However, the impedance phase measured at frequencies between 40 and 180 kHz was better at distinguishing between different tissue states. A 112% increase in measured impedance on occlusion was reported in [28], which was reduced to 67% on reperfusion. In [29, 30] impedance measurements were used to predict tissue pH, and these predictions correlated up to 95% with actual tissue pH measurements. In

[29], the authors discuss that tissue impedance measurement can be more accurate than tissue pH measurements in ischemia detection. In [30] up to 10 kHz the real part of the impedance increases with ischemia duration from 175 Ω at $t = 0$ to 200 Ω at $t = 2$ h and 225 Ω at $t = 4$ h. At higher frequencies, the impedance is unaffected. The absolute value of the imaginary part of the impedance also increases with the ischemic duration from 10 to 15 and 20 Ω. An increase of 20% in the low-frequency real part was reported. According to [31], where impedance was used to differentiate between normal, ischemic, and scar tissue, the best parameters to characterize the evolution of ischemia in tissues are the low-frequency resistance (with a variation of 20–150%) and the characteristic frequency (from 10 to 50%). The characteristic frequency is defined as the frequency at which the imaginary part of the impedance reaches a maximum. The high-frequency resistance also behaves similarly; however, changes from 5 to 20% were reported. In [32], a 200% increase in measured impedance magnitude of the myocardium is reported after induced coronary occlusion (with the phase exhibiting a similar response). The authors were able to differentiate between normal, ischemic, and infarcted tissue by impedance measurements. They reported a maximum impedance magnitude change due to ischemia at 1 kHz, with other authors reporting a maximum change at 200 Hz. Other studies have reported a maximum change (×3) in impedance phase at 5 kHz. In [15], the impedance of the squamous epithelium, where cells are tightly packed, was not found to be high at low frequencies and rapidly decreased at high frequencies, while in columnar tissue the difference between low- and high-frequency impedance was significantly less. In this way, the two types of tissue were distinguishable with the proposed probe. Ischemic injury in intestinal mucosa was monitored with ileum impedance, tonometry (measurement of CO_2), and pH (Fig. 3.4a) in [18]. Tonometry and pH are early indicators of ischemia, but not of mucosal damage, and especially in prolonged ischemia they are incapable of estimating tissue damage [18]. According to [18], while pH measurements responded immediately to induced ischemia, once tissue becomes acidotic, intestinal pH cannot offer any additional information regarding the progression of ischemic injury as pH readings stabilize to a constant pH value. At the same time, the measured impedance began to change, indicating functional and structural changes in the tissue. In [33], a 1 kHz frequency was used for the measurements as this is considered a sufficiently low frequency for studying the extracellular space. Simultaneously, this frequency is high enough to avoid measurement errors induced by the electrode–electrolyte interface. The magnitude of the impedance at 1 kHz during induced ischemia via renal vascular occlusion changed by 25% within 60 min. Simultaneous measurements of pH and K^+ were performed. All monitored parameters changed immediately and rapidly upon occlusion. Impedance phase and pH decreased while impedance magnitude and K^+ increased. The variation of each measurement point of the measured impedance was significantly lower than that on the pH and K^+ measurements. Furthermore, the baseline impedance values were stabilized much faster. After reperfusion, pH and K^+ measurements returned to baseline levels, but impedance did not, indicating structural and functional damage to the tissue, which is not detectable with pH and K^+ measurements. These

observations further enhance the position of impedance measurements for ischemia monitoring. In [34], the authors highlight the need for simultaneous multi-frequency measurements. Greater changes in magnitude were achieved due to ischemia at low frequencies (100 Hz), while for phase at higher frequencies (100 kHz). Both parameters returned to baseline levels following reperfusion (at 50 min) indicating no tissue structural damage (no cell death or membrane disruption). In [35], measurements of pH, K^+, and impedance were again performed in the kidneys of anesthetized rats. An early, rapid rise in K^+ and impedance magnitude were reported, followed by a slower increase in impedance phase, while the pH decreased rapidly, reaching a plateau. Only impedance was capable of providing information regarding the effects of ischemic preconditioning to the tissue. The continuous increase of K^+ and impedance following the plateau in pH was attributed to passive ion and water shifts taking place after adenosine triphosphate (ATP) depletion, which do not influence pH. The needle probe used was similar to that of Fig. 3.4b.

According to [37], the magnitude of the measured impedance is mainly subject to variations in the extracellular environment, while the phase is mostly influenced by variations of the membrane of the cell and its integrity. Consequently, edema, hemorrhage and cell necrosis will modify the electrical properties of tissues. Edema and tissue necrosis have been shown to be associated with a reduction of tissue electrical impedance. The authors in [32] discuss that the reason ischemia results in changes in impedance is due to changes in the extracellular volume induced due to cell swelling (and thus reduction of the extracellular space). According to [38], the edema due to ischemia (which is established due to the osmotic water shift from extracellular space to intracellular space) increases the extracellular (low-frequency) resistance and decreases the intracellular (high-frequency) resistance. Metabolic processes that take place during ischemia produce ions and ion shifts, which further decrease the high-frequency intracellular resistance. Table 3.1 summarizes the experimental settings used in the referenced literature.

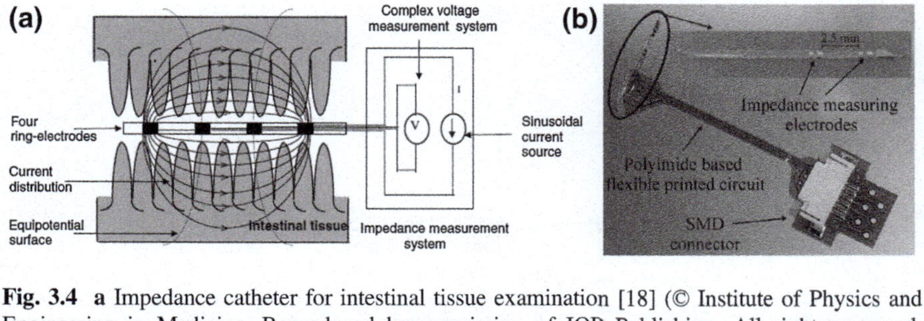

Fig. 3.4 **a** Impedance catheter for intestinal tissue examination [18] (© Institute of Physics and Engineering in Medicine. Reproduced by permission of IOP Publishing. All rights reserved. https://doi.org/10.1088/0967-3334/24/2/304). **b** A needle with an electrode array on a flexible (SU-8) substrate for kidney ischemia monitoring in rats (reprinted from [36], © 2009, with permission from Elsevier)

Table 3.1 Summary of published experimental settings for tissue ischemia monitoring via impedance

Paper	Frequency	Current	Impedance variation	Electrodes	Type
[25]	1 kHz, 10 kHz, 1 MHz	10 μA	100 Ω–1.5 kΩ	Bipolar	Iridium coated wires: active area 10 mm^2, diameter 0.2 mm
[37]	10, 30, 100, 300 kHz	n/a	10 Ω–6.7 kΩ	Tripolar	Catheter
[26]	5.5 kHz and 6 kHz	250 μA	n/a	Tetrapolar	Compound circular
[9]	5 kHz	5 μA (square)	400 Ω–1 kΩ	n/a	n/a
[39]	1–20 kHz (1, 8, 16 kHz)	5 μA	100 Ω–10 k Ω, [0–30°]	Tetrapolar	n/a
[27]	1 kHz–1 MHz	n/a	100–600 Ω, 0–30°	Tripolar and Tetrapolar	Tetrapolar: 0.4 mm diameter, 2.5 mm platinum wire needles
[28]	0.5–5 kHz	5 μA (square)	75 Ω–1.8 kΩ	Bipolar	Pacing electrodes
[29]	100 Hz–1 MHz	100 μA	10–200 Ω	Tetrapolar	10 mm separation stainless steel
[40]	5 kHz	100 μA	×4	Tetrapolar	Compound Circular
[14]	60 Hz–140 kHz	40 μA (square)	n/a	Tetrapolar	4 mm deep needles 3.8 mm^2, 2–3 mm apart
[31]	100 Hz–1 MHz	n/a	100–500 Ω	Tetrapolar	0.4 mm diameter platinum needles, 2.5 mm separation
[32]	250 Hz–50 kHz (1, 5, 15 kHz)	15 μA	100–700 Ω	Tetrapolar	Platinum pins: 3 mm separation, 3 mm insertion
[41]	100 Hz–100 kHz	11 mV	10 Ω–2 kΩ	Bipolar	2 cm separation, 0.79 cm^2
[38]	100 Hz–10 MHz	n/a	10 Ω–3.5 kΩ	Bipolar	1 × 4 mm^2, 10 mm separation
[30]	10 Hz–100 kHz	n/a	10 Ω–225 Ω	Tetrapolar	Ag/AgCl 1 mm diameter, 10 mm separation
[42]	1–100 kHz	n/a	n/a	Tetrapolar	n/a
[15]	9–650 kHz	10 μA	n/a	Tetrapolar	Au 0.8 mm diameter, 1.27 mm separation
[16]	n/a	n/a	n/a	Bipolar	4 mm wide, 1.6 mm separation
[17]	n/a	n/a	n/a	Bipolar	2 cm separation

(continued)

Table 3.1 (continued)

Paper	Frequency	Current	Impedance variation	Electrodes	Type
[18]	50 Hz–300 kHz	1.5 mA	n/a	Tetrapolar	n/a
[19]	n/a	n/a	n/a	Bipolar	2 cm separation
[33]	100 Hz–100 kHz 1 kHz	10 μA	n/a	Tetrapolar	Pt 300 μm × 300 μm 4 mm array length, 2.8 mm inner electrode 600 μm inner/outer distance
[34]	100 Hz–100 kHz	5, 50, 500 μA	2–9 kΩ	Tetrapolar	Pt 300 μm × 300 μm 4 mm array length, 2.8 mm inner electrode 600 μm inner/outer distance
[35]	1 kHz	10 μA	1–2.5 kΩ	Tetrapolar	Pt 300 μm × 300 μm 4 mm array length, 2.8 mm inner electrode 600 μm inner/outer distance
[36]	100 Hz–100 kHz	100 μA	1–4.5 kΩ	Tetrapolar	Pt 300 μm × 300 μm 2.5 mm inner electrode
[20]	1 kHz	6 μA	1–10 kΩ	Bipolar	Catheter, stainless steel

3.2.2 Bioimpedance Sensors

A wide range of probes for detecting and monitoring cancerous tissue have been reported in the literature. The authors in [11] reported on a bioimpedance probe (Fig. 3.5a) for the detection of cervical intraepithelial neoplasia (CIN), the second most common female cancer. The probe was evaluated in vivo with 166 patients referred to colposcopy due to an abnormal Papanicolaou smear test and was able to identify CIN with a sensitivity of 74% and a specificity of 53%. It was concluded that the probe achieved similar sensitivity and specificity to current screening tests, with the advantage of providing real-time results. In [12], a commercially available flexible electrode array (Fig. 3.5b) was used to detect intracranial neoplasms, with a particular interest in glioblastoma multiforme, the most frequent primary malignant brain tumor in adults. The electrode array consisted of 32 titanium nitride electrodes with a 30 μm diameter and 300 μm inter-electrode separation and two counter-electrodes. The system was used in vivo with four rat models. The system was capable of differentiating between cancerous and healthy brain tissue. The former resulted in considerably different impedance spectra at the 10–500 kHz frequency range, giving changes in the maximum relative impedance recorded of ~75%, the frequency at which this occurs of ~110%, and the relative impedance at 315 kHz of ~60%. A transrectal probe (Fig. 3.5c) for prostate cancer detection was reported in [13]. Prostate cancer is the most common cancer and the second most lethal male cancer in the USA and many western countries. Prostate-specific antigen (PSA) levels in blood serum of men serves as an initial screening tool for prostate cancer. Patients with elevated PSA levels are referred to a transrectal ultrasound (TRUS)-guided biopsy. However, TRUS is not prostate cancer-specific. Both hypo- and hyper-echoic regions may represent benign or malignant tumors. To address this, the authors modified a standard TRUS probe with additional electrodes on a flexible substrate to perform transrectal electrical impedance tomography. The ultrasound imaging data are used to obtain a priori anatomical information to aid the image reconstruction process. A large number of impedance measurement catheters have been developed for intraluminal measurements in applications such as intravascular monitoring for the detection of tissue proliferation induced by stent implantation (Fig. 3.5d) [43], Barrett's esophagus [15], gastrointestinal motility [16] and esophageal studies [20–24] (with simultaneous pH measurements) [17] and for monitoring the intestinal mucosa [18] and intestinal function [19]. At the smaller scale it is widely applied for monitoring cell cultures and interrogating individual cells and in biosensing applications for the detection of proteins [44], single stranded deoxyribonucleic acid (ssDNA), and bacteria [1].

In [45] the authors implemented their sensing electrodes on a complementary metal-oxide-semiconductor (CMOS) chip. Instead of using one of the metal layers, they implemented the electrodes using the top polysilicon layer. In contrast to the metal interconnect layers, polysilicon does not degrade when in contact with aqueous solutions. In addition, a passivating native oxide thin layer (~3 nm) spontaneously arises and stops further corrosion. Finally, a large number of

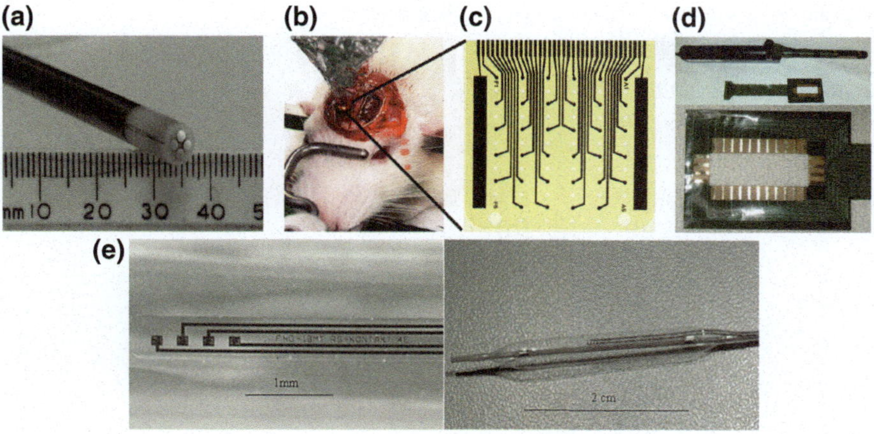

Fig. 3.5 **a** Impedance probes for CIN (from [11], with permission from Wolters Kluwer, http://journals.lww.com/ijgc/pages/default.aspx), **b** intracranial neoplasms (reprinted from [12], © 2013, with permission from Elsevier), **c** prostate cancer [13] (© Institute of Physics and Engineering in Medicine. Reproduced by permission of IOP Publishing. All rights reserved. https://doi.org/10.1088/0967-3334/31/8/s02) and **d** intravascular balloon catheter for the detection of neointimal proliferation following stent implantation (from [43], © Steinkopff Verlag 2005. With permission of Springer. Original figure caption: "Balloon catheter (Jomed Maestro 3.0 diameter, length 20 mm, Jomed, Rangendingen, Germany) with four integrated, microfabricated electrodes. The electrodes are arranged axially, 100 μm in diameter and spaced 333 μm apart)"

catheters for impedance measurements in the gastrointestinal tract have been developed. These are mainly a series (along the length of the catheter) of ring electrodes (e.g. Fig. 3.4a) [15–24]. The design of these appears to have been performed without considering the electric field properties of the electrode arrangements along the catheter length. Catheters and the relevant electronics for the combined measurement of impedance and pH have recently become commercially available (Digitrapper pH-Z recorder and VersaFlex catheter by Given Imaging, recently purchased by Covidien and subsequently by Medtronic). Finally, it is important to note that impedance spectroscopy has been used to assess the functionality of ion-selective electrodes [46].

3.2.3 Bioimpedance Sensor Modeling and Optimization

Commonly, issues related to the electrodes, electrode topologies, and their design are not discussed. Nevertheless, these are detrimental to the ability of the bioimpedance sensor to detect specific processes under investigation and to distinguish them from others. Electrode systems may consist of two (bipolar electrode system), three (tripolar), or four (tetrapolar) electrodes. In tissue bioimpedance applications, tetrapolar electrode systems are preferred. The reason for this is the inherent ability

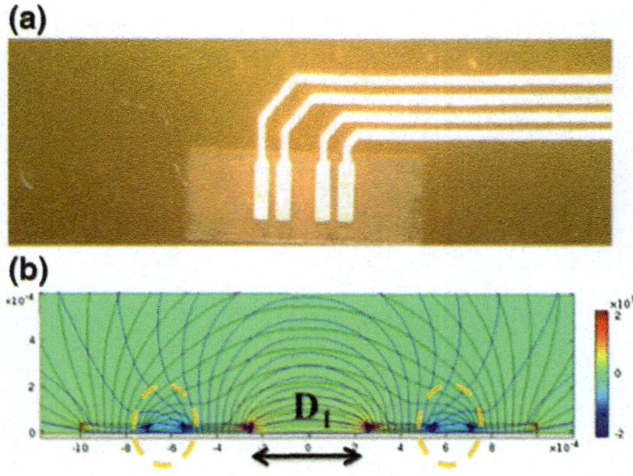

Fig. 3.6 A flexible impedance sensor and its 2D FEM modeling for the evaluation of S (© IEEE 2015. Reprinted with permission from [50])

of tetrapolar impedance measurements to diminish the influence of the polarization contact impedance formed at the electrode–tissue interface of the electrodes of the bioimpedance sensor used to inject a known current into the tissue [4, 47–50]. This interfacial impedance is typically significantly larger than the tissue impedance, and as the electrode size is reduced, this becomes increasingly an issue. This is achieved since the injected current does not flow through the measurement electrodes, as separate pairs are used for each. Consequently, there is no voltage drop across the measuring electrode interfacial impedance. If coplanar and parallel to each other electrodes are considered, as in Fig. 3.6, different injection/measurement electrode combinations can be implemented in a tetrapolar system as discussed in [48]. The most commonly used is the Wenner/Schlumberger (as it is known in the field of geophysical prospecting) arrangement, where injecting and measuring pairs correspond to the outer and inner electrodes respectively. Other arrangements include the dipole–dipole array and the cross array [48].

Of great importance is the understanding of the fundamental parameters of the physiological process being investigated and the analysis and optimization of the sensor geometry and the resulting electric field properties of the electrode array. Thus, the geometry of the electrode system is of high significance, as it is responsible for the electrical detection process. Important tools for achieving this are analytical (e.g. conformal mapping, CM [49]) and numerical modeling techniques (finite element method, FEM) [50]. Both commercially available (Ansys, Comsol, etc.) and custom FEM modeling packages allow the solution of complex problems in 2D (as in Fig. 3.6) and 3D. However, FEM is a computationally expensive and time-consuming approach. In addition, the accuracy of the solution depends on a number of parameters, such as the mesh and its quality and the solver used. In addition, FEM, being a numerical approach, does not allow an analytic equation to be obtained. Such an analytic model would be valuable, as it would directly elucidate the relationship between the property in question and the geometry. When the problem under examination is relatively simple or when an analytic expressions is required for optimization purposes, CM techniques can be used to obtain exact or approximate solutions to an electromagnetic problem [49]. However, the use of CM is restricted to 2D homogeneous and isotropic half-spaces. With CM, a problem imposed by a geometry that is challenging to analyze analytically is mapped to a geometry where the analysis of the problem is much simpler. In this way, the problem is first analyzed and solved in the new domain and is then back-transformed to the original geometry [49]. A common example is the transformation from a coplanar electrode geometry to a parallel plate geometry [49].

A vital tool for design optimization of bioimpedance sensors is the so-called Gezelowitz sensitivity theory [47–52]. The theorem is based on lead-field theory and the principle of reciprocity of four-port systems. According to this, the positioning of injecting and measuring leads can be switched without affecting the measured impedance (reciprocity). Thus, in the Schlumberger/Wenner array, let J_1 be the current density distribution when a current I_1 is injected between the outer pair of electrodes into a volume conductor with conductivity σ and permittivity ε and with the inner pair of electrodes being connected to a high impedance node. Also let J_2 be the reciprocal current density distribution, obtained within the same volume conductor when a current I_2 is injected between the inner pair of electrodes, with the outer pair of electrodes connected to a high-impedance node. These two reciprocal current densities are shown in Fig. 3.6b. Then, according to Geselowitz, the sensitivity field, S, is defined as the scalar field resulting from the vector dot product of J_1 and J_2, normalized by the product of the injected currents,

$$S = \frac{J_1 \cdot J_2}{I_1 I_2} \tag{3.15}$$

This is evaluated through the volume conductor under investigation, which may be a homogeneous isotropic medium or a more complex model, modeling different tissues, organs, and properties. S will have a different value in each infinitesimal volume within the model, indicating the aptitude of an electrode arrangement to detect change in impedance at that particular region. Being a vector dot product, S can have positive, negative, or zero values. This depends on the orientation of the current density lines in each minute area within the volume conductor. When the two vectors are perpendicular, S is zero. A volume with $S = 0$ indicates an area where the particular electrode arrangement is incapable of detecting an impedance change. When the angle between the two vectors is less than 90°, $S > 0$ and when it is greater than 90°, $S < 0$ [47–52]. Negative sensitivity means that if, for example, there is an impedance increase, the measured impedance will indicate a reduction in impedance (and vice versa). The magnitude of the sensitivity shows how much a region within the volume conductor contributes to the total measured impedance. In a bipolar electrode arrangement S is always positive. There are no negative sensitivity regions and S is equal to the square of the resulting current density. This is because the injecting and measuring leads are the same ($J_1 = J_2$). However, the tetrapolar system has both. Thus, when using a tetrapolar electrode system it is essential to examine the sensitivity characteristics of the electrode system, as it allows the assessment of its sensing ability and its optimization for a particular sensing application. The volume integral of S gives a measure of impedance.

$$Z = \int_{v} \frac{1}{(\sigma + \mathrm{j}\omega\varepsilon)} S\mathrm{d}v \tag{3.16}$$

From the above discussion, it is clear that the geometry of the electrode array defines its ability to interrogate different areas above the surface of the sensor and in the tissue. The separation between electrodes is the governing parameter [4, 47–52]. Hence, an optimum design requires a high and uniform sensitivity in the regions where a specific phenomenon inducing tissue impedance changes is to be detected. According to [40], ideally the array should demonstrate a high non-zero and constant sensitivity only in the desired volume of interest within the tissue. Negative sensitivity regions should be minimized, as changes taking place simultaneously at two locations with equal magnitude but opposite sign will show no change in measured impedance. The probe proposed by the authors reduces the negative sensitivity regions and achieves a depth within the volume that contributes significantly to the measured impedance equal to half the radius of the compound circular electrodes. In [15], the authors restricted 90% the sensitivity of the electrode topology within a height equal to the thickness of the tissue of interest. In [33], the authors used a simple analytic expression for a simple two-layer model to optimize an impedance probe for ischemia monitoring. Sensitivity theory was not considered; however, the distance between the inner pair electrodes was increased to maximize the voltage drop between these electrodes and hence the signal to noise ratio (SNR). This also increases the spatial resolution of the probe. In order to avoid making the sensor sensitive to the cellular structure of the tissue and rather to its homogeneous properties, a minimum distance between electrodes much larger than the cell size was used (300 μm). The platinum electrodes were coated with black platinum in order to increase the surface area of the electrodes and thus reduce the interfacial contact impedance. In order to enhance this, the electrodes were made as large as possible. The designed electrode topology has a spatial resolution of 4 mm. In [34], the authors improved their previous probe in [33] by replacing the Si substrate with a SiC substrate. This reduced current leakage between the substrate–solution interface, substantially improving the device performance for measurements up to 100 kHz. In [53] the authors used the probe in [33] to discriminate between healthy cold preserved organs for transplantation and those that had previously suffered ischemia. In [36], the authors replaced the Si and SiC substrates with a flexible SU-8 substrate (Fig. 3.4b). They argue that although biocompatible (Si and SiC) and of low current leakage (SiC), these substrates are highly resistant to mechanical stress. This rigidity causes a mechanical mismatch between the tissues and the probe, leading to tissue damage and subsequent scar tissue formation, which will hinder the functionality of the device and the quality of the measurements. Thus, they propose the use of flexible polymer substrate probes as a means to improve their biocompatibility. The SU-8 substrate, in contrast to previous ones, is not a semiconductor but an insulator, leading to a further reduction in current leakage through the substrate. In [50], the authors designed a bioimpedance probe (Fig. 3.6) for monitoring the mucosal layer in the gastro-intestinal (GI) tract and can thus be used for ischemia, Barrett's esophagus, and cancer detection. The gold-plated electrode sensor on polyimide (PI) was fabricated using commercially available flexible printed circuit board (PCB) fabrication. It was optimized using FEM such that 90% of the total volume S is confined within a height above the electrode surface that would focus the measurement in the mucosal layer of the tissue. FEM models incorporating the various tissue layers and their appropriate thickness and material properties further supported the optimization. The distance between the current injecting and voltage measuring electrodes was minimized close to the minimum allowed by the design rules of the process to reduce the negative sensitivity regions, and the 90% confinement of S within the volume conductor was calculated from FEM simulations using

$$\text{Percentage } S = \frac{\int_0^y \int_{-x_1}^{x_1} S \, \mathrm{d}x\mathrm{d}y}{\int_{\mathrm{S}} S \, \mathrm{dS}} 100\%, \tag{3.17}$$

where, as discussed in [50], y is the height of interest over the electrode surface, x_1 is the x-axis dimension of the 2D model and the denominator is the total surface (S) integral of S, i.e. the cell constant of the array.

3.2.4 *Bioimpedance Instrumentation*

As mentioned earlier, the focus in this chapter is on tissue impedance measurements and thus galvanostatic measurements. Such a bioimpedance measurement system requires a number of individual subsystems performing different functions. Here the basic building blocks are discussed. These include a voltage signal excitation, followed by a voltage-to-current converter, which injects a current to the tissue via two electrodes. This is followed by a voltage recording system via two other electrodes and subsequent signal processing steps. This is summarized in Fig. 3.7.

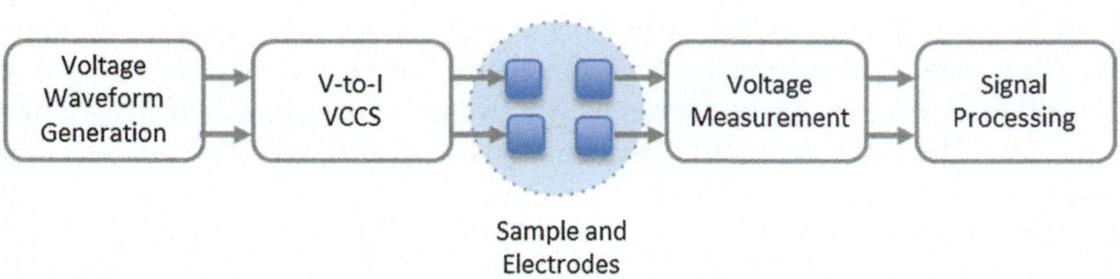

Fig. 3.7 Simplified block diagram of a typical galvanostatic impedimetric system for tissue bio-impedance applications

3.2.4.1 Voltage Excitation

As mentioned previously, different tissues and processes exhibit characteristic impedance variations at different frequency bands. Thus, the measurement of bioimpedance requires tissue interrogation and measurements across a bandwidth of interest (from a few Hz to several tenths of MHz, depending on the application). The majority of measurement systems employ swept frequency sine signals. A discrete set of frequencies, which are linearly or logarithmically distributed and with constant amplitude throughout the bandwidth, are typically used to excite the tissue. The voltage excitation system is responsible for setting and generating the frequencies used and their voltage amplitude. Commercially available benchtop signal generators [54], analog output data acquisition cards, or custom systems using commercially available microcontrollers, field programmable gate arrays (FPGA) [55] and direct digital synthesis (DDS) chips [50, 56] can be used to generate the signals of interest. Digital solutions, however, typically require high-speed digital-to-analog converters (DAC), large memory, additional control circuitry, and extensive filtering, leading to large, power-hungry implementations [55]. Alternatively, analog approaches can be used using traditional discrete op-amp-based oscillators and wave-shaping circuits [2], while filtering is used in most of the above-mentioned solutions if the waveforms generated are not sinusoidal and in general to reduce the spectral content of the signal.

Nevertheless, a wide range of applications (e.g. lung and cardiac monitoring and flow cytometry) involve transient and fast processes. These require very fast measurements across a wide bandwidth, such that the instantaneous impedance at each frequency is obtained. There are a number of different multi-frequency signals applicable to rapid impedance measurements. They can be aperiodic functions (e.g. white noise) or periodic excitations. The latter includes maximum length binary sequence (MLBS), chirp, discrete interval binary sequence (DIBS), Walsh function, and multi-sine signals [2, 54, 55]. From these, the multi-sine allows the greatest degree of flexibility for the generation of custom spectra. They can be comprised of a discrete set of specific frequencies, each with a preferred amplitude and phase [2, 54]. Such excitations may also lead to a higher SNR in contrast to MLBS and chirp excitations, as discussed in [54]. The amplitude of each frequency component in the MLBS spectrum decreases with increasing frequency and thus the SNR will vary across the bandwidth. Furthermore, MLBS will contain frequencies that are not of interest, thus wasting energy on those. MLBS, however, is often attractive because it can be very easily generated using a linear feedback shift register [54, 57]. Using fast logic circuits these simple circuit topologies can generate signals in the GHz range [54]. A chirp excitation is essentially a swept sine excitation. DIBS is a binary sequence, where the majority of the power is focused in the desired frequency range. As discussed in [54], DIBS and multi-sine excitations lead to more accurate measurements.

For a high SNR, it is necessary to employ a high amplitude for each frequency component. Ideally, this should be constant throughout the bandwidth. Nevertheless, biological systems, electrodes, and electronics demonstrate nonlinear

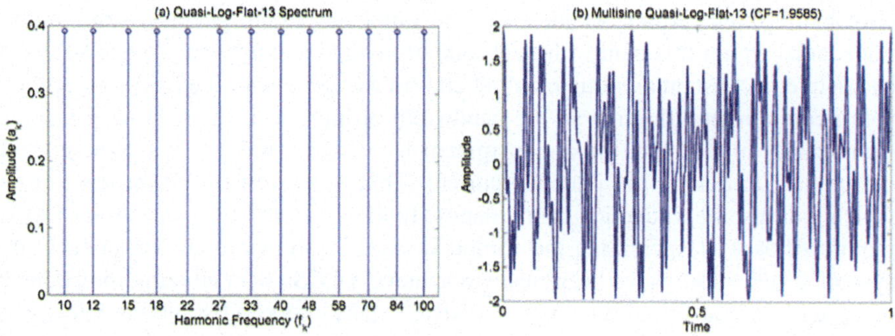

Fig. 3.8 A multi-sine with a quasi-logarithmic frequency distribution and a flat amplitude spectrum optimized in [59] with a CF = 1.9585 (© Institute of Physics and Engineering in Medicine. Reproduced by permission of IOP Publishing. All rights reserved. https://doi.org/10.1088/0967-3334/36/5/895)

behaviors when subject to signal levels beyond certain levels. In addition, the amplitude of the generated signal is one of the parameters controlling the injected current. As mentioned previously, this must also be within specific limits for safety purposes, which is governed by published guidelines. Consequently, the amplitude of each frequency component and the total amplitude of the composite multi-frequency signal need to be well-controlled and limited. This is especially important in implantable applications, where ultra-low power systems are necessary, and thus the power rails are severely limited. In addition, this is also important to avoid tissue stimulation, as is the case in neuroprosthetic applications. Signals can be characterized by what is known as the crest factor (CR). This is defined as the ratio of the signal's peak value and its root mean square (RMS) value and demonstrates how much of the total allowable amplitude is used for a specific power level. In addition, the larger the CF the lower the SNR and the more bits are required in the analog to digital conversion step for a given level of accuracy [54]. Thus it is important to minimize the composite signal's CF and this is done by an appropriate selection of phases for each frequency component. This is a subject that has recently attracted considerable attention and a number of solutions have been proposed using analytical and numerical approaches for waveform CF optimization [54, 55, 58, 59]. In addition, even nonlinearities cause an overestimation of the impedance at the fundamental frequencies. To avoid this, it is advised to use a log-odd frequency distribution as proposed in [54]. Figure 3.8 demonstrates a multi-sine excitation with a CR = 1.9585, which implements a constant amplitude for each frequency component and a quasi-log frequency distribution.

3.2.4.2 Voltage Controlled Current Source (VCCS)

For the implementation of the current source a VCCS is typically used, which, as the name suggests, is a circuit that converts an input sinusoidal voltage into an output sinusoidal current. An ideal current source has an infinite bandwidth, output impedance within this bandwidth and output voltage compliance. This is so that it can deliver a constant amplitude output current irrespective of the load connected at the circuit's output, throughout the bandwidth of interest. If the load impedance is significantly higher than the output impedance of the source, then the circuit will not be able to deliver the intended current. This will lead to measurement errors, which will also be a function of frequency. In this case, the current will need to be monitored as well, increasing the complexity, size, and power consumption of the system. Thus, efforts have been made to achieve VCCSs with an output impedance of up to 1 MΩ at frequencies up to 1 MHz. Many proposed solutions effectively measure the current across a sense resistor in order to use this information with feedback to regulate the current. However, parasitic capacitances intrinsic to the VCCS structure and at the VCCS output node reduce the output impedance of the circuit with increasing frequency. The voltage generated at the output node of the current source due to the injected current and the load impedance must be within the voltage compliance of the source, otherwise the signal will be distorted. This depends on the integrated circuit's output node characteristics, governed by the design, and the supply voltage used. The voltage signal can also be kept within the linear range of the system by appropriate selection of the maximum output current amplitude. This in turn depends on the transconductance of the VCCS and on the input voltage excitation waveform, as discussed in the previous section. Finally, the harmonic content of the output signal should be minimal; thus the VCCS should not distort the signal. Typically, decoupling capacitors in series with the output current are used (1–10 μF) to block DC currents from being injected into the tissue for safety purposes.

Many VCCSs based on discrete components have been proposed over the years. Solutions based on classical amplifiers include load-in-the-loop circuits, as in Fig. 3.9a [60]. The load is thus connected between the op-amp output through a small resistor (R_{safe}) and the virtual ground at the negative input terminal, and thus the load is floating. Provided that the amplifier has infinite input impedance, the output impedance will be infinite. The output current is equal to $I_{out} = V_{in}/R_{in}$. The authors obtained an output impedance of 150 kΩ at 1 MHz, claiming this to be the highest reported output impedance for a VCCS based on discrete components. The fact that the output impedance of this topology does not rely on resistor matching and a floating load can be realized with a single op-amp makes this topology very attractive. Other solutions are based on current-mode circuits, such as operational transconductance amplifiers (OTA), current conveyors, and current feedback amplifiers, as in the topology of Fig. 3.9b. This was recently used in [55] and is based on a current feedback amplifier with an accessible internal high-impedance node, allowing its use as a current conveyor. The DC feedback loop (gain of 10 V/V) ensures transconductance accuracy and avoids the VCCS saturating due to residual DC currents, which ideally are blocked from the in-series output capacitors. As in the load-in-the-loop circuit, the transconductance of the topology is again set by a single resistor. This topology was compared with a modified Howland circuit in [61]. The authors demonstrated that both circuits achieve an output impedance of

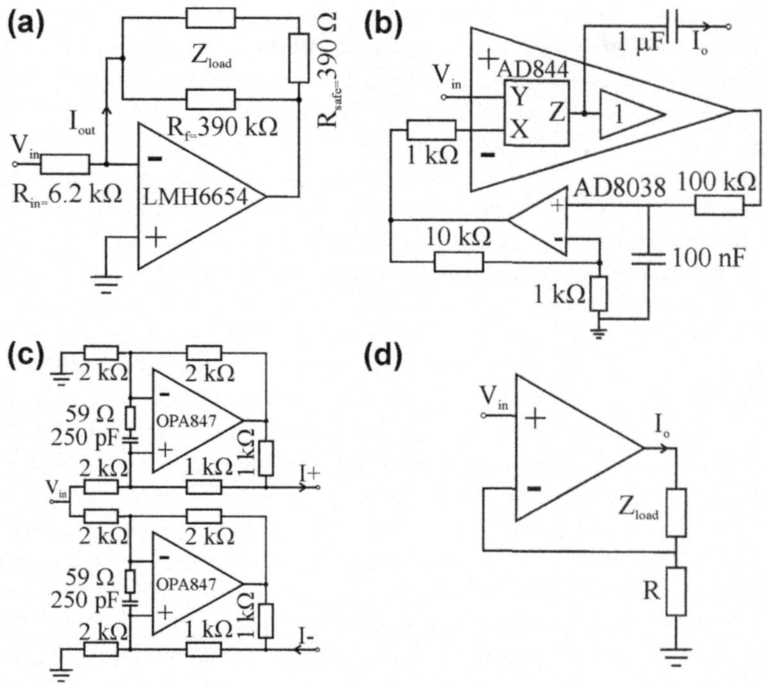

Fig. 3.9 **a** Load-in-the-loop VCCS topology proposed in [60]. **b** The current conveyor/current feedback amplifier-based topology used in [55]. **c** The symmetrical Howland VCCS implemented in [50]. **d** Conventional current source with feedback control of the current

70 kΩ at 1 MHz. The Howland circuit (Fig. 3.9c) is one of the most common topologies used in bioimpedance applications. What is known as the modified or improved Howland circuit was examined in [62] and more recently in [50, 56]. Nevertheless, all Howland circuits are challenged by a number of issues limiting their performance. One is the need for highly matched resistors (0.01% tolerance). As discussed in [56, 62], the pole location of the amplifier used is also very important. A final issue that needs attention is the PCB layout, as all tracks need to be as small as possible to minimize parasitic capacitances between the internal and external nodes, which may hinder the high-frequency output impedance of the current source. In [62] the authors report an output impedance of 61 kΩ at 1 MHz. A final approach for a VCCS is to fix the current through the load via the feedback topology shown in Fig. 3.9d. According to [63], this topology should be avoided for a number of reasons: (a) it is prone to instability; (b) when not connected to the sample, the amplifier output saturates to a voltage close to the supplies, and when it is eventually connected to the sample before the output stabilizes to the desired current levels, for a transient period large unsafe currents may be injected to the tissue unless a voltage clamping methodology is implemented, which may load the output further; and (c) the resistor connected to ground, R, must be small, otherwise a large common-mode voltage will be generated [63].

Single-ended VCCSs are referred to ground and have thus large common-mode voltages. The use of symmetrical (known also as differential, mirrored, or bipolar) current sources, where the load is essentially floating, allows the minimization of the common mode voltage at the load. Consequently, this reduces the voltage measurement errors at the front-end voltage recording electronics due to the inadequate common-mode rejection ratio (CMRR) of the recording amplifier. A symmetrical floating Howland VCCS was recently reported in [50] and is based on the recent single-ended design of [56] with a transconductance of 1 mA V^{-1}. This is essentially two Howland circuits differentially driven, thus producing anti-phase current outputs. Mismatches between the two VCCSs and imperfections may lead to a large common mode differential DC voltage appearing at the output (due to e.g. voltage offsets) leading to a DC current being injected to the tissue, which can be blocked using capacitors in series with the outputs of the two VCCSs.

The output impedance can be improved further by using active shields and guards (triax wiring), impedance converters to compensate the parasitics, minimizing the distance between sensor and the VCCS (active electrodes), and measuring the current and varying the input voltage. Apart from the active electrode scenario, which necessitates the need for encapsulation of the device, miniaturization, and application-specific integrated circuits (ASIC), all other options will introduce parasitics at the output due to the input capacitance of the amplifiers used or will cause instabilities and hence oscillation in the case of the impedance converters. Impedance converters are used to compensate for parasitic capacitances and improve the VCCS performance at high frequencies. However, this can be achieved only within a narrow frequency band, since these can be used to compensate the VCCS only at a single frequency. Thus for spectroscopic measurements a large number of such circuits would be required for each frequency component of interest. It is obvious from the above that the design of a VCCS is not a trivial task.

3.2.4.3 Voltage Readout Circuit Techniques

Following the injection of the current into the sample, the resulting voltage needs to be measured. Many different approaches have been proposed for the front-end electronics. Typically, first an instrumentation amplifier is used to provide accurate differential measurements and to suppress common-mode interferences. This should provide a common-mode rejection ratio above 90 dB within the bandwidth of interest. Unfortunately, the majority of commercially available discrete instrumentation amplifiers cannot provide such specifications; e.g. the AD8421 with a gain of 1 V/V has a CMRR of 60 dB and with 100 V/V, 100 dB at 100 kHz. If the recording electronics are implemented in an ASIC (as opposed to discrete components) analog signal processing offers significant power consumption and size reduction compared to digital signal processing. Nevertheless, for accurate measurements the imperfections of analog circuits need to be addressed. This will be discussed in more detail in Chap. 5. As mentioned previously, electrical impedance can be described by its magnitude and phase or, equivalently, by its real and imaginary parts. In the following section, analog and mixed signal techniques are discussed which can calculate the impedance using these two techniques.

Magnitude/Phase (MP) Measurement

Magnitude/phase measurement is the simplest technique for the measurement of bioimpedance. The magnitude of the recorded sinusoidal voltage signal across the load can be obtained with a peak detector or a full-wave rectifier. In the latter case, the signal needs to be subsequently filtered by a low-pass filter (LPF). To calculate the phase of the impedance under investigation a reference signal is required, which is in phase (0°) with the signal applied to the impedance being measured. This can be obtained through a current sense resistor. Converting this into a square wave and comparing it with a square wave version of the signal measured across the impedance allows the calculation of the phase. This can be performed with a phase detector, which can be an XOR gate, or a switching or analog demodulator (a multiplier). One such system was recently proposed in [64]. A simplified block diagram of the method is shown in Fig. 3.10.

It is obvious from this discussion that two different measurement channels are required, each with potentially dissimilar error sources and thus needing different compensation techniques to address them. One source of error is associated with synchronization in the phase channel due to the need for the in-phase signal reference. Amplifier and comparator offsets and unmatched delays will lead to errors in the phase measurement. In the magnitude channel, amplifier offsets are important, since the output of the full-wave rectifier or the peak detector are DC signals. If a synchronous rectifier is used in the system of Fig. 3.10 as in [64], the phase delay of the comparator generating the demodulation signal will also introduce errors. All of these become gradually more pronounced with increasing frequency. A disadvantage of this system is that for multi-frequency measurements a bank of high Q-factor bandpass filters (BPF) is necessary following the instrumentation amplifiers for each frequency component. In [65] an MP system based on discrete component was presented. The authors used a commercially available gain-phase detector (AD8302) IC from Analog Devices. The AD8302 performs division in the

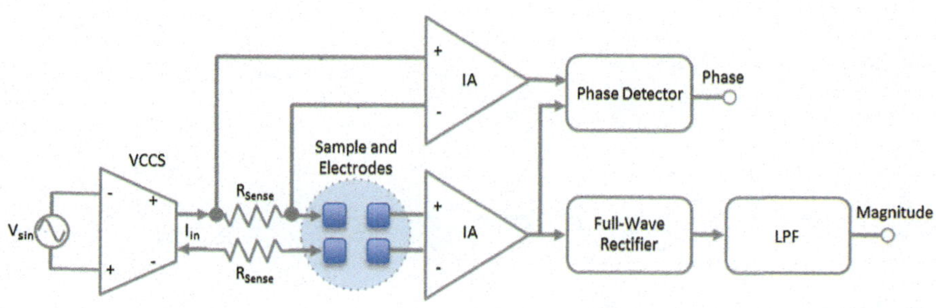

Fig. 3.10 The MP impedance recording front-end proposed in [64]

log-domain by subtraction of the signal across a known resistor and the signal across unknown impedance to obtain the magnitude. However, if the VCCS used can deliver a constant current at the frequency of interest for the load impedances under investigation, there is no need to divide the recorded voltage with the current, as this is just another gain factor within the system.

Real and Imaginary Component Measurement: Synchronous Detection (SD)

Real and imaginary signal component measurement is more complicated. Synchronous detection (SD, also known as lock-in, phase sensitive, and quadrature demodulation) is an established technique that allows this to be performed. If we consider a single frequency scenario, then two reference signals are required to perform this: an in-phase (I, 0°) and a quadrature (Q, 90°) signal of the same frequency with respect to the signal injected into the tissue. The multiplication of the measured sinusoidal voltage signal across the sample under test, with these two reference signals, allows the calculation of the real and imaginary component values, respectively [66], as shown in Fig. 3.11. This is a demodulation process, and if sine waves are used with an analog multiplier (e.g. a Gilbert cell), then this will output a DC voltage and a double frequency component. The latter is filtered out with an LPF and the former is a value proportional to the real or imaginary component of the signal. If the recorded voltage at the readout has an amplitude A (which is a function of the injected current, the frequency, the amplifier gain, and the impedance), a phase delay φ_z (due to the impedance), and an added DC voltage (C) due to electrode interface potentials and this is demodulated with an in-phase to the excitation signal of the same frequency $f_d = \omega_d/(2\pi)$ and a unity amplitude, then the demodulator output is

$$V_{\mathrm{SD\_Re}} = \frac{A}{2}\left[\cos(\varphi_z) - \cos(2\omega_d t - \varphi_z)\right] + \mathrm{C}\sin(\omega_d t). \tag{3.18}$$

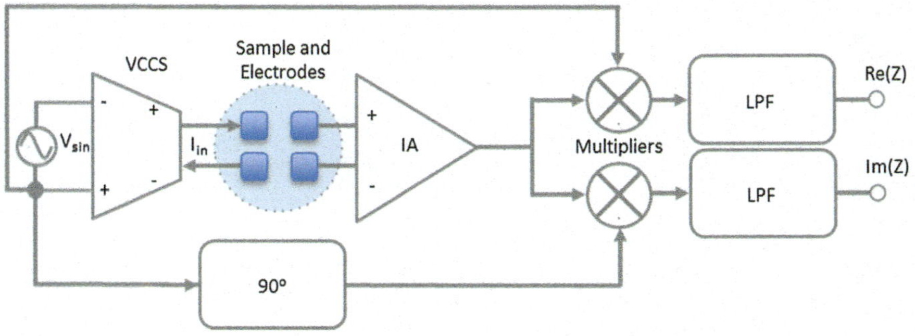

Fig. 3.11 Single-frequency SD [64]

If a Q demodulating signal is used, then

$$V_{\mathrm{SD\_Im}} = \frac{A}{2}\left[\cos\left(\varphi_z + \frac{\pi}{2}\right) - \cos\left(2\omega_{\mathrm{d}}t - \varphi_z + \frac{\pi}{2}\right)\right] + \mathrm{C}\sin\left(\omega_{\mathrm{d}}t + \frac{\pi}{2}\right). \quad (3.19)$$

All frequency components are subsequently removed by the LPF and the system output for each demodulating channel is a DC voltage proportional to the impedance.

$$V_{\mathrm{LPF\_Re}} = \frac{A}{2}\cos(\varphi_z), \quad (3.20)$$

and

$$V_{\mathrm{LPF\_Im}} = \frac{A}{2}\cos\left(\varphi_z + \frac{\pi}{2}\right). \quad (3.21)$$

Alternatively, a switching multiplier can be used with square wave demodulating signals as in [66]. Due to the multi-frequency content of the square wave, the demodulator output will have a complex frequency content. These are again filtered out with the LPF.

An important advantage of SD is that the multiplication and the LPF together impose a frequency selective property to the system, i.e. the system is inherently a high *Q*-factor BPF; the readout locks to a specific frequency component, which is demodulated to DC. Any other frequency components present, either in the demodulating signal or in the recorded signal (due to interference or because a multi-frequency excitation has been used) are modulated to higher frequencies (including electrode DC voltages). Thus, no additional BPFs are required as in the MP system for multi-frequency measurements. The farther the interfering signals are from the demodulating frequency, the greater their attenuation is. The ability of the system to reject these is quantified by the series or normal mode rejection ratio (SMRR, NMRR) [67]. If the systems output signal at the demodulating frequency is $v_O(f_d)$ and to an interfering frequency component $v_O(f_i)$, then the SMRR is defined as their ratio in decibels [67].

$$\mathrm{SMRR} = 20\log\left|\frac{v_O(f_{\mathrm{d}})}{v_O(f_{\mathrm{i}})}\right| \quad (3.22)$$

From the above equations, it is clear that in order to have a high SMRR, one option is for the frequency of interest and the interfering signal frequencies to be substantially different, $f_{\mathrm{i}} \ll f_{\mathrm{d}}$. This is not always controllable and it will be application-specific. The alternative is to have a high-order LPF. As discussed in [67], a second-order low-pass filter with the same cutoff frequency as a first-order filter provides an SMRR twice as large. Depending on the application, the two components may need to be recorded simultaneously. This is typically the case. In this scenario, similarly to the MP system, two channels are necessary. However, these are identical and their performance is governed by the same limitations. For accurate IQ measurements, the two channels should be accurately matched (e.g. delays and offsets). In applications where the measured impedance is not expected to change rapidly, a single channel may be used, in which case the demodulation signal in the single channel must be switched between 0° and 90°. Thus, the issue of matching the channels is resolved and the system can consume lower power and space. The cutoff frequency and order of the LPF are important and can be selected by considering the settling-time required, which may be governed by the application, and the lowest frequency of interest for the measurement of the sample's impedance. These are defined by the biological system being investigated and the application.

Apart from the matching issues, offset and synchronization are important for the same reasons as in the MP system. Matching is important when more than one SD channel is implemented for simultaneous real and imaginary component measurements and/or in multi-frequency systems. DC voltage offsets are also detrimental as the outputs of SD are DC voltages. The generation of accurate quadrature demodulating signals is challenging. Typically these are generated at the voltage excitation subsystem as in [2] or using quadrature analog oscillators or DDS systems [50]. A source of error, which is typically ignored, is due to the need of this system for precise synchronization between the injecting and measuring stages. The VCCS will introduce phase delays, which will lead to errors in the demodulation process at high frequencies, as the demodulation signals will no longer be in phase and in quadrature to the injected current. Generating these following the VCCS as in the MP system of Fig. 3.10, would significantly increase the area and power footprint of the system, particularly in multi-frequency applications. If a switched modulator is used the demodulation signals need to be square waves and if these are generated via squaring sine waves with comparators, the offset and delay of these will further hinder proper demodulation. These will result in an erroneous DC component in both real and imaginary outputs at high frequencies. This can be especially detrimental in swept or multi-frequency measurements, as the error will differ in value at each injected frequency component. The use of additional phase shifting circuitry can provide correction, implemented as a feedforward or feedback topology, at the expense of a small increase in system complexity and power consumption as in [2]. A recent development [66] of an ASIC SD system for multi-frequency applications demonstrated accurate measurements up to 1 MHz.

Synchronous Sampling (SS) and Other Techniques

Another technique for attaining the real and imaginary components of impedance is SS. Since the excitation frequency content, amplitude, and phase of each frequency component injected are typically known, there is no need to perform a fast Fourier transform (FFT) [55]. As the name implies this technique is based on the sampling of the measured voltage. This is performed at the exact instances where the input frequency components reach their peak and mid-point value (e.g. zero in a split supply system), allowing the real and imaginary components to be obtained [8, 68]. If such samples are obtained in both half cycles of a period (thus satisfying also Nyquist) and averaged, voltage offsets are eliminated [8]. Thus real and imaginary component measurement can be performed using a single channel, offsets are eliminated and the signal is simultaneously digitized. A zero-order hold (ZOH) circuit is typically used. According to [67], SS opens many spectral windows to noise and a ZOH has no response to these windows apart from DC. Thus, typically the signal is filtered prior to SS demodulation to preserve the SNR either by a BPF [68, 69] or a high-pass filter (HPF) [70]. This technique was extended to multi-frequency signals in [8]. If the measurements are averaged within the full period of a multi-frequency signal, then the calculation of the real and imaginary components at each frequency using a single channel can be performed [8], as in Fig. 3.12, which is a significant advantage over the other techniques. However, accurate synchronization for obtaining the samples is essential, which can be challenging, particularly at high frequencies, as is the generation of the appropriate signals for the sampling.

Apart from the AD8302 gain/phase chip used in [65], there are two commercially available systems on-chip for the measurement of impedance. The first is the AD5933, which, however, requires additional components in order to allow it to perform galvanostatic impedance measurements, as in [71], and its frequency measurement is limited to 100 kHz. Essentially, it does not have a VCCS. The second is the recently released AFE4300 by Texas Instruments. This does not need any external active components to perform impedance measurements, but it does need a number of passives. It is intended for weight scales which include body composition analysis. It is recommended for measurements up to ~80 kHz (due to an internal cutoff frequency of 150 kHz). Internally, it has two impedance measurement channels. The first measures only the magnitude of the impedance using a full wave rectifier and the second is an SD system, which, however, is limited to the measurement frequencies. This chip comes with multiplexers at the front end, allowing multiple electrode arrangements to be interrogated in a switched fashion. Both solutions have DDS systems on chip for waveform generation.

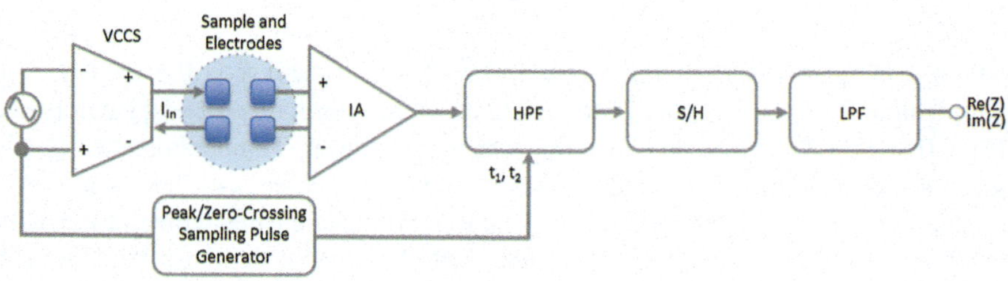

Fig. 3.12 System-level diagram of a single-frequency SS system [64]

3.3 Temperature Sensors

Temperature measurement is important in biomedical applications. Deviation of body temperature from normal, which is centered around 36.6 °C, is an indication of medical complications. Tissue temperature is also elevated in the vicinity of wounds and may indicate infection [72] and skin temperature, in association with other measurements, provides information regarding cardiovascular health, cognitive state, and malignancy [73]. According to [74], an implanted temperature sensor needs to sense temperatures between 35 and 42 °C, and temperature changes (due to physiological conditions or induced, for example, by the power dissipation of the implant device itself) of only 2 °C, can induce protein denaturation. In addition, many other sensors, such as ion-selective electrodes (ISE), ion-sensitive field-effect transistors (ISFET) and metallic-based sensors, such as strain and pressure sensors, are affected by temperature changes and thus compensating for this variation requires the measurement of temperature. For example the IrO_x pH sensors [75] had a temperature dependence of -1.6 ± 0.02 mV °C^{-1}. This corresponds to about 0.02 pH for a 1 °C change in temperature [75]. An example for the need of implanted temperature sensors is in orthopedic implants (discussed in more detail in the following section). Due to friction between the head and cup of a hip prosthesis, a temperature increase of up to 3.5 °C may be induced [76]. This is important to know, as it may imply a damaged or misaligned prosthesis that may eventually lead to further complications for the patient. The majority of temperature sensors used in biomedical applications are contact temperature sensors, where the sensor and the measurand are in thermal equilibrium, i.e. there is no thermal flow between them [77]. The simplest temperature sensor is a thin metal wire, typically Pt, Au, Ni, or Cu, the resistance of which will increase approximately linearly with increasing temperature. This temperature dependence is due to the scattering of electrons by lattice vibrations (phonons). At higher temperatures, phonons vibrate with higher amplitudes [78]. At low temperatures, e.g. −300 °C, there are very few phonons and the resistance is independent of temperature. Such devices are known as resistance temperature detectors (RTDs) or as resistance thermometers. Their operation relies on the temperature coefficient of resistance (TCR) of these metallic thin, narrow structures defined by

$$\alpha(T) = \frac{1}{R_0}\frac{\mathrm{d}R}{\mathrm{d}T} = \frac{1}{R_0}\frac{R - R_0}{T - T_0}, \tag{3.23}$$

where α is the TCR, R_0 is the resistance at T_0 = 0 °C and R is the resistivity at temperature T. Typically the TCR is defined as the normalized resistance change from 0 to 100 °C, provided that this change is linear. Temperature sensors in general can have either a positive (the resistance decreases with increasing temperature) or negative (the resistance increases with increasing temperature) temperature coefficient. Pt is often the material of choice as it demonstrates the most stable and linear temperature dependence over the widest temperature range and the greatest repeatability and stability over long periods of time. Pt typically has a TCR of $\sim 3.85 \times 10^{-3}$ °C^{-1} at room temperature [78]. Cu is used between −100 and 100 °C and has a TCR of 3.9×10^{-3} °C^{-1}. Ni has a high TCR (6×10^{-3} °C^{-1}), but is nonlinear and strain sensitive. The TCR of Au is 2.4×10^{-3} °C^{-1} and that of Ag is 4×10^{-3} °C^{-1}. Ideally the material used should be chemically stable and inert in oxidation, reduction, and corrosion [78]. Similarly to what was discussed earlier for impedance sensors, greater RTD measurement accuracy can be achieved using three- or four-point measurements, in order to eliminate the effect of the connecting leads. Commercially available RTDs come in many forms in addition to thin metal films. These include wire-wound topologies, where a metal wire is wrapped around an insulating core, and coiled topologies, where the wire coil is not limited by the core and can thus expand freely and without strain. In supported RTDs, it is important that the thermal expansion of the sensing material is matched to that of the substrate. Depending on the material, they may have a sensitivity of e.g. 0.3–0.4% from their room temperature value for every 1 °C [78] and can be used for temperatures between −200 and 500 °C and even up to 900 °C [78]. Pt RTDs are used as fluid flow velocity sensors, an application known as hot wire anemometry. Metal thin film RTDs have been used in large flexible and stretchable sensor arrays for epidermal/wearable [73, 79, 80] application as well as in sensor arrays to monitor entire organs [75] and in balloon catheters [81]. A linear response with a sensitivity of 23 mK was reported in [75] with the sensor as shown in Fig. 3.13 with an average response of 1.23 Ω °C^{-1}. The sensor was fabricated by depositing 40 nm thick Au on 5 nm Cr using an electron-beam evaporator. Blood flow occlusion via a pressure cuff on the upper arm led to measurable changes of skin temperature on the wrist of a human subject. The temperature decreased due to no blood flowing through and due to dissipation of tissue heat to the environment. With the use of a thermal model of the human wrist, measured temperature changes where converted into changes in blood flow rate.

Apart from RTDs, thermocouples and thermistors are two more families of temperature sensors. Thermistors are more sensitive to small temperature changes (resolution of 1 Ω/0.01 °C [82]) and, being suitable for achieving a high resolution within such small temperature ranges, have a faster response time but are less stable

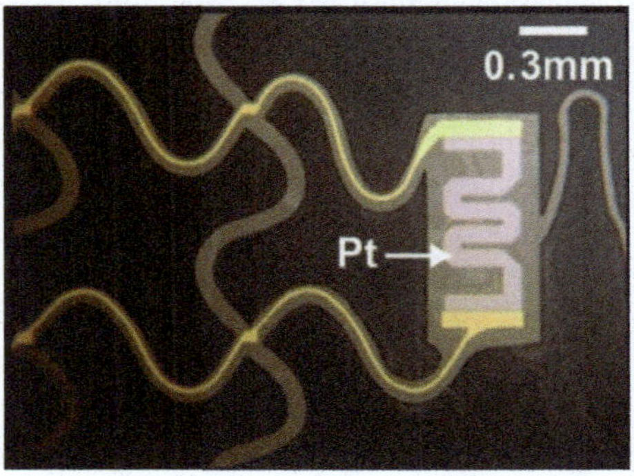

Fig. 3.13 A platinum resistor temperature sensor developed by depositing 50 nm Pt on 5 nm thick Ti by an electron-beam evaporator. The same structure was used in [81] (From [79]. Reprinted with permission from AAAS)

than RTDs. Their TCR is in the range of −0.03 °C^{-1}, ten times larger than that of RTDs. Thermistors also have a smaller temperature range (e.g. up to 250 °C or lower, typically up to 130 °C), which is, however, not important for biomedical applications. They are also highly nonlinear, but they are repeatable. Thermistors are made using ceramic or polymer materials and often from pressed semiconducting materials, e.g. sintered metal oxides of iron, cobalt, copper, magnesium, manganese, titanium, and nickel [78, 82]. The temperature dependence of thermistor resistance can be approximated by

$$R = R_0 \exp\left[1 - B\left(\frac{1}{T} - \frac{1}{T_0}\right)\right], \tag{3.24}$$

where B is a constant characteristic of the material used [82]. Because both thermistors and RTDs require a current flow through them, the possibility of self-heating of the sensing element needs to be considered. Thermistors have been used for detecting respiratory events during sleep [83, 84] and for evaluating the circadian rhythm [85] using thermistors positioned in the vicinity of the nostrils, while as discussed in [77] they have also been used in pulmonary artery catheters for thermal dilution measurement of cardiac output, rectal probes, esophageal catheters inserted though the mouth or nostrils during anesthesia, and in bladder catheters.

Thermocouples exploit the thermoelectric effect. The junction between two dissimilar metals at different temperatures gives rise to an open circuit voltage, which is a function of the junction temperature. Thus, no excitation signal is necessary, as in thermistors and RTDs. This is known as the Seebeck effect. Certain metal combinations must be used, which are welded together to create a thermocouple, and there are over 150 combinations, some of which are tabulated in [82]. These can be categorized depending on the materials used and a standardized letter designation is used to identify each, e.g. noble metal (Type B: Platinum/Platinum–30% Rhodium, Type R: Platinum/Platinum–13% Rhodium), base metals (Type E: Nickel–10% Chromium/Nickel–45% Copper), non-metals (using e.g. carbon, boron, carbide and boride), high temperature and refractory metals (Type G: Tungsten/Tungsten–26% Rhenium) [82]. The choice depends on the application. Thermocouples drift within a few hours of use and are less accurate than RTDs (accuracies bellows 1 °C are difficult to achieve), but can be used for higher temperatures than RTDs, which is important for industrial rather than biomedical applications, and have a much faster response time (ms vs. s). Their accuracy is further hindered by intermediate metals connecting to measuring equipment which creates parasitic junctions and drifts in other parts of the circuit [78]. Another disadvantage is that the open circuit potential developed across the junction that is to be measured is very small and in the range of a few tens of μV °C^{-1} and in the range of the potential formed at parasitic junctions. The latter can be compensated by measuring the temperature of the parasitic junction (e.g. using a thermistor or a RTD) and compensating for it. This is known as cold junction compensation, which, however, complicates the system topology, cost, power consumption, and size of the system. The use of a pair of thermocouples, one with a positive open circuit potential and one with a negative, can be combined to increase the recorded potential. The response of thermocouples is nonlinear, described by a complex polynomial equation, which can be as high as of ninth-order depending on the materials used [77]. Their need for frequent recalibration, the parasitic junction measurement system necessary, and their limited accuracy render this type of sensor unsuitable for implantable applications. Very thin wires are typically used for their realization, giving rise to large resistance, and consequently these are sensitive to noise, while the input impedance of the recording electronics needs to be high.

An alternative presented in [73], uses PIN diodes, the turn-on voltage of which is temperature-dependent. The PIN structures are formed by patterned doping of Si nano-membranes and the individual device dimensions are 100 μm × 200 μm. The authors use both the PIN diodes and the metallic structures as temperature sensors, as local microscale heaters, or both, simultaneously. The sensors did not demonstrate any hysteresis and demonstrated sensitivities of 14–23 mK, which is comparable to the 24 mK sensitivity of commercially available infrared cameras.

However, in applications where active electronics are involved, it is more convenient to exploit the temperature dependence of various on-chip electronic components and thus combine electronics and sensor on the same chip. This is typically achieved using transistor-level techniques and components available in CMOS technologies, and the bibliography on the subject is vast. These techniques originate from thermal management applications where there is a need to monitor temperature variations due to the environment or due to integrated circuit operation, which could affect system reliability and performance [86]. The majority of CMOS temperature sensors exploit bipolar transistors. Two types of bipolar transistors are typically available in CMOS processes: lateral and substrate. Due to a lower sensitivity to stress and a more ideal behavior, substrate bipolar junction transistors (BJT) are preferred. Figure 3.14a illustrates how such pnp BJTs are formed in a typical n-well CMOS process with a p^+ drain-source diffusion in an n-well. One disadvantage, which is evident from this figure, is that the BJT's collector is formed by the substrate; as a result, this terminal will be held at a constant potential, e.g. ground, limiting design flexibility in such processes. Another disadvantage is the low current gain of these devices. These, however, are not limiting the use of these devices in temperature-sensing applications [87]. If two such BJTs are operated, each with a different current, the difference in their base-emitter voltages (V_{BE}) is almost linearly proportional to absolute temperature (PTAT) [87]. Typically the sensitivity is -2 mV K^{-1} for individual BJTs. If the differential circuit mentioned is used, as described in more detail in [87], the ΔV_{BE} is PTAT and independent of process parameters and the absolute value of the collector currents and typical circuit designs lead to sensitivities of 198 μV K^{-1}. Using such technique measurements with an absolute accuracy of 0.1 °C between −50 and 130 °C can be achieved according to [87]. The group from the Delft University of Technology has produced a vast number of publications on the subject [87–89] and in recent publications they have focused on offset reduction and batch calibration to improve fabrication yield and minimize calibration complexity. With batch calibration [88], the authors achieved a system consuming 65 μW and an inaccuracy of ±0.25 °C from −70 to 130 °C, while with individual system calibration an inaccuracy better than ±0.1 °C was achieved. Typically a PTAT circuit is used with a complementary to absolute temperature (CTAT) current-generating circuit. While the PTAT increases with increasing temperature, the CTAT decreases. CMOS transistors operated in weak inversion, an operational region where the device VI characteristic follows an exponential behavior similar to a BJT and which is useful for low-power, low-frequency applications, can be exploited for the generation of PTAT and CTAT signals, as in [86]. The temperature-dependent voltages are converted to delays and these are used to generate a temperature modulated output pulse width. The sensor consumes 119 nW of power in this way; however, its operating range is not within that required for biomedical applications. In [74], a temperature-insensitive reference voltage was compared to a PTAT and a multilevel low-power comparator immediately digitized the signal for an implantable application, but consuming 37.3 μW. A system with a 10 μW power consumption that utilizes a PTAT with transistors in weak inversion, capable of measuring between 10 and 80 °C with an accuracy of 2.8 °C was presented in [90]. The system generates a clock output, the frequency of which is PTAT; however, in order to do so it requires an external temperature-insensitive clock. Another voltage to frequency converter system with a power consumption of 1.8 μW, an accuracy of 1 °C, a sensitivity of 442 Hz $°C^{-1}$ between −40 and 120 °C and nominal output frequency of 100 kHz was presented in [91]. Instead of using the BJT or subthreshold metal-oxide-semiconductor field-effect transistor (MOSFET or just FET) temperature-dependent characteristics, the authors used a beta multiplier with an on-chip 200 kΩ high-resistivity polysilicon resistor with temperature-dependent properties. This together with the temperature-dependent mobility of the MOSFETs used gave the temperature sensitivity of the topology. An alternative, based on ring

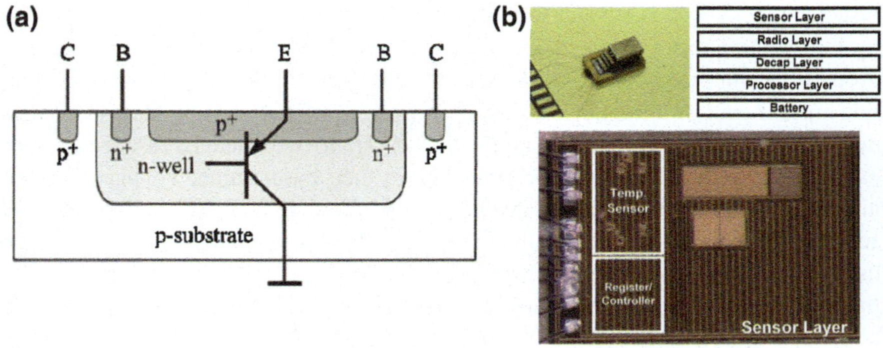

Fig. 3.14 **a** Substrate pnp bipolar transistor in an n-well CMOS technology (© IEEE 2004. Reprinted with permission from [87]). **b** A temperature-sensing node with 3D integration of multiple dies. The temperature-sensing die is located on the top of the ASIC stack (© IEEE 2014. Reprinted with permission from [93])

Table 3.2 Comparison of temperature sensing systems based on CMOS technology

Paper	Sensor type	Technology (μm)	Area (mm^2)	Supply (V)	Resolution	Calibration	Error (°C)	Temperature range (°C)	Power (μW)**
[86]	FET	0.18	0.0416	0.5, 1	0.14–0.21 °C/LSB	2-point	−0.8/+1	−10 to 30	0.119
[74]	n/a	0.35	n/a	2.5	±0.67 °C	n/a	n/a	36–40	37.3
[90]	FET	0.35	0.08	2.2	0.1 °C	1-point	−1.8/+1	10–80	10
[88]	BJT	0.7	4.5	2.5	0.025 °C	1-point	(±0.25, ±0.1)*	−55 to 125	62.5
[89]	BJT	0.16	0.08	1.5	0.02–0.005	1-point	±0.15	−55 to 125	5.1
[91]	FET, R	0.18	0.02175	1.2	442 Hz $°C^{-1}$	1-point	1	−40 to 120	1.8
[92]	FET, R	0.35	0.084	n/a	0.035	2-point	±0.1	35–45	0.11
[93]	FET	0.18	0.09	1.2	0.3	2-point	−1.4/+1.5	0–100	0.071

*Batch calibration, individual device calibration

**Calculated using the minimum supply voltage and supply current, where mentioned as such oscillators, the oscillating frequency of which is temperature-dependent, is presented in [92]. The system had a resolution of 0.035 °C, an inaccuracy of 0.1 °C between 35 and 45 °C following a two-point calibration step, and a power consumption of 110 nW. A cascaded bootstrapped current source was used, which generates a temperature-dependent output current due to the temperature dependence of the resistors and the threshold voltage (V_{Th}) of the transistors used. The authors of [93] developed a novel MOSFET-based sensing element comprised of two cascaded NMOSs with their gates connected to their common node. The topology eliminates process/temperature dependencies of V_{Th} and μ, and thus demonstrates low process variability, while being linear and consuming only 8 pW. These generate a voltage which, following a number of steps and stages, generates two voltages used in a voltage-controlled ring oscillator. The same is performed using a temperature-insensitive structure. The two generated oscillator frequencies are fed into two counters where frequency to digital conversion takes place. The overall system consumes only 71 nW for measurement between 0 and 100 °C, with an error of −1.4 °C/+1.5 °C and a resolution of 0.3 °C, and was used as a wireless sensor node as shown in Fig. 3.14b. The above are summarized and compared in Table 3.2. More temperature-sensing modalities, such as capacitance, quartz and noise thermometers, and fiber optic probes are discussed in [82].

3.4 MEMS-Based Motion Sensors

Micro-electro-mechanical systems (MEMS)-based technologies play an important role in the development of biomedical sensors [94] for measuring physical properties such as pressure, strain, motion (displacement, acceleration, and velocity), deformation, moment, torque, and other indices [95]. Accelerometers, gyroscopes, and magnetometers are the driving force in the development of MEMS sensors, with a plethora of commercially available options available. Nowadays, every smartphone has such sensors, as well as most fitness/wellness wearable devices. These sensors have been exploited for a wide range of applications, including wellness monitoring, rehabilitation applications (e.g. gait analysis), and monitoring of the elderly (e.g. fall detection and balance). Physical activity produced by skeletal muscles results in energy expenditure, which overall relates to health status in areas including cardiovascular disease, obesity, and diabetes mellitus [96]. Advances in micro/nano-fabrication have allowed the fabrication of MEMS-based sensing devices on the same substrate with the necessary CMOS sensor recording electronics, allowing high miniaturization, sensitivity, low-power operation, and low cost [97, 98]. While there is a significant interest in such devices for aerospace/military (e.g. navigation of missile systems) and consumer electronics applications (wearable fitness monitoring devices, gaming consoles, smartphone applications, and navigation), they also play an important role in biomedical applications. For example, they can be used to estimate the severity of symptoms in Parkinson's disease patients and the efficacy of treatments. Freezing of gait (FOG) is a common symptom in Parkinson's disease, related to falls, while dyskinesia is a side-effect of Parkinson's disease medication, and thus such devices can be used to detect falls and to assess and regulate drug intake. It is obvious that they are generally invaluable to neurological tremor patients. Tremor is a common movement disorder affecting 4% of people above 65 years old, and while it is not life-threatening it significantly affects the quality of life of patients [99]. Detecting falls with such devices is important for safety monitoring of high-risk patients, such as the elderly [100], as a means to tackle fall-related injuries. Another high-risk group that can benefit from such devices are children with cerebral palsy, a movement disorder that appears early in childhood. Another application discussed in [100] is the detection of epileptic seizures, where in primary and secondary epileptic crises patients experience a sudden loss of consciousness, being unable to interact with the environment. Stroke patients often experience reduced control of body segments due to paralysis. This takes place usually on the opposite side of the brain damaged by the stroke and can affect any part of the body. Paralysis on one side is known as hemiplegia, while one-sided weakness is called hemiparesis. Rehabilitation and therapy can help stroke patients improve. Motion sensors can be used to assess this process and to detect future strokes. The use of MEMS motion sensors to monitor neonates and assess their neuro-motor development and detect brain dysfunction was reviewed in [101]. Another example is the application of MEMS-based motion sensors to assess the stimulation parameters and location of implant electrodes in deep brain stimulation for tremor suppression. From the above, it is obvious that the information collected by motion sensors can be used as a feedback signal for rehabilitation, either in the form of changing e.g. drugs or drug dosage, or physiotherapy/exercise routines and methods, or as an electrical signal regulating stimulation platforms. This is important, as such stimulation platforms can be implantable neural stimulation systems regulating tremor by deep brain stimulation (DBS), suppressing epileptic episodes, or replicating the operation of the vestibular system. Such implanted neural interfacing devices are typically batteryless, communicating with the external world and being powered via inductive telemetry. Consequently, the MEMS sensors do not need to be implanted, but can be part of the external interrogation device and worn on the body. Nevertheless, these would still be a fundamental part of the implanted system and its operation. The vestibular system is a particularly interesting example. It is located in each ear and is responsible for the sense of balance, the head's motion, and spatial orientation, and helps stabilize vision through the vestibulo-ocular reflex (VOR) [102, 103]. It is comprised of two components: three semicircular canals orthogonal to each other (horizontal, anterior, and posterior), which are sensitive to radial acceleration, and the two otolith organs (saccule and utricle), which are sensitive to linear accelerations relative to gravity in two planes, approximately orthogonal to each other. Each is positioned to be parallel to the horizontal and vertical planes. Vestibular disorders include vestibular neuritis, labyrinthitis, Ménière's disease, vertigo, loss of balance, blurred vision, instability in locomotion, and nausea. The use of MEMS sensors emulating the operation of the vestibular system and implantable electronics for the treatment of vestibular disorders was examined in [102], where a commercial MEMS sensor was used, while in [104] an implantable bio-inspired MEMS-based microfluidic approach was used to develop a low-power MEMS device mimicking the operation of the natural vestibular system.

Accelerometers sense the linear acceleration in one or more directions, while gyroscopes sense the angular motion on one or more axes. Typically, the operation of an accelerometer is based on a proof mass (or seismic mass), which is attached to a mechanical suspension system. The latter is attached to a stable reference point [96]. Consequently, an inertial force applied to the system due to acceleration or gravity will force the mass to deflect, satisfying Newton's second law (force = mass × acceleration). This deflection can be assessed by a variety of methods, which include piezoelectric, piezoresistive, capacitive, electromagnetic induction, or optical techniques [94–96]. In piezoresistive MEMS, a cantilever is typically used and piezoresistive elements are fabricated on the cantilever and the proof mass. Bending of the cantilever due to the applied mechanical forces induces corresponding changes in resistance, typically measured in a Wheatstone bridge arrangement. Such systems can measure constant acceleration, e.g. gravity, but are temperature-sensitive and the change of resistance to be measured is small. Piezoresistive sensors and readout circuits will be discussed in more detail in Sect. 3.5. Piezoresistive MEMS cantilevers have attracted interest in the development of biosensors, where e.g. antibody-antigen binding events on the cantilever will induce mechanical forces on the cantilever due to increased mass [97, 98]. In [98] CMOS ASICs using a silicon-on-insulator (SOI) process were post-processed with a SOI micromachining process to create 200 μm × 50 μm and 150 μm 40 μm cantilever beams with a thickness of 0.6 μm. Piezoresistive structures on the cantilever were realized by boron ion implantation. An N^+ polysilicon layer was used to implement the piezoresistor in [97]. In piezoelectric sensors, the applied acceleration induces a force and consequently a displacement of the seismic mass, resulting in an output voltage signal proportional to the acceleration. Materials such as barium titanate, lead zirconate-titanate (PZT), zinc oxide (ZnO), aluminum nitride (AlN), and single-crystal quartz generate a potential difference across them as a function of applied mechanical stress or strain [94]. This is also proportional to the sensor thickness, and as a result they tend to lose sensitivity as the size becomes smaller. Nevertheless, they demonstrate a better SNR and in contrast to piezo-resistive sensors, which are typically used to sense strain in a single direction, piezoelectric sensors can detect strains in all directions simultaneously [105]. On the other hand, they do not respond to the constant component of acceleration [96]. As discussed in [94], the piezoelectric effect is described by the constitutive equation as

$$D_i = e_{ij}S_j + \varepsilon_0 \varepsilon_{ik}^S E_k, \tag{3.25}$$

where D_i is the electric displacement in the ith orthogonal direction (i = 1, 2, 3) due to six mechanical strain components S_j (three normal strain and three shear strain components, thus j = 1, 2, 3, 4, 5, 6) and three orthogonal electric field components E_k (k = 1, 2, 3), e is the piezoelectric strain coefficient and ε_{ik}^S is the relative permittivity of the material between different pairs of directions (i and k) at a constant mechanical strain S. As before, ε_0 is the permittivity of free space. MEMS-based piezoelectric structures usually adopt a parallel plate capacitor structure with the piezoelectric material between the two parallel plates of an area A carrying a charge Q. As discussed in [105], it is assumed that since MEMS sensors are essentially 2D structures, the polar axis is in the out-of-plane direction and only D_3 can be measured, in which case

$$D_3 = \frac{Q}{A}. \tag{3.26}$$

If there is no external applied electric field, and since $e_{31} = e_{32}$, the above may be rewritten as

$$D_3 = e_{31}(S_1 + S_2) + e_{33}S_3. \tag{3.27}$$

As will be discussed in Sect. 3.5, capacitive sensors have a high sensitivity and a low temperature drift, while they are easily implemented in most mechanical structures. Consequently, the displacement of the proof mass can be assessed capacitively. As shown in [106], differential capacitive accelerometers offer a higher SNR, reduce the effects of offsets and $1/f$ noise, and consequently have higher sensitivity and are sensitive to the constant component of accelerations. Such a structure can be implemented by stacking two capacitive structures on top of each other to form a capacitive potential divider, with a common capacitor plate implemented by a cantilever as in [106]. The induced forces will increase one capacitance while reducing the other. A MEMS motion sensor was presented in [107], where two coplanar electrodes formed a capacitive element. A suspended grounded seismic mass was formed above the capacitive structure. This affected the fringing electric field formed between the two coplanar electrodes as it moved above them as a function of motion. A three-axis capacitive accelerometer using a single proof mass developed by a post-processed CMOS chip was presented in [108]. The device demonstrated sensitivities of 0.53 mV G^{-1} along the x-axis, 0.28 mV G^{-1} along the y-axis and 0.2 mV G^{-1} along the z-axis within 0.8–6 G. The single proof mass design allows reduction of size and cost and does not require complex packaging procedures, as would the assembly of three identical individual accelerometers orthogonally to each other on a single substrate [109]. Sensing of each axis acceleration was based on three differential capacitive comb-finger structures, where mechanical motion induced changes in the gap between the fingers. The orientation and spring design of the structure allowed three separate structures each to respond to mechanical motion along one particular axis. The x- and y-axis springs have the same serpentine structures, while the z-axis spring is a V-shaped structure. The recording electronics are integrated on the same chip. A similar approach was presented in [110] by the same group, where improvements in the sensing elements lead to an increase in the sensitivities at each axis to 14.7 mV g^{-1} along the x-axis, 15.4 mV G^{-1} along the y-axis and

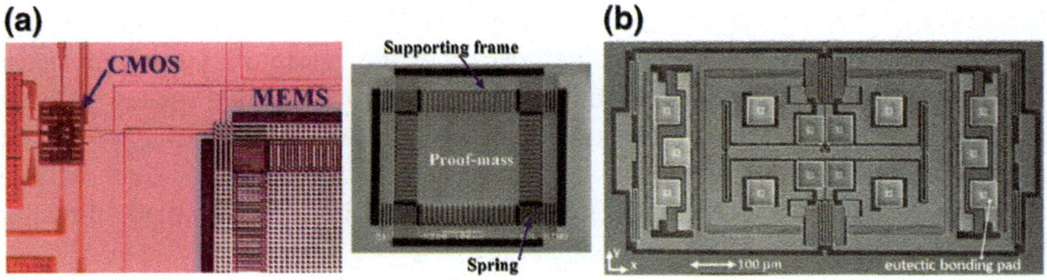

Fig. 3.15 **a** Left: Close-up showing the CMOS processing circuits implemented in the vicinity of the MEMS device. Right: Scanning electron microscope (SEM) image of a three-axis MEMS accelerometer (© IEEE 2012. Reprinted with permission from [110]). **b** SEM top image of a *z*-axis gyroscope with piezoresistive nanogauges (© IEEE 2015. Reprinted with permission from [114])

14.6 mV G^{-1} along the *z*-axis within a measurement range of 0.01–1 G. The developed MEMS device and CMOS electronics are shown in Fig. 3.15a. To improve the temperature stability of the accelerometer, the same group developed the system proposed in [111], where the mechanical structure was implemented by stacked pure oxide layers and the metal layers were limited to the sensing electrodes and electrical routings. This suppressed the thermal deformation of the suspended structures due to mismatch of the coefficients of thermal expansion between metal and oxide layers, while parasitic capacitances were reduced as well as noise, improving sensitivity and performance further. As argued in [109], while the use of a single proof mass leads to a compact design, it results to cross-axis sensitivity, hindering the use of such designs in high-performance applications.

Another disadvantage of various structures is the use of non-identical sensing elements, resulting in different performances for each axis. The proposed structure of [109] addresses the above shortcomings by implementing three individual differential capacitive accelerometers where the orthogonal alignment of the accelerometers is inherently achieved by the fabrication process. The device achieved significantly reduced noise, cross-axis sensitivity, and nonlinearity compared to the other designs and a range of ±10 G in the lateral and +12/−7.5 G in the vertical directions. A nano-G in-plane nano-photonic resonant optical accelerometer based on multilayer sub-wavelength diffraction nano-gratings was presented in [112]. Two sets of polysilicon nano-gratings separated by an air gap modulate an incident light source's polarization and near-field intensity according to their relevant lateral motion. As in other examples, differential sensing was employed to improve sensitivity. The advantages of such topologies lie in an improved sensitivity and performance close to the Brownian noise limits of the structure. The device achieved a sensitivity of 590 V G^{-1}. The devices demonstrated very high sensitivities to very low accelerations, requiring specialized packaging to address their fragility.

With regard to MEMS gyroscopes, there are three basic types: spinning rotor, ring laser, and vibrating mass. The most common type is vibratory gyroscopes. The subject of MEMS gyroscopes was recently reviewed in [113]. Such devices are typically composed of three elements: the proof mass and the sensing and drive elements. Sensing is typically performed via the capacitive measurement of the angular frequency through the Coriolis force applied along the sense direction at the suspended proof mass, which is kept in motion with a controlled amplitude along the drive direction [114]. This mode of operation is known as amplitude modulation (AM) mode. Generally speaking, if the proof mass moves according to $x = A_x \cdot \cos(\omega_x t)$, where x is the proof mass displacement, A_x is the amplitude of the movement, and ω_x is the angular speed of the driving system, then when an angular speed Ω is applied along the z-axis, a Coriolis force, F_y, along the y-axis will be applied according to [113]

$$F_y = 2m\Omega v = -2mA_x\omega_x \sin(\omega_x t), \tag{3.28}$$

where v is the linear moving speed of the proof mass. A minimum of two degrees of freedom motion capability is typically permitted to achieve Coriolis-induced transfer energy between the drive and sense resonance modes [115]. For example, in [115] the suspended drive and sense frames vibrate in only one direction, thus having one degree of freedom, while the proof mass vibrates along two orthogonal directions, having two degrees of freedom. This allows Coriolis coupling to be achieved between the drive and sense frames. When the resonant frequencies of these are matched, the rate sensitivity of the sense mode is amplified and maximized by its mechanical quality factor. However, when this is true it is challenging to achieve low noise, large bandwidth, long-term stability, and low power consumption simultaneously [114]. In any case, manufacturing tolerances will lead to mismatched mode frequencies. The drive and sense modes should ideally be decoupled; however, manufacturing tolerances will lead to undesirable mechanical crosstalk between the two. As discussed in [116], this coupling will lead to a measurement offset translating to an error as large as $1000° \text{ s}^{-1}$. Using various techniques, e.g. phase-sensitive detection, this can be minimized by two orders of magnitude, but not entirely eliminated. It is thus one of the dominant factors limiting the operating range, linearity, and bias instability of such devices [116]. Due to this coupling there is an induced double frequency component. When differential sensing is used, this appears as a common-mode signal, which can be greatly suppressed. However, in reality, due to manufacturing tolerances, this will not be entirely suppressed, since the geometry will not by exactly balanced and matched. Limiting to one the degree of freedom on the sense and drive modes reduces the coupling, while proper mechanical design to increase the rigidity of the sense mode electrodes reduces the double frequency components, unfortunately reducing the Coriolis response [116]. A piezoresistive sensing approach as an alternative to the common capacitive sensing methodology was examined in [114], in an effort to reduce the impact of noise. The z-axis MEMS device developed is shown in Fig. 3.15b.

MEMS multi-axis accelerometers and gyroscopes are often combined with MEMS magnetometers to create inertial measurement units and compasses, such as those found in smartphones [117]. In [118], for example, a z-axis magnetometer was combined with a three-axis gyroscope. There is a wide range of techniques that can be used to make a magnetometer. These include those based on anisotropic magnetoresistance (AMR), giant magnetoresistance (GMR), magnetostriction, the magnetoresistive effect, flux gate, the Hall effect, and the Lorentz force [119]. The measurement range of various techniques was discussed in [120]: Hall effect sensors are capable of measurements from the upper nT to T range; GMR sensors, which are based on the principle of anti-ferromagnetic coupling, are sensitive in the sub-microtesla range; and fluxgate sensors, which are based on the principle of magnetic saturation, are sensitive in the upper pT range, but are large power-hungry devices. Finally, superconducting quantum interference devices (SQUID) can be used in the fT range, but require cryogenic cooling and are sensitive to electromagnetic interferences. According to [120], the most sensitive reported magnetometer is a device based on scalar atomic magnetometry [121]. The most popular technique is the Lorentz force method, as it can be implemented as a MEMS implementation allowing integration with other MEMS-based sensors and thus the realization of compact, low-cost, low-power simple (when compared to other approaches) devices developed using common processes. In addition, they avoid the use of magnetic materials, do not pose an intrinsic sensitivity versus full-scale tradeoff as with AMR devices, and overcome the achievable noise densities of Hall sensors [122], while according to [119] they are hysteresis-free, in contrast to magnetoresistive and flux gate topologies. These devices are based on the measurement of the Lorentz force induced on a structure due to an applied magnetic field, B. In the presence of this field, a driving current I in a suspended structure orthogonal to B determines the force in a direction orthogonal to both B and I. This is described in more detail in [118]. The amplitude of this force is very small, which is particularly an issue when resolutions in the μT and below are needed. As before, the force is sensed via piezoresistive or capacitive transduction. In order to improve the sensitivity, high force sensors and/or structural mechanical force amplification mechanisms are required or to exploit the high Q-factor of microscale resonant structures [120]. As in gyroscopes, an AC signal is injected to the device at its resonant frequency. This amplifies the induced Lorentz force motion with a high quality factor. Similarly to what was described earlier for gyroscopes, this is an AM scheme and there is a tradeoff between operating in resonance and bandwidth. A sub-400 $\mathrm{nT}/\sqrt{Hz}$ resolution z-axis Lorentz force magnetometer implemented with a tuning fork combined with a separate custom signal processing ASIC was presented in [122] achieving a power consumption of 775 μW. A 27 pT magnetometer was presented in [119]. This was intended for brain imaging applications, and in particular for magneto-encephalography (MEG) recordings to map brain activity, where a minimum detectable signal of 10 fT is required. A magnetoelectric nano-plate resonator structure was presented in [117], based on a magnetostrictive/piezoelectric magnetoelectric heterostructure, comprised of a $(FeGaB/Al_2O_3)$ 10 multilayer with a high magnetostrictive coefficient, and with an AlN piezoelectric nano-plate resonator body. This structure is deposited on top of a Pt interdigital transducer excited by an AC signal. In the presence of a magnetic field, the Young's modulus of the structure changes and this translates in a change of resonant frequency. The proposed structure of [120], achieves a 2.8 $\mathrm{pT}/\sqrt{\mathrm{Hz}}$ sensitivity by using a thermal-piezoresistive amplification technique, where a DC bias current is used to drive two suspended resonator mass plates connected through a piezoresistive beam.

3.5 Strain and Pressure Sensors

A very important class of sensors is those assessing mechanics. Within this class, we include strain and stress, pressure, velocity, gyroscopes, and acceleration. MEMS technology and simple thin film planar and flexible electronics technologies play an important role in the development of such sensors. In this section the chapter focuses on stress/strain and pressure sensors.

3.5.1 *Strain*

The most common means of sensing strain is via changes in resistance. The classical resistive strain gauge is a thin film metal wire (copper-nickel or nickel-chrome alloys) with a serpentine geometry similar to that used in many RTDs. When a tensile force, F, is applied to a material with an unstressed cross-sectional area of A, the stress, σ_{str}, on the material is defined as the applied force per unit of cross-sectional area

$$\sigma_{str} = \frac{F}{A} \tag{3.29}$$

The suffix ‘str’ is used here to differentiate these parameters from the electrical ones defined earlier in the chapter. The length of the wire increases (axial elongation) and its diameter decreases (transversal contraction), thus decreasing A, proportionally to the stress. The larger the force, the larger the change. Thus strain has two components: the longitudal strain $(\varepsilon_{str-long})$ and the transverse strain $(\varepsilon_{str-trans})$. According to (3.2), this leads to an increase of resistance. If the initial length of the device, L_0, becomes L after a force is applied to it, then the ratio of the elongation to the original length is called the tensile strain, ε_{str}, and is given by

$$\varepsilon_{str-long} = \pm\frac{(L - L_0)}{L_0}. \tag{3.30}$$

The sign depends on whether the force is tensile (positive) or compressive (negative). Equivalently, for the transverse strain, if the original diameter is D_0 and the diameter following the applied strain is D, then

$$\varepsilon_{str-trans} = \mp\frac{(D - D_0)}{D_0}. \tag{3.31}$$

The sign depends on whether the force is tensile (negative) or compressive (positive). Strain is a nondimensional quantity. Stress and strain are linearly related in elastic materials and their relationship is described by Hooke's Law, according to which

$$\sigma_{str} = E_{str}\varepsilon_{str}, \tag{3.32}$$

where E_{str} is the elastic modulus of the material, also known as Young's modulus of elasticity, assumed to be a constant for a given material. The ratio of the absolute values of the longitudal and transverse strains is known as Poisson's ratio, given by

$$\nu_{str} = \left|\frac{\varepsilon_{str-trans}}{\varepsilon_{str-long}}\right|, \tag{3.33}$$

The above hold provided that the deformation is within the elasticity limits of the materials, such that they do not break or deform permanently. If not, then Hooke's law is no longer valid and the material behaves irreversibly and is considered to be in the plastic deformation region, where the relationship between stress and strain is no longer linear. Apart from the geometry of a material, under stress its conductivity may also change. This is due to crystal lattice deformations. The energy band curves of the material change, and consequently its effective mass. Conductivity is a function of effective mass. Nevertheless, in metals this is not a significant change and the effects of the geometrical changes dominate the response. The sensitivity of a strain sensor to the applied strain is defined by the gauge factor, *GF*, given by:

$$GF = \frac{R - R_0}{\varepsilon_{str}R_0}, \tag{3.34}$$

where R_0 is the resistance before the force is applied to the device and R is the resistance after deformation. Commercially available thin-film serpentine strain gauges typically have a *GF* of about 2. The direction of orientation of the conductive lines of the sensor defines the main orientation of deformation sensitivity. Instead of a single straight line, a serpentine structure of multiple parallel straight lines is used, which is effectively a longer wire. According to (3.2), the longer the conductor, the larger the resistance. Thus small deformations can result in larger resistance changes. It is evident from the discussion of RTDs that strain sensors are also temperature-sensitive, and vice versa. This can be dealt with by appropriate material selection and by having a dummy sensor that remains unstrained but in close proximity to the sensor, i.e. an RTD. Semiconductor strain gauges based on silicon or germanium, for example, offer a larger *GF*, e.g. n-type doped silicon may have a *GF* as large as 200 and an n-type of −140. In these devices, the *GF* is a function of both the doping material and its concentration, they demonstrate very small hysteresis and because the dopants are randomly oriented within the crystal, the device response is typically not sensitive to orientation. Lightly doped devices have a higher resistivity and a larger *GF*, but also greater temperature sensitivity. Nevertheless, such polycrystalline materials are very brittle and reach their elastic limit faster than metals, while their *GF* is nonlinear.

In [123] the authors controlled the deposition of platinum nanoparticles to form 2D layers of various densities using standard lithographic processing at room temperature. At low densities the films were semiconducting; with increasing nanoparticle density they became more metallic. The highest sensitivities were obtained for medium-density films leading to sensitivities of one order of magnitude higher than those of continuous metallic films, which typically have a *GF* of 2 [124, 125]. Gold interdigitated electrodes were developed with a gap of 10 μm and the nanoparticles were deposited between the electrodes. Recent advances in nanotechnology and micro/nano-fabrication and the widespread availability of metal nanoparticle-based, conductive polymer and other inks for inkjet and screen printing have opened the door for the development of lower cost strain gauge devices. The advantage of inkjet printing over screen printing is the fact that the former is a maskless technique, leading to fewer processing steps and significant cost reduction, as reduced ink quantities are used. In addition it is a contactless technology. PI (such as DuPont Kapton) and PET (polyethylene terephthalate, such as Mylar) films of various thicknesses are commonly used in flexible electronics applications. Such substrates have been used for inkjet-printed stain sensors such as the ones presented in [124–126]. In [123, 124] commercially available regular office inkjet printers and ink cartridges and nozzles were used, together with commercially available silver nanoparticle inks. Classic serpentine structures were implemented. Typically, three superimposed layers are printed [124] to achieve a good conductivity and good geometric features followed by nanoparticle sinterization via sample heating at temperatures in the region of 150 °C for half an hour (depending on the ink) [125, 127]. According to [124, 127], conventional strain

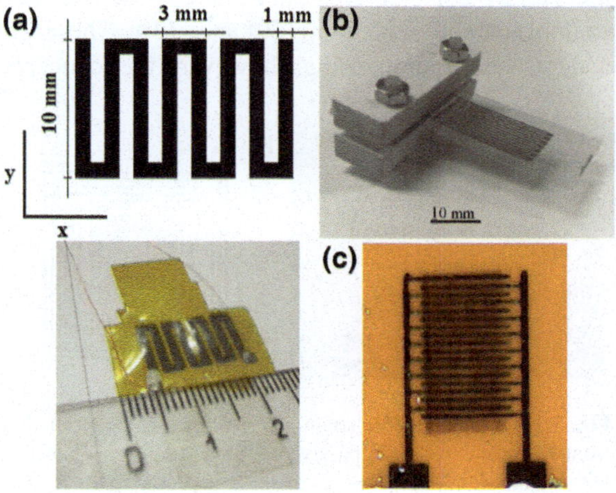

Fig. 3.16 The inkjet-printed strain sensors of **a** [124] (© IEEE 2014. Reprinted with permission from [124]) **b** [126] (© IEEE 2013. Reprinted with permission from [126]) and **c** [125] (© IOP Publishing. Reproduced with permission. All rights reserved, https://doi.org/10.1088/0964-1726/22/10/105028)

gauges made of metal foils and semiconductor slabs are larger and less suitable for attachment on curvilinear surfaces or to measure large deformations. *GF*s of 6.2 were achieved with the structure of Fig. 3.16a, while in [126], with track widths of 200 μm, a spacing of 300 μm, and a length of 1 cm, a *GF* of 21 was reported (Fig. 3.16b). In [125], similar classic serpentine structures were reported and were compared with interdigital strain sensors, where the main contributor to strain sensitivity was either an additional layer of poly(3,4-ethylenedioxythiophene) (PEDOT) or a layer of TIPS-pentacene, deposited on top of the structure's fingers, as shown in Fig. 3.16c. A versatile commercially available inkjet printer (Fujifilm Ditmatix) was used with printing nozzles and cartridges from the same company. Track widths of 40 μm and separations of 90 μm where achieved for the classic serpentine structure and an average nominal resistance of 830 Ω and a *GF* of 0.35. In the interdigital structure the separation between tracks was 30 μm and the track width was 40 μm. With the addition of the PEDOT layer these piezoresistive sensors led to an average nominal resistance of 1.42 kΩ and a *GF* of 2.48, while with TIPS-pentacene 100 kΩ. The authors used the PEDOT interdigital structure to create a 2D array of such strain sensors. Interceptions of lines with columns were prevented by using an insulating layer to prevent their short-circuiting. These strain sensors were used in smart hip prosthesis applications.

Carbon-based devices utilizing carbon nanotubes, graphite, and graphene have in recent years attracted significant attention, especially for flexible electronics and sensor applications. Strain sensors made using these materials and the ones previously mentioned have the ability to measure deformations up to 150% and demonstrate *GF*s up to 30 [127]. Screen-printed graphite strain sensors fabricated using a commercially available graphite ink on a 125 μm thick poly (ethylene naphthalate) (PEN) substrate were demonstrated in [127]. The devices were simple tracks, with additional silver screen-printed tracks used as interconnections. The authors reported a *GF* of 19.3 for the device of Fig. 3.17a. In a recent review, the use of carbon nanotubes for strain sensing was discussed in detail in [128], which reports *GF*s typically up to 22.4 for such devices. By combining strain sensors with inductors, passive devices can be realized such as the one shown in Fig. 3.17b. Such examples are presented in [129] and [130]. These were designed for environmental and structural monitoring; however, the concept is applicable to biomedical

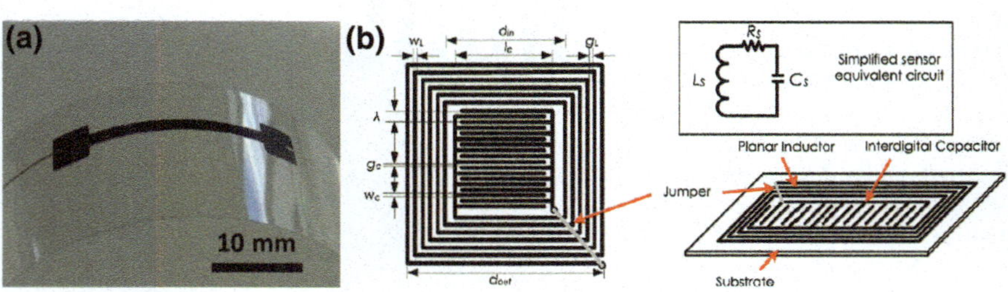

Fig. 3.17 **a** The graphite strain gauge from [127] (reprinted from [127], © 2014, with permission from Elsevier) and **b** the passive wireless strain sensor of [129] (© IOP Publishing. Reproduced with permission. All rights reserved, https://doi.org/10.1088/0957-0233/17/11/002)

applications as well. In [129], a capacitive strain sensor was presented, based on an interdigital structure. The capacitance of this topology is highly dependent on geometry; thus electrode deformation and change in the separation between the finger gaps will lead to a change in capacitance. This capacitive change will lead to a change in the resonant frequency of the *LC* circuit formed with the interdigital capacitor and the surrounding inductors, which is detectable. Similarly, in [130] a capacitive parallel plate strain sensor was presented. The capacitor plates were formed by single-walled carbon nanotube-based (SWCN) films with a 127 μm PET film serving both as the structure substrate and the capacitor dielectric. The authors extended this concept to pH sensing by developing SWCN-based pH-sensing films which demonstrated a varying resistance with pH. A passive strain sensor for orthopedic applications which was ultrasonically interrogated was presented in [131]. According to the authors, this allows the use of conventional commercially available ultrasound scanners to be used as the external readout system.

All the above can allow patients who have undergone such procedures to monitor in real time the loading and strains imposed during their daily activities on their implantable orthopedic device and to be alerted in a timely manner should any complications, excessive loadings, stresses, or movements take place. As suggested by [132], such implants can be used to develop patient-specific post-operative activity modification and tailor-made exercise routines for rehabilitation following orthopedic surgery.

3.5.1.1 Orthopedic, Dental, and Other Applications

In addition to implantable devices for monitoring metabolic/biochemical processes, an important class of implantable devices is orthopedic implants, replacing fractured/diseased/damaged bones in humans. Many such implants fail over time. Mechanical sensors are required in this case for two main reasons. The first is to understand the mechanics involved in the system under study, such that all possible movements and resulting forces to the implant are assessed. This is important in order to validate computational models and improve the design of these devices. For proper design and implant success, the influence that loads have on the implant need to be investigated and fragile regions have to be identified. In addition, such studies assist in the improvement of the alignment of the prosthesis during surgery [133]. The second reason is to examine over time the quality of the implant such that prompt intervention takes place (e.g. degradation of implant due to loosening, increased friction, high local temperature due to friction, misalignment, local fractures). Misalignment may lead to imbalance (e.g. in knee implants used for total knee arthoplasty [133]).

In [76], the authors presented an orthopedic implant comprised of a commercially available hollow shaft titanium hip endoprosthesis (Fig. 3.18). This was modified by inserting within the hollow shaft two telemetry units assembled on a ceramic substrate each comprising an ASIC and 18 passive components with a power consumption of 10 mW and dimensions of 7 × 14 mm. Each of these units was hermetically encapsulated using a metal cylinder. An additional power coil was used. Three strain gages were positioned within the hollow neck. The authors discuss that temperature changes up to 3.5 °C have been reported in such implants during 45 min walking, and this was attributed to friction. These temperature changes may lead to tissue damage and implant loosening. Hence, in addition to the strain sensors, the implant incorporated eight temperature sensors. These were negative temperature coefficient (NTC) thermistors, which are typically made from semiconducting materials such as metal oxides. An external device with a coil is fixed to the patient's belt. The designed implant is shown in Fig. 3.18.

A nine-channel telemetry system with an ASIC was presented in [134] for the interrogation of up to six strain sensors, a temperature sensor, and the telemetrically supplied rail voltage. The six commercially available strain sensors (with a nominal resistance of 350 Ω) allowed the calculation of the *x*, *y*, and *z* forces and moments. Shoulder endoprostheses, vertebral body replacements, and tibial trays were instrumented with the proposed device, as shown in Fig. 3.19.

A microfabricated strain sensing device for artificial knee implants (Fig. 3.20) was presented in [133]. A commercially available knee implant was used as an exemplar device for the sensor, the most affected component of which is a polyethelene insert. FEM modeling was used to identify locations exhibiting high compressive strain to pinpoint the ideal location for the sensors in this implant component. The device comprised four strain sensors developed using standard microfabrication techniques. Two were positioned in regions of high stress (thus termed active sensors) and two in regions of zero stress (termed passive sensors);

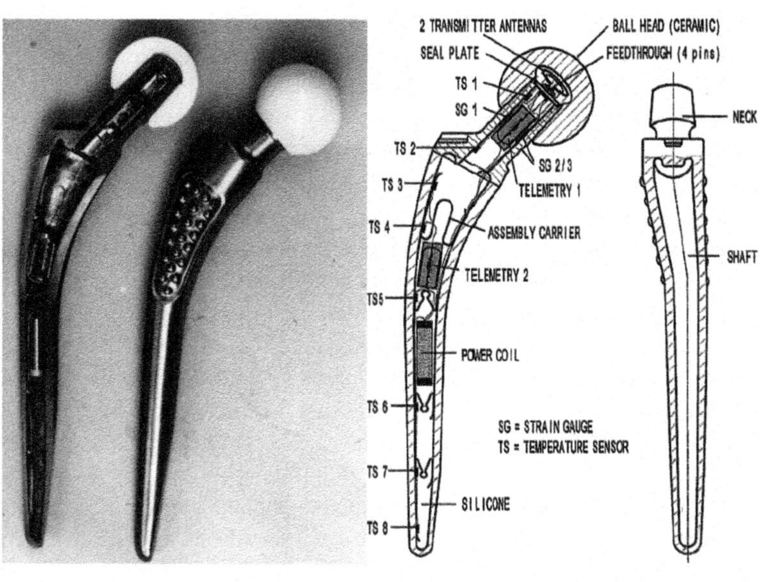

Fig. 3.18 An instrumented hip endoprosthesis with strain and temperature sensors and telemetry (reprinted from [76], © 1999, with permission from Elsevier)

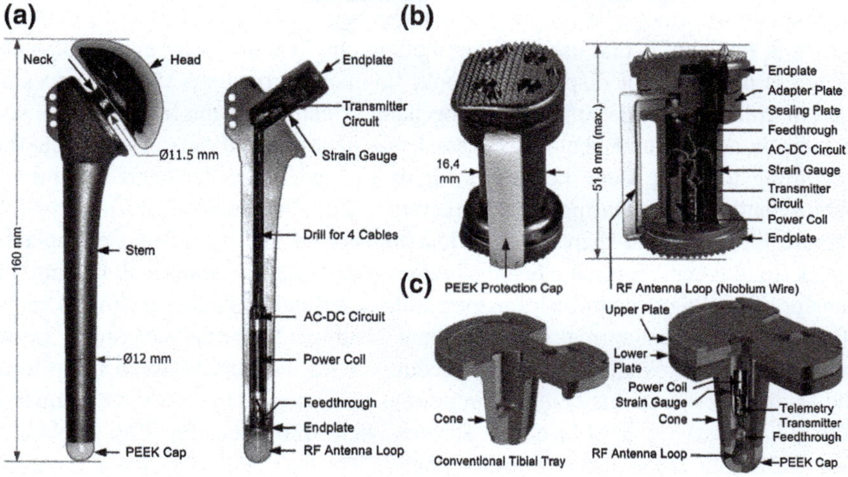

Fig. 3.19 **a** Shoulder, **b** vertebral body, and **c** tibial tray orthopedic implant with embedded strain and temperature sensors, readout electronics, and telemetry for powering and communication (© IEEE 2007. Reprinted with permission from [134])

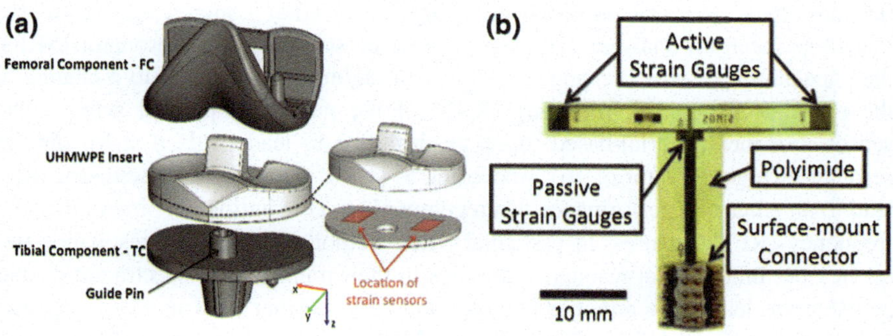

Fig. 3.20 **a** The commercially available artificial knee implant used and modified with the custom strain gauge system shown in **b** in [133]

the latter were used to compensate for temperature variations. The sensors were PI-metal-PI structures with the metal layer being an arrangement of thin-metallic wires. The resistance of the sensor was set to 3.2 kΩ to minimize power consumption. PI was used due to its chemical and thermal stability, biocompatibility, and low water uptake. The thickness of the sensor was equal to 10 μm and a standard Wheatstone bridge circuit was used with resistors of similar resistance to that of the sensor. The device demonstrated a linear response for forces between 0 and 1500 N (the force values typically applied to the knee during walking), with an average device sensitivity of 2.5 $\mu V\ N^{-1}$.

Apart from orthopedic implants replacing bones and joints, another important family of implants is that used in bone fracture applications, where an implantable device is used to treat displaced or multi-fragmented fractures. Such devices are used in order to provide sufficient biomechanical stability to the fractured site, such that the bone's natural healing process leads to a clinically successful outcome. According to [135], bone healing is highly dependent on the degree of implant stability and loading. During normal fracture healing the mechanical stability of the fracture increases and there is a gradual transfer of loading from the implanted device to the native tissue, healing callus and bone. In abnormal healing this transfer of loading is disturbed. Imaging techniques and manual assessments cannot distinguish between normal and abnormal healing. Abnormal healing is treated using salvage or revision surgical procedures, with the time between the primary and secondary operations being an important parameter in the clinical outcome. In [135–137], passive MEMS-based sensors were proposed for this application (Fig. 3.21a). In [135], in vivo results (Fig. 3.21b) from an ovine model were presented. The animal model employed both normal and delayed healing, which the proposed sensor was able to differentiate 21 days following the fracture. This is significantly earlier than when radiographic observations of calcified tissue can be acquired [135]. The sensor was adhered to modified commercially available six-hole locking fracture fixation plates. The sensor was 8 mm per side square and 0.8 mm thick and was interrogated using an external commercially available benchtop spectrum analyzer. The sensor is based on a split ring resonator (SRR), the capacitance of which changes via physical deformation leading to a change in the resonance response frequency (RRF) of the passive implanted sensor. The sensor is a metamaterial-based topology designed to lead to deep sharp dips at resonance (high Q-factor of 76). According to [136], metamaterial-based structures lead to an enhanced performance when compared to conventional (spiral coil) radio frequency (RF) structures. In particular SRRs demonstrate a lower resonance frequency per unit area, at resonance the dips in their transmission spectra are deeper and sharper, leading to a higher Q-factor and thus a higher SNR and consequently higher linearity. In [135, 136] a nested SRR topology was proposed featuring multiple SRRs (Fig. 3.21c) in a compact architecture. This enhanced in [136] the strain sensitivity of the device (0.0369 kHz/microstrain as opposed to 0.0259 kHz/microstrain and 1.09 vs. 0.723 kHz kgf^{-1}) and decreased its operating resonance frequency (506.2 MHz against 529.8 MHz) when compared to standard SRR topologies in free-space experiments. Soft tissue experiments indicated that these changed to 474.2 MHz and 4 kHz kgf^{-1} for the nested topology. The authors discuss that this change is due to the very high dielectric constant of the tissue, which improves the focusing of the incident electromagnetic waves [136]. The strain sensitivity of the device is due to the decrease in the gaps in the SRR structure under applied load, leading to an increase in the capacitance between the fingers of the device. Simultaneously the dielectric area between substrate and metal is increased and the capacitance between the two is also increased. Consequently, these lead to a decrease in resonant frequency. The nested topologies in [135] and [136] are the same, but the geometric parameters are different. Thus in [135] the

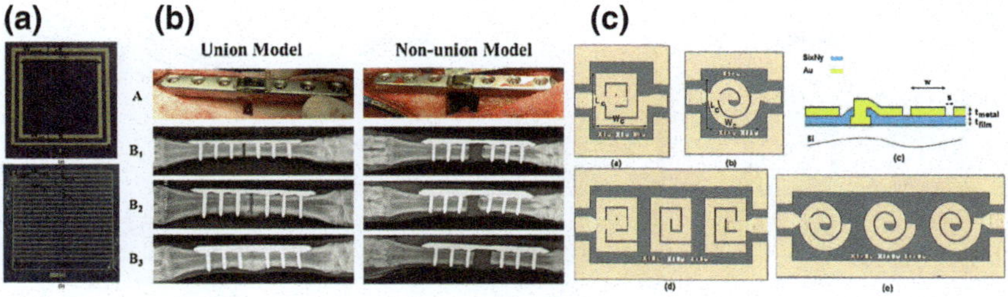

Fig. 3.21 **a** The classical SRR topology and the nested SRR structure proposed in [136] (© IEEE 2009. Reprinted with permission from [136]). **b** In vivo images obtained during surgery in [135] with different osteotomy gaps to model union and non-union healing conditions. Radiographic images at 7, 24, and 45 days (reprinted with permission from John Wiley and Sons). **c** Single and triplet rectangular and circular resonant devices from [137] and the device side view (license: https://creativecommons.org/licenses/by/3.0/, part **d** of the original figure not used)

overall device is 22.2 mm × 22.2 mm, while in [136] it is 8 mm × 8 mm. In [136], rectangular and circular single and triplet devices were presented for the same application. In the triplet topologies the middle device serves as the sensor and the other two devices are the receiving and transmitting antennas. The rectangular single device demonstrated a 330 MHz resonance shift, while the triplet a 170 MHz shift for a load of 3920 N. The circular devices demonstrated (for the same device parameters and applied load) a shift of 500 MHz for the single device and 260 MHz for the triplet and thus were more sensitive than the rectangular ones. The devices were 340 μm × 340 μm each, with two turns, a 60 μm wide Au metal, and 10 μm spacing between turns. The metal thickness was 100 nm and the dielectric film (Si_xN_y) was also 100 nm thick.

Another relevant application is stress monitoring in dental implants. As discussed in [138], such implants are fixed in the location of interest using a bridge, which is in turn fixed into its location using titanium implants into the bone and abutments, as in Fig. 3.22. Typically, the number of implants and abutments may vary between two and six in a patient. The abutments are made out of ceramics and other synthetic materials that provide low shrinkage during molding and low permeability for liquids [138]. It is necessary to understand the biomechanics involved in order to minimize the number of failed implants and to enhance the understanding of the bone regrowth process in the vicinity of the implanted titanium [138, 139]. To assess these, an external unit connected with wires to strain gauges through the patient's mouth is employed. These systems are non-portable and can only be employed in the hospital/clinic environment. In addition, due to the wires, the patient's chewing behavior is altered during measurements and thus errors in the measurements are introduced, while bruxing and clenching, which many patients do while sleeping, cannot be assessed. An ASIC for this application was presented in [138], where the electronics were molded in the prosthesis using an epoxy. With three strain gauges used in each abutment and up to six abutments possible, the ASIC could monitor up to 18 strain gauges. According to the same paper, the dental

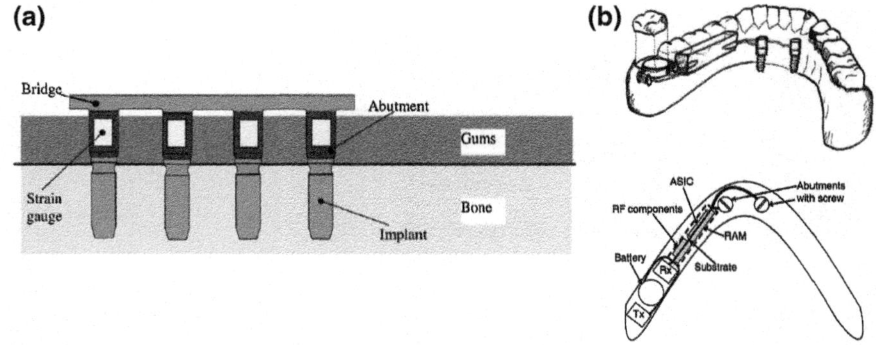

Fig. 3.22 **a** A schematic illustration of a dental implant with four abutments used to fix into position dental implants and **b** a demonstration of how the proposed implanted system of [139] is used (reprinted from [139], © 2002, with permission from Elsevier)

force signals have a bandwidth of 50 Hz and a measuring accuracy of 20 microstrain is required. This corresponds to an axial force accuracy of 19.4 N and a bending moment accuracy of 1.6 N m, with estimated maximum/minimum strains as a result of excessive biting of ±2000 microstrain.

According to [140], currently commercially available strain sensors are in the region of 2–5 mm, which is considered too large in some applications, where high resolution may be required. Their large size also hinders their use for implants due to their irregular geometries. In such applications, small, thin, flexible gauges that can conform to bone geometries and have greater sensitivities are required to obtain high resolution and accurate results. In the same paper, the authors discuss that polydimethylsiloxane (PDMS) can only form good contact with bone if a silicone-based adhesive is used [140]. This type of adhesive requires a day to cure at room temperature, thus making it undesirable for in vivo applications where such PDMS-based sensors have to be directly glued on a bone surface. To address the above, the authors present a strain sensor with a 2.5 times greater sensitivity (a gauge factor of $GF = 2.56$) than commercially available sensors. The devices consist of a thin-film metal layer (120 nm gold with a 10 nm chromium adhesion layer), encapsulated between two layers of parylene-C and fabricated with standard microfabrication techniques. The resistance of the sensor was ~180 Ω. The serpentine structure created had a length of 350 and 500 μm, a track width pitch between tracks of 10 μm, and 10 turns in total.

Apart from orthopedic replacements and bone fractures, articular cartilage osteochondral scaffold implants used as a support for cartilage regeneration can also be instrumented with strain sensors to monitor joint loading. Such scaffolds are made from biodegradable materials and with surface features which encourage bone attachment. These are used as carriers for cartilage maturation and development prior to implantation [141], as in Fig. 3.23. In [141], three strain gauges were attached to such scaffolds and according to [132] such implantations with additional readout electronics and telemetry have proven adept at detecting small changes in

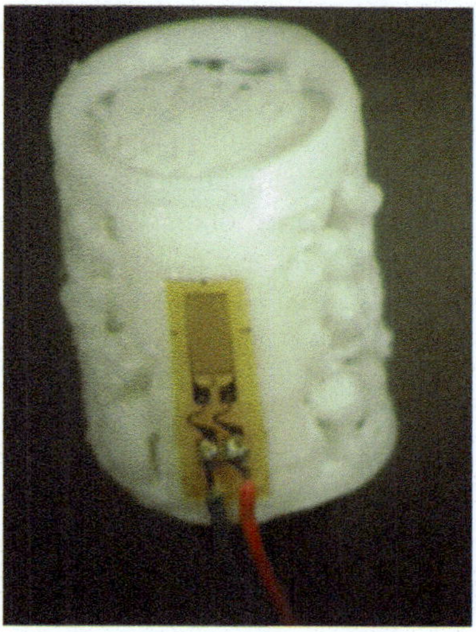

Fig. 3.23 An implantable osteochondral cartilage regeneration scaffold instrumented with strain gauges to measure cartilage loads in vivo (reprinted from [132], © 2012, with permission from Wolters Kluwer Health, Inc., journal; URL: http://journals.lww.com/jaaos/pages/default.aspx)

the forces applied to the joint and have indeed confirmed that loads applied to the system increase during healing. According to [132], such implants have been used to verify results obtained from gait and motion analysis studies and to verify and improve analytical models of strain loading. These studies have also been used to improve tissue engineering processes, where various loading procedures are used in cartilage bioreactors.

3.5.2 *Pressure*

Apart from applications related to strain, which is of interest in orthopedics, another related application is pressure monitoring. Pressure is a force applied perpendicular to a surface that results in strain, and it is used typically with fluids (liquids and gases). Similarly to stress, which is used with solids and can be both perpendicular and parallel to a surface, it is defined as the force per unit area as in (3.29); however, in contrast to stress, pressure is a scalar quantity. Pressure in organs such as the brain, eye, heart, and bladder is highly regulated. Monitoring its value locally can thus provide physicians with important information regarding a patient's health. In many cases, if pressure regulation is diminished or lost, organ function may be compromised or may lead to patient death. Typically, pressure is measured indirectly using ex vivo methods such as sphygmomanometry and ultrasound. These, however, are unsuccessful in delivering reliable and localized data for many applications [142]. Catheter-based pressure sensors have been used as a means to address the issue of localization and accuracy. Nevertheless, these often require the patient to be sedated, and as a result the condition being monitored may be affected, leading in turn to measurement uncertainties. The actual process of pressure measurement in the clinic and the stress experienced by patients has been found to affect measurement accuracy [143]; this is known as the white coat syndrome or white coat hypertension. An additional issue with current techniques is that they provide a snapshot in time and thus they may miss important events, peaks, or troughs. Thus, implantable pressure monitoring devices are necessary.

The majority of pressure sensors use a membrane and a sealed cavity. The membrane deflects under applied pressure. Resistance or capacitance changes are typically used as the transduction mechanism. In resistance-based detection, a piezoresistive material is deposited on the membrane, or the membrane itself is fabricated out of a piezoresistive material. In capacitance-based detection an electrode is positioned on the top and bottom of the cavity with one of the two being the deflectable membrane and the second typically being rigid, forming together a parallel plate capacitor. Capacitive sensing typically demonstrates higher pressure sensitivity and low temperature drift in contrast to piezoresistive devices [143]. For high accuracy measurements, differential measurements are preferred. Atmospheric pressure can be used as the baseline; however, this is dependent on atmospheric temperature, temperature, and time of day, leading to a baseline drift. Alternatively, the difference in the pressure measurement between two points can be used (e.g. the pressure difference between the inside and outside of the bladder [143]), where the pressure of one of the two can be constant and known. For absolute pressure measurements differential measurements against a perfect vacuum are necessary. However, creating a perfect vacuum that is also stable over extended periods of time is challenging.

3.5.2.1 Sphincter of Oddi

One such application where localized measurements are essential for diagnosis is sphincter of Oddi (SO) dysfunction, which leads to abdominal pain and where catheter-based SO manometry (SOM) is typically used. SOM can be performed with two techniques. The first uses three piezoelectric solid state sensors oriented radially 90° from each other at the tip of catheter. The second type, which is the most common, is water-perfused. Such catheter-based solutions for SO may lead to pancreatitis. This risk has been found to depend upon the type of catheter used (perfusion catheter with an aspiration port vs. a standard perfusion catheter) as well as the measurement technique used (solid state vs. perfusion) [144].

3.5.2.2 Bladder Pressure

Urinary incontinence is another application where there is a need for implantable pressure monitoring. Catheter-based methods are currently common, where a pressure sensor is intraurethrally used to access the bladder. This is a painful process that takes place in the clinic for short periods of time, especially due to infection and stone formation in the bladder. An implanted solution that can be used

Fig. 3.24 **a** The fully packaged implanted bladder pressure catheter system presented in [145] (© Springer Science + Business Media, LLC 2008. With permission of Springer; original source caption: "The implanted device after fully being fully packaged. The pressure sensor is housed at the end of a 7.5 French catheter, which is implanted directly into the bladder or peritoneal cavity. The other end of the catheter is connected into the sensor node, which consists of the dot mote (microcontroller and wireless transmitter), the amplifying electronics, and battery. The device is wrapped in LDPE film and molded in medical-grade PDMS"). **b** The tethered sensor used in [146] with a diameter of 13.6 mm and a thickness of 2 mm (© IEEE 2011. Reprinted with permission from [146])

together with neuromodulation devices in a closed-loop system as a means of treating incontinence would improve the quality of life of millions of people. In [145] an active system was presented that was comprised of active electronics, wireless transmitter, sensor, catheter, and battery (Fig. 3.24a). The system was capable of continuous measurements over three days in porcine models, and had a resolution of 0.02 psi and a detection range of 1.5 psi. A commercially available piezoresistive pressure sensor was used, which is configured into a temperature-compensated Wheatstone bridge. A bladder pressure sensing system was also presented in [146], where the authors once again used a commercially available sensor (a piezoresistive pressure sensor mounted on an 8 mm × 8 mm ceramic substrate by Merit Sensor Systems). The devices were silicone encapsulated (Fig. 3.24b) via a two-part molding process and were placed in the bladder wall. Compared to an intravesical sensor with a transurethral catheter, the implanted ones detected earlier bladder contractions evoked via pelvic nerve electrical stimulation.

3.5.2.3 Blood Pressure

In the developed world, cardiovascular diseases (CVD) are the leading cause of death. Blood pressure monitoring provides information regarding the cardiovascular system. Thus, the predominant method of diagnosis of these is whether the patient has hypertension. According to [147], 10–55% of Europeans suffer from arterial hypertension, with 20% demonstrating systolic blood pressure levels higher than 160 mmHg. Such high levels increase the risk of stroke heart attack, coronary heart disease, aortic aneurism, and angina pectoris, among others. External cuff-based devices and catheter-based solutions once again face the same disadvantages as in other applications (white coat effect, patient discomfort, large size).

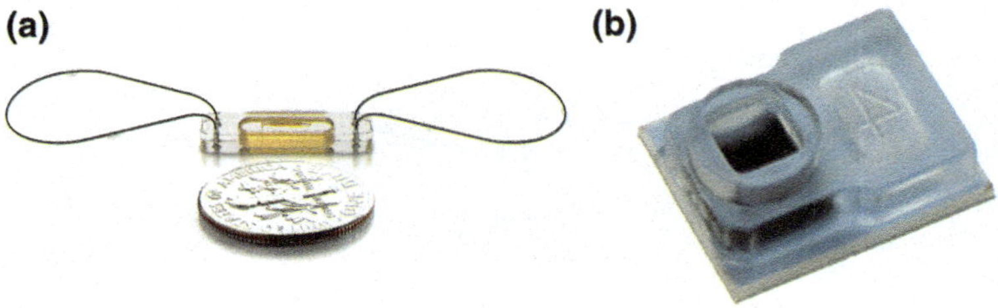

Fig. 3.25 **a** The passive EndoSensor/CardioMEMS HF from St Jude Medical [148] for the detection of heart failure within the pulmonary artery (image printed with permission from [149]). **b** The Merit Sensor Systems blood pressure sensor. This sensor is not intended for implantable applications [150] (image printed with permission from Merit Sensor Systems)

Commercially available solutions do exist. Such examples include the Remon ImPressure/Remon CHF (now owned by Boston Scientific) for measurements within an aneurism sac and the EndoSensor/CardioMEMS HF from St Jude Medical for the detection of heart failure within the pulmonary artery. The latter is a passive telemetric device 3 mm × 15 mm composed of a pressure-sensitive capacitor and an inductor, shown in Fig. 3.25a. An external device with an external inductor interrogates the implanted device.

Merit Sensor Systems produce a wide range of pressure sensors for industrial applications and also for various medical applications, but they are not implantable. Their solution uses a piezoresistive Wheatstone bridge with a chemically etched silicon diaphragm, the mechanical deflection of which due to exerted pressure leads to a measurable electrical signal change. It utilizes a dielectric gel barrier protection, is temperature-compensated, and can measure pressure between −30 and 300 mmHg. Its dimensions are 10.5 mm × 8.1 mm and a thickness of 4.1 mm and is shown in Fig. 3.25b. Implanted blood pressure monitoring was demonstrated in

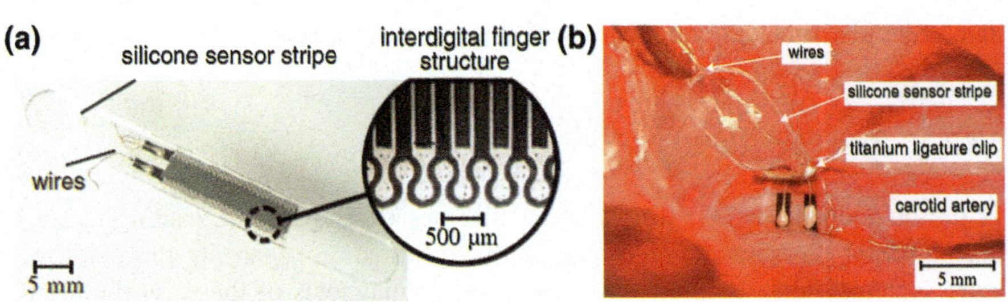

Fig. 3.26 **a** A flexible capacitive strain gauge for blood pressure monitoring (© Springer Science + Business Media, LLC 2012. With permission of Springer from [147]. original source caption: "Silicone sensor strip with the dimensions of 40 × 5 × 0.5 mm^3"). **b** The in vivo evaluation of this sensor [147]. The ligature clip holds the sensor in place (© Springer Science + Business Media, LLC 2012. With permission of Springer from [147]. Original source caption: "In vivo test with a sensor strip wrapped around a carotid artery (Ø 5.4 mm) of a domestic pig. The sensor strip is fixed and closed by a titanium ligature clip")

[147], where a flexible capacitive strain gauge was presented (Fig. 3.26a). The device was wrapped around an arterial blood vessel as in Fig. 3.26b. Changes in the diameter of the vessel were obtained in this way, which are related to blood pressure. Silicone was used as both the substrate and encapsulant. The conductive tracks were formed by PEDOT:PSS. These were patterned without masks, via a 1064 nm 3.6 W Nd:YAG laser system. The gaps and track width of the interdigital structure were 95 and 200 μm respectively for all 95 finger pairs. Fabricated sensors demonstrated a sensitivity of −0.18% mmHg^{-1} and could be elongated by up to 15% without being damaged.

A parallel plate capacitor concept was used in [151], where two metal electrodes were placed in parallel, with the artery and blood being the dielectric, as in Fig. 3.27a. A change in blood pressure would change the distance between the two plates and thus the capacitance. The authors used a flexible sensor node approach realized through a PI substrate, thinned silicon chips, and an inductor for telemetry. The flexible silicon chips were realized by chemical etching to make them thinner than 50 μm. Thermal compressive bonding was used to mount these on the PI substrate and electroplating was used to form electrical connections between the two due to the fragility of the thinned silicon. The system could be attached on surfaces with a curvature of 10 mm or less. The resonant frequency of the *LC* circuit formed was thus a function of blood pressure demonstrating a linear response where for changes between 0 and 800 mmHg it changed from ~151 MHz to ~153 MHz. A capacitive diaphragm MEMS pressure sensor with active recording electronics and telemetry is presented in [152]. The sensor chip and the ASIC were 3D stacked and assembled on a flexible PCB with additional external passive components and coil. This is shown in Fig. 3.27b. A silicone

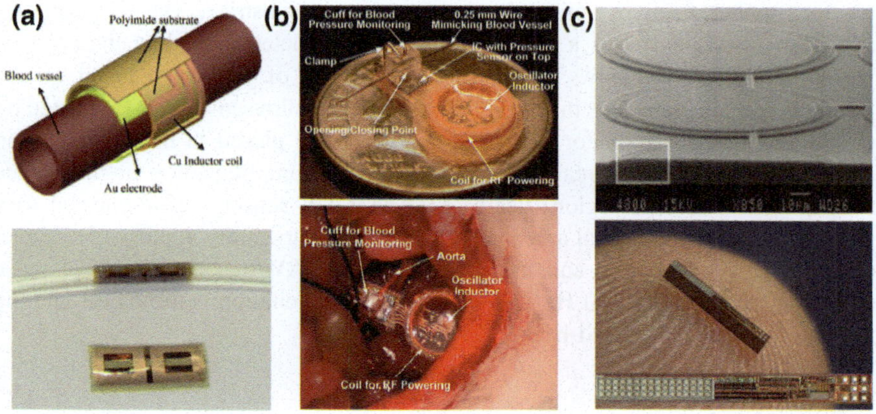

Fig. 3.27 **a** Illustration of a pressure-sensing system wrapped around a blood vessel and its realization (reprinted from [151], © 2005, with permission from Elsevier). **b** The pressure monitoring system developed for mice in [152] and its implantation (© IEEE 2009. Reprinted with permission from [152]). **c** SEM image of an on-chip pressure sensor and its integration with the associated signal processing electronics (© IEEE 2012. Reprinted with permission from [153])

rubber cuff is instrumented with the sensor and wrapped around the blood vessel. The cuff is filled with a low-viscosity insulating fluid, silicone oil, together with a silicone sensing membrane which is in contact with the vessel. This couples the blood pressure waveform from the vessel to the MEMS sensors. The system demonstrated a nonlinearity below 2.5%, and a hysteresis of 1% for the measurement of pressures between 750 and 950 mmHg.

An alternative to the aforementioned approaches is to position the implantable sensor within the vessel. In [153] such an approach is presented, where the implant is positioned within the femoral artery. The system measures temperature and pressure, the measurement range of which is 30–300 mmHg with a ±1 mmHg error and a power consumption of 300 μW. The system (22 cm long) is telemetrically powered and interrogated and consists of two parts, the sensor tip (comprised of a pressure sensing chip in a metal capsule with a 1 mm outer diameter) and data cable (length of 20 cm) and the transponder unit (2 cm diameter and 4 mm thick). The sensor is a monolithic capacitive pressure sensor comprised of 48 membranes, as shown in Fig. 3.27c, half of which are not pressure-sensitive and are used for temperature compensation. The temperature sensor is realized by a parasitic bipolar transistor. The sensor chip is 350 μm thick and 5.6 mm × 0.7 mm in size. The telemetry operates at 133 kHz and the telemetry chip with its additional external passive components and rigid PCB are molded into an epoxy. All materials in contact with blood are covered with an antithrombogenic coating. The system was tested in vivo in a sheep model and measurements correlated with reference measurements performed with a commercially available catheter system.

The authors of [154] presented a diaphragm sensor for the measurement of blood pressure within the left ventricle of the heart. Instead of capacitance or resistance the authors used surface acoustic wave (SAW) sensing as the transduction mechanism. SAW sensors are produced by patterning metallic interdigital structures together with reflective gratings on a piezoelectric substrate. The resonant frequency of such devices is a function of many parameters including the acoustic wave propagation velocity. The latter is a function of stresses induced on the piezoelectric substrate. The thin membrane of the pressure sensor was fabricated by quartz and a SAW resonator was patterned on it and placed in the middle of an air-filled cavity. With increasing pressure, the resonant frequency of the device (which is in the MHz region) decreases. The SAW sensor was connected to a pseudo normal mode helical antenna with circular polarization. A pulsed RF signal at 868 MHz is generated and used to excite the SAW resonator, with a pulse repetition frequency of 100 Hz. After 1 μs, the RF excitation is turned off and the resonator energy is radiated back.

3.5.2.4 Nerve Cuff Electrode Applications

Cuff electrodes are common for neural signal recordings. If a pressure larger than 20 mmHg is applied to the nerve via the cuff electrode, the nerve may become damaged. In [155], the authors propose the use of pressure sensors to monitor the applied pressure and protect the nerve. The proposed topology was a parallel plate capacitive pressure sensor formed by two gold electrodes and PDMS as the dielectric on PI. Plasma treatment of PDMS was necessary due to its poor adhesion to gold and PI. The final device was 7 mm × 7 mm. A similar parallel plate topology was presented in [156], where the PI was replaced with Teflon and the gold was replaced by a conductive PDMS/graphene nanoplatelet nanocomposite. The device was fabricated using layer by layer spin coating and demonstrated a gauge factor of 3. A passive implantation of a similar topology was presented in [157], where it was combined with an inductor.

3.5.2.5 Intraocular Pressure (IOP)

Intraocular pressure (IOP) monitoring is another application of great importance, as it is crucial for the diagnosis and management of glaucoma, the second leading cause of blindness in the world. IOP is determined by the production rate of aqueous humor and the drainage rate of aqueous humor (primarily through the trabecular meshwork located in the anterior chamber angle). An increase in the IOP may be due to drainage ducts being blocked or due to narrowing or closure of the angle between the iris and cornea. IOP is dynamically changing, and current techniques (e.g. the Goldmann tonometer) provide only a snapshot. Glaucoma will lead to intense eye pain, blurred vision, mid-dilated pupil and photophobia, redness, headaches, nausea, and vomiting. If untreated, glaucoma will lead to optical nerve damage and eventually sight loss. There is no cure, but early diagnosis and treatment slows down the evolution of the disease and may even eliminate it. Glaucoma patients have an IOP of 20 mmHg compared to a 10 mmHg in healthy ones. Implanted devices would allow clinicians to monitor IOP continuously and assess treatment efficacy, but must address challenges related to continuous eye movement and must not block the patient's sight. Such examples are presented in [158–160], where flexible parylene-based passive *LC* topologies are demonstrated for IOP measurement (Fig. 3.28). These are implanted in the anterior chamber and fixed to the iris. The authors presented two sensors in [159], one is based on a parallel plate capacitor, the top electrode of which is part of a membrane that is deformed with changing IOP, while in the second version both the capacitance and the inductance are IOP dependent. These structures are shown in Fig. 3.28a. The total device dimension of the former is 4 mm × 2 mm and demonstrated a sensitivity of 468 kHz $mmHg^{-1}$ and the latter 4 mm × 1 mm and a sensitivity of 1.14 MHz $mmHg^{-1}$; both devices were linear between 0 and 30 mmHg. The IOP dependence on these leads to resonant frequency shifts detectable by an external circuit. This sensor-interrogating device was integrated into a pair of glasses.

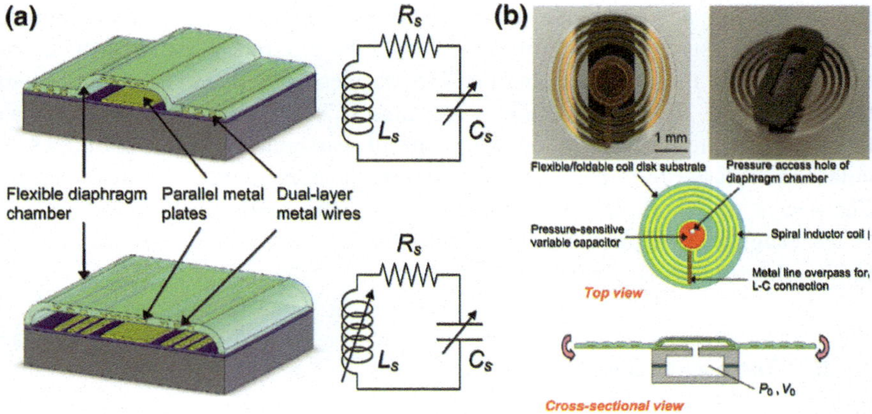

Fig. 3.28 **a** The passive *LC* IOP sensors of [159] (© IEEE 2008. Reprinted with permission from [159]). In the top structure only the capacitor is variable with IOP, while in the bottom both C and L are variable. **b** The flexible capacitive IOP sensor of [160] (© IEEE 2010. Reprinted with permission from [160])

3.5.2.6 Intracranial Pressure (ICP)

ICP monitoring is another important application. Brain injury and tumors, meningitis and aneurysms may lead to intracranial hypertension. Cerebrospinal fluid drainage for the relief of the fluid pressure via a lumbar puncture or through an external ventricular drain (EVD), are common techniques used. In hydrocephalus patients, catheters are used to drain and thus relieve pressure into the abdominal

Fig. 3.29 **a** The Codman Microsensor ICP. The Raumedic ICP sensors: **b** NeuroVent P-C, **c** NeuroVent P and **d** Neurodur (image from [161], license: https://creativecommons.org/licenses/by/4.0/

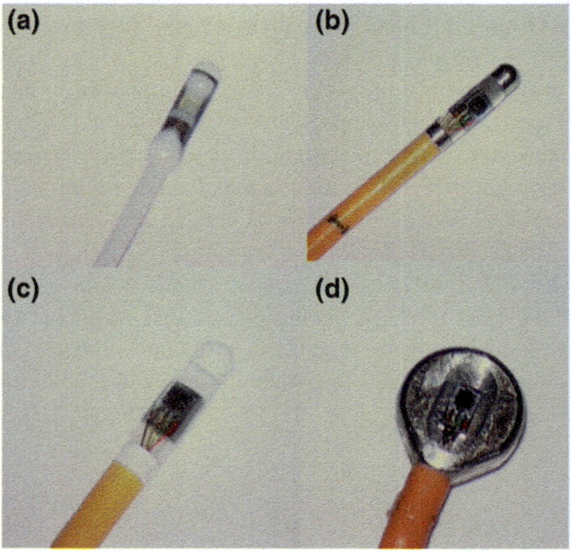

Application	Pressure Range (mmHg)	Pressure Resolution (mmHg)	Frequency Bandwidth (Hz)	Packaging Considerations
ICP	−10 to 50	1	0–30	Shunt integration, catheter
IOP	12–22, >22 abnormal	2	0–30	Needle delivery, contact lens
Blood Pressure	50–180	1	0–200	Catheter, stents
Bladder Pressure	10–70, ~150 during voiding	1	3–5	Intra-urethral Catheter
Intra-abdominal Pressure	0.2–16.2	--	0–15	--

Fig. 3.30 Sensor and electronics design specifications and packaging considerations as summarized in [143] (license: https://creativecommons.org/licenses/by/3.0/. Used with no changes)

cavity. Such shunts often fail, and thus it is important to monitor intracranial pressure. Commercially available solutions which combine EVD catheters with pressure sensing include the Codman Microsensor ICP (a miniature strain gauge on a titanium case at the tip of a flexible nylon tube shown in Fig. 3.29a) [161] (Codman, Johnson & Johnson, Rayham, MA, USA) and the Integra LifeSciences Coorporation (Plainsboro, New Jersey, United States) Camino (a fiber optic sensor) [143, 163–165]. Raumedic AG (Munchberg GE) also has a number of ICP sensor solutions shown in Fig. 3.29. These, however, must be used in a clinical environment, and long-term monitoring may lead to infection.

Figure 3.30 summarizes the ranges of pressure necessary for different applications, the frequency content of the recorded signals, and the necessary pressure resolution and packaging considerations.

3.5.3 *Readout Circuit for Resistive Sensors*

In this section, classical topologies for interrogating resistive sensors such as RTDs and strain sensors are discussed. The Wheatstone bridge is the classical circuit topology used for precisely measuring small changes of resistance. This is shown in Fig. 3.31. Two resistive potential dividers are used with the same resistor ratios.

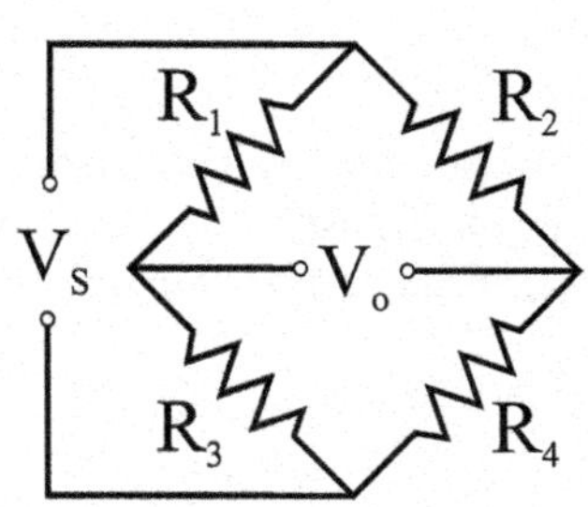

Fig. 3.31 The Wheatstone bridge circuit

The same potential is applied across both potential dividers and the potential difference of the two potential divider outputs is equal to zero. A high input impedance instrumentation amplifier is used to measure this. When one of these resistors is an RTD or a strain gauge, changes in its value with lead to measurable changes in the potential difference between the two output voltages of each potential divider. This is known as a quarter bridge circuit.

The output voltage, V_O, is given by

$$V_O = V_S \frac{R_1 R_4 - R_2 R_3}{(R_1 + R_3)(R_2 + R_4)}. \tag{3.35}$$

If resistor R_4 is e.g. a strain gauge, as in Fig. 3.32a, and a force leads to a resistance change of δR_4, then

$$V_O = V_S \frac{R_1 (R_4 + \delta R_4) - R_2 R_3}{(R_1 + R_3 + \delta R_4)(R_2 + R_4)}. \tag{3.36}$$

Since the bridge is initially balanced and $R_1 R_4 = R_2 R_3$, and if the change in resistance is very small ($\delta R_4 / R_4 \ll 1$), as is typically the case with metal foil strain gauges, then

$$V_O \approx V_S \frac{\delta R_4 R_1}{(R_1 + R_3)(R_2 + R_4)}. \tag{3.37}$$

Using (3.34), where $R - R_O = \delta R_4$ and $R_O = R_4$, then

$$\varepsilon_{str} \approx \frac{1}{GF} \frac{V_O}{V_S} \frac{(R_2 + R_4)^2}{R_2 R_4}. \tag{3.38}$$

If two of the resistors, R_3 and R_4, are replaced with strain gauges, then the circuit is called a half-bridge. If these are placed in the front and back of a body experiencing a force perpendicular to the sensors, such that one is experiencing a tensile strain and the other a compressive strain, as in Fig. 3.32b, and both sensors have equal sensitivities and are of the same sign (positive or negative) then the equation above becomes

$$\varepsilon_{str} \approx \frac{1}{2GF} \frac{V_O}{V_S} \frac{(R_2 + R_4)^2}{R_2 R_4}. \tag{3.39}$$

As can be seen, for the same applied strain, the output voltage is doubled and thus also the sensitivity of the circuit. If all four resistors are replaced with sensors, the circuit is called a full bridge circuit, and then

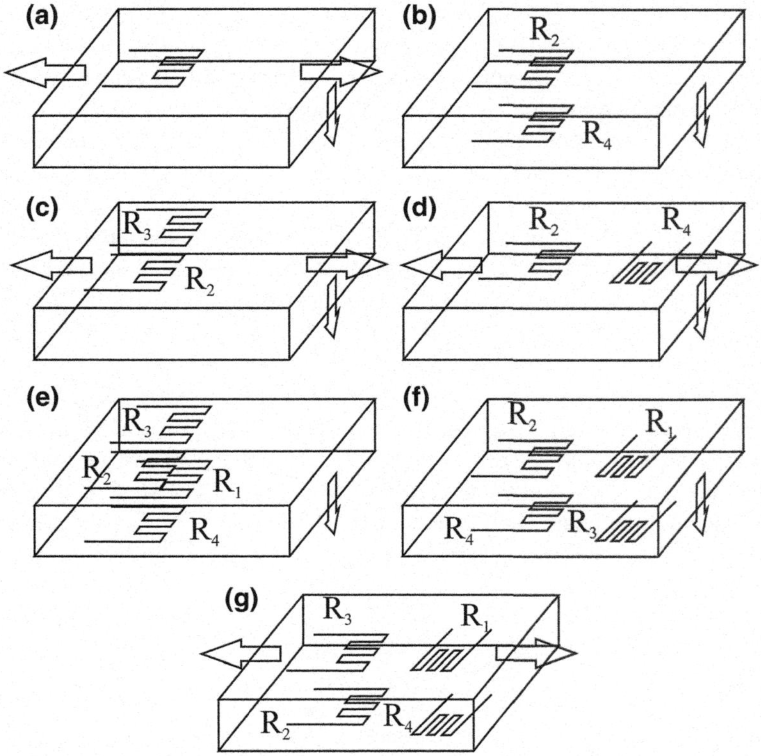

Fig. 3.32 Various strain sensing arrangements: **a** The quarter bridge circuit for the measurement of axial and perpendicular strain. **b** Half bridge topology for temperature compensation and bending strain measurement. No axial strain sensitivity. If resistors R_1 and R_4 are used instead, then the circuit is sensitive only to axial strains but not temperature. **c** Bending and axial strain measurement using two sensors, with double sensitivity, but no temperature compensation. **d** If R_4 is not bonded, this topology compensates for temperature. If bonded it measures both axial and perpendicular strain and compensates for the Poisson effect. **e** Full bridge circuit with four times greater sensitivity than the quarter bridge circuit and temperature compensation for bending strain measurement. **f** Full bridge for bending strain measurement and with both temperature and Poisson effect compensation. **g** Full bridge for axial measurements with temperature and Poisson effect compensation

$$V_0 \approx V_S \frac{R_2 R_4}{(R_2 + R_4)^2} \left(\frac{\delta R_1}{R_1} - \frac{\delta R_2}{R_2} - \frac{\delta R_3}{R_3} + \frac{\delta R_4}{R_4} \right). \tag{3.40}$$

The above shows why in the half bridge it was R_3 and R_4 (or equivalently R_2 and R_4) that were used for the two sensors facing each other. Both their corresponding terms in the equation and their change in resistance have opposite signs, and consequently they are not cancelling each other out. The advantage of this topology is also that temperature will affect both sensors the same. As a result, this common mode signal will be cancelled out, as it will be subtracted, compensating for temperature. This topology is also insensitive to axial strain. If the two sensors are used to measure the same strain, e.g. they are next to each other, as in Fig. 3.32c, then e.g. R_1 and R_4, or R_3 and R_2, can be used for the two sensors, since their change of resistance has the same sign and they thus do not cancel out. Another approach for temperature compensation is to use e.g. R_2 as the second sensor, which, however, remains unstrained; it is positioned for example at a right angle to the principal axis of strain, is not bonded to the surface, and is in close proximity and in thermal contact as in Fig. 3.32d. Then a temperature change will affect both sensors the same and if they are identical its effect will be cancelled out. This topology compensates for temperature and remains sensitive to axial strain. If the second sensor is also bonded to the material under strain, the axial strain will cause a reduction in resistance to R_2 due to contraction, which, however, will be added. This will thus compensate for the Poisson effect and this arrangement is known as Poisson's gauge. In a full bridge configuration, R_1 and R_4 can be positioned next to each other such that both can sense compressive strain, and R_2 and R_3 can be placed on the other side to sense the tensile strain, as in Fig. 3.32e. Such an arrangement compensates for temperature changes and measures the bending strain, while being insensitive to axial strain. It can be shown that for such full bridge, the sensitivity increases by four times, according to

$$\varepsilon_{str} \approx \frac{1}{4GF} \frac{V_O}{V_S} \frac{(R_2 + R_4)^2}{R_2 R_4}. \tag{3.41}$$

If R_2 and R_4 are placed facing each other and R_1 and R_3 are placed perpendicular to the axis of strain and also facing each other, as in Fig. 3.32f, then both temperature and the Poisson effect are compensated for when measuring the bending strain. If R_1 and R_4 both measure the compressive Poisson effect and R_2 and R_3 the tensile strain, as in Fig. 3.32g, then the axial strain can be measured, with both the Poisson effect and temperature effects being compensated for. The effectiveness of the above topologies will be hindered among others, by device mismatch (temperature coefficients, *GF*, resistance), bridge symmetry (e.g. unmatched lead resistances), device alignment and positioning, temperature differentials between device locations, and the anisotropic properties of the materials being studied under stress, which would change the strains sensed by the sensors. In addition to the above, strain gauges can be arranged at angles when, for example, both the magnitude and the direction of the force need to be evaluated. In the preceding discussion, small resistive changes were assumed and approximate relationships were consequently discussed. If high accuracy is required and if large changes are induced, the above equations for the quarter bridge are not adequate, as the response is nonlinear. Feedback to the bridge can be used to linearize the system. As discussed throughout the chapter, capacitive sensors are typically used to control the frequency of oscillation of a signal, providing a simple readout circuit with a quasi-digital output.

3.6 Neural Interfaces

In the electrophysiological research field electrical neural recordings and stimulations are performed in excitable tissue such as brain, muscle, peripheral nerves, or sensory systems. The activity of the neurons is recorded as an action potential known as the extracellular potential. Large numbers of microelectrodes are developed for chronically implanted devices, which requires understanding of the electrochemical mechanism that underlies the behavior of the neural simulation and recording electrodes [166]. Recording microelectrodes should be no larger than approximately 2000–4000 μm^2 and the action potentials should be recorded close to the target neurons to maintaining consistent neural recording, as these will affect the quality of the recording. Different types of material are used for the recording electrodes, such as stainless steel, tungsten, platinum, platinum-iridium alloys, iridium oxide, titanium nitride, and PEDOT.

Other important factors are the electrode impedance, where the thickness and composition of the connective tissue surrounding the electrode could vary after implantation. Important parameters, such as SNR, signal distortion and built up voltage are influenced by the electrode-electrolyte impedance. Adverse tissue response can be the result of insertion trauma [167] due to foreign-body response to the electrode material [168, 169] or micromotion of the electrode [170, 171], which may induce further trauma.

Some strategies for lowering the impedance involve the development of iridium oxide modified platinum microelectrodes [172]. Iridium oxide is a material that can undergo Faradaic electrochemical reactions, which leads to a much larger charge capacity, but there are issues related to long-term stability at high current injection levels, limiting its potential for chronic implantation.

A potential improvement can be achieved by additional deposition with conductive polymers [173]. Different types of conductive polymers such as PPy and PEDOT doped with different anion dopants such as CLO_4^- and poly(styrene sulphonate) are deposited on different metals. These structures have Faradaic and non-Faradaic charge transfer at the electrode/neural interface, which is related to the rough surface and the large electroactive area and charge capacity.

Carbon-based materials present a very high surface area, resulting in large charge capacity. CNT-based fibers are presented in [174], where effective stimulation, long-term stability, and biocompatibility are discussed. Liquid crystal polymer (LCP) could be used instead of polyimide as an electrode substrate for long-term implatantion due to its low water absorption rate [175].

In [174] biocompatible CNT fiber (capable of bidirectional interactions with neural activity) microelectrodes for neural stimulation and recording are presented. The CNT fibers are used without additional surface modification, and high charge density, low impedance, and flexible microelectrodes for stable interface manipulation of the activity of neural ensembles are achieved.

Over the years, a functional response by depolarizing the membranes of excitable cells is initated with electrode stimulation. Depolarization occurs by the flow of ionic current between two or more electrodes, one of which is close to the target tissue. Neural stimulation electrodes in general fall into one of the following categories: macroelectrodes with high charge/phase threshold and low charge density threshold; microelectrodes with low charge/phase threshold and high charge density threshold. In the latter case the charge density is high and so are electrode degradation and tissue damage. The narrow safe window for penetration for the brain is $\leq 2000\ \mu m^2$.

Glass pipettes, placed next to or inside the target neurons, are extensively used, and this type of electrode is easy to make with a stable electrode/neural interface. However, they are quite fragile, can be blocked by the tissue, and have a low bandwidth, and the best application for this type of sensor is for single cell simulation and recording experiments [176]. The need for new ideas supporting the understanding of the molecular and cellular mechanism related to brain activity is driving the next generations of developments in neural interfaces, combining electrical and optical modalities for fundamental research and human clinical applications. Very interesting are multimodal fibers with a $\sim$10–50 μm diameter (for single cell targeting) that can record and manipulate neural activity, as reported in [177] (Fig. 3.33).

Classical detection technologies such as microdialysis are limited by their low temporal resolution (such as sampling rate), which is why additional development to detect the concentration of neurotransmitters, metabolites, glucose, hormones, neuropeptides, enzymes, and pH could help for more detailed analysis of neural circuits [178]. Multifunctional fibers with the ability for simultaneous electrical, optical, and chemical interrogation of neural circuits in vivo have been developed [179]. Millisecond precision is required for probing and manipulating neurons. Some probes use materials exceeding the elastic moduli of the neural tissues, where long-term experiments showed decrease in SNR. Different factors are related to the decreased SNR, such as breach of the blood–brain barrier, tissue damage when micromotion is observed of the implant with respect to the brain, and disruption of the glial network. Different types of printing technologies are reported, such as

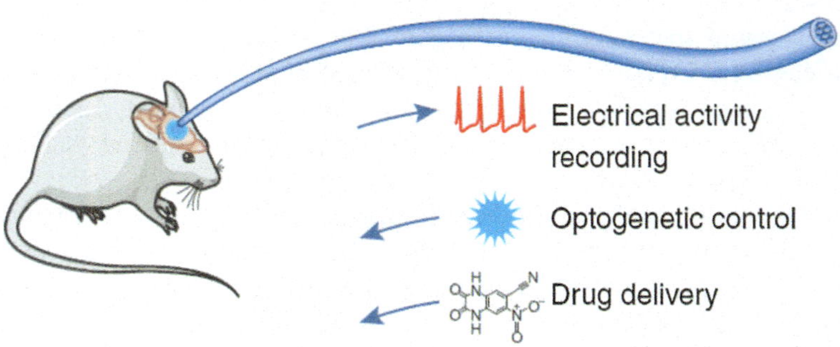

Fig. 3.33 Optical-fiber based microprobe implanted in the skull, allowing free movement and simultaneous recording of neural activity, optogenetic control, and local drug delivery. Reprinted by permission from Macmillan Publishers Ltd: Nature Biotechnology, from [177], © 2015

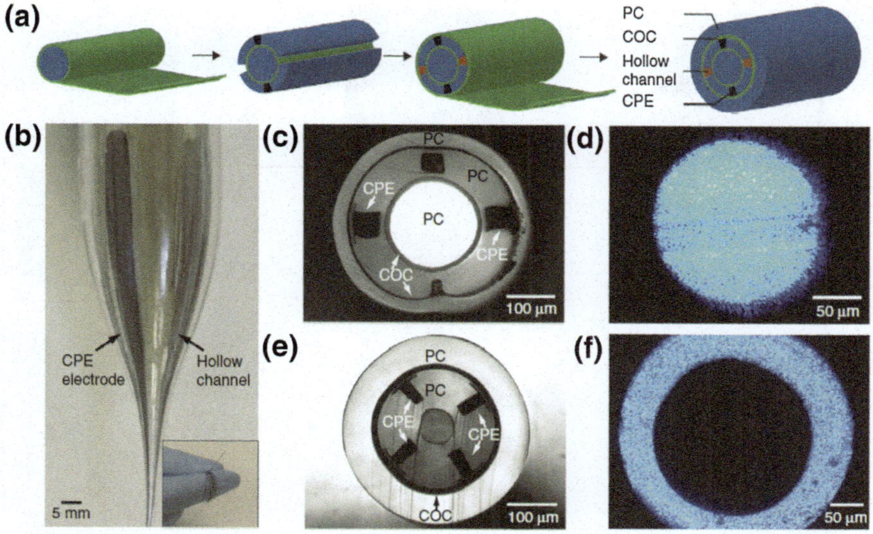

Fig. 3.34 Fiber probe fabrication and characterization. **a** Preform for multimodal fiber probes. **b** Transition stage. Inset: fiber around finger. **c**, **e** Cross-sectional optical images. **d**, **f** Cross-sectional optical images of a $\lambda = 473$ nm light source coupled into the two designs. Reprinted by permission from Macmillan Publishers Ltd: Nature Biotechnology, from [179], © 2015

microcontact and micromachining for development of neural probes on flexible substrates, but these techniques are limited for the use of integrated multiple electrodes or drug delivery systems without further increase in overall dimensions. [179] presented a neural probe with feature sizes as small as 5 μm (probe diameters ranging from 70 to 700 μm) fabricated from metal, composite, and polymer material by a thermal drawing process. The fibers were able to perform simultaneous optogenetic stimulation (further explanation of optical detection can be found in Chap. 6), neural recording, and drug delivery. This is shown in Fig. 3.34. The limitation of these probes is their recording capabilities, because of the nature of the formulation technology.

Nanomaterials are used for neural prostheses and the advances in electroactive materials (including carbon nanotubes, graphene, silicone nanowires, conductive polymers, and hybrid organic-inorganic nanomaterials) are reviewed in [180]. Some examples of implantable electrodes are shown in Fig. 3.35. A parylene cuff electrode system which is self-closed and used for peripheral nerve recording was designed, fabricated, and characterized in [195]. The mechanical structure is minimized compared with conventional cuff electrodes and the precise curls and closing of the cuff are achieved through control of the different parylene layers during heat treatment. Electrodeposition of iridium oxide is done on the sites to increase the charge capacity and for reduced interfacial impedance. Figure 3.36 shows the 100 μm diameter parylene electrodes.

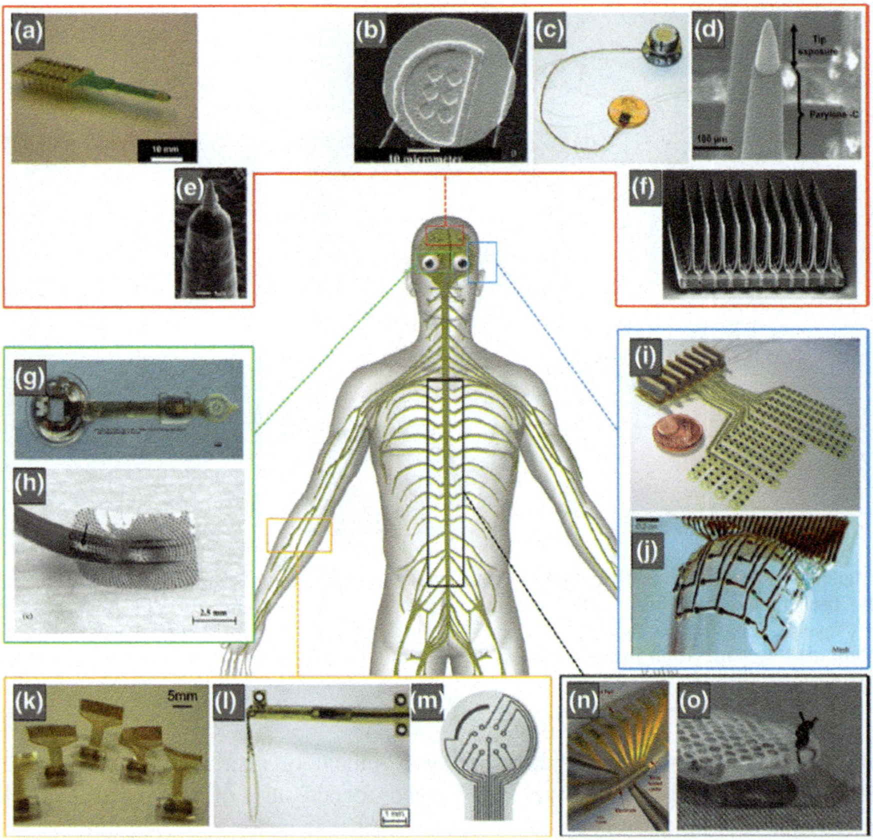

Another self-locking system was demonstrated in [196], where the cuff diameter can be adjusted through the self-locking mechanism. The proposed design was tested for over 11 weeks of implantation and it induced no damage to the axons within the rat sciatic nerve. Figure 3.37 shows that the electrode sites of the cuff electrodes are placed on the surface of the nerve, the intrafascicular electrodes penetrate the epineurium around the entire nerve and the perineurium around the fascicles [197]. The parylene cuff electrode is designed as shown in Fig. 3.38, where there is a parylene strip with a guide tongue at the end, a gold locking loop, gold ratchet teeth at the edges, pads for external connection, and a parylene ribbon cable. After full release from the silicon substrate the cuff electrode weights 8 mg.

The main drawback of the nerve cuff electrode is the lack of spatial selectivity within the nerve: recording and stimulation from a specific fascicle or pathway is difficult. In [198] a matrix cuff was developed to investigate different positions and number of contacts affecting the selectivity. Intrafascicular devices [199, 200] and micro-electrode arrays [197, 201] are more invasive but more spatially selective.

The development of optoprobes is a research field attracting significant attention [202]. An optoprobe is defined as a device having one of the following modalities:

◂**Fig. 3.35** Reproduced with permission from [180], © Wiley 2014. **a** Eight-channel silicon substrate Michigan electrode array. Reproduced with permission from [181], © Elsevier 2008. **c** SEM of a Michigan electrode. Reproduced with permission from [182], © Elsevier 2003. **d** BrainGate microelectrode (i.e. Utah array) connected by a 13 cm ribbon cable. Reproduced with permission from [183], © Nature Publishing Group 2006. **e** High-magnification image of the Utah array. Reproduced with permission from [184], © Elsevier 2010. **g** SEM of parylene coated Au tip. Reproduced with permission from [185], © IOP Publishing 2012. **h** SEM image of 100 microelectrodes of Utah electrode array. Reproduced with permission from [183], © Nature Publishing Group 2006. **i** An epi-retinal vision prosthesis, implant with parylene-C and silicone rubber encapsulation. Reproduced with permission from [186], © IOP Publishing 2009. **j** Heat molded and annealed retinal electrode array. Reproduced with permission from [187] © Elsevier 2008. **k** Electrode array with a 16 mm diameter. Reproduced with permission from [188], © IOP Publishing 2009. **l** Optical image of polyimide electrode arrays with silk support wrapped onto a glass hemisphere. Reproduced with permission from [189], © Nature Publishing Group 2010. **m** Cross section showing the structure of the neutral plane at the electrode array. Reproduced with permission from [190], © Springer 2010. **n** Transversal intrafascicular multichannel electrode. Reproduced with permission from [191], © Wiley 2011. **o** Enlarged view of the regenerative electrodes. Reproduced with permission from [192], © Elsevier 2004. **p** PDMS-substrate wrapped around a wire of similar diameter (2 mm). Reproduced with permission from [193], © Springer 2008. **q** Paddle-style epidural electrode. Reproduced with permission from [194], © Congress of Neurological Surgeons 2011, published by Wolters Kluwer Health

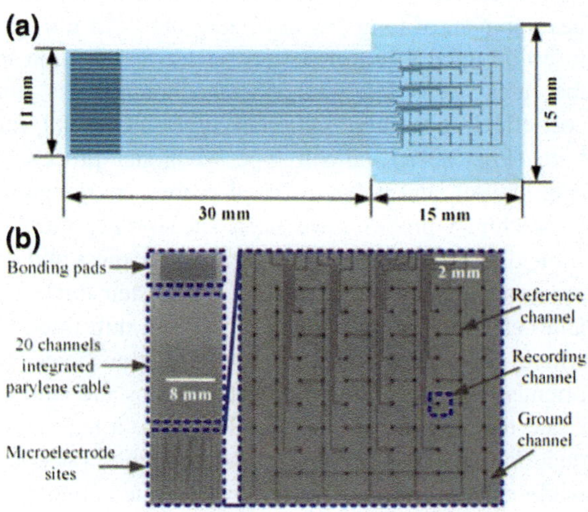

Fig. 3.36 **a** Parylene planar electrode topology and **b** photographs of the parylene planar electrodes (© IEEE 2015. Reprinted with permission from [195])

(i) delivery of light to specific site(s); (ii) electrical recording and delivery of light; (iii) optical detection and delivery of light; and (iv) different types of recording techniques (such as electrical, optical, chemical, or magnetic) and light delivery. The light is delivered to the soft tissue or the neural tissue using totally implantable devices, portable devices or tethered systems. Tethered devices are connected to the outside world by wires or optical waveguides inserted into the tissue.

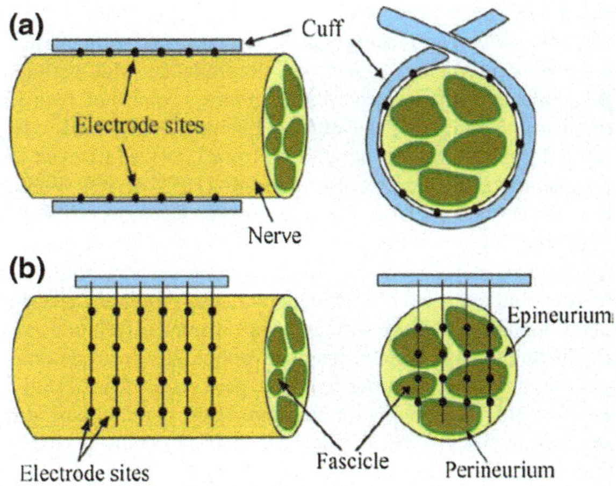

Fig. 3.37 Cuff electrode **a** and intrafascicular electrode **b** from both side and front (© IEEE 2014. Reprinted with permission from [196])

Several considerations must be taken into account with regard to the functionality, usability, long-term implantation, and reliability of a neural probe. These include probe material selection, different microfabrication processes, and interface biocompatibility. The desire to go deeper into the brain, for example, needs the development of long (>5 mm) and thin (<50 μm) probes, but in this case we have insertion axial forces, reaction forces, and tension forces. Further development in the area of CMOS-MEMS technologies is the basis for mass-produced cost-effective probes. Different types of neural microelectrodes, such as metal wire (Fig. 3.39a), silicon type Michigan probes (Fig. 3.39b) and Utah probes (Fig. 3.39c), and polymer based probes are reviewed in [203].

Michigan probes have been successfully employed. Nevertheless, they have some disadvantages, related to probe thickness and durability, since the wet etching process used in their fabrication limits their thickness to the region of 15 μm. The Utah electrode array is made from 1.83 mm thick boron-doped silicon substrates, where polyimide is used as the insulation layer. The length of the horizontally oriented probes in the array is limited by the silicon wafer thickness used to fabricate them, and consequently they have a maximum length of 1.5 mm. The production process suffers from low production rates, unsuitable for batch fabrication, while each probe shaft can only facilitate a single recording site [203]. The metal interconnection wires at the back of the array, used to connect each electrode with pads, are stiff and unsuitable for chronic implantations [203]. Well-known microneedle-like electrodes are Michigan probe [204] and Utah array [205] electrodes, whereas representatives of sheet electrodes are the cuff electrode [206] and the flexible multiplexed electrode array [207]. Non-invasive sheet electrodes are preferable for in vivo testing because this type of probe does not cause reactive glial sheath encapsulation around the neural interface, which could disturb the neural stimulation and recording [208].

Further developments in neural technology combined with direct laser lithography (good for small production lines) are reported in [209]. Different types of

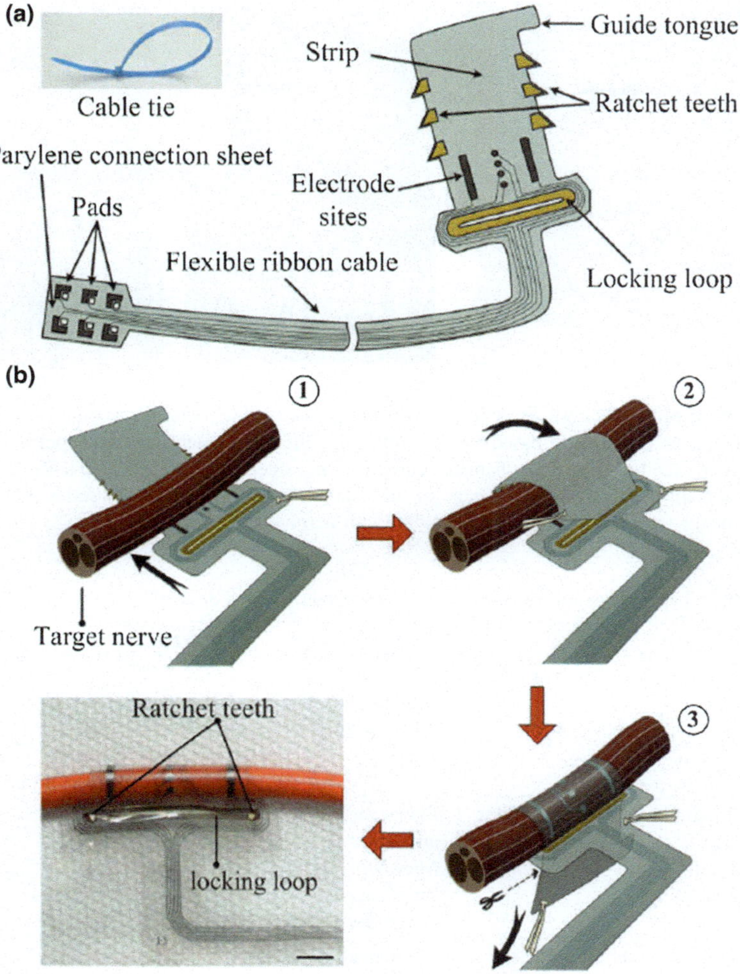

Fig. 3.38 **a** Parylene cuff electrode with self-locking mechanism and flexible parylene cable. **b** Illustration of the cuff electrode implantation procedure. The cuff electrode is wrapped around a hook-up wire (1.3 mm in diameter) to demonstrate the end result. The scale bar is 2 mm (© IEEE 2014. Reprinted with permission from [196])

polymers are used, such as polyimide and parylene-C to cover the metal and silicon region to form a biocompatible interface. A drawback of this type of electrode is that the electrodes are not sufficiently stiff to pierce brain tissue. In some cases an additional support layer is used for increased stiffness [210], such as a 5–10 μm thin silicon support layer. Implantable neural electrode arrays based on polymers, manufactured with MEMS technology, have been reported [211, 212], where smoother coupling with the neural tissue is achieved with flexible devices. Recent

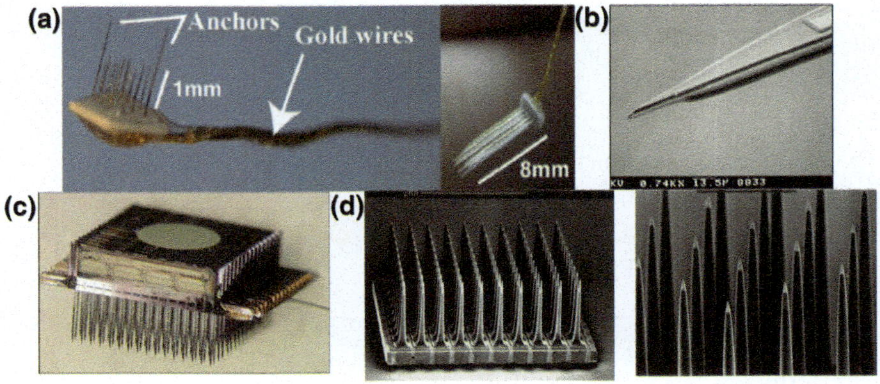

Fig. 3.39 **a** Multi-electrode array (reprinted from [214], © 2007, with permission from Elsevier), **b** SEM images of a silicon-based neural probe (© IEEE 2004. Reprinted with permission from [204]). **c** A 3D 1024-site 128-channel neuroelectronic interface (© IEEE 2004. Reprinted with permission from [204]). **d** SEM image of the Utah array and insulated electrodes with exposed platinum tips (reprinted from [215], © 1999, with permission from Elsevier)

Table 3.3 Types of neural recording electrodes (table adapted from [213])

Electrode type	Electrode design	References
Penetrating cortical electrodes	Microwires Cortical microelectrodes	[216–218] Thin needle type electrodes [219] Michigan electrodes [220] Utah electrode arrays [205, 221]
Non-penetrating cortical electrodes	Planar electrode arrays Microwires Intrafascicular electrodes	[207] [222] Longitudinal type [223] Transverse type [224]
Non-penetrating peripheral nerve electrodes	Microelectrode arrays Regenerative interfaces Cuff electrodes	Utah slanted electrode arrays [225, 226] Cylinder type with a slit [227] Helical electrodes [228]

Reprinted with permission from Springer, © 2016

trends in the field of therapeutic interventions with electrical stimulation showing different types of neural recording electrodes are summarized in Table 3.3 [213].

Pulse signals with an appropriate magnitude and time duration are applied to the tissue for electrical stimulation via conductive electrode interfaces. The properties of the excited tissues, electrode placement in relation to the target tissues, the types of electrodes used, and stimulus waveform parameters are important for the efficacy of the electrical stimulation. The stimulated area is influenced by monopolar and bipolar stimulation. In the case of monopolar configuration a single reference or

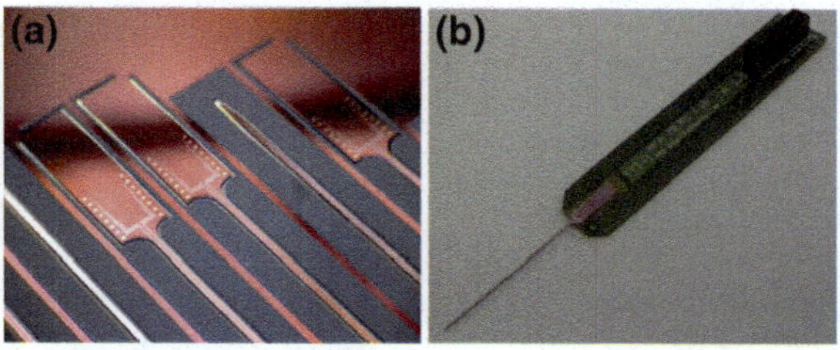

Fig. 3.40 **a** 16-channel microelectrodes on a silicon wafer. **b** Silicon-based neural electrode on a PCB with shaft length 3 cm (© IEEE 2013. Reprinted with permission from [230])

return electrode and several stimulation electrodes are used, which affects a large area of tissue because of the distant reference electrode that tends to spread the electric field. In a bipolar configuration, a localized stimulation area is achieved because of the stimulation pulses that are applied between two closely spaced electrodes. In tripolar configuration, multiple return paths for spreading/adjusting current paths are established in proximity to the electrodes. The current density and the electric field intensity are affected by the shape of the stimulating electrode. Planar type electrodes induce a lower electric field at the tip compared to sharp-tipped electrodes. When a highly focused stimulation is required tipped electrodes are used. For better tissue contact, depending on the field strength, pillar or bumpy electrode structures are adopted.

All of the components of an implant must be either nontoxic to cells or encapsulated within nontoxic materials that can serve as a diffusion barrier to prevent elution into the body. Conducting materials for interconnects and electrodes should have little tendency to create corrosion products in a physiological environment without a voltage bias and during stimulation with current pulses. The devices must have a biomechanically safe type of design with respect to geometry, flexibility, and weight [229]. Deep reactive ion etching is used for formation of long Si-based neural microelectrodes with good mechanical stability, robust implantation, and targeting capability, such as those shown in Fig. 3.40.

Two- and three-dimensional electrode arrays can be tethered to subcutaneous electronics using highly flexible ribbon cables [231], as in Fig. 3.41. For chronic studies, further optimization of the electrode–tissue interface will play a dramatic role, as requirements are driving the size of the electrode towards cellular dimensions and below. In addition to electrode technology, substantial progress has been made in the site selection, cables, amplification circuitry, neural signal processors, and wireless interfaces needed for chronic investigations in neuroscience and for neural prostheses. Some of these advances are discussed in Chap. 5.

A multi-channel, flexible polyimide-type C-shaped neural interface electrode device with protruding metal contacts for signal recording has been designed and fabricated as a non-invasive system in [232] consisting of a polyimide–metal–polyimide sandwich structure. Electrodes for neural stimulation apply a specific pattern of electrical current and transduce the external commands for motor or sensory controls into nerve signals. It is important that the provided stimulation should remain below the electrode's charge-carrying capacity to avoid an irreversible electrochemical process and nerve damage, and to maintain low impedance

Fig. 3.41 A neural microsystem consisting of 2D and 3D arrays of implanted electrodes in the cortex with ribbon type cables connecting them to subcutaneous electronics (© IEEE 2008. Reprinted with permission from [231])

over long periods of time so as to withstand erosion in the physiological environment [233, 234].

Based on their nerve invasiveness, electrode designs can be [213, 232]: (1) extraneural electrodes, for example cuff electrodes or similar which encircle nerve surfaces [235–237]; (2) interfascicular electrodes, where the active contact is in between the nerve bundles [238]; (3) intrafascicular electrodes, such as longitudinal or transverse intrafascicular electrodes [224, 239, 240], where the active contacts are inside the nerve fascicles and which penetrate the epineurium; and (4) regenerative sieve electrodes [213]. More selective communication with individual nerve fascicles is possible if the invasiveness is higher, but nerve damage is more serious [241]. Various examples of different types of neural interfaces are compared in Table 3.4. For the electrode fabrication biocompatible materials such as gold (Au), platinum (Pt), platinium black, and platinium/iridium (Pt/Ir) are used for the contact sites and silicon and polymers as substrate materials [242]. Polyimide is used as the substrate material for development of multi-channel, non-invasive, flexible C-shaped neural interface electrode devices with metal contacts for signal recording. Common polymer materials used to fabricate flexible neural interfaces include

Table 3.4 Comparison of neural interfaces (table adapted from [257])

		Microneedle-like (invasive)	Sheet-like (non-invasive)
For central nervous system	Rigid	Neural silicon probe integrated with sensors [258]	Microelectrode glass array [259]
	Flexible	Flexible ultra-thin neural probe [260]	Graphene-based electrode array [261]
For peripherial nervous system	Rigid	Slanted Utah array [225]	Metal wire monopolar electrode [262]
	Flexible	Transverse intrafascicular multichannel electrode [224]	Cuff electrode [263]

Reproduced with permission from Wiley, © 2015

SU-8 [243], parylene [244], polyimide [245], liquid crystal polymer [246], silk [189], and benzocyclobutene (BCB) [247].

An improved version of the flexible cuff electrode with silicone rubber is discussed in [248], while the electrical parameters are optimised for better communication with the peripheral nerve system [249]. Sinkjaer et al. improved signal recording with cuff electrodes in a very noisy environment using subspace separation algorithms [250, 251]. N. Donaldson et al. worked on a cuff electrode with tripolar recording, reducing the interferences from the electromyography (EMG) and the recorded electroneurogram (ENG) signals are enhanced [252, 253]. Struijk and Andreasen reported an optimized configuration and closure of the cuff electrode which contributes to the neural signal acquisition [254]. Stieglitz et al. [255] developed technology for development of micromachined, multichannel, flexible, lightweight cuff electrodes. The blood pressure is shown to be controlled by the cuff electrode through selective stimulation of the vagal nerve [256].

There are a few challenges related to the cuff electrode: (1) to record signals from nerves with different diameters and (2) communication capability with very fine nerves at the distal end (e.g. the diameter is less than 300 μm). Cuff electrodes are technically flexible devices, but the platinum wires are embedded inside the polydimethylsiloxane (PDMS) sleeves which makes the flexible cuff electrodes a bit stiff,

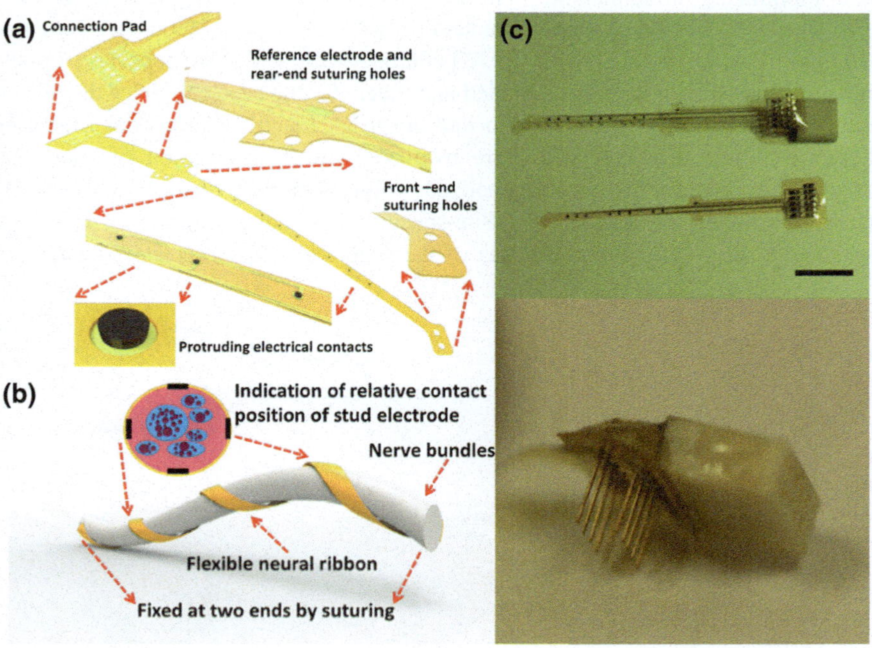

Fig. 3.42 **a**, **b** Neural ribbon electrode and its application. **c** Optical images of final device. The scale bar is 5 mm (reprinted with permission from Wiley and Sons, from [257], © WILEY-VCH Verlag GmbH & Co. KGaA 2015, Weinheim)

and thus prevents the cuff electrode from fully conforming to the small nerve. A flexible and biocompatible neural ribbon electrode that can achieve self-adaptation to nerves with various diameters communicates with small nerves is demonstrated in [257]. The device can match with nerves of different diameters because of the spiral wrapping mechanism. This mechanism ensures good contact and communication between the nerve bundles and the points of contact. A polyimide device with a thickness of 10 μm is integrated with 200 nm electrical tracing.

In Fig. 3.42 a strip-like device is shown proposed in [257]. The neural ribbon electrodes are wrapped around the nerves with fixation by sutures at the two ends and have been successfully implanted on rat nerves of various diameters (300–600 μm).

A nerve electrode with minimal tissue damage (preventing inflammation) during regeneration of the nervous system has been developed using coated nerve electrodes with PEG hydrogel that contains poly (lactic-co-glycolic) acid (PLGA) microspheres (MS) loaded with anti-inflammatory cyclosporin A (CsA) [264]. Micro-wells play a role in the increased surface area, and additional chemical treatment of the surface with aminopropylmethacrylamide is done to improve the adhesive interface between the electrode and the hydrogel. For improved nerve signal quality, a conductive polymer, poly(3,4-ethylenedioxythiophene)-poly(styrenesulfonate) (PEDOT/PSS), is used. In [265] a polyimide nerve cuff electrode is reported with a controllable drug loading/release function. Electro-spinning method is used for development of stable recording of peripheral nerve signals, stimulation, and minimizing inflammation (anti-inflammatory drug loading/release, dexamethasone (DEX)-loaded poly L-lactic acid (PLLA) and/or poly lactic-co-glycol acid (PLGA). Polyethylene glycol (PEG) patterned with UV was deposited on the functional nerve cuff with DEX-loaded nanofibers for the acceleration of the release rate of the drug (increased diffusion rate of DEX). Electrical properties (decreaed because of the PEG patterning) are recovered by the deposition of additional conductive layer of electro-polymerized poly(3,4-ethylenedioxythiophene)-poly (styrenesulfonate) (PEDOT:PSS). Site-specific drug administration is achieved, which significantly lowers the doses of the drug and minimizes the potential for

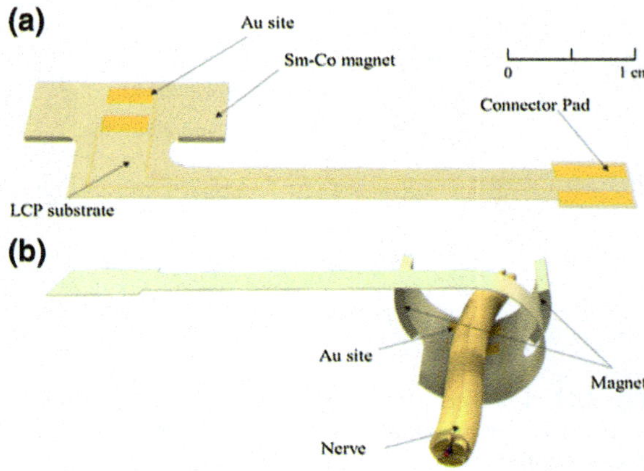

Fig. 3.43 **a** LCP-based MENCE device in its natural state. **b** Curled MENCE before installation on a nerve [266] (© IOP Publishing. Reproduced with permission. All rights reserved. https://doi.org/10.1088/1741-2560/13/6/066014)

significant side-effects. Adhesion between the Ti/Pt surface and the deposited nanofibers is improved by UV-patternable PEG, which are bonded with van der Waals forces, and the initial burst of the DEX release rate is increased by controlling the diffusion rate.

Another approach is to use a pair of magnets fully embedded and encapsulated in a liquid crystal polymer (LCP) substrate to utilize magnetic force in order to replace the conventional installing techniques of cuff electrodes as proposed in [266] and shown in Fig. 3.43. The advantage of using a magnet-embedded neural cuff electrode (MENCE) is that it can be reusable in acute experiments, has good cost-effectiveness, and can be easily repositioned during surgery, which would be helpful in both acute and chronic cases.

Fabrication of stimulating nerve cuff electrodes with Pt, IrOx, poly (3,4-ethylenedioxythiophene) (PEDOT), and platinum black (Pt black) on polyimide

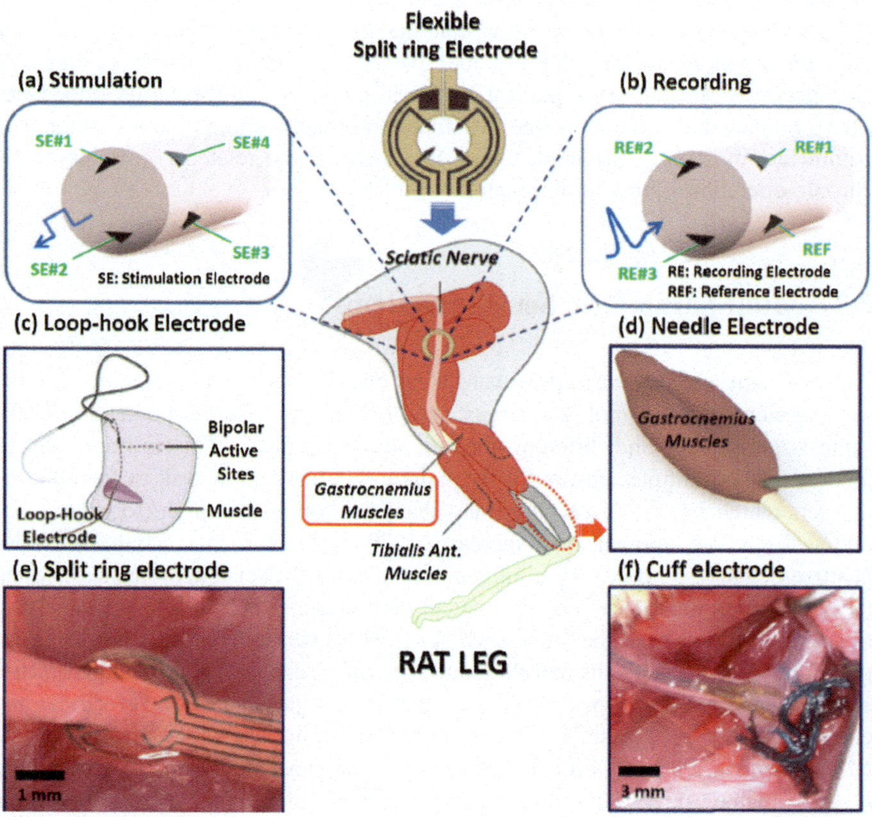

Fig. 3.44 Implanted flexible split ring electrode on the static nerve for **a** stimulation and **b** recording. **c** Intramuscular loop-hook bipolar electrode. **d** Subdermal electrical stimulation using a needle electrode. Picture of implanted **e** flexible split ring electrode **f** commercial cuff electrode (Microprobe Inc., USA) on the static nerve in rats (reprinted from [268], © 2017, with permission from Elsevier)

for use as a stimulating electrode was discussed in [267]. Stimulating nerve cuff electrodes with Pt black had the highest charge delivery capacity (62 times higher than Pt), the highest charge injection capacity (six times higher than Pt), and the lowest interfacial impedance (2.9 times lower than Pt). The lowest interfacial impedance of Pt black (followed by PEDOT, Pt, and IrOx) could be associated with the surface roughness of the electrode, its reactivity, and capacitance in the solution. The greater charge injection capacity could minimize the nerve damage, including low inflammatory cell infiltration and less fibrous tissue infiltration. Stimulating nerve cuff electrode with Pt black is expected to provide biocompatible and stable electrical stimulation without electrode dissolution and tissue damage.

Flexible split ring electrodes quickly and easily implanted at any desired position on peripheral nerves are presented in [268]. Selective stimulation demonstrated that gastrocnemius (GM) and tibialis anterior (TA) muscles were selectively activated by different pairs of active electrodes on a split ring electrode using a transverse bipolar configuration and stimulation currents of 200 μA (Fig. 3.44).

From the above it can be seen that the general trend is towards electrode miniaturization, increased flexibility, multiplexed electrode arrays of large numbers, and electrode surface functionalization for improved electrode–tissue impedance [213]. Implantable stimulators needs to function reliably, and numerous factors and parameters must be considered, taking into account the related constrains of the chronic operation following implantation [269].

3.7 Multi-parametric Sensing Systems

An important and natural step forward from classical systems employing a single sensing element is to employ a variety and plethora of sensors for detection of the same target marker and different markers and phenomena. This is important, as readings from multiple sensors for the same target markers as well as for different markers which relate to each other may change in a linked fashion (for example if one is decreased another may increase) [270, 271], and this can reduce false positive/negatives, improving confidence and clinical decision robustness. Many metabolites are interdependent and patient-dependent, so it also allows spatial and temporal relationships to be established [272]. However, each sensing element must provide independent results and they must not affect each other. For amperometric sensors, there are two types of crosstalk between adjacent sensors: electrical and chemical. Electrical crosstalk is due to capacitive coupling, while chemical crosstalk is due to overlapping diffusion layers between adjacent working electrodes and is unavoidable, since the sensors must consume analyte in order to perform a measurement [272]. Center-to-center distances between electrodes less than 100 μm are reported to be problematic [272].

Multi-sensing systems (systems with many sensors, not necessarily different) have already been discussed in this and the previous chapter, and many of these are multi-parametric sensing systems [73, 75, 76, 79–81, 273–293]. The active and

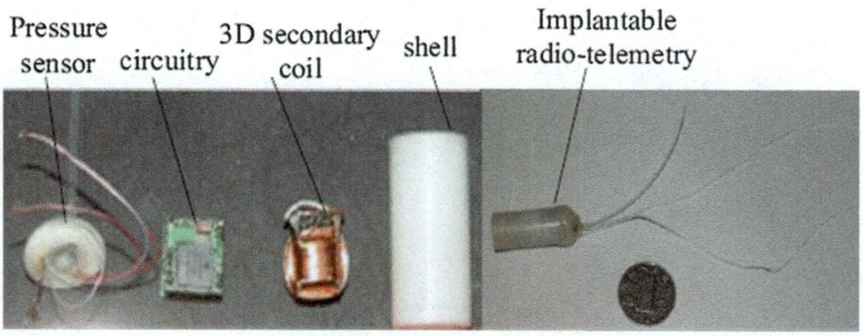

Fig. 3.45 Implantable capsule for measurement of pressure and temperature using commercially available technologies (reprinted from [295], © 2015, with permission from Elsevier)

passive flexible/stretchable sensing systems of Rogers et al., discussed previously, currently set the state of the art, incorporating tactile, optical, electrochemical and electrical sensors on the same platform.

There is a wide range of proposed implantable systems using commercially available electronics. Many currently commercially available microcontrollers incorporate basic analog blocks, such as amplifiers. Many also come with compact device packaging using e.g. ball grid arrays (BGA), or are available as unpackaged devices, i.e. as dies. Simultaneously, widely available regular rigid and flexible PCB fabrication technologies can achieve minimum geometrical features down to a few tenths of a micron. All the above have encouraged researchers to develop implantable systems using such technologies. The advantage of this is the faster development times and the ability to focus on the development of the specialized sensors, since many of the electronics required are standard. In [294] a ZigBee wireless platform for in vivo measurement in pigs of aortic and ventricular pressure and temperature was presented. The system consumed 13.33 μW of power. Using a switching architecture the analog front end is reconfigurable, such that different types of sensors can be interrogated with a small number of components. The complete readout system before packaging was 48 mm × 33 mm × 15 mm. A similar system for measurement of pressure, temperature, and biopotentials in rabbits was presented in [295]. The capsule was 13 mm in diameter and 30 mm in height. Two silver wires were used for biopotential recording and a small plastic tube was used to interface with the pressure sensor. A 3D secondary coil was used for telemetry. This is shown in Fig. 3.45.

A team from Ecole Polytechnique Federale de Lausanne has developed a number of systems for implantable chemical sensing [293, 296–298]. In [296] they used commercially available electronics to develop a potentiostat (Fig. 3.46a) and electrodes. The electrodes were functionalized with a multi-walled carbon nanotube (MWCNT)-chloroform solution and an enzymatic probe (glucose, lactate, and glutamate oxidases). ATP sensors were prepared by a mixture of glucose oxidase (GOD) and hexokinase (HEX). The system presented in [297] used five electrodes for electrochemical sensors, one temperature sensor, a pH sensor, and an ASIC (0.18 μm technology, power consumption of 530 μW) for chrono-amperometry

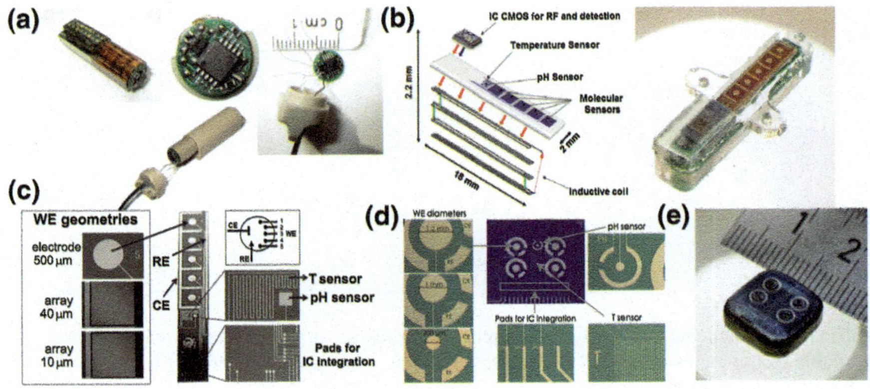

Fig. 3.46 Implantable multi-parametric sensing systems: **a** A system using of-the-shelf components (© IEEE 2013. Reprinted with permission from [296]). **b** A fully implantable encapsulated sensing system with ASIC, inductive telemetry, and multi-parametric monitoring using the sensors shown in **c** (© IEEE 2015. Reprinted with permission from [297]). **d** The electrochemical and temperature sensors of the miniature telemetrically operated multi-parametric sensing system shown in **e** [298] (© IEEE 2014. Reprinted with permission from [298])

and cyclic voltammetry powered by an inductive telemetric link. This multi-parametric platform (2.2 mm × 1.5 mm) was fabricated on a silicon substrate with a native oxide and a Ti layer followed by Pt to form all the metallization and a biocompatible top Al_2O_3 passivation. The temperature sensor was a Pt wire, the pH sensor was an anodic iridium oxide film on Pt, and the electrochemical sensors were comprised of a common Pt counter and RE and five Pt working electrodes. The latter were functionalized with carbon nanotubes (CNT) and enzymes entrapped in a chitosan matrix and sealed behind a porous polycarbonate membrane for biocompatibility. The implant was coated with parylene-C and an outer silicone soft shell. These are shown in Fig. 3.46b, c. Biocompatibility tests in vitro and in vivo demonstrated the ability of the packaging to prevent copper corrosion, leaking, and cytotoxicity for more than seven days. They also demonstrated that the sensor membranes did not induce cytotoxicity in the short term (there are concerns regarding the biocompatibility of CNTs). In [298] the same group developed a further miniaturized multi-parametric platform for freely moving rats. The ASIC and peripheral electronics interrogated a temperature sensor (based again on a Pt wire with a 34 kΩ resistance at 20 °C for lower power), a pH sensor (anodic iridium oxide), and four three-electrode systems for cyclic voltammetry and chrono-amperometry (working electrodes of 0.5, 1 and 1.2 mm diameter), all made with Pt (Fig. 3.46d). The system consumed about 1.6 mW. Two separate channels were used for data transmission (loop antenna at 869 MHz, 40 cm away) and remote powering of the implant (13.56 MHz, 3 cm). The sensors were used for the chrono-amperometric detection of glucose and the voltammetric detection of etoposide and mitoxantrone, two anti-cancer drugs. The electrode platform was fabricated following a similar protocol as before. An epoxy adhesive was used to assemble the system, which was encapsulated by parylene-C, and a biocompatible

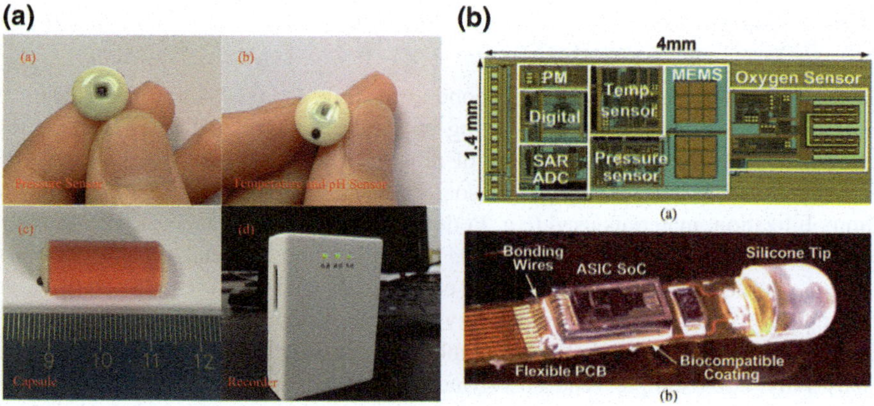

Fig. 3.47 **a** An implantable GI tract monitoring capsule for pH, temperature and pressure operating up to 136 h (© IEEE 2015. Reprinted with permission from [300]). **b** The intracranial neuromonitoring catheter of (© IEEE 2014. Reprinted with permission from [301])

membrane as in Fig. 3.46e. In vitro and in vivo assessment demonstrated the biocompatibility of the system (for up to 35 and 30 days respectively).

Monitoring of the gastrointestinal tract with multimodal platforms was recently presented in [299, 300]. The system was capable of monitoring temperature (negative temperature coefficient thermistor), pH (ISFET) and pressure, with an ASIC and was packaged in a capsule measuring 11 mm × 25 mm as shown in Fig. 3.47a. The system had a battery life of 136 h. The data were wirelessly transferred to an external data recorder and the system was evaluated in vivo with 20–22 human subjects. In [301] a multi-parametric sensing platform for intracranial monitoring of pressure, oxygen, and temperature was presented. The system was based on a 1.4 mm × 4 mm ASIC with a power consumption of 166 μW. The pressure sensor is a MEMS capacitive sensor fabricated on the CMOS chip, the temperature sensor is an on-chip sensor based on parasitic BJTs, and the O_2 sensor is a three-electrode chrono-amperometric system comprised of 16 working electrodes, a common counter, and RE all made out of Pt, which is then covered by Nafion. The sensitivity of the temperature sensor was 2 mV $°C^{-1}$, of the O_2 sensors 194 pA/mmHg, and of the pressure sensor 18.6 aF $mmHg^{-1}$. The complete system was fitted on a catheter, as shown in Fig. 3.47b.

3.8 Sensor Metrics

In this section, we define a number of widely used metrics used to define sensor performance. Examples are given, using various sensing modalities discussed in this and the previous chapter, as well as challenging issues and solutions related to these.

3.8.1 Sensitivity

The amount by which a sensor responds to a unit change of the target event is referred to as its sensitivity. Depending on the sensor type and its characteristics, sensitivity may depend on a variety of different factors. For example, in amperometric sensors it is a function of sensor geometry, enzyme concentration, enzyme immobilization, materials used (e.g. Pt demonstrates the highest sensitivity to H_2O_2 oxidation among inert materials) [270, 272], the inner polymer layers (either for enzyme immobilization or interference elimination), and the outer membranes used to improve issues related to limited oxygen or to minimize biofouling/leaching. It has been reported [302] that such membranes can lead to a 60–75% reduction in sensitivity, for example. The absorbance of the products of the electrochemical reaction and of some analytes such, as oxygen, on the surface of the working electrode also leads to a reduction of sensitivity. Applying high potentials to clean the electrode surface can re-establish the sensor's sensitivity, at the cost of a reduction in selectivity.

3.8.2 Selectivity

The ability of a sensor to distinguish and detect only a single varying target event when multiple other changing events and potential targets exist and take place in the vicinity of the sensor is termed the selectivity of the sensor. In the ideal case, the recognition process should have high specificity, such that the sensor is sensitive towards one particular ion, atom, or molecule. In reality, this is rarely achieved, and in general, the recognition system will react to a range of analytes, but with low sensitivity. In electrochemical sensors, target analyte specific molecular recognition elements are employed. In amperometric sensors, these are primarily enzymes. In implantable applications reagentless operation is required, and thus oxidases are typically used. For example, in amperometric enzymatic sensors a high potential (0.6–0.7 V vs. Ag/AgCl reference electrode (RE)) is used to oxidize enzymatically generated hydrogen peroxide (H_2O_2). However, this high potential leads to the oxidation of endogenous species such as nitric oxide (NO), ascorbic acid (AA), dopamine, and uric acid (UA), among others [270, 272]. This will affect the selectivity of the sensor, as the sensor will provide erroneous results. One solution is to use selective membranes such as Nafion [270, 271], which, however, leads to reduced sensitivity and increased response time [272]. Alternatively, a dual electrode system can be used, where the second electrode is blank, such that the background current can be established and cancelled out. Another approach is to reduce the oxidation potential required. Modification of the working electrodes with carbon nanotubes has led to the reduction of this to −0.2 V, for example [270]. At such low potentials few endogenous species are electroactive, thus improving the selectivity of the sensor. The use of multi-enzyme systems to provide differential measurements and thus cancel out the effect of interfering species has also been proposed [271].

3.8.3 *Limit of Detection (LOD)*

The limit of detection (LOD) of a sensor refers to the lowest/smallest change in the target event that can be detected by a change in the sensor output signal. This change needs to be statistically different from the baseline signal recorded in the absence of the target event/analyte. This is a very important parameter when detecting minute changes. In amperometric sensors, the use of nano/micro-electrodes which exhibit hemispherical diffusion profiles has been shown to provide lower detection limits.

3.9 Conclusion

Implantable sensors offer a direct means of measuring in vivo mechanical and functional changes of patient physiology and pathophysiology. Implantable sensors need to be small, minimally invasive, wireless, and low power. They need to be highly selective and linear, with limits of detection appropriate to the target application, and they need to provide repeatable and stable measurements with a rapid response and low drift and hysteresis. In addition, they need to have a long shelf life. They need limited sample preparation or none at all, if possible, and they need to be low cost, simple to use, and provide a clear improvement in the patient's quality of life.

Electrical (action potential recording, imperimetric) and tactile (strain and pressure) sensors are all necessary for different implantable sensing applications, which include surgical site infection (SSI) monitoring, diabetes, neuroprosthesis, ischemia, cardiovascular diseases, and orthopedic applications. Electrical bioimpedance is a powerful method for interrogating tissues, with strong diagnostic capabilities, which can be used in combination with electrochemical sensors to yield higher diagnostic accuracy and specificity. The design of bioimpedance sensors requires the use of electromagnetic modeling techniques to optimize the sensitivity of the sensors for a particular application. Strain and pressure are important, particularly for orthopedic applications involving human motion and orthopedic implants. Strain and pressure sensors based on resistive, capacitive, and inductive changes are discussed in this chapter, along with metamaterial-based approaches. Both electrochemical sensors and strain sensors require temperature recordings to compensate for their inherit temperature dependence. However, temperature can be compensated for using differential measurements. Various types of temperature sensors exist, with the vast majority being serpentine metal foil structures on flexible substrates and semiconductor-based devices. Low-power miniaturized electronics with high accuracy are necessary to interrogate the sensors. The Wheatstone bridge for resistive sensors, such as strain gauges and RTDs, is a standard topology, with many variations depending on the application. For strain sensors, different orientations and combinations of sensors to implement quarter, half and full bridge sensors were discussed. Bioimpedance instrumentation was also discussed in detail, with a discussion of three different front-end systems and current sources and their relative merits and disadvantages. It is obvious from the above and from discussions throughout this chapter and others that multi-parametric sensing systems are essential, with a large number of such examples presented in the literature.

References

1. F. Lisdat, D. Schäfer, The use of electrochemical impedance spectroscopy for biosensing. Anal. Bioanal. Chem. **391**(5), 1555–1567 (2008)
2. P. Kassanos, I.F. Triantis, A CMOS multi-sine signal generator for multi-frequency bioimpedance measurements, in *2014 IEEE International Symposium on Circuits and Systems (ISCAS)*, (2014), pp. 249–252
3. W.R.B. Lionheart, J. Kaipio, C.N. McLeod, Generalized optimal current patterns and electrical safety in EIT. Physiol. Meas. **22**(1), 85–90 (2001)
4. S. Grimnes, O.G. Martinsen, *Bioimpedance and Bioelectricity Basics*, 1st edn. (Academic Press, Suffolk, UK, 2000)
5. C. Gabriel, S. Gabriel, E. Corthout, The dielectric properties of biological tissues: I. Literature survey. Phys. Med. Biol. **41**(11), 2231–2249 (1996)
6. O.G. Martinsen, S. Grimnes, H.P. Schwan, *"Interface phenomena and dielectric properties of biological tissue", in Encyclopedia of Surface and Colloid Science* (Marcel Dekker, New York, 2002), pp. 2643–2652
7. H.P. Schwan, Electrical properties of body tissues and impedance plethysmography. IRE Trans. Med. Electron. PGME **3**, 32–46 (1955)
8. M. Min, T. Parve, A. Ronk, P. Annus, T. Paavle, Synchronous sampling and demodulation in an instrument for multifrequency bioimpedance measurement. IEEE Trans. Instrum. Meas. **56**(4), 1365–1372 (2007)
9. C.L. del Rio et al., Early time course of myocardial electrical impedance during acute coronary artery occlusion in pigs, dogs, and humans. J. Appl. Physiol. **99**(4), 1576–1581 (2005)
10. A. McEwan, J. Tapson, A. van Schaik, D.S. Holder, Code-division-multiplexed electrical impedance tomography spectroscopy. IEEE Trans. Biomed. Circuits Syst. **3**(5), 332–338 (2009)
11. S. Abdul, B.H. Brown, P. Milnes, J.A. Tidy, The use of electrical impedance spectroscopy in the detection of cervical intraepithelial neoplasia. Int. J. Gynecol. Cancer **16**(5), 1823–1832 (2006)
12. H.-G. Jahnke et al., Impedance spectroscopy—an outstanding method for label-free and real-time discrimination between brain and tumor tissue in vivo. Biosens. Bioelectron. **46**, 8–14 (2013)
13. Y. Wan, R. Halter, A. Borsic, P. Manwaring, A. Hartov, K. Paulsen, Sensitivity study of an ultrasound coupled transrectal electrical impedance tomography system for prostate imaging. Physiol. Meas. **31**(8), S17–S29 (2010)

14. J. Harms, A. Schneider, M. Baumgartner, J. Henke, R. Busch, Diagnosing acute liver graft rejection: experimental application of an implantable telemetric impedance device in native and transplanted porcine livers. Biosens. Bioelectron. **16**(3), 169–177 (2001)
15. C.A. González-Correa et al., Virtual biopsies in Barrett's esophagus using an impedance probe. Ann. N. Y. Acad. Sci. **873**(1), 313–321 (1999)
16. H.N. Nguyen, J. Silny, S. Matern, Multiple intraluminal electrical impedancometry for recording of upper gastrointestinal motility: current results and further implications. Am. J. Gastroenterol. **94**(2), 306–317 (1999)
17. A.J. Bredenoord, B.L.A.M. Weusten, D. Sifrim, R. Timmer, A.J.P.M. Smout, Aerophagia, gastric, and supragastric belching: a study using intraluminal electrical impedance monitoring. Gut **53**(11), 1561–1565 (2004)
18. C.A. González, C. Villanueva, S. Othman, R. Narváez, E. Sacristán, Impedance spectroscopy for monitoring ischemic injury in the intestinal mucosa. Physiol. Meas. **24** (2), 277–289 (2003)
19. H. Imam, C. Sanmiguel, B. Larive, Y. Bhat, E. Soffer, Study of intestinal flow by combined videofluoroscopy, manometry, and multiple intraluminal impedance. Am. J. Physiol. Gastrointest. Liver Physiol. **286**(2), G263–G270 (2004)
20. J.L. Gonzalez-Guillaumin, D.C. Sadowski, O. Yadid-Pecht, K.V.I.S. Kaler, M.P. Mintchev, Multichannel pressure, bolus transit, and pH esophageal catheter. IEEE Sens. J. **6**(3), 796–803 (2006)
21. A.J. Bredenoord, B.L.A.M. Weusten, R. Timmer, A.J.P.M. Smout, Minimum sample frequency for multichannel intraluminal impedance measurement of the oesophagus. Neurogastroenterol. Motil. **16**(6), 713–719 (2004)
22. A. Al-Zaben, V. Chandrasekar, Computation of intraluminal impedance. Physiol. Meas. **25** (1), 61 (2004)
23. J. Fass et al., Measuring esophageal motility with a new intraluminal impedance device: first clinical results in reflux patients. Scand. J. Gastroenterol. **29**(8), 693–702 (1994)
24. R. Tutuian, M.F. Vela, S.S. Shay, D.O. Castell, Multichannel intraluminal impedance in esophageal function testing and gastroesophageal reflux monitoring. J. Clin. Gastroenterol. **37**(3), 206–215 (2003)
25. F. Mellert et al., Detection of (reversible) myocardial ischemic injury by means of electrical bioimpedance. IEEE Trans. Biomed. Eng. **58**(6), 1511–1518 (2011)
26. J. Wtorek et al., Monitoring of myocardium state during off-pump coronary artery by-pass grafting. Physiol. Meas. **29**(6), S393–S405 (2008)
27. Y. Salazar, R. Bragos, O. Casas, J. Cinca, J. Rosell, Transmural versus nontransmural in situ electrical impedance spectrum for healthy, ischemic, and healed myocardium. IEEE Trans. Biomed. Eng. **51**(8), 1421–1427 (2004)
28. R. Dzwonczyk, C. del Rio, D.A. Brown, R.E. Michler, R.K. Wolf, M.B. Howie, Myocardial electrical impedance responds to ischemia and reperfusion in humans. IEEE Trans. Biomed. Eng. **51**(12), 2206–2209 (2004)
29. S. Kun, B. Ristic, R.A. Peura, R.M. Dunn, Algorithm for tissue ischemia estimation based on electrical impedance spectroscopy. IEEE Trans. Biomed. Eng. **50**(12), 1352–1359 (2003)
30. B. Ristic, S. Kun, R.A. Peura, Muscle tissue ischemia monitoring using impedance spectroscopy: quantitative results of animal studies, in *Proceedings of the 19th Annual International Conference of the IEEE Engineering in Medicine and Biology Society*, (1997), pp. 2108–2111
31. O. Casas et al., In vivo and in situ ischemic tissue characterization using electrical impedance spectroscopya. Ann. N. Y. Acad. Sci. **873**(1), 51–58 (1999)
32. M.A. Fallert et al., Myocardial electrical impedance mapping of ischemic sheep hearts and healing aneurysms. Circulation **87**(1), 199–207 (1993)
33. A. Ivorra et al., Minimally invasive silicon probe for electrical impedance measurements in small animals. Biosens. Bioelectron. **19**(4), 391–399 (2003)
34. R. Gómez et al., A SiC microdevice for the minimally invasive monitoring of ischemia in living tissues. Biomed. Microdevices **8**(1), 43–49 (2006)

35. A. Sola et al., Multiparametric monitoring of ischemia-reperfusion in rat kidney: effect of ischemic preconditioning. Transplantation **75**(6), 744–749 (2003)
36. M. Tijero et al., SU-8 microprobe with microelectrodes for monitoring electrical impedance in living tissues. Biosens. Bioelectron. **24**(8), 2410–2416 (2009)
37. J. Cinca et al., Changes in myocardial electrical impedance in human heart graft rejection. Eur. J. Heart Fail. **10**(6), 594–600 (2008)
38. M. Schäfer, C. Schlegel, H.-J. Kirlum, E. Gersing, M.M. Gebhard, Monitoring of damage to skeletal muscle tissues caused by ischemia. Bioelectrochem. Bioenerg. **45**(2), 151–155 (1998)
39. A. Yufera, A. Rueda, J.M. Munoz, R. Doldan, G. Leger, E.O. Rodriguez-Villegas, A tissue impedance measurement chip for myocardial ischemia detection. IEEE Trans. Circuits Syst. Regul. Pap. **52**(12), 2620–2628 (2005)
40. J. Wtorek, L. Jozefiak, A. Polinski, J. Siebert, An averaging two-electrode probe for monitoring changes in myocardial conductivity evoked by ischemia. IEEE Trans. Biomed. Eng. **49**(3), 240–246 (2002)
41. E. Marzec, K. Wachal, The electrical properties of leg skin in normal individuals and in patients with ischemia. Bioelectrochem. Bioenerg. **49**(1), 73–75 (1999)
42. S. Kun, R.A. Peura, Tissue ischemia detection using impedance spectroscopy, in *Proceedings of the 16th Annual International Conference of the IEEE Engineering in Medicine and Biology Society, 1994. Engineering Advances: New Opportunities for Biomedical Engineers*, (1994), pp. 868–869
43. T. Süselbeck et al., In vivo intravascular electric impedance spectroscopy using a new catheter with integrated microelectrodes. Basic Res. Cardiol. **100**(1), 28–34 (2004)
44. P. Kassanos, R.K. Iles, R.H. Bayford, A. Demosthenous, Towards the development of an electrochemical biosensor for hCGβ detection. Physiol. Meas. **29**(6), S241–S254 (2008)
45. F. Segura-Quijano, J. Sacristán-Riquelme, J. García-Cantón, M.T. Osés, A. Baldi, Towards fully integrated wireless impedimetric sensors. Sensors **10**(4), 4071–4082 (2010)
46. A. Radu et al., Diagnostic of functionality of polymer membrane—based ion selective electrodes by impedance spectroscopy. Anal. Methods **2**(10), 1490–1498 (2010)
47. P. Kassanos, A. Demosthenous, R.H. Bayford, Towards an optimized design for tetrapolar affinity-based impedimetric immunosensors for lab-on-a-chip applications, in *IEEE Biomedical Circuits and Systems Conference, 2008. BioCAS*, (2008), pp. 141–144
48. P. Kassanos, A. Demosthenous, R.H. Bayford, Comparison of tetrapolar injection-measurement techniques for coplanar affinity-based impedimetric immunosensors, in *IEEE Biomedical Circuits and Systems Conference, 2008. BioCAS*, (2008), pp. 317–320
49. P. Kassanos, A. Demosthenous, R.H. Bayford, Optimization of bipolar and tetrapolar impedance biosensors, in *Proceedings of 2010 IEEE International Symposium on Circuits and Systems (ISCAS)*, (2010), pp. 1512–1515
50. P. Kassanos, H.M.D. Ip, G.-Z. Yang, A tetrapolar bio-impedance sensing system for gastrointestinal tract monitoring, in *2015 IEEE 12th International Conference on Wearable and Implantable Body Sensor Networks (BSN)*, (2015), pp. 1–6
51. S. Grimnes, Ø.G. Martinsen, Sources of error in tetrapolar impedance measurements on biomaterials and other ionic conductors. J. Phys. Appl. Phys. **40**(1), 9–14 (2007)
52. B.H. Brown, A.J. Wilson, P. Bertemes-Filho, Bipolar and tetrapolar transfer impedance measurements from volume conductor. Electron. Lett. **36**(25), 2060–2062 (2000)
53. M. Genescà et al., Electrical bioimpedance measurement during hypothermic rat kidney preservation for assessing ischemic injury. Biosens. Bioelectron. **20**(9), 1866–1871 (2005)
54. B. Sanchez, G. Vandersteen, R. Bragos, J. Schoukens, Basics of broadband impedance spectroscopy measurements using periodic excitations. Meas. Sci. Technol. **23**(10), 105501 (2012)
55. B. Sanchez, X. Fernandez, S. Reig, R. Bragos, An FPGA-based frequency response analyzer for multisine and stepped sine measurements on stationary and time-varying impedance. Meas. Sci. Technol. **25**(1), 015501 (2014)

56. A.S. Tucker, R.M. Fox, R.J. Sadleir, Biocompatible, high precision, wideband, improved Howland current source with lead-lag compensation. IEEE Trans. Biomed. Circuits Syst. **7** (1), 63–70 (2013)
57. T. Sun, S. Gawad, C. Bernabini, N.G. Green, H. Morgan, Broadband single cell impedance spectroscopy using maximum length sequences: theoretical analysis and practical considerations. Meas. Sci. Technol. **18**(9), 2859–2868 (2007)
58. J. Ojarand, M. Min, P. Annus, Crest factor optimization of the multisine waveform for bioimpedance spectroscopy. Physiol. Meas. **35**(6), 1019–1033 (2014)
59. Y. Yang, F. Zhang, K. Tao, B. Sanchez, H. Wen, Z. Teng, An improved crest factor minimization algorithm to synthesize multisines with arbitrary spectrum. Physiol. Meas. **36** (5), 895–910 (2015)
60. F. Seoane, R. Macías, R. Bragós, K. Lindecrantz, Simple voltage-controlled current source for wideband electrical bioimpedance spectroscopy: circuit dependences and limitations. Meas. Sci. Technol. **22**(11), 115801 (2011)
61. P. Bertemes-Filho, B.H. Brown, A.J. Wilson, A comparison of modified Howland circuits as current generators with current mirror type circuits. Physiol. Meas. **21**(1), 1–6 (2000)
62. H. Hong, M. Rahal, A. Demosthenous, R.H. Bayford, Comparison of a new integrated current source with the modified Howland circuit for EIT applications. Physiol. Meas. **30** (10), 999–1007 (2009)
63. A.C. Ivorra, Contributions to the measurement of electrical impedance for living tissue ischemia injury monitoring—OpenThesis, Universitat Politécnica de Catalunya, 2005
64. P. Kassanos, I.F. Triantis, A. Demosthenous, A CMOS magnitude/phase measurement chip for impedance spectroscopy. IEEE Sens. J. **13**(6), 2229–2236 (2013)
65. Y. Yang, J. Wang, G. Yu, F. Niu, P. He, Design and preliminary evaluation of a portable device for the measurement of bioimpedance spectroscopy. Physiol. Meas. **27**(12), 1293–1310 (2006)
66. P. Kassanos, L. Constantinou, I.F. Triantis, A. Demosthenous, An integrated analog readout for multi-frequency bioimpedance measurements. IEEE Sens. J. **14**(8), 2792–2800 (2014)
67. R. Pallás-Areny, J.G. Webster, *Analog Signal Processing* (Wiley, New York, 1999)
68. R. Gonzalez-Landaeta, O. Casas, R. Pallas-Areny, Heart rate detection from plantar bioimpedance measurements. IEEE Trans. Biomed. Eng. **55**(3), 1163–1167 (2008)
69. R. Pallas-Areny, O. Casas, A novel differential synchronous demodulator for AC signals. IEEE Trans. Instrum. Meas. **45**(2), 413–416 (1996)
70. R. Pallas-Areny, J.G. Webster, Bioelectric impedance measurements using synchronous sampling. IEEE Trans. Biomed. Eng. **40**(8), 824–829 (1993)
71. C. Margo, J. Katrib, M. Nadi, A. Rouane, A four-electrode low frequency impedance spectroscopy measurement system using the AD5933 measurement chip. Physiol. Meas. **34** (4), 391–405 (2013)
72. N. Mehmood, A. Hariz, R. Fitridge, N.H. Voelcker, Applications of modern sensors and wireless technology in effective wound management. J. Biomed. Mater. Res. B Appl. Biomater. **102**(4), 885–895 (2014)
73. R.C. Webb et al., Ultrathin conformal devices for precise and continuous thermal characterization of human skin. Nat. Mater. **12**(10), 938–944 (2013)
74. A.C. Paglinawan, Y.-H. Wang, S.-C. Cheng, C.-C. Chuang, W.-Y. Chung, CMOS temperature sensor with constant power consumption multi-level comparator for implantable bio-medical devices. Electron. Lett. **45**(25), 1291–1292 (2009)
75. L. Xu et al., 3D multifunctional integumentary membranes for spatiotemporal cardiac measurements and stimulation across the entire epicardium. Nat. Commun. **5**, 3329 (2014)
76. F. Graichen, G. Bergmann, A. Rohlmann, Hip endoprosthesis for in vivo measurement of joint force and temperature. J. Biomech. **32**(10), 1113–1117 (1999)
77. D.P. Jones, *Biomedical Sensors* (Momentum Press, New York, 2010)
78. D. Li (ed.) Resistance thermometers, in *Encyclopedia of Microfluidics and Nanofluidics* (Springer, US, 2008), pp. 1790–1790
79. D.-H. Kim et al., Epidermal electronics. Science **333**(6044), 838–843 (2011)

80. W.-H. Yeo et al., Multifunctional epidermal electronics printed directly onto the skin. Adv. Mater. **25**(20), 2773–2778 (2013)
81. D.-H. Kim et al., Materials for multifunctional balloon catheters with capabilities in cardiac electrophysiological mapping and ablation therapy. Nat. Mater. **10**(4), 316–323 (2011)
82. P.R.N. Childs, J.R. Greenwood, C.A. Long, Review of temperature measurement. Rev. Sci. Instrum. **71**(8), 2959–2978 (2000)
83. A. BaHammam, Comparison of nasal prong pressure and thermistor measurements for detecting respiratory events during sleep. Respiration **71**(4), 385–390 (2004)
84. J. Fei, I. Pavlidis, Thermistor at a distance: unobtrusive measurement of breathing. IEEE Trans. Biomed. Eng. **57**(4), 988–998 (2010)
85. E. Jovanov, D. Raskovic, R. Hormigo, Thermistor-based breathing sensor for circadian rhythm evaluation. Biomed. Sci. Instrum. **37**, 493–497 (2001)
86. M.K. Law, A. Bermak, H.C. Luong, A Sub-μW embedded CMOS temperature sensor for RFID food monitoring application. IEEE J. Solid-State Circuits **45**(6), 1246–1255 (2010)
87. M.A.P. Pertijs, G.C.M. Meijer, J.H. Huijsing, Precision temperature measurement using CMOS substrate pnp transistors. IEEE Sens. J. **4**(3), 294–300 (2004)
88. A.L. Aita, M.A.P. Pertijs, K.A.A. Makinwa, J.H. Huijsing, G.C.M. Meijer, Low-power CMOS smart temperature sensor with a batch-calibrated inaccuracy of ±0.25 °C (±3σ) from −70 °C to 130 °C. IEEE Sens. J. **13**(5), 1840–1848 (2013)
89. K. Souri, Y. Chae, K.A.A. Makinwa, A CMOS Temperature Sensor With a Voltage-Calibrated Inaccuracy of 0.15 °C (3) From 55 to 125 °C. IEEE J. Solid-State Circuits **48**(1), 292–301 (2013)
90. K. Ueno, T. Asai, Y. Amemiya, Low-power temperature-to-frequency converter consisting of subthreshold CMOS circuits for integrated smart temperature sensors. Sens. Actuators Phys. **165**(1), 132–137 (2011)
91. C. Azcona, B. Calvo, N. Medrano, S. Celma, CMOS quasi-digital temperature sensor for battery operated systems. Electron. Lett. **49**(21), 1338–1340 (2013)
92. A. Vaz et al., Full passive UHF tag with a temperature sensor suitable for human body temperature monitoring. IEEE Trans. Circuits Syst. II Express Briefs **57**(2), 95–99 (2010)
93. S. Jeong, Z. Foo, Y. Lee, J.-Y. Sim, D. Blaauw, D. Sylvester, A fully-integrated 71 nW CMOS temperature sensor for low power wireless sensor nodes. IEEE J. Solid-State Circuits **49**(8), 1682–1693 (2014)
94. F. Khoshnoud, C.W. de Silva, Recent advances in MEMS sensor technology-biomedical applications. IEEE Instrum. Meas. Mag. **15**(1), 8–14 (2012)
95. F. Khoshnoud, C.W. de Silva, Recent advances in MEMS sensor technology-mechanical applications. IEEE Instrum. Meas. Mag. **15**(2), 14–24 (2012)
96. C.-C. Yang, Y.-L. Hsu, A review of accelerometry-based wearable motion detectors for physical activity monitoring. Sensors **10**(8), 7772–7788 (2010)
97. Y.J. Huang et al., A CMOS cantilever-based label-free DNA SoC with improved sensitivity for Hepatitis B virus detection. IEEE Trans. Biomed. Circuits Syst. **7**(6), 820–831 (2013)
98. X. Yu, Y. Tang, H. Zhang, T. Li, W. Wang, Design of high-sensitivity cantilever and its monolithic integration with CMOS circuits. IEEE Sens. J. **7**(4), 489–495 (2007)
99. G. Grimaldi, M. Manto, Neurological tremor: sensors, signal processing and emerging applications. Sensors **10**(2), 1399–1422 (2010)
100. S. Patel, H. Park, P. Bonato, L. Chan, M. Rodgers, A review of wearable sensors and systems with application in rehabilitation. J. NeuroEngineering Rehabil. **9**, 21 (2012)
101. H. Chen, M. Xue, Z. Mei, S. Bambang Oetomo, W. Chen, A review of wearable sensor systems for monitoring body movements of neonates. Sensors **16**(12), 2134 (2016)
102. T.G. Constandinou, J. Georgiou, A micropower tilt-processing circuit. IEEE Trans. Biomed. Circuits Syst. **3**(6), 363–369 (2009)
103. H. Zeng, Y. Zhao, Sensing movement: microsensors for body motion measurement. Sensors **11**(1), 638–660 (2011)
104. C.M. Andreou, Y. Pahitas, J. Georgiou, Bio-inspired micro-fluidic angular-rate sensor for vestibular prostheses. Sensors **14**(7), 13173–13185 (2014)

105. S. Kon, R. Horowitz, A high-resolution MEMS piezoelectric strain sensor for structural vibration detection. IEEE Sens. J. **8**(12), 2027–2035 (2008)
106. Y.M. Wang, P.K. Chan, H.K.H. Li, S.E. Ong, A low-power highly sensitive capacitive accelerometer IC using auto-zero time-multiplexed differential technique. IEEE Sens. J. **15** (11), 6179–6191 (2015)
107. G. Langfelder, A.F. Longoni, A. Tocchio, E. Lasalandra, MEMS motion sensors based on the variations of the fringe capacitances. IEEE Sens. J. **11**(4), 1069–1077 (2011)
108. C.M. Sun, M.H. Tsai, Y.C. Liu, W. Fang, Implementation of a monolithic single proof-mass tri-axis accelerometer using CMOS-MEMS technique. IEEE Trans. Electron Devices **57**(7), 1670–1679 (2010)
109. S. Tez, U. Aykutlu, M.M. Torunbalci, T. Akin, A Bulk-micromachined three-axis capacitive MEMS accelerometer on a single die. J. Microelectromechanical Syst. **24**(5), 1264–1274 (2015)
110. M.H. Tsai, Y.C. Liu, W. Fang, A three-axis CMOS-MEMS accelerometer structure with vertically integrated fully differential sensing electrodes. J. Microelectromechanical Syst. **21** (6), 1329–1337 (2012)
111. M.H. Tsai, Y.C. Liu, K.C. Liang, W. Fang, Monolithic CMOS-MEMS pure oxide tri-axis accelerometers for temperature stabilization and performance enhancement. J. Microelectromechanical Syst. **24**(6), 1916–1927 (2015)
112. U. Krishnamoorthy et al., In-plane MEMS-based nano-g accelerometer with sub-wavelength optical resonant sensor. Sens. Actuators Phys. **145–146**, 283–290 (2008)
113. G. Zhanshe, C. Fucheng, L. Boyu, C. Le, L. Chao, S. Ke, Research development of silicon MEMS gyroscopes: a review. Microsyst. Technol. **21**(10), 2053–2066 (2015)
114. S. Dellea, F. Giacci, A.F. Longoni, G. Langfelder, In-Plane and out-of-plane MEMS gyroscopes based on piezoresistive NEMS detection. J. Microelectromechanical Syst. **24**(6), 1817–1826 (2015)
115. S. Sonmezoglu, S.E. Alper, T. Akin, An automatically mode-matched MEMS gyroscope with wide and tunable bandwidth. J. Microelectromechanical Syst. **23**(2), 284–297 (2014)
116. S.E. Alper, Y. Temiz, T. Akin, A compact angular rate sensor system using a fully decoupled silicon-on-glass MEMS gyroscope. J. Microelectromechanical Syst. **17**(6), 1418–1429 (2008)
117. Y. Hui, T. Nan, N.X. Sun, M. Rinaldi, High resolution magnetometer based on a high frequency magnetoelectric MEMS-CMOS oscillator. J. Microelectromechanical Syst. **24**(1), 134–143 (2015)
118. G. Langfelder, C. Buffa, A. Frangi, A. Tocchio, E. Lasalandra, A. Longoni, Z-axis magnetometers for MEMS inertial measurement units using an industrial process. IEEE Trans. Ind. Electron. **60**(9), 3983–3990 (2013)
119. K. Sinha, M. Tabib-Azar, 27 pT silicon nitride MEMS magnetometer for brain imaging. IEEE Sens. J. **16**(17), 6551–6558 (2016)
120. V. Kumar, A. Ramezany, M. Mahdavi, S. Pourkamali, Amplitude modulated Lorentz force MEMS magnetometer with picotesla sensitivity. J. Micromechanics Microengineering **26** (10), 105021 (2016)
121. D. Sheng, S. Li, N. Dural, M.V. Romalis, Subfemtotesla scalar atomic magnetometry using multipass cells. Phys. Rev. Lett. **110**(16), 160802 (2013)
122. P. Minotti, S. Brenna, G. Laghi, A.G. Bonfanti, G. Langfelder, A.L. Lacaita, A Sub–400-nT/sqrt(HZ), 775-uW, multi-loop MEMS magnetometer with integrated readout electronics. J. Microelectromechanical Syst. **24**(6), 1938–1950 (2015)
123. J.L. Tanner, D. Mousadakos, K. Giannakopoulos, E. Skotadis, D. Tsoukalas, High strain sensitivity controlled by the surface density of platinum nanoparticles. Nanotechnology **23** (28), 285501 (2012)
124. M. Borghetti, E. Sardini, M. Serpelloni, Preliminary study of resistive sensors in inkjet technology for force measurements in biomedical applications, in *2014 11th International Multi-Conference on Systems, Signals Devices (SSD)*, (2014), pp. 1–4

125. V. Correia, C. Caparros, C. Casellas, L. Francesch, J.G. Rocha, S. Lanceros-Mendez, Development of inkjet printed strain sensors. Smart Mater. Struct. **22**(10), 105028 (2013)
126. B. Ando, S. Baglio, All-inkjet printed strain sensors. IEEE Sens. J. **13**(12), 4874–4879 (2013)
127. A. Bessonov, M. Kirikova, S. Haque, I. Gartseev, M.J.A. Bailey, Highly reproducible printable graphite strain gauges for flexible devices. Sens. Actuators Phys. **206**, 75–80 (2014)
128. O. Kanoun et al., Flexible carbon nanotube films for high performance strain sensors. Sensors **14**(6), 10042–10071 (2014)
129. Y. Jia, K. Sun, F.J. Agosto, M.T. Quiñones, Design and characterization of a passive wireless strain sensor. Meas. Sci. Technol. **17**(11), 2869–2876 (2006)
130. K.J. Loh, J.P. Lynch, N.A. Kotov, Inductively coupled nanocomposite wireless strain and pH sensors. Smart Struct. Syst. **4**(5), 531–548 (2008)
131. F. Umbrecht et al., Wireless implantable passive strain sensor: design, fabrication and characterization. J. Micromechanics Microengineering **20**(8), 085005 (2010)
132. E.H. Ledet, D. D'Lima, P. Westerhoff, J.A. Szivek, R.A. Wachs, G. Bergmann, Implantable sensor technology: from research to clinical practice. J. Am. Acad. Orthop. Surg. **20**(6), 383–392 (2012)
133. W. Hasenkamp et al., Design and test of a MEMS strain-sensing device for monitoring artificial knee implants. Biomed. Microdevices **15**(5), 831–839 (2013)
134. F. Graichen, R. Arnold, A. Rohlmann, G. Bergmann, Implantable 9-channel telemetry system for in vivo load measurements with orthopedic implants. IEEE Trans. Biomed. Eng. **54**(2), 253–261 (2007)
135. K.C. McGilvray et al., Implantable microelectromechanical sensors for diagnostic monitoring and post-surgical prediction of bone fracture healing. J. Orthop. Res. **33**(10), 1439–1446 (2015)
136. R. Melik et al., Nested metamaterials for wireless strain sensing. IEEE J. Sel. Top. Quantum Electron. **16**(2), 450–458 (2010)
137. R. Melik, E. Unal, N. Kosku Perkgoz, C. Puttlitz, H.V. Demir, Circular high-Q resonating isotropic strain sensors with large shift of resonance frequency under stress. Sensors **9**(12), 9444–9451 (2009)
138. W. Claes, W. Sansen, R. Puers, A 40-μA/channel compensated 18-channel strain gauge measurement system for stress monitoring in dental implants. IEEE J. Solid-State Circuits **37** (3), 293–301 (2002)
139. W. Claes, R. Puers, W. Sansen, M.D. Cooman, J. Duyck, I. Naert, A low power miniaturized autonomous data logger for dental implants. Sens. Actuators Phys. **97–98**, 548–556 (2002)
140. G.Y. Yang, G. Johnson, W.C. Tang, J.H. Keyak, Parylene-based strain sensors for bone. IEEE Sens. J. **7**(12), 1693–1697 (2007)
141. C.P. Geffre, P.R. Finkbone, C.L. Bliss, D.S. Margolis, J.A. Szivek, Load measurement accuracy from sensate scaffolds with and without a cartilage surface. J. Investig. Surg. Off. J. Acad. Surg. Res. **23**(3), 156–162 (2010)
142. E.L. Tan, B.D. Pereles, B. Horton, R. Shao, M. Zourob, K.G. Ong, Implantable biosensors for real-time strain and pressure monitoring. Sensors **8**(10), 6396–6406 (2008)
143. L. Yu, B.J. Kim, E. Meng, Chronically implanted pressure sensors: challenges and state of the field. Sensors **14**(11), 20620–20644 (2014)
144. P.R. Pfau et al., Sphincter of Oddi manometry. Gastrointest. Endosc. **74**(6), 1175–1180 (2011)
145. R. Tan et al., Development of a fully implantable wireless pressure monitoring system. Biomed. Microdevices **11**(1), 259–264 (2008)
146. J. Melgaard, N.J.M. Rijkhoff, Detecting the onset of urinary bladder contractions using an implantable pressure sensor. IEEE Trans. Neural Syst. Rehabil. Eng. **19**(6), 700–708 (2011)
147. P. Bingger, M. Zens, P. Woias, Highly flexible capacitive strain gauge for continuous long-term blood pressure monitoring. Biomed. Microdevices **14**(3), 573–581 (2012)

148. "St. Jude Medical CardioMEMS™ HF System." [Online]. Available: http://www.sjm.com/cardiomems
149. "St. Jude Medical CardioMEMS HF System Prompts Changes That Improve Heart Failure Management and Reduce Hospitalizations | Business Wire," 04-Apr–2016. [Online]. Available: http://www.businesswire.com/news/home/20160404005737/en/St.-Jude-Medical-CardioMEMS-HF-System-Prompts. Accessed 07 Feb 2017
150. "Merit Sensor BP SERIES." [Online]. Available: https://meritsensor.com/products/bp-series/ . Accessed 28 Jan 2016
151. K.-H. Shin, C.-R. Moon, T.-H. Lee, C.-H. Lim, Y.-J. Kim, Flexible wireless pressure sensor module. Sens. Actuators Phys. **123–124**, 30–35 (2005)
152. P. Cong, N. Chaimanonart, W.H. Ko, D.J. Young, A wireless and batteryless 10-bit implantable blood pressure sensing microsystem with adaptive RF powering for real-time laboratory mice monitoring. IEEE J. Solid-State Circuits **44**(12), 3631–3644 (2009)
153. N.J. Cleven et al., A novel fully implantable wireless sensor system for monitoring hypertension patients. IEEE Trans. Biomed. Eng. **59**(11), 3124–3130 (2012)
154. O.H. Murphy et al., Continuous in vivo blood pressure measurements using a fully implantable wireless SAW sensor. Biomed. Microdevices **15**(5), 737–749 (2013)
155. C.-C. Chiang, C.-C.K. Lin, M.-S. Ju, An implantable capacitive pressure sensor for biomedical applications. Sens. Actuators Phys. **134**(2), 382–388 (2007)
156. M.K. Filippidou, E. Tegou, V. Tsouti, S. Chatzandroulis, A flexible strain sensor made of graphene nanoplatelets/polydimethylsiloxane nanocomposite. Microelectron. Eng. **142**, 7–11 (2015)
157. K.G. Ong, C.A. Grimes, A resonant printed-circuit sensor for remote query monitoring of environmental parameters. Smart Mater. Struct. **9**(4), 421–428 (2000)
158. J.C.-H. Lin, Y. Zhao, P.-J. Chen, M. Humayun, Y.-C. Tai, Feeling the pressure: a parylene-based intraocular pressure sensor. IEEE Nanotechnol. Mag. **6**(3), 8–16 (2012)
159. P.-J. Chen, D.C. Rodger, S. Saati, M.S. Humayun, Y.-C. Tai, Microfabricated implantable parylene-based wireless passive intraocular pressure sensors. J. Microelectromechanical Syst. **17**(6), 1342–1351 (2008)
160. P.-J. Chen, S. Saati, R. Varma, M.S. Humayun, Y.-C. Tai, Wireless intraocular pressure sensing using microfabricated minimally invasive flexible-coiled LC sensor implant. J. Microelectromechanical Syst. **19**(4), 721–734 (2010)
161. P.K. Eide, A. Bakken, The baseline pressure of intracranial pressure (ICP) sensors can be altered by electrostatic discharges. Biomed. Eng. OnLine **10**(1), 75 (2011)
162. "Codman Neuro ICP EXPRESS® Monitoring System." [Online]. Available: https://www.depuysynthes.com/hcp/codman-neuro/products/qs/ICP-EXPRESS-Monitoring-System. Accessed 28 Jan 2016
163. "Integra Camino Intracranial Pressure Monitor." [Online]. Available: http://www.integralife.com/index.aspx?redir=detailproduct&Product=757&ProductName=Integra%AE%20Camino%AE%20Intracranial%20Pressure%20Monitor%20%28CAM02%29&ProductLineName=ICP%20Monitoring&ProductLineID=11&PA=neurosurgeon. Accessed: 28 Jan 2016
164. "Integra Camino Intracranial Pressure Monitor Brochure."
165. "Integra Camino Intracranial Pressure Monitoring Kit Brochure."
166. S.F. Cogan, Neural stimulation and recording electrodes. Annu. Rev. Biomed. Eng. **10**(1), 275–309 (2008)
167. D.J. Edell, V.V. Toi, V.M. McNeil, L.D. Clark, Factors influencing the biocompatibility of insertable silicon microshafts in cerebral cortex. IEEE Trans. Biomed. Eng. **39**(6), 635–643 (1992)
168. D.H. Szarowski et al., Brain responses to micro-machined silicon devices. Brain Res. **983**(1–2), 23–35 (2003)
169. R. Biran, D.C. Martin, P.A. Tresco, Neuronal cell loss accompanies the brain tissue response to chronically implanted silicon microelectrode arrays. Exp. Neurol. **195**(1), 115–126 (2005)

170. J.K. Niparko, R.A. Altschuler, J.A. Wiler, X. Xue, D.J. Anderson, Surgical implantation and biocompatibility of central nervous system auditory prostheses. Ann. Otol. Rhinol. Laryngol. **98**(12), 965–970 (1989)
171. Y.-T. Kim, R.W. Hitchcock, M.J. Bridge, P.A. Tresco, Chronic response of adult rat brain tissue to implants anchored to the skull. Biomaterials **25**(12), 2229–2237 (2004)
172. M.B.A. Fontes, Electrodes for bio-application: recording and stimulation. J. Phys: Conf. Ser. **421**(1), 012019 (2013)
173. A.R. Harris, S.J. Morgan, J. Chen, R.M.I. Kapsa, G.G. Wallace, A.G. Paolini, Conducting polymer coated neural recording electrodes. J. Neural Eng. **10**(1), 016004 (2013)
174. F. Vitale, S.R. Summerson, B. Aazhang, C. Kemere, M. Pasquali, Neural stimulation and recording with bidirectional, soft carbon nanotube fiber microelectrodes. ACS Nano **9**(4), 4465–4474 (2015)
175. Z. Lertmanorat, F.W. Montague, D.M. Durand, A flat interface nerve electrode with integrated multiplexer. IEEE Trans. Neural Syst. Rehabil. Eng. **17**(2), 176–182 (2009)
176. K. Najafi, Solid-state microsensors for cortical nerve recordings. IEEE Eng. Med. Biol. Mag. **13**(3), 375–387 (1994)
177. C.G. Herrera, A.R. Adamantidis, An integrated microprobe for the brain. Nat. Biotechnol. **33** (3), 259–260 (2015)
178. A.N. van den Pol, Neuropeptide transmission in brain circuits. Neuron **76**(1), 98–115 (2012)
179. A. Canales et al., Multifunctional fibers for simultaneous optical, electrical and chemical interrogation of neural circuits in vivo. Nat. Biotechnol. **33**(3), 277–284 (2015)
180. P. Fattahi, G. Yang, G. Kim, M.R. Abidian, A review of organic and inorganic biomaterials for neural interfaces. Adv. Mater. **26**(12), 1846–1885 (2014)
181. M.R. Abidian, D.C. Martin, Experimental and theoretical characterization of implantable neural microelectrodes modified with conducting polymer nanotubes. Biomaterials **29**(9), 1273–1283 (2008)
182. X. Cui, D.C. Martin, Electrochemical deposition and characterization of poly (3,4-ethylenedioxythiophene) on neural microelectrode arrays. Sens. Actuators B Chem. **89**(1–2), 92–102 (2003)
183. L.R. Hochberg et al., Neuronal ensemble control of prosthetic devices by a human with tetraplegia. Nature **442**(7099), 164–171 (2006)
184. S. Negi, R. Bhandari, L. Rieth, R. Van Wagenen, F. Solzbacher, Neural electrode degradation from continuous electrical stimulation: comparison of sputtered and activated iridium oxide. J. Neurosci. Methods **186**(1), 8–17 (2010)
185. A. Goryu, R. Numano, A. Ikedo, M. Ishida, T. Kawano, Nanoscale tipped microwire arrays enhance electrical trap and depth injection of nanoparticles. Nanotechnology **23**(41), 415301 (2012)
186. T. Stieglitz, Development of a micromachined epiretinal vision prosthesis. J. Neural Eng. **6** (6), 065005 (2009)
187. D.C. Rodger et al., Flexible parylene-based multielectrode array technology for high-density neural stimulation and recording. Sens. Actuators B Chem. **132**(2), 449–460 (2008)
188. B. Rubehn, C. Bosman, R. Oostenveld, P. Fries, T. Stieglitz, A MEMS-based flexible multichannel ECoG-electrode array. J. Neural Eng. **6**(3), 036003 (2009)
189. D.-H. Kim et al., Dissolvable films of silk fibroin for ultrathin conformal bio-integrated electronics. Nat. Mater. **9**(6), 511–517 (2010)
190. S.P. Lacour et al., Flexible and stretchable micro-electrodes for in vitro and in vivo neural interfaces. Med. Biol. Eng. Comput. **48**(10), 945–954 (2010)
191. C. Hassler, T. Boretius, T. Stieglitz, Polymers for neural implants. J. Polym. Sci., Part B: Polym. Phys. **49**(1), 18–33 (2011)
192. P. Negredo, J. Castro, N. Lago, X. Navarro, C. Avendaño, Differential growth of axons from sensory and motor neurons through a regenerative electrode: A stereological, retrograde tracer, and functional study in the rat. Neuroscience **128**(3), 605–615 (2004)

193. K.W. Meacham, R.J. Giuly, L. Guo, S. Hochman, S.P. DeWeerth, A lithographically-patterned, elastic multi-electrode array for surface stimulation of the spinal cord. Biomed. Microdevices **10**(2), 259–269 (2008)
194. J.D. MacDonald, K.J. Fisher, Technique for steering spinal cord stimulator electrode. Oper. Neurosurg. **69**(1), ons83–ons87 (2011)
195. X. Kang, J.Q. Liu, H. Tian, B. Yang, Y. Nuli, C. Yang, Self-closed parylene cuff electrode for peripheral nerve recording. J. Microelectromechanical Syst. **24**(2), 319–332 (2015)
196. H. Yu, W. Xiong, H. Zhang, W. Wang, Z. Li, A parylene self-locking cuff electrode for peripheral nerve stimulation and recording. J. Microelectromechanical Syst. **23**(5), 1025–1035 (2014)
197. A. Branner, R.B. Stein, R.A. Normann, Selective stimulation of cat sciatic nerve using an array of varying-length microelectrodes. J. Neurophysiol. **85**(4), 1585–1594 (2001)
198. J. Zariffa, M.K. Nagai, Z.J. Daskalakis, M.R. Popovic, Influence of the number and location of recording contacts on the selectivity of a nerve cuff electrode. IEEE Trans. Neural Syst. Rehabil. Eng. **17**(5), 420–427 (2009)
199. S.M. Lawrence, G.S. Dhillon, K.W. Horch, Fabrication and characteristics of an implantable, polymer-based, intrafascicular electrode. J. Neurosci. Methods **131**(1–2), 9–26 (2003)
200. S.M. Lawrence, G.S. Dhillon, W. Jensen, K. Yoshida, K.W. Horch, Acute peripheral nerve recording Characteristics of polymer-based longitudinal intrafascicular electrodes. IEEE Trans. Neural Syst. Rehabil. Eng. **12**(3), 345–348 (2004)
201. K. Warwick et al., The application of implant technology for cybernetic systems. Arch. Neurol. **60**(10), 1369–1373 (2003)
202. M.T. Alt, E. Fiedler, L. Rudmann, J.S. Ordonez, P. Ruther, T. Stieglitz, Let there be light - optoprobes for neural implants. Proc. IEEE **105**(1), 101–138 (2017)
203. M. HajjHassan, V. Chodavarapu, S. Musallam, NeuroMEMS: neural probe microtechnologies. Sensors **8**(10), 6704–6726 (2008)
204. K.D. Wise, D.J. Anderson, J.F. Hetke, D.R. Kipke, K. Najafi, Wireless implantable microsystems: high-density electronic interfaces to the nervous system. Proc. IEEE **92**(1), 76–97 (2004)
205. E.M. Maynard, C.T. Nordhausen, R.A. Normann, The Utah intracortical electrode array: a recording structure for potential brain-computer interfaces. Electroencephalogr. Clin. Neurophysiol. **102**(3), 228–239 (1997)
206. G.G. Naples, J.T. Mortimer, A. Scheiner, J.D. Sweeney, A spiral nerve cuff electrode for peripheral nerve stimulation. IEEE Trans. Biomed. Eng. **35**(11), 905–916 (1988)
207. J. Viventi et al., Flexible, foldable, actively multiplexed, high-density electrode array for mapping brain activity in vivo. Nat. Neurosci. **14**(12), 1599–1605 (2011)
208. A.B. Schwartz, X.T. Cui, D.J. Weber, D.W. Moran, Brain-controlled interfaces: movement restoration with neural prosthetics. Neuron **52**(1), 205–220 (2006)
209. M. Kindlundh, P. Norlin, U.G. Hofmann, A neural probe process enabling variable electrode configurations. Sens. Actuators B Chem. **102**(1), 51–58 (2004)
210. K.-K. Lee et al., Polyimide-based intracortical neural implant with improved structural stiffness. J. Micromechanics Microengineering **14**(1), 32–37 (2004)
211. P.J. Rousche, D.S. Pellinen, D.P. Pivin, J.C. Williams, R.J. Vetter, D.R. Kipke, Flexible polyimide-based intracortical electrode arrays with bioactive capability. IEEE Trans. Biomed. Eng. **48**(3), 361–371 (2001)
212. A. Mercanzini et al., Demonstration of cortical recording using novel flexible polymer neural probes. Sens. Actuators Phys. **143**(1), 90–96 (2008)
213. A.C. Patil, N.V. Thakor, Implantable neurotechnologies: a review of micro- and nanoelectrodes for neural recording. Med. Biol. Eng. Comput. **54**(1), 23–44 (2016)
214. S. Musallam, M.J. Bak, P.R. Troyk, R.A. Andersen, A floating metal microelectrode array for chronic implantation. J. Neurosci. Methods **160**(1), 122–127 (2007)
215. R.A. Normann, E.M. Maynard, P.J. Rousche, D.J. Warren, A neural interface for a cortical vision prosthesis. Vision. Res. **39**(15), 2577–2587 (1999)

216. A. Jackson, E.E. Fetz, Compact movable microwire array for long-term chronic unit recording in cerebral cortex of primates. J. Neurophysiol. **98**(5), 3109–3118 (2007)
217. G. Lehew, M.A.L. Nicolelis, State-of-the-art microwire array design for chronic neural recordings in behaving animals, in *Methods for Neural Ensemble Recordings, Second Edition* (CRC Press, Boca Raton, FL, 2007), pp. 1–20
218. D.A. Schwarz et al., Chronic, wireless recordings of large-scale brain activity in freely moving rhesus monkeys. Nat. Methods **11**(6), 670–676 (2014)
219. T.D.Y. Kozai et al., Ultrasmall implantable composite microelectrodes with bioactive surfaces for chronic neural interfaces. Nat. Mater. **11**(12), 1065–1073 (2012)
220. D.R. Kipke, R.J. Vetter, J.C. Williams, J.F. Hetke, Silicon-substrate intracortical microelectrode arrays for long-term recording of neuronal spike activity in cerebral cortex. IEEE Trans. Neural Syst. Rehabil. Eng. **11**(2), 151–155 (2003)
221. P.J. Rousche, R.A. Normann, Chronic recording capability of the Utah intracortical electrode array in cat sensory cortex. J. Neurosci. Methods **82**(1), 1–15 (1998)
222. Å.B. Vallbo, K.-E. Hagbarth, Activity from skin mechanoreceptors recorded percutaneously in awake human subjects. Exp. Neurol. **21**(3), 270–289 (1968)
223. N. Lago, K. Yoshida, K.P. Koch, X. Navarro, Assessment of biocompatibility of chronically implanted polyimide and platinum intrafascicular electrodes. IEEE Trans. Biomed. Eng. **54** (2), 281–290 (2007)
224. T. Boretius et al., A transverse intrafascicular multichannel electrode (TIME) to interface with the peripheral nerve. Biosens. Bioelectron. **26**(1), 62–69 (2010)
225. H.A.C. Wark et al., A new high-density (25 electrodes/mm^2) penetrating microelectrode array for recording and stimulating sub-millimeter neuroanatomical structures. J. Neural Eng. **10**(4), 045003 (2013)
226. H. Xu et al., Conductive PPY/PDLLA conduit for peripheral nerve regeneration. Biomaterials **35**(1), 225–235 (2014)
227. P.B. Yoo, D.M. Durand, Selective recording of the canine hypoglossal nerve using a multicontact flat interface nerve electrode. IEEE Trans. Biomed. Eng. **52**(8), 1461–1469 (2005)
228. W.M. Grill, Jr., M.D. Tarler, J.T. Mortimer, Implantable helical spiral cuff electrode, US5505201 A, 09 Apr 1996
229. T. Stieglitz, M. Schuetter, K.P. Koch, Implantable biomedical microsystems for neural prostheses. IEEE Eng. Med. Biol. Mag. **24**(5), 58–65 (2005)
230. G. Márton et al., In vivo measurements with robust silicon-based multielectrode arrays with extreme shaft lengths. IEEE Sens. J. **13**(9), 3263–3269 (2013)
231. K.D. Wise, A.M. Sodagar, Y. Yao, M.N. Gulari, G.E. Perlin, K. Najafi, Microelectrodes, microelectronics, and implantable neural microsystems. Proc. IEEE **96**(7), 1184–1202 (2008)
232. N. Xue et al., Polymeric C-shaped cuff electrode for recording of peripheral nerve signal. Sens. Actuators B Chem. **210**, 640–648 (2015)
233. S.B. Brummer, L.S. Robblee, F.T. Hambrecht, Criteria for selecting electrodes for electrical stimulation: theoretical and practical considerations. Ann. N. Y. Acad. Sci. **405**(1), 159–171 (1983)
234. G.E. Loeb, R.A. Peck, Cuff electrodes for chronic stimulation and recording of peripheral nerve activity. J. Neurosci. Methods **64**(1), 95–103 (1996)
235. K.H. Polasek, H.A. Hoyen, M.W. Keith, D.J. Tyler, Human nerve stimulation thresholds and selectivity using a multi-contact nerve cuff electrode. IEEE Trans. Neural Syst. Rehabil. Eng. **15**(1), 76–82 (2007)
236. M.A. Lebedev, M.A.L. Nicolelis, Brain–machine interfaces: past, present and future. Trends Neurosci. **29**(9), 536–546 (2006)
237. L. Kenney et al., An implantable two channel drop foot stimulator: initial clinical results. Artif. Organs **26**(3), 267–270 (2002)
238. D.J. Tyler, D.M. Durand, A slowly penetrating interfascicular nerve electrode for selective activation of peripheral nerves. IEEE Trans. Rehabil. Eng. **5**(1), 51–61 (1997)

239. A. Kundu, K.R. Harreby, K. Yoshida, T. Boretius, T. Stieglitz, W. Jensen, Stimulation selectivity of the 'thin-film longitudinal intrafascicular electrode' (tfLIFE) and the 'transverse intrafascicular multi-channel electrode' (TIME) in the large nerve animal model. IEEE Trans. Neural Syst. Rehabil. Eng. **22**(2), 400–410 (2014)
240. S. Raspopovic, M. Capogrosso, S. Micera, A computational model for the stimulation of rat sciatic nerve using a transverse intrafascicular multichannel electrode. IEEE Trans. Neural Syst. Rehabil. Eng. **19**(4), 333–344 (2011)
241. X. Navarro, T.B. Krueger, N. Lago, S. Micera, T. Stieglitz, P. Dario, A critical review of interfaces with the peripheral nervous system for the control of neuroprostheses and hybrid bionic systems. J. Peripher. Nerv. Syst. **10**(3), 229–258 (2005)
242. L.A. Geddes, R. Roeder, Criteria for the selection of materials for implanted electrodes. Ann. Biomed. Eng. **31**(7), 879–890 (2003)
243. S.H. Cho, H.M. Lu, L. Cauller, M.I. Romero-Ortega, J.B. Lee, G.A. Hughes, Biocompatible SU-8-based microprobes for recording neural spike signals from regenerated peripheral nerve fibers. IEEE Sens. J. **8**(11), 1830–1836 (2008)
244. S. Takeuchi, D. Ziegler, Y. Yoshida, K. Mabuchi, T. Suzuki, Parylene flexible neural probes integrated with microfluidic channels. Lab Chip **5**(5), 519–523 (2005)
245. S. Takeuchi, T. Suzuki, K. Mabuchi, H. Fujita, 3D flexible multichannel neural probe array. J. Micromechanics Microengineering **14**(1), 104–107 (2004)
246. S.E. Lee et al., A flexible depth probe using liquid crystal polymer. IEEE Trans. Biomed. Eng. **59**(7), 2085–2094 (2012)
247. K. Lee, J. He, R. Clement, S. Massia, B. Kim, Biocompatible benzocyclobutene (BCB)-based neural implants with micro-fluidic channel. Biosens. Bioelectron. **20**(2), 404–407 (2004)
248. P.E.K. Donaldson, Aspects of silicone rubber as an encapsulant for neurological prostheses. Med. Biol. Eng. Comput. **29**(1), 34–39 (1991)
249. N.D. Donaldson, P.E.K. Donaldson, When are actively balanced biphasic ('Lilly') stimulating pulses necessary in a neurological prosthesis? I historical background; Pt resting potential; Q studies. Med. Biol. Eng. Comput. **24**(1), 41–49 (1986)
250. J.O. Larsen, M. Thomsen, M. Haugland, T. Sinkjær, Degeneration and regeneration in rabbit peripheral nerve with long-term nerve cuff electrode implant: a stereological study of myelinated and unmyelinated axons. Acta Neuropathol. (Berl.) **96**(4), 365–378 (1998)
251. B. Upshaw, T. Sinkjaer, Digital signal processing algorithms for the detection of afferent nerve activity recorded from cuff electrodes. IEEE Trans. Rehabil. Eng. **6**(2), 172–181 (1998)
252. I.F. Triantis, A. Demosthenous, N. Donaldson, On cuff imbalance and tripolar ENG amplifier configurations. IEEE Trans. Biomed. Eng. **52**(2), 314–320 (2005)
253. A. Demosthenous, J. Taylor, I.F. Triantis, R. Rieger, N. Donaldson, Design of an adaptive interference reduction system for nerve-cuff electrode recording. IEEE Trans. Circuits Syst. Regul. Pap. **51**(4), 629–639 (2004)
254. L.N.S. Andreasen, J.J. Struijk, Signal strength versus cuff length in nerve cuff electrode recordings. IEEE Trans. Biomed. Eng. **49**(9), 1045–1050 (2002)
255. T. Stieglitz, H. Beutel, M. Schuettler, J.-U. Meyer, Micromachined, polyimide-based devices for flexible neural interfaces. Biomed. Microdevices **2**(4), 283–294 (2000)
256. D.T.T. Plachta et al., Blood pressure control with selective vagal nerve stimulation and minimal side effects. J. Neural Eng. **11**(3), 036011 (2014)
257. Z. Xiang et al., Progress of flexible electronics in neural interfacing—a self-adaptive non-invasive neural ribbon electrode for small nerves recording. Adv. Mater. **28**(22), 4472–4479 (2016)
258. E.G.R. Kim et al., 3D silicon neural probe with integrated optical fibers for optogenetic modulation. Lab Chip **15**(14), 2939–2949 (2015)
259. A.B. Kibler, B.G. Jamieson, D.M. Durand, A high aspect ratio microelectrode array for mapping neural activity in vitro. J. Neurosci. Methods **204**(2), 296–305 (2012)

260. Z. Xiang et al., Ultra-thin flexible polyimide neural probe embedded in a dissolvable maltose-coated microneedle. J. Micromechanics Microengineering **24**(6), 065015 (2014)
261. D.-W. Park et al., Graphene-based carbon-layered electrode array technology for neural imaging and optogenetic applications. Nat. Commun. **5**, 5258 (2014)
262. D.J. Tyler, D.M. Durand, Functionally selective peripheral nerve stimulation with a flat interface nerve electrode. IEEE Trans. Neural Syst. Rehabil. Eng. **10**(4), 294–303 (2002)
263. C. Veraart, W.M. Grill, J.T. Mortimer, Selective control of muscle activation with a multipolar nerve cuff electrode. IEEE Trans. Biomed. Eng. **40**(7), 640–653 (1993)
264. D.N. Heo et al., Multifunctional hydrogel coatings on the surface of neural cuff electrode for improving electrode-nerve tissue interfaces. Acta Biomater. **39**, 25–33 (2016)
265. S.J. Park et al., Functional nerve cuff electrode with controllable anti-inflammatory drug loading and release by biodegradable nanofibers and hydrogel deposition. Sens. Actuators B Chem. **215**, 133–141 (2015)
266. J. Seo, J.H. Wee, J.H. Park, P. Park, J.-W. Kim, S.J. Kim, Nerve cuff electrode using embedded magnets and its application to hypoglossal nerve stimulation. J. Neural Eng. **13** (6), 066014 (2016)
267. Y.J. Lee, H.-J. Kim, S.H. Do, J.Y. Kang, S.H. Lee, Characterization of nerve-cuff electrode interface for biocompatible and chronic stimulating application. Sens. Actuators B Chem. **237**, 924–934 (2016)
268. S. Lee et al., Selective stimulation and neural recording on peripheral nerves using flexible split ring electrodes. Sens. Actuators B Chem. **242**, 1165–1170 (2017)
269. S. Nag, N.V. Thakor, Implantable neurotechnologies: electrical stimulation and applications. Med. Biol. Eng. Comput. **54**(1), 63–76 (2016)
270. S. Vaddiraju, I. Tomazos, D.J. Burgess, F.C. Jain, F. Papadimitrakopoulos, Emerging synergy between nanotechnology and implantable biosensors: a review. Biosens. Bioelectron. **25**(7), 1553–1565 (2010)
271. G.S. Wilson, M. Ammam, In vivo biosensors. FEBS J. **274**(21), 5452–5461 (2007)
272. G.S. Wilson, R. Gifford, Biosensors for real-time in vivo measurements. Biosens. Bioelectron. **20**(12), 2388–2403 (2005)
273. H. Cao et al., An implantable, batteryless, and wireless capsule with integrated impedance and pH sensors for gastroesophageal reflux monitoring. IEEE Trans. Biomed. Eng. **59**(11), 3131–3139 (2012)
274. H.-J. Chung et al., Stretchable, multiplexed pH sensors with demonstrations on rabbit and human hearts undergoing ischemia. Adv. Healthc. Mater. **3**(1), 59–68 (2014)
275. E. Lindner et al., In vivo and in vitro testing of microelectronically fabricated planar sensors designed for applications in cardiology. Fresenius J. Anal. Chem. **346**(6–9), 584–588 (1993)
276. V.V. Cosofret, E. Lindner, T.A. Johnson, M.R. Neuman, Planar micro sensors for in vivo myocardial pH measurements. Talanta **41**(6), 931–938 (1994)
277. V.V. Cosofret, M. Erdosy, T.A. Johnson, R.P. Buck, R.B. Ash, M.R. Neuman, Microfabricated sensor arrays sensitive to pH and K^+ for ionic distribution measurements in the beating heart. Anal. Chem. **67**(10), 1647–1653 (1995)
278. S.A.M. Marzouk, S. Ufer, R.P. Buck, T.A. Johnson, L.A. Dunlap, W.E. Cascio, Electrodeposited iridium oxide pH electrode for measurement of extracellular myocardial acidosis during acute ischemia. Anal. Chem. **70**(23), 5054–5061 (1998)
279. J.L. Gonzalez-Guillaumin, D.C. Sadowski, K.V.I.S. Kaler, M.P. Mintchev, Ingestible capsule for impedance and pH monitoring in the esophagus. IEEE Trans. Biomed. Eng. **54** (12), 2231–2236 (2007)
280. I.B. Tahirbegi, M. Mir, J. Samitier, Real-time monitoring of ischemia inside stomach. Biosens. Bioelectron. **40**(1), 323–328 (2013)
281. I.B. Tahirbegi, M. Mir, S. Schostek, M. Schurr, J. Samitier, In vivo ischemia monitoring array for endoscopic surgery. Biosens. Bioelectron. **61**, 124–130 (2014)
282. M. Mir, R. Lugo, I.B. Tahirbegi, J. Samitier, Miniaturizable ion-selective arrays based on highly stable polymer membranes for biomedical applications. Sensors **14**(7), 11844–11854 (2014)

283. M. Kubon et al., A microsensor system to probe physiological environments and tissue response, in *2010 IEEE Sensors*, (2010), pp. 2607–2611
284. A. Weltin, B. Enderle, J. Kieninger, G.A. Urban, Multiparametric, flexible microsensor platform for metabolic monitoring. IEEE Sens. J. **14**(10), 3345–3351 (2014)
285. G. Urban et al., Miniaturized multi-enzyme biosensors integrated with pH sensors on flexible polymer carriers for in vivo applications. Biosens. Bioelectron. **7**(10), 733–739 (1992)
286. A. Guiseppi-Elie, S. Brahim, G. Slaughter, K.R. Ward, Design of a subcutaneous implantable biochip for monitoring of glucose and lactate. IEEE Sens. J. **5**(3), 345–355 (2005)
287. A.R.A. Rahman, G. Justin, A. Guiseppi-Elie, Towards an implantable biochip for glucose and lactate monitoring using microdisc electrode arrays (MDEAs). Biomed. Microdevices **11**(1), 75–85 (2008)
288. A.R.A. Rahman, G. Justin, A. Guiseppi-Wilson, A. Guiseppi-Elie, Fabrication and packaging of a dual sensing electrochemical biotransducer for glucose and lactate useful in intramuscular physiologic status monitoring. IEEE Sens. J. **9**(12), 1856–1863 (2009)
289. C.N. Kotanen, A. Guiseppi-Elie, Characterization of a wireless potentiostat for integration with a novel implantable biotransducer. IEEE Sens. J. **14**(3), 768–776 (2014)
290. M. Stanacevic, K. Murari, A. Rege, G. Cauwenberghs, N.V. Thakor, VLSI potentiostat array with oversampling gain modulation for wide-range neurotransmitter sensing. IEEE Trans. Biomed. Circuits Syst. **1**(1), 63–72 (2007)
291. M. Mollazadeh, K. Murari, G. Cauwenberghs, N. Thakor, Wireless micropower instrumentation for multimodal acquisition of electrical and chemical neural activity. IEEE Trans. Biomed. Circuits Syst. **3**(6), 388–397 (2009)
292. M. Roham et al., A wireless IC for wide-range neurochemical monitoring using amperometry and fast-scan cyclic voltammetry. IEEE Trans. Biomed. Circuits Syst. **2**(1), 3–9 (2008)
293. G. Massicotte, S. Carrara, G. Di Micheli, M. Sawan, A CMOS amperometric system for multi-neurotransmitter detection. *IEEE Trans. Biomed. Circuits Syst.*, **10**(3), 731–741 (2016)
294. P. Valdastri et al., An implantable ZigBee ready telemetric platform for in vivo monitoring of physiological parameters. Sens. Actuators Phys. **142**(1), 369–378 (2008)
295. H. Liu et al., An implantable radio-telemetry system for detecting multiple bio-parameters of a small animal based on wireless energy transmission. Mechatronics **28**, 18–26 (2015)
296. S. Carrara et al., Remote system for monitoring animal models with single-metabolite bio-nano-sensors. IEEE Sens. J. **13**(3), 1018–1024 (2013)
297. A. Cavallini et al., A subcutaneous biochip for remote monitoring of human metabolism: packaging and biocompatibility assessment. IEEE Sens. J. **15**(1), 417–424 (2015)
298. C. Baj-Rossi et al., Full fabrication and packaging of an implantable multi-panel device for monitoring of metabolites in small animals. IEEE Trans. Biomed. Circuits Syst. **8**(5), 636–647 (2014)
299. F. Xu, G. Yan, K. Zhao, L. Lu, J. Gao, G. Liu, A wireless capsule system with ASIC for monitoring the physiological signals of the human gastrointestinal tract. IEEE Trans. Biomed. Circuits Syst. **8**(6), 871–880 (2014)
300. L. Lu, G. Yan, K. Zhao, F. Xu, An implantable telemetry platform system with ASIC for in vivo monitoring of gastrointestinal physiological information. IEEE Sens. J. **15**(6), 3524–3534 (2015)
301. W.P. Chan et al., A monolithically integrated pressure/oxygen/temperature sensing SoC for multimodality intracranial neuromonitoring. IEEE J. Solid-State Circuits **49**(11), 2449–2461 (2014)
302. C.N. Kotanen, A. Guiseppi-Elie, Monitoring systems and quantitative measurement of biomolecules for the management of trauma. Biomed. Microdevices **15**(3), 561–577 (2013)

Chapter 4
Sensor Embodiment and Flexible Electronics

P. Kassanos, S. Anastasova, C. M. Chen and Guang-Zhong Yang

List of Acronyms

3D-MIN	3D multifunctional integumentary membrane
ASIC	Application-specific integrated circuit
CAD	Computer aided design
CDG	Chemically derived graphene
CINE	Combination of interconnects and electronics
CMOS	Complementary metal-oxide-semiconductor
CNT	Carbon nanotubes
CP	Conductive polymer
CPC	Conductive polymer composite
CVD	Chemical vapor deposition
DAC	Digital to analog converter
DoD	Droplets-on-demand
ECG	Electrocardiogram
EEG	Electroencephalogram
EMG	Electromyogram
EOG	Electro-oculogram
FBR	Foreign body response
FDA	Food and Drug Administration
FEM	Finite element method
FET	Field-effect transistor
FLEPS	Flexible letterpress stamping
FLG	Few layer graphene
FPC/FPCB	Flexible printed circuit board
FR-4	Flame retardant type four
GI	Gastrointestinal
GO_x	Glucose oxidase
GO	Graphene oxide

P. Kassanos (✉) · S. Anastasova · C. M. Chen · G.-Z. Yang
The Hamlyn Centre, Imperial College London, London, UK
e-mail: p.kassanos@imperial.ac.uk

G.-Z. Yang (ed.), *Implantable Sensors and Systems*
https://doi.org/10.1007/978-3-319-69748-2_4

h-BN	Hexagonal boron nitride
HF	Hydrogen fluoride
HMDS	Hexamethyldisiloxane
IC	Integrated circuit
ICP	Intrinsically conductive polymer
ISO	International Organization for Standardization
LCP	Liquid crystal polymer
LED	Light emitting diode
μCP	Micro contact printing
MEMS	Micro-electro-mechanical-system
MMD	Multilevel matrix deposition
MMIC	Micromolding in capillaries
MOEMS	Micro-opto-electro-mechanical system
MOS	Metal-oxide-semiconductor
MOSFET	Metal-oxide-semiconductor field-effect transistor
MRI	Magnetic resonance imaging
MSM	Metal-semiconductor-metal
MTTF	Mean-time-to-failure
NIL	Nanoimprint lithography
NM	Nano-membrane
NO	Nitric oxide
NP	Nanoparticle
OFET	Organic field-effect transistor
OLED	Organic light emitting diode
OTFT	Organic thin-film transistors
OVPD	Organic vapor phase deposition
PA	Polyacetylene
PANI	Polyaniline
PAR	Polyarylate
PBS	Phosphate buffered saline
PC	Polycarbonate
PCB	Printed circuit board
PCL	Polycaprolactone
PDGF	Platelet-derived growth factor
PDMS	Polydimethylsiloxane
PEDOT	Poly (3,4-ethylene dioxythiophene)
PEDOT:PSS	Poly(3,4-ethylenedioxythiophene) polystyrene sulfonate
PEEK	Polyetheretherketone
PEG	Polyethylene glycol
PEN	Poly(ethylene naphthalate)
PEO	Polyethylene oxide
PES	Polyethersulphone
PF	Polyfuran
PI	Polyimide

PLGA	Poly(lactic-co-glycolic acid)
PMMA	Poly (methylmethacrylate)
PNB	Polynorbornene
PPG	Photoplethysmography
PPV	Poly(p-phenylene-vinylene)
PPy	Polypyrrole
PTh	Polythiophene
PVA	Polyvinyl alcohol
PVD	Physical vapor deposition
RF	Radio frequency
RH	Relative humidity
rGO	Reduced graphene oxide
SAB	Surface activated bonding
SAM	Self-assembled monolayer
SEM	Scanning electron microscope
SMP	Shape-memory polymer
SOI	Silicon on insulator
SPR	Surface plasmon resonancel
TMDs	Transition metal dichalcogenides
TPU	Thermoplastic polyurethane
TRM	Tissue response modifier
UBM	Under bump metallization
UV	Ultraviolet
UVO	Ultraviolet/ozone
VEGF	Vascular endothelial growth factor
VTE	Vacuum thermal evaporation

4.1 Introduction

Sensor embodiment and packaging are particularly important for implantable systems. One key element is the development of flexible electronics. Traditional electronics, based on rigid silicon technologies, is associated with a number of intrinsic disadvantages. The inherent brittleness of inorganic semiconductors and stiffness of Si wafer-based devices represent a major issue when interfaced with tissues. This is because our internal organs are complex and they have innate responses to reject foreign bodies. Furthermore, tissues are soft, and they undergo constant motion and deformation. In this chapter, we will discuss current progress in flexible printed circuit board (FPC/FPCB) technologies and provide a review of new fabrication techniques and materials for making soft devices and interconnects suitable for implantable applications. Issues related to geometrical designs for mechanically resilient flexible devices, hermetic packaging, biocompatibility and encapsulation are addressed.

4.1.1 Commercial Flexible Printed Circuit Board (FPC) Technology

FPC or FPCB technologies have been commercially available for a number of decades. Most companies employ this technology in order to embed the electronics into a specific shape of enclosure of their products. In commercially available technologies, copper is the most common conductive material for high-volume, low-cost, mass-produced FPCs. Instead of using glass-reinforced flame retardant epoxy (FR-4, a rigid material) as a substrate, which is the material typically used for regular rigid PCBs, manufacturers use soft materials such as polyimide (PI, Kapton), polyester (Mylar), aramid, and laminated glass epoxy. These materials provide different electrical, mechanical, chemical, and thermal properties that make PCBs more flexible/bendable.

Polyester and PI are materials commonly used by virtue of their wide temperature range (polyester: −60 to 105 °C, PI: −200 to 300 °C), higher tensile strength (polyester: 22–28, PI: 25–30 PSI * 1000), and organic and good dielectric strength (both materials are in a range of 3400–3600 V mil^{-1}).

Figure 4.1a shows the general process of making commercial FPCs. First of all, the PI substrate and conductive layer (copper) must be properly attached to each other to form a sheet and must be chemically pre-cleaned. Following this process, the circuit artwork is patterned on a photoresist layer and after exposure to ultraviolet (UV) light and wet etching the circuit is patterned. The next step involves the use of laser cutting equipment to drill small diameter holes (vias), which are then plated with copper. After copper plating, a cover layer has to be aligned and tacked on to the panel sheet prior to a lamination step. Some designs may require an extra step in the procedure to stiffen certain locations, such as gold fingers, vias, and sensors. Finally, a tested panel sheet is punched into several circuits by highly precise punch tooling or a laser-cutting machine. An example of a two-layered FPC cross-section is shown in Fig. 4.1b.

The advantages of using commercial FPC protocols include cost-effective, large volume production and consistent electrical properties. A disadvantage is its low tensile strength due to a limited selection of available materials. Nevertheless, commercially available FPC technologies have been widely used for medical devices. One example is the bendable oxygenation sensor shown in Fig. 4.1c. Other examples include an implantable microelectrode array for extracellular stimulation of the invertebrate locomotory system presented in [1] and an implantable bioimpedance sensor presented in [2].

To extend beyond the current FPCs for flexible, stretchable, and transient electronics and devices, ultra-thin platforms with high tensile strengths have been developed in recent years. These will be discussed in the following sections.

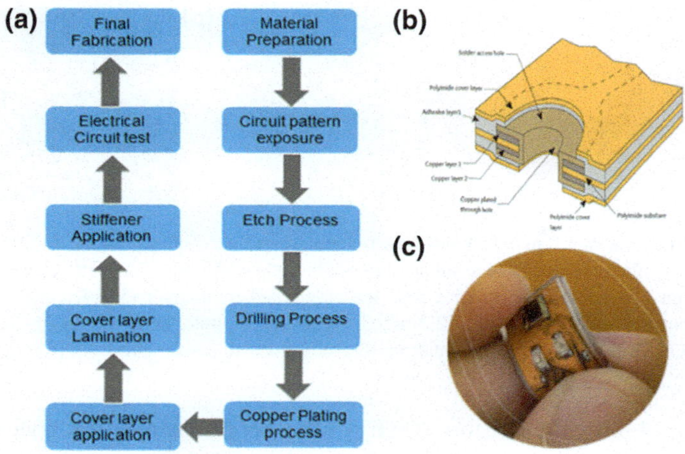

Fig. 4.1 **a** A general fabrication process of commercial FPCs. **b** A general commercial two-layered FPC process side view. **c** An example of a bendable FPC showing the Hamlyn oxygenation monitoring sensor

4.1.2 Biomedical Flexible Devices

Thus far, many flexible or semi-flexible devices have been developed for biomedical applications. They can be placed directly on the skin, subcutaneously, or on organs in the human body. In [3], a rigid PCB based microcontroller board integrated with a flexible strain gauge pressure sensor is used for respiration monitoring for newborns. A multi-sensor platform is developed in [4] for disabled patients, which uses a headband with textile electrodes to record electro-oculograms (EOG) related to eye movements and a glove integrated with flex-resistive sensors and an accelerometer to detect hand gestures. Other examples include the development of flexible photoplethysmography (PPG) for blood oxygen saturation monitoring [5, 6] and miniature ultrasound transducers [7] (Fig. 4.2).

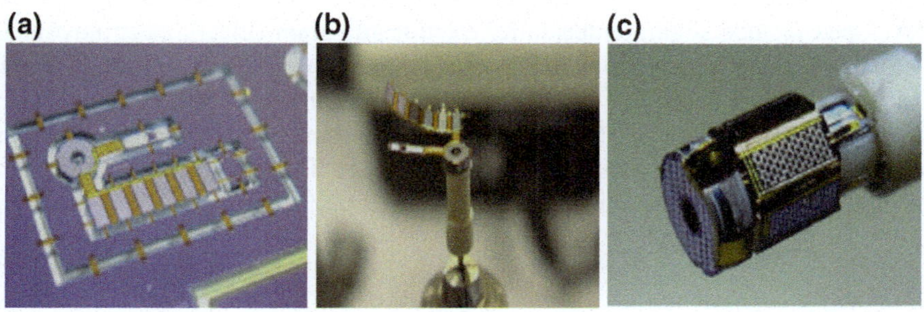

Fig. 4.2 **a** F2R technology. **b**, **c** Segmented 360° ultrasound transducer assembled on an instrument. Reprinted with permission from Ref. [7]. © IEEE 2012

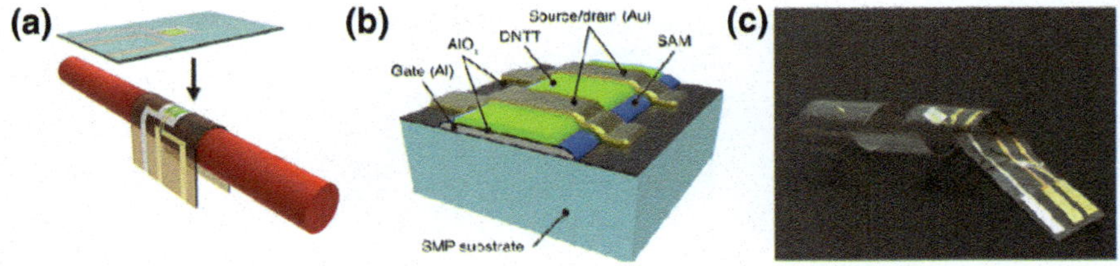

Fig. 4.3 **a** Illustration of an OTFT device wrapped around a cylindrical shape. **b** Structure of the OTFT device illustrating different substrates: SMP, gate: aluminum; dielectric layer: aluminum oxide and an n tetradecyl phosphoric acid self-assembled monolayer (SAM) combined with an organic semiconductor—dinaphtho[2,3-b:2′,3′-f]thieno[3,2-b]thiophene (DNTT); drain/source: Au. **c** The OTFT with a helical deformation. Reproduced with permission from [8]. © Wiley 2014

Organic thin-film transistors (OTFT) have been integrated on shape-memory polymer (SMP) substrate implantable devices [8]. They demonstrate shape changing and soft conformability on 3D surfaces following a temperature stimulus. Figure 4.3a illustrates the device being wrapped round a cylintrical shape, while (b) shows a cross-section of the OTFT and (c) illustrates how the OTFT is bent into a helix. Another example of flexible implant is a blood flow ultrasonic Doppler sensor from Cook Medical, which is used to monitor both arterial and venous blood flows after transplant surgery. Figure 4.4 shows the sensor, which is tethered to an external system.

4.1.2.1 Epidermal Electronics

Due to its relative ease of access, there have been many applications of flexible electronics for epidermal sensing. For example, a team at the University of Illinois at Urbana-Champaign [10–20] has developed a variety of sensing methodologies (electrical, chemical, and optical sensors) using flexible substrates. Sufficiently thin layers (~ 30 μm) and devices (<7 μm) that are flexible and stretchable are used to achieve conformal contact to the skin with sufficient adhesion by relying on van der Waals forces [11, 12]. Water-soluble polymer sheets serve as a temporary support to transfer the devices to the skin, similarly to temporary tattoos. The sensors developed include strain, temperature, electrocardiogram (ECG), electromyogram (EMG), electroencephalogram (EEG, alpha rhythms), and light emitting diodes (LED) and photodetectors for oximetry. Other work in the field includes an ultraflexible photonic skin [17] with three-color polymer LEDs, and organic photodetectors for reflective pulse oximetry and display (Fig. 4.5). The total thickness of the device (3 μm) was thinner than the epidermal layer of the skin.

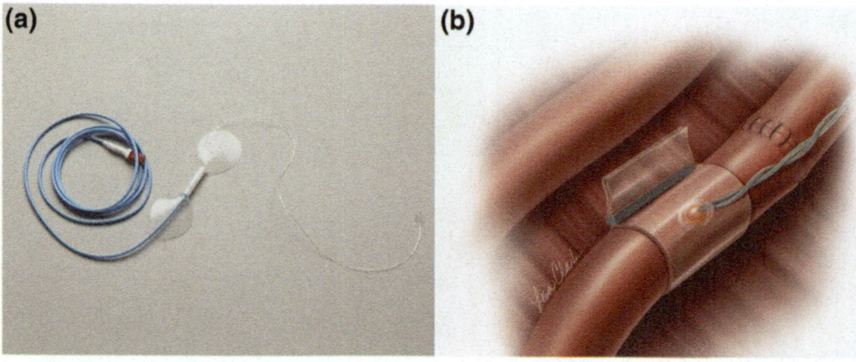

Fig. 4.4 **a** An ultrasonic Doppler device from Cook Medical. **b** Illustration of the sensor wrapped around a blood vessel after reconstruction (permission for use granted by Cook Medical, Bloomington, Indiana [9])

4.1.2.2 Catheter-Based Devices

Medtronic, Boston Scientific, and others offer commercially available and widely used clinical catheters, which allow radiofrequency ablation, laser or cryo-ablation, and mapping of cardiac electrical activities. Traditional catheters require point-by-point manual placement of the point electrodes to different regions of the tissue for ablation or electrophysiological measurements. Balloon-based catheters are a promising alternative. Similarly to epidermal electronic devices, such devices can be attached to cardiac and angioplasty balloon catheters. Balloon inflation flexes and stretches the device on the surface of the catheter to allow the sensors on its surface to achieve stable contact intraluminally. Physiological measurements can be performed during ablation or other procedures, allowing the physician to adjust and optimize the treatment during the procedure in real time. Attachment of the balloon surface to the tissue can be assessed in real time via bipolar electrodes for impedance measurement, as discussed in [21]. Multiplexing techniques can significantly reduce the wiring required. Post-processing of complete integrated circuits fabricated with commercial silicon on insulator (SOI) technologies allows transfer of the devices onto flexible substrates [15]. Such catheters have been used for epicardial recordings (electrophysiological, temperature, tactile) and ablation in rabbit hearts [16].

4.1.2.3 Large Area, Full Organ Engulfing Multisensory Systems

Apart from epidermal and catheter-based applications, similar devices have been demonstrated as multi-sensory mapping systems covering large areas of organs such as the brain [14] and heart [10, 21, 22]. These devices can conformably wrap around organs and establish good contact without the need for sutures, adhesives, or

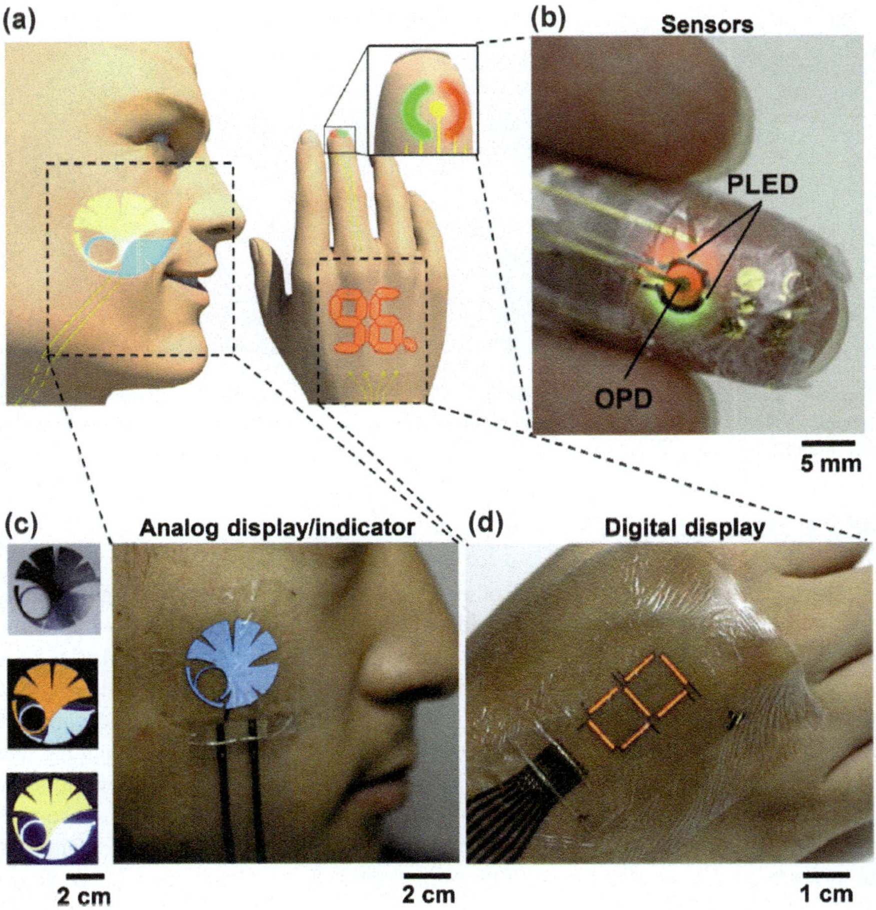

Fig. 4.5 **a** Illustration of the optoelectronic skin system. **b** The organic optical sensor attached to a finger. **c** The color-changing indicator attached to the cheek of a subject and **d** the seven-segment digital flexible skin display on a hand. From Science Advances [17]. This work is licensed under BY-NC (http://creativecommons.org/licenses/by-nc/4.0/)

fixtures [21]. In [14], the use of two metal oxide semiconductor (MOS) transistors per electrode (one for buffering, one for multiplexing) allowed the fabrication of a dense array with thousands of electrodes to sample large areas of the brain with high spatial resolution (Fig. 4.6). This subdural active electrode array was used in vivo on a cat brain to record brain activity during seizures and visual-evoked responses. 3D elastic sheets matching the geometry of the epicardium with arrays of optical, electrical, temperature, and tactile sensors were used to completely envelop a rabbit heart in [10] and were capable of moving synchronously with the underlying cardiac tissue (Fig. 4.7). This system allowed pacing of the heart and was capable of high-density electrophysiological detection of ischemia and reperfusion,

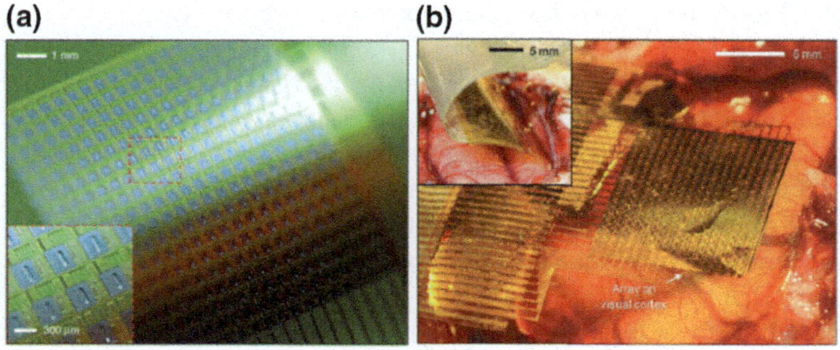

Fig. 4.6 **a** A flexible, high-resolution multiplexed electrode array of [14], comprised of 360 channels. The electrodes were 300 μm × 300 μm with a spacing of 500 μm. Inset: a closer view demonstrating a number of units. **b** The electrode array on the visual cortex of a feline model, and in the inset, the same array inserted into the interhemispheric fissure. Reprinted by permission from Macmillan Publishers Ltd on behalf of Cancer Research UK: Nature Neuroscience [14]. © 2011

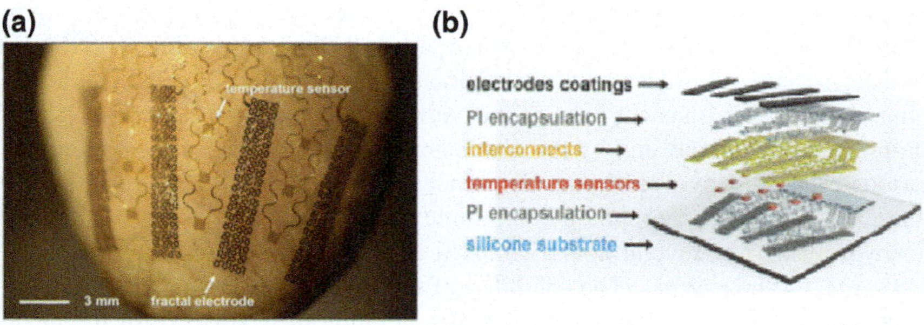

Fig. 4.7 **a** A stretchable 3D-MIM placed on a Langendorff-perfused rabbit heart. **b** The structure of the 3D-MIN. Reproduced with permission from [22]. © Wiley 2015

as well as tachycardia and temperature monitoring during cold perfusion and ablation. Similar multi-sensory systems could be fabricated for monitoring of the lungs and other organs.

Similarly to [10], in [22] 3D multifunctional integumentary membranes (3D-MIN) were presented. The fabricated device had eight electrodes that can be used to record ECG or to deliver stimulus pulses. The electrodes are electrochemically modified to reduce their contact impedance with PEDOT:PSS or nanotextured Pt–Ir. Co-integrated in the same device are nine temperature sensors/heaters for monitoring ablation procedures. These devices are essentially resistors, which can elevate the temperature to 70.8 °C to perform point contact ablation (Fig. 4.7b). A novel fractal design Greek cross (discussed in Sect. 4.4) of the electrode area allows 20% uniaxial stretchability in the horizontal direction and 15% biaxial stretchability, which permits operation of the device during the natural cycle of beating.

4.1.2.4 Transient Resorbable Electronics and Drug Release

For implantable sensing, most efforts are focused on electronics durability and extended lifetime, and in biomedical applications on biocompatibility (see Sect. 4.5.2) of the outermost layers of the device and packaging hermeticity (see Sect. 4.5.1). However, in many applications, a device may only need to be implanted for a few weeks or months. It would thus be preferable if the device could be removed without the need for a surgical operation. For certain applications (e.g. the GI tract), the device can be removed via natural orifices. With the appropriate choice of materials, devices can also be designed such that following a specific period of time and given the appropriate conditions, they disintegrate and dissolve after they have served their purpose. This is particularly important for green electronics applications, but especially useful in biomedical applications. When an implant has fulfilled its purpose the question is to whether leave it within the body or to remove it to mitigate potential future risks.

In [18], Mg, Si, and MgO and SiO_2 were used as conductive, semiconducting, and insulating materials, respectively, while silk (which is water-soluble and enzymatically degradable) served as the substrate and packaging material. Mg and Ti were used as adhesion promoters where needed. Si nano-membranes (NMs) are used for transistors, diodes, photodetectors, strain gauges, solar cells, and temperature sensors. The transient behavior of various components can be controlled by the protective materials and by the dimensions and structuring of the devices. The various processes involved in the transience of these materials depend on temperature, pH, thickness, morphology, and ionic content. Mg electrodes are consumed via hydrolysis over several hours, while Si and SiO_2 in phosphate buffered saline (PBS, pH 7.4) take weeks. The solubility of silk is programmable by controlling its crystallinity. There are two main timescales defining the transience of the device: the transience of the encapsulation and that of the Mg. For example, a transistor fabricated with the proposed process and materials of [18] can be fabricated with 1 μg of Si and can be dissolved in 30 μl of biofluid. The authors demonstrated the technology by fabricating a passive device consisting of an inductor and a resistor, which served as a heater. This was implanted under the skin of a rat and served as a thermal therapy device for surgical site infections as a transient bactericidal device. Inductive coupling powered the device, inducing, through the resistor, local temperature changes of up to 4 °C. Controlled by the crystallinity of the silk, the device transience was 15 days, during which the device sterilized the wound. Si NMs are bioresorbable via their dissolution, which involves hydrolysis and depends on pH. Experiments with such Si NM at body temperature demonstrate a dissolution of 4.5 nm day^{-1} [18]. Metal-oxide-semiconductor field-effect transistors (MOSFET) formed with such Si NM SiO_2 gate dielectrics, Mg electrodes, and MgO and crystalized silk encapsulations demonstrated relatively unchanged device operation for up to 90 h [18].

In order to achieve fully functional devices, which in many cases require processing steps that are incompatible with biodegradable substrates, the authors separated deposition, etching, and lithography from the biodegradable substrates.

The devices are fabricated on initial carrier substrates (diluted polyimide on poly (methylmethacrylate) (PMMA) on a Si wafer) and subsequently transferred to the biodegradable substrates using transfer printing techniques.

Biodegradable substrates studied include also poly(lactic-co-glycolic acid) (PLGA), polycaprolactone (PCL) and rice paper. Rice paper rapidly absorbs water and swells, finally disintegrating. The authors demonstrated transient impedance electrodes for skin hydration monitoring applications related to wound healing. The device used phosphorous-doped silicon for the electrodes, Mg for the interconnects, and SiO_2 as a dielectric. The authors also demonstrated two passive devices consisting of a coil and a resistor as a nonantibiotic transient therapy device for controlling surgical site infection. A temperature increase of up to 5 °C was reported with a device comprised of two sets of a coil connected to a resistor, each with a different resonant frequency and with 15-day functional transience [18]. A similar device using Mg for a coil and resistive heating element (300 Ω) on a silk substrate was presented in [13]. With an excitation at 80 MHz and a power of 500 mW, a temperature of up to 20 °C was achieved in in vivo studies where a device comprised of a coil and a resistor was implanted subcutaneously in a mouse. In vitro experiments in bacterial cultures were performed and temperature changes of up to 28 °C were achieved with 100 mW at 80 MHz. Bacteria inhibition zones were formed corresponding to the location of heat treatment. In vivo experiments with bacterial infected mice with a similar device implanted at the infection site, demonstrated induced temperatures of up to 42 and 49 °C, with 100 and 500 mW at 80 MHz, respectively with 10 min heat treatments [13]. The detectable bacterial count was reduced in mice treated with the device. Analysis of tissue samples 15 days after implantation showed Mg levels close to physiological concentrations. The authors expanded this work, by loading silk layers encapsulating the device with antibiotics (ampicillin). Thus, the wireless temperature control via the inductor and heating element and the boiling time of the raw silk fibers and the water annealing time of the silk films, controlled the release kinetics of the drug for a set silk crystallinity level [13].

Implantable resorbable passive optic devices have also been demonstrated. Silk is an optically clear material, which can be molded into microstructured optical elements such as microprism arrays. This simple device allowed the increase of near infrared light returning to an external detector. This in turn improved deep-tissue imaging and contrast enhancement. In vivo experiments demonstrated a threefold improvement. Doping the silk with Au-nanoparticles (NP) allowed the absorbance of green laser irradiation, which is subsequently converted into heat. Such a device demonstrated a localized temperature increase of 5 °C in vivo and was also used for the elimination of bacteria. Alternatively, the silk can be doped with therapeutic substances, such as doxorubicin [20], used for the treatment of various cancers. The transience of the silk microprism controls the release of the dopant, while the change in the reflected optical signal depends on the degradation of the device and thus gives also a measure of the drug being released. Histological assessment following implantation of such devices [18, 20, 23] from two to four weeks revealed no significant inflammatory response, encapsulation, or fibrosis,

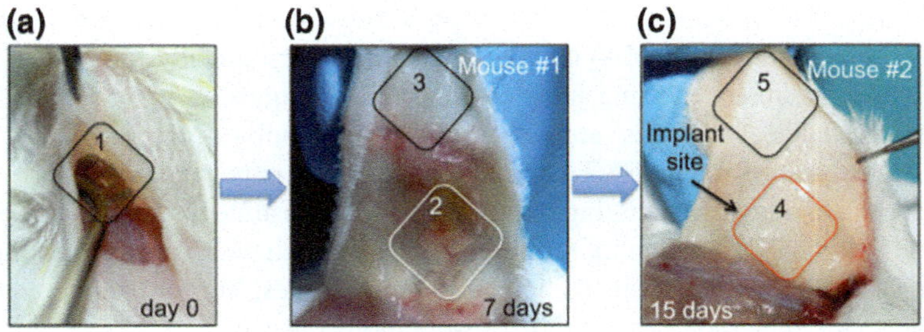

Fig. 4.8 a–c Show the devices implanted underneath subdermal region of two BALB/c mice on day 0, day 7, and day 15. Reproduced with permission from [13]

while only faint residues of the devices were detectable. Reintegration into the subdermal layer and revascularization were evident via histological examination. Figure 4.8 [13] shows one such device implanted underneath the skin of BALB/c mice. Mouse #1 was examined after seven days and the device was moderately degraded, while examination of mouse #2 demonstrated that the device was fully degraded after 15 days.

4.2 Fabrication Methods for Advanced Flexible Interconnects and Devices

4.2.1 Microfabrication Technology

Microfabrication technology is a family of processes classically used for the fabrication of semiconductor devices and integrated circuits. This typically includes processes such as photolithography, doping, polishing, deposition techniques, and etching. Silicon wafers are often used as a substrate because of their wide availability, low cost, and surface flatness, but other materials, such as glass, quartz, and plastics, are also common. Over the past three decades, this technology has been extended beyond the manufacturing of integrated circuits (IC) and into the development of small-scale sensors for both 2D and 3D miniaturized devices. Conductors and insulators are deposited on the wafer by techniques such as physical vapor deposition (PVD), e.g. sputtering, or chemical vapor deposition (CVD) technology. Alternatively, spin coating can be used to deposit on the wafer initially thin films of e.g. polydimethylsiloxane (PDMS) or PI prior to other material deposition. Photolithography is then used to introduce thin film masking patterns deposited on the top. A positive (Novolac/diazide photoresist or polymethyl-methacrylate (PMMA)) or a negative (poly chloromethyletyrene, AZ5206 and SU-8) photoresist (silicon or other elastic substrates) is first deposited using spin coating and is prebaked. This is then followed by UV light exposure to transfer the geometric pattern of a photomask onto the resist and removal (NaOH,

KOH, and NH_4OH) of unwanted resist by a solution called the developer, following a post-exposure bake step. The wafer is then often hard-baked (up to 180 °C) to make the resist resilient to subsequent processing steps. Etching is then used to remove areas on the wafer that are not protected by the photoresist. Different forms of etching exist, such as wet (anisotropic, isotopic, HF, and electrochemical) or dry (vapor and plasma) etching. Finally the remaining resist is removed using liquid chemistry or oxygen plasma. This family of fabrication technologies has been extensively applied to fabricate micro-engineering surfaces, sensors and transistors, and analog, digital, and mixed-signal integrated circuits, as well as micro-opto-electro-mechanical systems (MOEMS), micro-electro-mechanical-systems (MEMS) [24–27], and they are also used extensively for the fabrication of flexible and stretchable electronic devices.

4.2.1.1 Photolithography Transfer Patterning

The process of making very thin and high tensile strain flexible interconnects is rapidly advancing. Those applications are not only suitable for electronic system integration, but also widely used in sensor applications, such as tactile sensing. Several publications have proven that with soft substrates the thickness of designs can be thinner than that of commercial FPC.

The fabrication process followed by [28] for the fabrication of Au-patterned interconnects is shown in Fig. 4.9a. The first step involved spinning a layer of 70 μm PDMS on a silicon wafer with a recommended mixed weight ratio (10:1) and curing at 70 °C for 12 h, and subsequently deposition of a 2 μm Cr adhesion layer on top of the PDMS. This was then coated by a 40 nm thick Au layer. A 0.45 μm thick film layer of positive photoresist-AZ 1505 on the metallic conductor was then patterned via UV light exposure using a contact mask aligner. Subsequently, etching allowed the conductive pattern to be exposed. In order to encapsulate the conductive layer with a very thin cover, another PDMS layer was diluted with hexane [28] on a mixed ratio (1:4) and spun at 6000 rpm for half a minute and then cured at 70 °C for 12 h. The final step involved depositing an 80 nm mask layer of Al on top of a second PDMS layer, and that was wet etched by a solution of H_2PO_4:H_2O:CH_3OOH:HNO_3 (16:2:1:1%v/v). An image of the patterned interconnects encapsulated with PDMS is shown in Fig. 4.9b. The results shown in Fig. 4.9c are electrical resistance measurements of two different widths of Au conductors on PDMS compared with the expected results.

Figure 4.10 shows another novel fabrication process termed “CINE” (combination of interconnects and electronics) [29]. The first step in CINE is fabricating the Au interconnects with PI encapsulation on a 4″ diameter Si substrate. The Si wafer was coated first with hexamethyldisiloxane (HMDS) to promote strong adhesion with PMMA. In order to create gold contact pads, the PI layer was etched. PMMA and HMDS were dissolved with acetone, as shown in Fig. 4.10c. The next step in the process was to attach an adhesive tape on top of the array of interconnects apply pressure by hand, and gradually remove the Si substrate. This was

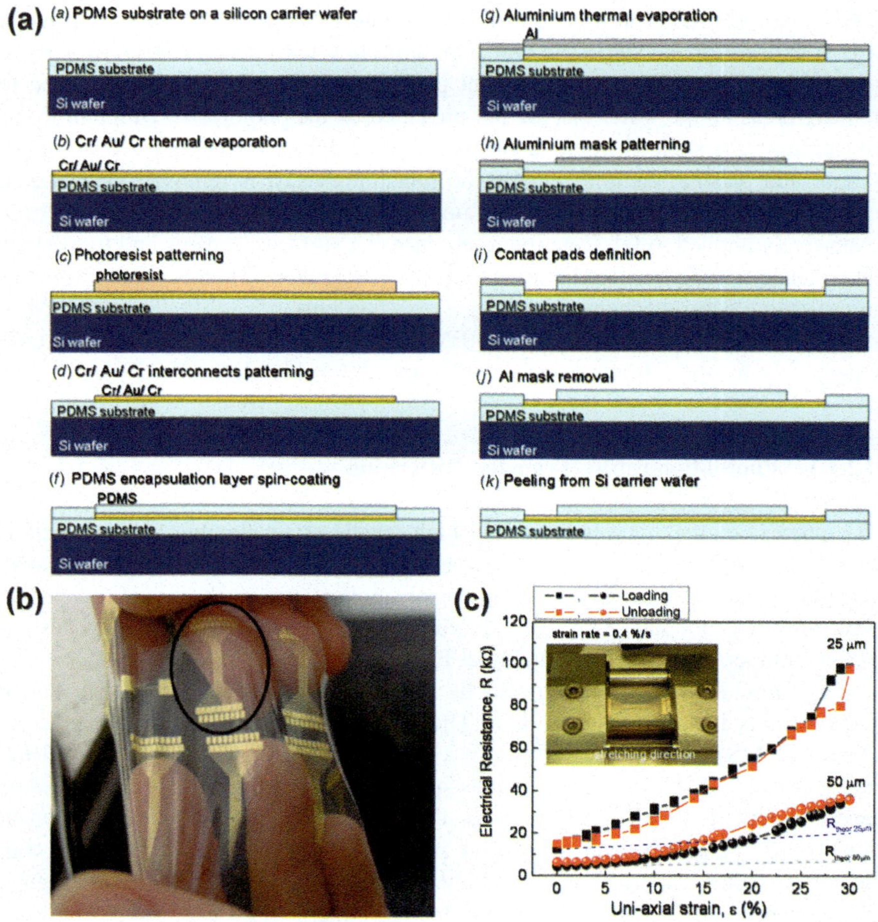

Fig. 4.9 **a** Fabrication process of making Au-patterned PDMS-encapsulated interconnects. **b** Photo of stretchable gold-patterned interconnects and **c** the electrical properties of two different tracks as a function of strain, compared with theoretical results. Reproduced with permission from [28].

followed by a process of exposing the commercial 1 mm Ecoflex soft substrate. The tape and interconnect array were then O_2 plasma-treated to enhance bonding. Finally, the tape and array were put into contact and pressed onto the Ecoflex by applying 5 kPa pressure for 10 min.

The results in Fig. 4.11 show that the electrical properties changed slightly after 1000 and 10,000 stretching cycles. The stretch reliability test was tested from 0% strain to 100% strain. Spectra recorded from a red LED source (Fig. 4.11b) were

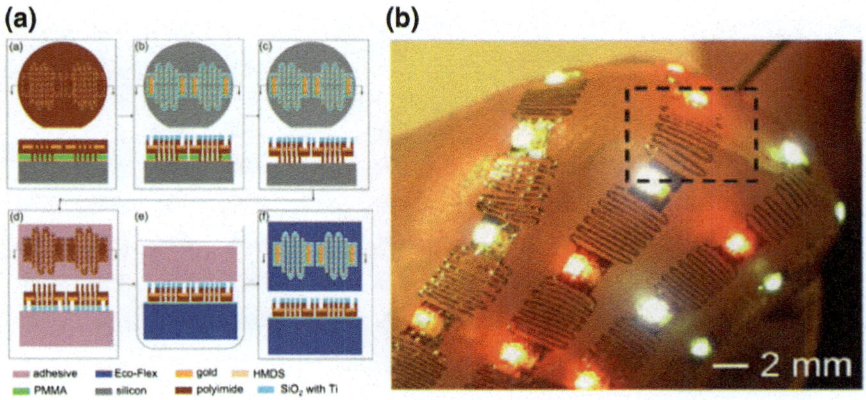

Fig. 4.10 **a** CINE process. **b** Photo of LED array bonded on gold interconnects. Reproduced with permission from [29]. © Wiley 2011

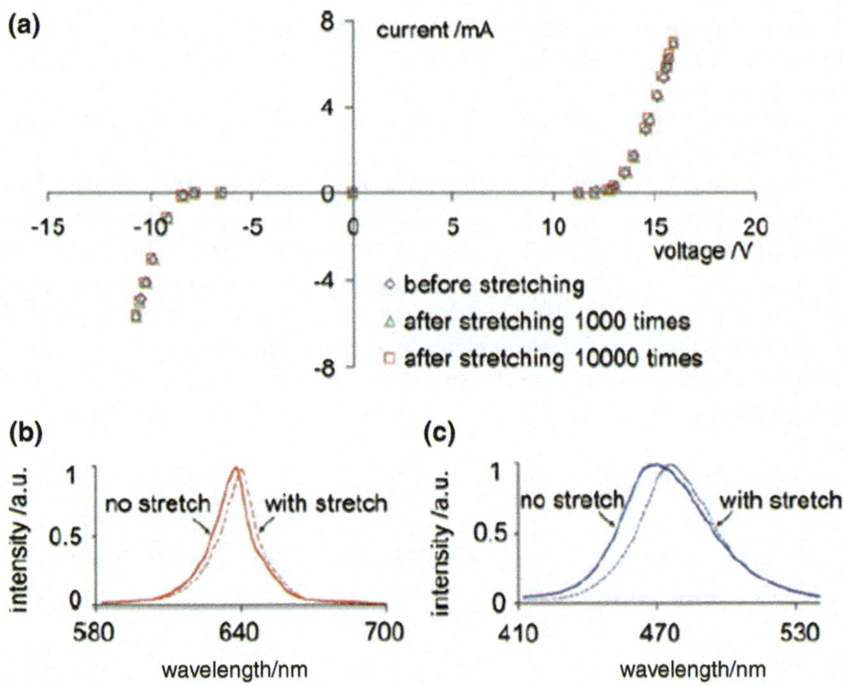

Fig. 4.11 **a** Results of electrical testing after stretching 1000–10,000 times. **b** Red LED light spectrum intensity between no stretch and stretch. **c** Blue LED light spectrum. Reproduced with permission from [29]. © Wiley 2011

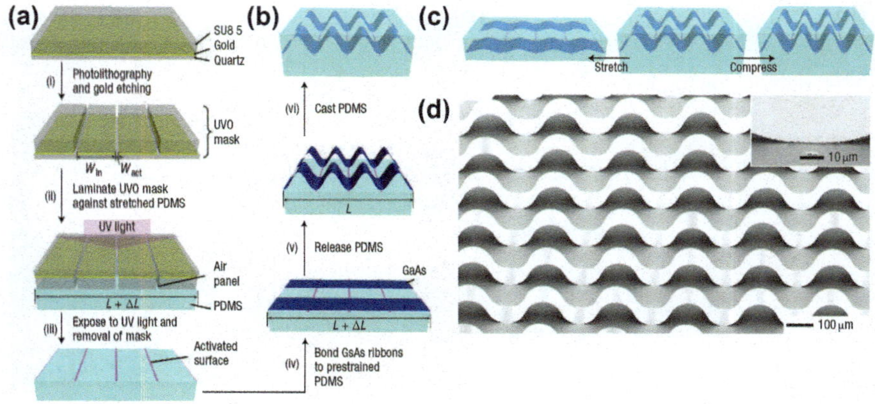

Fig. 4.12 Fabrication steps for the development of stretchable nanoribbons. Reproduced with permission from Macmillan Publishers Ltd: Nature Nanotechnology [30]. © 2006

measured at tensile strains of 0 and 67% and the spectral peaks only shifted 2.3 nm. Compared to the red LED, the spectral peak shift of a blue LED at 0 and 67% tensile strains was 6.8 nm (Fig. 4.11c).

Another approach, presented in [30], is the development of 3D shapes in nanoribbons. This is achieved by using a ultraviolet/ozone (UVO) mask in order to pholithographically selectively modify the surface chemistry of the substrate, such that the adhesion sites of semiconducting materials such as Si and GaAs are controlled and by prestretching the substrate (PDMS). The PDMS substrate is subsequently released. After this process, the GaAs (or Si) nanoribbons take on a wavy shape. Finally, a top PDMS layer is deposited for encapsulation. This technique has been used to implement flexible stretchable interconnects and also devices such as metal-semiconductor-metal (MSM) photodetectors [30]. Stretchability up to 100% and compressibility up to 25% were demonstrated. These are summarized in Fig. 4.12.

4.2.2 Transfer Printing, Imprint Lithography, and Related Technologies

Transfer printing is a family of techniques with high resolution (<10 nm) that can be used with a wide variety of materials for large-area assembly of devices. Small molecules, carbon nanotubes (CNT) and graphene [31, 32], polymers, DNA, photoresists, oxides, metals, and inorganic semiconductors can all be developed in the form of an ink suitable for transfer printing [33]. Transfer printing was recently reviewed in [33]. The three basic variations of transfer printing are demonstrated in Fig. 4.13, and here we will consider these in terms of cold welding [34], a form of transfer printing typically used for patterning metals [35–37]. Patterning is performed by bonding two surfaces together on contact or with the application of moderate pressure (1 atm) [36]. In all three, the first step is the creation of the stamp, which is embossed with the desired pattern. In the additive transfer mode, Fig. 4.13a, the raised regions of the stamp correspond to regions of the final device where material is deposited. The stamp is typically made from a soft material such as PDMS. Before coating the stamp (with the specific desired thickness required for the final device) with the material (e.g. noble metals such as Au, as they do not become oxidized, which would prevent bonding) to be patterned, an adhesion reduction layer is deposited on the stamp. This ensures that the material will be easily transferred from the stamp to the device substrate. A thin layer (~5 nm) of the same material is deposited on the final device substrate. This is known as the strike layer [36]. When bringing the two into contact under small pressure, cold welding takes place and the material from the stamp is transferred to the substrate. The leftover thin residual strike layer left between the transferred patterns is easily removed using sputtering or gentle etching using a reactive gas.

In the subtractive transfer mode, as shown in Fig. 4.13b, a layer of material is deposited on the device substrate at the thickness required for the final device. The stamp, which in this case is a hard stamp, is coated with an adhesion enhancement layer and then by a thin layer of the same material to be patterned. The raised regions of the stamp in this case correspond to areas of the device where material is to be removed. The two are then brought into pressure contact, as before, and a cold weld is formed. A higher pressure will cause the material at the edge of the stamp to fracture. This allows the material to be lifted off the substrate. It is important to note that in the additive process no fracturing is required, and it is thus a simpler method.

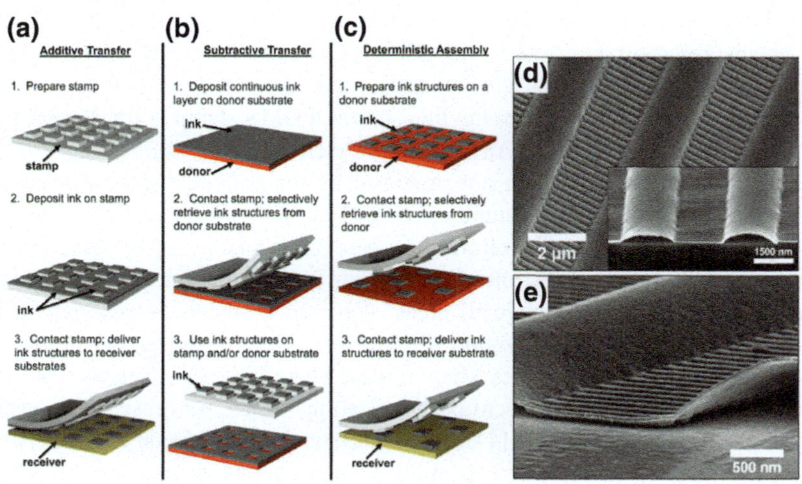

Fig. 4.13 **a** Additive mode, **b** subtractive mode, and **c** deterministic mode of transfer printing. Reprinted with permission from [33]. © Springer 2013. **d** and **e** scanning electron microscope (SEM) images of Au microchannels connected with arrays of nanochannels fabricated using transfer printing. Reprinted with permission from [39]. © American Chemical Society 2003

The last form of transfer printing is known as deterministic transfer (Fig. 4.13c). This is a process combining (a) and (b) and it can be used for example to pick up pre-made electrodes from a donor substrate and then place them on a receiver substrate. The latter is valuable because it separates material growth and processing from the device substrate and the stamp. This is important, as it significantly expands the materials capable of being processed and patterned.

In transfer printing in general, thermal, mechanical, and chemical strategies need to be employed in order to facilitate high quality and successful transfer printing. Important, for example, are the adhesion forces and the chemistry between materials. Using such transfer printing techniques, a flexible organic transistor was sequentially printed on a polyethylene terephthalate (PET) substrate in [38], an organic LED (OLED) in [35], and graphene transistors in [32], while in [39] transfer printing was used to print Au micro- and nanochannels and lines on top of each other (Fig. 4.13b). The combination of photolithographic, deposition, and other techniques and transfer printing has been used to fabricate complete electronic and sensing flexible and stretchable devices. In [40, 41], doped silicon micro/nanoribbons were transfer printed onto PMMA or PI substrates and used to form complementary metal-oxide-semiconductor (CMOS) inverters, ring oscillators, differential amplifiers, and sensors.

A similar technique to transfer printing, known as microcontact printing (μCP) [42, 43], uses a hard (Si) photolithographically patterned master mold to create a soft elastomeric stamp, typically made using PDMS. The stamp is saturated typically with a thiol (e.g. methyl-terminated thiols with long alkyl chains, polar group-terminated thiols, or aliphatic dithiocarboxylic acids) or silane ink by immersing the stamp in the ink or by applying some ink to the patterned section of the stamp by drop-casting. Only apolar inks were initially used because of the hydrophobic nature of PDMS [42]. Ink diffuses into the bulk of the PDMS stamp and an ink reservoir is created [42]. The thiol ink solvent, e.g. ethanol, is then evaporated using nitrogen and the stamp is dried. When the stamp is then brought into contact with an Au surface, the thiol on the PDMS stamp is transferred to the Au surface. The thiol-based SAM acts as an etch resist. A wet-etching step then takes place, where Au sections not protected by the patterned thiol are removed. This is the basic μCP process, summarized by Fig. 4.14, where μCP is followed by etching. Instead, μCP can be followed by electroless plating or by selective electropolymerization [44]. Electroless plating can take place on the SAM-covered areas or on the non-SAM-covered areas, depending on the process parameters. Area-selective electropolymerization is performed using metalized substrates, with the SAM layer defining the areas where it takes place. The SAM acts as a barrier to electron transfer, preventing electropolymerization, which, however, takes place on the non-SAM-covered regions [44]. Other variations of this (such as positive μCP to generate an identical replica of the master rather than its inverse) are discussed in detail in [42]. It is important to note also that the PDMS stamp hydrophobicity can be modified, by e.g. oxygen plasma and UV ozone treatment, to allow the use of polar inks.

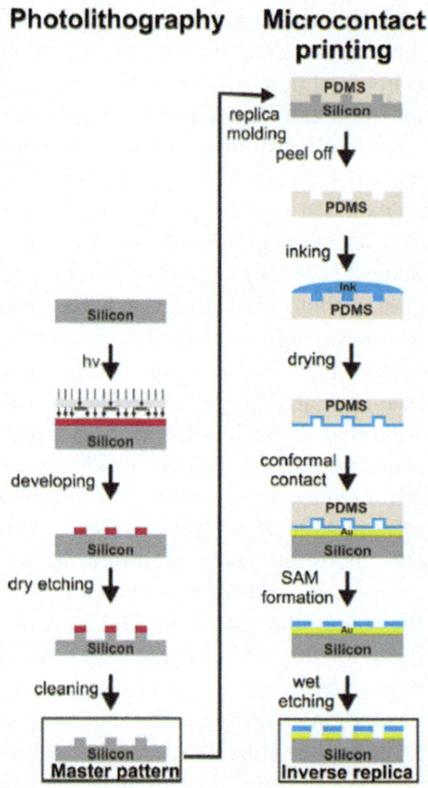

Fig. 4.14 Schematic illustration demonstrating the photolithographic process used for fabricating the patterned mold, the creation of the stamp from the mold, and the μCP process for patterning Au on a substrate. Reprinted with permission from [42]. © Wiley 2009

Another technique is soft contact lamination [45], where metal-coated soft (PDMS) rubber stamps are brought into contact with a substrate to establish electrical contacts. No pressure or adhesives are applied and the contacts are reversible and formed at room temperature. Stamp masters are created on silicon using photolithographically patterned resists, which are used to cast the stamps. Ti (1 nm) and Au (15 nm) are deposited on the raised and recessed regions of the stamps (not the sidewalls) using electron beam deposition. Bringing this into contact with a flat smooth surface leads to "wetting", leading to atomic scale contact establishing electrical contacts where necessary. This was used to form a top contact back-gated organic field-effect transistor (OFET) with a pentacene channel and SiO_2 dielectric, the gate being the Si substrate, with the channel and drain electrodes formed by the stamp, with channel lengths from 250 to 150 nm [45].

Imprint lithography is another family of techniques based on compression molding [46–48]. First, a mold, made of metals, dielectrics, or semiconductors, e.g. SiO_2 molds on Si substrate, is patterned using e.g. electron beam lithography. The resolution of the technique is limited only by the resolution the techniques used to create the mold [49]. Then, a thin resist is cast on a substrate, and subsequently the mold is pressed into the resist, while being heated if, for example, the resist is a thermoplastic polymer, to soften it relative to the mold. Thus, in this case, the resist is not modified by changing its chemical structure with radiation. Alternatively, UV radiation can be used with different resists, but this requires the UV radiation to reach the resist either through the substrate or through the mold (Fig. 4.15). The mold is then removed and anisotropic etching removes any resist remaining in the compressed regions. Metal can then be deposited on the resist and on the sample. Subsequently dissolving the resist leaves behind the metal on the substrate of interest [46]. The compression molding technique eliminates many issues related to classic photolithography, such as scattering issues and diffraction, and can be used to pattern large areas with high throughput and a resolution below 10 nm [46, 48]. In the first demonstration of the technique to fabricate electronic devices [48], nanoimprint lithography (NIL) was used to pattern quantum dot, wire (100 nm wide and 2 μm long), and ring transistor channels, while polysilicon gates and source and drain metal electrodes were patterned with photolithography. NIL was also used to pattern with high resolution small molecule and polymer light-emitting materials for OLED devices (Fig. 4.15) [50]. Methods combining photolithography and imprinting and customized for roll-to-roll fabrication on plastic flexible substrates have also been presented [51–53], where it was used in [54] as a means of fabricating the gate in an OFET. It is clear from the above that there are various forms of imprint lithography. These include thermal and UV (discussed above), photo, step-and-flash, and roller nanoimprinting [43]. In all of the above techniques a common challenge is the mold alignment and most importantly accurate control of the applied pressure of the mold to the device/substrate, the mold fabrication process, and materials used in conjunction with the inks and materials and the interfacial surface chemistries between these, which all have to be optimized. Another challenge is associated with the accuracy of the etching of the residual

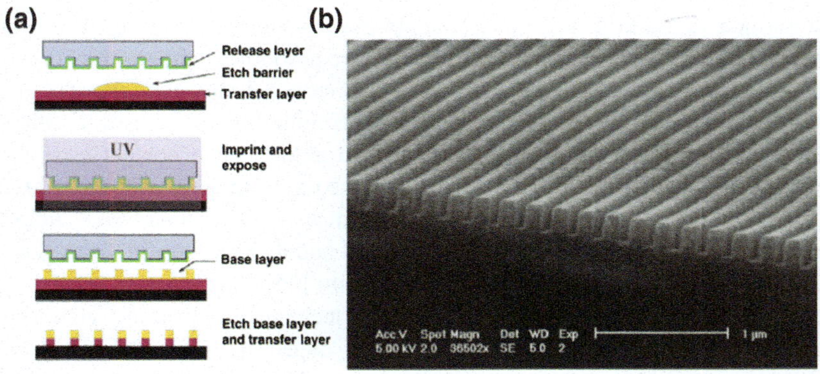

Fig. 4.15 **a** A schematic representation of a UV-based imprint lithography process. Reprinted from [49]. © 2005, with permission from Elsevier. **b** NIL-patterned light emitting structures. Reprinted from [50], with the permission of AIP Publishing

resist/strike layer. Challenges also include the swelling of the soft polymer stamp used for transferring SAM inks, which leads to enlarged patterning, and the deformation of the PDMS stamp, due to its soft nature, leading to pattern distortion.

4.2.3 Stencil and Screen Printing

So far several methodologies have been introduced that can be used to fabricate flexible/stretchable interconnects and devices with a resolution within the μm range and below. Stencil and screen printing are simple fabrication processes that can be used outside the clean room to realize devices and interconnects typically with a resolution of a few tens of microns and above [55]. Figure 4.16a shows a copper foil, which was wet-etched to produce a stencil mask to pattern conductive patterns on PDMS. Lines 150 μm wide with a 100 μm spacing (Fig. 4.16b) were achieved using 100 μm thick stencils. The ink used was created by mixing 99.9% silver powder with the PDMS elastomer. These were mixed and degassed for 3 and 1 min with spin speeds of 2000 and 2200 rpm respectively. A squeegee was used to push the ink through the mask, which generated the conductive traces shown in

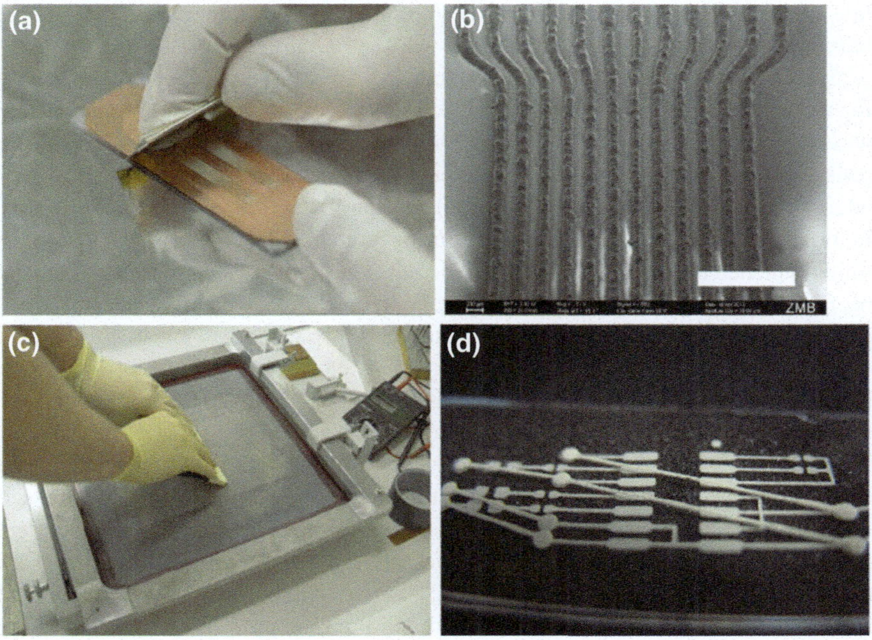

Fig. 4.16 **a** Stencil printing on PDMS using a copper stencil. **b** SEM image of stencil printed conductive tracks. **c** Screen printing using a mesh screen and the same ink. **d** A double-sided printed circuit with vias (scale bar: 10 mm). Reprinted by permission from Scientific Reports [55]. © Macmillan Publishers 2014

Fig. 4.16b. Stencils can be easily fabricated using wet etching or laser cutting, and for resolutions of tens of μm, commercially available PCB manufacturing processes can be exploited for low cost fabrication using industry-standardized processes.

A problem with stencil printing is that closed lines cannot be printed. This is not an issue with screen printing (Fig. 4.16c), which is the most mature technology for printed flexible electronics. It is a simple, fast, reliable, high-throughput, low-cost technique suitable for mass production and roll-to-roll fabrication. Screen printing uses a stainless steel or polyester mesh screen typically with 200–400 threads per inch (tpi). A thin emulsion layer (e.g. 13 μm) defines the pattern to be printed. Because there is a mesh with openings between threads, there are no continuous open paths, as in stencil printing. Consequently, closed lines can be printed. Similarly to stencil printing, a squeegee is used to push the ink through the mesh to print the desired patterns. The quality of the printing is influenced by factors such as the printing speed, ink viscosity, squeegee geometry, material, pressure and angle against the screen, and the surface tension of the substrate. In addition, the material, tpi, and strength of the screen also play an important role. In contrast to inkjet inks, screen printing inks are highly viscous in order to be properly dispensed to the substrate through the mesh of the screen. The screen can be used multiple times. One issue is the potential drying of residual ink within the screen's mesh, following repeated use of the screen, which leads to distortions in the printed patterns. Screen printing is considered a non-contact printing technique, as the screen is typically a few mm above the substrate. Following printing, the patterned substrate is then baked.

Screen printing was also used to pattern the Ag-loaded PDMS discussed earlier for stencil printing in the same paper [55]. Results show that a high conductivity was achieved with an Ag loading of the PDMS of 25 vol.%, (600 S m^{-1}) and resistance increase insignificantly by changing to 22, 19, and 16%. However, adding more Ag beyond 19 vol.% fragilized the material and decreased the maximum achievable elongation at rupture during stretching tests. Vias were formed by filling punched holes with this elastomeric ink (Fig. 4.16d). Screen printing has also been used to fabricate all layers in parallel-plate flexible pressure sensors using piezoresistive and piezoelectric composites [56, 57], electrochemical sensors such as ion selective electrodes (ISE) [58], passive electronic components (resistors, inductors, capacitors) [59], humidity sensors [60], OLEDs [61], and passive electronic components, sensors. and interconnects on fabrics [62, 63]. In [63], a lamination of the printed fabric with a thermoplastic polyether elastomer meltdown coating was used to protect the printed structures during washing and daily usage. Special nonwoven fabrics were used based on polyester, nylon, and polyethylene fibers. Finally, in [64] screen printing was used to print high-gain amplifiers and digital to analog converters (DAC) on flexible substrates (Fig. 4.17). A polycarbonate (PC) substrate was used and the inks used to realize the devices were all commercially available, apart from the semiconductor layer (TIPS-pentacene, 6,13-Bis(triisopropylsilylethynyl)pentacene), which was also the only layer not printed by screen printing, but with a slot die coater. Carrier mobilities of 1.5 cm^2 V^{-1} s^{-1} were reported.

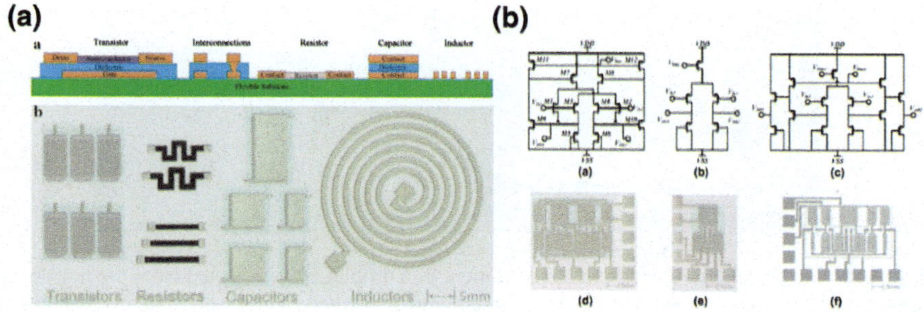

Fig. 4.17 Screen printed **a** electronic components and **b** differential amplifiers. Reprinted from [64]. © 2014, with permission from Elsevier

4.2.4 Direct-Write Techniques: Inkjet Printing and Others

Inkjet printing has attracted significant interest over the last decade for flexible electronics fabrication. The technology is essentially the same as that used for commercial personal printers for printing graphics on paper. Inkjet printing is a direct-write technique, requiring no masks, in contrast to all previous methods. The process of designing and fabricating masks and the issues related to photolithography are thus avoided. In addition, ink is dispensed only where it is required. Thus, substantially less material is wasted. This is also true since no resists or etchants are required for patterning. Low-viscosity inks are printed though a nozzle, which is piezoelectrically, thermally, ultrasonically, or electrohydrodynamically controlled to dispense droplets on demand (DoD). Thermally actuated printing nozzles are avoided with organics, because heat may damage the functional organic material [44]. Thus small droplets of 1–100 pL of ink are injected one by one though an orifice, while moving the printed head in two or three dimensions. Usually, the lateral dimension is kept constant and to a minimum (about 1–3 mm above the printing substrate) to avoid air currents, which may influence the droplet trajectories. Typical nozzle diameters are 10–80 μm in diameter. Initially, before printing, ink is flushed through the system and the static pressure at the nozzle is adjusted. The ink is held in place at the orifice by surface tension and a meniscus is formed. The pressure is adjusted such that this meniscus is flat and stable. An important first step in the optimization of the printing parameters is related to droplet formation and ejection. Figure 4.18 shows a droplet being ejected and created. As the drop falls towards the substrate, a long tail is formed, which is subsequently converted into a leading droplet and a number of smaller satellite droplets. These satellites may catch up and merge with the leading droplet and can thus be ignored. If not, they are problematic and the print parameters need to be altered such that these satellites are eliminated. This is because these might follow different trajectories, affecting the print quality. Parameters such as the static pressure at the nozzle, ink temperature (if possible), and the nozzle excitation waveform have to be adjusted. These will also control the drop size and the ejection velocity to some extent.

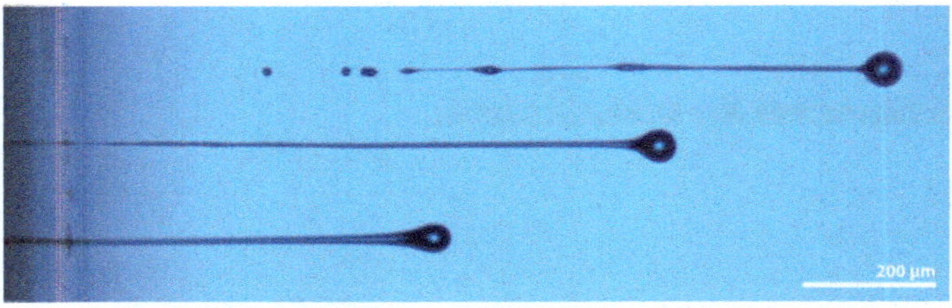

Fig. 4.18 Image captured by a high-speed camera of a droplet as it is ejected from the nozzle and formed prior to impact with the substrate. A column is formed, which subsequently becomes a main droplet and additional smaller satellite droplets. Reproduced with permission from [65].

The minimum printable feature size (typically a minimum resolution of 20–50 µm) is a function of many parameters, including nozzle diameter, substrate properties (such as hydrophobicity and flatness), ink temperature, droplet speed, nozzle excitation parameters, printing speed, and the air pressure applied to the system. These would define the diameter of a single droplet on a substrate. According to [66], the drop footprint on the substrate is a linear function of the droplet diameter in flight and it increases with decreasing droplet/substrate contact angle. For a contact angle of 10°, the drop footprint is approximately three times the drop diameter in flight [66]. Typically, however, lines and other shapes are printed, not individual drops. To achieve this, a series of droplets must be printed with optimized droplet spacing. Thus, drops come in contact and interact with other drops. This will influence the liquid flow and spreading, as will the drying of the droplets. The droplets are usually pinned in position, and as the droplet evaporates at a greater rate towards its edges, fluid will flow from the center of the drop to replace the ink that evaporated at the contact line [66]. This leads to an increased deposition of ink particles towards the edge of the droplet at the initial contact line (Fig. 4.19a). This will continue until the droplet solidifies, leading to the formation of a ring rather than a homogeneous circular spot. This is known as the coffee ring or coffee stain effect, illustrated and shown in Fig. 4.19a, and apart from taking place during the printing process, it will also occur during the solidification process [66–68]. This also depends on the aforementioned range of parameters. If the spacing between droplets is not properly optimized the printed patterns will not form the required geometry. Figure 4.19b shows printed lines when the droplets are either too separated or too closely spaced. This again depends on the parameters mentioned above, particularly droplet spacing, printing speed, temperature, droplet velocity, and substrate/ink contact angle. More information can be found in [69–72].

It is obvious from the above that establishing the optimum printing parameters for a particular ink and substrate, resolution, and device is a complex process. A number of techniques have been proposed in order to achieve better quality and smaller features. One approach is to prepattern the printing substrate photolithographically with e.g. trenches or to selectively change the surface chemistry of the

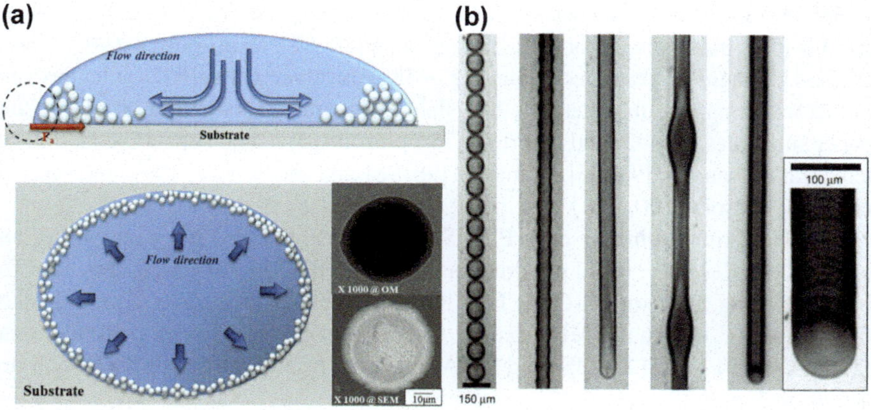

Fig. 4.19 **a** Illustration and optical and scanning microscope images of the coffee ring effect, resulting from ink droplet spreading and particle concentration towards the droplet edges. Reprinted from [67], with permission from Elsevier. **b** Examples of printed patterns with various droplet spacings. Using the terminology, from left to right these can be described as individual drops, scalloped, uniform and bulging patterns, and stacked coins [68]. Reprinted with permission from [68]. © American Chemical Society 2008

substrate and the contact angle [66]. However, such approaches can complicate and increase the cost of the printing process further. Apart from the complexity of the process to print good quality features, other disadvantages of inkjet printing include clogging of the nozzles and slow printing speed (and thus low throughput), rendering the technique unsuitable for mass fabrication [43]. Nozzle clogging is particularly common with thermal and piezoelectric nozzles [44]. Office inkjet printers can be modified for inkjet printing of devices, but commercial systems are also available (Microdrop Autodrop, FujiFilm Dimatix, MicroFab Jetlab).

The final step in the process is the solidification of the printed ink. The use of e.g. Ag NP inks to print conductive tracks is common. Such inks, prior to and following drying at room temperature, however, are not conductive. A sinterization step must thus follow in order to merge the NPs together and thus make the printed structures conductive. At the curing temperature of NP inks, the organic shell of the particles is lost, allowing contact between adjacent particles and at higher temperatures sinterization takes place, where NPs merge to form a continuous percolating network [73]. Typically, inks require baking temperatures in the region of 150 °C and there is continuous effort to bring sintering temperatures down in order to decrease the manufacturing cost and to make inks compatible with more substrates, materials, and processes. An alternative is to use lasers to selectively sinter printed NP [74–77]. Laser irradiation follows the printed structures, sintering the central part of the printed line without damaging the substrate. Unsintered material is washed away. This has the potential to lead to narrower conductive features; however, it is a very slow process (0.2 mm s^{-1}) [73]. Another approach is to use microwave radiation, which is reported to reduce the sintering time by a factor of 20 [73], or IR incandescent lamps [78]. With other inks, solvent evaporation, cooling, polymer gelling, or a chemical step may be required [66].

Inkjet printing has been used to produce a wide range of devices, such as printed circuits [79], inductors, capacitors, and resistors [80–82], an all-polymer *RC* filter in [83, 84], magnetic resonance imaging (MRI) receiver coils [85], antennas and inductors on paper substrates [86–88], field-effect transistors (FET) on plastic and paper substrates [89–92] and inverters [75], piezoresistive and capacitive tactile and temperature sensors [93–95], humidity [96–98] and gas sensors [99], and electrochemical sensors [100, 101].

Ag NPs printed directly onto PDMS can be easily peeled off without any encapsulation. Recently, [102] presented a modified MPTMS-PDMS process used for inkjet printing a sensor on-chip. This has the potential to lead to a new field of fabrication of stretchable electronics. Optical images of Ag NPs printed onto (a) PDMS, (b) MMD PDMS, and (c) Ag patterns printed on to MMD PDMS are shown in Fig. 4.20.

Another direct write technique is laser-induced forward transfer [103, 104]. In this technique, solid or liquid materials and inks from a donor substrate are transferred to the device (receiving) substrate with high special resolution by pulsed laser irradiation. With the appropriate laser irradiation power, pulsing, and scanning speed, material is ejected from the donor substrate, creating a jet similar to inkjet printing in the case of liquid inks as shown in Fig. 4.21. This technique has been used to print Ag NP inks [105], OLEDs [106], and polymer/CNT composites for chemical sensors [107], and for the fabrication of biosensors, where the liquid solvent transports biomolecules (such as human IgG in a PBS and single-stranded DNA) to the receiver substrate.

With the rapidly growing market of 3D printing technology and additive manufacturing, probe-based electrochemical deposition provides a reliable and repeatable solution to fabricate interconnects [108, 109]. A probe-based system consists of (1) a 3D micropipette (filled with electrolyte −0.5 M $CuSO_4$ aqueous solution and a Cu wire inserted as the anode) embedded in a long travel Z piezo-stage that can control the fine direction of travel path; (2) a high-sensitivity electrometer that monitors the ionic current; and (3) an XY piezo-stage. Freestanding nanoscale copper wires can be fabricated using this technique by finely moving the XYZ piezo-stages. Figure 4.22a shows the steps in interconnect fabrication by using the

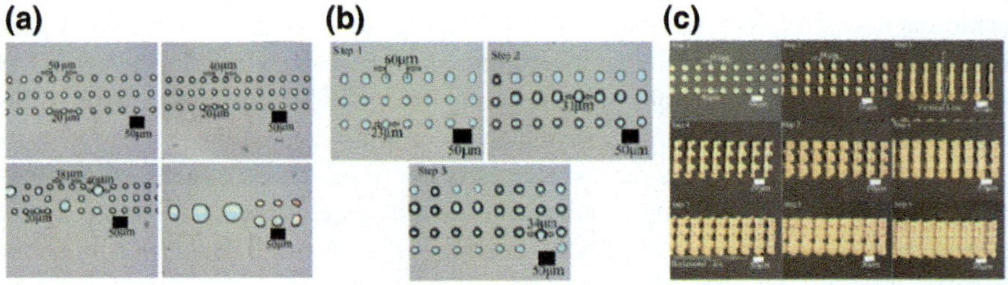

Fig. 4.20 **a** Silver droplets on untreated PDMS. **b** Silver droplet on MMD PDMS. **c** Silver patterns on MMD PDMS. Reproduced from [105] with permission of The Royal Society of Chemistry

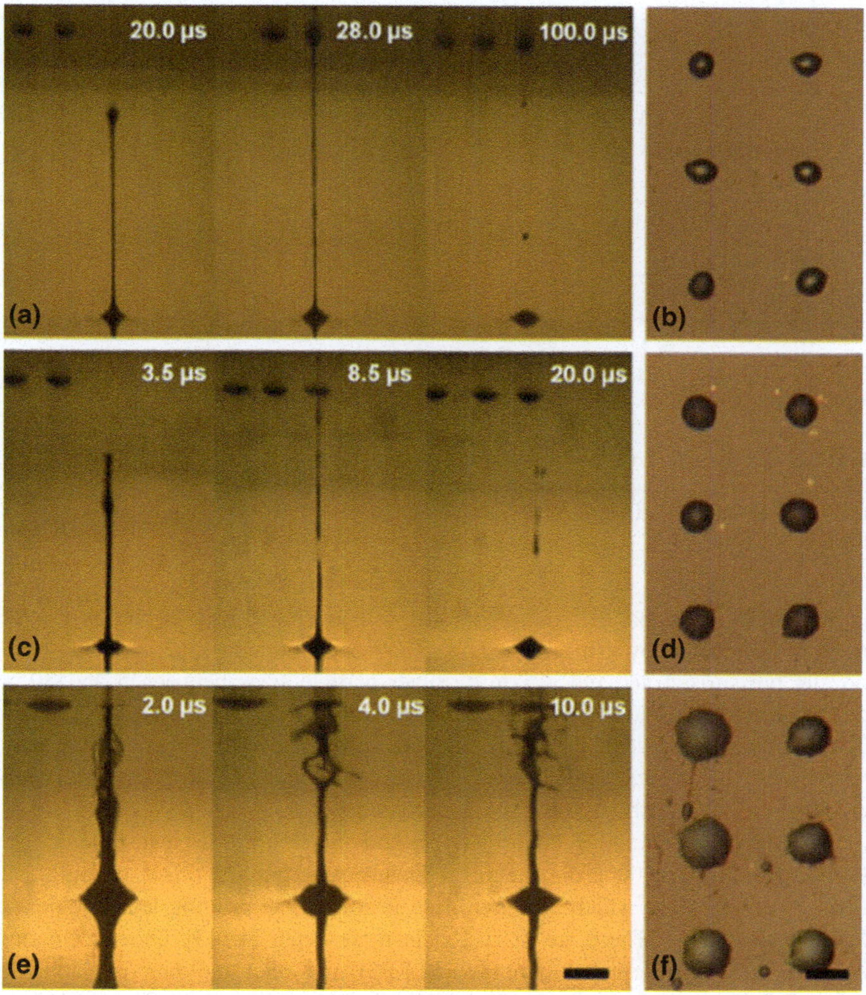

Fig. 4.21 Lift of Ag NP ink using various laser fluences and the resulting deposited droplets. The donor substrate is a 1 mm thick quartz plate coated with a 40 nm laser-absorbing Ti layer. This is at the bottom of the setup and Ag NP ink is deposited on it. The donor is illuminated with the laser from the bottom of the arrangement (not shown). The NP jet produced following laser irradiation impinges on a poly(ethylene naphthalate) (PEN) substrate located above the donor substrate. **a**, **b** 100 mJ cm^{-2} (ejection threshold laser fluence); **c**, **d** 200 mJ cm^{-2} (smooth jetting); and **e**, **f** 230 mJ cm^{-2} (turbulent jetting). Scale bar: 50 μm. Reproduced from [105]. © Springer 2013

3D meniscus-confined eletrodeposition process, while Fig. 4.22b, c are SEM images demonstrating 20 nanowires deposited on a central pad and the uniform quality of the bonds. Figure 4.22d shows the IV characteristics obtained when manufacturing a ~740 nm diameter Cu wire and the inset shows a broken wire, caused when applying a current that is too high [109].

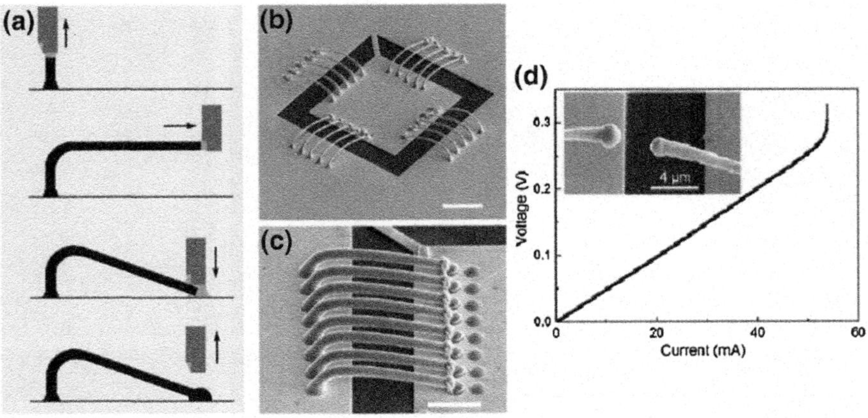

Fig. 4.22 **a** 3D wire bonding using meniscus-confirmed electrodeposition. **b** Cu nanowires bonding on a central pad of area 50 µm × 50 µm. **c** The SEM close-up demonstrating the uniformity of the bonds. **d** Linear IV curve of fabricating the nanowire. From [109]. Reprinted with permission from AAAS

An omni-directional printing technique was developed in [110] to print structures in 2D and 3D by using silver NPs. The system contained a syringe with ink, nozzle, pressurized air, and a fixture. Silver particle sizes between 5 and 50 nm demonstrated an optimal flow behaviour through nozzles between 1 and 30 µm. The technique was used to fabricate planar array microelectrodes with a minimum width of 2 µm, multilayered microelectrodes printed with up to 11 layers, and out-of-plane (3D) arch-shaped interconnects.

In addition to the above fabrication methods, techniques used in flexible electronics also include slot die coating, gravure and gravure-offset printing, and flexographic printing which are often used in roll-to-roll printing technologies for flexible electronics. These are discussed in more detail in [43]. Other techniques include flexible letterpress stamping (FLEPS) and micromolding in capillaries (MMIC) [44].

With 3D printing technology continually evolving, structural electronics circuit design is another option that can be exploited to fit electronic designs and components into different shapes of enclosures. Several versions of a novelty six-sided gaming dice were presented in [111], embedded with various circuits and components, arranged and interconnected in 3D and fabricated by employing an enhanced 3D printing technology. The team laid the circuitry out by using electronic computer-aided design (CAD) software and then deformed the 2D circuitry into a 3D geometry in order to import the design into mechanical CAD software for structural electronics interconnects pre-processing. Cavities and interconnects were formed by laser-cured liquid photopolymers for placing electronic components together with adhesives (UV curable) and conductive inks. The results showed that the enhanced 3D printing technology combined with a stereolithrography process could save up to 75% of the manufacturing time of a traditional flexible PCB

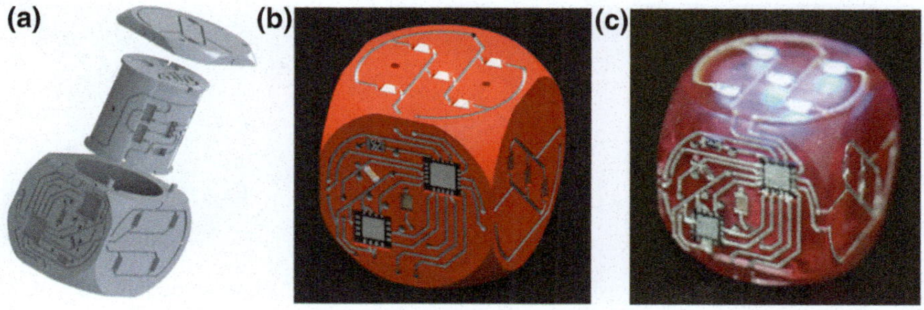

Fig. 4.23 **a** Exploded CAD illustration of the 3D structural electronics platform. **b** CAD illustration of the assembled device and **c** the final device. Reprinted with permission from [111]. © IEEE 2014

embedded into an inkjet mold. The proposed structural electronics manufacturing process took 30 h to complete, while a traditional FPC manufacturing and mold process took 120 h. Figure 4.23a, b show the final 3D structural circuit design with and without components on CAD software, while (c) shows the actual fabricated gaming dice.

Another interesting approach related to textile electronics using conductive threads was recently presented in [112]. Threads were used and functionalized to realize strain, pH, and glucose sensors, conductive interconnects, electrodes, and microfluidic networks. These were used to interface with tissues in 3D as sutures.

4.3 Materials for Flexible/Stretchable Electronics

The use of unconventional materials (organic materials, conductive polymers, and graphene, among others) for flexible electronics has attracted extensive interest in recent years to enhance the capabilities of electronic devices. Changes in protocols in terms of materials and fabrication processes and parameters have led to novel and unique devices to meet the challenging specifications of different applications.

If the device is to be integrated into a soft, biological structure, then an important feature would be the stretchability of the material [113]. Different materials are investigated for such applications, such as CNT platforms for flexible and stretchable electronics [114] and CNT networks as a channel semiconductor in flexible thin-film transistors, and as stretchable conductors in various sensors. A recent example is the incorporation of CNTs for the development of flexible logic gates (Fig. 4.24a) and in electronic skins, where OLEDs and pressure-sensitive rubber are connected such that the OLEDs will light up according to localized pressure (Fig. 4.24b, c). An important material for implantable biomedical devices is graphene, which has been shown to work as a protective, antibacterial, conductive, and bioactive overlayer [115]. Promising materials that can be used in

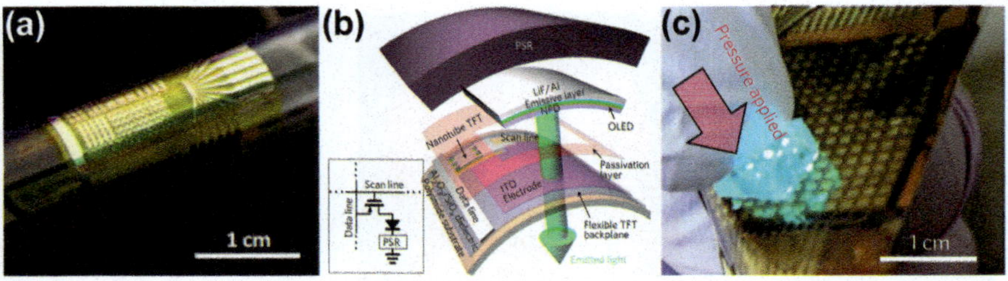

Fig. 4.24 Flexible logic gates using CNT-TFTs: **a** Optical image showing the flexibility of the circuits; the sample is wrapped around a glass tube with a diameter of 10 mm. Reprinted with permission from [117]. © American Chemical Society 2012. **b** User interactive e-skin using CNT-TFT, an OLED and a layer of PSR integrated on a PI substrate. **c** Operation of the e-skin showing OLEDs turned locally where pressure is applied. Reprinted by permission from Macmillan Publishers Ltd: Nature Materials, © 2013 [118]

biomedical devices for different levels of stretchability [116] are inorganic materials on soft elastomers, which, however, limit large deformations of the elastomer. This will be discussed again in Sect. 4.4.

4.3.1 Substrates for Flexible and Stretchable Electronics

Ideally, substrate materials used for flexible and stretchable electronics to replace standard rigid materials should possess specific properties, depending on the application and device/system functionality and required performance. These include properties such as a low coefficient of thermal expansion (CTE), thermal stability, solvent resistance, and dimensional stability. In addition, they must be resistant to the absorption of gases and humidity. Some of these, such as dimensional stability, are by default not possible with stretchable substrates. It is important to note that while these may not be necessary for the operation of a device, they will most probably be critical in the fabrication process. Properties related to solvents, particularly temperature, are especially important, as in many cases these dictate the fabrication processes and device materials used for a specific substrate material. For example, Si typically requires processing temperatures exceeding 500 °C, which is significantly higher than the glass transition temperature of common plastic substrates, e.g. PI, which has one of the highest at 270 °C [43]. Resistance to humidity and solvent absorption are important, as these may significantly influence the substrate's properties, e.g. its electrical resistance. Typically, substrates are used as insulating materials. Localized changes in the substrate's resistance would be problematic, as they would create short circuits, potentially leading to device failure. Other important substrate material properties include the material's ease of handling, if it is rollable (e.g. for roll-to-roll fabrication processes) or foldable, flatness (crucial for the quality of deposited thin films), weight, and conformability [43].

Plastic polymeric materials are dominant over others (e.g. metal, thin glass), due to their transparency (depending on the application) and bendability. Typically such materials are characterized by a low glass transition temperature. A major research topic has thus been the development of materials, inks, and fabrication methods that require low processing temperatures (below 200 °C). According to [43], polymers can be divided into three main categories: (1) semi-crystalline, (2) amorphous, and (3) solution cast amorphous polymers. Group (1) includes PET, polyetheretherketone (PEEK) and PEN. Group (2) includes polyethersulphone (PES) and PC. These are non-crystalline thermoplastics, which are solvent cast or melt-extruded [43]. Other amorphous polymer materials used in flexible electronics are polyarylate (PAR), polynorbornene (PNB), and (most importantly) PI. Depending on the processing temperatures during device fabrication or device operation, the CTE difference and the induced thermal stresses between substrate material and device materials may be important. For light-emitting and display applications, where light emission takes place through the substrate, the optical clarity of the substrate will also be important. PI, for example, is transparent, but it is yellow.

PDMS is a polymer that has been widely used for microfluidics and biosensing due to its biocompatibility, ease of fabrication, and ability to be bonded to many materials to form complex multilayer systems. A promising route for development of compliant MEMS and microelectronics is using stretchable bioelectronics through patterning of Au conductors embedded in PDMS films [28]. Metal patterns were fabricated on PDMS substrate by deposition of an intermediate parylene layer [119]. Another commonly used biocompatible polymer is PMMA, which has been used e.g. as a coating for intravascular medical devices [120]. These (PMMA and PDMS) were compared and were both suggested for use in implantable devices in the bladder [121]. One of the limitations of PDMS is that it has low impact resistance.

Different flexible polymers, such as PI, parylene, and liquid crystal polymers (LCP), have been used in implantable electrode arrays, structural membranes, and telemetric coils [122]. LCP-based materials, because of their lower water absorption, lower coefficient of thermal expansion, and biocompatibility, have been shown to work reliably in the warm, moist, and challenging environment of the human body [123, 124]. Using self-assembly of non-ionic surfactant (N,N-bis (2-hydroxyethyl) dodecanamide), liquid crystals showed excellent biocompatibility and high conductivity. The conductivity of the system was adjusted with change in the humidity and temperature [125].

Other substrates used include fabrics and transient bioresorbable/biodegradable substrates, such as rice paper and silk, discussed in Sect. 4.1.2.4. A material is termed biodegradable if it has the ability to be broken down and dissolved without the release of toxic byproducts [126–128]. Rogers and co-workers have reported electronic devices based on Si nanomembranes that are stretchable and also soluble under physiological conditions [18]. A very exciting development is the fabrication of all-organic circuits from components that are all biodegradable [129]; semiconductors, electrodes, and substrates could be tailored by synthesis to degrade. In an early example of a biodegradable organic electronic device, Bettinger et al.

fabricated a thin-film transistor on a resorbable PLGA substrate; another substrate for biodegradable organic devices is poly(lactic acid) [130]. Silk fibroin is both highly elastic and biodegradable, as discussed earlier with respect to Fig. 4.8 [131]. Liu et al. produced an organic solar cell where the substrate used was biodegradable [132]. Melanin, one naturally occurring organic semiconductor polymer in the body, is degraded into harmless byproducts. The drawback is that melanin has a low mobility of charge carriers and is mechanically brittle [126, 133]. Ideally, good charge transport properties should be combined with programmable degradation, as discussed in [113, 134–136].

4.3.2 Conductive Polymers

Conductive polymers (CP) represent an important class of materials for flexible electronics and in particular for biomedical applications [137]. They find application in many areas [138], such as tissue engineering (biocompatible, biodegradable scaffolds containing stimuli to enhance tissue regeneration), neural probes (implantable electrodes for recording or stimulating neurons), biosensors (devices containing biomolecules as sensing elements), drug delivery (devices for storage and controlled release of drugs [139, 140]), and bio-actuators (devices to create mechanical force that could be used as "artificial muscle"-type actuators). CPs are interesting materials with important applications in biology and medicine because they interact with the biological samples but still maintain good biocompatibility [141]. Different types of CPs can be generated, depending on the dopant used. Consequently, there are CPs with p-type dopants (oxidants) [142, 143] and n-type dopants (reducing agents) [144].

Over the last decade, growing interest has been observed in the development of tactile sensing for artificial skins [145]. The focus is on the development of such systems combining simple, low-cost preparation, high flexibility, and low power consumption. Depending on the physical nature of the transduction method there are piezoelectric [146], optical [147], magnetic [148], ultrasonic [149], resistive, and capacitive [150, 151] materials. Materials with piezoresistive properties are reported [145] to be the best candidates for artificial sensing "skins" which can emulate the sensing of the human skin and simultaneously fit the shape of the robot structure or artificial limb. CPs have an important role in these developments.

CPs possess very attractive mechanical, optical, electronic, magnetic, wetting, and microwave-absorbing properties and can be designed to have high electrical conductivity and good environmental stability, which are quite important. Their electrical properties can be tailored to approach those of metals and inorganic semiconductors. At the same time, however, they can retain a number of polymer properties, including material flexibility and ease of processing and synthesis. In the field of nanoscience and nanotechnology various conductive polymers, such as polyacetylene (PA), polypyrrole (PPy), poly(p-phenylene-vinylene) (PPV), poly (3,4–ethylene dioxythiophene) (PEDOT), polyfuran (PF), and polythiophene (PTh), have drawn significant attention [152]. There are different types of CPs based on their type of electrical charge, such as delocalized pi electrons, ions, and conductive nanomaterials.

In the field of bioelectronics, a new class of materials has emerged known as conductive polymer composites (CPC). These are comprised of one or more non-conductive polymers and a conductive filler material, which is distributed inside the polymer matrix. CNTs and metal NPs are often used as filler materials. Composite nanofibers with high surface area produced using electrospinning have also been used for mixing CPs with polymer solutions. Percolation theory is used to determine the conductivity of the network of filler formed within the matrix at a given weight percentage. A percolation threshold is observed when the conductivity is decreased because of interrupted paths of electron transport. Another new class of CPs is known as intrinsically conductive polymers (ICP). These consist of monomers that can be reduced or oxidized such that they can be made conductive, e.g. polyaniline (PANI), PA, PPy, PTh. There are also redox polymers which are less conductive compared to ICPs due to localized redox sites. In ICPs the conductivity is achieved through electron flow. Their use in electrochemical sensing is limited because of their very low ionic conductivity at room temperature. ICPs have a very flexible structure, which can be modified, and efficient transfer of electrons is observed. They are commonly used in biosensor development as intermediate layers and in electronics for signal readout. This type of polymer is biocompatible, which makes them good for entrapment of biomolecules [153–155].

Conjugated CPs contain delocalized electron networks, due to alternating single and double bonds. Recently, there has been significant interest in the development of composite-material sensors by combining Si and polymers. Such materials are used for embedded Si-based sensing elements in a polymer skin [156] and packaging of silicone-based elements in a protective polymer [157]. In the case of curved surfaces, it is challenging to use Si and other inorganic materials, because of their brittleness. Flexible sensors containing multi-wall CNTs in PDMS and liquid metal interconnects as an elastic conductive material are also used [158]. Another development is the highly twistable tactile sensing array intended for artificial skin applications (Fig. 4.25). This was made by dispersing a CP, which served as the tactile sensing element, on a grid of sensing electrodes [159]. The stretchable sensing electrodes and scanning traces were helical and were implemented by winding copper wires around elastic nylon fibers. PDMS was used as the base material. Pressure and thermal sensors have also been prepared by processing PI or PEN and organic transistor-based electronic circuits [8]. Cross-talk is eliminated by using a grid of Cu electrodes [160].

A major challenge has been the immobilization of the CP-matrix on the substrate for CP-based electrochemical sensors [141]. Some of the techniques used, such as photolithography and e-beam lithography, are not preferred because of possible induced changes in the polymer composites. Other methods proposed include soft lithography, electrochemical deposition, and inkjet printing. Chemical conjugation

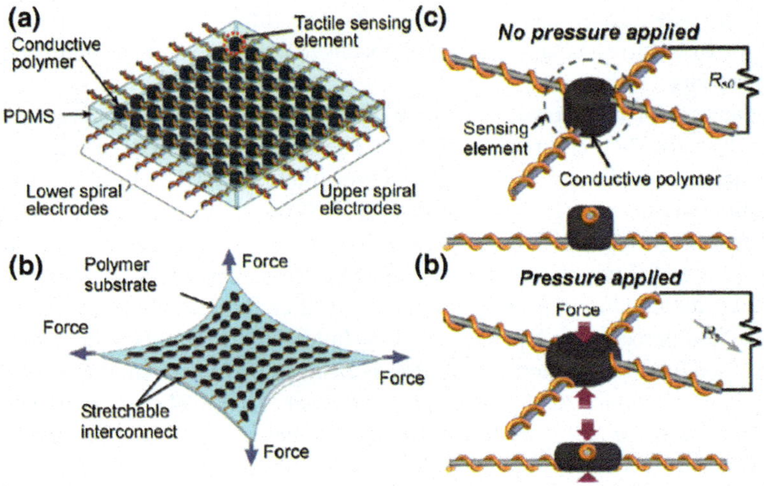

Fig. 4.25 **a** Illustration of the tactile sensing array. **b** The same device under stretching. **c** A sensor pixel without and with **d** applied pressure. Reprinted from [159]. © 2011, with permission from Elsevier

of the CP to the substrate surface [161, 162] is another possible form of integration. MEMS technology can contribute towards minimizing the sensing elements such that a large number of sensing elements can be deployed per unit area [145].

4.3.3 Graphene

Carbon is the fourth most common element and is the fundamental building block of life. It is commonly found in its elemental forms as coal, diamond, and graphite, and is used in a wide range of applications, including carbon electrodes for the production of silicon and other materials and in batteries (zinc-carbon batteries), as carbon-based materials can withstand large currents and temperatures [163]. Graphite is a good electrical and thermal conductor, but the conductivity is anisotropic, differing along the different crystal axes, i.e. it is different along a graphene sheet and different along stacked graphene sheets, while it demonstrates no inherent electrochemistry [163]. Graphene is a 2D monolayer of sp^2-bonded carbon atoms in a honeycomb lattice and a flexible transparent conductor with a large surface-area-to-volume ratio, mechanical flexibility, high stability in electrochemical environments, and good electronic and optical properties [164]. Similarly to graphite, pristine graphene does not demonstrate inherent electrochemistry. Single-layer graphene sheets can be wrapped into 0D fullerenes, rolled into 1D nanotubes and stacked into 3D graphite. According to [165], the electronic structure of graphene evolves rapidly as the number of stacked layers increases, approaching the 3D limit of graphite with 10 layers. Single-layer and, to some approximation, two-layer graphene are zero-gap semiconductors with one type of electron and one type of hole. With 3–10 layers, a number of different charge carriers appear and conduction and valence bands start to overlap, complicating the electronic spectra of the material [165]. Structures with more than 10 layers are considered as graphite. It is clear from the above that the term *graphene* is used to describe a range of different materials, often fabricated using different methodologies and consequently with considerably different electrical, electrochemical, and other properties. This has been addressed and discussed in detail in [163]. Few layer graphene (FLG) and its use in making electronic devices was first demonstrated in 2004 [166]. Since then, a wide range of electronic devices have been proposed based on this material, such as the one presented in [167], where a stack of different metallic graphene layers, insulating hexagonal boron nitride layers and various semiconducting monolayers, was used to implement an LED. The remarkably high carrier mobilities and high transconductances achieved with graphene allow its use in radio frequency (RF) applications, where analog transistors with cutoff frequencies in the range of hundreds of gigahertz have been demonstrated [168]. Graphene has no bandgap, which makes it is good for other applications, but its use for digital transistors (used as switches) is not advantageous. This is because a low off-state current is not achievable. Figure 4.26 illustrates the cross-section of a graphene-based transistor with a back-gate.

A number of different flexible and stretchable graphene transistors have been presented in recent years [170, 171]. In [171] an all-graphene-based FET array was presented on a stretchable rubber substrate (Fig. 4.27). Graphene was used as the channel and the source and drain electrodes, and no metals were used. The insulating gate material was an ion gel dielectric (poly(styrenemethyl methacrylatestyrene) (PSPMMA-PS) triblock copolymer and 1-ethyl-3-methylimidazolium bis (trifluoromethylsulfonyl)imide ([EMIM][TFSI]) ionic liquid in ethyl acetate solvent (weight ratio of 0.1:0.9:9)), and the gate electrode was made out of PEDOT:PSS, both printed using aerosol jet printing. The devices demonstrated stable operation when stretched up to 5% following 1000 stretching cycles.

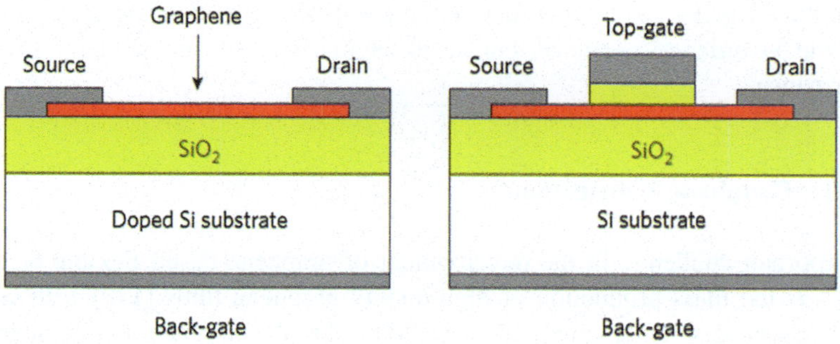

Fig. 4.26 Illustration of graphene-based transistor with a back-gate. Reprinted by permission from Macmillan Publishers Ltd: Nature Nanotechnology [169]. © 2010

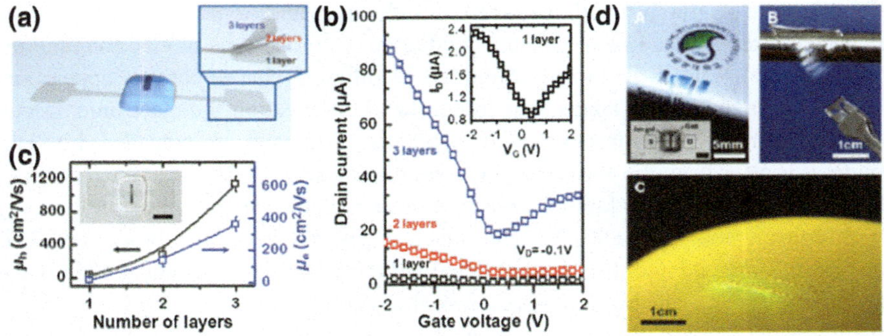

Fig. 4.27 **a** Illustration of the fabricated device developed using one, two, or three graphene layers. **b** The IV characteristics of the devices for different graphene layers and (inset) enlarged single layer device response. **c** The hole and electron mobilities of the devices as a function of the number of graphene layers. The inset demonstrates a fabricated device on a PDMS substrate, with the scale bar equal to 300 μm. **d** Examples of fabricated devices on different substrates: (*A*) PET, (*B*) PDMS, (*C*) balloon. Reprinted with permission from [171]. © American Chemical Society 2011

Apart from its use in transistors, graphene is also used for photodetectors [172–177]. The gapless band structure of graphene allows it to exhibit a broad range of photodetection ranging from UV to terahertz. This is not possible with any other material. Among other unique properties, it is also characterized by high carrier mobility, which allows high-speed conversion of photons or plasmons into electrical currents or voltages [177]. Several different mechanisms may be responsible for photodetection in graphene, including the photovoltaic effect, the bolometric effect, the plasma-wave assisted mechanism, the photo-thermoelectric effect and the photogating effect. These are discussed in more detail in [177]. As discussed in the same review paper, there are many different classes of graphene photodetectors. One classic structure is the metal-graphene-metal structure. It has been demonstrated that photo-generated carriers exist in narrow regions of the order of 200 nm at the interface of graphene with the metal electrodes due to the induced band bending from the charge transfer at the interface. In [174], the authors used an interdigitated metal electrode structure, where different metals were used for each electrode in order to create an enhanced asymmetric electric field for improved performance.

4.3.3.1 Graphene Fabrication

An important challenge in the development of graphene-based flexible electronic devices is the mass production of high-quality graphene films [178] with controllable and well-defined features. The method used to development the graphene film has a direct impact to the film's properties [163, 179, 180]. According to [163], fabrication methods can be separated into two categories: (1) top-down approaches typically involve the exfoliation of graphite using mechanical, chemical, or electrochemical methods or the opening/unzipping of CNTs. (2) Bottom-up approaches assemble small organic molecular building blocks into graphene. This is achieved using catalytic (CVD) or thermal (SiC decomposition) methods, or chemically via organic synthesis. In terms of bottom-down approaches, chemically derived graphene (CDG) and graphene oxide (GO) have attracted significant attention [163, 180, 181]. Individual sheets of GO are essentially graphene sheets decorated with oxygen functional groups. CDG (reduced GO, rGO) is another 2D form of carbon, but with additional chemical moieties, rendering it different from graphene. Chemical synthesis of graphite oxide sheets, exfoliated chemically into individual GO sheets, which are subsequently reduced (rGO), offers a simple method for CDG synthesis [181]. As discussed in [178], liquid phase exfoliation and reduction of GO has advantages in quantity, yield, and cost and is thus very promising for high-yield manufacturing. Reduction of GO can be performed by a variety of methods, which include exposure to chemicals such as hydrazines and hydrides in strongly alkaline environments, thermally above 200 °C in reducing or inert environments, electrochemically, and photocatalytically. However, none of these methods leads to complete removal of oxygen [163], while thermal reduction leads to the creation of vacancies, complicating the electrical spectra of the material further. GO has a C to O ratio of 2 while rGO can have ratios from 3 to 15 [163]. GO is thus insulating with a sheet resistance of $10^{12}\ \Omega\ \text{sq}^{-1}$ and reduction of GO substantially reduces the sheet resistance with reported values between 340 and 28.6 $\text{k}\Omega\ \text{sq}^{-1}$ [181]. The sheet resistance of rGO can be fine-tuned over 12 orders of magnitude, depending on the reduction process used [178]. These are higher than that of ideal graphene, but can be further reduced significantly with doping. Residual oxygen functional groups from the fabrication process used (e.g. in rGO) give such graphene sheets an inherent redox activity, since some of these redox processes are reversible. The solubility of GO has allowed the development of GO and rGO dispersions (the latter e.g. via chemical reduction of GO in solution), allowing large area film deposition using techniques such as drop-casting, dip coating, spin coating, and spaying.

In terms of bottom-up approaches, single- and few-layer graphene are typically grown epitaxially using CVD. Large-scale production of continuous high-quality graphene films for practical applications have proven challenging. This was achieved recently on rigid substrates and on flexible substrates for roll-to-roll production using CVD on metal foils [182, 183], as well as wafer-scale production on SiO_2/Si wafers [184]. CVD is currently considered the most promising method for high-quality industrial-scale fabrication of graphene. CVD graphene does not demonstrate inherent electrochemistry. Nevertheless, inherent electrochemistry can be demonstrated due to the exposed regions of metal substrates used, edges, and defects. Epitaxial growth on SiC is another method capable of producing wafer-scale graphene, which is, however, difficult to detach and transfer to flexible substrates [178] and expensive due to the prohibitively high cost of single-crystal SiC wafers. It also has poor electrocatalytic properties [163]. A problem common to all fabrication techniques lies in the transfer of the graphene films to flexible substrates. These could be grown, for example, via CVD on Cu or Ni and transferred to PET or PDMS. Typically, wet etching of supporting metals (Ni, Cu, Pt) is used [178, 182, 185, 186]. This step in the fabrication process, however, damages and contaminates the graphene, while its cost is forbidding [178]. As highlighted in [178], the bubbling method reported in [184] is a promising alternative. The various graphene preparation methods are summarized in [163]. Regarding electrochemical applications, one of the main issues is that graphene sheets deposited on conducting surfaces restack, hindering the electrochemical behaviour of devices. This is due to strong sheet-to-sheet van der Waals interactions [187], which essentially lead to a reduction of the specific surface area and thus capacitance. Well-defined porous 3D graphene structures address this issue. They can serve as an individual sensing platform or they can be decorated with catalytic NPs (Pt, Pd) to enhance oxygen reduction, or with metal dichalcongenides for enhanced hydrogen evolution [163]. 3D graphene is usually grown via CVD on metal substrates (Cu or Ni) and demonstrates similar electrochemistry to CVD graphene. It can also be prepared via the graphitization of 3D carbon, leading to a multi-layered graphitic material with numerous defects [163].

4.3.4 Hexagonal Boron Nitride (h-BN)

Hexagonal boron nitride (*h-BN*) is an inorganic material used in many different technologies, generally deposited by CVD. It is a good electrical insulator and can provide a direct bandgap [188]. In 1990, researchers from Japan [189] found that deposition of an AlN layer on a sapphire substrate before growing a GaN nitride film would lead to a direct bandgap of 3.39 eV. Researchers found that *h-BN* has a larger direct bandgap energy than GaN and that a single-crystal *h-BN* shows a dominant absorption band at 215 nm in high-pressure and high-temperature

Table 4.1 Experimental bandgap energy values for *h-BN*

Method	Eg (eV)	Wavelength (nm)
X-ray emission spectra	3.6	344
X-ray emission spectra	3.85 ± 0.05	322 ± 4
Optical absorption spectra	3.9	318
UV absorption spectra	4.3	288
Reflection spectra	4.5	275
Optical reflectivity	5.2 ± 0.2	238 ± 9
Multiband luminescence spectra	≥ 5.5	225
Optical spectra	5.89	210
Reflection, absorption and photo-conductivity spectra	5.8	214
Absorption spectra	5.83	213
Temperature dependence of electrical resistivity	7.1 ± 0.1	175

conditions [190]. It is thus a promising material for UV lasers. A comparison of bandgap energy of graphite-like *h-BN* is reproduced and shown in Table 4.1 [188].

Apart from the large direct bandgap, *h-BN* is described by a number of additional properties, which make it an important material for graphene-based devices. The substrate material supporting graphene and the quality of the interface between the two play a significant role in the performance of the device [191]. Graphene-based devices are typically fabricated on an SiO_2 substrate. The SiO_2 surface roughness and scattering induced due to charge surface states and impurities lead to graphene carrier mobilities inferior to those theoretically expected. Other oxides have similar limitations. *h-BN*, on the other hand, has an atomically smooth planar surface that suppresses rippling of graphene, with a lattice very similar to it (lattice mismatch 1.7%). It is insulating with dielectric properties similar to SiO_2 and is relatively free of dangling bonds and surface charge traps [192]. Consequently, graphene-based devices on *h-BN* substrates have demonstrated improved performance with mobilities an order of magnitude better than devices on SiO_2 substrates [192]. It can thus be used as a transistor gate dielectric, as demonstrated in [193–195] and Figs. 4.28 and 4.29. A comparison of the graphene transistors of Fig. 4.28b from [195] fabricated on SiO_2 and *h-BN* substrates demonstrated the superior performance of the device on the *h-BN* substrate with a small-signal transconductance and an effective carrier mobility that were 8.5 and 4 times greater, respectively. In [194] atomic layers of graphene/*h-BN* were peeled off and transferred to flexible substrates, such as PDMS and PMMA, while in [196] a graphene/*h-BN* transistor on a flexible PI film was presented (Fig. 4.29).

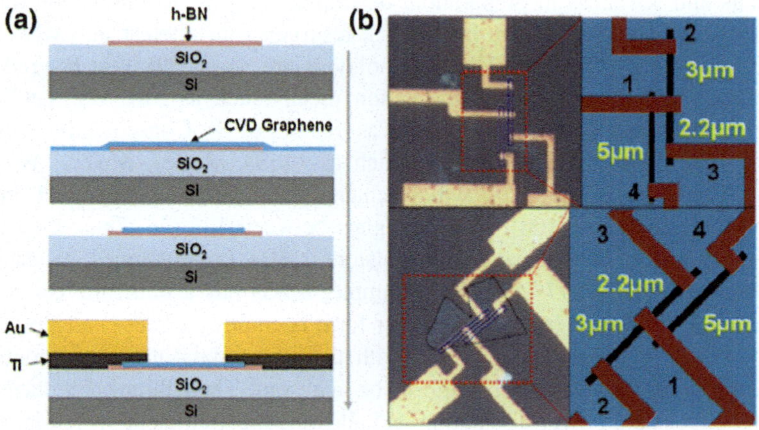

Fig. 4.28 **a** Fabrication process of the graphene FET on h-BN. **b** Top: graphene FET on SiO_2. Bottom: graphene FET on h-BN. The numbers illustrate that there are three different devices with a common source terminal and with different gate lengths, but the same gate width of 500 nm. Reprinted from [195], with the permission of AIP Publishing

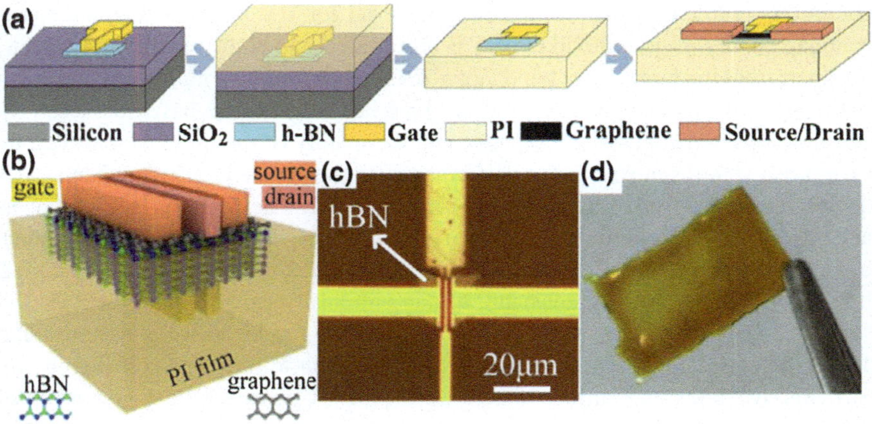

Fig. 4.29 **a** The fabrication process of the graphene/*h-BN* transistor on PI. **b** A fabricated device illustration. **c** Image of a fabricated device. **d** The device on the PI substrate. Reprinted with permission from [196]. © IEEE 2013

4.3.5 Transition Metal Dichalcogenides (TMDs)

Although graphene-based transistors have a high carrier mobility, they typically have a low current on/off ratio and cannot be effectively switched off due to the lack of a bandgap [168]. TMDs are emerging as the potential alternatives to graphene. They have been studied for decades, but following the growing interest in 2D materials and advances in preparation, detection, analysis, manipulation, and fabrication of 2D materials, there has been a resurgence in interest in TMDs and in particular in their 2D forms. These monolayers are atomically thin layered materials, composed of a layer of a transition metal (such as Mo, W), sandwiched between two layers of a calcogen (such as Te, Se, or S). Consequently, semiconducting 2D TMDs include materials such as MoS_2, $MoSe_2$, $MoTe_2$, WS_2, and WSe_2. These have bandgaps which change from indirect to direct in their single-layer 2D forms [168], with bandgaps between 1–2 eV. Thus, they can be used to implement devices such as transistors, LEDs, photovoltaics, batteries, and photodetectors [168, 197–200]. For example, MoS_2 has a bulk indirect bandgap equal to 1.3 eV, which in its single layer form becomes a direct bandgap equal to 1.8 eV [168]. The bandgap possible with these materials allows, in contrast to graphene, a high on/off current ratio to be achieved. Their thinness reduces short channel-effects and power dissipation and allows a more efficient switching control [168]. They have a mobility index comparable to that of Si. TMDs are also compatible with flexible opto/electronics. Single-layer MoS_2 is 30 times stronger than steel and is deformable up to 11% before breaking [168]. Similar methods to those used for graphene fabrication can also be used for TMDs. These include mechanical, chemical, or ultrasonic exfoliation or CVD. The above, together with their relatively high Earth abundance, justify the recent interest in TMDs.

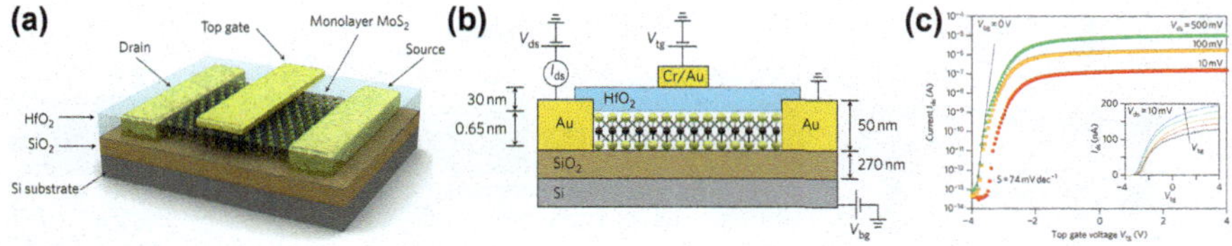

Fig. 4.30 **a** Structure of MoS_2 FET, **b** cross-sectional view of MoS_2 FET with electrical connections, and **c** IV characteristic curves of the MoS_2. Reprinted by permission from Macmillan Publishers Ltd on behalf of Cancer Research UK: [201]. © 2011

A monolayer MoS_2 FET with a high current on/off ratio ($\sim 10^8$), subthreshold slope of 74 mV per decade, n-type conduction, and a mobility at room temperature of >200 $cm^2\ V^{-1}\ S^{-1}$, was presented in [201]. Its structure and its IV characteristics are shown in Fig. 4.30. The $MoTe_2$-based transistor presented in [202] achieved an on/off ratio greater than 10^6, a subthreshold swing of $\sim$140 mV per decade and mobilities in the region of 10 $cm^2\ V^{-1}\ s^{-1}$. A flexible transparent FET fabricated on a polymeric substrate (PEN) was demonstrated in [203]. This combined a number of 2D materials that were mechanically stacked together. In detail, semiconducting MoS_2 was used as the channel, insulating *h-BN* as the dielectric and conductive graphene as the gate electrode. The devices demonstrated a mobility of 45 $cm^2\ V^{-1}\ s^{-1}$ and an unchanged performance up to a strain of 1.5%. More information regarding 2D-based material electronics can be found in [204].

The direct bandgaps make TMDs highly attractive for optoelectronic applications. A flexible TMD/polymer band-selective photodetector on a filter paper substrate was recently presented in [205]. A novel fabrication method was demonstrated with the mixing of a ratio (5:5) of two TMD materials, $MoSe_2$ and MoS_2. This resulted in a photodetector that can perform detection in both the invisible and visible spectra (at wavelengths from 532 to 1064 nm with light intensities of 63 and 230 mW cm^{-2}, respectively). More information regarding 2D material-based photodetectors can be found in [177].

TMD nanosheets can also be embedded in one of the layers as a hole injection layer (called bulk TMDs) in organic light-emitting diodes (OLED) [206]. By treating with an UVO process, it was assumed that the UVO introduced oxygen into the lattices of the TMDs' nanosheets. The luminance performance was increased and the operating voltage was down to 4.4 V [206]. The structure of the OLED and an image of the OLED stability over time without encapsulation are shown in Fig. 4.31.

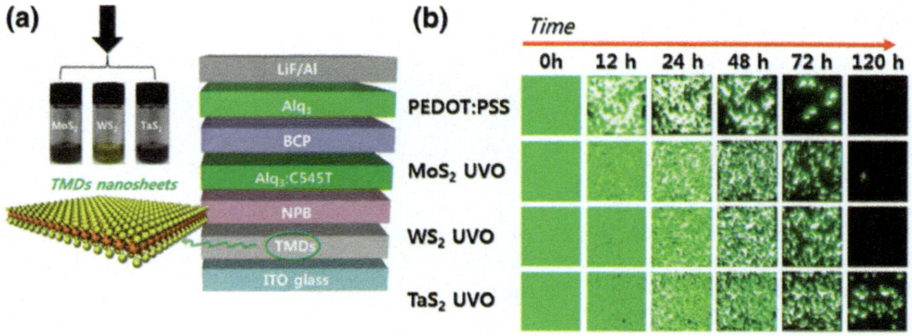

Fig. 4.31 **a** The structure of the OLED and **b** the stability of four different OLEDs over time without encapsulation and when exposed in air. Reprinted with permission from [206]. © Wiley 2015

4.3.6 Organic Materials

Organic materials and their use in opto/electronics have received significant interest in recent decades. A wide variety of devices have been developed using such materials, such as OLEDs, OTFTs, and organic solar cells. The term "organic electronic materials" is used to refer to carbon-rich compounds, which can be categorized into three main groups: small molecules, polymers, and biological materials [36]. Thus, many such devices have already been discussed in previous sections. Organic materials have been found to be a useful and low-cost solution in many optical and electronic applications, such as large area displays, photovoltaics, and biosensors [207, 208]. In contrast to silicon-based devices, which require processing temperatures exceeding 200 °C, typically above 800 °C, organic-based devices can be fabricated near room temperature. Aromatic and conjugated materials are typically used as the active semiconducting layer in OFETs. The term "small molecules" is used to describe organic compounds with well-defined molecular weight [36]. Semiconducting small molecules such as tetracene, pentacene, sexithiophene, copper phthalocyanine, and rubrene are often used. Pentacene is the most common organic small molecule semiconducting material used for organic transistors, because it generally provides the largest carrier mobilities [209], with hole mobilities of 5 $cm^2\ V^{-1}\ s^{-1}$ and even up to 35 $cm^2\ V^{-1}\ s^{-1}$ reported [210]. Pentacene transistors and ring oscillators on a paper substrate (hot-pressed cotton-fiber paper) were presented in [211]. The devices had a carrier mobility of 0.2 $cm^2\ V^{-1}\ s^{-1}$ and an on/off ratio of 10^6. A flexible pressure sensing array with pentacene transistors is shown in Fig. 4.32. A disadvantage of pentacene is that it is easily oxidized when exposed to oxygen and thus the carrier mobility is decreased.

A challenge in the development of organic transistors has been the development of n-type devices. Materials such as the fullerene C_{60} and hexadecafluorocopper-phthalocyanine ($F_{16}CuPc$) have been used successfully to develop n-type devices,

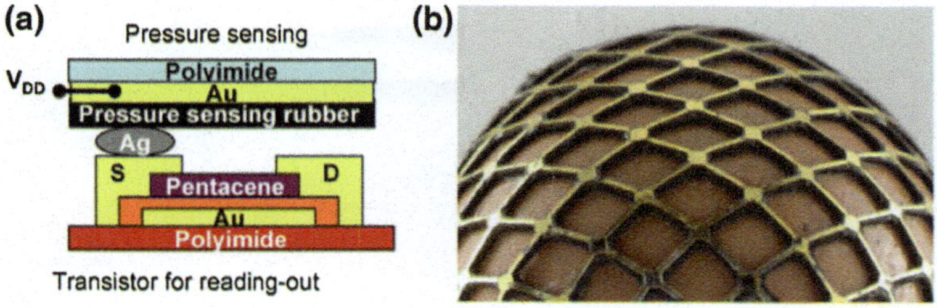

Fig. 4.32 **a** A pressure sensor comprised of a pressure sensing rubber layer and a pentacene-based organic FET. This device forms the unit cells used in the realization of the active matrix pressure sensing array illustrated in **b** being used to cover an egg. The device was processed with a CO_2 laser to pattern it. Reprinted with permission from [214]. © Wiley 2010

with the latter allowing air-stable devices [209]. In addition to small molecules, polymers such as PPV, poly(3-hexylthiophene) (P3HT) and polydiacetylene are also commonly used. In contrast to small molecules, polymers are long-chain molecules with an "indeterminate number of molecular repeat units" [36]. PPV is an emissive semiconductor often used in OLEDs. An OFET developed using biocompatible and flexible materials was used in [212] for bidirectional stimulation and recording of primary neurons. The development of OTFTs and OLEDs together on the same bendable plastic substrate has recently been achieved by managing to keep the temperatures throughout the whole fabrication process below 150 °C [213]. The luminance of the presented 2.5" full-color, top-emission display did not significantly degrade within a storing period of over two months.

A number of organic biocompatible and biodegradable materials, that are also environmentally friendly, were presented in [135, 136]. Materials such as caramelized glucose, plastics based on potato and corn starch (Ecoflex by BASF), and hard edible gelatin can be used as substrates. Materials such as nucleobases (adenine and guanine), DNA, sugars (glucose, sucrose, and lactose) and caffeine can be used as dielectrics, while dyes used as colorings in the textile, cosmetic, food, and other industries, such as indanthrene yellow G, indanthrene brilliant orange RF, perylene diimide, p-type beta-carotene, and n-type indigo [134], can be used as semiconductors. The authors presented an edible OFET based on an aurin-coated gelatin substrate, a guanine and adenine gate dielectric, perylene diimide as the organic semiconductor, and gold for the source, drain, and gate electrodes. The device demonstrated a mobility of 0.02 $cm^2\ V^{-1}\ S^{-1}$ and a gate capacitance per unit area of 5.1 $nF\ cm^{-2}$.

Fabrication methods commonly used with organic materials include solution deposition methods (spin-on or spray-on, e.g. polymers which are solution processed in general), inkjet printing, vapor-phase deposition (such as vacuum thermal evaporation (VTE) or organic vapor phase deposition (OVPD), typically used for small molecules), thermal imaging dry transfer (using laser ablation of materials from a donor substrate in contact with the device substrate), and imprint lithography, self-assembly techniques and cold welding using stamps [36].

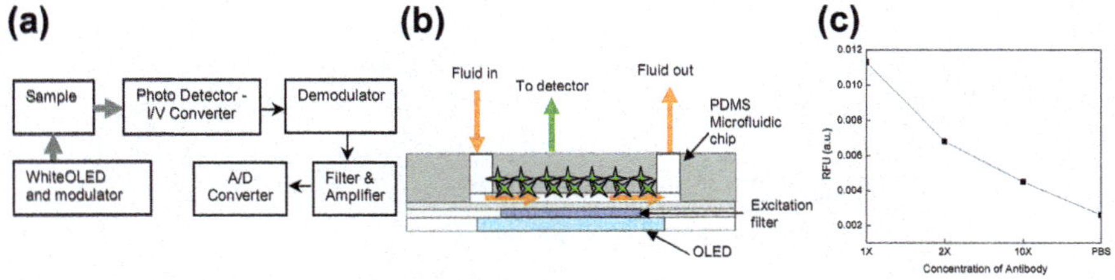

Fig. 4.33 **a** Block diagram of the white OLED system. **b** Side view of the microfluidic-based optical biosensing system and **c** correlation of fluorescence with different antibody concentrations. Reprinted with permission from [215]. © IEEE 2008

A microfluidic chip integrated with organic photodetectors and light-emitting devices was presented in [215]. A system block diagram and a side view of the microfluidic chip integrated with a biosensor are shown in Fig. 4.33a, b. OLED-based photodetectors detected fluorescence changes from a 0.012 μl volume chamber containing 3 ng of donkey anti-sheep IgG (immunoglobulin G) mixed with Alexafluor 488 fluorophores. Eight detectors were fabricated around the white OLED and the microfluidic biosensor on a glass substrate. The white OLED emitted the whole spectrum from 350 to 750 nm, while by using different organic materials, photodetectors with sensitivities to different wavelengths were fabricated. Figure 4.33c shows a fluorescence correlation at different antibody concentrations [215]. A wide range of flexible/stretchable devices based on organic materials were recently discussed in [214].

4.4 Trace Patterns for Flexible/Stretchable Electronics

Apart from fabrication protocols and the selection of substrate and device materials, another method of enhancing the tensile strength and thus the flexibility and stretchability of devices is via their geometrical design [216–218]. Figure 4.34 demonstrates the difference between different substrates and various scenarios for interconnect design for stretchable electronics. In coplanar-type devices, the geometry of the interconnect has been optimized such that deformations are coped with in 2D. In non-coplanar type designs, deformations are coped with in 3D, while in compound type devices, the substrate itself forms part of the device by incorporating within the polymer various materials (e.g. CNTs, Ag nanowires, or metal NPs [219], e.g. Fig. 4.17) [219, 220].

Thus far, a major focus in stretchable electronics has been on the optimization of the interconnect geometry, to form what is called an island-bridge layout. In this approach, active devices reside in the island and are connected with each other via bridges formed by the interconnects [221]. In this way, interconnects deform upon stretching and the active devices undertake minimal deformations. Some such devices were discussed and illustrated in previous sections of this chapter (Figs. 4.7 and 4.10). One way to achieve this is to use a prestretched PDMS substrate with

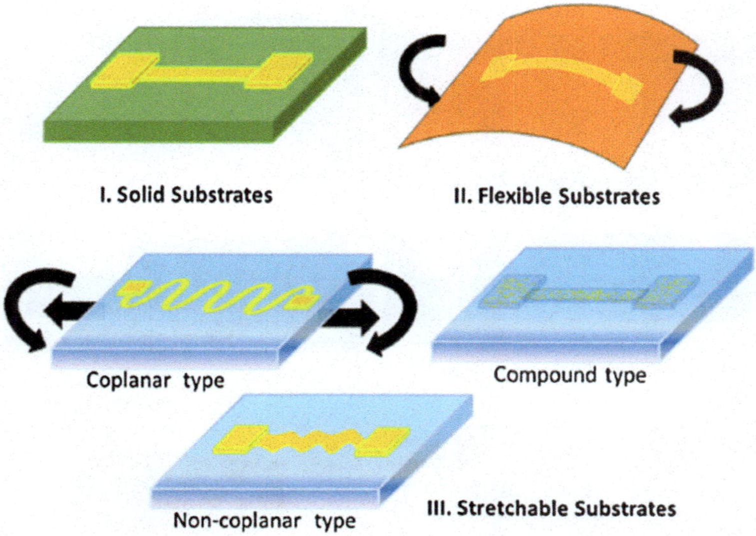

Fig. 4.34 Types of substrates and methodologies for stretchable interconnects [219]

patterned trenches and to transfer print the fabricated device onto it [222]. After releasing the substrate, the interconnects form arc bridges that hang over the PDMS trenches and below the device plane (Fig. 4.35). Alternatively, flat elastomers can be used, in which case upon release the arced interconnects become buckled and deform above the island plane [116]. To achieve this, the PDMS substrate needs to be treated (with ozone) such that strong bonds are formed with the islands and weak bonds with the interconnects, as discussed with respect to Fig. 4.12. It must be highlighted that all inorganic semiconductors are brittle, fracturing with strains as low as 1%. Thus, a major challenge is to enhance, via optimized mechanical design, the stretchability of these materials in particular, which are crucial for device development. Inorganic films become highly flexible to out-of-place bending when their thickness is in the range of 100 nm [116].

Direct deposition of materials on pre-stretched substrates [116, 223, 224] is another method of creating flexible interconnects and devices. Releasing the substrate leads to a wave structure. Allowable strains of up to 28% have been reported using this technique [219].

Another approach is to use serpentine interconnects that can be fully bonded, partially bonded, or completely non-bonded on the substrate [221]. Fully bonded serpentine interconnects on prestrained substrates are discussed in detail in [221]. A silicone rubber substrate (Ecoflex, Smooth-On, Inc.) was used with a transfer-printed 50 μm wide serpentine interconnect composed of a 0.3 μm thick copper tracks sandwiched between 1.2 μm thick PI layers. The serpentine design is composed of semicircles connected with straight lines with a length and distance of 500 μm. Supporting Cu wires with PI, rather than directly placing the metal on the

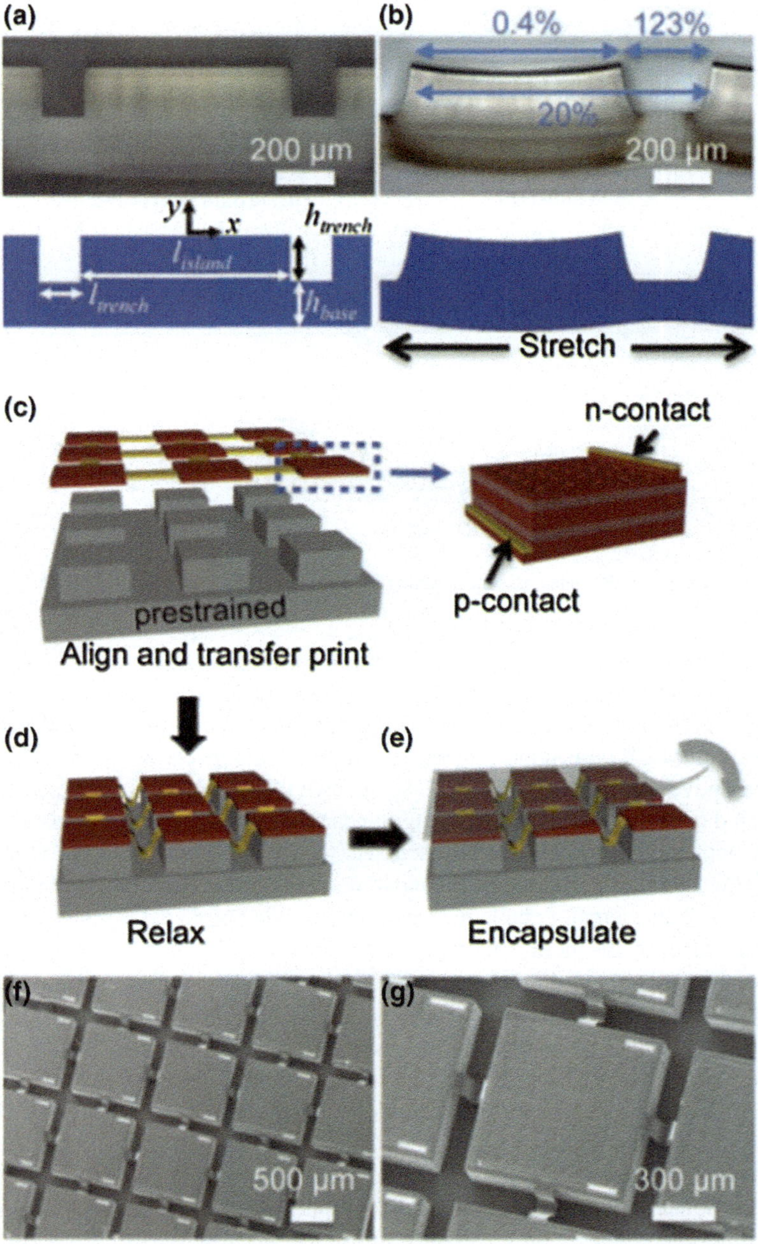

Fig. 4.35 Images and FEM model validation of **a** a relaxed and **b** a stretched elastomeric substrate and **c–e** fabrication steps of a stretchable GaAs photovoltaic device in an island-bridge topology; **f**, **g** are SEM close-ups of the islands and bridges. Reprinted with permission from [222]. © Wiley 2011

elastomer, has been shown to enhance the stretchability and durability of stretchable interconnects, with improved elongation of up to 30% larger according to [225]. Experiments demonstrated that the elastic stretchability of the serpentine interconnect is 53% if no prestretching is used, which becomes 185% when 85% prestretching is used. Total stretchability changed from ~137 to 283%. It is important to note that for highly prestretched samples, buckling takes place when releasing them [221, 226]. This is out-of-plane bending and twisting, particularly of the semicircular regions, as the straight lines stay vertical when compressed, experiencing less bending. Detailed discussion and FEM simulation of horseshoe and other interconnect designs can be found in [225–230]. The authors discuss that a similar approach can be used with horseshoe and self-similar serpentine [231] interconnects. It is important to highlight the results of [232], where the effect of the distance between adjacent parallel interconnects on stretchability was evaluated. The strain experienced by parallel interconnects with distances greater than 2.5 mm was found to be the same as that of a single interconnect. Smaller distances, however, led to higher strains. Simulation presented in [219] based on the geometrical parameterization shown in Fig. 4.36 demonstrated that for a fixed *W*/*R* ratio, the maximum possible applied strain increases linearly up to $\theta = 115°$. For greater angles, stretchability is significantly reduced and remains essentially constant. The effect of changes of the *L*/*R* ratio depends on the other parameters and may have different effects and should thus be carefully tailored. Keeping θ constant at 90° and changing the *W*/*R* ratio demonstrated that for ratios of 0.5, 0.1, 0.033, and 0.025 the maximum applied strains were 14, 30, 52, and 48% respectively. Thus, reducing this ratio makes the interconnect more stretchable. Nevertheless, this will also increase the resistance of the track and reduce the maximum allowable current through the track. Breaking down a track into multiple parallel tracks of smaller *W* has the advantage of decreasing *W*/*R* and thus also the strain on the interconnect [229], while creating parallel current paths to reduce the total track resistance and increase the maximum current [233]. The thickness of the track is

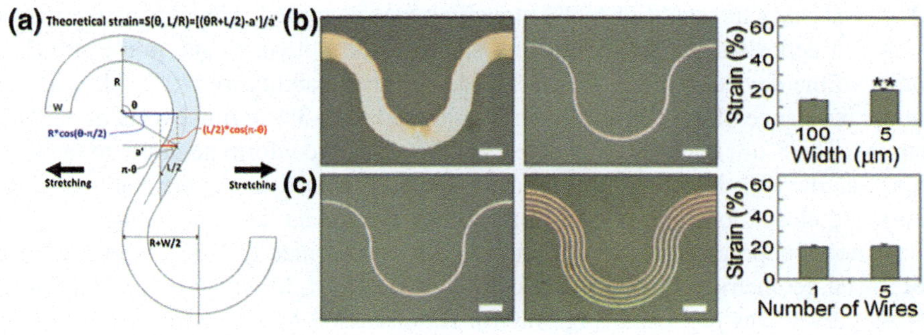

Fig. 4.36 **a** Illustration of a generic serpentine structure, demonstrating the geometrical design parameters of the structure, which are the radius (*R*), angle (*θ*), length of the straight section (*L*), and width of the track (*W*) [219]. The effect of **b** *W*/*R* ratio and **c** number of parallel conductors on the stretchability of the interconnect. Reprinted with permission from [233]. © Wiley 2004

also important. Greater thickness provides greater stretchability, as it reduces the effect of vertical compressive stresses [219].

In [218], 2D diamond-shaped interconnects were presented, which were optimized using the finite element method (FEM). A liquid alloy was used to cover and enhance points in the design, which had a greater possibility of breaking during stretching. Deposition, wet etching, and lithography were used to pattern Au on PDMS and PDMS reservoirs. Screen printing was also used to pattern a liquid alloy in these reservoirs. From calculations, the strain in the metal was found to be 18.6% by assuming a 50% strain in the elastomer. Results also showed that liquid-alloy-covered joints improved the tensile strength of the device by up to 60%, with a resistance variation of 30%.

In [217], a 400 tpi mesh screen, and an Ag paste were used to screen print conductive tracks on a 50 μm thick thermoplastic polyurethane (TPU) substrate which was stretched by 20%. The TPU was baked at 70 °C while being stretched, and the ink was sintered at 150 °C. Straight lines, rectangular, zigzag, and horseshoe structures were printed. Circuits sintered the longest (up to 60 min) with a horseshoe structure demonstrated the largest extendibility (8%).

In [216], photoablation was used for the fabrication of stretchable interconnects. The authors exploit the fact that the threshold fluence for photoablation of metals is much greater than that of polymers, in order to use photolithographically patterned metals on PI as a photoablation mask to pattern the PI. This was used to create various meandering and rectilinear two-layer interconnects. FEM simulations and resistance measurements were used to assess the stretchability of the structures. The electrical resistance comparison between rectilinear and meandering patterns demonstrated that the former fails at a uniaxial stretching of 7% and the latter at 50%, with a change in resistance of 5%.

In [234], the authors used standard microfabrication techniques to create a highly stretchable patch for adaptive thermotherapy. The interconnects were of a serpentine-type geometry and interconnected an array of thermal elements in an island-bridge approach. The interconnects demonstrated a stretchability of 800% (Fig. 4.37a). However, such high stretchabilities were not reversible. A revesibilility threshold of 600% stretching was identified. The conductive tracks were Cu deposited on PI. The heating elements did not contribute to the stretchability of the device, limiting the overall device stretchability to 270%. The change in the resistance of the thermal patch was limited to 2–3% during cycles of applied strains. The 8 μm thickness of the device allowed it to conform naturally to the skin via van der Waals forces. The heaters were also used as temperature sensors, allowing closed-loop operation of the system to regulate the temperature.

A novel approach to interconnect design was presented in [235], where the use of fractal geometries, such as von Koch, Peano, Hilbert, Moore, Vicsek, and Greek cross curves was proposed (Fig. 4.37b). As discussed in [116], fractal curves are described by self-similarity, where small sections of the geometry resemble the whole pattern. Such design can be optimized to withstand increased strains along specific dimensions and to support various deformations (e.g. radial and biaxial). It has been shown [116] that increasing the level of self-similarity increases the

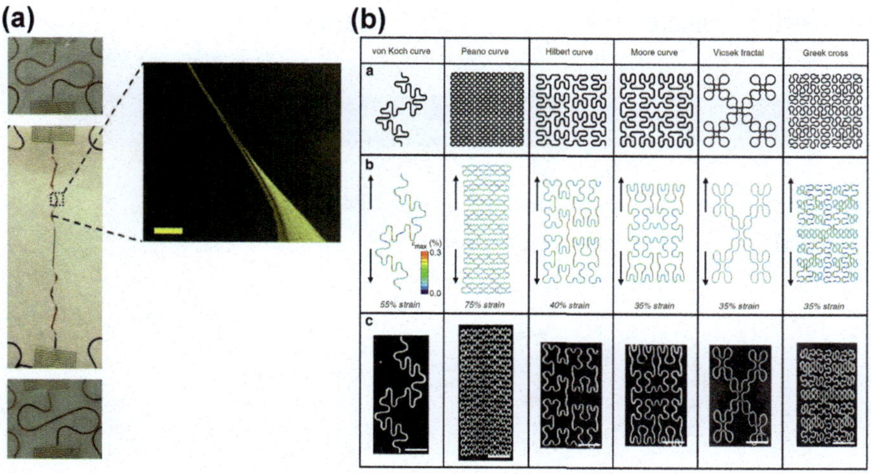

Fig. 4.37 **a** Stretchable interconnect. Top: before elongation, middle: 800% elongation; bottom: after 600 cycles of 600% elongation. The scale bar is 2 cm. Reprinted with permission from [234]. © Wiley 2014. **b** Fractal-based types of curves for flexible interconnects, electrodes, and devices. Reprinted by permission from Macmillan Publishers Ltd: Nature Communication [235]. © 2014

stretchability of an interconnect. PI-covered metal and Si-nanomembranes on elastomeric substrates using various such designs were demonstrated and were used to implement antennas, temperatures sensors, heating elements, and electrodes for electropotential recordings. Figure 4.37b [235] summarizes the tensile strain achieved by six different interconnect patterns tested under similar conditions.

The above approaches allow reduction of the stiffness of devices such that they are comparable to or smaller then the surface on which the devices are to be mounted (e.g. tissues or skin). When this is achieved, the devices can be used to perform their intended operation unobtrusively and without the risk of being damaged.

4.5 Packaging, Encapsulation, and Biocompatibility

Unfortunately, the challenges in the development of implantable medical devices extend beyond those in the design of interconnects, sensors, readout electronics, and powering. Once the aforementioned components are put in place, they need to be appropriately packaged prior to implantation. Important aspects of this fundamental and challenging step in the manufacturing of implantable medical devices are associated with the biocompatibility and hermeticity of the encapsulation of the final device. The accompanying challenges have a dramatic impact on the performance of these systems and have prevented their widespread use. According to [236], packaging materials must be non-inflammatory, non-toxic, non-allergenic,

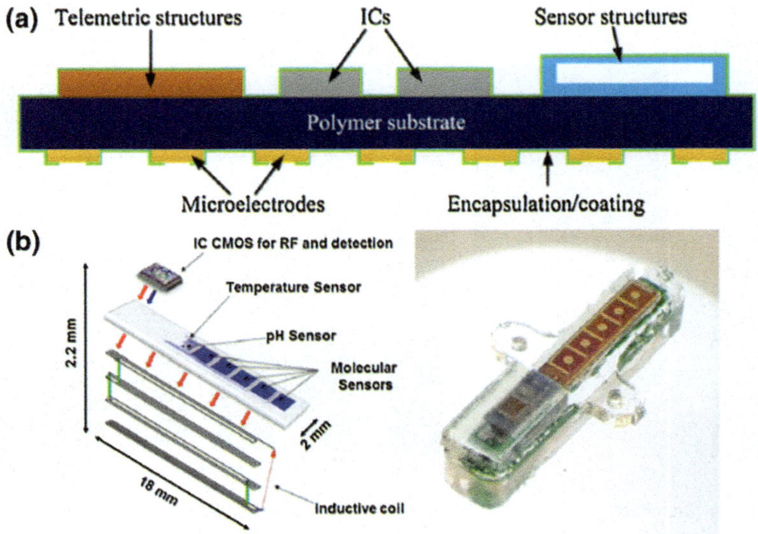

Fig. 4.38 **a** Schematic of packaging components in an implantable sensor. Reprinted from [123]. © 2014, with permission from Elsevier, **b** an implantable sensing platform. Left: assembly; right: final packaged device. Reprinted with permission from [239]. © IEEE 2015

non-carcinogenic, and non-thrombogenic (when located within the bloodstream). Therefore, the package must be biocompatible and hermetic and should not outgas, as in the case of moisture/gas penetration the device will not function [237]. These requirements will be discussed in the following sections. The issue of polymer encapsulation and packaging is discussed in detail and it is also reviewed in detail in [123]. Currently, existing technologies do not address all the challenges regarding in vivo use [238]. Figure 4.38 demonstrates examples of packaging of components and implantable platforms.

4.5.1 Encapsulation and Hermeticity

The long-term stable packaging of implantable devices is a challenging task. This is due to the limitations imposed on the task by the application itself. These devices are placed within biological tissues and organs and submerged in biological fluids which contain proteins, enzymes, cells, and other organic and inorganic materials which create a corrosive, reactive, and hostile environment for implanted devices [240]. Much work on this subject has been carried out by groups led by Najafi [240–243], Stieglitz [244–247], and Donaldson [248–250]. According to [241], size, material properties, mechanical structure and rigidity, biocompatibility, required lifetime, and maximum allowable temperature, are some of the parameters that need to be considered. The packaging and encapsulation must be mechanically durable and chemically stable and have a dual role; they need to protect the implant from the harsh biological environment and protect the tissue from the implant itself [236, 242]. These needs must be satisfied for periods from a few days or months to decades (e.g. pacemakers) depending on the application, while cables and their connections to the main device need to withstand several hundred million bending movements [236].

One of the biggest challenges is preventing corrosion. Corrosion will lead to ionic currents, which will lead to erroneous sensor recordings and eventually sensor destruction and potential bio-contamination. Thus, all implant components need to be protected against moisture and ions; water decomposition from voltage-carrying leads must also be prevented [244]. Thus, the hermeticity of the packaging is important. This is defined as a helium leakage rate below 10^{-9} mbar s^{-1} [244]. A number of approaches have been followed, which include the use of a capsule/ enclosure bonded to the device substrate, the use of thin-film protective materials, and combinations of the two approaches. The device cannot be isolated completely from its environment, as a direct interaction is required selectively and reliably between it and specific implant constituents, i.e. the sensors. Thus, openings in the packaging are often required. Typically, the sensor will be external to the main package containing the electronics and a reliable connection is required between the two. Another important aspect is that of assembly, which becomes important when a number of different systems (e.g. sensors and electronics) need to be combined [241, 242]. In addition to bio/chemical biocompatibility, which will be thoroughly discussed below, a final important aspect is that of structural biocompatibility, i.e. the mechanical interaction between the implanted device and the host environment, influenced by "sharp edges and corners as well as micromotions and mismatched Young's moduli" [244].

In terms of interfacing classical device substrates used for electronics and sensors with tissues, using Si as a substrate comes with the advantages of ease of processing, low cost, material abundance, and high biocompatibility. However, some disadvantages have been reported. During kidney tissue impedance measurements for ischemia detection, for example, AC current leakage across the substrate at high frequencies (i.e. above 5 kHz) was reported [251–253]. It was shown that at these frequencies, the capacitance of the substrate–solution interface was the main obstacle for measuring tissue impedance, as the substrate partially short-circuits the system, with injected current leaking across the substrate and generating a false image of decreasing measured impedance. Deposition of isolating passivation layers could possibly partially stop the leaks across the Si substrate, but it could be tricky if there are edge shapes, as in needles. Different coatings, such as silicone, parylene, and other organic polymers can be used to prevent current leaks across device edges. SiC is a semiconductor material mostly used in high-power, high-frequency, and high-temperature applications that can be processed in clean-room environments using standard semiconductor processes. In addition, SiC possesses several properties that make it a suitable substrate for the production of biomedical microdevices. SiC is transparent, chemically inert, and refractory, and has been shown to be a more hemocompatible material than Si and related biomaterials. Moreover, SiC is more resistant to mechanical stress than Si and possesses a wide bandgap and a high electrical breakdown constant.

Regarding hermetic capsule enclosures, Ti (metal) and alumina (Al_2O_3, a ceramic) are the most popular choices, while glass is less common. The vast majority of commercially available implantable devices use these types of encapsulation and these materials [244]. The use of Ti cans with glass-sealed feedthroughs for connection to external leads has been proven to be the most effective way of packaging implantable devices by the widespread success of cardiac pacemakers [236, 241, 242].

Bonding of the capsule to a substrate is a challenging step in implant packaging, and a number of different approaches have been proposed in the literature, with anodic (also known as electrostatic) and eutectic bonding being the most popular options [240–248]. Anodic bonding is typically applied between a glass structure (made of Pyrex 7740 glass) and a polished Si substrate [236, 240–243]. The two are brought in contact and heated to 300–400 °C. The glass is softened and its electrical resistance is reduced, and a high voltage (800–1500 V) is applied across the interface such that a permanent, irreversible, and hermetic bond is formed between the two. One such example is illustrated in Fig. 4.39. This packaging methodology was tested in accelerated conditions, where it demonstrated mean-time-to-failure (MTTF) of >100 years. In saline solutions under accelerated conditions, soaked in salt water for nine years, and implanted in animals (in the abdominal cavity, on the surface of the dura in the skull, and in the thigh in contact with the muscle in guinea pigs) for periods up to two years, it proved its effectiveness for implanted devices [240–243]. Biopsies and tissue examination demonstrated that there was no connective tissue attached to the devices implanted in the abdominal cavity and it had the least effect on the animal in contrast to other implant locations. Dura implants led to a greater reaction, demonstrating afocal fibrosis in the vicinity of the device. Finally, thigh implants led to a greater reaction by the host, leading to a large fibrous capsule around the device and to an atrophied muscle. The devices were sterilized prior to implantation using ethylene oxide gas. Anodic bonds can also be

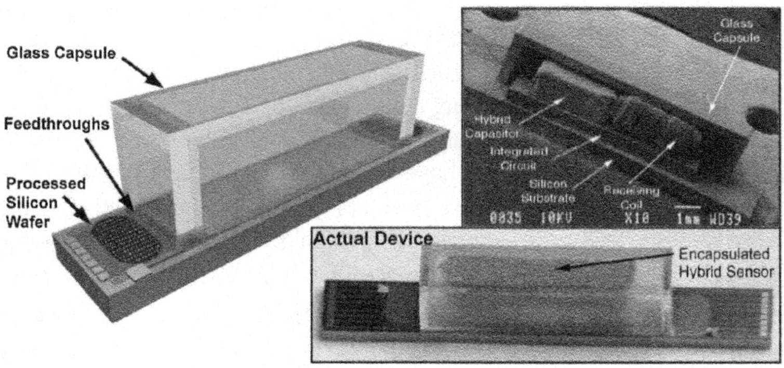

Fig. 4.39 Anodically bonded packaging. Reprinted with permission from [240]. © IEEE 2005

formed between Pyrex and metal films commonly found on ICs; however, the strength of the bond is reduced [236].

In eutectic bonding, a solder of a suitable material can be formed in the bonding area. The temperature is raised (300–400 °C) until the solder flows and creates a bond. One of the most common methods is the eutectic bonding of Au and Si. Once again a voltage across the two materials is applied while they are bonded together [241, 242]. This is typically achieved by having an Au bonding ring on the glass package and an Si bonding ring on the chip or common substrate, or vice versa.

Both anodic and eutectic bonding techniques can be applied not only to individual chips, but can also be performed at the wafer level, such that multiple chips can be packaged together, thus reducing packaging time and cost. One such recent example is presented in [248], where eutectic bonding of Au with Si was used to bind an application specific integrated circuit (ASIC) to a cap wafer leading to a hermetic gas-filled capsule. The cap wafer was directly bonded to the ASIC wafer by a wafer-bonding process before dicing. Au is not CMOS-compatible and thus the wafers had to be post-processed. The top metal layer (Al) of the process was used to define a metal ring surrounding the active area of the ASIC (Fig. 4.40). Similarly to bond pads, these were not passivated. The Au was deposited using electroless under bump metallization (UBM). To do this, an Ni layer was deposited using electroless plating followed by electroless plating with Au. Finally, the electroless Au was electroplated to form a thick (4.5 μm) Au layer. The cap wafer was patterned to define the Si ring and cavities were formed via an etching process, since the surface of CMOS chips is not flat. Patterns were etched on the top of the cap wafer to aid in the dicing process. The eutectic point of Au–Si is 363 °C, which is below the 450 °C that CMOS processes are known to withstand, and only a small amount of force (150–250 N) is required to avoid mechanical damage to the wafers. The amount of the material contributing to the reaction also needs to be optimized to reduce spillage of the solder.

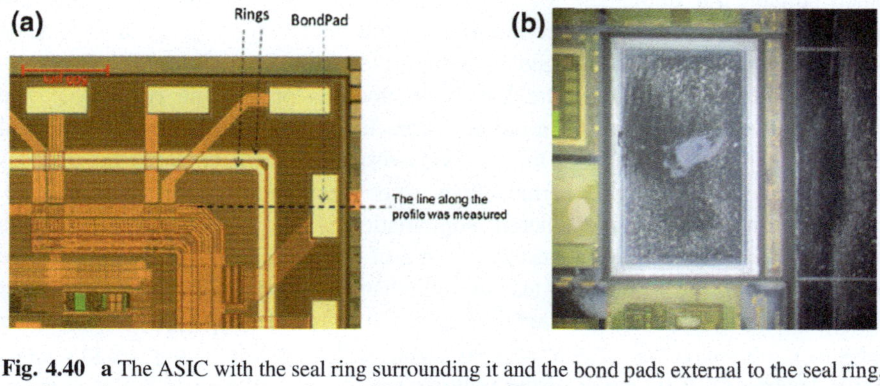

Fig. 4.40 **a** The ASIC with the seal ring surrounding it and the bond pads external to the seal ring, and **b** a micropackage on top of an ASIC, with surrounding ASICs exposed. Reprinted with permission from [248]. © IEEE 2005

An alternative to capsules and enclosures are thin film coatings. These can occupy comparatively a much smaller area and can easily take any shape or form. Nevertheless, materials that can be used to obtain such films are either not hermetic or may fail easily by being too thin. Thin-film materials can be organic or inorganic. Organic materials include epoxies, silicones, and polymers, such as PIs, polyurethanes, and parylene-C. Some of their advantages include that most can be deposited at low temperatures; they can easily conform to any geometry, since many are initially in a liquid state before being cured; and finally, their characteristics can be adjustable to suit the application of interest. The main disadvantages of these include the fact that they do not lead to a hermetic package; many can be penetrated by water over time (some in fact are actually used as humidity sensing films, e.g. PI; this will be discussed in the following section) and they can easily degrade in harsh environments. In general, the use of polymers is associated with a high moisture absorption rate, high-frequency electrical loss, and low mechanical stability. In some cases, polymers are used in combination with metals at high bonding temperatures and pressures, which could change their material and interfacial properties. The use of various bonding adhesives and physical/chemical deposition methods may lead to non-hermetic, permeable, optically opaque, and poorly conductive interfaces. Even so, such organic materials have found widespread use, as they can be used in applications where the device does not need to be implanted for prolonged periods of time or in cases where the local conditions are controlled or the specifications are relaxed. Thus, polymer-based packaging using polymers with improved characteristics and the development of low-pressure and temperature integration techniques are critical for packaging [123]. A number of technologies are using LCPs or surface-activated bonding (SAB) methodologies. Polymer–polymer/metal bonding is done using SAB technologies, which give reliable, miniaturized, high-performance packages for implantable sensors [132, 254, 255].

As an outer packaging material, medical grade silicone is used to create a soft shell suitable for implantation [239]. It is the most widespread option and has excellent adhesion to many common materials and exhibits very low leakage currents, while it is characterized by a high electrical resistivity and low permeability to ions [236, 241]. The low bulk modulus of silicone rubber, i.e. the material's resistance to uniform compression, leads to minute stress on electrical bonds [236]. It causes a mild foreign body response and it is stable for decades [244]. However, it is highly permeable to water vapor and thus moisture-sensitive. A very promising adhesive/adherent combination is silicone rubber with alumina ceramic, demonstrating an adhesion lifetime of 30 years [236]. Polyurethanes are stiffer, tougher, and with higher tear strength than silicone rubber, leading to thinner encapsulations. However, they are susceptible to stress cracking [236] and the ether groups in polyether polyurethane are oxidized by the metal ions they are in contact with, leading to degradation of the polymer.

Parylene-C is hydrophobic and is deposited via vapor deposition at room temperature. It leads to very conformal thin (hundreds of angstroms) coatings with no pinholes [241, 242] and demonstrates low water vapor absorption and permeability to water, gases, and ions [236, 244], while it has a high electrical resistivity and dielectric strength and is flexible. However, it demonstrates poor adhesion to common materials [236, 241, 242, 244]. Its adhesion is satisfactory to Au, Pt, and silicon nitride, while it is very low to PI [244]. To tackle this, an intermediate adhesion layer is deposited. Parylene-C [256] is an inert and biocompatible polymer typically used as a diffusion barrier to prevent any leaking of potentially hazardous substances in the body and to avoid corrosion of the electronic components. It is commonly used as a barrier medium for implantable microdevices, such as pacemakers, catheters, chemical sensors, probes, and microelectrodes.

PMMA is a transparent synthetic thermoplastic with good biocompatibility for in vivo applications and has an impact strength higher than glass. It has poor resistance to solvents and other chemicals like acids, which make it unsuitable for devices intended to be used in the stomach. Teflon is another promising material, as is PI, which can be used for periods up to a few years, but is unsuitable for longer periods. The main hurdle with such materials is that they require high temperatures for reasonable deposition rates, and some result in films that are not conformal enough. As a result these are not popular choices. Porous PC membranes are used in microdialysis filters in commercial biosensors. The choice of an outer membrane is critical to the proper operation and stability of the device following implantation. Different types of membranes, such as collagen, PC, cellulose acetate, Nafion, and polyurethane, can also be used. It is reported that the most reactive material is PI and the least reactive is polytetrafluoroethylene.

The use of cage systems for housing implantable devices was investigated in [257]. These cylinders were made out of porous polyvinyl alcohol (PVA) sheets with an 80 mm pore size and/or stainless steel meshes with 0.54 mm openings and were implanted subcutaneously in rats for five weeks. Upon explantation, continuous fibrotic capsules formed around all of the cage types were evident, with fluid inside the cages. Metal cages allowed more growth of tissue inside the cages, due to their greater porosity, whereas those covered by PVA did not. Diffusion across the fibrous tissue of analytes, such as Na^+, K^+, and urea, which is crucial for chronic kidney disease patients, was examined and the analytes were found to be able to diffuse through the fibrotic tissue. Diffusion of urea was found to be slower than the other solutes, possibly due to its larger molecule size. An absence of tissue growth inside the PVA sponge-wrapped metal cage was observed, indicating that this is a promising approach for implantable biosensors and therapeutic devices.

As mentioned in previous sections, biocompatible, mechanically flexible, light-weight polymer-based materials have emerged as alternatives to metals for electronics, e.g. microelectrode arrays and telemetric coils. In the case of implantable electrodes, the operating environment, type of polymer substrate, electrode materials, and surface/interfacial adhesion layers all affect their performance [258].

An example of an inorganic non-conducting material is silicon nitride, which is extensively used in the semiconductor industry as a passivation layer for the protection of ICs, as it is impervious to moisture. The quality of thin films produced with silicon nitride depends on the fabrication method followed (see also discussion on ISFETs), and it should be a dense film with no pinholes. Temperatures above 400 °C are required, with the highest quality films produced using low-pressure CVD at 800 °C. Depending on the voltages applied, the thickness of the passivation needs to be adjusted [236]. Silicon carbide, polycrystalline diamond and tantalum oxide are alternatives, with the former two requiring temperatures above 400 °C [241, 242]; these have led to stable hermetic packages [244]. Metal thin films (e.g. Au) represent another alternative. Provided they are inert (corrosion free), they offer an excellent barrier to moisture. For obvious reasons, an intermediate insulating layer is required. This can be any of the insulating materials mentioned earlier, provided that good adhesion is achieved. The thickness of this intermediate film and its dielectric constant are important, as the parasitic capacitance from the circuit to the package must be minimized. It must also cure at a higher temperature than the subsequently deposited materials. PI has been used for example, together with gold, which demonstrated a MTTF of 35 years [241, 242]. Some hybrid examples of hermetic and non-hermetic encapsulation are illustrated in Fig. 4.41 [244].

A simple low-cost alternative similar to eutectic binding is presented in [245–247] and is shown in Fig. 4.42. The authors used a ceramic alumina substrate and multiple layers of screen printing inks (two Pt/Au metal, glass, and dielectric paste), which they patterned using laser ablation to form interconnects and circuit connections. The first metal layer allows electrical connections between the internal and external areas of the hermetic package. The second metal layer was used in order to solder a red brass metal capsule using a low melting point metallic solder

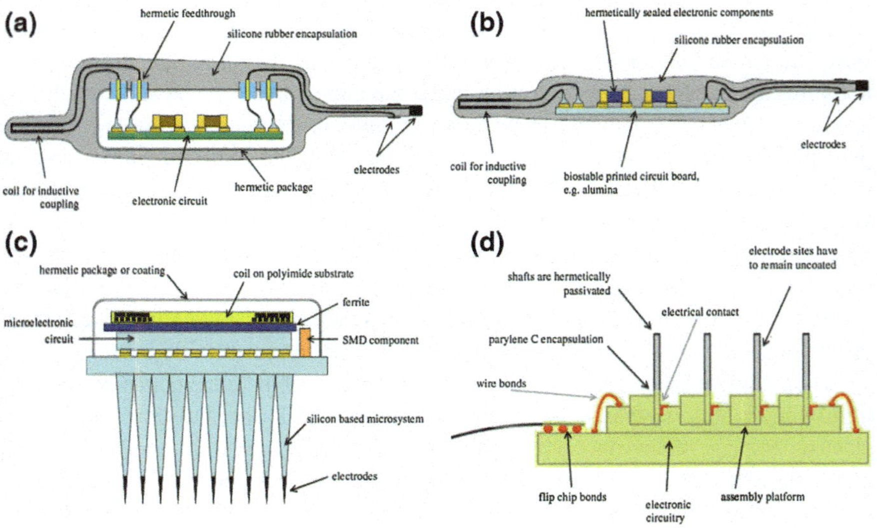

Fig. 4.41 **a** A hermetic encapsulation with feedthroughs for connection to external electrode sensors and coil, and a final polymer encapsulation. **b** Use of hermetically sealed components with a non-hermetic system level encapsulation. **c** Hermetic encapsulation of all electronics and sensors without leads. **d** Non-hermetic encapsulation at the system level using hermetically packaged components similarly to (**b**). Reprinted from [244] with permission from Springer. © 2010

[245, 246]. Alternatively, glass solder can be used to minimize the effect on inductive coupling telemetry [247]. The glass layer ensures good adhesion with the silicone rubber encapsulant. A hole on the top of the metal capsule allowed vapour to leave the package in a controlled environment at elevated temperatures and in weak vacuum before being filled with helium gas. Finally, the hole was sealed using solder. A minimum lifetime to water-induced failure of a few hundred years was reported [245]. The packaging process reported in [245] requires temperatures between 125 and 350 °C, while that of [247] requires temperatures up to 1500 °C. In addition to the above techniques, [246] recommends the use of desiccants, also known as water getters. These are materials that can bind water molecules either chemically or physically. Chemical binding desiccants can be implemented using various metal oxides (typically CaO) and are based on their reaction with oxygen, which typically leads to a non-reversible exothermal process. Physically binding desiccants lead to reversible adsorption of water molecules. Examples of such materials are silica gel, molecular sieves, and activated clay [245]. Silica gel is a highly porous silicon oxide which thus has a very large water-adsorbing area. It is useful at 30% relative humidity (RH) and higher at room temperature. Molecular sieves may be based on glasses, clays, charcoals, and other materials. Similarly to silica gel, they have a large surface area; however the porosity is optimized to adsorb molecules of specific polarity and size. They are effective over large temperature ranges and for humidities of 8% RH or larger. Activated clay desiccants use layers of silicate, and water molecules are adsorbed between them. It is most effective for high RH. The interested reader may refer to Chapter 4 in [259] for a thorough discussion on this and other related techniques.

4.5.1.1 Humidity Sensing

It is clear from the preceding section that a hermetically sealed packaged implant is essential for long-term operation. Periodically assessing the hermeticity of the package, i.e. the humidity, is thus essential to ensure proper implant operation. According to [246], apart from the water in the environment of the packaged implant (i.e. tissue), there are water sources that can be found internally within a hermetically sealed implant. Water can be generated within the device by chemical

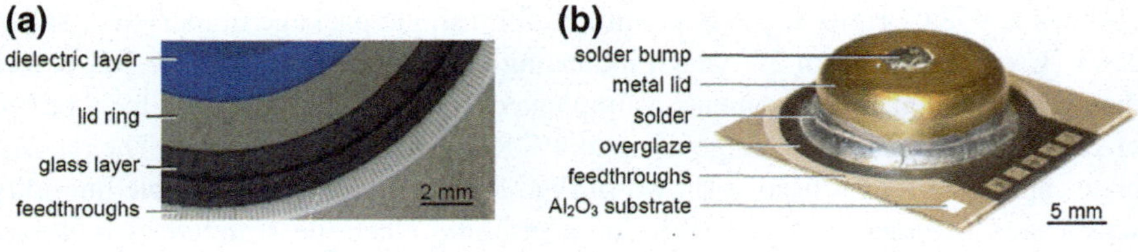

Fig. 4.42 **a** The different screen-printed and laser-patterned layers on the alumina substrate. **b** The hermetically packaged implant prior to silicone rubber encapsulation. The height is 5 mm and the substrate dimensions are 25 mm × 25 mm. Reprinted with permission from [245]. © IEEE 2010

reactions, particularly if high voltages are associated with the application, and water may have been adsorbed to the components within the package before its hermetic sealing. The limit for acceptable humidity in a hermetically sealed implanted device is 5000 ppm [246]. In an ambient temperature of 37 °C and with atmospheric pressure within the package this translates to an RH of 8.1% according to [246]. A minimum of 27% RH (corresponding to 17,000 ppm) is required for thin-film Al corrosion progression. The majority of humidity sensors available in the literature are capacitive sensors. Advantages of capacitive sensing include low power consumption, a high output signal amplitude that can be easily achieved, a wide operating range, low temperature dependence, and simple readout electronics [249]. However, they do suffer from the low humidity sensitivity of the polymer dielectrics typically used. In [260], the authors presented a number of PI-based humidity sensors. The sensor was comprised of a bottom metal plate followed by a PI layer with a thickness of 300–1200 Å and a top perforated metal electrode. The top electrode structure was comprised of fingers to enable moisture to reach the PI layer. The larger the gap between the fingers the easier it is for the moisture to reach or leave the sensing membrane. Different geometries of the same basic structure were examined. A structure with 25 μm width fingers with a spacing of 15 μm and a 1×1 mm^2 area led to a sensitivity of 0.86 pH/% RH for the 1200 Å thickness and 3.4 pF/% RH for 300 Å. With this topology, sensitivity can be increased by increasing the area or by reducing the PI thickness. The capacitance of the sensor is given by the typical equation for capacitance, except that the dielectric permittivity of the film is humidity-dependent.

In [243], the authors developed a passive humidity sensor using one of the sensors of [260]. The capacitive sensor was $1 \times 1 \times 0.5$ mm^3 and its capacitance changed from 180 to 240 pF between 20 and 80% RH. This was connected to a coil wound around a ferrite substrate (20 μH). This formed an *LC* tank circuit, the resonant frequency of which depended on the capacitance of the sensor. The device was telemetrically interrogated via magnetic induction. Using an external loop antenna (400 nH) the authors measured the resonance frequency shift as a change in the load impedance reflected back to the antenna. This varied between 4.5 and 5.1 MHz, giving a sensitivity of 8 kHz/% RH. The resolution of the system was ± 2.5% RH. The sensor was validated using hermetic and non-hermetic packages in humidity chambers, saline baths, and animal models. In [240] the authors monitored the humidity of micropackages (using a PI-based humidity sensor) soaking in saline for a period of 9.9 years at 23 °C, 321 days at 85 °C, and 70 days at 95 °C. Humidity was monitored for a single package immersed in saline at 97 °C over a period of 2.7 years, indicating a change of less than 5% RH before the package failed. Measurements of implanted devices in a guinea pig for a period of 22 months indicated a change of up to 4% RH thoughout the period. The parallel plate capacitive sensor used is shown in Fig. 4.43a. A CMOS-compatible humidity sensor was demonstrated in [249]. An interdigital electrode structure was implemented using the top metal layer, and the PI top passivation layer available in the CMOS process used was exploited as a humidity-sensing layer. Analytical conformal mapping techniques and FEM modeling were used to optimize the geometry

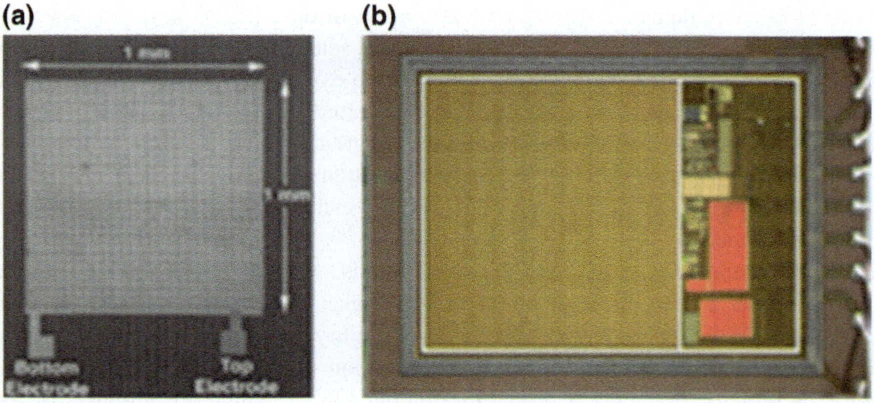

Fig. 4.43 **a** The parallel plate capacitive humidity sensor. Repinted with permission from [260]. © IEEE 2001. **b** A CMOS interdigital humidity sensor (marked (**a**) on the figure) with integrated electronics (marked (**b**) on the figure). Reprinted with permission from [261]. © IEEE 2013

of the sensor given the geometric constraints of the CMOS process, i.e. the PI layer thickness is fixed and equal to ~2.8 μm, the minimum allowable track width is 1.2 μm, and the track gap is 1 μm. According to these the electrode separation was set to 3.5 μm and the track width to 1.5 μm. The sensor was $2 \times 2\ mm^2$ and resulted in a sensitivity of 0.03 pF/% RH. One such sensor, together with the necessary electronics, is shown in Fig. 4.43b. In [250] the authors combined a humidity sensor with corrosion and temperature sensors. The latter two were single metal layer structures. The resistance of these structures was monitored, as it increases due to corrosion.

4.5.2 Foreign Body Response (FDR), Inflammation, and Implantable Sensor Performance

The in vivo reliability and lifetime of implantable sensors dictates the success of such devices. An implantable sensor's biocompatibility is assessed by two main outcomes: (a) the host's reaction to the implant—the device must not adversely affect the function of the host—and (b) examining whether the sensor performance has been negatively affected by the host environment [254, 255, 262, 263]. For implants to be used in vivo, an important feature is their chemical and biological inertness. However, in some cases biological activity is wanted, as in the case of bone regeneration or in the case of minimizing undesirable activities such as infection or blood clotting. It is incorrect to talk about a "biocompatible" sensor using materials assumed to be biocompatible, since the size and morphology of the implant has an important and sometimes dominant effect on the tissue response. The majority of implantable sensors are unfortunately incapable of performing reliable in vivo measurements for more than a few hours or days [264]. Within a relatively short period of time, it is common for implanted sensors to exhibit a 50% decrease in their sensitivity, for example [262, 263]. Once explanted, sensors which showed diminished performance in vivo have been reported to revert to their normal performance [264]. As a result, this performance degradation is attributed to the in vivo environment. Sensors are typically implanted transcutaneously, but can also be positioned within cavities of the human body, such as vascular, esophageal, and intestinal lumens.

In transcutaneous sensor implant applications, soon after implantation, tissue inflammation and the foreign body response take place. This is summarized in Fig. 4.44. Tissue inflammation is due to tissue injury occurring during implantation and its continuous presence in the specific section of the body. Small molecules, such as protein fragments (e.g. serum albumin [263]), which are products of the acute inflammatory response, will physically block analyte diffusion to the sensing element [262]. Shortly after implantation, a protein and (phagocytic) cell layer with a thickness of 10–100 μm is formed on the implanted device in an attempt to destroy the implant. This forms a barrier, preventing target analytes from diffusing towards the vicinity of the sensor, which affects the local concentration of analytes due to the consumption of oxygen and glucose by certain phagocytic cells and reactive oxygen species (H_2O_2, O^{2-}, NO, OH^-, $OONO^-$) and enzymes are generated to degrade the implant [263–266]. This process (e.g. its intensity and timing) depends on many factors, such as the size, shape and chemistry of the implant [263]. This acute response takes place within seconds of implantation and lasts for about 3–4 days [262, 263, 267]. It is followed by a chronic inflammatory response or a modified version of the healing process [263] due to the continual presence of the implanted device [268]. During this stage, the tissue surrounding the implant becomes continuously avascular in an attempt to isolate the foreign device [266]. This is essentially the foreign body response (FBR), which is the natural response of the body to confine the device within fibrous tissue such that it does not interact with the surrounding tissue. Eventually, the device is enclosed by a vascular collagenous fibrotic capsule with a thickness of 50–200 mm [268], isolating it from the surrounding tissue. This is a process that takes 1–2 weeks following implantation [263]. A detailed discussion regarding the above can be found in [268]. Many sensors, such as (oxidase) enzymatic glucose amperometric sensors, require oxygen as part of their transduction process. The foreign body response will lead to a significant change in the local concentration of oxygen and other analytes, thus further degrading sensor performance. According to [264], foreign body response leads to a substantially higher reduction in device performance and is far more challenging to tackle than electrical fouling, enzyme degradation, or electrode shielding [264].

The main concern where blood is associated, especially in intravascular implants, is thrombosis. Intravascular implantation is the most desirable implant location for most target analytes and applications. Immediately after implantation, blood proteins (e.g. fibrinogen) adsorb onto the device surface, triggering coagulation. Subsequently, platelets bind with these proteins and are then activated. Their

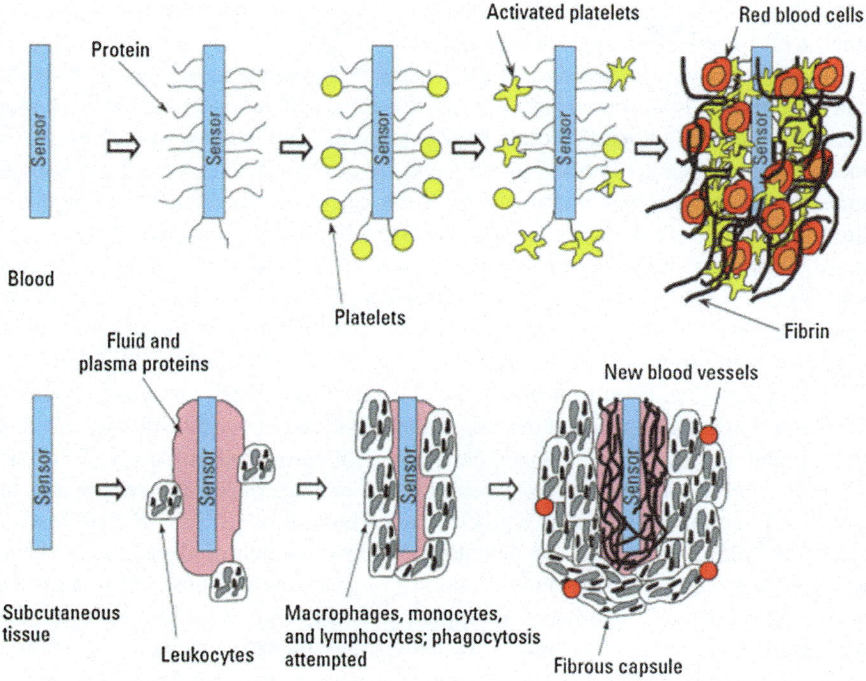

Fig. 4.44 Top: the sequence of events that leads to thrombus formation on the surface of implanted sensors within the vascular lumen. Bottom: the sequence of events that leads to the formation of fibrous capsules around transcutaneously implanted sensors. Reprinted with permission from [267]. © American Chemical Society 2006

activation leads to the release of chemicals (such as additional fibrinogen), which accelerate clotting. This process is illustrated in Fig. 4.44. Essentially a thrombus is created which can potentially block the sensor surface, degrading the sensor's response. This could also become an embolus and occlude the circulation, in which case it would not allow the sensor to obtain realistic readings [262] and would potentially risk the patient's health due to the risk of thromboembolism. The adherence of active cells on the sensor surface creates a local microenvironment which is different from that of the bulk. These metabolically active cells will consume oxygen and glucose, producing CO_2, and will thus change the local concentration of these, while this will also lead to changes in pH due to the increase in CO_2 [266, 267]. It is clear that these will lead to unrealistic sensor readings. The adhesion and activation of platelets depend on many factors, such as the coagulation status of the patient, the blood flow in the vessel, and its size. Administration or use within the device encapsulation of anticoagulants like heparin will not prevent platelet adhesion and activation, but can reduce the size of the thrombus formed in the device [267]. One measurable parameter indicating that the foreign body response of the organism has been established is an observed higher electrode interfacial impedance [269, 270].

In addition to inflammation and the foreign body response, another process that can compromise long-term device operation is the calcification of implanted devices [268]. In this case mineral calcium builds up on the device. Due to factors such as mechanical damage, connective tissue cells are devitalized. As a result, membrane phosphorus cannot work and reacts with the calcium present in the extracellular fluid. This process eventually leads to the formation of calcium phosphate crystals [268]. While this issue has mainly been reported in heart valves [268] and breast implants, it is also an issue with implantable sensors [271]. Similarly to previous solutions, drugs such as dexamethasone or levamisole can be used [268]. Other approaches are discussed in [268] and are mainly related to heart valve implants.

A final issue with sensors implanted for extended periods of time is bacterial adhesion. Implantable medical devices are intended to improve the quality of life of people and to save lives. Nevertheless, implant-associated infection [272] still remains a serious complication. Accumulation of bacteria on the sensor and the resulting infection are related to a process called biofouling [273, 274]. This leads to infection. During this process an increase in the level of neutrophils and white blood cells is observed, which normally function to ingest foreign substances. After 24–48 h, the neutrophils gradually disappear and are replaced by monocytes and macrophages. Toxic peroxynitrite ($OONO^-$) and hydroxyl radicals are formed during the reaction of the superoxide produced and NO. The macrophages attempt to ingest the implant, which is not possible because it is much larger, which leads to endocytosis and the formation of giant cells. Subsequently, different markers of inflammation are released, such as cytokines, leukotrienes, growth factors, and proteases [275]. This once again leads eventually to the formation of a foreign body capsule. Attention has also been paid to the role of cytokines and chemokines in activating the inflammatory and wound-healing processes [276]. One consequence of the function of the macrophages is the decrease in the local pH due to oxidative processes. Despite sterilization, bacterial infection remains a significant problem in the widespread deployment of implantable sensing. The issue with sterilization is that in many cases it damages the sensor (e.g. it may degrade polymer and enzymatic layers of a sensor). In addition, it does not eradicate the risk of bacterial infection [266]. Bacteria attached to the surface of a device will form microcolonies, which secrete a polymeric shield. These biofilms are physiological and chemical barriers resistant to phagocytosis and antibiotics [266]. Sterilization may be performed using UV or gamma irradiation, autoclaving, gaseous sterilization (ethylene oxide), use of antiseptics such as alcohol, and oxygen plasma treatment of the device.

Nevertheless, measurement discrepancies may not only be due to biocompatibility and/or biofouling issues. During transcutaneous implantation, the tissue is damaged. As a result, damaged cells may not be able to function normally, thus giving non-realistic sensor readouts. In addition, the wound-healing process will lead to increased glucose and oxygen consumption, which in the case of glucose sensors will further affect the local analyte levels (since oxygen is required for the sensor to function) and thus sensor reading accuracy.

4.5.2.1 Strategies for Enhancing Biocompatibility of Implanted Devices

Many approaches have been proposed in the literature to tackle these issues. Materials which mimic normal tissue function and morphology and that release active agents are some of the most successful approached and will be discussed below. The use of biocompatible coatings such as Nafion, PVA hydrogels, and humic acids has been proposed [264, 267]. These membranes must achieve an unperturbed and uninterrupted fluidity of the target analyte in the vicinity of the sensing element; they must reduce protein adsorption and allow the implanted device to be assimilated with the surrounding tissue [262–264]. In addition, they should be thin and porous, such that they allow the implanted sensor to respond rapidly to analyte concentration changes. According to [263, 277], the most common characteristics between the functional groups most resistant to protein adsorption are "(1) lack of hydrogen bond donors, (2) hydrogen bond accepting groups, (3) polar functional groups, (4) uncharged groups; while polarity, conformal mobility and wettability did not correlate with protein resistance". However, as mentioned earlier, the non-toxicity of the encapsulation materials to tissue is not a sufficient parameter guaranteeing unhindered implant sensor operation, since even such materials will trigger immune responses. Using a patterned encapsulation to promote vascularity and angiogenesis instead of a smooth surface has been shown to improve sensor performance.

More recent is the use of tissue response modifiers (TRMs) to tackle inflammation and the foreign body response. One approach is to embed biodegradable microspheres (commonly made from PLGA) containing such TRMs within biocompatible anti-biofouling polymers, the degradation of which may be triggered by changes in pH, other ion concentrations, or electrical stimulus [264]. Examples of TRFs are anti-inflammatories (e.g. dexamethasone and superoxide dismutase mimic), growth factors (vascular endothelial growth factor, VEGF, and PDGF), and vasodilator agents like nitric oxide (NO) to suppress the inflammatory response. The release kinetics of the TRMs from the encapsulation is important, however: once these have been exhausted the immune response will be triggered. Thus this is suppression with a limited period of effectiveness. The use of angiogenesis-promoting factors, such as VEGF, has triggered concerns associated with cancers [263]. The use of VEGF delivered from a coating on a sensor has been applied to biosensors. Release of an anti-inflammatory, dexamethasone, has also been used as a means of modulating the foreign body response when combined with VEGF release [282, 283]. VEGF appears to increase inflammation, while dexamethasone appeared to decrease neuro–vascularization.

Anticoagulants and antibiotics are not preferable [266]. The former, while useful in preventing thrombus formation, may lead to uncontrollable bleeding, especially in patients suffering from hemophilia and renal failure [266]. The latter would be used to tackle bacterial infections; however, antibiotics are continually proven to be less effective due to the antibiotic resistance developed by many pathogens and their associated side-effects [266]. The advantage of the active release of antibiotics is that high local doses of a tailored selection of antibiotics can be released against specific pathogens without exceeding toxicity levels and avoiding harmful side-effects in other regions of the body [274]. Commonly used antibiotics in such applications are vancomycin, amoxicillin, and gentramicin. [274]. The efficacy of the drug depends on the delivery mechanism, and in particular on the rate and manner in which it is released. Biocompatible polymer coatings, such as polyurethane and silicone rubber, or biodegradable polymers such as collagen, cancellous bone grafts, polyorthoesters, and PLGA, can be used for controlled delivery. Significant work is directed towards the synthesis and surface modification of biomaterials, such as polyethylene glycol (PEG) and polyethylene oxide (PEO), with the purpose of minimizing protein adsorption and cell adhesion [278]. The strategy to minimize protein adsorption applies not only to the acute inflammatory response, but also to processes that lead to the attachment of cells to the implant. PEG has demonstrated excellent anti-adhesive properties due to the low deposition of plasma proteins [255]. PEG can also be deposited on a surface as patterning in a brush-like configuration that sterically hinders microorganism attachment [275]. These PEG coatings may additionally utilize immobilized reactive groups for potential functionalization of the surface.

Surface modifications to prevent biofilm growth include the use of modified chitosan, zwitterionic coatings, formulations of Ag, immobilized enzymes, and various combinations of these [254, 255, 273, 274, 277, 278]. It has been shown that reduced protein adsorption and reduced foreign body capsule formation are achieved by materials with phospholipids. Zwitterionic coatings have also become a recent area of interest [197, 266]. Zwitterions are molecules that contain both positive and negative charges while having an overall neutral charge balance. Cationic molecules are known to have antimicrobial effects [277, 279]. Switchable zwitterionic materials have also been developed, which have been coated on material surfaces to facilitate bactericidal properties in devices. In another report, the zwitterionic polymer coating was combined with Ag NPs to increase the effect of the coating [266].

Another approach against bacterial infection is the use of Ag. As mentioned previously, the degradation products of the latter are common metabolites, and thus toxicity is not an issue [274]. Many antibiotics are specific for certain bacteria and are ineffective against other. An alternative to antibiotics is Ag, a nontoxic antibacterial agent with an antibacterial efficacy against a wide range of pathogens commonly found in implantable applications, such as *Escherichia coli*, *Staphylococcus aureus*, *Staphylococcus epidermitis* and *Pseudomonas aeruginosa* [274]. An additional advantage of Ag is that bacteria cannot develop resistance to it. However, it must be noted that it is the oxidized form of silver (Ag^+) that is antibacterial. Metallic Ag coated surfaces have failed to decrease bacterial adhesion. On the other hand, polymers actively releasing Ag^+ show strong antibacterial behavior [274].

A final alternative for combating bacterial infection is the controlled release of antibodies. Polyclonal IgG released controllably from a hydrogel matrix has been shown to reduce implant infection in vivo in a mouse model. Up to 90% of the IgG was released the first nine hours, with lower levels released over the following 21 h following implantation, when most of the IgG loaded into the hydrogel was released. This significantly increased the survival rate of the mice against a certain strain of *P. aeruginosa* (after 10 days 70% against 0% for a control group with the same hydrogels without the IgG antibodies) [274]. The same experiment with a different strain of the same bacterium resulted in a 20% against a 100% survival rate.

Many authors particularly highlight the use of NO [266, 267, 274, 280]. It is a "potent inhibitor of platelet adhesion and activation" in blood vessels and is "important in wound healing" [267]. According to [267], vascular endothelial cells continuously release NO, which diffuses into the blood vessel lumen and into the surrounding muscle tissue. It plays a vital role in maintaining normal blood pressure and preventing platelet activation in healthy blood vessels. Thus, artificial NO release by the vascular implant to its local environment can inhibit platelet adhesion on its surface and subsequent activation. As a result, this will minimize vasoconstriction in the vicinity of a vascular implant. In addition, NO has a short half-life, which is less than one second when glucose and hemoglobin are present, which is always the case in vessels. Thus, NO released by an implant will be consumed almost immediately and within the vicinity of the implant. According to [267], implanted glucose, CO_2, and oxygen sensors have been developed such that their sensitivity is not influenced by NO release or by the NO release mechanism, and that these do not contribute to the electrochemical analytical signal recorded. Implantation for up to 23 and 15 h (Fig. 4.45) of NO-releasing catheters has been shown to be superior to controls [267]. Another approach, according to [267], is, instead of implementing a transient controlled release of NO, to use device coatings that can locally generate NO from endogenous substances available in the blood [266, 267]. This can lead to NO release for longer periods of time. NO is also useful in transcutaneous environments, as it plays a significant role in wound-healing. It can reduce the inflammatory response, as it down-regulates inflammatory response mediators [267]. In addition NO promotes angiogenesis [266, 267]. The vascularity in the vicinity of the implant is important, as it allows more nutrients to reach the implantation site, promotes healing and enhances the mass transfer of (target) analytes in the vicinity of the sensor. NO-releasing materials have also been shown to reduce bacterial colonization and growth [266, 267]. The technology of NO release for implanted sensors is discussed in detail in [266, 267, 274]. The same group has produced blood glucose implantable enzymatic amperometric sensors utilizing NO release. This is illustrated in Fig. 4.45. The sensor is comprised of a Pt electrode modified by a xerogel layer with immobilized glucose oxidase (GO_x), a polyurethane layer used to protect the GO_x from diffusion of NO, a polyurethane layer with NO-releasing hydrogen particles, and a polyurethane outer film acting as a barrier membrane [266]. Although the beneficial effects on sensor performance and lifetime have been reported and highlighted, their influence on the local tissue and the host organism has not been fully examined; these could potentially be hazardous in the long-term [284]. It should also be noted that NO is electroactive at the potential corresponding to the oxidation of peroxide.

An example of an implantable smart tattoo glucose sensor is presented in [281]. The authors developed alginate microspheres for active release of dexamethasone and diclofenac sodium, which are anti-inflammatory drugs. This approach allowed drug delivery for 30 days at a controlled rate and suppression of the inflammation. In [282], PLGA microspheres dispensed in PVA hydrogels are presented. The PLGA were loaded with dexamethasone, which was slowly released to block inflammation and fibrous encapsulation and the PVA allowed rapid diffusion of the target analytes to the sensor. The tissue in the vicinity of the implant remained normal, with no inflammatory reaction or fibrous encapsulation during the period of the study (one month).

Nanostructured surfaces have different contact angles from smooth surfaces, which may hinder protein adsorption, while they may demonstrate reduced time constants for electron exchange between proteins and an implant surface, thus achieving slower clotting [264]. Nanopatterned nanoporous titania, Si, and carbon have been proposed as solutions. Nanoporous materials, the permeability of which is a function of pH, temperature, analyte concentrations, and magnetic and electric fields, have also been proposed. Such materials can expand and collapse such that they desorb proteins from the implant surface and thus allow proper analyte transference toward the sensing element. Another approach is the use of functional materials, e.g. proteases, embedded within polymers. These, for example, degrade adsorbed proteins.

Several possible ways of modifying the surface area before implantation have been reported. In [277], SAMs were exploited and surface plasmon resonance (SPR) was used to investigate protein binding following SAM deposition [277]. Another possible way to promote selective cell adhesion is to use peptides, such as arginineglycine-aspartic acid [285], that are known to interact with matrix proteins such as fibronectin, laminin, collagen, and vitronectin. It could be an issue if the

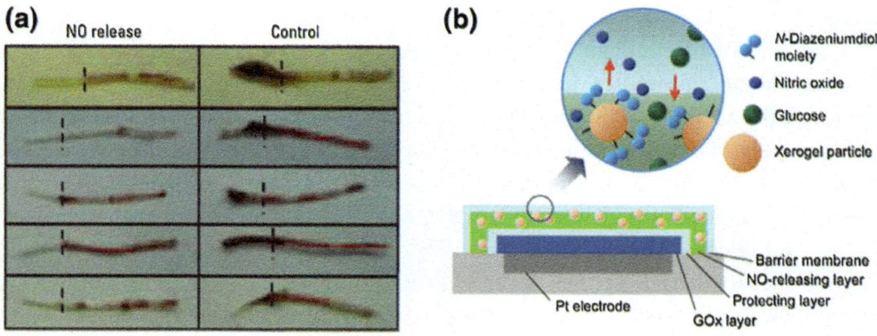

Fig. 4.45 **a** Thrombus formation on distal tips of electrochemical oxygen-sensing catheters of sensors with NO-releasing PDMS layers and controls implanted in porcine arteries of the same animals for at least 15 h. The section of the sensor to the left of the dotted line indicated the area exposed to flowing blood. Reprinted with permission from [267]. © Americal Chemical Society 2006. **b** Enzymatic glucose amperometric sensor employing NO release. Reprinted with permission from [283]. © American Chemical Society 2004

protein is absorbed on modified, non-functional confirmations [286]. Other surface materials used are hydrogels, which are chemically cross-linked multivinyl monomers such as methacrylate and acrylate, as well as additional cross-linked components of poly(ethyleneglycol)-co-poly(lactic acid) or PVA, or natural polymers, such as hyaluronic acid and chondroitin sulfate [286]. Some of these were mentioned earlier as well.

The extent of the foreign body response depends on the tissue damage imposed during implantation. As a result, the size of the implant is a significant parameter. Thus the sensing elements, the associated drive, recording, and power data transmission electronics, and the overall system packaging and encapsulation need to be extremely small. With regard to the sensors, 10–50 μm diameter wires are used; carbon nanofibers and nanotubes are used to reduce their size. Transistors as sensors and part of the transduction process have also been proposed. CNTs, nanorods, nanowires, and polymers have been used as discussed earlier as transistor channel materials to perform detection by monitoring changes in transistor transconductance, gate conductance, or threshold voltage. The use of polymers reduces the potential of tissue damage, being soft and flexible. In addition in many applications the implantable sensing system may need to adapt and conform to the tissue/organ geometry. In such applications, the use of polymeric materials as the sensor's substrate and sensing elements provides a significant advantage over traditional rigid silicon-based technologies. Many of these have already been discussed in detail.

When using such nanomaterials, it is important to consider their biocompatibility. NPs used may enter the human body, and whether these are life-threatening or not is currently debated in the scientific community, and more research is needed to quantify and establish their pharmacokinetics, toxicity, and immunogenicity. In addition, each nanomaterial is different and such studies need to be performed for each one of them. In any case, care must be taken such that this does not occur. Even if these are buried deep in the device, leaching may cause them to contact the surrounding tissue. An important family of materials is that of biodegradable materials for transient sensors and electronics. These materials can dissolve naturally and controllably over time, such that implanted devices can biodegrade at the end of their useful lifetime. Such devices were also discussed earlier. This is similar to biodegradable sutures used routinely in surgical procedures.

4.5.3 Biocompatible Standards for Medical Devices

In previous sections, we have outlined key technological advances for sensor embodiment. It is important also to highlight that such devices need to be safe and avirulent, such that they can be used on various locations of the human body (Fig. 4.46). ISO 10993 and 14971 are the international standards that define guidelines for the biological evaluation of medical devices and the application of risk management for such devices [287]. Figure 4.46 shows the locations on the

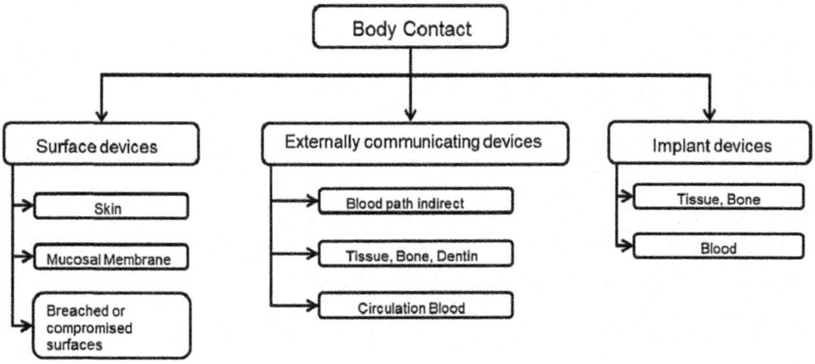

Fig. 4.46 Test categories of ISO 10993 international standard of biological evaluation of medical devices and food and drug administration (FDA) modified matrix

body which need to use biocompatible materials when medical devices are attached, either in the hospital or at home. The first column includes surface devices, and in particular devices that need to be placed on skin, mucosal membranes, and breached/compromised areas. The second column includes cases of externally communicating devices where the devices are in contact with the blood path indirectly: tissues, bone, dentin, and circulation blood. The third case includes implant devices on target sites such as tissue, bone, and blood. The testing items include cytotoxicity, sensitization, irritation, acute toxicity, subchronic toxicity, genotoxicity, heamocompatibility, chronic toxicity, and carcinogenicity. The testing durations are (1) ≥ 24 h, (2) 1–30 days, and (3) >30 days, depending on the application. Discrepancies of test items are shown in Table 4.2 [287].

4.6 Conclusions

The design requirements of wearable and, particularly, implantable medical devices, are stricter, more diverse and limiting, and greater in number than those imposed in traditional portable electronics and sensing applications. Biomedical devices must seamlessly integrate with the body such that they don't cause discomfort, damage tissues, or affect measurements. This is not possible with traditional rigid electronics. Their geometric conformity is crucial. Furthermore, the additional burden on healthcare systems due to an aging population and to obesity, the need of cost-effectiveness and increase in life expectancy by reducing the percentage of readmissions and hospital stay, have generated a need for a new generation of medical devices and changes in healthcare practices. Together with advances in materials science, fabrication, and nanotechnology, the above needs have led to the development of flexible, stretchable, and transient bioresorbable electronic devices.

Table 4.2 Test items of ISO 10993 and FDA modified matrix ("X" corresponds to required test items and "O" corresponds to items which may be considered as a required test item)

Body-contact	Contact duration	Cytotoxicity	Sensitization	Irritation	Acute toxicity	Sub-chronic toxicity	Genotoxicity	Implant	Hemo-compatibility	Chronic toxicity	Carcino-genicity
Skin	≤ 24 h	X	X	X							
	1–30 days	X	X	X							
	>30 days	X	X	X							
Mucosal membrane	≤ 24 h	X	X	X							
	1–30 days	X	X	X	O	O		O			
	>30 days	X	X	X	O	X	X	O		O	
Breached or compromised surface	≤ 24 h	X	X	X	O						
	1–30 days	X	X	X	O	O		O			
	>30 days	X	X	X	O	X	X	O		O	
Blood path indirect	≤ 24 days	X	X	X	X				X		
	1–30 days	X	X	X	X				X		
	>30 days	X	X	X	X	X	X	O	X	O	O
Tissue, bone, dentine	≤ 24 days	X	X	X	O						
	1–30 days	X	X	X	X	X	X	X			
	>30 days	X	X	X	X	X	X	X		O	O
Circulation blood	≤ 24 h	X	X	X	X		O		X		
	1–30 days	X	X	X	X	X	X	X	X		
	>30 days	X	X	X	X	X	X	X	X	O	O
Tissue, bone	≤ 24 h	X	X	X	O						
	1–30 days	X	X	X	X	X	X	X			
	>30 days	X	X	X	X	X	X	X		O	O
Blood	≤ 24 h	X	X	X	X	X		X	X		
	1–30 days	X	X	X	X	X	X	X	X		
	>30 days	X	X	X	X	X	X	X	X	O	O

Instead of Si wafers, polymeric substrates are preferred, such as PI and PET for flexibility and PDMS for stretchability, and even rice paper [78] and silk [13] substrates for transient resorbable electronic devices. To enhance the flexibility and stretchability inherent to such materials and an unhindered device performance under mechanical stresses, the geometry of interconnects and devices need to be optimized. Serpentine, horseshoe, fractal geometries, and wavy nanoribbons are just a few of the approaches that can be implemented to achieve this on elastomeric substrates. Prestretching of these prior to material deposition, device transfer, and assembly enhances the mechanical robustness of the devices further.

Unconventional materials, such as graphene, CNTs, organic semiconductors, TMDs, and nanopatterned inorganic semiconductors have led to a new generation of devices suitable for flexible/stretchable electronics. Soft substrates require low-temperature processing, which further motivates the use of such materials. These, however, cannot match the performance of traditional electronics (e.g. high carrier mobility). Either because of the nature of these materials or because of the processing temperatures of various fabrication technologies, fabrication techniques such as imprint lithography, transfer, and inkjet and screen printing have come into play, allowing the processing and deposition of new materials on soft substrates at low temperature and low cost.

Medical devices employing the above design concepts have been fabricated to address a number of challenges related to wearable and implantable electronics, such as multiplexed arrays of high density, employing a wide range of sensors such as PPG, ECG, strain, and pH sensors. These are important, for example, for monitoring the homeostasis of the organism and providing information that indicates early signs of adverse events, such as surgical site infection or organ failure.

References

1. A. Bozkurt, A. Lal, Low-cost flexible printed circuit technology based microelectrode array for extracellular stimulation of the invertebrate locomotory system. Sens. Actuators Phys. **169**(1), 89–97 (2011)
2. P. Kassanos, H.M.D. Ip, G.-Z. Yang, A tetrapolar bio-impedance sensing system for gastrointestinal tract monitoring, in *2015 IEEE 12th International Conference on Wearable and Implantable Body Sensor Networks (BSN)* (2015), pp. 1–6
3. C. Kallmayer, E. Simon, Large area sensor integration in textiles, in *2012 9th International Multi-Conference on Systems, Signals and Devices (SSD)* (2012), pp. 1–5
4. A. Nelson et al., Wearable multi-sensor gesture recognition for paralysis patients, in *2013 IEEE SENSORS* (2013), pp. 1–4
5. R.G. Haahr et al., An electronic patch for wearable health monitoring by reflectance pulse oximetry. IEEE Trans. Biomed. Circuits Syst. **6**(1), 45–53 (2012)
6. C.M. Chen, R. Kwasnicki, B. Lo, G.-Z. Yang, Wearable tissue oxygenation monitoring sensor and a forearm vascular phantom design for data validation, in *2014 11th International Conference on Wearable and Implantable Body Sensor Networks* (2014), pp. 64–68
7. R. Dekker et al., Living chips and chips for the living, in *2012 IEEE Bipolar/BiCMOS Circuits and Technology Meeting (BCTM)* (2012), pp. 1–9

8. J. Reeder et al., Mechanically adaptive organic transistors for implantable electronics. Adv. Mater. **26**(29), 4967–4973 (2014)
9. Doppler Blood Flow Monitor, Cook medical. [Online]. Available: https://www.cookmedical.com/products/33eecd89-149b-4d3c-955a-213541b21142/. Accessed: 22 Sep 2016
10. L. Xu et al., 3D multifunctional integumentary membranes for spatiotemporal cardiac measurements and stimulation across the entire epicardium. Nat. Commun. **5**, 3329 (2014)
11. D.-H. Kim et al., Epidermal electronics. Science **333**(6044), 838–843 (2011)
12. W.-H. Yeo et al., Multifunctional epidermal electronics printed directly onto the skin. Adv. Mater. **25**(20), 2773–2778 (2013)
13. H. Tao et al., Silk-based resorbable electronic devices for remotely controlled therapy and in vivo infection abatement. Proc. Natl. Acad. Sci. **111**(49), 17385–17389 (2014)
14. J. Viventi et al., Flexible, foldable, actively multiplexed, high-density electrode array for mapping brain activity *in vivo*. Nat. Neurosci. **14**(12), 1599–1605 (2011)
15. S.P. Lee et al., Catheter-based systems with integrated stretchable sensors and conductors in cardiac electrophysiology. Proc. IEEE **103**(4), 682–689 (2015)
16. D.-H. Kim et al., Materials for multifunctional balloon catheters with capabilities in cardiac electrophysiological mapping and ablation therapy. Nat. Mater. **10**(4), 316–323 (2011)
17. T. Yokota et al., Ultraflexible organic photonic skin. Sci. Adv. **2**(4), e1501856 (2016)
18. S.-W. Hwang et al., A physically transient form of silicon electronics. Science **337**(6102), 1640–1644 (2012)
19. M.S. Mannoor et al., Graphene-based wireless bacteria detection on tooth enamel. Nat. Commun. **3**, 763 (2012)
20. H. Tao et al., Implantable, multifunctional, bioresorbable optics. Proc. Natl. Acad. Sci. **109** (48), 19584–19589 (2012)
21. D.-H. Kim et al., Electronic sensor and actuator webs for large-area complex geometry cardiac mapping and therapy. Proc. Natl. Acad. Sci. **109**(49), 19910–19915 (2012)
22. L. Xu et al., Materials and fractal designs for 3D multifunctional integumentary membranes with capabilities in cardiac electrotherapy. Adv. Mater. **27**(10), 1731–1737 (2015)
23. G. Park et al., Immunologic and tissue biocompatibility of flexible/stretchable electronics and optoelectronics. Adv. Healthc. Mater. **3**(4), 515–525 (2014)
24. M.A. Deeds, P.A. Sandborn, MOEMS chip-level optical fiber interconnect. IEEE Trans. Adv. Packag. **28**(4), 612–618 (2005)
25. M. Lapisa, G. Stemme, F. Niklaus, Wafer-Level Heterogeneous Integration for MOEMS, MEMS, and NEMS. IEEE J. Sel. Top. Quantum Electron. **17**(3), 629–644 (2011)
26. R. Özgün, B.J. Jung, B.M. Dhar, H.E. Katz, A.G. Andreou, Silicon-on-insulator (SOI) integration for organic field effect transistor (OFET) based circuits, in *2011 IEEE International Symposium of Circuits and Systems (ISCAS)* (2011), pp. 2253–2256
27. M. Luo, A.W. Martinez, C. Song, F. Herrault, M.G. Allen, A Microfabricated wireless RF pressure sensor made completely of biodegradable materials. J. Microelectromechanical Syst. **23**(1), 4–13 (2014)
28. T. Adrega, S.P. Lacour, Stretchable gold conductors embedded in PDMS and patterned by photolithography: fabrication and electromechanical characterization. J. Micromechanics Microengineering **20**(5), 055025 (2010)
29. X. Hu, P. Krull, B. de Graff, K. Dowling, J.A. Rogers, W.J. Arora, Stretchable inorganic-semiconductor electronic systems. Adv. Mater. **23**(26), 2933–2936 (2011)
30. Y. Sun, W.M. Choi, H. Jiang, Y.Y. Huang, J.A. Rogers, Controlled buckling of semiconductor nanoribbons for stretchable electronics. Nat. Nanotechnol. **1**(3), 201–207 (2006)
31. M.J. Allen et al., Soft transfer printing of chemically converted graphene. Adv. Mater. **21** (20), 2098–2102 (2009)
32. X. Liang, Z. Fu, S.Y. Chou, Graphene transistors fabricated via transfer-printing in device active-areas on large wafer. Nano Lett. **7**(12), 3840–3844 (2007)

33. A. Carlson, A.M. Bowen, Y. Huang, R.G. Nuzzo, J.A. Rogers, Transfer printing techniques for materials assembly and micro/nanodevice fabrication. Adv. Mater. **24**(39), 5284–5318 (2012)
34. C. Kim, P.E. Burrows, S.R. Forrest, Micropatterning of organic electronic devices by cold-welding. Science **288**(5467), 831–833 (2000)
35. C. Kim, S.R. Forrest, Fabrication of organic light-emitting devices by low-pressure cold welding. Adv. Mater. **15**(6), 541–545 (2003)
36. S.R. Forrest, The path to ubiquitous and low-cost organic electronic appliances on plastic. Nature **428**, 911–918 (2004)
37. C. Kim, M. Shtein, S.R. Forrest, Nanolithography based on patterned metal transfer and its application to organic electronic devices. Appl. Phys. Lett. **80**(21), 4051–4053 (2002)
38. D.R. Hines, V.W. Ballarotto, E.D. Williams, Y. Shao, S.A. Solin, Transfer printing methods for the fabrication of flexible organic electronics. J. Appl. Phys. **101**(2), 024503 (2007)
39. J. Zaumseil et al., Three-dimensional and multilayer nanostructures formed by nanotransfer printing. Nano Lett. **3**(9), 1223–1227 (2003)
40. J.-H. Ahn et al., Bendable integrated circuits on plastic substrates by use of printed ribbons of single-crystalline silicon. Appl. Phys. Lett. **90**(21), 213501 (2007)
41. D.-H. Kim et al., Materials and noncoplanar mesh designs for integrated circuits with linear elastic responses to extreme mechanical deformations. Proc. Natl. Acad. Sci. **105**(48), 18675–18680 (2008)
42. A. Perl, D.N. Reinhoudt, J. Huskens, Microcontact printing: limitations and achievements. Adv. Mater. **21**(22), 2257–2268 (2009)
43. S. Khan, L. Lorenzelli, R.S. Dahiya, Technologies for printing sensors and electronics over large flexible substrates: a review. IEEE Sens. J. **15**(6), 3164–3185 (2015)
44. R. Parashkov, E. Becker, T. Riedl, H.H. Johannes, W. Kowalsky, Large area electronics using printing methods. Proc. IEEE **93**(7), 1321–1329 (2005)
45. J. Zaumseil, T. Someya, Z. Bao, Y.-L. Loo, R. Cirelli, J.A. Rogers, Nanoscale organic transistors that use source/drain electrodes supported by high resolution rubber stamps. Appl. Phys. Lett. **82**(5), 793–795 (2003)
46. S.Y. Chou, P.R. Krauss, P.J. Renstrom, Imprint lithography with 25-nanometer resolution. Science **272**(5258), 85–87 (1996)
47. S.Y. Chou, P.R. Krauss, P.J. Renstrom, Nanoimprint lithography. J. Vac. Sci. Technol. B **14** (6), 4129–4133 (1996)
48. S.Y. Chou, P.R. Krauss, W. Zhang, L. Guo, L. Zhuang, Sub-10 nm imprint lithography and applications. J. Vac. Sci. Technol. B **15**(6), 2897–2904 (1997)
49. D.J. Resnick, S.V. Sreenivasan, C.G. Willson, Step & flash imprint lithography. Mater. Today **8**(2), 34–42 (2005)
50. J. Wang, X. Sun, L. Chen, S.Y. Chou, Direct nanoimprint of submicron organic light-emitting structures. Appl. Phys. Lett. **75**(18), 2767–2769 (1999)
51. P. Maury et al., Roll-to-roll UV imprint lithography for flexible electronics. Microelectron. Eng. **88**(8), 2052–2055 (2011)
52. J. Han, S. Choi, J. Lim, B.S. Lee, S. Kang, Fabrication of transparent conductive tracks and patterns on flexible substrate using a continuous UV roll imprint lithography. J. Phys. Appl. Phys. **42**(11), 115503 (2009)
53. S.H. Ahn, L.J. Guo, High-Speed roll-to-roll nanoimprint lithography on flexible plastic substrates. Adv. Mater. **20**(11), 2044–2049 (2008)
54. H. Gold et al., Self-aligned flexible organic thin-film transistors with gates patterned by nano-imprint lithography. Org. Electron. **22**, 140–146 (2015)
55. A. Larmagnac, S. Eggenberger, H. Janossy, J. Vörös, Stretchable electronics based on Ag-PDMS composites. Sci. Rep. **4** (2014)
56. S. Khan, W. Dang, L. Lorenzelli, R. Dahiya, Flexible pressure sensors based on screen-printed P(VDF-TrFE) and P(VDF-TrFE)/MWCNTs. IEEE Trans. Semicond. Manuf. **28**(4), 486–493 (2015)

57. S. Khan, S. Tinku, L. Lorenzelli, R.S. Dahiya, Flexible tactile sensors using screen-printed P (VDF-TrFE) and MWCNT/PDMS composites. IEEE Sens. J. **15**(6), 3146–3155 (2015)
58. J. Ping, Y. Wang, Y. Ying, J. Wu, Application of electrochemically reduced graphene oxide on screen-printed ion-selective electrode. Anal. Chem. **84**(7), 3473–3479 (2012)
59. D. Numakura, Advanced screen printing 'practical approaches for printable & flexible electronics', in *2008 3rd International Microsystems, Packaging, Assembly Circuits Technology Conference* (2008), pp. 205–208
60. R. Soukup, A. Hamáček, J. Řeboun, Organic based sensors: Novel screen printing technique for sensing layers deposition, in *2012 35th International Spring Seminar on Electronics Technology* (2012), pp. 19–24
61. G.E. Jabbour, R. Radspinner, N. Peyghambarian, Screen printing for the fabrication of organic light-emitting devices. IEEE J. Sel. Top. Quantum Electron. **7**(5), 769–773 (2001)
62. Y. Kim, H. Kim, H.J. Yoo, Electrical characterization of screen-printed circuits on the fabric. IEEE Trans. Adv. Packag. **33**(1), 196–205 (2010)
63. B. Karaguzel et al., Flexible, durable printed electrical circuits. J. Text. Inst. **100**(1), 1–9 (2009)
64. J. Chang, X. Zhang, T. Ge, J. Zhou, Fully printed electronics on flexible substrates: high gain amplifiers and DAC. Org. Electron. **15**(3), 701–710 (2014)
65. G.D. Martin, S.D. Hoath, I.M. Hutchings, Inkjet printing—the physics of manipulating liquid jets and drops. J. Phys: Conf. Ser. **105**(1), 012001 (2008)
66. B. Derby, Inkjet printing of functional and structural materials: fluid property requirements, feature stability, and resolution. Annu. Rev. Mater. Res. **40**(1), 395–414 (2010)
67. Y.-F. Liu, W.-S. Hwang, Y.-F. Pai, M.-H. Tsai, Low temperature fabricated conductive lines on flexible substrate by inkjet printing. Microelectron. Reliab. **52**(2), 391–397 (2012)
68. D. Soltman, V. Subramanian, Inkjet-printed line morphologies and temperature control of the coffee ring effect. Langmuir **24**(5), 2224–2231 (2008)
69. J. Stringer, B. Derby, Formation and stability of lines produced by inkjet printing. Langmuir **26**(12), 10365–10372 (2010)
70. E. Fribourg-Blanc, D.M.T. Dang, C.M. Dang, Characterization of silver nanoparticle based inkjet printed lines. Microsyst. Technol. **19**(12), 1961–1971 (2013)
71. T.H.J. van Osch, J. Perelaer, A.W.M. de Laat, U.S. Schubert, Inkjet printing of narrow conductive tracks on untreated polymeric substrates. Adv. Mater. **20**(2), 343–345 (2008)
72. B.J. Kang, J.H. Oh, Geometrical characterization of inkjet-printed conductive lines of nanosilver suspensions on a polymer substrate. Thin Solid Films **518**(10), 2890–2896 (2010)
73. J. Perelaer et al., Printed electronics: the challenges involved in printing devices, interconnects, and contacts based on inorganic materials. J. Mater. Chem. **20**(39), 8446–8453 (2010)
74. A. Chiolerio et al., Inkjet printing and low power laser annealing of silver nanoparticle traces for the realization of low resistivity lines for flexible electronics. Microelectron. Eng. **88**(8), 2481–2483 (2011)
75. S. Chung, S.O. Kim, S.K. Kwon, C. Lee, Y. Hong, All-inkjet-printed organic thin-film transistor inverter on flexible plastic substrate. IEEE Electron Device Lett. **32**(8), 1134–1136 (2011)
76. I. Theodorakos, F. Zacharatos, R. Geremia, D. Karnakis, I. Zergioti, Selective laser sintering of Ag nanoparticles ink for applications in flexible electronics. Appl. Surf. Sci. **336**, 157–162 (2015)
77. S.H. Ko, J. Chung, H. Pan, C.P. Grigoropoulos, D. Poulikakos, Fabrication of multilayer passive and active electric components on polymer using inkjet printing and low temperature laser processing. Sens. Actuators Phys. **134**(1), 161–168 (2007)
78. D. Tobjörk et al., IR-sintering of ink-jet printed metal-nanoparticles on paper. Thin Solid Films **520**(7), 2949–2955 (2012)
79. P.J. Smith, D.-Y. Shin, J.E. Stringer, B. Derby, N. Reis, Direct ink-jet printing and low temperature conversion of conductive silver patterns. J. Mater. Sci. **41**(13), 4153–4158 (2006)

80. S. Jeong, H.C. Song, W.W. Lee, Y. Choi, B.-H. Ryu, Preparation of aqueous Ag Ink with long-term dispersion stability and its inkjet printing for fabricating conductive tracks on a polyimide film. J. Appl. Phys. **108**(10), 102805 (2010)
81. G. McKerricher, J.G. Perez, A. Shamim, Fully inkjet printed RF inductors and capacitors using polymer dielectric and silver conductive ink with through vias. IEEE Trans. Electron Devices **62**(3), 1002–1009 (2015)
82. J.F. Salmerón et al., Properties and printability of inkjet and screen-printed silver patterns for RFID antennas. J. Electron. Mater. **43**(2), 604–617 (2013)
83. Y. Liu, T. Cui, K. Varahramyan, All-polymer capacitor fabricated with inkjet printing technique. Solid-State Electron. **47**(9), 1543–1548 (2003)
84. B. Chen, T. Cui, Y. Liu, K. Varahramyan, All-polymer RC filter circuits fabricated with inkjet printing technology. Solid-State Electron. **47**(5), 841–847 (2003)
85. D. Mager et al., An MRI receiver coil produced by inkjet printing directly on to a flexible substrate. IEEE Trans. Med. Imaging **29**(2), 482–487 (2010)
86. B.S. Cook, A. Shamim, Inkjet printing of novel wideband and high gain antennas on low-cost paper substrate. IEEE Trans. Antennas Propag. **60**(9), 4148–4156 (2012)
87. L. Yang, A. Rida, R. Vyas, M.M. Tentzeris, RFID tag and RF structures on a paper substrate using inkjet-printing technology. IEEE Trans. Microw. Theory Tech. **55**(12), 2894–2901 (2007)
88. R. Vyas et al., Inkjet printed, self powered, wireless sensors for environmental, gas, and authentication-based sensing. IEEE Sens. J. **11**(12), 3139–3152 (2011)
89. H.-Y. Tseng, V. Subramanian, All inkjet-printed, fully self-aligned transistors for low-cost circuit applications. Org. Electron. **12**(2), 249–256 (2011)
90. S. Chung, J. Jang, J. Cho, C. Lee, S.-K. Kwon, Y. Hong, All-inkjet-printed organic thin-film transistors with silver gate, source/drain electrodes. Jpn. J. Appl. Phys. **50**(3S), 03CB05 (2011)
91. T. Kawase, T. Shimoda, C. Newsome, H. Sirringhaus, R.H. Friend, Inkjet printing of polymer thin film transistors. Thin Solid Films **438–439**, 279–287 (2003)
92. H. Sirringhaus et al., High-resolution inkjet printing of all-polymer transistor circuits. Science **290**(5499), 2123–2126 (2000)
93. D.J. Lichtenwalner, A.E. Hydrick, A.I. Kingon, Flexible thin film temperature and strain sensor array utilizing a novel sensing concept. Sens. Actuators Phys. **135**(2), 593–597 (2007)
94. H. Al-Chami, E. Cretu, Inkjet printing of microsensors, in *IEEE 15th International Mixed-Signals, Sensors, and Systems Test Workshop 2009 (IMS3TW '09)* (2009), pp. 1–6
95. P. Alpuim, V. Correia, E.S. Marins, J.G. Rocha, I.G. Trindade, S. Lanceros-Mendez, Piezoresistive silicon thin film sensor array for biomedical applications. Thin Solid Films **519**(14), 4574–4577 (2011)
96. F. Molina-Lopez, D. Briand, N.F. de Rooij, All additive inkjet printed humidity sensors on plastic substrate. Sens. Actuators B Chem. **166–167**, 212–222 (2012)
97. M.V. Kulkarni, S.K. Apte, S.D. Naik, J.D. Ambekar, B.B. Kale, Ink-jet printed conducting polyaniline based flexible humidity sensor. Sens. Actuators B Chem. **178**, 140–143 (2013)
98. A. Rivadeneyra, J. Fernández-Salmerón, M. Agudo, J.A. López-Villanueva, L.F. Capitan-Vallvey, A.J. Palma, Design and characterization of a low thermal drift capacitive humidity sensor by inkjet-printing. Sens. Actuators B Chem. **195**, 123–131 (2014)
99. U. Altenberend et al., Towards fully printed capacitive gas sensors on flexible PET substrates based on Ag interdigitated transducers with increased stability. Sens. Actuators B Chem. **187**, 280–287 (2013)
100. P. Sjöberg et al., Paper-based potentiometric ion sensors constructed on ink-jet printed gold electrodes. Sens. Actuators B Chem. **224**, 325–332 (2016)
101. N. Komuro, S. Takaki, K. Suzuki, D. Citterio, Inkjet printed (bio)chemical sensing devices. Anal. Bioanal. Chem. **405**(17), 5785–5805 (2013)
102. J. Wu et al., Inkjet-printed microelectrodes on PDMS as biosensors for functionalized microfluidic systems. Lab Chip **15**(3), 690–695 (2015)

103. C.L. Kang, Y. Xu, K.L. Yung, W. Chen, Laser Induced Forward Transfer. Adv. Mater. Res. **591–593**, 1135–1138 (2012)
104. P. Serra, J.M. Fernandez-Pradas, M. Colina, M. Duocastella, J. Dominguez, J.L. Morenza, Laser-induced forward transfer: a direct-writing technique for biosensors preparation. J. Laser MicroNanoengineering **1**(3), 236–242 (2006)
105. C. Boutopoulos, I. Kalpyris, E. Serpetzoglou, I. Zergioti, Laser-induced forward transfer of silver nanoparticle ink: time-resolved imaging of the jetting dynamics and correlation with the printing quality. Microfluid. Nanofluidics **16**(3), 493–500 (2013)
106. R. Fardel, M. Nagel, F. Nüesch, T. Lippert, A. Wokaun, Laser-induced forward transfer of organic LED building blocks studied by time-resolved shadowgraphy. J. Phys. Chem. C **114** (12), 5617–5636 (2010)
107. C. Boutopoulos, C. Pandis, K. Giannakopoulos, P. Pissis, I. Zergioti, Polymer/carbon nanotube composite patterns via laser induced forward transfer. Appl. Phys. Lett. **96**(4), 041104 (2010)
108. A.P. Suryavanshi, M.-F. Yu, Probe-based electrochemical fabrication of freestanding Cu nanowire array. Appl. Phys. Lett. **88**(8), 083103 (2006)
109. J. Hu, M.-F. Yu, Meniscus-confined three-dimensional electrodeposition for direct writing of wire bonds. Science **329**(5989), 313–316 (2010)
110. B.Y. Ahn et al., Omnidirectional printing of flexible, stretchable, and spanning silver microelectrodes. Science **323**(5921), 1590–1593 (2009)
111. E. Macdonald et al., 3D printing for the rapid prototyping of structural electronics. IEEE Access **2**, 234–242 (2014)
112. P. Mostafalu, M. Akbari, K.A. Alberti, Q. Xu, A. Khademhosseini, S.R. Sonkusale, A toolkit of thread-based microfluidics, sensors, and electronics for 3D tissue embedding for medical diagnostics. Microsyst. Nanoeng. **2**, 16039 (2016)
113. T.F. O'Connor, K.M. Rajan, A.D. Printz, D.J. Lipomi, Toward organic electronics with properties inspired by biological tissue. J. Mater. Chem. B **3**(25), 4947–4952 (2015)
114. L. Cai, C. Wang, Carbon nanotube flexible and stretchable electronics. Nanoscale Res. Lett. **10**(1), 1–21 (2015)
115. J. Li et al., Graphene film-functionalized germanium as a chemically stable, electrically conductive, and biologically active substrate. J. Mater. Chem. B **3**(8), 1544–1555 (2015)
116. S. Wang, Y. Huang, J.A. Rogers, Mechanical designs for inorganic stretchable circuits in soft electronics. IEEE Trans. Compon. Packag. Manuf. Technol. **5**(9), 1201–1218 (2015)
117. C. Wang, J.-.C. Chien, K. Takei, T. Takahashi, J. Nah, A.M. Niknejad, Extremely bendable, high-performance integrated circuits using semiconducting carbon nanotube networks for digital, analog, and radio-frequency applications. Nano Lett. **12**(3), 1527–1533 (2012)
118. C. Wang, D. Hwang, Z. Yu, K. Takei, J. Park, T. Chen, User-interactive electronic skin for instantaneous pressure visualization. Nat Mater **12**, 899–904 (2013)
119. N. Chou, S. Yoo, S. Kim, A largely deformable surface type neural electrode array based on PDMS. IEEE Trans. Neural Syst. Rehabil. Eng. **21**(4), 544–553 (2013)
120. S.J. Kim et al., The potential role of polymethyl methacrylate as a new packaging material for the implantable medical device in the bladder. Biomed. Res. Int. **2015**, 852456 (2015)
121. W. J. Bae et al., AB222. Comparison of biocompatibility between PDMS and PMMA as packaging materials for the intravesical implantable device: changes of macrophage and macrophage migratory inhibitory factor. Transl. Androl. Urol. **3**(Suppl 1) (2014)
122. B. Rubehn, T. Stieglitz, In vitro evaluation of the long-term stability of polyimide as a material for neural implants. Biomaterials **31**(13), 3449–3458 (2010)
123. Y. Qin, M.M.R. Howlader, M.J. Deen, Y.M. Haddara, P.R. Selvaganapathy, Polymer integration for packaging of implantable sensors. Sens. Actuators B Chem. **202**, 758–778 (2014)
124. M.M.R. Howlader, M. Iwashita, K. Nanbu, K. Saijo, T. Suga, Enhanced Cu/LCP adhesion by pre-sputter cleaning prior to Cu deposition. IEEE Trans. Adv. Packag. **28**(3), 495–502 (2005)

125. Y. Zhang, D. Li, Y. Li, S. Zhang, M. Wang, Y. Li, High electric conductivity of liquid crystals formed by ordered self-assembly of nonionic surfactant N,N-bis (2-hydroxyethyl)dodecanamide in water. Soft Matter **11**(9), 1762–1766 (2015)
126. C.J. Bettinger, J.P. Bruggeman, A. Misra, J.T. Borenstein, R. Langer, Biocompatibility of biodegradable semiconducting melanin films for nerve tissue engineering. Biomaterials **30** (17), 3050–3057 (2009)
127. C.L.E. Nijst et al., Synthesis and characterization of photocurable elastomers from poly (glycerol-co-sebacate). Biomacromol **8**(10), 3067–3073 (2007)
128. M. Strange, D. Plackett, M. Kaasgaard, F.C. Krebs, Biodegradable polymer solar cells. Sol. Energy Mater. Sol. Cells **92**(7), 805–813 (2008)
129. A. Campana, T. Cramer, D.T. Simon, M. Berggren, F. Biscarini, Electrocardiographic recording with conformable organic electrochemical transistor fabricated on resorbable bioscaffold. Adv. Mater. **26**(23), 3874–3878 (2014)
130. G. Mattana, D. Briand, A. Marette, A. Vásquez Quintero, N.F. de Rooij, Polylactic acid as a biodegradable material for all-solution-processed organic electronic devices. Org. Electron. **17**, 77–86 (2015)
131. D.-H. Kim et al., Dissolvable films of silk fibroin for ultrathin conformal bio-integrated electronics. Nat. Mater. **9**(6), 511–517 (2010)
132. Y. Liu et al., Highly flexible and lightweight organic solar cells on biocompatible silk fibroin. ACS Appl. Mater. Interfaces. **6**(23), 20670–20675 (2014)
133. C.J. Bettinger, Z. Bao, Organic thin-film transistors fabricated on resorbable biomaterial substrates. Adv. Mater. **22**(5), 651–655 (2010)
134. M. Irimia-Vladu et al., Indigo—a natural pigment for high performance ambipolar organic field effect transistors and circuits. Adv. Mater. **24**(3), 375–380 (2012)
135. M. Irimia-Vladu, N.S. Sariciftci, S. Bauer, Exotic materials for bio-organic electronics. J. Mater. Chem. **21**(5), 1350–1361 (2011)
136. M. Irimia-Vladu et al., Biocompatible and biodegradable materials for organic field-effect transistors. Adv. Funct. Mater. **20**(23), 4069–4076 (2010)
137. T.K. Das, S. Prusty, Review on conducting polymers and their applications. Polym.-Plast. Technol. Eng. **51**(14), 1487–1500 (2012)
138. N.K. Guimard, N. Gomez, C.E. Schmidt, Conducting polymers in biomedical engineering. Prog. Polym. Sci. **32**(8–9), 876–921 (2007)
139. S. Kim, J.-H. Kim, O. Jeon, I.C. Kwon, K. Park, Engineered polymers for advanced drug delivery. Eur. J. Pharm. Biopharm. **71**(3), 420–430 (2009)
140. J.K. Oh, R. Drumright, D.J. Siegwart, K. Matyjaszewski, The development of microgels/nanogels for drug delivery applications. Prog. Polym. Sci. **33**(4), 448–477 (2008)
141. S. Nambiar, J.T.W. Yeow, Conductive polymer-based sensors for biomedical applications. Biosens. Bioelectron. **26**(5), 1825–1832 (2011)
142. H. Shirakawa, E.J. Louis, A.G. MacDiarmid, C.K. Chiang, A.J. Heeger, Synthesis of electrically conducting organic polymers: halogen derivatives of polyacetylene, (CH)x. J. Chem. Soc. Chem. Commun. **16**, 578–580 (1977)
143. C.K. Chiang et al., Electrical conductivity in doped polyacetylene. Phys. Rev. Lett. **39**(17), 1098–1101 (1977)
144. C.K. Chiang, S.C. Gau, C.R.F. Jr, Y.W. Park, A.G. MacDiarmid, A.J. Heeger, Polyacetylene, (CH)x: n-type and p-type doping and compensation. Appl. Phys. Lett. **33** (1), 18–20 (1978)
145. S. Stassi, V. Cauda, G. Canavese, C.F. Pirri, Flexible tactile sensing based on piezoresistive composites: a review. Sensors **14**(3), 5296–5332 (2014)
146. C. Li, P.M. Wu, S. Lee, A. Gorton, M.J. Schulz, C.H. Ahn, Flexible dome and bump shape piezoelectric tactile sensors using PVDF-TrFE copolymer. J. Microelectromechanical Syst. **17**(2), 334–341 (2008)
147. V. Maheshwari, R.F. Saraf, High-resolution thin-film device to sense texture by touch. Science **312**(5779), 1501–1504 (2006)

148. T. Nelson, R. vanDover, S. Jin, S. Hackwood, G. Beni, Shear-sensitive magnetoresistive robotic tactile sensor. IEEE Trans. Magn. **22**(5), 394–396 (1986)
149. G.M. Krishna, K. Rajanna, Tactile sensor based on piezoelectric resonance. IEEE Sens. J. **4** (5), 691–697 (2004)
150. S.C.B. Mannsfeld et al., Highly sensitive flexible pressure sensors with microstructured rubber dielectric layers. Nat. Mater. **9**(10), 859–864 (2010)
151. J.A. Dobrzynska, M.A.M. Gijs, Flexible polyimide-based force sensor. Sens. Actuators Phys. **173**(1), 127–135 (2012)
152. Y. Zhou, C.-W. Chiu, H. Liang, Interfacial structures and properties of organic materials for biosensors: an overview. Sensors **12**(11), 15036–15062 (2012)
153. J. Janata, M. Josowicz, Conducting polymers in electronic chemical sensors. Nat. Mater. **2** (1), 19–24 (2003)
154. M. Gerard, A. Chaubey, B.D. Malhotra, Application of conducting polymers to biosensors. Biosens. Bioelectron. **17**(5), 345–359 (2002)
155. D.D. Borole, U.R. Kapadi, P.P. Mahulikar, D.G. Hundiwale, Conducting polymers: an emerging field of biosensors. Des. Monomers Polym. **9**(1), 1–11 (2006)
156. C.-C. Wen, W. Fang, Tuning the sensing range and sensitivity of three axes tactile sensors using the polymer composite membrane. Sens. Actuators Phys. **145–146**, 14–22 (2008)
157. B.J. Kane, M.R. Cutkosky, G.T.A. Kovacs, A traction stress sensor array for use in high-resolution robotic tactile imaging. J. Microelectromechanical Syst. **9**(4), 425–434 (2000)
158. H. Hu, K. Shaikh, C. Liu, Super flexible sensor skin using liquid metal as interconnect, 2007 in *IEEE Sensors* (2007), pp. 815–817
159. M.-Y. Cheng, C.-M. Tsao, Y.-Z. Lai, Y.-J. Yang, The development of a highly twistable tactile sensing array with stretchable helical electrodes. Sens. Actuators Phys. **166**(2), 226–233 (2011)
160. Y.-J. Yang et al., An integrated flexible temperature and tactile sensing array using PI-copper films. Sens. Actuators Phys. **143**(1), 143–153 (2008)
161. B. Dong, M. Krutschke, X. Zhang, L. Chi, H. Fuchs, Fabrication of polypyrrole wires between microelectrodes. Small **1**(5), 520–524 (2005)
162. B. Dong, D.Y. Zhong, L.F. Chi, H. Fuchs, Patterning of conducting polymers based on a random copolymer strategy: toward the facile fabrication of nanosensors exclusively based on polymers. Adv. Mater. **17**(22), 2736–2741 (2005)
163. A. Ambrosi, C.K. Chua, A. Bonanni, M. Pumera, Electrochemistry of graphene and related materials. Chem. Rev. **114**(14), 7150–7188 (2014)
164. A. Nathan et al., Flexible electronics: the next ubiquitous platform. Proc. IEEE. **100**(Special Centennial Issue), May 2012, pp. 1486–1517
165. A.K. Geim, K.S. Novoselov, The rise of graphene. Nat. Mater. **6**(3), 183–191 (2007)
166. K.S. Novoselov et al., Electric field effect in atomically thin carbon films. Science **306** (5696), 666–669 (2004)
167. F. Withers et al., Light-emitting diodes by band-structure engineering in van der Waals heterostructures. Nat. Mater. **14**(3), 301–306 (2015)
168. Q.H. Wang, K. Kalantar-Zadeh, A. Kis, J.N. Coleman, M.S. Strano, Electronics and optoelectronics of two-dimensional transition metal dichalcogenides. Nat. Nanotechnol. **7** (11), 699–712 (2012)
169. F. Schwierz, Graphene transistors. Nat. Nanotechnol. **5**(7), 487–496 (2010)
170. S.-K. Lee et al., All graphene-based thin film transistors on flexible plastic substrates. Nano Lett. **12**(7), 3472–3476 (2012)
171. S.-K. Lee et al., Stretchable graphene transistors with printed dielectrics and gate electrodes. Nano Lett. **11**(11), 4642–4646 (2011)
172. S. Riazimehr et al., Spectral sensitivity of graphene/silicon heterojunction photodetectors. Solid-State Electron. **115**(Part B), 207–212 (2016)
173. M.C. Lemme et al., Gate-activated photoresponse in a graphene p–n junction. Nano Lett. **11** (10), 4134–4137 (2011)

174. T. Mueller, F. Xia, P. Avouris, Graphene photodetectors for high-speed optical communications. Nat. Photonics **4**(5), 297–301 (2010)
175. X. Gan et al., Chip-integrated ultrafast graphene photodetector with high responsivity. Nat. Photonics **7**(11), 883–887 (2013)
176. A. Pospischil et al., CMOS-compatible graphene photodetector covering all optical communication bands. Nat. Photonics **7**(11), 892–896 (2013)
177. F.H.L. Koppens, T. Mueller, P. Avouris, A.C. Ferrari, M.S. Vitiello, M. Polini, Photodetectors based on graphene, other two-dimensional materials and hybrid systems. Nat. Nanotechnol. **9**(10), 780–793 (2014)
178. D.-M. Sun, C. Liu, W.-C. Ren, H.-M. Cheng, A review of carbon nanotube- and graphene-based flexible thin-film transistors. Small **9**(8), 1188–1205 (2013)
179. L. Buglione, E.L.K. Chng, A. Ambrosi, Z. Sofer, M. Pumera, Graphene materials preparation methods have dramatic influence upon their capacitance. Electrochem. Commun. **14**(1), 5–8 (2012)
180. A. Bonanni, M. Pumera, High-resolution impedance spectroscopy for graphene characterization. Electrochem. Commun. **26**, 52–54 (2013)
181. G. Eda, M. Chhowalla, Chemically derived graphene oxide: towards large-area thin-film electronics and optoelectronics. Adv. Mater. **22**(22), 2392–2415 (2010)
182. S. Bae et al., Roll-to-roll production of 30-inch graphene films for transparent electrodes. Nat. Nanotechnol. **5**(8), 574–578 (2010)
183. T. Kobayashi et al., Production of a 100-m-long high-quality graphene transparent conductive film by roll-to-roll chemical vapor deposition and transfer process. Appl. Phys. Lett. **102**(2), 023112 (2013)
184. L. Gao, G.-X. Ni, Y. Liu, B. Liu, A.H. Castro Neto, K.P. Loh, Face-to-face transfer of wafer-scale graphene films. Nature **505**(7482), 190–194 (2014)
185. C. Yan, J.H. Cho, J.-H. Ahn, Graphene-based flexible and stretchable thin film transistors. Nanoscale **4**(16), 4870–4882 (2012)
186. K.S. Kim et al., Large-scale pattern growth of graphene films for stretchable transparent electrodes. Nature **457**(7230), 706–710 (2009)
187. M.F. El-Kady, V. Strong, S. Dubin, R.B. Kaner, Laser Scribing of High-Performance and flexible graphene-based electrochemical capacitors. Science **335**(6074), 1326–1330 (2012)
188. V.L. Solozhenko, A.G. Lazarenko, J.-P. Petitet, A.V. Kanaev, Bandgap energy of graphite-like hexagonal boron nitride. J. Phys. Chem. Solids **62**(7), 1331–1334 (2001)
189. H. Amano, T. Asahi, I. Akasaki, Stimulated emission near ultraviolet at room temperature from a GaN film grown on sapphire by MOVPE using an AlN buffer layer. Jpn. J. Appl. Phys. **29**(Part 2, No. 2), L205–L206 (1990)
190. K. Watanabe, T. Taniguchi, H. Kanda, Direct-bandgap properties and evidence for ultraviolet lasing of hexagonal boron nitride single crystal. Nat. Mater. **3**(6), 404–409 (2004)
191. M. Wang et al., A platform for large-scale graphene electronics—CVD growth of single-layer graphene on CVD-grown hexagonal boron nitride. Adv. Mater. **25**(19), 2746–2752 (2013)
192. C.R. Dean et al., Boron nitride substrates for high-quality graphene electronics. Nat. Nanotechnol. **5**(10), 722–726 (2010)
193. K.K. Kim et al., Synthesis and characterization of hexagonal boron nitride film as a dielectric layer for graphene devices. ACS Nano **6**(10), 8583–8590 (2012)
194. Z. Liu et al., In-plane heterostructures of graphene and hexagonal boron nitride with controlled domain sizes. Nat. Nanotechnol. **8**(2), 119–124 (2013)
195. E. Kim, T. Yu, E.S. Song, B. Yu, Chemical vapor deposition-assembled graphene field-effect transistor on hexagonal boron nitride. Appl. Phys. Lett. **98**(26), 262103 (2011)
196. J. Lee et al., High-performance current saturating graphene field-effect transistor with hexagonal boron nitride dielectric on flexible polymeric substrates. IEEE Electron Device Lett. **34**(2), 172–174 (2013)

197. R. Coehoorn, C. Haas, J. Dijkstra, C.J.F. Flipse, R.A. de Groot, A. Wold, Electronic structure of $MoSe_2$, MoS_2, and WSe_2. I. Band-structure calculations and photoelectron spectroscopy. Phys. Rev. B **35**(12), 6195–6202 (1987)
198. C. Feng, J. Ma, H. Li, R. Zeng, Z. Guo, H. Liu, Synthesis of molybdenum disulfide (MoS_2) for lithium ion battery applications. Mater. Res. Bull. **44**(9), 1811–1815 (2009)
199. C. Lee et al., Frictional characteristics of atomically thin sheets. Science **328**(5974), 76–80 (2010)
200. J.A. Wilson, A.D. Yoffe, The transition metal dichalcogenides discussion and interpretation of the observed optical, electrical and structural properties. Adv. Phys. **18**(73), 193–335 (1969)
201. B. Radisavljevic, A. Radenovic, J. Brivio, V. Giacometti, A. Kis, Single-layer MoS_2 transistors. Nat. Nanotechnol. **6**(3), 147–150 (2011)
202. N.R. Pradhan et al., Field-effect transistors based on few-layered α-$MoTe_2$. ACS Nano. **8**(6), 5911–5920 (2014)
203. G.-H. Lee et al., Flexible and transparent MoS2 field-effect transistors on hexagonal boron nitride-graphene heterostructures. ACS Nano. **7**(9), 7931–7936 (2013)
204. G. Fiori et al., Electronics based on two-dimensional materials. Nat. Nanotechnol. **9**(10), 768–779 (2014)
205. D. B. Velusamy et al., Flexible transition metal dichalcogenide nanosheets for band-selective photodetection. Nat. Commun. **6**, 8063 (2015)
206. C. Kim, T.P. Nguyen, Q.V. Le, J.-M. Jeon, H.W. Jang, S.Y. Kim, Performances of liquid-exfoliated transition metal dichalcogenides as hole injection layers in organic light-emitting diodes. Adv. Funct. Mater. **25**(28), 4512–4519 (2015)
207. A.C. Arias, J.D. MacKenzie, I. McCulloch, J. Rivnay, A. Salleo, Materials and applications for large area electronics: solution-based approaches, Chem. Rev. **110**(1), 3–24 (2010)
208. S.H. Chang, C.H. Chiang, F.S. Kao, C.L. Tien, C.G. Wu, Unraveling the enhanced electrical conductivity of PEDOT:PSS thin films for ITO-free organic photovoltaics. IEEE Photonics J. **6**(4), 1–7 (2014)
209. H. Klauk, Organic thin-film transistors. Chem. Soc. Rev. **39**, 2643–2666 (2010)
210. T.W. Kelley et al., Recent progress in organic electronics: materials, devices, and processes. Chem. Mater. **16**(23), 4413–4422 (2004)
211. F. Eder, H. Klauk, M. Halik, U. Zschieschang, G. Schmid, C. Dehm, Organic electronics on paper. Appl. Phys. Lett. **84**(14), 2673–2675 (2004)
212. V. Benfenati et al., A transparent organic transistor structure for bidirectional stimulation and recording of primary neurons. Nat. Mater. **12**(7), 672–680 (2013)
213. M. Katsuhara et al., 44.2: Distinguished Paper: a reliable flexible OLED display with an OTFT backplane manufactured using a scalable process. SID Symp. Dig. Tech. Pap. **40**(1), 656–659 (2009)
214. T. Sekitani, T. Someya, Stretchable, large-area organic electronics. Adv. Mater. **22** (2010)
215. A.S. Azam, M. Boukadoum, R. Izquierdo, A. Acharya, M. Packirisamy, Integrated multifunctional fluorescence biosensor based on OLED technology, in *2008 Joint 6th International IEEE Northeast Workshop on Circuits and Systems and TAISA Conference (NEWCAS-TAISA 2008)* (2008), pp. 173–176
216. K.L. Lin, K. Jain, Design and fabrication of stretchable multilayer self-aligned interconnects for flexible electronics and large-area sensor arrays using excimer laser photoablation. IEEE Electron Device Lett. **30**(1), 14–17 (2009)
217. K.-S. Kim, K.-H. Jung, S.-B. Jung, Design and fabrication of screen-printed silver circuits for stretchable electronics. Microelectron. Eng. **120**, 216–220 (2014)
218. H.J. Kim, T. Maleki, P. Wei, B. Ziaie, A Biaxial Stretchable interconnect with liquid-alloy-covered joints on elastomeric substrate. J. Microelectromechanical Syst. **18** (1), 138–146 (2009)
219. H. Hocheng, C.-M. Chen, Design, fabrication and failure analysis of stretchable electrical routings. Sensors **14**(7), 11855–11877 (2014)

220. T. Cheng, Y. Zhang, W.-Y. Lai, W. Huang, Stretchable thin-film electrodes for flexible electronics with high deformability and stretchability. Adv. Mater. **27**(22), 3349–3376 (2015)
221. Y. Zhang et al., Experimental and theoretical studies of serpentine microstructures bonded to prestrained elastomers for stretchable electronics. Adv. Funct. Mater. **24**(14), 2028–2037 (2014)
222. J. Lee et al., Stretchable GaAs photovoltaics with designs that enable high areal coverage. Adv. Mater. **23**(8), 986–991 (2011)
223. C. Yu, H. Jiang, Forming wrinkled stiff films on polymeric substrates at room temperature for stretchable interconnects applications. Thin Solid Films **519**(2), 818–822 (2010)
224. S.P. Lacour, J. Jones, Z. Suo, S. Wagner, Design and performance of thin metal film interconnects for skin-like electronic circuits. IEEE Electron Device Lett. **25**(4), 179–181 (2004)
225. Y.Y. Hsu, M. Gonzalez, F. Bossuyt, J. Vanfleteren, I.D. Wolf, Polyimide-enhanced stretchable interconnects: design, fabrication, and characterization. IEEE Trans. Electron Devices **58**(8), 2680–2688 (2011)
226. Y. Zhang et al., Buckling in serpentine microstructures and applications in elastomer-supported ultra-stretchable electronics with high areal coverage. Soft Matter **9** (33), 8062–8070 (2013)
227. Y.-Y. Hsu, M. Gonzalez, F. Bossuyt, F. Axisa, J. Vanfleteren, I. De Wolf, The effects of encapsulation on deformation behavior and failure mechanisms of stretchable interconnects. Thin Solid Films **519**(7), 2225–2234 (2011)
228. Mario Gonzalez, Fabrice Axisa, Frederick Bossuyt, Yung-Yu. Hsu, Bart Vandevelde, Jan Vanfleteren, Design and performance of metal conductors for stretchable electronic circuits. Circuit World **35**(1), 22–29 (2009)
229. M. Gonzalez, F. Axisa, M.V. Bulcke, D. Brosteaux, B. Vandevelde, J. Vanfleteren, Design of metal interconnects for stretchable electronic circuits. Microelectron. Reliab. **48**(6), 825–832 (2008)
230. O. van der Sluis, Y.Y. Hsu, P.H.M. Timmermans, M. Gonzalez, J.P.M. Hoefnagels, Stretching-induced interconnect delamination in stretchable electronic circuits. J. Phys. Appl. Phys. **44**(3), 034008 (2011)
231. S. Xu et al., Stretchable batteries with self-similar serpentine interconnects and integrated wireless recharging systems. Nat. Commun. **4**, 1543 (2013)
232. Y.-Y. Hsu, M. Gonzalez, F. Bossuyt, F. Axisa, J. Vanfleteren, I.D. Wolf, The effect of pitch on deformation behavior and the stretching-induced failure of a polymer-encapsulated stretchable circuit. J. Micromechanics Microengineering **20**(7), 075036 (2010)
233. D.S. Gray, J. Tien, C.S. Chen, High-conductivity elastomeric electronics. Adv. Mater. **16**(5), 393–397 (2004)
234. A.M. Hussain, E.B. Lizardo, G.A. Torres Sevilla, J.M. Nassar, M.M. Hussain, Ultrastretchable and flexible copper interconnect-based smart patch for adaptive thermotherapy. Adv. Healthcare Mater. **4**(5), 665–673 (2015)
235. J.A. Fan et al., Fractal design concepts for stretchable electronics. Nat. Commun. **5**, 3266 (2014)
236. L. Bowman, J.D. Meindl, The packaging of implantable integrated sensors. IEEE Trans. Biomed. Eng. **BME-33**(2), 248–255 (1986)
237. D.F. Williams, *Biocompatibility of Clinical Implant Materials* (CRC Press, Boca Raton, 1981)
238. G. Jiang, D.D. Zhou, Technology advances and challenges in hermetic packaging for implantable medical devices, in *Implantable Neural Prostheses 2*, ed. by D. Zhou, E. Greenbaum (Springer, New York, 2009), pp. 27–61
239. A. Cavallini et al., A subcutaneous biochip for remote monitoring of human metabolism: packaging and biocompatibility assessment. IEEE Sens. J. **15**(1), 417–424 (2015)

240. T.J. Harpster, S.A. Nikles, M.R. Dokmeci, K. Najafi, Long-term hermeticity and biological performance of anodically bonded glass-silicon implantable packages. IEEE Trans. Device Mater. Reliab. **5**(3), 458–466 (2005)
241. K. Najafi, Packaging of implantable microsystems, in *2007 IEEE Sensors* (2007), pp. 58–63
242. K. Najafi, in *Micropackaging technologies for integrated microsystems: applications to MEMS and MOEMS* (2003), pp. 1–19
243. T.J. Harpster, S. Hauvespre, M.R. Dokmeci, K. Najafi, A passive humidity monitoring system for in situ remote wireless testing of micropackages. J. Microelectromechanical Syst. **11**(1), 61–67 (2002)
244. T. Stieglitz, Manufacturing, assembling and packaging of miniaturized neural implants. Microsyst. Technol. **16**(5), 723–734 (2010)
245. M. Schuettler, J.S. Ordonez, T.S. Santisteban, A. Schatz, J. Wilde, T. Stieglitz, Fabrication and test of a hermetic miniature implant package with 360 electrical feedthroughs, in *2010 Annual International Conference of the IEEE Engineering in Medicine and Biology Society (EMBC)* (2010), pp. 1585–1588
246. M. Schuettler, A. Schatz, J.S. Ordonez, T. Stieglitz, Ensuring minimal humidity levels in hermetic implant housings, in *2011 Annual International Conference of the IEEE Engineering in Medicine and Biology Society, EMBC* (2011), pp. 2296–2299
247. J.S. Ordonez, M. Schuettler, M. Ortmanns, T. Stieglitz, A 232-channel retinal vision prosthesis with a miniaturized hermetic package, in *2012 Annual International Conference of the IEEE Engineering in Medicine and Biology Society (EMBC)* (2012), pp. 2796–2799
248. N. Saeidi, M. Schuettler, A. Demosthenous, N. Donaldson, Technology for integrated circuit micropackages for neural interfaces, based on gold–silicon wafer bonding. J. Micromechanics Microengineering **23**(7), 075021 (2013)
249. N. Saeidi, J. Strutwolf, A. Marechal, A. Demosthenous, N. Donaldson, A capacitive humidity sensor suitable for CMOS integration. IEEE Sens. J. **13**(11), 4487–4495 (2013)
250. N. Saeidi, A. Demosthenous, N. Donaldson, J. Alderman, Design and fabrication of corrosion and humidity sensors for performance evaluation of chip scale hermetic packages for biomedical implantable devices, in *European Microelectronics and Packaging Conference 2009 (EMPC 2009)* (2009), pp. 1–4
251. A. Ivorra et al., Minimally invasive silicon probe for electrical impedance measurements in small animals. Biosens. Bioelectron. **19**(4), 391–399 (2003)
252. R. Gómez et al., A SiC microdevice for the minimally invasive monitoring of ischemia in living tissues. Biomed. Microdevices **8**(1), 43–49 (2006)
253. M. Tijero et al., SU-8 microprobe with microelectrodes for monitoring electrical impedance in living tissues. Biosens. Bioelectron. **24**(8), 2410–2416 (2009)
254. A.A. Sharkawy, B. Klitzman, G.A. Truskey, W.M. Reichert, Engineering the tissue which encapsulates subcutaneous implants. I. Diffusion properties. J. Biomed. Mater. Res. **37**(3), 401–412 (1997)
255. A.A. Sharkawy, B. Klitzman, G.A. Truskey, W.M. Reichert, Engineering the tissue which encapsulates subcutaneous implants. III. Effective tissue response times. J. Biomed. Mater. Res. **40**(4), 598–605 (1998)
256. T. Trantidou, D.J. Payne, V. Tsiligkiridis, Y.-C. Chang, C. Toumazou, T. Prodromakis, The dual role of Parylene C in chemical sensing: acting as an encapsulant and as a sensing membrane for pH monitoring applications. Sens. Actuators B Chem. **186**, 1–8 (2013)
257. G.S. Prihandana et al., Solute diffusion through fibrotic tissue formed around protective cage system for implantable devices. J. Biomed. Mater. Res. B Appl. Biomater. **103**(6), 1180–1187 (2015)
258. M.M.R. Howlader, A.U. Alam, R.P. Sharma, M.J. Deen, Materials analyses and electrochemical impedance of implantable metal electrodes. Phys. Chem. Chem. Phys. **17** (15), 10135–10145 (2015)
259. M. Schuettler, T. Stieglitz, Microassembly and micropackaging of implantable systems, in *Implantable Sensor Systems for Medical Applications* (Woodhead Publishing, Oxford, 2013), pp. 108–149

260. M. Dokmeci, K. Najafi, A high-sensitivity polyimide capacitive relative humidity sensor for monitoring anodically bonded hermetic micropackages. J. Microelectromechanical Syst. **10** (2), 197–204 (2001)
261. D. Cirmirakis, A. Demosthenous, N. Saeidi, N. Donaldson, Humidity-to-Frequency sensor in cmos technology with wireless readout. IEEE Sens. J. **13**(3), 900–908 (2013)
262. G.S. Wilson, M. Ammam, In vivo biosensors. FEBS J. **274**(21), 5452–5461 (2007)
263. G.S. Wilson, R. Gifford, Biosensors for real-time *in vivo* measurements. Biosens. Bioelectron. **20**(12), 2388–2403 (2005)
264. S. Vaddiraju, I. Tomazos, D.J. Burgess, F.C. Jain, F. Papadimitrakopoulos, Emerging synergy between nanotechnology and implantable biosensors: a review. Biosens. Bioelectron. **25**(7), 1553–1565 (2010)
265. C.N. Kotanen, A. Guiseppi-Elie, Monitoring systems and quantitative measurement of biomolecules for the management of Trauma. Biomed. Microdevices **15**(3), 561–577 (2013)
266. J.H. Shin, M.H. Schoenfisch, Improving the biocompatibility of in vivo sensors via nitric oxide release. Analyst **131**(5), 609–615 (2006)
267. M. Frost, M.E. Meyerhoff, In vivo chemical sensors: tackling biocompatibility. Anal. Chem. **78**(21), 7370–7377 (2006)
268. Y. Onuki, U. Bhardwaj, F. Papadimitrakopoulos, D.J. Burgess, A review of the biocompatibility of implantable devices: current challenges to overcome foreign body response. J. Diabetes Sci. Technol. **2**(6), 1003–1015 (2008)
269. L.A. Geddes, R. Roeder, Criteria for the selection of materials for implanted electrodes. Ann. Biomed. Eng. **31**(7), 879–890 (2003)
270. A. Radu et al., Diagnostic of functionality of polymer membrane – based ion selective electrodes by impedance spectroscopy. Anal. Methods **2**(10), 1490–1498 (2010)
271. R.C. Mercado, F. Moussy, In vitro and in vivo mineralization of Nafion membrane used for implantable glucose sensors. Biosens. Bioelectron. **13**(2), 133–145 (1998)
272. S.R. Shah, A.M. Tatara, R.N. D'Souza, A.G. Mikos, F.K. Kasper, Evolving strategies for preventing biofilm on implantable materials. Mater. Today **16**(5), 177–182 (2013)
273. J.D. Patel, M. Ebert, R. Ward, J.M. Anderson, S. epidermidis biofilm formation: effects of biomaterial surface chemistry and serum proteins. J. Biomed. Mater. Res. A **80A**(3), 742–751 (2007)
274. E.M. Hetrick, M.H. Schoenfisch, Reducing implant-related infections: active release strategies. Chem. Soc. Rev. **35**(9), 780–789 (2006)
275. J.M. Anderson, Biological responses to materials. Annu. Rev. Mater. Res. **31**(1), 81–110 (2001)
276. J.A. Jones et al., Proteomic analysis and quantification of cytokines and chemokines from biomaterial surface-adherent macrophages and foreign body giant cells. J. Biomed. Mater. Res. A **83A**(3), 585–596 (2007)
277. E. Ostuni, R.G. Chapman, R.E. Holmlin, S. Takayama, G.M. Whitesides, A survey of structure—property relationships of surfaces that resist the adsorption of protein. Langmuir **17**(18), 5605–5620 (2001)
278. N. Tirelli, M.P. Lutolf, A. Napoli, J.A. Hubbell, Poly(ethylene glycol) block copolymers. Rev. Mol. Biotechnol. **90**(1), 3–15 (2002)
279. I.A. Silver, R.J. Murrills, D.J. Etherington, Microelectrode studies on the acid microenvironment beneath adherent macrophages and osteoclasts. Exp. Cell Res. **175**(2), 266–276 (1988)
280. A. Koh, S.P. Nichols, M.H. Schoenfisch, Glucose sensor membranes for mitigating the foreign body response. J. Diabetes Sci. Technol. **5**(5), 1052–1059 (2011)
281. R.D. Jayant, M.J. McShane, R. Srivastava, In vitro and in vivo evaluation of anti-inflammatory agents using nanoengineered alginate carriers: towards localized implant inflammation suppression. Int. J. Pharm. **403**(1–2), 268–275 (2011)
282. Y. Wang, F. Papadimitrakopoulos, D.J. Burgess, Polymeric 'smart' coatings to prevent foreign body response to implantable biosensors. J. Controlled Release **169**(3), 341–347 (2013)

283. J.H. Shin, S.M. Marxer, M.H. Schoenfisch, Nitric oxide-releasing sol–gel particle/polyurethane glucose biosensors. Anal. Chem. **76**(15), 4543–4549 (2004)
284. G.S. Wilson, M.A. Johnson, In-vivo electrochemistry: what can we learn about living systems?. Chem. Rev. **108**(7), 2462–2481 (2008)
285. D.L. Hern, J.A. Hubbell, Incorporation of adhesion peptides into nonadhesive hydrogels useful for tissue resurfacing. J. Biomed. Mater. Res. **39**(2), 266–276 (1998)
286. B.D. Ratner, S.J. Bryant, Biomaterials: where we have been and where we are going. Annu. Rev. Biomed. Eng. **6**(1), 41–75 (2004)
287. Use of International Standard ISO 10993-1, 'Biological evaluation of medical devices—Part 1: Evaluation and testing within a risk management process': Guidance for Industry and Food and Drug Administration Staff. U.S. Department of Health and Human Services, Food and Drug Administration, Center for Devices and Radiological Health, 16 June 2016

Chapter 5
Ultra-Low Power Application-Specific Integrated Circuits for Sensing

P. Kassanos, H. Ip and Guang-Zhong Yang

List of Acronyms

AAP	Axon action potentials
AC	Alternating current
ACFB	Autocorrelation feedback
ADC	Analog-to-digital converter
AGC	Automatic gain control
APS	Active pixel sensors
ASIC	Application specific integrated circuit
AZ	Auto-zeroing
BIST	Built-in self-test
BJT	Bipolar junction transistor
BPF	Bandpass filter
CDS	Correlated double sampling
CHS	Chopper stabilization
CMFB	Common-mode feedback
CMOS	Complementary metal-oxide semiconductor
CMRR	Common-mode rejection ratio
CMS	Correlated multiple sampling
CNS	Central nervous systems
CORDIC	Coordinate rotation digital computer
CS	Compressed sensing
CTAT	Complementary-to-absolute-temperature
CT ΔΣ	Continuous-time ΔΣ-modulator
CV	Cyclic voltammetry
DAC	Digital-to-analog converter
DBS	Deep brain stimulation
DC	Direct current
DDA	Differential difference amplifier

P. Kassanos (✉) · H. Ip · G.-Z. Yang
The Hamlyn Centre, Imperial College London, London, UK
e-mail: p.kassanos@imperial.ac.uk

G.-Z. Yang (ed.), *Implantable Sensors and Systems*
https://doi.org/10.1007/978-3-319-69748-2_5

DDS	Direct digital synthesizer
DNA	Deoxyribonucleic acid
DSP	Digital signal processing
DT ΔΣ	Discrete-time ΔΣ-modulator
ECoG	Electrocorticography
ECG	Electrocardiography
EEG	Electroencephalography
EIT	Electrical impedance tomography
EMG	Electromyography
EMI	Electromagnetic interference
ENG	Electroneurography
EOG	Electrooculography
ESD	Electrostatic discharge
FBVA	Feedback voltage attenuation
FET	Field-effect transistor
FFCA	Feedforward current attenuation
FFVA	Feedforward voltage attenuation
FGMOS	Floating gate MOS
FIR	Finite impulse response
FPAA	Field-programmable analog array
FPGA	Field-programmable gated array
FS	Frequency span
FSCV	Fast scan cyclic voltammetry
FSK	Frequency shift keying
HD3	Third order harmonic distortion
HPF	High-pass filter
IA	Instrumentation amplifier
IC	Inversion coefficient
IF-ADC	Integrate and fire ADC
IIR	Infinite impulse response
KCL	Kirchhoff's current law
LFP	Local field potential
LNA	Low noise amplifier
LPF	Low-pass filter
LSB	Least significant bit
LUT	Lookup table
MEMS	Micro-electro-mechanical system
MOSFET	Metal-oxide field effect transistor
MSB	Most significant bit
NEF	Noise efficiency factor
OOS	Output offset storage
OSR	Oversampling ratio
op-amp	Operational amplifier
OTA	Operational transconductance amplifier

PEF	Power efficiency factor
PGA	Programmable gain amplifier
PND	P+ non-salicide diffusion
PNS	Peripheral nervous systems
PSD	Power spectral density
PSRR	Power supply rejection ratio
PTAT	Positive-to-absolute-temperature
PWV	Pulse wave velocity
Q	Quality factor
RMS	Root-mean-square
ROM	Read-only memory
S	Slope
SAR	Successive approximation register
SC	Switched capacitor
SCCMOS	Super cutoff CMOS
SD	Synchronous detection
SFDR	Spurious-free dynamic range
SNR	Signal to noise ratio
SoC	System-on-chip
TC	Temperature coefficient
TCN	Negative temperature coefficient
TCP	Positive temperature coefficient
TDC	Time-to-digital converter
TEM	Time-encoding machine
THD	Total harmonic distortion
TI	Transimpedance
TIA	Transimpedance amplifier
TSC	Triangle-to-sine converter
TSV	Through-silicon-vias
VCCS	Voltage-controlled current source
VFC	Voltage-to-frequency converter
VVR	Voltage variable resistor

5.1 Introduction

In the quest for ever-reducing system size and increased integration and functionality, application-specific integrated circuit (ASIC) technology plays a pivotal role in modern implants, where custom circuits designed at transistor and device levels are replacing off-the-shelf commercial chips and bulky benchtop systems. Recently, commercial system-on-chip (SoC) devices encompassing digital microcontrollers, radio, and analog–digital converters, as well as reconfigurable amplifier circuits, are widely available. Despite this, further development of ASIC-specific implantable systems is required, particularly in the area of multi-channel array sensor interfaces, ultra-low power data acquisition, and circuits that work with specialized micro-sensors for implants. ASICs designed to focus on a particular application have given designers the freedom to optimize power consumption for a set task, unlike general-purpose SoCs that have to cater for a wide range of applications and hence typically consume more power. In this chapter, we begin with a survey on the latest development of ASICs and related integrated systems from literature. This is followed by an overview of technological trends in integrated circuit/sensor fabrication and fusion. The rest of the chapter focuses on a number of engineering aspects related to ultra-low power ASIC circuits appropriate for implantable sensors and sensor front-ends, covering bioimpedance, neural and electrochemical sensor measurement circuits, as well as low-power analog-to-digital converter design and architectures.

5.1.1 ASIC Integrated Sensing—Overview and Examples

Components of an SoC built for medical implant and biological applications can be broadly classified into sensing, powering, and stimulation circuits. Sensing is the most common component, with ASICs designed to take measurements from a wide range of physical, optical, and biochemical sensors, some monolithically integrated on the ASIC. We cover some of the crucial sensing circuits in later parts of this chapter, while power circuits are discussed in Chap. 7.

Figure 5.1 illustrates the typical architecture of a sensing SoC. The analog front-end interfaces directly with the sensors and mostly comprises signal amplifiers and filters. Sensors can be divided into two categories depending on the interfacing circuits required. "Direct measurement" sensors transduce sensing target signals into electrical signals that can be detected directly (e.g. photodiode for converting light into charge), whereas "excitation measurement" sensors require the sensing

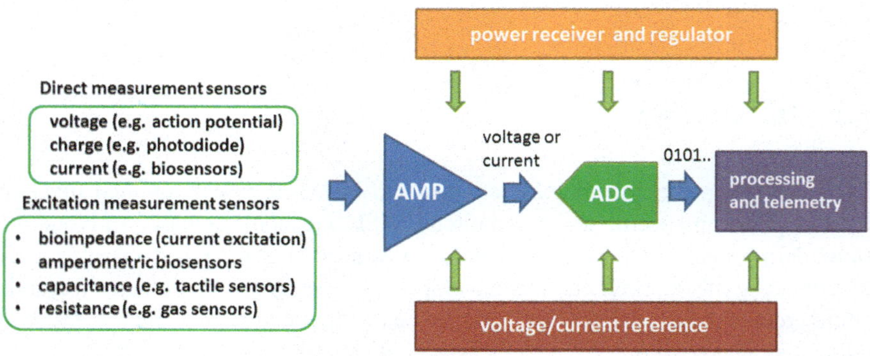

Fig. 5.1 Typical building blocks of a sensing SoC

circuits to exert an excitation (or "bias") signal on the sensor for correct operation. Examples of the second type of sensors includes bioimpedance sensing that requires controlled alternating current (AC) current injection into the tissue while a voltage is measured. This will be covered extensively later in this chapter. The type of sensor used thus determines the analog front-end setup, taking into account input signal magnitude and bandwidth requirements. Typically, for voltage measurements instrumentation amplifiers (IA) can be used to provide high-impedance buffering and differential signal gain. For on-chip current measurements, transimpedance amplifiers (TIA) with either capacitive (average current integration) or pseudo-resistor feedback are commonly used. Charge measurements are usually carried out with the same capacitive feedback amplifier or captured and amplified with a switched capacitor (SC) circuit taking advantage of the linear Q-V relationship across a capacitor. For excitation sources, a voltage source can be driven directly from a buffered digital-to-analog converter (DAC), whereas current sources are commonly derived from V-I converters. The front-end amplifiers of an ASIC are typically band-limited to a minimum speed necessary to process the input signal, this restricts out-of-band noise, improving the signal-to-noise ratio (SNR). Additional filters can be placed at this stage to further improve signal integrity. Common filters used are continuous-time transconductance-C (G_m-C) or current-mode filters. Both types exploit the tunability of transistor bias currents to realize a range of cutoff frequencies. This is particularly advantageous for low-frequency filters, as this approach eliminates the need for large value (taking up large chip area) resistors and/or capacitors to realize a large *RC* time constant. Following amplification and filtering, the analog signal is to be digitized by an analog–digital converter, giving an output of two binary states. The two most common types of analog–digital converters (ADC) used in implants and biological SoCs are successive approximation register (SAR) ADCs and time-encoding ADCs. The output of a SAR ADC is the typical pulse coded modulation with "0" and "1" commonly coded by two voltage levels. The time-encoding approach represents the two binary states as timing (e.g. duty cycle) variations on a waveform and can be particularly robust given component imperfections. We cover in detail in this chapter both types of ADCs in later parts of this chapter. The SoC can be powered remotely for batteryless operation or wireless charging of a small battery. Wireless powering can be carried out with inductive, ultrasonic, or RF devices. These are covered extensively in Chap. 7. All the analog circuits in an SoC will require steady reference voltage/currents to operate. These need to be nano-power consumption references suitable for use in nano-micro power SoCs. For the rest of this section, we survey some of the latest developments in SoC for implants and biological applications.

One of the major advantages of developing an ASIC for sensing is the shear density of electronic integration. For neuroscience experiments, it is clear that large arrays of electrodes are advantageous to capture neural network activities. In [1], a 52-recording channel ASIC is integrated with a 455-electrode probe measuring 100 μm × 10 mm × 50 μm. All electrodes and readout electronics are implemented with an all complementary metal-oxide-semiconductor (CMOS) process

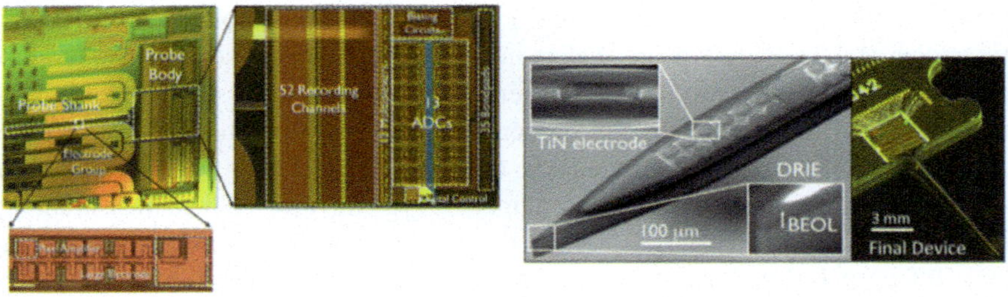

Fig. 5.2 52 channel neural probe on all-CMOS process (© IEEE 2014. Reprinted with permission from [1])

(see Fig. 5.2). For neural recordings, the typical bandwidth is from DC (direct current, e.g. 0 Hz) to 10 kHz, covering average local field potentials (LFP, 1–300 Hz) and action potentials with frequency components up to 10 kHz. The signal levels are in the µV–mV range (see Fig. 5.8). Placing the amplifier circuitry in the immediate vicinity of the sensor electrode thus minimizes the effect of noise and disturbance pickup for carrying a small signal over lengths of wirings. A fixed gain of 20 V/V is thus implemented for the amplifiers underneath the sensing electrodes, followed by a programmable gain amplifier (PGA) and filter and 10-bit SAR ADC at the probe body (see Fig. 5.2). The sensing electrodes are realized with titanium nitride deposition during chip post-processing, which also involves thinning the entire wafer to 50 µm. The unique dimension of the all-CMOS neural probe reported in [1] results in 784 devices available on a 200 mm wafer of the TSMC 0.18 µm technology.

Yin et al. [2] presented a 100 channel neuro-recording ASIC as part of an implant (Fig. 5.3). A 3.2 GHz frequency shift keying (FSK) data link has been integrated as part of the system for transmitting the multi-channel data as well as a 150 kHz inductive charging circuit for the 200 mAh lithium ion battery. Neural recordings of low signal levels require high-performance low-noise pre-amplifiers for signal conditioning. By adopting a full custom design and limiting the amplifier bandwidth on-chip, while appropriately sizing the input transistors of the amplifiers,

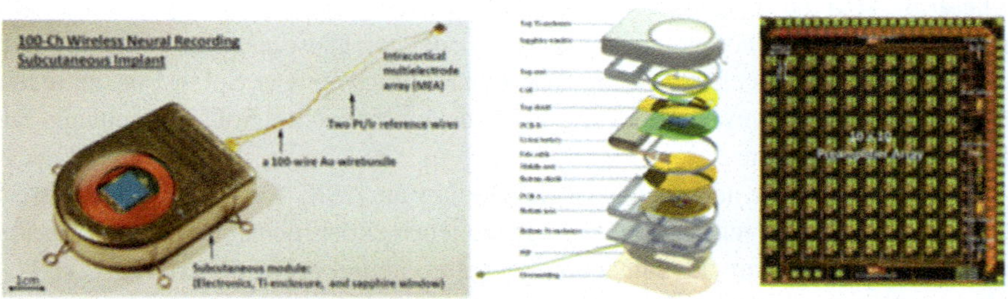

Fig. 5.3 100 channel neuro-recording system with a 100 amplifier channel ASIC (© IEEE 2013. Reprinted with permission from [2])

energy-efficient yet low-noise pre-amplifiers can be realized. The amplifiers reported in [2] have an input-referred root-mean-square (RMS) noise of 2.83 μV_{rms} over the 1 Hz–10 kHz signal bandwidth. Inputs are AC-coupled to each of the 100 pre-amplifiers via capacitors. For on-chip amplifiers, capacitor feedback is generally preferred to set the gain value compared with using resistors, as capacitors do not draw DC current and have relatively controlled matching characteristics (as in Fig. 5.9a). The use of capacitors thus lowers the overall power consumption as well as simplifying amplifier designs, with the single output capacitor determining the amplifier bandwidth. The AC-coupling filters a significant amount of low-frequency noise, and specially designed pseudo-resistors achieving GΩ resistance, consisting of transistors and DC current bias, are used in conjunction with the input capacitor to set the high-pass frequency cutoff for the amplifier. A gain of 46 dB has been set by capacitor ratios of 10 pF/0.05 pF.

Typical on-chip interconnect densities of ASICs are only limited by technology and signal integrity and speed. On-chip vias in the latest fabrication technologies are of the order of <1 μm^2. On-chip components such as analog amplifiers and digital circuits can thus be highly integrated. Off-chip connections between ICs are enhanced by the availability of through-silicon-vias (TSVs) and dedicated silicon interposers for dense routing between chips. As a result, a 3D stack of dies can be realized. Apart from the dramatic reduction of system size, one of the direct consequences of tight IC integration is the short signal path and improved signal integrity. Three horizontal layers of chips are integrated in [3] as a high-density electrocorticography (ECoG) system. The three horizontal layers consist of a micro-probe array of 24 × 24 electrodes, followed by an interposer and finally three readout ASICs in the last horizontal layer. The ASICs respectively carry out analog signal amplification, digitization, wavelet transform, and control signal generation. The front-end amplifiers employ capacitors and pseudo-resistors in the feedback loop to stabilize gain and minimize area and power consumption, whereas the ADC implemented is an SAR ADC (see Sect. 5.3). The wavelet transform follows a lifting algorithm that essentially decomposes the filters into more efficient operations.

The advent of microfabrication technology has enabled a plethora of miniature devices, including micro-electro-mechanical systems (MEMS) and microfluidic chips, some of which leverage the IC fabrication processes to achieve miniaturization. System integration methodologies have enabled the combination of ASICs and other microfabricated devices in mm-scale packages, with the ASIC designed to interface with the micro-sensors or energy harvesting units, carrying out signal processing, data acquisition, and energy conversion. Figure 5.4 shows a 27 mm^2 assembly of a power management ASIC together with a hermetically sealed piezoelectric MEMS device that is capable of harvesting ambient vibration energy and storing it in an ultra-capacitor [3]. The device is suitable for powering wireless sensors that consume up to 100 μW. The MEMS device consists of a 300 mg tungsten mass, which generates a voltage and charge at its electrodes when displaced. The ASIC is responsible for converting the MEMS-generated time-varying voltage to a steady supply for charging the ultra-capacitor, while consuming

Fig. 5.4 Integrated system for vibrational energy harvesting (© IEEE 2014. Reprinted with permission from [4])

minimum power to maximize efficiency. As we show in Chap. 7, the typical architecture of an energy-harvesting ASIC consists of a rectifier converting AC inputs to a DC voltage. This is followed by a regulator to provide a smoothed output. A trickle charger also provides the appropriate charging waveforms for the ultra-capacitor. Auxiliary circuits to prevent the capacitor from discharging backwards into the ASIC are also present.

A recent example of ASIC technology transforming benchtop instruments to a mm-scale SoC is given in [5]. Impedance spectroscopy or dielectric spectroscopy has been demonstrated to have numerous applications in the area of biomedical science, including sample analysis of bodily fluids and tissues (see Sect. 5.2.5). The mostly label-free, non-destructive, and real-time nature of impedance measurements makes it particularly suitable for a number of applications. To carry out impedance spectroscopy measurements, a setup with variable sinusoidal excitation followed by measurement circuits is required. Typical excitation and measurement frequencies depend on the samples being investigated and can vary from a few Hz to the GHz range. The traditional setup requires a benchtop vector analyzer, which is typically bulky and expensive to deploy. This measurement setup has been integrated into an SoC of 3 mm × 3 mm in [4]. We discuss more details of measurement circuits for bioimpedance measurements in the later parts of this chapter. A multi-target sensing platform for implantable blood monitoring featuring glucose and lactate sensors and an ASIC to carry out cyclic voltammetry (CV) measurements has been reported in [6] (Fig. 5.5). The biosensors are microfabricated with a silicon substrate where metalized electrodes of titanium and platinum were created by evaporation. The electrodes are subsequently functionalized by drop-casting glucose oxidase/lactate oxidase and multi-walled carbon nanotubes. Enzymatic biosensors, such as those used in [6], require sensor control and readout circuits known as potentiostats. The task of a potentiostat is to set up the electrical operating conditions for the sensors in the form of a stable voltage, and concurrently measure the current delivered by the

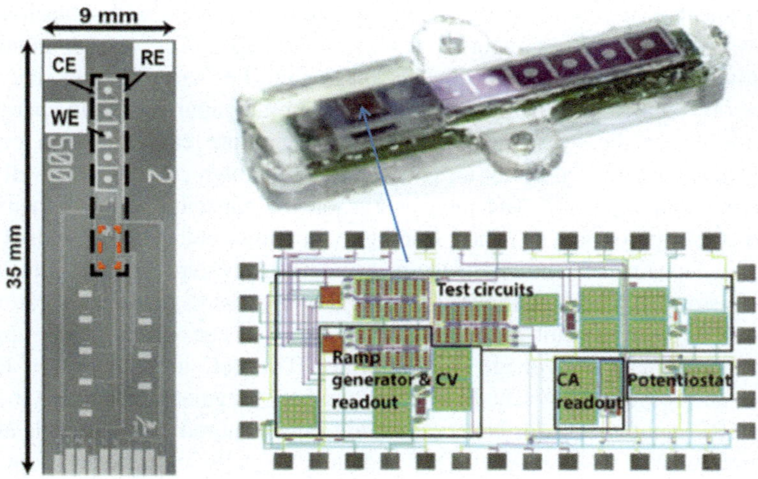

Fig. 5.5 Integrated biochemical sensing with CV ASIC (© IEEE 2014. Reprinted with permission from [6])

sensor. CV is carried out by applying a ramp voltage to the sensor electrodes over time. CV is typically used to study a range of redox processes and can determine the reduction potential of an analyte, useful in setting up a chronoamperometry experiment to measure target concentration with a biosensor. The potentiostat ASIC thus consists of the potentiostat as well as control circuitry for producing the ramp and timing parameters for the CV. The challenge of carrying out CV with on-chip potentiostats is the typically small time constants of the measurements and the high sensitivity required to measure the low input currents presented by the sensors, some down to the pA region, demanding very low-noise amplifiers. The low-power potentiostat presented in [6] consumes 220 μW from a 1.8 V supply. The IC is fabricated by 0.18 μm technology and the ramp waveform is generated with digital circuits (direct digital frequency synthesizer (DDFS), more commonly known as a direct digital synthesizer, DDS) and a 9-bit digital-to-analog converter. The potentiostat consists of folded cascode two-stage amplifiers with class AB input stage to maximize signal range.

Huang et al. [7] introduced an ASIC with surface modifications to incorporate a MEMS capacitor with glucose-sensitive hydrogel, an N-type poly-Si nanowire field-effect transistor (FET) for Troponin-I protein detection, and ion-sensitive field-effect transistors (ISFET) (see Chap. 2) as a pH sensor, as well as resistor- and temperature-independent biasing circuits for temperature measurements. The IC can be powered by a 1 MHz RF signal and/or a solar cell supplying <1 mW of power. Radio transmission of sensor data is facilitated by a 403 MHz radio link. The overall system footprint, including the RF coil and solar cell and lens is 20 mm × 25 mm on a miniature PCB. The analog front-end amplifier reported in [7] acts as a generic block for voltage, current, capacitance, and resistance measurements. The circuit is based on differential amplifiers with capacitor feedback, similar to the ones reported widely in literature. For voltage inputs, the capacitive divider configures the fixed voltage gain. For current inputs, the circuit acquires average current measurement by allowing the input current to charge up a capacitor where the voltage is measured. To measure capacitance, a constant voltage is applied to a capacitor array consisting of known capacitors as well as the unknown input capacitor. The total charge on the capacitor array is then redistributed with appropriate switches, leading to an output voltage proportional to the change in capacitance. The resistance measurements are carried out by placing the input resistor in a bridge configuration with known resistors. Upon application of a known voltage, the circuit essentially carries out R-V conversion before applying the voltage signal to a voltage gain amplifier. The SoC incorporates an RF-DC rectifier and respective switched-capacitor and boost regulator to convert the solar cell and RF input energy for battery charging. The acquired sensor signals are first converted with a 10-bit SAR ADC, and then digitally processed before being transmitted off-chip by a simple ring-oscillator generated carrier and a power amplifier.

While it is advantageous to have a large number of parallel data acquisition channels on a sensing SoC, the large amount of data originating from the sensor arrays creates strain on the telemetric system. The data rate is often increased to accommodate this at the expense of power consumption. To reduce the amount of data for transmission, compressed sensing (CS) can be used with many examples of on-chip implementations. CS is based on an alternative sampling criterion for sparse signals. Traditionally, analog signals are sampled at frequencies greater than twice the signal bandwidth—the Nyquist sampling criterion. The sampled signal then can be perfectly reconstructed with an ideal low-pass filter (LPF). Equivalently, fitting sinc functions at each of the sampled points in the time domain retrieves the original signal. The Nyquist sampling theorem assumes LPF reconstruction and allows any signal band-limited to $0.5 \cdot f_{samp}$ to be reconstructed. For more restricted signals that can be expressed as a limited sum of orthonormal basis functions (e.g. having only a few frequency components), we can relax the sampling to sub-Nyquist intervals $<2 \cdot f_{BW}$. An example of a sparse signal in the frequency domain with a sinusoidal basis is an amplitude-modulated sine wave. According to the Nyquist sampling criterion, a sampling frequency of at least twice the bandwidth is required for perfect reconstruction with an LPF. However, since the signal is sparse, with only a few frequency components, we can reconstruct the signal with only the constituent components. In CS, the main task is to estimate the constituent components with optimization. The end result is a sampling scheme with moderate quality and a large reduction in data rate. A four-fold reduction in data rate has been reported in [8], with an intracranial electroencephalography (EEG) recording SoC consuming 10.5 μW per channel. The CS scheme starts with acquiring a number of random sensor measurements and carrying out summation over the random samples. In [8], random sampling is carried out by controlling a sampling switch with a feedback register-based pseudorandom number generator, whereas the summation is realized with SC circuits. This simple processing in the analog domain shifts the power consumption from digital circuits. In [9], a different approach is taken, with random sampling carried out in the digital domain, achieving a 10% compression ratio. Deep brain stimulation (DBS) has been clinically demonstrated to improve essential tremor and Parkinson's disease symptoms. A closed-loop system comprises neuro-recording electrodes for measuring abnormal signs (such as change in energy) in the LFP signals in combination with a stimulation setup to determine the most effective set of stimulation parameters, such as signal amplitude, pulse width, and repetition. A wirelessly powered closed-loop stimulation SoC has been presented in [10]. Four parallel channels of low-noise amplifier, each giving a gain of 54 dB and a second-order bandpass roll-off implemented by cascaded fully differential amplifiers with AC coupling and pseudo-resistor/capacitor feedback are available. The acquired neural signal is processed logarithmically, resulting in reduced bit resolution in the digital domain. The digitization is carried out by a 100 sps 8 bit pipeline ADC to accommodate the multi-channel neural signal data rate requirements. Whereas the reduction in the number of bits improves power efficiency, the processing of data in the log domain also eliminates the need for complex multiplier circuits. The stimulation circuits are comprised of eight channels of balanced current generators as cascaded current mirrors driven by parallel 6 bit DACs. The self-powered chip consumes 0.5 mW for processing and supplies up to 4.18 mA for current stimulation, and is powered remotely by a 915 MHz RF source with a 2.4 GHz backscattering telemetry link.

5.1.2 Monolithic Sensor Integration with ASICs

Due to the versatility of the silicon substrate, and for the benefit of a more compact system with improved signal integrity, sensors are sometimes integrated on the same die as the ASIC. Here we briefly describe a number of post-processing cases for monolithically integrating sensors on ASICs, with a focus on post-processed ICs. An in-depth description of various sensor fabrication processes has already been provided in Chaps. 2–4. A large number of the processing steps of MEMS microfabrication coincide with IC fabrication. As a result, "MEMS-CMOS" processes are available for co-fabrication of transistor circuits and MEMS devices. Post-processing of fabricated ASICs to realize MEMS devices after all the active circuits have been realized is also possible [11]. Generally, the silicon substrate for CMOS ASICs allows the fabrication of a limited range of MEMS devices with moderate performance compared to dedicated MEMS processes, whereas the typical CMOS-MEMS process facilitates the construction of MEMS devices at the expense of circuit performance. The realization of both MEMS sensors and ASICs on the same chip, however, reduces the amount of integration required and improves signal integrity, with the sensors placed in close proximity to the readout circuits. In [11], an SoC with the complete chain of signal acquisition including wireless telemetry has been modified to include cantilevers for deoxyribonucleic acid (DNA) detection. The cantilever consists of a native polysilicon layer (and metal layers) that acts as a piezoresistor. This is fabricated by dry etching down to the metal layer followed by gold deposition. Dry etching of the silicon substrate creates a suspended cantilever. The deposited gold layer is used for secure attachment of probe DNA structures which will bind to specific DNA targets. This binding modifies the mass load on the suspended cantilever, hence bending the piezoresistive polysilicon strip. This change in resistance is detected by the analog front-end circuitry of the ASIC as a change in oscillation frequency and transmitted off-chip through a 433 MHz radio link.

As described in [11], the typically offered aluminum and copper metallization in standard CMOS processes is not suitable to use in sensing electrodes in the majority of the cases. In [12], modifications from the top metal layer of an ASIC are again carried out to facilitate the deployment of heterogeneous protein-based biosensors in situ. The system takes advantage of the circuit integration by realizing a large number of readout channels on-chip, vastly reducing footprint and power consumption. The processing described in [12] starts with selectively etching away the SiO_2 passivation layer (see Fig. 5.6) of the ASIC, then gold or silver electrodes are subsequently deposited on the exposed top IC metal layer. The silver electrodes are then chlorinated to form silver chloride reference electrodes, followed by a glass frit-mimicking Nafion membrane layer, whereas the gold electrodes are functionalized by attaching membrane proteins or enzyme interfaces on the gold electrodes. Carbide, sulphide, and ammonia gas sensors have been integrated on the top metal layer of a readout ASIC [13] by polymer-carbon-nanotube deposition for the rapid diagnosis of ventilator-associated pneumonia. The polymer provides gas selectivity and expands upon exposure to the appropriate gas targets. This expansion results in a change in resistance, subsequently detected by interdigitized electrodes connected to a voltage amplifier while a known current is injected into the sensor. As well as the sensor front-ends, the ASIC incorporates a SAR ADC followed by a RISC processor implementing a statistical neural network for on-chip classification of bacteria detection.

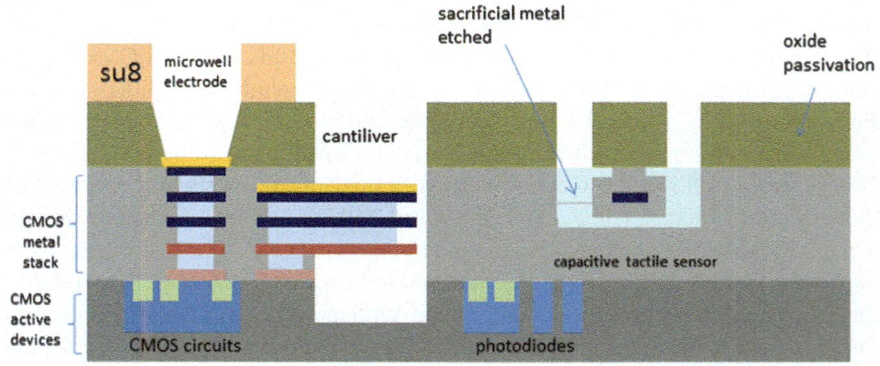

Fig. 5.6 Illustration of monolithic sensor–ASIC integration

CMOS imagers are generally suited for high-speed acquisition. CMOS photodiodes can be realized with native CMOS technologies and a number of commercially available processes offer well-characterized photodetector cells. These are known as active pixel sensors (APS), consisting of a reverse-biased diode and a reset switch with the diode connected to a small amplifier. Upon receiving light, the diode charges up a capacitor (typically this is the parasitic input capacitor of the amplifier) and produces a voltage at the output. In [14], a 32 K pixel array capturing images at 58 Hz has been reported. Each of the pixels measures $7.5 \times 7.5\ \mu m^2$. The entire construct is used for blood flow imaging in a free-moving rat. Tactile sensors can be integrated on-chip with minimal post-processing steps following standard commercially available ASIC fabrication technologies [15]. Tactile sensors created this way are capacitors with a moving upper plate. When the upper plate is displaced by pressure applied on the surface, the capacitance of the device changes. This is subsequently detected by the on-chip readout circuits. The sensors reported in [15] are discs of 24 μm radius, consisting of the top and penultimate metal layer of the CMOS chip. The top metal layer together with oxide insulation makes up the top moving plate of the capacitor. The entire top plate is made into a suspending structure by wet etching of sacrificial metal routings (see Fig. 5.6). The tactile sensor produces a displacement of 750 μN/100 nm for incoming pressure with change in capacitance of the order of a few fF. The capacitance-to-voltage converter is based on switched-capacitor circuits and incorporates a reference capacitor array for calibration against process variations during fabrication.

5.1.3 Prosthetics and Integrated Circuit Technology

One of the major drivers of implant technology is prosthetics. A particular example is the area of cochlear implants, where a range of products are commercially available. A typical setup for cochlea prosthesis consists of an implanted device that interfaces with an electrode array to provide direct stimulus to the auditory nerve. An external sound processing device equipped with a microphone transmits power and compressed stimulus data to the implant via an inductive link. Recently [16], researchers reported a cochlear implant system that is designed to be completely implanted, using the inner ear canal as a microphone. Here we highlight some of the SoC designs presented in [16]. For speech recognition purposes, it has been demonstrated that only a few spectral channels are required. For eight channels of stimulation, the total power consumption on the reported stimulation circuits was kept below 0.5 mW in [16]. With a total power consumption of less than 1 mW, the chip can be used in conjunction with a reasonably sized ultra-capacitor. The ultra-capacitor is suitable for implant use, as it can withstand a large number of recharging cycles, far exceeding that of conventional rechargeable batteries. The drawback of the ultra-capacitor is the relatively low energy density, which can be addressed by low-power circuit techniques that are covered in later parts of this chapter. The implant chip interfaces with a piezoelectric sensor to measure the sound induced motion of the umbo in the middle ear. This consists of an *RC* feedback operational amplifier (op-amp) configured as a charge amplifier followed by a PGA (resistor switching) and an ADC driver. The input sound is digitized by a 9-bit/16 ksps ADC running on a 0.6 V supply. The low supply voltage is crucial for low-power ADC and sound processor operation. The sound processor and the digital arbitrary waveform generator produce interleaved multichannel stimulus signals. Up to eight channels of stimulation can be produced with reconfigurable digital filter banks based on multi-rate processing. The processor output is used to control a current switch matrix consists of multiple current mirrors. A single current source with voltage-controlled resistors (MOS switches) is used for interleaved electrode stimulation. Charge balancing is achieved by shorting electrodes after each pulse (see Sect. 5.2.9).

High sensor density, customizable power consumption, and tiny footprint are typical requirements of retinal prosthesis for sight restoration. These requirements are mostly met by a SoC implementation. The aim of a retinal prosthetic is to replace the damaged photoreceptor and neural circuitry with an image sensor and neural stimulator. The acquired image thus needs to be processed to form a spatiotemporal stimulation pattern suitable for the nerve interface. An external camera can be used to capture the images, which is subsequently wirelessly relayed to the implanted neural stimulator IC [17]. Depending on the implant location, these are termed epiretinal (in front/on the retina), sub-retinal (behind the retina), and supra-choroidal implants (between the choroidal and the sclera). To apply sufficient current (of the order of 100 s of mA) to the electrode tissue interface (impedance of $\sim$30 kΩ and increasing with reduced electrode area to match biological nerve density), ASICs for retinal implants are fabricated with high-voltage technologies. High-voltage devices are typically implemented with thicker gate oxides. The ASIC reported in [17] consists of 1024 stimulation channels and separate power and data links, with the power transmitted more efficiently through the 2 MHz lower frequency inductive link, whereas data is transmitted over a 20 MHz DPSK channel. Bandwidth reduction techniques, including address redundancy, are used to reduce the data rate required for 60 fps image transmission at 614 kbps. The power inductive coils are center tapped to recover a number of $\pm$ supply voltages to power the stimulation front-end and logic circuits separately. This minimizes power consumption by utilizing a high supply voltage ($\pm$10 V) only when required, leaving the rest of the circuits operating in low-voltage/low-power mode. Stimulation waveforms are biphasic, with a balance between the two phases to achieve zero overall charge injection (see Sect. 5.2.9). This balanced stimulus avoids electrolysis and electrode corrosion as well as preventing tissue damage. The entire chip occupies 37.6 mm^2 and is designed to be connected to external electrode arrays made of platinum with Teflon insulations. The correct retinal stimulation waveform is still an area of active research: while charge balancing is an essential characteristic, waveform symmetry in the two phases is not necessary. In [18], an epiretinal system with an ASIC stimulator that also supports arbitrary stimulation waveforms of 4 bits/ 109.2 μs resolution has been reported. This is shown in Fig. 5.7. The stimulation ASIC supports a 20 Mbps data rate with a 160 MHz carrier, while power is supplied remotely through a 10 MHz inductive link. The stimulation front-end consists of current mirrors and p/nMOS drain terminals for high-impedance outputs. The imbalance in charge due to the unmatched p/nMOS push-pull stage is released periodically to avoid build-up. Having the imager in the retinal prosthesis system placed externally on a separate device means that the data link between the imager and the implant has to accommodate a high data rate for high-resolution stimulation. One way to mitigate this is to have image sensing in situ on the implant. The external device thus only acts as a remote power source with a small data rate link for housekeeping. A sub-retinal implant ASIC with integrated imager is reported in [19]. The ASIC is tested in vitro with degenerative retinal tissues from mice. A gold micro-electrode array on SU-8 substrate was used for the experiment. The photodetection circuit reported in [19] is in the classic "3T" configuration with a reverse biased photodiode supplying charge to a parasitic capacitor at the input of an open drain common-source amplifier. An extra switch allows the pixel to be placed in reset since the subsequent amplifier is in open loop. The common-source amplifier drives the gate of a source follower with a resistor placed at the source to carry out voltage-current conversion. Current stimulation is provided through two balanced terminals, with the two current phases (in and out) going through the same type of (n and p) transistors. This provides a better charge balance compared to the push-pull output stages presented in [17, 18].

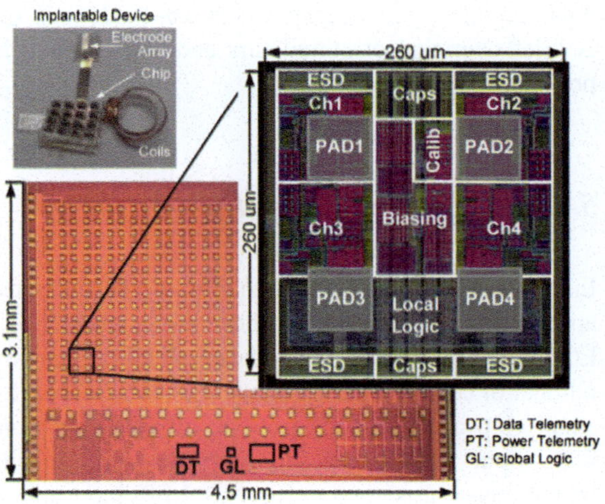

Fig. 5.7 Die micrograph and implantable epiretinal prosthesis device (© IEEE 2013. Reprinted with permission from [18])

5.2 Analog Circuit Techniques for Sensor Interfacing and Signal Processing

In this section, analog front-end electronics interfacing with sensors will be discussed. An overview of the main signal processing and conditioning blocks will be provided. A common task in various biomedical applications involves the recording of voltages. Consequently, the discussion will begin with an overview of common recording amplifiers used for applications such as electrocardiography (ECG), electromyography (EMG), electroneurography (ENG), and bioimpedance. Within this context, however, we will not discuss standard amplifier topologies at the transistor level, such as the design of telescopic, folded-cascode, or balanced operational transconductance amplifiers (OTAs), as there are many textbooks dedicated to this subject.

A task common to all acquisition systems is filtering. Linearized OTAs for G_m-C filter applications will be discussed, which are typically preferred over other filter types in biomedical applications. Sensor systems must provide accurate measurements irrespective of manufacturing tolerances in the ASIC fabrication. A discussion of offset compensation techniques is also provided, with a focus on chopping, as it is more suitable for continuous-time recordings and provides unique features in terms of noise, which are crucial for the recording of minute biomedical signals buried in interferences. Leakage in switches in deep sub-micron technologies is an issue and solutions to this will also be discussed, as in many cases topologies using switches may be interfaced with sensors. As will be discussed later, switches play a crucial role in all methods useful for compensating circuit imperfections. The discussion will then focus on various sensing methodologies, such as bioimpedance, voltammetry and amperometry, and voltage-to-frequency converters (VFC), a technique applicable to any sensing method, capable of producing quasi-digital outputs which can be directly transmitted to a data aggregator. A brief overview of stimulators and capacitive sensor interfacing circuits is also provided.

5.2.1 *Front-End Recording Amplifiers*

In this section, common amplifier topologies for sensor interfacing and implantable applications are discussed. These are intended for various applications, such as ECG, EEG, ENG, or bioimpedance. In all these biomedical applications, there are a number of common design challenges, which apart from low-power, low-noise, and small chip area include electrode offset elimination, amplifier offset elimination, high input impedance, and (according to the application) specific bandwidth requirements, as shown in Fig. 5.8. LFPs have an amplitude of 20 μV–5 mV, with a frequency content between 1 and 500 Hz, and action potentials are characterized with amplitudes of 50–500 μV, with frequencies 100 Hz–10 kHz [20–22].

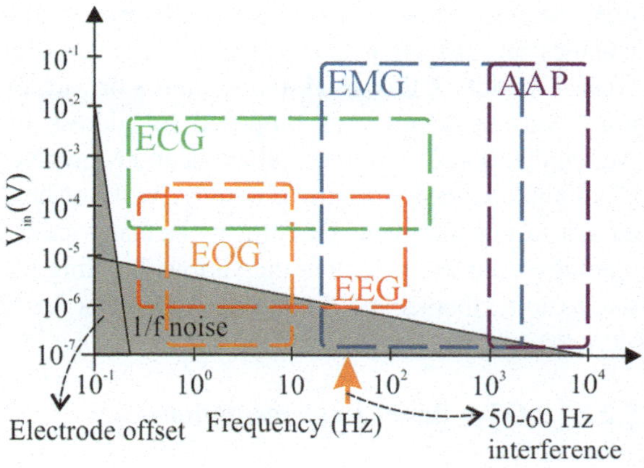

Fig. 5.8 A not-to-scale plot of the amplitude versus frequency characteristics of biopotential signals (EEG, ECG, electrooculogram (EOG), axon action potentials (AAP), and EMG) and interfering (noise, offset, and mains coupling) signals. Compiled with information from [25] and [26]

Other electrophysiological signal sources are shown in Fig. 5.8. Consequently, because of the small amplitude of these signals, noise and interference can significantly affect the recording quality. The total noise in the recorded signal contains contributions from the thermal and flicker (1/*f*) noise of the amplifier and the thermal noise of the recording electrodes [22, 23]. The latter, together with background noise, is in the region of 20 μV for an electrode with an impedance of 1 MΩ and a bandwidth of 5 kHz [24]. pMOS transistors exhibit lower 1/*f* noise (one or two orders of magnitude less than nMOS, provided that $|V_{GS}|$ does not greatly exceed V_T [20]), and are thus typically preferred for implementing the input pair transistors of the bio-amplifier. A large gate area will also lead to lower noise, as 1/*f* noise is inversely proportional to gate area [20, 22]. Chopping, as will be discussed later in more detail, is a commonly used technique for reducing 1/*f* noise. As discussed in Chap. 2, electrochemical reactions taking place at the electrode interface with the biology will lead to a double-layer impedance, which is a function of the electrochemistry at the electrode surface, the electrode surface properties, the ions present, and their activity, temperature, pH, and the contact the electrode maintains with the biology (pressure, motion). The formation of the double layer is due to charge redistribution at the interface, which will lead to a potential drop across it, i.e. a DC voltage. Differences in the interfaces of electrode pairs leads to different electrode DC potentials, which will lead to a differential input DC potential seen at the inputs of the recording amplifier. This DC offset voltage is typically 1–10 mV, but can be as large as 50 mV [22] and even 1–2 V [20]. Since high gains are needed to amplify the small μV range electrophysiological signals, these offsets will saturate the recording amplifier and lead to signal distortion. A number of different sensors (e.g. resistive or capacitive) and signal processing methods (e.g. synchronous detection (SD), used in bioimpedance measurements and also known as synchronous demodulation, lock-in, or quadrature demodulation/detection) have a DC output. Thus, random and systematic amplifier voltage offsets need to be eliminated for accurate sensor readout, preferably automatically, without manual trimming on an ASIC-by-ASIC basis. Another issue with large input DC levels from the electrodes, when transistors are directly connected to them, is related to the voltages that the input transistors can tolerate according to the process specifications. To address this, thick-oxide transistors are used [22]. In the following sections, various approaches proposed to address these issues will be discussed. Although crucial, as biomedical applications involve typically the recording of very small signals, designing for low noise will not be discussed in the examples discussed in the following sections, as in many cases this depends on the particular topology used to implement the various building blocks, and the reader is referred to classic textbooks and the referenced papers for further details.

5.2.1.1 OTA-Based Implementations

In this section we will discuss low-noise amplifiers intended for recording of neural signals. The AC-coupled OTA-based capacitive feedback amplifier of Fig. 5.9a was originally proposed by Harrison and Charles in [20] as a low-noise amplifier (LNA) for implantable neural recording applications. Since then it has been used in a wide range of neural recording and sensor applications, where instead of being used as the first block in the signal processing steps as originally intended, it is often used in subsequent steps as a PGA, as in [25–32]. The midband gain of the circuit is set by the capacitive ratio $A_m = C_1/C_2$. C_1 capacitively couples the input signal, thus discarding any electrode DC offset voltages. This capacitor should be significantly smaller than the electrode impedance to reduce signal attenuation [33]; nevertheless, these tend to be large (e.g. 20 pF) and occupy large chip area. As discussed in [34], a "capacitance less than 100 pF ensures negligible signal attenuation at the electrode". C_1, together with parasitic capacitors, forms a capacitive voltage divider attenuating the input signal. Due to mismatch between the two C_1s

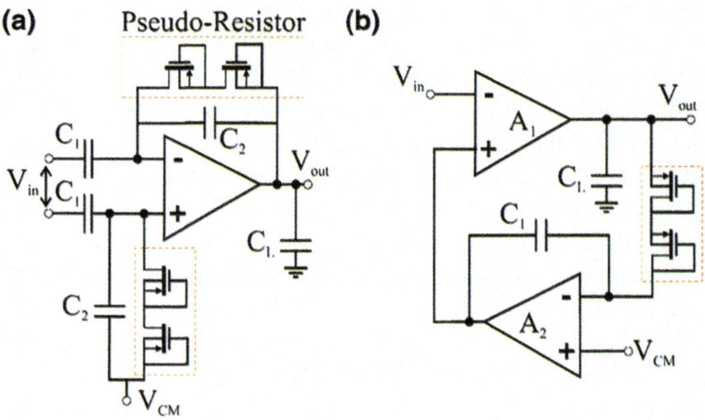

Fig. 5.9 **a** The AC-coupled neural recording amplifier with pMOS pseudo-resistors, proposed in [20]. **b** The DC-coupled amplifier with active low-frequency suppression, proposed in [21]

and the parasitics, a different voltage attenuation will be applied to each input branch, which in turn reduces the common-mode rejection ratio (CMRR) [35]. A solution is to use large C_1s—much larger than the mismatch in the parasitics [35] —as large capacitors are easier to match, improving to some extent the CMRR. The large size of these capacitors may prevent the integration of many such amplification blocks in multichannel systems [22]. The capacitors also affect the input referred noise of the amplifier ($\overline{v^2_{\mathrm{ni,amp}}}$). This is related to the input referred noise ($\overline{v^2_{\mathrm{ni,OTA}}}$) and input capacitance (C_{in}, in the region of a few fF) of the OTA according to [20]:

$$\overline{v^2_{ni,amp}} = \left(\frac{C_1 + C_2 + C_{IN}}{C_1}\right)\overline{v^2_{ni,OTA}}. \tag{5.1}$$

The input impedance of this topology is equal to $Z_{\mathrm{in}} = 1/s(C_1 + C_{\mathrm{p}})$, where C_{p} is the parasitic capacitance at the input node due to ASIC packaging, PCB, and external leads. Hence, a large input capacitance will reduce the input impedance, which is of course frequency-dependent. Consequently, a tradeoff can be seen, from the above equation and discussions, between input impedance, chip area, CMRR, and gain and noise, as a large capacitor C_1 for a large gain and CMRR and low noise will lead to a reduced input impedance and large chip real estate occupation. In [20], a detailed design procedure for low noise and low power is provided. A useful metric originally proposed in [36] and used in [20] and other sources is the noise efficiency factor (NEF) of an amplifier, which demonstrates the tradeoff between noise and power:

$$\mathrm{NEF} = V_{\mathrm{ni,rms}}\sqrt{\frac{2I_{\mathrm{tot}}}{4\pi U_{\mathrm{t}}kT \cdot BW}}, \tag{5.2}$$

where $V_{\mathrm{ni,rms}}$ is the input-referred RMS noise voltage, I_{tot} is the total current in the amplifier, and BW is the amplifier bandwidth. The power efficiency factor (PEF) is another metric, defined as the square of the NEF multiplied by the supply voltage, to assess the efficiency with respect to the supply [37, 38]:

$$\mathrm{PEF} = \mathrm{NEF}^2(V_{\mathrm{DD}} - V_{\mathrm{SS}}). \tag{5.3}$$

The low-pass cutoff frequency is set by the midband gain, the transconductance of the OTA and the load capacitance, C_{L}. If C_1 and C_{L} are significantly smaller than C_2, then the bandwidth is to a first approximation equal to $g_{\mathrm{m}}/(A_{\mathrm{m}}C_{\mathrm{L}})$, where g_{m} is the OTA transconductance. The series connection of the pMOS transistors implements a very high value incremental resistance (greater than 10^{11} Ω) for setting the DC operating point. Consequently, the cutoff frequency of the input high-pass filter (HPF) can be easily set in the mHz range. As discussed in [39], this cutoff must be as low as possible to eliminate low-frequency noise introduced by the parallel combination of C_2 and the pseudo-resistors, which rolls off as $1/f^2$ and appears at the gate of the input transistors of the amplifiers and can thus be amplified. This is inherently achieved by the extremely large resistance of the pseudo-resistors. This is set by C_2 and the pseudo-resistors.

These pMOS pseudo-resistors (Fig. 5.10a) were originally proposed in [40]. With a negative V_{GS} each transistor operates as a diode-connected pMOS. When V_{GS} is positive, the parasitic pnp bipolar junction transistor (BJT) formed between the transistor's source, well, and drain, is activated and the transistors act as diode-connected BJTs [40]. Two devices in series are typically used as shown to reduce the distortion for large output signals [20]. These pseudo-resistors, however, suffer from large random variations due to process variations and are temperature dependent (up to an order of magnitude). In addition, they are nonlinear for large output voltage swings [20, 22]. This is because the resistance will be large for small signals and vice versa [41]. In addition, the junction leakage current from the n-type well to the p-substrate, which will be in the region of pA or greater, can bias the common high-impedance bulk node in the pseudo-resistor, affecting further the resistance of the element [26]. Finally, the value of this element is not programmable. According to [41], there is a tradeoff between noise and distortion with regard to the number of pseudo-resistor elements connected in series. Using more pseudo-resistors decreases nonlinearity (distortion decreases approximately linearly with the number of elements, as the voltage across each element is smaller), but increases noise.

A number of approaches have been proposed to address the issues of the pseudo-resistor. One is to use pMOS transistors biased in weak inversion, using a gate bias voltage to control the equivalent resistance with the bulks shorted together to their source or drains, as in Fig. 5.10b [42, 45]. While this structure is programmable, it remains nonlinear and asymmetric around the output common-mode

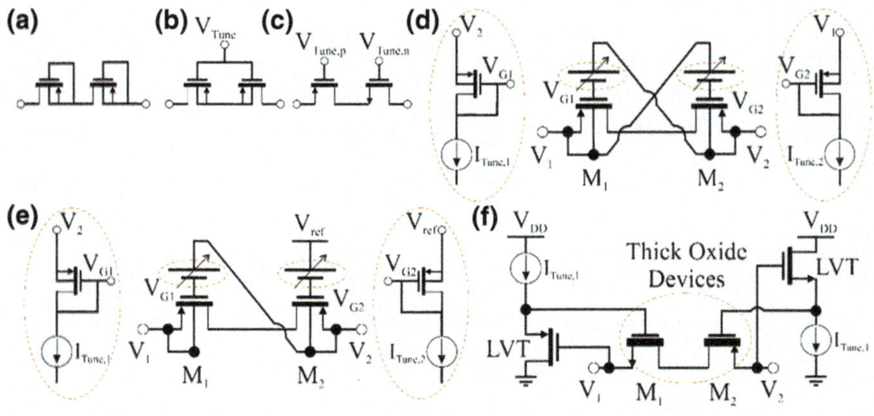

Fig. 5.10 **a** The pseudo-resistor structure proposed in [40] and used in the neural recording amplifier of [20]. **b** The tunable pseudo-resistor implemented in [42]. **c** The complementary structure proposed in [43]. **d** The cross-coupled structure of [26]. **e** The pseudo-cross-coupled approach proposed in [44]. **f** The fixed V_{GS} pseudo-resistor proposed in [24]

voltage [44]. Furthermore, for small V_{GS}s, where the resistance is high, although the linearity is good for large swings it can lead to leakage issues, and when used in the feedback of the OTA the resulting drift will change the input common-mode voltage and eventually saturate the amplifier [24]. It was shown in [24] that the total harmonic distortion (THD) of this structure was below 1% for frequencies below 1 kHz, and it can be very low if thick-oxide transistors are used. The complementary topology of Fig. 5.10c [43], where the two subthreshold transistors are controlled separately, is symmetric around the common-mode output voltage level and thus linear over a wide input range, while it does not add significant parasitics or noise [44]. However, it is highly susceptible to process, common-mode, and offset variations [24, 44]. In [26], the fully balanced pseudo-resistor structure of Fig. 5.10d was presented, where each transistor functions as an active resistor for half the cycle of a sine wave input signal. The resistor is tuned by the current sources and achieves symmetric resistance variation around the quiescent point and across process corners, with good linearity [44]. Nevertheless, the response remains partially asymmetric, leading to an increase in THD, while it adds excess noise to the amplifier [44]. In addition, the DC current flowing through the topology leads to a difference between the common-mode voltage at the input and output nodes, degrading resistor linearity. To address the DC current issue, the authors proposed the pseudo-cross-coupled resistive structure of Fig. 5.10e, where V_{ref} is biased at the DC level of V_A. Consequently, the current flowing through the topology is zero and hence there is no DC voltage drop across it. This leads to a symmetrical characteristic around the quiescent output common-mode voltage and greater linearity, but only for 0.2 V. Since the DC current through the two transistors is zero, they have a very small transconductance and hence they add very little noise to the amplifier. The authors demonstrated an improvement in THD and they highlight the sensitivity of the structure to OTA offsets. However, the technology leakage current will not allow the resistor to be tuned by more than a few hundred GΩ to achieve the low cutoff frequencies necessary (mHz–Hz). A composite structure was thus proposed where the classic tunable pseudo-resistor (Fig. 5.10b) was combined with the pseudo-cross-coupled structure (Fig. 5.10e). The former was used as a switch for coarse tuning. A final approach for the implementation of a pseudo-resistor was proposed in [24] and [46]. The authors proposed a pseudo-resistor topology based on transistors, where their V_{GS} is kept constant irrespective of the voltage across it to make the device resistance constant regardless of the imposed voltage swings. This is achieved by dynamically maintaining the gate voltages of each transistor with a source follower, as in Fig. 5.10f. An nMOS with a pMOS connected in series and implemented with thick-oxide transistors demonstrated the best performance when compared with other combinations. The transistors are kept in subthreshold and feedback ensures that their V_{GS} is constant even for large voltage swings. The resistance of the element can be tuned by the bias current of the source follower. A large current increases $|V_{GS}|$ and reduces the resistance. Low V_T transistors were used to allow wide $|V_{GS}|$ tunability with a small device. Resistance variation of less than 16% was achieved with this approach for swings between 0.3 and 0.9 V. More accurate biasing can be achieved when used in the feedback of the OTA, as leakage issues are overcome. A THD improvement of 10–15 dB was reported [24, 46] against the traditional pseudo-resistor when used in the amplifier topology of Fig. 5.9a. Programmability of the total pseudo-resistance via a gated approach of serially connected elements was presented in [41].

A number of authors have examined alternatives to the pMOS pseudo-resistors. The use of linearized G_m topologies in the feedback was proposed in [37] to achieve programmable cutoff frequencies. In [47], an SC resistor was implemented. A series-to-parallel charge-sharing scheme was used to reduce the charge transfer by a factor of 10, as shown in Fig. 5.11a. This allowed the realization of large resistances (R_{SC}) with a high switching frequency (f_{SC}) and large capacitors ($R_{SC} = 10/(Cf_{SC})$). Duty-cycled resistors were proposed in [34] as in Fig. 5.11b, where if the switch is driven by a clock with a duty-cycle factor D, then the average resistance of the element at low input signal frequencies is equal to R/D. Hence, if $R = 1$ MΩ and $D = 1/40{,}000$, this leads to an equivalent resistance (R_{eq}) of $R_{eq} = 40$ GΩ. This element achieves high linearity and can be tuned by D over wide ranges.

In the original design of [20], a simple balanced OTA topology was used with a theoretical NEF ≈ 2.9. In another implementation, presented in [39], the authors used a folded-cascode topology with source degenerated current sources to achieve a NEF of 2.067, with the theoretical limit of the topology being 2.02. A large gain is required with this topology for low noise. This consequently leads to a large input capacitance and a tradeoff between large chip area and gain. The use of the clamped T-capacitor network in the feedback proposed in [48] addresses this tradeoff, as it reduces the effective feedback capacitor value and achieves the same midband gain with reduced input capacitance for the same high-pass pole and thus smaller chip area and higher input impedance. As highlighted in [22], a higher feedback resistance to maintain the same high-pass pole increases noise and area overhead. The area reduction achieved by [48] nevertheless is not considerable, since extra capacitors are needed to form the T-network. Lower NEF and power consumption can be achieved in amplifier arrays using the partial OTA-sharing scheme proposed in [49], where telescopic cascade OTAs were used, providing the best noise–power tradeoff. Nevertheless, this topology to be properly biased needs different input and output common-mode voltages. This was achieved by two additional in-series

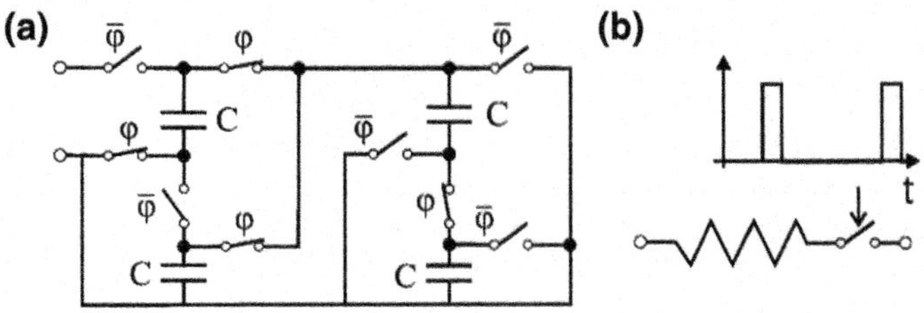

Fig. 5.11 **a** The SC resistive element proposed in [47]. **b** Duty-cycled resistor for large average resistance, proposed in [34]

diode-connected pMOS transistors at the virtual ground, with their sources shorted and used to set the input common-mode voltage.

A number of designs propose the use of DC-coupled amplifiers, where there are no input DC-blocking capacitors and DC offsets are removed via feedback. These can be either analog or digital methods. The DC-coupling approach of [21] is similar to common topologies used for ECG recordings and implements active low-frequency suppression via feedback, again using pMOS pseudo-resistors to achieve a low high-pass cutoff as in Fig. 5.9b. No input capacitors are required, but it needs two amplifiers, increasing power consumption. The time constant of the Miller integrator formed by A_2 is defined by C_1 and the pseudo-resistor; the mid-band gain and low-pass cutoff frequency are set by A_1. The high-pass corner is also dependent on the gain of A_1. The common-mode voltage level is defined by V_{ref}. The inverting Miller integrator low-pass filters the signal at the output of A_1 and applies a negative signal to the positive input of A_1. Consequently, the feedback remains negative. Thus the DC and low-frequency noise are tracked and eliminated. In addition, the topology also eliminates internal offsets due to mismatches and process variations in both amplifiers, while drift from electrodes and amplifiers is also attenuated. A large open loop gain is required for A_2 to attenuate non-ideal effects in the integrator (see Fig. 5.23). Another disadvantage of the topology is that the resistance of the pseudo-resistor is a function of the amplifier's output node voltage swing. This will affect the high-pass pole of the topology, which is also sensitive to the open-loop gain of the feedback amplifier [22].

Apart from the topologies discussed above, other recently proposed solutions include the amplifier suggested in [50], where feedback is employed to match the input and neutralization capacitances to compensate for the input leakage current and thus increase input impedance. A novel design approach for an overall high CMRR was presented in [51].

A few proposed designs implement a digital DC servo loop to eliminate input DC voltages, as in [22, 52, 53]. Approaches for eliminating the electrode DC offset and increasing the input impedance through feedback will be discussed in Sect. 5.2.2.1 within the context of chopping, since chopping will decrease the input impedance, while electrode DC offset reduction loops are often implemented in systems using chopping.

5.2.1.2 Current Feedback

In most biomedical applications, high CMRR, low noise, programmable gain, low power, high integration and high input impedance, among other parameters, are essential at the front-end. An IA is thus typically employed. The classic three-op-amp IA with resistive feedback is not suitable for such applications. The CMRR depends on accurate resistor matching, which is hard to achieve and requires laser trimming, which is not available in standard CMOS. The topology requires operational amplifiers with low output impedances and large drive capabilities to drive the feedback resistors. This translates to high currents and thus large power consumption, while the use of op-amps, which are typically internally compensated for stability, will impose further constraints with regard to the frequency response of the topology. Current feedback IAs address these issues. An example is shown in Fig. 5.12 and is a CMOS implementation of [54] first demonstrated in [55]. The CMOS implementation is composed of an input transconductance stage and an output transimpedance (TI) stage. The input transconductors of each stage are resistively degenerated. Each stage has a feedback loop around each transconductor and these are isolated from each other. Consequently, this topology is classified as a local current feedback IA [56]. The input transistors provide high input impedance and impose the input differential voltage across resistor R_1. For a differential input signal the output currents of the input transconductance stage will become unbalanced and amplifier A will balance the drain current of the input pair transistors via active current feedback, provided that the current feedback gain is large. In this way the input voltage appears across R_1, and consequently the input stage performs as a unity gain buffer. High-gain amplifier B balances the currents at the output stage in a similar manner. The output voltage is equal to the potential drop across resistor R_2 plus the reference voltage. The current at the output branch is mirrored from the input. Thus the currents are matched and $i_{R1} = i_{R2}$. Consequently,

$$i_{R_1} = \frac{(v_{in+} - v_{in-})}{R_1}, \tag{5.4}$$

$$i_{R_2} = \frac{(v_o - v_{ref})}{R_2} \tag{5.5}$$

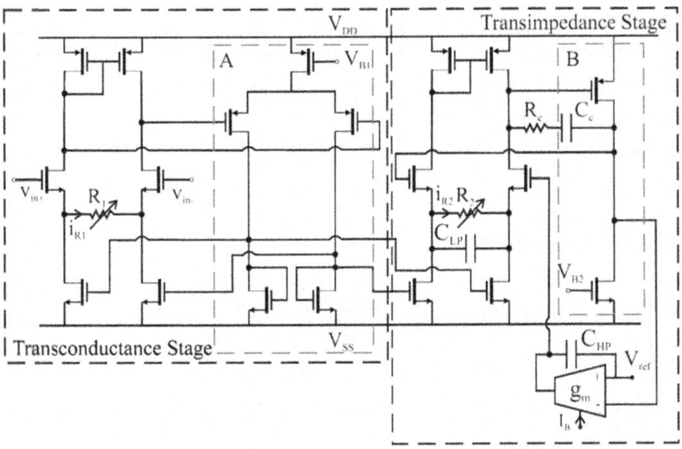

Fig. 5.12 The current feedback IA topology used in [28, 55–60] for bioimpedance, neural recording, and EEG applications

and

$$\frac{(v_o - v_{ref})}{(v_{in+} - v_{in-})} = \frac{R_2}{R_1}. \tag{5.6}$$

Hence, the gain is set only by the two resistors and the CMRR is not dependent on resistor matching, but rather on device matching, as discussed in detail in [28, 56]. The fact that there are only two resistors reduces the chip real estate required, while since there is no global feedback a single high-impedance node simplifies the frequency compensation required for stability. In the input stage, there is no need for compensation, but at the output stage, because the loop gain is high, a compensation network $R_C C_C$ can be added for stability. Various implementations for B have been proposed [55, 57, 58]. The inclusion of a capacitor across R_2 introduces a low-pass pole to the amplifier at $f_{LP} = 1/(2\pi R_2 C_{LP})$, which can be adjusted according to the application, while a G_m-C LPF between output node and one of the inputs of the TI stage introduces a high-pass pole to the topology, if needed, for active low-frequency suppression (similarly to [21]), where the pole is given by $f_{HP} = g_m/(2\pi C_{HP})$. The noise is dominated by that of the input transconductor [28]. For low noise, R_1 must be kept small, the input transistor W/L must be made large, and the length of the load transistors should be larger than that of the input pair. This is discussed in detail in [28]. This topology has been used for EEG recording [55], bioimpedance measurements [57–60] (because of the ability of the topology, due to current feedback, to sustain a high CMRR at high frequencies—a 90 dB CMRR at 2 MHz has been reported [56]), and neural recordings [28].

A current-balancing IA has been proposed in [25] which has been used in a number of EEG, ECG, and bioimpedance systems [29, 61–63]. A simplified schematic is shown in Fig. 5.13. As in the previous example, the differential input is copied across R_1 and the current through R_2 is a copy of that through R_1 creating the output voltage. Hence, once again, the gain is given by the ratio of two resistors, R_2/R_1, provided that certain conditions are met [25]. Capacitor C_C can be used to set the low-pass pole of the topology, which is given by [25]:

$$f_{LP} = \frac{1}{2\pi R_2 C_2 \alpha}, \tag{5.7}$$

where

$$\alpha = \left(1 + \frac{1}{g_{m,M_1} R_1}\right)\left(1 + \frac{1}{g_{m,M_2} R_2}\right). \tag{5.8}$$

The CMRR of the topology depends on the DS conductance mismatch of the input pair transistors and of the current sources I_1. To increase the CMRR the input transistors $M_{1,2}$ are replaced by an input stage that makes the V_{SD} of the transistor insensitive to the input DC level. This input stage also reduces the input capacitance of the IA [35]. Source followers are added to the drains of the input transistors to

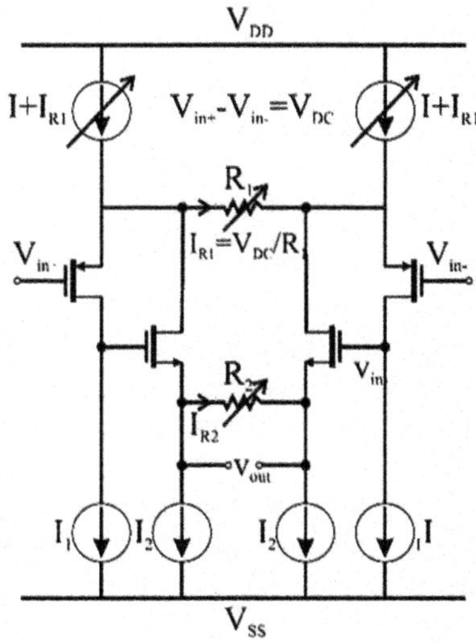

Fig. 5.13 The current-balancing IA proposed in [25, 62]

drive the gates of $M_{2,3}$ to level shift the signal and increase the voltage swing of the IA. A similar approach to Fig. 5.13 was proposed in [32, 35, 64], in that there is a separate input transconductance and an output TI stage.

5.2.1.3 Differential Difference Amplifier (DDA) Topologies

A commonly used amplifier topology at the front-end and/or at subsequent processing stages is the differential difference amplifier (DDA). Originally proposed in [65], it has been used as the front-end amplifier for EEG/ECG recordings with chopping [66], for bioimpedance measurements for electrical impedance tomography (EIT) [67], as a back-end PGA with chopping in a bioimpedance measurement system [68, 69], and as a C-V measurement system for MEMS accelerometers [70]. A DDA uses at the input stage a number (≥ 2) of differential pairs. This is illustrated in Fig. 5.14a for n = 2. Thus, in the case when there are two pairs, these are transconductance elements converting the differential input voltages ($V_{\mathrm{pp}} - V_{\mathrm{pn}}$) and ($V_{\mathrm{np}} - V_{\mathrm{nn}}$) of each into two differential output currents (Δi_{p} and Δi_{n}). These can then be subtracted without additional circuits, in the current domain, by shorting the appropriate nodes together. An appropriate active load converts this into a voltage, where, if A_{OL} is the DDA's open-loop gain and all non-idealities are ignored, this leads to [71]:

$$V_o = A_{OL}\left(\left(V_{pp} - V_{pn}\right) - \left(V_{np} - V_{nn}\right)\right). \tag{5.9}$$

As discussed in [65], a number of different circuit operations with the need for multiple op-amps and passive components can be performed with a DDA. For example, a floating input comparator realized with a single DDA would require

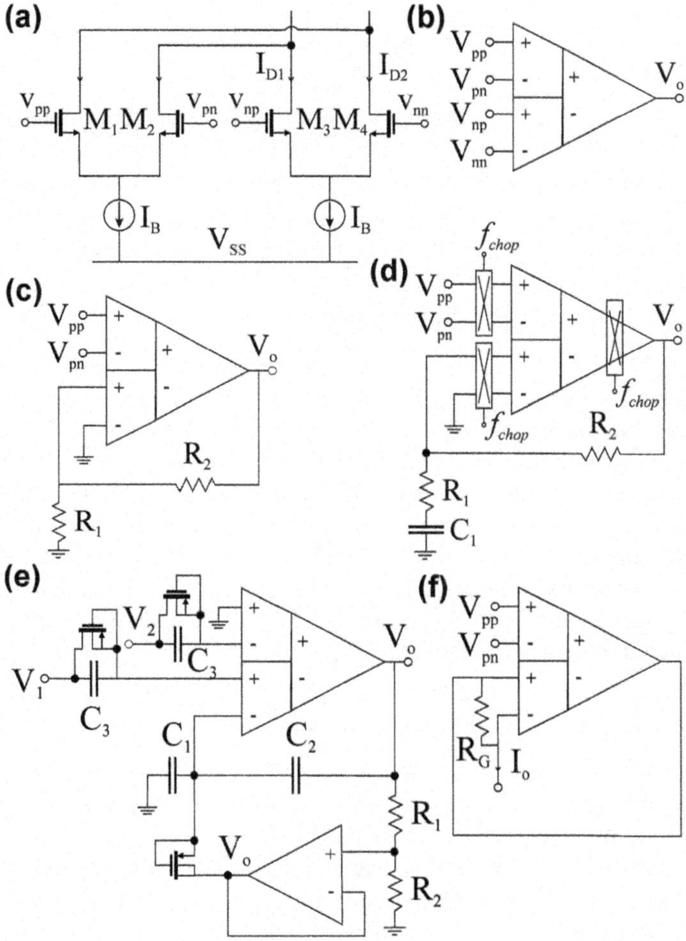

Fig. 5.14 **a** The input stage of a DDA with two input pairs. **b** The symbol of a complete DDA. **c** The DDA configured as an IA [65]. **d** The chopped DDA-based IA proposed in [66] for input DC offset removal. **e** The DDA-based amplifier proposed in [67]. **f** The DDA configured as a VCCS [65]

three op-amps, and a level shifter would require three op-amps and three resistors instead of a single DDA, while a voltage inverter can be implemented without external resistors. More important to the focus of this chapter is the use of the DDA to implement an IA, as shown in Fig. 5.14c, where a single DDA can be used with two resistors setting the IA gain (similarly to the current feedback approach) instead of the classic three op-amp design, where the CMRR is dependent on accurate resistor matching. The gain (A_{IA}) is equal to

$$A_{IA} = 1 + \frac{R_2}{R_1}. \tag{5.10}$$

However, the use of the resistors will lead to higher power consumption. This IA approach was implemented in [66], where pMOS input pair transistors were used for reduced $1/f$ noise. Chopping was used to reduce $1/f$ noise further and for offset elimination. A capacitor, C_1, is added to the circuit to remove the input DC offset voltage, as in Fig. 5.14d. This modifies the transfer function of the circuit to

$$A_{IA}(s) = \frac{R_2 C_1 s}{R_1 C_1 s + 1} + 1. \tag{5.11}$$

Consequently, for DC signals, the gain is 1 V/V and the circuit does not saturate because of electrode offset voltages, due to the high-pass cutoff introduced by the feedback LPF with a pole at $1/2\pi R_1 C_1$. This RC time constant has to be big for a very low pole position. As discussed in [72], an advantage of the chopped DDA is that the effects of charge injection and clock feedthrough from the chopping switches (discussed in detail in Sect. 5.2.4) at each input pair counteract those at the other. A hybrid approach between [20] and [66] was implemented in [67], although chopping was not used, and is shown in Fig. 5.14e. When parasitic capacitances are not considered, the gain is given by

$$A_{CL} = 1 + \frac{C_1}{C_2}. \tag{5.12}$$

As discussed in [67], when $C_0 = C_1 + C_2$, the effect of the parasitics is cancelled. pMOS pseudo-resistors were used for DC biasing of the input nodes. A DC loop fixed the DC gain to 2 V/V. In this way, the output DC level was twice that of the input. By setting the input common-mode voltage equal to half the required supply voltage, this fixed the output supply to the required level. The DDA input stage was followed by a common-source and a third common-drain stage.

Another interesting application of the DDA is for the implementation of a voltage-controlled current source (VCCS) as in Fig. 5.14f, where the transconductance of the circuit is defined by $G_m = 1/R_G$. The additional input terminals also allow the elimination of output offset voltages or input offset voltages due to mismatched electrodes, for example, using a similar approach to [21]. Other DDA-based implementations of analog signal processing blocks were presented in [71], including four quadrant multipliers, grounded MOS resistors, and active filters. As discussed and implemented in the paper, for large signal-handling capabilities, the transconductance input stages should be linearized. This is because the input stages of the DDA, in contrast to an op-amp, do not have the property of a virtual short circuit similar to an OTA, as there will be no feedback mechanism forcing each pair of terminals to the same potential. Early fully differential implementations of the DDA were first proposed in [71, 73].

5.2.2 Dynamic Offset and Noise Compensation

For high accuracy and sensitivity measurements, issues arising from the non-ideal effects of integrated amplifiers need to be compensated for. These include input-referred DC offset voltage and thermal and flicker noise. Thermal noise is a wideband signal, while flicker noise is a narrowband signal (decreasing by 1/*f*). Compensating for these increases the dynamic range of the amplifier, which is particularly important in low-power sensor applications where the output swings are limited. These should first be minimized by the proposed transistor-level design and layout techniques (common centroid layout, keeping devices close to each other, and ensuring the same orientation and surrounding environment). Offset is a DC parameter, however, and this can drift over time due to temperature, ageing, and packaging stress. Variation in the drain current (I_D) of transistors can be caused by mismatches in threshold voltage (V_T) and the transconductance (gain) factor of a transistor ($\beta = \mu_{n,p} C_{ox} W/L$). According to [74], these are equal to

$$\frac{\delta I_D}{\delta \beta} \approx \frac{I_D}{\beta}, \tag{5.13}$$

in both strong and weak inversion, and

$$\frac{\delta I_D}{\delta V_{th}} = -g_m = \begin{cases} -2\sqrt{\beta I_D}, & \text{in strong inversion} \\ -\frac{I_D}{nU_T}, & \text{in weak inversion} \end{cases}, \tag{5.14}$$

where $\mu_{n,p}$ is the mobility the charge carrier of a transistor, C_{ox} is the gate capacitance of the transistor ($\mu_{n,p}$ and C_{ox} are both determined by the technology), W and L are the width and length of the transistor (design variables), and U_T is the thermal voltage, equal to $kT/q = 0.02585$ V, where $k = 1.38064852 \times 10^{-23}$ J K^{-1} is Boltzmann's constant, $q = 1.602176565 \times 10^{-19}$ C is the charge of an electron and T is the temperature in kelvins ($T = 300$ K at room temperature). These have been extensively discussed in [75, 76] and recently summarized in [74]. Factors β and V_T have a stochastic variation due to mismatch. The standard deviation (σ) of V_T is given by

$$\sigma^2(V_T) = \frac{A_{V_T}^2}{WL} + S_{V_T}^2 D^2, \tag{5.15}$$

where D is the distance between two transistors and A_{V_T} and S_{V_T} are process-related parameters. The standard deviation of β is given by

$$\frac{\sigma^2(\beta)}{\beta} = \frac{A_W^2}{W^2 L} + \frac{A_L^2}{WL^2} + \frac{A_{C_{ox}}^2}{WL} + \frac{A_{\mu_{n,p}}^2}{WL} + S_\beta^2 D^2 \approx \frac{A_\beta^2}{WL} + S_\beta^2 D^2, \tag{5.16}$$

where $A_\beta^2 = A_{C_{ox}}^2 + A_{\mu_{n,p}}^2$ (typically provided by the process) and A_W, A_L, $A_{C_{ox}}$, $A_{\mu_{n,p}}$, and S_β are, as before, process-related parameters. It is obvious from the above that mismatches are inversely proportional to device area; however, large devices introduce larger parasitics, and in applications requiring high integration using device dimensions as a means of reducing offset may not be feasible. In addition, as mentioned earlier, 1/*f* noise is inversely proportional to device area. The power spectral density (PSD) of the low-frequency input referred noise of an amplifier can be written as [77]

$$S_n = S_0\left(1 + \frac{f_f}{|f|}\right), \tag{5.17}$$

as it contains both white and 1/*f* noise, where S_0 is the white noise PSD and f_f is the 1/*f* noise corner frequency, where the 1/*f* noise equals the white noise floor. An alternative approach involves the use of circuit techniques to remove the effect of mismatches. There are two main families of techniques proposed to address these: chopper stabilization and (CHS) and auto-zeroing (AZ). These also affect noise. These will be discussed in the following section and the reader is also referred to [77], upon which the core of our discussion is based, and also to [74]. A discussion of more complex approaches (such as ping-pong-pang, nested chopping, combinations of auto-zero and chopping, etc.) than those discussed here, but still based on auto-zero and chopping, can be found in [74].

5.2.2.1 Chopper Stabilization

The principle of chopping is illustrated in Fig. 5.15a. Before amplification, the input signal is modulated by a signal with a much higher frequency (f_{chop}) than

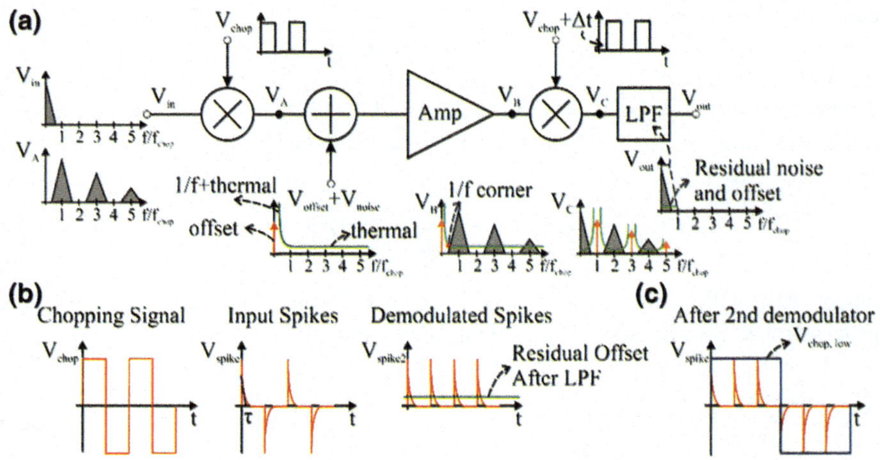

Fig. 5.15 **a** The chopping principle. **b** Residual offset due to switching spike transient at the input modulator (due to switch non-idealities), which are amplifier and demodulated. **c** The effect of nested chopping on the spiking transient, leading to a zero average after the output LPF

those in the input signal and is transposed to its odd harmonics. f_{chop} should be twice the maximum frequency of the input signal to avoid aliasing. Modulation is usually performed with a square wave signal. Thus, the polarity of the offset voltage is periodically switched at a high frequency. This is then amplified and filtered by the transfer characteristics of the amplifier. In applications where $1/f$ noise is important, the clock frequency is selected such that the modulated input signal is at a frequency higher than f_f. The signal is then demodulated using the same frequency signal. This is achieved by using an input and an output modulator (also known as choppers). Thus, on average this process cancels DC offset and low-frequency noise, since the recorded signal is modulated and demodulated back to baseband, while offset and noise are modulated only once and remain at the odd harmonics of the chopping signal. A subsequent LPF can remove these and the signal is ideally recovered, amplified with no offset and low-frequency noise content. Chopping and its use in DDAs were also discussed in [72], while an improved chopping techniques was proposed in [78]. Similarly to [79–81], in a number of papers [82], while the input chopper is positioned before the input transistors, the output chopping takes place at low impedance nodes internal to the amplifier structure before the dominant pole at the output terminal, so that the amplifier bandwidth does not limit the chopping frequency. Instead of using regular switches or transmission gates, bootstrapped switches were used in [27, 83, 84], since in reduced power supply systems, the on resistance of switches increases, increasing the thermal noise of the modulator. Chopper offset-stabilization is a technique with a secondary signal path, where an auxiliary amplifier compensates for the offset of the main amplifier. The auxiliary amplifier needs to be offset-free, which can be achieved with chopping. Such topologies are discussed in more detail in [74]. Alternatively, AZ can be used, which is discussed latter. This technique works only in negative feedback amplifiers.

The finite amplifier bandwidth and phase delay imposed on the input signal by the amplifier will be problematic, as they introduce frequency components around the even harmonics of f_{chop} and will affect the overall gain of the amplifier, respectively. These will appear as switching glitches at the output [74]. The imposed delay to the signal at high frequencies will necessitate demodulation with an equally delayed clock [85].

In contrast to AZ, chopping does not lead to aliasing of the broadband noise. As discussed in [77], the chopper-modulated baseband PSD is approximately constant, approaching asymptotically and being less than the input white noise, and if the amplifier corner frequency is much larger than the chopping frequency, then the output PSD is equal to the white noise PSD. In addition, the PSD of the chopper modulated baseband flicker noise can be approximated by a white noise component. The total residual noise, S_C, according to [77] can then be approximated for $\left|ff_{chop}^{-1}\right| \leq 0.5$ and $f_c f_{chop}^{-1} \gg 1$, by

$$S_C \cong S_0\left(1+\frac{0.8525f_f}{f_{chop}}\right), \tag{5.18}$$

where f_c is the amplifier cutoff frequency.

Any residual offset is mainly attributed to chopper switch non-idealities, and in particular to the input chopper, as spikes caused by them will be amplified and demodulated. This is illustrated in Fig. 5.15b. These are attributed to clock feed-through and charge injection. As discussed in [86], these spikes are due to a certain charge ΔQ, which is injected into the input capacitance of the amplifier, C_{in}, which in turn causes a spike with a peak amplitude of $V_{inj} = \Delta Q/C_{in}$. Differentially, this also depends on the mismatch between different elements comprising the chopper and also mismatches in the input impedance at each input node, giving rise to a differential input spike signal. This discharges exponentially with a time constant $\tau = 2R_sC_{in}$, where R_s is the switch on resistance and the source resistance, which may vary, and is periodic with a frequency f_{chop}. This time constant is much smaller than $1/(2f_{chop})$ and most of the energy of these spikes will be at frequencies higher than f_{chop}. For $\tau \ll 1/2\mathrm{f}_{chop}$, the residual offset ($V_{off,res}$) is equal to [77]

$$V_{off,res} = 2\tau f_{chop} V_{inj}. \tag{5.19}$$

This offset can be reduced significantly if the amplifier bandwidth is equal to $2f_{chop}$, with a resulting decrease to the overall DC gain of the amplifier ($A_v = 0.81A_0$), but with the residual offset being equal to [77]

$$V_{off,res} = \left(2\tau f_{chop}\right)^2 V_{inj}. \tag{5.20}$$

The residual offset is thus proportional to f_{chop} [74]. From the above equation, it can be seen that there is a tradeoff between noise and offset. An $f_{chop} = f_c$ is a good compromise according to [77, 87].This is equivalent to making the main amplifier selective to certain frequencies such that these spikes are filtered out before being demodulated by the output chopper. A second-order LPF with a cutoff frequency f_{chop} or preferably a bandpass filter (BPF) with a center frequency f_{chop} provides an optimal solution [77, 83, 86, 88]. This is due to the phase characteristics of the filter [86]. However, matching the filter center frequency to f_{chop} is challenging. Since the residual offset is a function of the amplitude of the digital clock driving the choppers, limiting the clock amplitude to a level below the supply levels, but which is capable of switching the chopper, can reduce the residual offset [74]. Another solution is to use a phase delay such that the demodulated spike signal will have an average value of zero, or to use a delay between the positive and negative half cycles of the chopping waveform [74]. A final solution for the elimination of the residual offset due to these spikes is the use of a spike filter prior to the output LPF. This was implemented in [25, 35, 62, 84]. This is a track and hold stage, the hold phase of which is very brief and coincides with the switching of the input chopper. In this way the output is sampled to the capacitor before the spike appears, and during the spike duration the held voltage at the capacitor is used; thus chopping glitches are not passed to subsequent stages. Using small well-matched transistors to implement the switches in the choppers will reduce the residual charge [89]. Current noise in chopper amplifiers was discussed in detail in [90]. This, when combined with high source impedances, is converted into a voltage noise. This can significantly contribute to the total input-referred noise. This noise has a white PSD with an amplitude proportional to f_{chop}. To reduce chopper noise the authors suggest the use of minimum-sized transistors to reduce charge injection, a low f_{chop} higher than the 1/*f* corner, a low swing clock driver, and the use of bootstrapped input choppers to drive the switches with a constant overdrive.

Residual offset can be reduced by using nested chopping [74, 89, 91], where two additional choppers are added, one at the input and one at the output, driven by a clock of lower frequency. The inner choppers are driven by a frequency greater than the 1/*f* corner frequency for an optimum noise performance. This, however, leads to a significant residual offset, as this frequency will need to be high, especially if nMOS input transistors are used, since residual offset is proportional to chopping frequency. The two external choppers modulate the residual offset at a lower frequency, reducing the effect of charge injection due to the choppers driven at a higher frequency, as shown in Fig. 5.15c. The lower frequency should be chosen according to the frequency content of the input signal. Nested chopping was used in [89, 91], achieving an offset below 100 nV in [89].

The output LPF is important, as it removes all offset signals which have been modulated to the odd harmonics of the chopping signal. An important question is how this should be implemented. Using an active topology will result in additional power consumption and, most importantly, additional offset. Passive solutions have been proposed but implemented with external components, as the large time constants required translate to a large chip area. This was followed in [89], where a passive *RC* filter with a cutoff of 3 Hz was implemented off-chip. In many applications requiring a high level of integration (e.g. multichannel systems) off-chip LPF implementations may be undesirable [92]. Increasing f_{chop} to relax LPF cutoff requirements will increase residual offset. As discussed by the authors, the filter can be part of an integrating ADC. An on-chip LPF was implemented in [93] using a simple passive SC LPF. The transfer function of such a filter exhibits zeros that were matched with the residual ripple frequency by driving it by the same chopping clock. Such sample-and-hold is easy to integrate and does not limit the bandwidth. An SC notch filter driven at f_{chop} was also used in [94] to remove the residual output ripple. This approach will lead to a certain amount of noise folding due to the sampling. In any case, if the output LPF order is not high enough and if its cutoff is not low enough relative to the modulated offset and noise signals, there will be a ripple signal at the output of the LPF with a frequency of f_{chop} and an amplitude proportional to the offset and the cutoff of the LPF [74]. This can be reduced further by using a large time constant, which is often implemented off-chip.

An alternative is to use feedback techniques to eliminate the modulated noise and offset which appears as a ripple on the output signal. A continuous-time AC-coupled ripple reduction loop was proposed in [92]. A sense capacitor converts the ripple at the output of the LPF to a current, with an amplitude proportional to the derivative of the output ripple. This is then synchronously demodulated and integrated to create a DC voltage proportional to the output ripple. This is converted into a current through a transconductor and is injected to the input to remove the ripple, and then used to cancel it at the input. This loop acts as a high-Q BPF with a center frequency at f_{chop}. Through the feedback, this translates into a notch, the width of which is governed by the unity-gain bandwidth of the loop gain. This leads to a 5 μV offset, a 1/*f* corner at 1 MHz, and a thermal noise PSD of 15 $nV/\sqrt{Hz}$. This was also performed in [95], where the feedback filter was implemented with an auto-zeroed SC filter. This topology is shown in Fig. 5.16a. A similar ripple reduction loop, termed an autocorrelation feedback (ACFB) loop, was proposed in [96]. The difference is that the loop is between the input and output of the output chopper, prior to the LPF. This allows the loop gain to be independent of the output load or the closed loop gain of the amplifier; the sense capacitor in [92] loads the output and should be small. A transconductor measures the ripple, which is then demodulated to DC with a chopper driven at f_{chop}, and an SC notch filter creates a nulling voltage, which is in turn applied to the input of the output chopper to eliminate the chopping ripple. A disadvantage of this is that since the ripple is sensed at the virtual ground node (the filter input) the signal will be low, requiring large gain. An alternative solution was proposed in [34, 97] where a number of passives were added, circumventing the need for feedback loops and additional active elements. The large resistances needed were implemented with transistors positioned within the feedback loop, where voltage swings are small and thus nonlinearity is not an issue.

The input impedance of AC-coupled amplifiers discussed in Sect. 5.2.1.1 can be improved using feedback from the output node to the input [34, 35, 82, 95]. This is particularly important when input capacitors are used in combination with chopping prior to these capacitors, as this will essentially appears as a SC and thus an equivalent input resistor with a resistance of $1/(2f_{chop}C_1)$, reducing the input impedance of the topology to a few MΩ for typically used chopping frequencies and input capacitors [95]. Positive feedback was used in [82, 95], where the output voltage is converted into a current and is injected back into the signal source. In [95] an additional pair of feedback capacitors was used in the feedback, as in Fig. 5.16a. However, this method is sensitive to the unknown parasitics at the input, which in combination with an input chopper will reduce the input impedance further. The positive feedback loop cannot compensate for this, and as a result it can increase the impedance only by 2.5–10 times. The feedback capacitor (C_{fb}) can be made programmable to fine tune this further. This approach was also followed in [82]. In this case, however, chopping was performed after the input capacitors and the feedback loop path consisted of an AC-coupled inverting amplifier with a gain of C_4/C_3 (set to 1 V/V) and a programmable capacitor array as in Fig. 5.16b. The input impedance of the topology without the feedback, as discussed earlier, is approximately equal $1/s(C_1 + C_p)$. The additional loop increases the input impedance by a factor β:

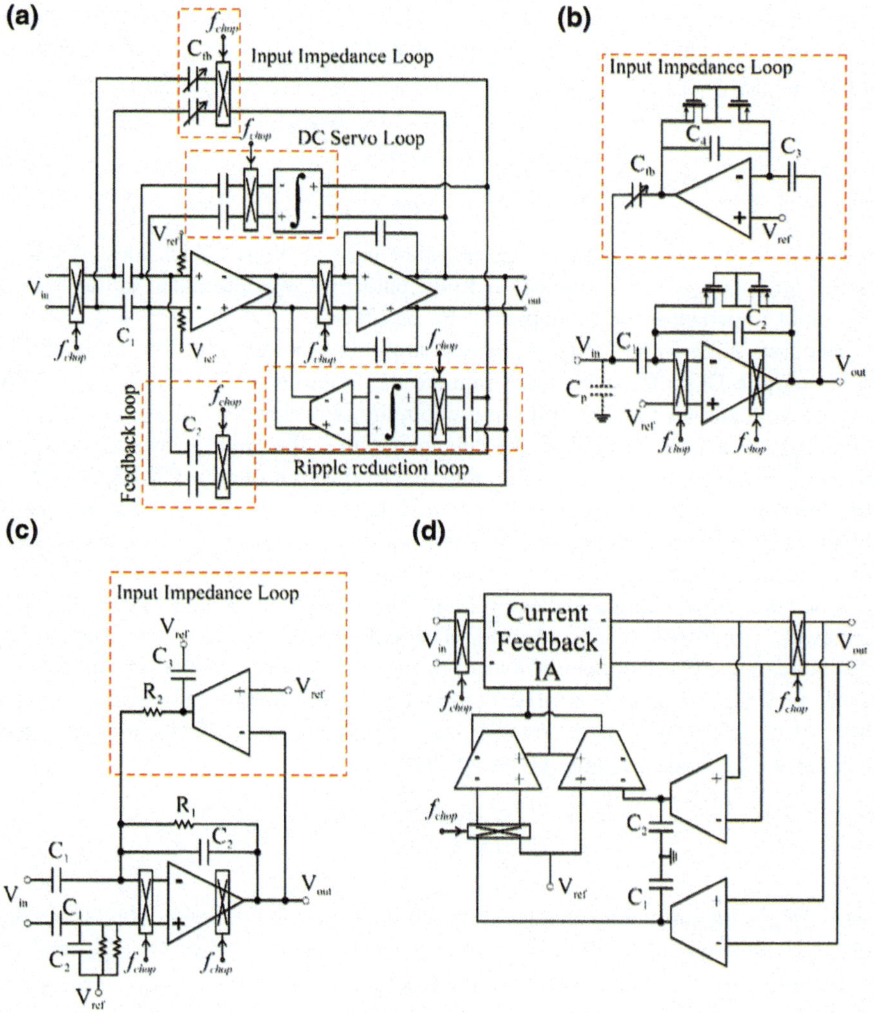

Fig. 5.16 **a** The capacitively-coupled chopper amplifier proposed in [95], employing a ripple reduction loop, an input impedance enhancement positive feedback loop, and a DC servo loop. The capacitive feedback together with the input capacitors set the gain as in Fig. 5.9a. **b** The input impedance-boosting loop implemented in [82]. **c** The amplifier topology proposed in [47]. **d** The loops implemented in the current feedback amplifier of [25] for eliminating DC electrode and IA offsets

$$\beta = \frac{C_1 + C_p}{C_1 + C_p - A_v C_{fb}}. \tag{5.21}$$

The feedback capacitor cannot be made arbitrarily large, as it may influence the stability of the circuit, and must be such that $\beta > 0$, with a maximum value given by [82]

$$C_{fb} = \frac{C_1 + C_p}{\left(\frac{C_1 C_4}{C_2 C_3} - 1\right)} \approx \frac{C_1 + C_p}{A_v}, \tag{5.22}$$

where A_v is the gain and C_{fb} is the capacitor used in the feedback loop. An alternative approach, which is insensitive to the parasitics, was presented in [34], where the input impedance can be improved by 25 times.

Sections 5.2.1.1 and 5.2.1.2 discussed amplifier topologies for the elimination of the electrode DC offset which can saturate the amplifier with large gains and low supply voltages. In [34, 95, 98] input chopping is performed before the input DC blocking capacitors. This has the advantage that the input capacitors will be dynamically matched through chopping, increasing the overall CMRR; however, the input electrode DC offset will be modulated to f_{chop} and will no longer be blocked by the input capacitors. This will then be processed by the amplifier and demodulated at the output appearing as an output DC signal. This input DC offset can be suppressed using a DC servo loop. This was implemented in the capacitive feedback amplifier of [34, 95, 98]. The output DC signal is amplified, then up-modulated by a chopper and fed back to the amplifier's virtual ground though capacitors C_{hp}, to cancel any DC components at the output, subtracting it at the input, similarly to what is done in [21]. Hence the loop acts as an HPF. The servo loop creates a high-pass pole given by [95]

$$f_{hp} = \frac{C_{hp}}{C_2} f_{int}, \tag{5.23}$$

where f_{int} is the unity gain frequency of the integrator in the servo loop. The value of C_{hp} is determined by the maximum expected electrode offset ($V_{eo,max}$) and the maximum possible output voltage of the feedback integrator ($V_{int\_out,max}$) according to [95]

$$V_{int_{out,max}} = \frac{C_1}{C_{hp}} V_{eo,max}. \tag{5.24}$$

The feedback integrator was implemented with a very large time constant SC integrator.

However, as highlighted in [47], servo loops for active cancellation of the DC input offset voltage limit the minimum supply voltage of the amplifier. In contrast to [34, 95, 98], in [47] chopping is performed after the input capacitors at the amplifier virtual ground nodes, sacrificing CMRR for a high input impedance and lower power supply. Hence, as when no chopping is used, the electrode DC offset voltage is rejected passively at the input and is not processed by the amplifier, and consequently there is no risk of amplifier saturation for a large electrode offset and limited power supply and amplifier output voltage swing. Furthermore, the chopper will no longer load the electrodes at the input. As in the previous case, where chopping combined with C_1 lead to an SC resistor reducing the input capacitance, in this case the chopping frequency combines with the input capacitance of the amplifier C_{in} to introduce a parasitic SC resistance at the input, which is, however, in the GΩ region. This parasitic introduces a current path between the two input terminals of the amplifier, and consequently any amplifier offset voltage appearing at the input will cause a current to flow through this parasitic, which can saturate the amplifier through the large feedback resistor. To address this issue, the authors used a DC servo loop implemented with a G_m-C filter with an off-chip >10 μF capacitor to integrate the amplifier output and generate a current proportional to the output error to cancel this current. As before, this leads to a high-pass characteristic. Nevertheless, the feedback resistors are still needed to cancel a zero in the transfer function. In this topology, using chopping increases the noise, requiring very large C_1s, which also reduces the input impedance, since the input referred noise is inversely proportional to C_1 and proportional to f_{chop} according to [22]

$$\left|V_{n,in}^2\right| = \left|V_{n,OTA}^2\right| \left(\frac{C_1 + C_2}{C_1} + \frac{2\pi f_{chop} C_{in}}{sC_1}\right)^2. \tag{5.25}$$

DC servo loops were also implemented in the current-balancing amplifiers in [25, 32, 35, 62], as in Fig. 5.16d. Two loops are implemented in [25], one to remove electrode offset voltages and the second to remove IA offsets. Thus the main purpose of chopping in this case is to remove low-frequency noise. For the first loop the amplifier output is integrated before the output chopper, and for the second loop the output chopper output is integrated. G_m-C first-order integrators are used and simple differential pairs are used to convert the signal into a current to be injected across resistor R_1 in Fig. 5.13. Two loops were used for the removal of the electrode offset in [62]: a coarse digital one and a fine analog one. Nevertheless, these voltage-to-current servo loops can only compensate for a few millivolts of electrode offset, as their capability is limited by noise and the maximum current provided by the feedback loop [64]. A voltage-to-voltage alternative was presented in [64] to extend the electrode offset compensation capability up to ±350 mV.

5.2.2.2 Auto-Zero (AZ)

The AZ technique is a sampling-based technique and in its basic implementation it is not suitable in applications requiring continuous-time operation. Nevertheless, as will be discussed further, this can still be facilitated. The AZ can take a number of forms. The unwanted signal (offset and noise) are sampled and stored and then subtracted from the recorded signal at the input of the amplifier, at its output, or at an intermediate node. In this way the DC offset can be completely removed.

In contrast to offset, which is constant, noise is time-varying and random. This is particularly true for wideband thermal noise. The correlation between the sample of the noise stored in the capacitors and the instantaneous noise will determine the efficiency in noise cancellation. AZ is more efficient in reducing $1/f$ noise over thermal noise as the autocorrelation of two $1/f$ samples decreases more slowly over time than that thermal noise [77]. If the sampling frequency is greater than f_f, the collected sample will also be highly correlated with this, and low-frequency random noise, such as $1/f$, will be greatly reduced. The latter is achieved, however, with an increase in the noise floor because of the aliasing of the wideband noise (and to a lesser extent of the $1/f$ noise as well) as a result of the sampling process [74, 77]. As discussed, in AZ there are two phases: the sampling (φ_S) and amplification phases (φ_A). The selection of the different phases is determined by switches. During φ_S, offset and noise are sampled. During this time period, the amplifier cannot be used for signal processing. During φ_A, the stored samples of offset and noise are subtracted from the instantaneous signal at the input of the amplifier. An auxiliary nulling input port is often used to perform the subtraction, which could, for example, be a second input differential pair, as in a DDA. Sampling can be performed in the analog domain with a sample-and-hold circuit (S/H) or in the digital domain [77]. During φ_S, errors may be introduced affecting the stored voltage, such as charge injection due to the switches. This will lead to a residual offset. The AZ technique is well suited for SC implementations, where the signals are sampled at each phase and held. During the hold period of SC implementations the amplifier can be auto-zeroed. Correlated double sampling (CDS) is another method, which can be considered as a special case of AZ. CDS is essentially an AZ phase followed by an S/H; the amplifier output is hence sampled twice per clock cycle and the difference between these two samples is taken. Offset and $1/f$ noise will be approximately constant between these two cycles due to the fast sampling clock. Consequently, they are highly correlated and subtraction of the two samples eliminates them. The sampling frequency must be higher than f_f (typical ratio of 10) [99]. Unfortunately, thermal noise is doubled due to the double sampling [99]. An extension to AZ and CDS is correlated multiple sampling (CMS), where CDS is combined with averaging. M samples are averaged at a reset stage and after a recording event and subsequently the two averages are differentiated. More information can be found in [85].

AZ can be applied by using open-loop or closed-loop approaches [77, 85]. The open-loop method is shown in Fig. 5.17a. During φ_S the S_1 and S_2 are grounded, then S_2 is opened and the amplifier offset times its gain is stored at the capacitor. During φ_A, S_1 connects the amplifier back to the input signal. As is evident, this approach, also called output offset storage (OOS), can only be used if the amplifier and/or the input-referred offset voltage are small, such that the output offset remains within the saturation limits of the amplifier, otherwise the amplifier will saturate. OOS can possibly be applied before the second stage of an amplifier which provides a large gain in order to avoid saturation due to the overall large gain. In addition, the subsequent stages connected to the capacitor should be of large input impedance, otherwise the capacitor may discharge.

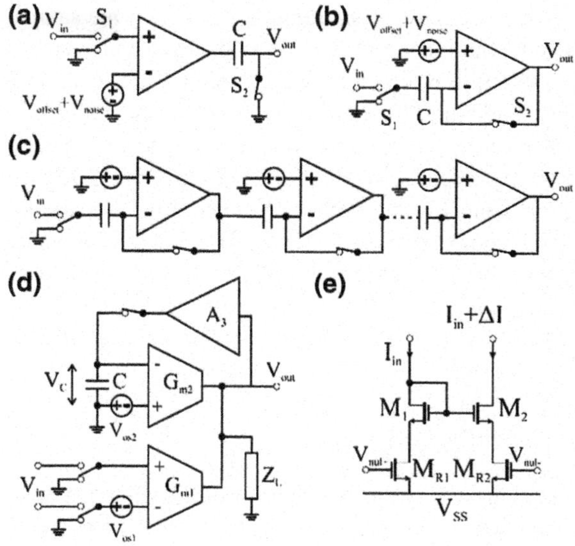

Fig. 5.17 **a** Open loop AZ. **b** Closed loop AZ. **c** Multistage amplifier closed loop AZ operation. **d** AZ operation with an additional nulling input. **e** Current mirror with triode transistors acting as auxiliary inputs for offset trimming

In high-gain amplifiers, the closed-loop approach is preferred. This is shown in Fig. 5.17b. In this case, the amplifier is configured as a unity-gain buffer, the input is shorted to a known potential (as before, e.g. ground) and the offset and noise are stored across *C*. If the amplifier gain is large the voltage stored across the capacitor is approximately equal to the voltage offset. This remains stored across the capacitor, as the CMOS amplifier will have a large input impedance and will thus not discharge.

In both cases, the residual charge will be dominated by the charge injection charge due to the switch imperfections. In a fully differential configuration this can be further reduced, as the residual offset will then be the differential of charge injection mismatch. These techniques can also be applied to multi-stage amplifiers, as discussed in [77]. If the switches at each stage are opened sequentially, then the residual offset of each stage adds to the input-referred offset of the next and can thus be eliminated. Consequently, the residual offset will only be determined by the final stage and the total residual offset can thus be made very small, as when referred to the input it will be divided by the total gain of the whole topology. Closed-loop AZ in a multi-stage amplifier is shown in Fig. 5.17c.

As mentioned before, the AZ function can also be performed by introducing an auxiliary input port, as in a DDA. This is shown in Fig. 5.17d. Thus, there are two input transconductors, G_{m1} and G_{m2}, and one (G_{m2}) is only used for offset elimination. A buffer or amplifier (A_3) can be used between the output and the input of the second transconductor for isolation from the storage capacitor and to provide a large loop gain and reduce the residual offset further [74, 77]. The ratio of the transconductance of the two pairs needs to be properly selected, as $V_c = -(G_{m1}/G_{m2})V_{os}$, such that the voltage stored across the capacitor is much larger than the input-referred offset of the compensation loop formed by the second transconductor and the feedback amplifier, avoiding at the same time saturation of G_{m2}. The residual offset can be minimized by increasing the loop gain (A_2A_3), while the residual offset due to charge injection can also be minimized if a number of conditions are satisfied as discussed in [77]. Another approach for the nulling input realization involves the use of linear region transistors as resistive degeneration elements in current mirrors [58, 77, 100] and is shown in Fig. 5.17e. AZ and CDS have been applied to various signal processing blocks, such as SC amplifiers for capacitive sensor readouts [99, 101, 102], comparators [103], and S/H circuits [104, 105]. It is important also to note the possibility of subtracting offset voltages using floating gate transistors [106]. In any case, caution must be taken with regard to charge injection and leakage current, which may discharge the offset/noise storage capacitor.

Continuous-Time AZ

As mentioned in the previous section, since the amplifier needs to be disconnected from its input signal to perform the AZ function, AZ is not suitable for continuous-time applications. Nevertheless, ping-pong AZ is a method alleviating this issue [100, 107–109], as shown in Fig. 5.18a. Here there are two copies of the same amplifier and at each cycle only one is used for signal processing, while the second is performing the AZ function. In this time-shared operation, AZ can thus be utilized for continuous-time measurements. However, as is self-evident, this requires twice the chip area and power consumption. In addition, during the handover between each amplifier, there will be switching transients appearing at the output, because the outputs of the two amplifiers are uncorrelated [107]. Nevertheless, ping-pong can be applied in cases where chopping cannot be applied, as in the case of offset compensation of continuous-time filters, as chopping requires an LPF for eliminating the modulated offset unless it is passively performed. A residual offset less than 100 μV and a noise of 225 nV/$\sqrt{\text{Hz}}$ were reported in the ping-pong AZ op-amp scheme of [109]. Ping-pong AZ was combined with chopping in [108] to compensate a current feedback IA. The input stage was ping-pong auto-zeroed to remove the offset and attenuate $1/f$ noise and chopping was applied to remove the low-frequency noise. Low frequency noise was reduced to white-noise levels (27 nV/$\sqrt{\text{Hz}}$), offset was 2.8 μV and a CMRR of 140 dB was achieved. Settling phases were added in the AZ process in order to prevent switching transients reaching the output. A chopping frequency half that of the AZ process was used, to prevent the introduction of switching transients from chopping. However, this was a general-purpose design consuming 1.7 mA from a 3–5 V supply and a large chip area. A programmable first-order G_m-C filter with an input-referred offset less than 210 μV and handover glitches between the ping-pong cycles less than 40 mV was presented in [107] (Fig. 5.18a). Large storage capacitors were used to reduce clock feedthrough and charge injection (see 5.26).

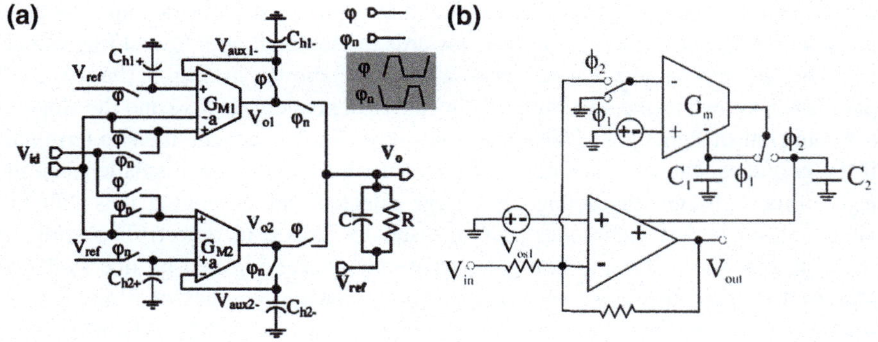

Fig. 5.18 **a** A ping-pong AZ G_m-C LPF (© IEEE 2007. Reprinted, with permission, from [107]). **b** Continuous-time AZ offset-stabilization

In another continuous-time implementation of AZ, the main amplifier is never disconnected from the input signal and processes the input continuously. This is known as AZ offset stabilization. A second amplifier is used that first performs AZ on itself during φ_1 to become offset-free (using C_1) and then samples the offset of the main feedforward amplifier during φ_2 (using C_2) to compensate it as well, as shown in Fig. 5.18b. Nulling input ports are needed in both the auxiliary and main amplifiers. This approach can only be applied if the main amplifier is used in closed-loop negative feedback and the voltage differential at the input is approximately equal to the main amplifier offset. This feedforward AZ method, however, suffers from intermodulation effects between the AZ clock and the input signal, which leads to distortion [107]. The output signal has a lower switching transient glitch content than ping-pong AZ, as there is no handover involved [107]. This approach is analyzed in detail in [77, 110, 111].

5.2.3 *Low Leakage Analog Switches*

Metal-oxide field-effect transistor (MOSFET) switches are important for a wide range of applications, such as in discrete-time analog circuits (SC circuits) or in the sampling of analog signals in e.g. pipeline ADCs using sample-and-hold circuits, for the storage of voltages, e.g. offsets in AZ, in the development of multiplexers, and in the development of switch modulators, e.g. in chopping. However, the classic simple single-transistor switch has a non-zero and nonlinear on-resistance. In addition, mechanisms such as charge injection, clock feedthrough, sampled noise and current leakage degrade the performance of the switch [77].

Charge is stored in the channel when the transistor is on and the charge is divided and injected between source and drain when the transistor is turned off. This is known as charge injection. The degree with which the charge, q_{inj}, will be divided between the two terminals, depends on the ratio of the capacitance at the drain (parasitic and hold capacitance), C_h, and at the total parasitic capacitance at the source, C_p, as the charge will flow towards the terminal with lowest impedance, i.e. towards the largest capacitance. These are shown in Fig. 5.19a. A third parameter, known as the switching parameter, is dependent on R_{ON} and the slope of the gate control signal [77]. The charge will split equally between the two terminals if the two capacitances are equal or if the switching time is much smaller than the $R_{ON}C$ time constant. Predicting the charge injected and simulating it is difficult; nevertheless, circuit techniques can be used to reduce its effect. The simplest solution is to use a transmission gate, where the charge injected by one switch is absorbed by the other. However, matching the channel charges of n and p transistors is poor and signal-dependent [77]. The transistors will be driven by complementary clock signals. The phase jitter between the two will hinder the compensation of charge injection further. Alternatively, C_p can be made larger than C_h by adding extra capacitance and the clock can be made to have a slow switching transient to make the switching parameter less than one, such that the majority of the charge goes to C_p (Fig. 5.19b). However, this limits the maximum clock frequency, while the charge flow due to the overlap capacitance, C_{ov}, is not compensated for [77]. Another approach is to make $C_p = C_h$ such that the charge is equally divided and to use half-sized dummy switches at each terminal driven by complementary clock signals to compensate for the injected charge as in Fig. 5.19c. A third solution is to use the dummy switches as before and a clock signal with a very short transition time to split the current equally (Fig. 5.19d). The dummy switches in both scenarios, however, need to be well matched. A final approach is to use a fully differential topology (Fig. 5.19e), such that the injected charge

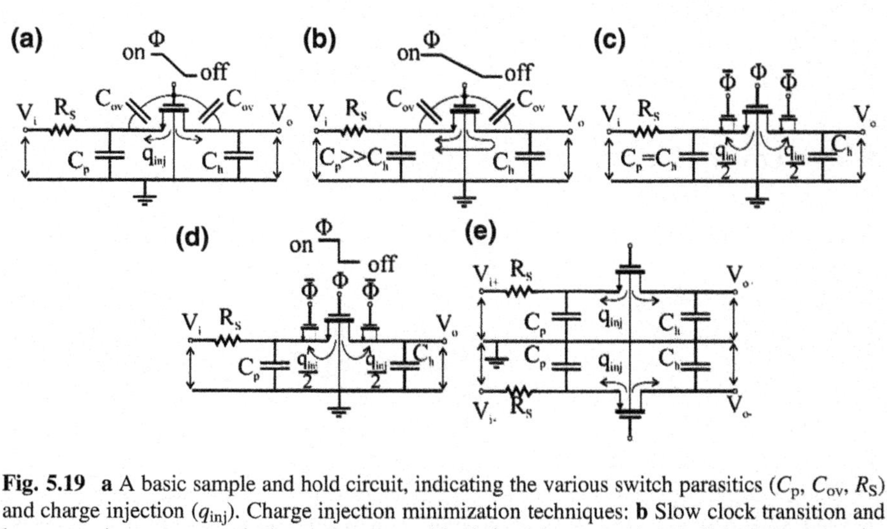

Fig. 5.19 **a** A basic sample and hold circuit, indicating the various switch parasitics (C_p, C_{ov}, R_S) and charge injection (q_{inj}). Charge injection minimization techniques: **b** Slow clock transition and large capacitor, **c** symmetrical capacitances and half-sized dummy switches, **d** half-sized dummy switches and short transition time on the clock, and **e** fully differential topology (adjusted from [77])

becomes a common-mode signal, which can then be rejected. Nevertheless, this requires accurate matching of q_{inj} between the two paths.

The term "clock feedthrough" is used to describe the coupling of the gate control signal to the output and input nodes, due to the C_{GS} and C_{GD} overlap parasitic capacitances. Not all of this charge will flow to C_{h} as some will flow to ground. An attenuation factor, $\alpha_{\mathrm{att}} < 1$, is thus used to described this charge. In addition, an error charge due to channel noise will also be stored in C_{h}. Finally, when the hold time is long, and/or the temperature is high, T_{h}, the leakage current, I_{leak}, through the drain-to-bulk junction should also be considered. Sub-micron CMOS technologies (e.g. 0.18 μm, 0.13 μm, and 90 nm) with low threshold voltages suffer from sub-threshold leakage. Ion implantation is used to lower the threshold voltage, and according to [112] this process leads to an effective leakage current of a minimum sized transistor in a modern CMOS process, in the range of 10 pA. Of particular importance is the reverse diode leakage current of the drain or source diffusions. This is particularly important for transistors used as switches together with a capacitor for temporary analog storage, e.g. to store a voltage for offset removal (e.g. AZ offset compensation) or for signal variables (e.g. sample-and-hold circuits). Storing a signal in this way is low power, simple, and compact, but it is only possible to do so for short timescales. This is because the current leakage of the switch during its off state will change the charge stored in C_{h}, and it will eventually discharge it. The total voltage error ΔV at C_{h} is thus equal to [77]

$$\Delta V = \frac{q_{\mathrm{inj}}}{C_{\mathrm{h}}} + \alpha_{\mathrm{att}} \frac{C_{\mathrm{GD}}}{C_{\mathrm{GD}} + C_{\mathrm{h}}} V_{\mathrm{G-clock}} + \sqrt{\frac{kT}{C_{\mathrm{h}}}} + \frac{I_{\mathrm{leak}} T_{\mathrm{h}}}{C_{\mathrm{h}}}, \tag{5.26}$$

As can be seen from the above equation, all parameters decrease with increasing C_{h}. In most cases the first term dominates.

Drain leakage is considered to be the summation of drain-to-bulk diode leakage (when the diode and bulk are biased at different potentials) and subthreshold conduction. The diode leakage in an nMOS can be modeled by the diode equation to a first-level approximation according to [113]

$$I_{DB} = I_S \left[\mathrm{e}^{-\frac{V_{DB}}{U_T}} - 1 \right], \tag{5.27}$$

where I_{DB} is the diode current, V_{DB} is the drain-to-bulk potential, and I_{S} is the diode saturation current. The latter is constant and proportional to the junction area and perimeter [113]. These parasitic diodes are shown in Fig. 5.20a. The subthreshold current is described by (5.72) and V_{T} by (5.130). As discussed in [113], to make subthreshold conduction negligible, V_{GS} has to be decreased such that $I_{\mathrm{DS}} \ll I_{\mathrm{DB}}$. The voltage at which subthreshold conduction can be ignored continuously decreases with decreasing threshold voltage in modern CMOS processes and goes below 0 V. In single supply systems, this will thus be less than the supply voltages [112, 113]. This is demonstrated in Fig. 5.20b. These corner simulation results from [112] of minimum sized transistors in a 0.35 μm technology for

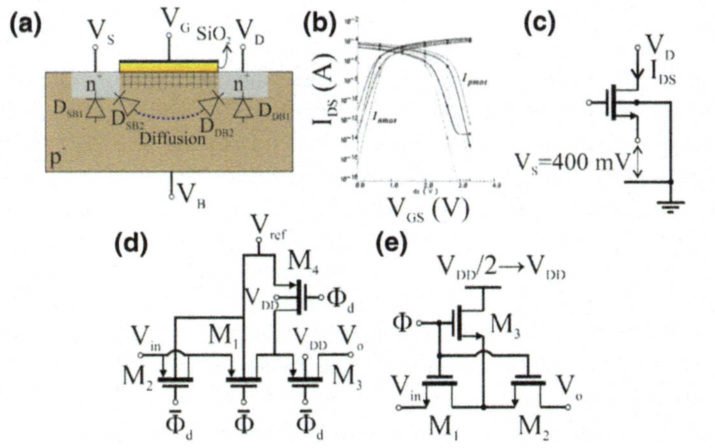

Fig. 5.20 **a** Accumulation-mode leakage source [113]. **b** Corner analysis simulation of minimum-sized nMOS and pMOS transistors in a 0.35 μm technology with $V_{DS} = V_{DD}/2$ [112]. **c** Source voltage shifting to minimize the subthreshold conduction current when the transistor is off [112]. **d** The low-leakage, low-charge injection switch proposed in [113]. M_1 is twice as big as the other transistors, which are minimum-sized. **e** The AT-switch for subthreshold leakage suppression [115]. In the nMOS configuration shown the drain of M_3 needs to be at a potential between $V_{DD}/2$ and V_{DD}. In the pMOS realization this needs to be set at a potential between V_{SS} and $V_{DD}/2$

$V_{DS} = V_{DD}/2 = 1.65$ V) show that for $V_{GS} = 0$ V for the nMOS and $V_{GS} = V_{DD}$ for the pMOS the current has not reached the diode reverse current limit and the off current is much larger. In the worst-case scenarios, these simulations indicate for the nMOS a current of 10 pA and for the pMOS 1 pA. Thus, one approach to minimize subthreshold conduction, for example, is, in an nMOS transistor, to bias the source at a potential where when $V_G = 0$ V leads to a $V_{GS} < 0$, with V_{GS} being equal to or less than the voltage where subthreshold conduction terminates and diode leakage dominates. In e.g. [112, 113] $V_{GS\_OFF} = -400$ mV was found to be a good choice for the scenarios examined. This is shown in Fig. 5.20c. The advantage of source shifting is that generating and supplying gate voltage outside the rails is not necessary to generate a negative V_{GS}. The disadvantage of this is that it limits the available voltage range and it requires additional circuitry to generate these bias voltages. Source shifting essentially increases V_T, changing V_{SB}. However, this increases I_{DB}. In a double-well process, however, the nMOS can sit inside its own well, in which case the bulk can be biased at the same potential as the drain, e.g. $V_{DD}/2$, in which case I_{DB} will also be reduced. In a transmission gate the difference of the two leakage currents from nMOS and pMOS transistors flows to the output node. Thus, for a given output voltage, the junction areas of the two transistors can be adjusted to cancel the current to some extent. However, this area ratio will be valid only for a specific output voltage and will depend on matching. Beyond weak inversion a transistor operates in a region known as accumulation mode of operation [113]. As shown in [113], within this region, for values of V_{SB} approaching

V_{DB} the drain current is independent of V_{GS} and constant to about 30 aA. It was concluded that the accumulation-mode current is mainly due to the coupling of the source-to-bulk diode current to the drain terminal and vice versa. This was attributed to carrier diffusion under the accumulation layer, as shown in Fig. 5.20a. Thus, when $V_{SB} = V_{DB}$ the accumulation source/drain interaction is minimized. The authors also discussed an additional leakage mechanism in transistors fabricated inside wells. In this case, the well-to-substrate junction interacts with the drain similarly to accumulation-mode coupling of the source-to-bulk junction to the drain. The well-to-substrate junction will be reverse-biased and the drain-to-well junction is not. Consequently, there will be a differential leakage current from the drain to the bulk. Considering the above, the authors proposed the topology of Fig. 5.20d. Transistor M_2 is a dummy transistor used to cancel charge injection, as in Fig. 5.19c. The bulks of M_1 and M_2 are biased by a voltage, V_{ref}, the same as the voltage of the non-inverting terminal of the op-amp used to realize a storage cell with this switch. V_{ref} is chosen to be much less than V_{DD}, such that M_1 is well in accumulation mode when off to minimize subthreshold conduction. Transistor M_3 is used to disconnect during the storage period the source of M_1 from the output node, such that M_4 can connect it to V_{ref}. In this way all terminals are at V_{ref} during the storage period and leakage is minimized. Two clock signals and their inverse are needed. One (Φ_d) is a delayed version of the other (Φ). It must be highlighted that the charge-injection cancellation with the dummy device needs M_1 to be twice as big, and this, together with the dummy device, leads to a four times larger total equivalent area. With the diode saturation current being proportional to the junction area, the leakage current will thus also be four times larger. With this topology the leakage current was measured to be in the region of 20–60 aA. Using a differential topology to eliminate charge injection instead of the dummy switch consequently decreased the leakage current to the range of 2–9 aA.

In many cases, switches are used in the inputs of signal recording amplifiers in applications requiring high input impedance and CMRR. As discussed in [114], leakage currents may be delivered to the tissue, and depending on the application and range of currents, this may have implications for the accuracy of a measurement in potentiostatic of galvanostatic systems or delivering charges to tissues when not needed, e.g. in stimulation applications. This is because when switches are turned off charge in the channel and parasitic capacitances is injected in the drain and source terminals [113].

Subthreshold leakage was also targeted in [115], where its significant effect in low V_T transistors is highlighted. The leakage current in a switched-capacitor integrator with 0.1 V V_T transistors was 500 times larger than when 0.3 V transistors were used, leading to an error of 8%. According to the authors, the leakage is nonlinearly dependent on the input signal. In the past the authors proposed a method called super cutoff CMOS (SCCMOS) to suppress leakage. However, this topology required gate voltages outside the power rails and thus required level shifting and charge-pumping circuits. This was to apply a negative V_{GS} and to bring the subthreshold conduction current to zero, as discussed earlier with regard to Fig. 5.20b, c. The proposed analog T-Switch (AT-Switch) shown in Fig. 5.20e used clock signals within the supply rail. This is achieved by shifting the analog ground to a higher voltage, e.g. $V_{DD}/2$. In this way the device is biased deep in the cutoff region in the off state. The advantage of this is that there is no need for signals outside the voltage rails as in SCCMOS and bootstrapped switches, and there is no need for dual-well processes to realize transmission-gate equivalents as in [113], since the technique does not require bulk biasing. The disadvantage is that, since the analog ground is shifted to $V_{DD}/2$, the dynamic range is limited. The switch consists of two nMOS transistors M_1, M_2, and a pMOS M_3. Another advantage of this topology is that the switch requires a single clock signal applied to the gates of the three transistors, which are shorted together. During the off-state ($V_G = V_{SS}$), node A is set to $V_{DD}/2$ through M_3. In this way V_{GS} is negative and M_2 is deeply cut off. Thus, this is essentially a source shifting method as with those discussed earlier. This topology was recently used in [114] to restrict current leaking into the tissue in a neural stimulator. A similar topology using only nMOS transistors was presented in [116]. In this case, however, two different clocks are necessary: one controls M_{1A} and M_{1B} and the second M_{1C}.

Another issue with conventional switches is related to low power supply operation. As the maximum power supply voltage is decreased, driving the switch with sufficient over-drive voltage is challenging, and for rail-to-rail operation control signals that exceed the power supply rail are required [117, 118]. This is particularly the case when low-V_T transistors are not available. The conductivity between source and drain terminals depends on their potentials against the channel potential. For V_{GS} less than V_T, the resistance increases abruptly. The conductance of a single MOSFET switch is thus strongly signal-dependent. A transmission gate (a nMOS and pMOS switch connected in parallel) has a region where the channel conductance is approximately signal independent. Nevertheless, when the supply voltage is reduced to less than the sum of the absolute values of the V_Ts of the two transistors, this vanishes and the switch does not conduct reliably [118]. Bootstrapping switch topologies overcome the issues of poor channel conduction and varying on-resistance [119]. These topologies ensure that in the off state, the gate is grounded and that in the on state, the gate to channel voltage is constant. To achieve this, a constant voltage is supplied between the gate and source, irrespective of the source input signal, which is generated and stored during the off state in a capacitor, establishing a constant conductance between drain and source, independent of the input signal. This stored voltage can exceed V_{DD}, depending on the input signal, and it is thus important that the gate dielectric does not fail. A number of bootstrapped topologies have been presented in the literature [117–122]. A simple topology to achieve this was presented in [118] and is shown in Fig. 5.21. During the off state of Φ, i.e. when $\overline{\Phi}$ is 1, the bootstrapped capacitor C_b is charged to $V_{DD} - V_{SS}$. And S_3 is on connecting the gate of M_1 to V_{SS} and the switch is off. In the on state, charged C_b acts as a floating battery which is connected between the gate and source terminals of M_1 via S_1 and S_2. This makes the conductance approximately constant if the parasitic capacitance and the body effect are ignored, and is equal to [118]

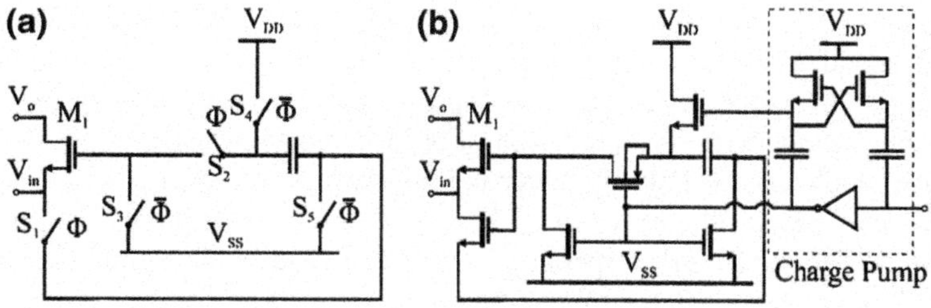

Fig. 5.21 **a** A simple bootstrapped switch topology. **b** Implementation of the bootstrapped switch

$$g_{DS} = \mu_n C_{ox}\frac{W}{L}\left(V_{GS} - V_{T,n}\right) = \mu_n C_{ox}\frac{W}{L}\left(V_{DD} - V_{SS} - V_{T,n}\right). \quad (5.28)$$

If $V_{DD} - V_{SS} > V_{T,n}$, the switch will conduct across the whole supply voltage range and the switch's conductance will be signal-independent. S_1, however, must also conduct rail-to-rail. Thus, this switch can also be bootstrapped. A simple realization of this topology is shown in Fig. 5.21b. S_3 and S_5 can be implemented by simple single-transistor switches, as their switching operation is trivial. S_4, however, cannot be implemented by a single pMOS, since when the capacitor is charged to voltages exceeding V_{DD} it would leak during the high state of Φ. To overcome this, the authors used an nMOS switch driven by a charge pump level-shifting the clock signal swinging between V_{DD} and $2V_{DD}$. S_2 is implemented with a pMOS. Its bulk potential must be at least as high as the potential across C_b. This can be addressed by connecting the bulk to the source or, preferably, by generating and biasing it to a constant $2V_{DD}$ voltage. S_2 cannot be an nMOS, as that would require large clock signals. Thus, M_1 cannot be compensated for the body effect. This has been addressed in other examples in [118] and it has also been addressed in [117]. The interested reader is referred to [117–122] for more discussions and other advanced topologies.

5.2.4 *Operational Transconductance Amplifiers (OTA), and G_m-C Filters and Linearized Transconductors*

5.2.4.1 OTAs

An OTA, as the name implies, is an analog building block transforming an input voltage ($V_i = V_1 - V_2$) into an output current (I_o), i.e. it is a VCCS. An OTA is essentially a simplified op-amp, in the sense that it does not include a low output impedance output stage and the related internal feedback for stability. An advantage of OTAs is that the transconductance (g_m) is programmable by a current (I_B). An ideal OTA has an infinite bandwidth, input and output impedances, SNR, and a linear input voltage range as large as the rails, i.e. a constant transconductance for all input signal amplitudes). Non-idealities include a non-zero input and output capacitances (C_i and C_o), a frequency-dependent transconductance ($g_m(f)$) and a finite output conductance (G_o) [123–125]. For low-frequency applications (i.e. most biomedical applications) only the latter is important, as it reduces the DC voltage gain and the filter quality (Q) factor. The high output impedance is particularly important, as for this circuit to operate as a VCCS it needs to be significantly higher than any load in order to deliver the required output current. Otherwise, the current will flow to ground through internal nodes. The symbol of an OTA is shown in Fig. 5.22a and its ideal small-signal equivalent circuit in Fig. 5.22b.

The output current is given by

$$I_o = g_m(V_1 - V_2), \tag{5.29}$$

where g_m is a function of I_B and parameters associated with the technology used, transistor geometries, transconductor topologies, and others. OTAs can be used to implement various functions required in analog signal processing. Due to the high output impedance (Z_o) of an OTA, when loaded with a resistor with a resistance smaller than Z_o the output current will flow through the resistor, leading to an output voltage. In the case of Fig. 5.22c a non-inverting amplifier is realized with a gain given by $V_o/V_i = g_m R_L$, while in the case of Fig. 5.22d, an inverting amplifier is realized, with a gain of $V_o/V_i = -g_m R_L$. In both cases, the output impedance of the circuit is defined by the load, i.e. $Z_o = R_L$. The circuit of Fig. 5.22e is a voltage variable resistor (VVR) [125], with a resistance equal to $Z_i = 1/g_m$. This is a very useful topology for the implementation of on-chip resistances. If the OTA is loaded with a capacitor, then the simplest first-order LPF, i.e., integrator, is realized as in Fig. 5.22f. This is a voltage variable integrator with a differential input and is a

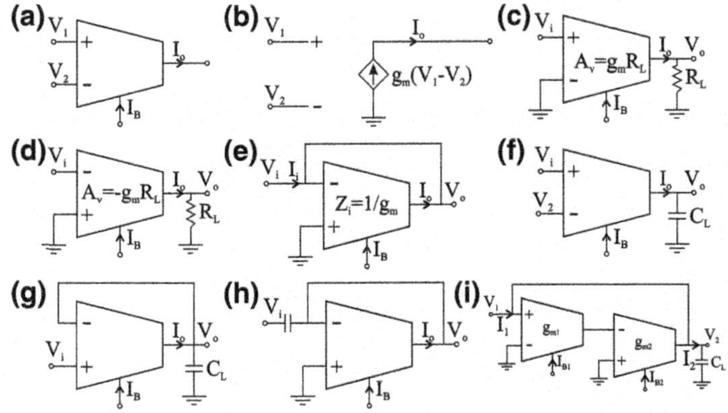

Fig. 5.22 OTA **a** symbol and **b** ideal equivalent circuit; the OTA as: **c** a non-inverting amplifier, **d** an inverting amplifier, **e** a voltage variable resistor (VVR), **f** a simple integrator, **g** another LPF, **h** a HPF, and **i** a grounded inductor

basic building block for many filters. The transfer function of this topology is given by

$$\frac{V_0}{(V_1 - V_2)} = \frac{g_m}{sC_L}. \tag{5.30}$$

The ideal magnitude and phase response of an integrator has a pole at the origin and a 90° phase shift at the unity-gain frequency. A high DC gain is an important characteristic of a good integrator [126]. A real integrator, however, has a finite DC gain (A) and a dominant pole shifted from the origin to ω_{ug}/A, where ω_{ug} is the unity-gain frequency, while there may be additional higher-frequency non-dominant poles. Due to the finite DC gain, there is a phase lead at ω_{ug}, while the non-dominant poles lead to excess phase at higher frequencies. This is shown in Fig. 5.23. These translate to a decrease in the Q-factor of the integrator and lead to errors in the pass-band of a filter implemented using the integrator, as they modify the filter transfer function [127, 128].

If a resistor R_L is connected in parallel with the capacitor, the integration becomes lossy; the gain is controlled by g_m and the loss is fixed by $R_L C_L$, according to

$$\frac{V_0}{(V_1 - V_2)} = \frac{g_m R_L}{sR_L C_L + 1}. \tag{5.31}$$

It is obvious from the above discussion that R_L can be implemented with the circuit of Fig. 5.22e, in which case the pole frequency can be adjusted by varying the transconductance (g_{mL}) of the VVR as in

$$\frac{V_o}{(V_1 - V_2)} = \frac{g_m}{sC_L + g_{mL}}. \tag{5.32}$$

The structure of Fig. 5.22g is also a lossy integrator, which, however does not have an adjustable gain, only an adjustable pole location and a single-ended input, with a transfer function defined by

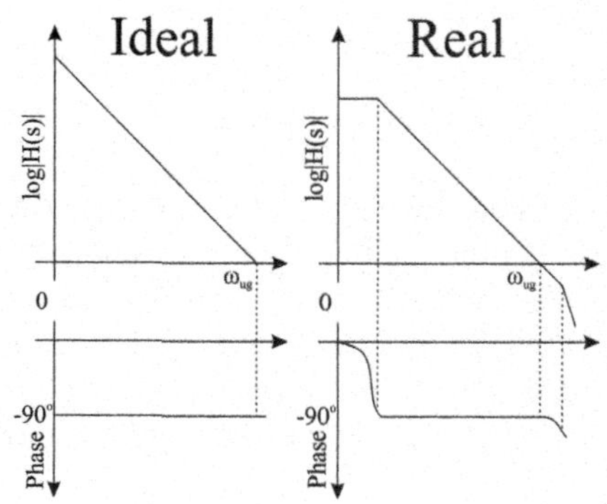

Fig. 5.23 Magnitude and phase Bode plots of an ideal and a real integrator

$$\frac{V_0}{V_i} = \frac{g_m}{sC_L + g_m}. \quad (5.33)$$

The topology of Fig. 5.22h is a simple first-order HPF with an adjustable pole location and constant gain (similarly to the previous circuit), as in

$$\frac{V_0}{V_i} = \frac{sC_L}{sC_L + g_m}. \quad (5.34)$$

The −3 dB cutoff frequency of the last two first-order circuits is given by

$$f_{-3dB} = \frac{g_m}{2\pi C_L}. \quad (5.35)$$

The last circuit (Fig. 5.22i) is a gyrator, i.e. a voltage variable impedance inverter. When such a circuit is loaded as a capacitor, C_L, as in Fig. 5.22i, it behaves as an inductor [125, 129]. This topology thus allows the emulation of inductors, which are otherwise challenging to implement on-chip and model with CAD. In this implementation, there are two current sources connected in a loop. The equations describing the two output currents of each transconductor are given by

$$I_1 = g_{m2}V_2 \quad (5.36)$$

and

$$I_2 = g_{m1}V_1. \quad (5.37)$$

Manipulating the above two equations, and if $Z_1 = V_1/I_1$ and the admittance $Y_2 = I_2/V_2 = sC_L$, leads to [123, 125]

$$\frac{V_1}{I_1} = \frac{1}{g_{m1}g_{m2}}\frac{I_2}{V_2} \Leftrightarrow Z_1 = \frac{sC_L}{g_{m1}g_{m2}} = sL_{eff} \quad (5.38)$$

Using the above basic building blocks and circuit analysis techniques allows the realization of high-order filters of all types.

5.2.4.2 G_m-C Filters

The implementation of filters using OTAs comes with several advantages, such as design simplicity, high integration, low component count, and programmability. This family of filters is typically preferred to over active-*RC* and MOSFET-C filters, especially in high-frequency telecommunications applications and in biomedical applications. The former is due to their low-impedance internal nodes allowing higher frequency operation, and no need for a high-speed power-hungry clock, as in discrete-time switched-capacitor filters. Since OTAs do not have an output stage, they do not have internal feedback compensating circuits (Miller compensation) for stability. The latter is due to the fact that biomedical applications typically require filters with large time constants and thus large resistors and/or capacitors to achieve this. The reason for this is demonstrated in Fig. 5.8.

As discussed in the previous section, in G_m-C filters, resistors are emulated with OTAs, which can easily realize very small transconductances and thus equivalent resistances ($g_m = 1/R$). In this way OTAs replace resistors in a manner that allows small on-chip capacitors in the range of a pF and lower to be used. G_m-C filters thus require a significantly reduced total component count and chip area, thus making them ideal for integration. Classic textbooks, such as [123, 124], and more recently [130], discuss the subject and the synthesis of filter topologies in detail and the reader is referred to these. Such filters have been used to synthesize biomimetic circuits, e.g. emulating human hearing, to implement implantable devices for the hearing impaired, i.e. artificial ears. A good and common example [58, 59, 77, 126, 131, 132] is the biquadratic filter of Fig. 5.24.

The transfer function ($H(s)$), the cutoff frequency (ω_0), and Q-factor (Q) of the LPF output are given by [131]

$$H_{LP}(s) = \left(\frac{g_{m1}}{g_{m2}}\frac{g_{m3}g_{m4}}{C_1C_2}\right) \Big/ \left(s^2 + \frac{g_{m2}}{C_1}s + \frac{g_{m3}g_{m4}}{C_1C_2}\right),$$
$$\omega_0^2 = \frac{g_{m3}g_{m4}}{C_1C_2}, \quad Q_{LP} = \frac{C_1}{g_{m2}}\omega_0. \tag{5.39}$$

For the bandpass output the transfer function becomes [126, 132]

$$H_{BP}(s) = \left(\frac{g_{m1}}{C_1}s\right) \Big/ \left(s^2 + \frac{g_{m2}}{C_1}s + \frac{g_{m3}g_{m4}}{C_1C_2}\right), \tag{5.40}$$

while the center frequency and Q-factor equations are the same as before [126]. The same transconductor can be used for all stages and the transconductance of each stage can be adjusted such that the various filter properties can be tuned accordingly. This biquad topology has been used to realize a filter structure with a widely

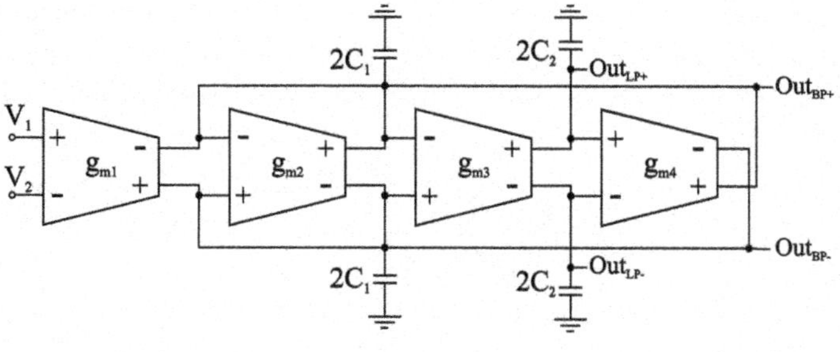

Fig. 5.24 Fully differential second-order biquadratic filter with low-pass and bandpass outputs

programmable center frequency (5.9–58 MHz) and Q-factor (0.36–10) [132] for field-programmable analog arrays (FPAA) applications, a 200 MHz seventh-order filter [133], a 1 V sixth-order filter for a 5.8 GHz transceiver [134], a 100 kHz to 20 MHz filter for telecommunications applications [135], and in bioimpedance applications [58, 59].

5.2.4.3 Linearized Transconductors

A typical transconductor has limited linear input voltage range. An important part of the development of G_m-C filters is the linearization of OTAs. In this section, both strong and weak inversion linearization techniques will be discussed, focusing only on differential voltage input topologies based on differential pairs (Fig. 5.25). Fully differential topologies, e.g. topologies with a differential output, can cancel out all even-order harmonics and should be preferred over single-ended topologies. However, in most cases these require an additional mechanism to control the common-mode output DC level (common-mode feedback, CMFB), which to some extent increases the overall transconductor footprint and power consumption. In some cases, depending on the filter topology, the CMFB of several transconductors may be combined. One such example in the filter structure of Fig. 5.24; this will be discussed at transistor level in the following sections. Sanchez-Sinencio classifies transconductor linearization techniques as methodologies (a) attenuating the input signal, (b) using several non-linear transconductors, (c) using negative feedback, or (d) using a nonlinear function and its inverse. The literature on linearization techniques is vast. In this section we discuss popular techniques and focus on those typically used in biomedical applications. As will be discussed, several techniques can be combined to achieve a wide input linear range.

Strong Inversion Linearization Techniques

The Differential Pair in Strong Inversion

The most basic differential input transconductor, which is the basis of all the following discussions, is the differential pair (also known as the long-tail pair or the source-coupled pair) shown in Fig. 5.25a. In strong inversion, where the device is operated above its threshold voltage, the current–voltage characteristics of the device, when the device is operated in saturation, are defined by the classic square law equation:

$$\begin{aligned} I_{\mathrm{D}} &= \frac{1}{2}\mu_{\mathrm{n,p}} C_{\mathrm{ox}} \frac{W}{L} (V_{\mathrm{GS}} - V_{\mathrm{T}})^2 (1 + \lambda V_{\mathrm{DS}}) \\ &= \frac{1}{2} K_{\mathrm{n,p}} \frac{W}{L} (V_{\mathrm{GS}} - V_{\mathrm{T}})^2 (1 + \lambda V_{\mathrm{DS}}), \end{aligned} \tag{5.41}$$

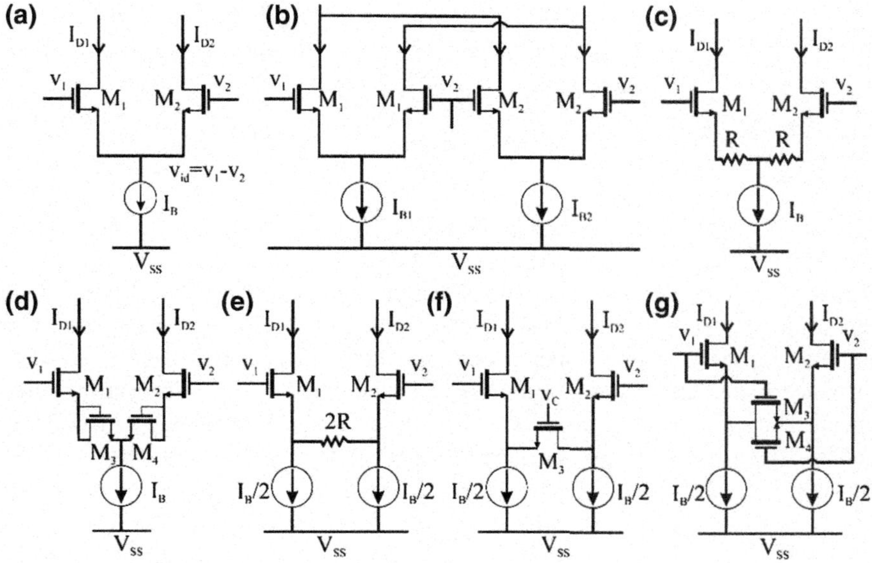

Fig. 5.25 **a** The simple differential pair transconductor, **b** the cross-coupled linearized transconductor; source degeneration with: **c** two resistors, **d** diode connected transistors, **e** single resistors and two current sources, **f** a transistor operated in the triode region, and **g** two varying bias triode transistors

where I_D is the drain current, μ is the carrier mobility, C_{ox} is the gate oxide capacitance, V_{GS} is the gate-to-source voltage, $K_{n,p} = \mu_{n,p}C_{ox}$, V_T is the threshold voltage, λ is the channel length modulation coefficient, and V_{DS} is the drain-to-source voltage. If channel length modulation effects are neglected (i.e. $\lambda = 0$), the entire tail current flows through transistor M_1 ($I_{D1} = I_B$ and $I_{D2} = 0$, in which case if $V_{G2} = 0$, $V_S = -V_T$) and we solve (5.41) for V_{GS}, then

$$V_{GS1} = V_T + \sqrt{\frac{2I_B}{K_n\frac{W}{L}}} = V_T + \sqrt{2}V_{ov}, \tag{5.42}$$

where V_{ov} is the overdrive voltage at the quiescent bias point, where $V_{GS1} = V_{GS2}$, $V_{id} = 0$ and $I_{D1} = I_{D2} = I_B/2$. The maximum input differential voltage range leading to a change in the output current is thus

$$-\sqrt{2}V_{ov} \leq V_{id} \leq \sqrt{2}V_{ov}. \tag{5.43}$$

Beyond this range, changes in V_{id} will always generate an output current of I_B, i.e. the response saturates. Thus, with an appropriate selection of tail current and input pair transistor dimensions, the input differential voltage range can be set. Using (5.41) to describe I_{D1} and I_{D2}, recognizing that $I_{D1} + I_{D2} = I_B$ and subtracting the two allows the calculation of the differential output current I_o as [136]

$$I_o = I_{D1} - I_{D2} = g_m V_{id}\sqrt{1 - \frac{K_n \frac{W}{L} V_{id}^2}{4I_B}}, \tag{5.44}$$

where the small-signal transconductance, g_m, is defined as

$$g_m = \sqrt{K_n \frac{W}{L} I_B} \tag{5.45}$$

It is obvious from the above equations that the response of the circuit is not linear due to the V_{id}^2 term. The response is linear only within a limited range, and as the input signal increases the response becomes increasingly nonlinear. The transconductance is defined as the first derivative of the current with respect to the input signal. Thus, for large signals, g_m, is non-constant and a function of V_{id}. The Taylor series expansion of I_0 gives a better understanding of the nonlinearity of the function.

$$I_o = g_m\left(V_{id} + 0 - \frac{1}{8}\frac{1}{V_{ov}^2}V_{id}^3 + 0 - \frac{1}{128}\frac{1}{V_{ov}^4}V_{id}^5 + \cdots\right). \tag{5.46}$$

The THD is a measure of the harmonic distortion present in a signal and it is defined as the sum of the powers of each harmonic component divided by the power of the fundamental. Even-order nonlinearities are cancelled due to the differential output and nonlinearities higher than third order decrease rapidly and can be neglected [137]. Thus the THD can be approximated by considering only the third-order harmonic distortion (HD_3), which is equal to

$$HD_3 = \left(K_n \frac{W}{L}\right)\frac{1}{32I_B}V_{id}^2 = \frac{1}{16V_{ov}^2}V_{id}^2. \tag{5.47}$$

Typically a response is considered linear if the THD is below 1%. According to [137] the following must then be met

$$-0.4\sqrt{2}V_{ov} \le V_{id} \le 0.4\sqrt{2}V_{ov}. \tag{5.48}$$

Thus, the linear range can be increased either by increasing I_B or L or by decreasing W. This, however, restricts the input common-mode range to allow sufficient headroom for the tail current source (I_B) to remain in saturation. This is thus a drawback for low-voltage applications.

Source Degeneration

Source degeneration is one of the most common linearization techniques used. It can take many forms, as depicted in Fig. 5.25. The topologies of Fig. 5.25c, e achieve the same transconductance. However, in (c) the noise of the current sink is divided across the two branches and appears as a common-mode noise contribution, while in (e) the noise from each sink appears as a differential noise current, and hence the noise of this circuit is higher. In addition, the common-mode input range in the topology of (c) is reduced due to the voltage drop across each resistor consuming voltage headroom ($I_B R/2$). Simply speaking, the way source degeneration works is by making the gain of the circuit a weaker function of the transconductance of the input pair transistors. A portion of the input voltage appears across the resistor rather than a V_{GS} drop, e.g. the applied signal swing across gate and source is reduced. Thus the signal swing applied across the gate and source terminals is reduced and the resulting output current is more linear [138]. If the transconductance of the input transistors is $g_{m\text{-}1,2}$, then the transconductance ($g_{m\text{-}sd}$) of the circuit is given by

$$g_{m-sd} = \frac{g_{m-1,2}}{1 + g_{m-1,2}R} \approx \frac{1}{R}. \tag{5.49}$$

with the far right part of the equation being true for $g_{m\text{-}1,2}R \gg 1$. This leads to an approximately constant V_{GS} for the input transistors. While the transconductance is reduced by $1 + g_{m\text{-}1,2}R$, the third-order harmonic is reduced by the square of the same factor [139]. A large resistor is required to achieve a wide linear range. Good quality resistors (high resistivity per unit length and linearity) are typically not an option on commercial CMOS processes, thus the implementation of the degeneration resistor on silicon is an issue.

Single Triode Transistor Source Degeneration

The resistor can be replaced by diode connected transistors as in Fig. 5.25d or by a transistor operating in the deep triode region as in (f) [140, 141]. In the triode region $V_{DS} < V_{ov}$ and the current voltage relationship of a FET is given by

$$I_D = K_n \frac{W}{L}\left[(V_{GS} - V_T)V_{DS} - \frac{1}{2}V_{DS}^2\right]. \tag{5.50}$$

And for $V_{DS} \ll V_{ov}$ (in deep triode) the squared term disappears and the drain current is approximately linear. Differentiating I_D with respect to V_{DS} then gives

$$r_{DS} = \frac{1}{K_n \frac{W}{L}(V_{GS} - V_T)} = \frac{1}{g_{DS}}, \tag{5.51}$$

where g_{DS} is the drain to source conductance. For large input swings, however, M_3 will not remain in the deep triode region and the resistance will substantially change. To ensure M_3 is in the ohmic region, a large V_C must be used, while this voltage allows tuning of the transconductance. The advantage of this is that since $g_{m\text{-}sd}$ is independent of I_B and $g_{m\text{-}1,2}$, tuning of $g_{m\text{-}sd}$ via V_C does not affect other circuit parameters, e.g. the output conductance [141]. If $(W/L)_{1,2}$ is large, then the V_{GS} of the input pair transistors is essentially constant [140]. A disadvantage of this topology is its high sensitivity to input common-mode signals [139]; V_C must follow the input common-mode level for M_3's resistance to be defined correctly [138].

Double Triode Transistor Source Degeneration

The circuit of Fig. 5.25g [142] addresses this, exhibiting a low sensitivity to common-mode input signals. For an input signal of zero, M_3 and M_4 are in the deep triode region. As V_1 increases M_3 stays in triode while M_4 becomes saturated. This is because in one case both drain and gate voltages rise, while in the other the drain rises while the gate and source voltages drop. Thus, as V_1 is increased the resistance of M_3 decreases. This decrease tries to counterbalance the typically decreasing transconductance with increased input voltage. Hence, even if one of the devices leaves the deep triode region, the transconductor remains relatively linearized. According to [142] the optimum relationship between the input and degeneration transistors is given by

$$\left(\frac{W}{L}\right)_{1,2} \approx 6.7\left(\frac{W}{L}\right)_{3,4}. \tag{5.52}$$

The transconductance of this topology is given by [143]

$$g_{\text{m-sd}} = \left(K_{\text{n}}^2\left(\frac{W}{L}\right)_{1,2}\left(\frac{W}{L}\right)_{3,4}\right)\sqrt{\frac{2I_{\text{B}}}{K_{\text{n}}\left(\frac{W}{L}\right)_{1,2}}}\Bigg/\left[\left(\frac{1}{2}K_{\text{n}}\left(\frac{W}{L}\right)_{1,2}\right)+2\left(K_{\text{n}}\left(\frac{W}{L}\right)_{3,4}\right)\right]. \tag{5.53}$$

Hence, $g_{\text{m-sd}}$ can be tuned by I_{B}.

Cross-Coupled Pairs

Cross-coupling two differential pairs (Fig. 5.25b) and appropriately scaling the input pair dimensions and bias currents such that the third-order nonlinear terms of each pair are matched, allows its cancellation [137]. To accomplish that it is obvious from (5.46) that the following must be true [137, 144]

$$\left[\left(\frac{W}{L}\right)_1\Big/\left(\frac{W}{L}\right)_2\right]^3 = \frac{I_{\text{B1}}}{I_{\text{B2}}}. \tag{5.54}$$

However, in order to avoid the linear terms being cancelled out as well, the following must also hold true

$$\left(\frac{W}{L}\right)_1 \neq \left(\frac{W}{L}\right)_2 \quad \text{and} \quad I_{\text{B1}} \neq I_{\text{B2}} \tag{5.55}$$

In which case, the remaining linear term defines the transconductance, which is simply the difference of the linear terms of each differential pair:

$$g_m = g_{m1} - g_{m2} = \sqrt{K_n \left(\frac{W}{L}\right)_1 I_{B1}} - \sqrt{K_n \left(\frac{W}{L}\right)_2 I_{B2}}. \quad (5.56)$$

Another popular approach is the cross-coupled quad cell of [145].

Combination of Techniques

As mentioned earlier, various techniques can be combined in order to achieve an even greater linearized input range. One example is presented in [144], where resistive source degeneration was combined with cross-coupling of differential pairs, as in Fig. 5.26a. It must be noted that using the methodology described for the cross-coupled linearization, complete cancellation of the third-order component cannot be achieved, since various non-idealities are not considered, in particular mobility degradation [144, 146]. This is more pronounced in submicron technologies, since as devices are reduced in size the average electric field per unit area applied at the gate increases. Due to this large vertical electric field, the carrier velocity is lowered by mobility degradation. The vertical field forces and confines the carriers in the channel within a narrower region towards the silicon and the gate oxide interface, where imperfections impede their movement, slowing them down. This reduces the current capability of the device and consequently its

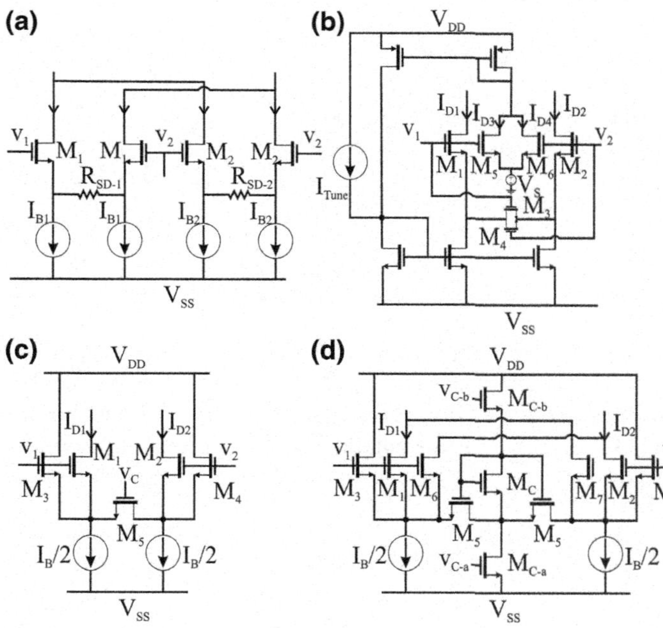

Fig. 5.26 **a** Combination of source degeneration and cross-coupling [144]. **b** Combination of adaptive biasing and source degeneration [148]. **c** Combination of source degeneration and current division [149–152] and **d** as in (**c**) with cross-coupling [149]

transconductance, and odd harmonics are introduced in I_D [138]. According to [144], when this is taken into account the classic square law transistor equation becomes

$$I_D = \frac{K_{n,p}\frac{W}{L}}{1+\theta(V_{GS}-V_T)}(V_{GS}-V_T)^2, \quad (5.57)$$

where θ is the mobility degradation factor, typically in the region of 0.1 V^{-1} [147]. As the thickness of the gate oxide (t_{ox}) is reduced, θ increases, because the field becomes stronger within the oxide. According to [144], θ can be interpreted as a degeneration resistance, $R_\theta = 2\theta/K$. Adding an actual degeneration resistor, R_{SD}, will add with R_θ. Substituting in (5.46) V_{id} with

$$V_{id} \rightarrow \frac{V_{id}}{(1+N_r)}, \quad (5.58)$$

where

$$N_r = g_m \frac{R_\theta + R_{SD}}{2} \quad (5.59)$$

is the degeneration factor. As a result the condition for third-order distortion cancellation becomes

$$\left[\frac{\left(\frac{W}{L}\right)_1}{\left(\frac{W}{L}\right)_2}\right]^3 = \frac{I_{B1}}{I_{B2}}\left(\frac{1+\sqrt{K_n\left(\frac{W}{L}\right)_1 I_{B1}}\frac{(R_{\theta1}+R_{SD-1})}{2}}{1+\sqrt{K_n\left(\frac{W}{L}\right)_2 I_{B2}}\frac{(R_{\theta2}+R_{SD-2})}{2}}\right)^6. \quad (5.60)$$

The overall transconductance is given by

$$g_m = \frac{g_{m1}}{(1+N_{r1})} - \frac{g_{m2}}{(1+N_{r2})}. \quad (5.61)$$

A similar circuit was presented in [140], where cross-coupling was used to eliminate the nonlinearities of the triode transistors used for degeneration. In another implementation source degeneration was combined with adaptive biasing (Fig. 5.26b) [148]. In this case, the adaptive tail bias current of the input differential pair is equal to

$$I_B = I_{Tune} + K_n\left(\frac{W}{L}\right)_{5,6}(V_{GS5,6}-V_T)^2 + \frac{1}{4}K_n\left(\frac{W}{L}\right)_{5,6}V_{id}^2. \quad (5.62)$$

Accordingly [148], the relationship between the different transistor dimensions should follow

$$\left(\frac{W}{L}\right)_{5,6} = \frac{\left(\frac{W}{L}\right)_{1,2}}{2m^2}, \tag{5.63}$$

where

$$m = 1 + \frac{\left(\frac{W}{L}\right)_{1,2}}{4\left(\frac{W}{L}\right)_{3,4}}, \tag{5.64}$$

The additional current I_{Tune} is used to tune the transconductor. According to the authors, good linearity is achieved for $1.5 \leq m \leq 1.75$.

In [149–152] source degeneration with current division and cross-coupling were combined to achieve very small transconductances for pole frequencies below 10 Hz for biomedical applications. When triode transistor source degeneration with current division is used, as in Fig. 5.26c, and when the conductance (g_{DS5}) of M_5 is much smaller than the $g_{\mathrm{m3,4}}$ of $M_{3,4}$, then the transconductance of the OTA is [151]

$$G_{\mathrm{m}} = \frac{2g_{\mathrm{m1,2}}g_{\mathrm{DS5}}}{2g_{\mathrm{DS5}} + (1+M)g_{\mathrm{m1,2}}}, \tag{5.65}$$

where

$$M = \frac{g_{\mathrm{m3,4}}}{g_{\mathrm{m1,2}}} \gg 1. \tag{5.66}$$

For $2g_{\mathrm{DS}} \ll (\mathrm{M} + 1)g_{\mathrm{m1,2}}$,

$$G_{\mathrm{m}} = \frac{2g_{\mathrm{DS5}}}{(1+M)}. \tag{5.67}$$

Thus, for low-frequency pole locations, G_{m} can be reduced by increasing M. By increasing M, however, the drain currents are reduced by the same factor. Thus for large M factors, the offset voltage because of leakage currents will increase [151]. With regard to noise, as discussed earlier, $1/f$ noise can be reduced by increasing the length of the input transistors, and the thermal noise by reducing $g_{\mathrm{m1,2}}$. According to [150], for reduced harmonic distortion

$$g_{\mathrm{m3,4}} \geq 5g_{\mathrm{DS5}}. \tag{5.68}$$

Using cross-coupling can reduced the transconductance of the circuit even further, at the cost of a higher sensitivity to mismatches and higher noise [149, 151]. If

$$N = \frac{g_{\mathrm{m6,7}}}{g_{\mathrm{m1,2}}}, \tag{5.69}$$

then

$$G_m = \frac{2(N-1)}{M+N+1} g_{DS5}. \tag{5.70}$$

If $N \approx 1$, a very small transconductance can be achieved. Using this topology, a 2.4 Hz LPF with a 60 dB dynamic range was demonstrated in [149]. The triode transistor was split into two transistors, and to reduce the dependence of its conductance to the common-mode voltage, the gate voltage is related to the common source voltage with the addition of the diode connected transistor M_C as in Fig. 5.26d. Controlling the drain–source voltage across M_C using two additional transistors ($M_{C\text{-}a}$ and $M_{C\text{-}b}$), allows the circuit's transconductance to be adjusted.

Weak Inversion Linearization Techniques

The weak-inversion operation domain of MOSFETs is particularly attractive for biomedical applications, in which energy consumption is one of the key metrics. Such applications are characterized by the need for a long battery life, periodic device activity and, in most cases, low speed. Bioelectrical (ECG, EEG, EMG) signals are typically recorded from subhertz frequencies (e.g. 0.05 Hz) up to 1 kHz [25] (see Fig. 5.8). However, depending on the application, measurements may extend to the 10 kHz range (ENG) [23]. Acquisition rates can be of the order of seconds to minutes, and thus subthreshold performance can be adequate. Typically, amplification precedes all other steps, with filtering being a secondary step in a typical analog signal processing chain, such that the signal is amplified sufficiently, unless these two functions are combined as in [20].

The Differential Pair in Weak Inversion

For a transistor operated in weak inversion to be in saturation the following condition must apply [153, 154]

$$V_{DS} \geq 4U_T \text{ to } 5U_T, \tag{5.71}$$

Thus V_{DS} must be approximately larger than 100–130 mV for the transistor to be in saturation. Weak inversion thus provides the minimum saturation voltage, which is also independent of V_T. This allows the power supply to be reduced significantly. The drain current of a transistor in weak inversion is given by

$$I_D = \pm I_{spec-n,p} e^{\left(\frac{V_G - V_T}{n_{n,p} U_T}\right)} \left(e^{\left(\frac{-V_S}{U_T}\right)} - e^{\left(\frac{-V_D}{U_T}\right)} \right). \tag{5.72}$$

As V_{DS} approaches 4–5U_T, the dependence of the drain voltage on the drain current can be neglected and the second term in the parenthesis disappears. Thus, in saturation I_D is given by

$$I_D = \pm I_{spec-n,p} e^{\left(\frac{V_G - V_T}{n_{n,p} U_T}\right)} e^{\left(\frac{-V_S}{U_T}\right)} = I_{D0-n,p} e^{\left(\frac{V_G}{n_{n,p} U_T}\right)} e^{\left(\frac{-V_S}{U_T}\right)}, \quad (5.73)$$

where I_{spec} is the specific current, defined as the current when $V_{GS} = V_T$. This is equal to

$$I_{spec-n,p} = 2n_{n,p} \mu_{n,p} C_{ox} U_T^2 \frac{W}{L} = I'_{spec-n,p} \frac{W}{L}. \quad (5.74)$$

I'_{spec} is the specific current of a square device and is technology dependent, and n is the subthreshold slope (different for nMOS and pMOS transistors), typically considered to be constant. Is some textbooks and papers, instead of n the parameter κ appears, where $n = 1/\kappa$ [153, 155–157]. The parameter n can be estimated by measuring the slope of the log-linear plot of I_D vs. V_G in the weak inversion region and can range between 1 and 1.5 or even up to 2 [153, 158–160]. It is defined as the required increase in gate voltage for the drain current to increase by one order of magnitude ($\times 10$ on a linear scale or by 1 on a logarithmic scale). I_{D0} is known as the leakage current and is defined as

$$I_{D0-n,p} = I_{spec-n,p} e^{\left(\frac{-V_T}{n_{n,p} U_T}\right)}. \quad (5.75)$$

In some textbooks, I_D is approximated by

$$I_D \approx I_{spec-n,p} e^{\left(\frac{V_{GS} - V_T}{n_{n,p} U_T}\right)}, \quad (5.76)$$

At this point it is important to introduce the inversion coefficient (IC). This is an indicator estimating a transistor's region of operation. For $IC > 10$, a device operates in strong inversion, for $IC < 0.1$ a device is in weak inversion, and for values of IC in the region of $0.1 < IC < 10$ a device is in moderate inversion. IC is defined as the ratio of the drain current (I_D) and the specific current (I_{spec}) [158, 160]

$$IC = \frac{I_D}{I_{spec-n,p}}. \quad (5.77)$$

If we use (5.73) to define the current across each transistor in a simple differential pair as in that of Fig. 5.25a and take their difference (I_0), then since $V_{S1} = V_{S2} = V_S$

$$I_0 = I_{D1} - I_{D2} = I_{D0-n,p}e^{\left(\frac{-V_s}{U_T}\right)}\left[e^{\left(\frac{V_{G1}}{n_n U_T}\right)} - e^{\left(\frac{V_{G2}}{n_n U_T}\right)}\right]. \tag{5.78}$$

Bearing in mind that, as before, $I_{D1} + I_{D2} = I_B$, then

$$I_B = I_{D1} + I_{D2} = I_{D0-n,p}e^{\left(\frac{-V_s}{U_T}\right)}\left[e^{\left(\frac{V_{G1}}{n_n U_T}\right)} + e^{\left(\frac{V_{G2}}{n_n U_T}\right)}\right]. \tag{5.79}$$

Dividing the two and recognizing that $V_{G1} = V_1$, $V_{G2} = V_2$ and that $V_{id} = V_1 - V_2$, then

$$\frac{I_0}{I_B} = \left[e^{\left(\frac{V_1}{nU_T}\right)} - e^{\left(\frac{V_2}{nU_T}\right)}\right] \Big/ \left[e^{\left(\frac{V_1}{n_n U_T}\right)} + e^{\left(\frac{V_2}{n_n U_T}\right)}\right]. \tag{5.80}$$

Multiplying both numerator and denominator by $e^{-\left(\frac{V_1+V_2}{2nU_T}\right)}$ [156] leads to a hyperbolic tangent, thus [153–157]

$$I_0 = \frac{I_B\left[e^{\left(\frac{V_{id}}{2n_n U_T}\right)} - e^{-\left(\frac{V_{id}}{2n_n U_T}\right)}\right]}{\left[e^{\left(\frac{V_{id}}{2n_n U_T}\right)} + e^{-\left(\frac{V_{id}}{2n_n U_T}\right)}\right]} = I_B \tanh\left(\frac{V_{id}}{2n_n U_T}\right). \tag{5.81}$$

A major drawback of the differential pair in weak inversion is its limited linear input range. This can be demonstrated with the differential transconductance evaluated by taking the first derivative of the differential output current with respect to the differential input voltage

$$G_m = \frac{\partial I_0}{\partial V_{id}} = \frac{I_B}{2n_n U_T}\left[\cosh\left(\frac{V_{id}}{2n_n U_T}\right)\right]^{-2} = \frac{I_B}{2n_n U_T}\left[1 - \tanh^2\left(\frac{V_{id}}{2n_n U_T}\right)\right]. \tag{5.82}$$

For $V_{id} = 0$, the nominal value of the transconductance is

$$G_m = \frac{I_B}{2n_n U_T}. \tag{5.83}$$

Thus, the transconductance can be adjusted linearly only via the tail bias current. If we define the linear range by 1% deviation from the nominal value of G_m, then the input linear range of the differential pair in weak inversion is constant and equal to 14.7 mV_{pp} [155]. If it is defined as 1% of THD, then the linear range is equal to 60 mV_{pp}. Distortion of the transconductor is due to the tanh function in the differential output current. The first three terms of the Taylor series expansion of tanh are

$$\tanh(x) = x - \frac{1}{3}x^3 + \frac{2}{15}x^5 - \cdots \tag{5.84}$$

Further analysis can be found in [155]. Another way of expressing the current through each transistor as a function of tail current is by recognizing again that $I_{D1} + I_{D2} = I_B$ and that $V_{S1} = V_{S2}$.

$$I_1 = I_B \frac{1}{1 + e^{\left(\frac{-V_{id}}{n_n U_T}\right)}} \tag{5.85}$$

and

$$I_2 = I_B \frac{1}{1 + e^{\left(\frac{V_{id}}{n_n U_T}\right)}} \tag{5.86}$$

For some applications the limited linear range of the differential pair in weak inversion (~ 15 mV$_{pp}$) may be sufficient. An example is presented in [161], where the authors developed a seventh-order BPF implementation of a continuous wavelet transform for the processing of EEG signals (1–50 μV, 1–70 Hz). The filter center frequencies were between 1–64 Hz and consumed 60 pW of power.

Source Degeneration

Similarly to the case of strong inversion, in weak inversion source degeneration is a popular linearization technique.

Diode-Connected Source Degeneration Transistors

A simple method to avoid the use of resistors is to use diode-connected transistors as in Fig. 5.25d. This circuit was proposed in [162] and used to implement a cascade of second-order filter section modeling auditory processing, i.e. a silicon cochlea. In this circuit the following hold true:

$$I_{D1} = I_{D0-1} e^{\left(\frac{V_1}{n_n U_T}\right)} e^{\left(\frac{-V_{S1}}{U_T}\right)} = I_{D0-1} e^{\left(\frac{V_1}{n_n U_T}\right)} e^{\left(\frac{-V_{G2}}{U_T}\right)}. \tag{5.87}$$

But I_{D1} is also equal to

$$I_{D1} = I_{D0-2} e^{\left(\frac{V_{G2}}{n_n U_T}\right)} e^{\left(\frac{-V_{S2}}{U_T}\right)}. \tag{5.88}$$

Equating the right-hand sides of (5.87) and (5.88) and solving for $e^{\left(\frac{-V_{G2}}{U_T}\right)}$ leads to the following equations for the current flowing across M_1 and M_3

$$I_{D1} = I_{D0-eff} e^{\left(\frac{V_1}{n_n(n_n+1)U_T}\right)} e^{\left(\frac{-n_n V_{S2}}{(n_n+1)U_T}\right)}. \tag{5.89}$$

where

$$I_{D0-eff} = \left(I_{D0-1}{}^{\frac{1}{n_n}} I_{D0-2}\right)^{\frac{n_n}{(n_n+1)}}. \tag{5.90}$$

Following the same process as before with I_{D1} and I_{D2} allows the calculation of the differential output current as [162]

$$I_0 = I_B \tanh\left(\frac{V_{id}}{2n_n(n_n+1)U_T}\right), \tag{5.91}$$

while the transconductance is [155]

$$G_m = \frac{I_B}{2n_n(n_n+1)U_T}\left[\cosh\left(\frac{V_{id}}{2n_n(n_n+1)U_T}\right)\right]^{-2}. \tag{5.92}$$

For $V_{id} = 0$, the nominal value of the transconductance is

$$G_m = \frac{1}{(n_n+1)}\frac{I_B}{2n_n U_T}. \tag{5.93}$$

In this way, the transconductance is reduced by a factor of $1/(n_n + 1)$. The linear range achieved with this method is approximately 35.8 $mV_{p\text{-}p}$ [155]. The authors in [162] propose the addition of additional diode connected transistors to increase the degeneration. Thus, for two diodes at each side the output current becomes

$$I_0 = I_B \tanh\left(\frac{V_{id}}{2n_n\left(n_n^2+n_n+1\right)U_T}\right), \tag{5.94}$$

while the transconductance is

$$G_m = \frac{I_B}{2n_n\left(n_n^2+n_n+1\right)U_T}\left[\cosh\left(\frac{V_{id}}{2n_n\left(n_n^2+n_n+1\right)U_T}\right)\right]^{-2}. \tag{5.95}$$

For $V_{id} = 0$, the nominal value of the transconductance is

$$G_m = \frac{1}{\left(n_n^2 + n_n + 1\right)} \frac{I_B}{2n_n U_T}. \tag{5.96}$$

As discussed with strong inversion designs, the disadvantage of using degeneration at that particular position is the reduction in the allowable input-common-mode range required to keep the tail current source in saturation, requiring an increased power supply for proper operation, while the thermal noise injected is also increased [162].

Single Diffusor

The degeneration method of Fig. 5.25f can also be used. Transistors operating in the weak-inversion triode (ohmic) region are usually termed "diffusors". The current of the transistor depends on the V_{DS} and is described by (5.73). It can be shown [153] that this equation can be rewritten as

$$I_D = \pm 2 I_{D0-n,p} e^{\left(\frac{V_G}{n_{n,p} U_T}\right)} e^{\left(\frac{-(V_D + V_S)}{2U_T}\right)} \sinh\left(\frac{V_{DS}}{2U_T}\right). \tag{5.97}$$

For a small V_{DS}, high-order terms (third-order terms and higher) can be neglected, leading to a Taylor series expansion equal to

$$I_D \approx \pm I_{D0-n,p} e^{\left(\frac{V_G}{n_{n,p} U_T}\right)} e^{\left(\frac{-(V_D + V_S)}{2U_T}\right)} \frac{V_{DS}}{U_T}. \tag{5.98}$$

From this the equivalent resistance can be obtained:

$$R_{DS} = \frac{U_T}{I_{D0-n,p}} e^{\left(\frac{(V_D + V_S)}{2U_T}\right)} e^{\left(\frac{-V_G}{n_{n,p} U_T}\right)}. \tag{5.99}$$

Thus the device can act as a linear resistor for a set gate voltage and common-mode voltage $(V_D + V_S)/2$. Transistors operated in such a regime can be used as pseudo-conductances to implement resistive networks if only currents are considered as input and output signals instead of voltages and all transistors are of the same type (nMOS or pMOS) [153, 154, 163]. In this current-mode approach, the transistor is linear for a wide range of currents in contrast to the voltage-mode approach, where the device is linear for a limited range of voltages. Defining the pseudo-voltages ($V^*_{D,S}$) and pseudo-conductance (G^*) as [153, 154]

$$V^*_{D,S} = \pm V_0 e^{\left(\frac{-V_{D,S}}{U_T}\right)}. \tag{5.100}$$

$$G^* = \frac{I_{D0-n,p}}{V_0} e^{\left(\frac{V_G}{n_{n,p} U_T}\right)} = \frac{1}{R^*}. \tag{5.101}$$

Then (5.72) becomes

$$I_{DS} = G^{*}\left(V_D^{*} - V_S^{*}\right). \tag{5.102}$$

Interconnecting transistors via their source and drain creates a network that is linear for current, but not for voltages. At each node the current splits linearly. This is applicable also in moderate and strong inversion; however, in these cases, the gate voltages of all transistors must be the same. In [164] the authors controlled the division via the *W*/*L* of the transistors. In weak inversion, however, each resistance in the network is programmable also via the gate voltage using (5.101). No voltages should be applied or measured in such current-mode circuits; thus V_0 in (5.100) and (5.101) is irrelevant [154]. Using the above to analyze the circuit of Fig. 5.25f and (5.72) for the current across M_3 and (5.73) for the currents across the matched input transistors, M_1 and M_2, of the differential pair, the current I_3 can we calculated by

$$I_3 = I_{D0,n} e^{\left(\frac{V_{G3}}{n_n U_T}\right)} \left(\frac{I_{D2}}{I_{D0,n} e^{\left(\frac{V_{G2}}{n_n U_T}\right)}} - \frac{I_{D1}}{I_{D0,n} e^{\left(\frac{V_{G1}}{n_n U_T}\right)}} \right). \tag{5.103}$$

For $V_{id} = 0$ and $V_{CM} = (V_1 + V_2)/2$, the nominal value of the current is

$$I_3 = I_{D0,n} e^{\left(\frac{V_{G3} - V_{CM}}{n_n U_T}\right)} \left(I_{D2} - I_{D1}\right). \tag{5.104}$$

Hence, the current flowing through the diffusor is proportional to $(I_{D2} - I_{D1})$ and the proportionality factor can be controlled via V_{CM} and V_{G3} [153]. Similarly to the strong inversion operation of the circuit, the degeneration depends on the input common-mode level. It can be shown [155] that the differential output current is equal to

$$I_0 = I_B \tanh\left(\frac{V_{id}}{2n_n U_T} - \tanh^{-1}\left[\sinh\left(\frac{V_{id}}{2n_n U_T}\right)/2m + \cosh\left(\frac{V_{id}}{2n_n U_T}\right)\right]\right). \tag{5.105}$$

The transconductance is then [155]

$$G_m = \frac{I_B}{\cosh^2\left(\frac{V_{id}}{2n_n U_T} - \tanh^{-1}\left[\frac{\sinh\left(\frac{V_{id}}{2n_n U_T}\right)}{2m + \cosh\left(\frac{V_{id}}{2n_n U_T}\right)}\right]\right)} \frac{\left(4m^2 + 2m\cosh\left(\frac{V_{id}}{2n_n U_T}\right)\right)}{\left(4m^2 + 4m\cosh\left(\frac{V_{id}}{2n_n U_T}\right) + 1\right)} \frac{1}{2n_n U_T}. \tag{5.106}$$

where

$$m = \left(\frac{W}{L}\right)_3 \Big/ \left(\frac{W}{L}\right)_{1,2}. \tag{5.107}$$

Setting the second derivative of G_m equal to zero returns a single positive root $m = 0.25$. The nominal transconductance for $V_{id} = 0$ and $m = 0.25$ is then

$$G_m = \frac{I_B}{2n_n U_T}\frac{2m}{1+2m} = \frac{1}{3}\frac{I_B}{2n_n U_T}. \tag{5.108}$$

The linear range with this circuit for 1% linearity is 116.8 $mV_{p\text{-}p}$. This is approximately eight times greater than a simple differential pair in weak inversion.

Double Diffusor

From the previous discussion and from (5.104) it is obvious that for the degeneration to be independent of the input signal, then V_{G3} must be equal to V_{CM}. This is achieved with the circuit of Fig. 5.25g. The differential output current in this case is [157]

$$I_0 = I_B \tanh\left(\frac{V_{id}}{2n_n U_T} - \tanh^{-1}\left[\frac{1}{4m+1}\tanh\left(\frac{V_{id}}{2n_n U_T}\right)\right]\right). \tag{5.109}$$

The transconductance is consequently given by [155]

$$G_m = \frac{I_B}{\cosh^2\left(\frac{V_{id}}{2n_n U_T} - \tanh^{-1}\left[\frac{1}{4m+1}\tanh\left(\frac{V_{id}}{2n_n U_T}\right)\right]\right)} \frac{\left((16m^2+8m)\cosh^2\left(\frac{V_{id}}{2n_n U_T}\right) - 4m\right)}{\left((16m^2+8m)\cosh^2\left(\frac{V_{id}}{2n_n U_T}\right) + 1\right)} \frac{1}{2n_n U_T}. \tag{5.110}$$

By once again setting the second derivative equal to zero, the only positive root occurs at $m = 0.5$, while according to simulations [154] suggests a value of 0.4. For $V_{id} = 0$ and $m = 0.5$, G_m becomes

$$G_m = \frac{I_B}{2n_n U_T}\frac{4m}{4m+1} = \frac{2}{3}\frac{I_B}{2n_n U_T}. \tag{5.111}$$

The achievable linear range with this topology in weak inversion is 58.4 $mV_{p\text{-}p}$, which is half that of the topology with the single diffusor.

The Multi-tanh Technique

An alternative and popular method borrowed from bipolar transistor technology [165–167] is the use of a plurality of differential pairs in weak inversion [154, 155, 157, 168, 169], a technique known as the multi-tanh method. This method is similar

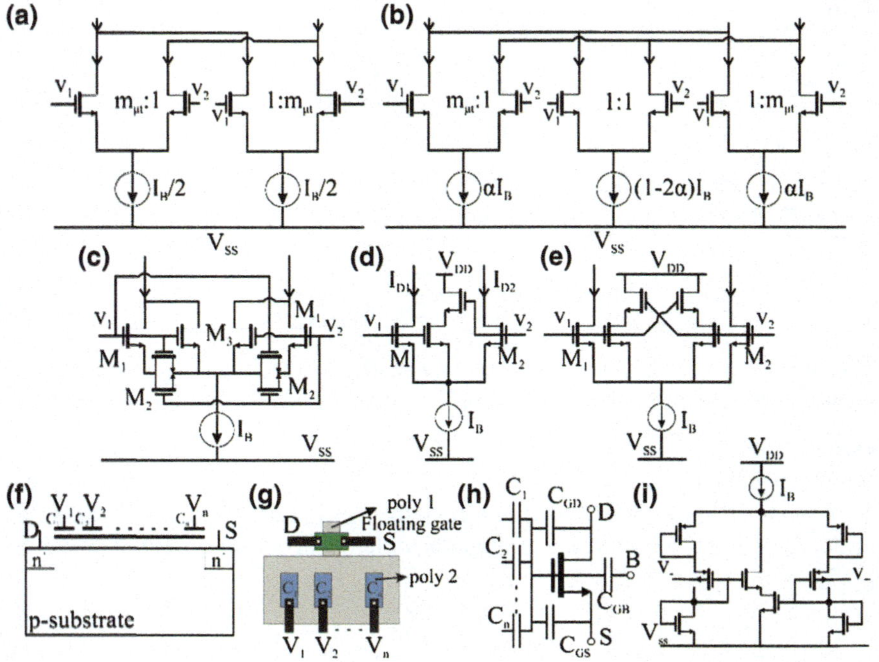

Fig. 5.27 **a** The multi-tanh doublet and **b** triplet transconductors. **c** The doublet with double diffusors. **d** Bump linearization. **e** Balanced bump circuit. **f** Schematic representation of a FGMOS, **g** its layout implementation, and **h** its equivalent circuit (adapted from [184]). **i** The weak-inversion low-power linearized transconductor of [182] employing source and gate degeneration, bump linearization, and bulk-driven input transistors

to the cross-coupling techniques in strong inversion discussed earlier. When two such differential pairs are used, biased with the same tail current ($I_B/2$), the circuit is referred to as a doublet. The topology uses the sum of currents of the differential pairs with controlled input offset voltages. These are defined by making one of the transistors in each pair $m_{\mu t}$ times larger than the other, as shown in Fig. 5.27a. This leads to an input offset voltage of

$$\Delta V_{\mathrm{id}} = n_{\mathrm{n,p}} U_{\mathrm{T}} \ln m_{\mu t}. \tag{5.112}$$

An alternative way of achieving this is by using floating gate transistors. Define

$$m_{\mathrm{\mu t}} = \frac{\left(\frac{W}{L}\right)_1}{\left(\frac{W}{L}\right)_2} = \frac{\left(\frac{W}{L}\right)_4}{\left(\frac{W}{L}\right)_3} = \mathrm{e}^{\ln m}. \tag{5.113}$$

Then the total differential output current, which is equal to the addition of the two differential output currents of each pair, is equal to [155, 157]

$$I_0 = \frac{I_B}{2}\left[\tanh\left(\frac{V_{id}}{2n_n U_T} + \frac{\ln m}{2}\right) + \tanh\left(\frac{V_{id}}{2n_n U_T} - \frac{\ln m}{2}\right)\right]. \quad (5.114)$$

The transconductance is then

$$G_m = \frac{I_B}{2n_n U_T}\left(\frac{1}{2\cosh^2\left(\frac{V_{id}}{2n_n U_T} + \frac{\ln m}{2}\right)} + \frac{1}{2\cosh^2\left(\frac{V_{id}}{2n_n U_T} - \frac{\ln m}{2}\right)}\right). \quad (5.115)$$

Equating the second derivative to zero leads to a positive root greater than unity equal to $m_{\mu t} = 2 + \sqrt{3} \approx 3.732$. A second root occurs at $m_{\mu t} = \frac{1}{(2+\sqrt{3})}$. For $V_{id} = 0$ and $m_{\mu t} = 3.732$

$$G_m = \frac{I_B}{2n_n U_T}\left(\frac{4}{m + 2 + \frac{1}{m}}\right) = \frac{2}{3}\frac{I_B}{2n_n U_T}. \quad (5.116)$$

The doublet has the same voltage-to-current relationship as the topology with the double diffusor and achieves the same linear range of 58.4 mV_{pp} [155]. A disadvantage of the topology when compared to the double diffusor topology is that the ideal scaling factor controlling the relative device geometries is not as easily realizable. As a result, a value of $m_{\mu t} = 4$ is typically used [154, 155]. The doublet linearization technique was recently used for the implementation of a 70 nW sixth-order BPF for measuring the acoustic signal from a wearable breathing detector [169]. The authors used $m_{\mu t} = 4$, as with this value the THD was shown to be a monotonically increasing function of V_{id}. For a linear input range of 1%, $m_{\mu t} = 5$ can be used satisfying the criterion even when mismatches are considered. This value was used in [168] with a tail current of 200 pA and a 2 V supply.

In [166], the combination of the doublet technique with various forms of resistive source degeneration in bipolar technology were proposed, demonstrating significantly improved performance. These, however, to the author's knowledge, have not been demonstrated in CMOS technology. A doublet with each pair incorporating double diffusors was proposed in [170]. This circuit was optimized numerically and an optimized linear range of 98.2 mV_{pp} was achieved for

$$m_1 = \left(\frac{W}{L}\right)_2 \Big/ \left(\frac{W}{L}\right)_1 = 0.2449. \quad (5.117)$$

and

$$m_2 = \left(\frac{W}{L}\right)_3 \Big/ \left(\frac{W}{L}\right)_1 = 0.1567. \tag{5.118}$$

The M_1, M_2, and M_3 transistors are shown in Fig. 5.27c. Such accuracy is challenging to achieve in reality. As will be discussed, this circuit achieves the same input linear range as the triplet.

A triplet implementation can also be implemented as in Fig. 5.27b [155, 165–167]. In the triplet, a normal differential pair with matched input transistors in added to the doublet. The doublet pairs, however, now have a tail current αI_B while the regular pair have a $(1–2\alpha)I_B$ tail current. The total differential output current is

$$\begin{aligned} I_0 = \alpha I_B &\left[\tanh\left(\frac{V_{id}}{2n_n U_T} + \frac{\ln m}{2}\right) + \tanh\left(\frac{V_{id}}{2n_n U_T} - \frac{\ln m}{2}\right)\right] \\ &+ (1 - 2\alpha)\tanh\left(\frac{V_{id}}{2n_n U_T}\right). \end{aligned} \tag{5.119}$$

The transconductance is given by

$$\begin{aligned} G_m = \alpha \frac{I_B}{2n_n U_T} &\left(\frac{1}{2\cosh^2\left(\frac{V_{id}}{2n_n U_T} + \frac{\ln m}{2}\right)} + \frac{1}{2\cosh^2\left(\frac{V_{id}}{2n_n U_T} - \frac{\ln m}{2}\right)}\right) \\ &+ (1 - 2\alpha)\frac{I_B}{2n_n U_T}\frac{1}{2\cosh^2\left(\frac{V_{id}}{2n_n U_T}\right)}. \end{aligned} \tag{5.120}$$

According to [155], setting both the second and fourth derivatives equal to zero leads to $m_{\mu t} = 4 + \sqrt{15} \approx 7.873$ and $\alpha = 25/66$. For these values and for $V_{id} = 0$ then

$$G_m = \frac{I_B}{2n_n U_T}\left(\frac{8\alpha}{m_{\mu t} + 2 + \frac{1}{m_{\mu t}}} + 1 - 2\alpha\right) = \frac{6}{11}\frac{I_B}{2n_n U_T}. \tag{5.121}$$

With these parameters this topology achieves an input linear range of 98.2 mV_{pp}. These values for $m_{\mu t}$ and α are, however, difficult to implement. In [154, 166] an $m_{\mu t} = 13$ and a tail current of $I_B/2$ for the doublet pairs was proposed, with the third normal pair with equal input transistors biased with a tail current of $3I_B/8$. Higher-order cells offer a small improvement for a considerable increase in complexity [166]. A recent example is a plurality of asymmetric cross-coupled pairs combined with source degeneration using diode-connected transistors, presented in [171]. The authors report an input linear range of ~400 mV_{pp} for a tail current of 24 nA.

Bump Linearization

Bump linearization is one more popular linearization technique for differential pairs in weak inversion [153, 170, 172–182]. The original circuit consists of a differential pair and an additional third branch composed of two cascaded transistors, each with their gates connected to one of the differential inputs as in Fig. 5.27d. The current through the middle branch is a bump-shaped function of V_{id} and is thus known as the bump output. The addition of I_1 and I_2 forms the anti-bump output [172, 183]. These two currents are complementary [172]. When $V_{id} = 0$ current flows through all three branches. As $|V_{id}|$ increases, the potential of the common-source node of the differential pair will follow the potential of the highest of the input nodes. This will eventually force the current through the middle branch to zero, as either of the two transistors will eventually switch off. Apart from a region near $V_{id} = 0$, where there is a current flowing through the bump transistors, the differential output current follows the typical tanh response. The three currents sum to I_B ($I_B = I_1 + I_2 + I_{mid}$), with [172, 183]

$$I_{mid} = \frac{I_B}{1 + \frac{4}{m}\cosh^2\left(\frac{V_{id}}{2n_n U_T}\right)}, \tag{5.122}$$

where

$$m = \frac{\left(\frac{W}{L}\right)_{mid}}{\left(\frac{W}{L}\right)_{1,2}}. \tag{5.123}$$

The width of the bump of I_{mid} depends on m, as m controls the amount of current flowing through the middle branch. Thus, the larger m is, the larger the current through the bump transistors. It can be shown that the differential output current is defined by [183]

$$I_0 = \frac{I_B \tanh\left(\frac{V_{id}}{2n_n U_T}\right)}{1 + \frac{m}{4}\operatorname{sech}^2\left(\frac{V_{id}}{2n_n U_T}\right)} \tag{5.124}$$

For $m = 0$ the above equation reduces to that of the simple differential pair, i.e. (5.81). Consequently the transconductance of this circuit is

$$G_m = \frac{I_B}{2n_n U_T}\left(\frac{\operatorname{sech}^2\left(\frac{V_{id}}{2n_n U_T}\right)\left(1 + \frac{m}{4}\operatorname{sech}^2\left(\frac{V_{id}}{2n_n U_T}\right)\right) + \left(\frac{m}{2}\right)\tanh^2\left(\frac{V_{id}}{2n_n U_T}\right)\operatorname{sech}^2\left(\frac{V_{id}}{2n_n U_T}\right)}{\left(1 + \frac{m}{4}\operatorname{sech}^2\left(\frac{V_{id}}{2n_n U_T}\right)\right)^2}\right). \tag{5.125}$$

As before, by differentiating and setting the second derivative of G_m to zero the optimum value of m can be found. This is for a value of $m = 2$ according to [182, 183]. The improved linearity is due to the adaptive current available to flow through the input pair transistors, since part of the tail current flows through the middle branch as a function of the input signal. The Taylor series expansion of the differential output current for $m = 2$ is [182, 183]

$$\frac{I_0}{I_B} = \frac{2}{3}\frac{V_{id}}{2n_n U_T} - \frac{8}{135}\frac{V_{id}^5}{2n_n U_T} - \cdots \tag{5.126}$$

Comparing this to (5.84) it is evident that the transconductance has been reduced by 2/3 and that the third-order nonlinearity has completely disappeared. This topology was expanded to a symmetrical structure by adding an additional branch with two more transistors with interchanged gate connections, as in Fig. 5.27e [170, 173, 179, 183]. According to [170] a maximally flat transconductance is achieved with $m = 1$ achieving a linear range of 58.4 mV_{pp}, which is identical to that of the doublet examined earlier.

Transconductance Reduction Techniques Applicable to All Inversion Regions

To scale the input signal to bring it within the linear range of the transconductor, the following approaches can be used.

Floating Gate Input Pair

The transconductance can also be reduced by using floating-gate techniques [152]. A multiple input floating gate device (FGMOS) can be built by using a regular MOSFET and by connecting its gate to n input capacitors in double-poly silicon technologies, as shown in Fig. 5.27f–h [184]. In this way the gate is capacitively coupled to its input(s). The voltage at the floating gate, V_{FG}, is thus given by [184]

$$V_{FG} = \frac{1}{C_T}\left(C_{GS}V_S + C_{GD}V_D + C_{GB}V_B + \sum_{i=1}^{n} C_i V_i\right) \tag{5.127}$$

where the total capacitance, C_T, is given by

$$C_T = \left(C_{GS} + C_{GD} + C_{GB} + \sum_{i=1}^{n} C_i\right), \tag{5.128}$$

where C_i and V_i are the input capacitances and voltages, and C_{GS}, C_{GD}, and C_{GB} are the parasitic capacitances from the gate to the other three terminals, as shown in Fig. 5.27h, with voltages V_S, V_D, and V_B, respectively. It is clear from the above that the voltage at the floating gate is a function of the parasitic capacitances. In weak inversion the greatest contribution is from the gate–bulk capacitance, C_{GB}, and since typically $V_B = 0$ in single-supply low-power circuits, its effect can be minimized [184]. Because of the voltage division at the input capacitors of a floating gate, attenuation of the input signal takes place, in this way keeping the signal within the linear range of the differential pair, thus effectively reducing the transconductance of the circuit. This is, however, achieved at the cost of increased input-referred noise [168]. If there are two inputs to the floating gate and thus two capacitors, C_1 and C_2, and current division is also incorporated provided that the parasitic capacitances between the floating gate and the other terminal of the input devices are comparatively negligible (C_1 and C_2 should be 5–10 times larger than the total parasitic seen at the floating gate, C_P), then the transcoductance of the circuit is given by [152, 185]

$$G_m = \frac{C_1}{C_1 + C_2 + C_P} g_{m-FG}, \tag{5.129}$$

Another advantage of FGMOS is that by using multiple input floating gates, these can be used to compensate for offsets. A disadvantage contributing to offsets is the trapped charge in the floating gate arising during the fabrication process. This is the same as in ISFETs examined in Chap. 2, which can be tuned with similar techniques (hot-electron injection or electron tunneling), as discussed in [186]. The techniques described can be used in this case as well. An alternative is to implement the simple layout technique proposed in [187, 188]. In [185], the authors combined resistive source degeneration, cross-coupling, floating-gate inputs, and current division via triode transistors in the moderate inversion region. In [184], the authors combined FGMOS input transistors with what is essentially a FGMOS version of the bump linearization technique achieving complete elimination of the cubic nonlinearity term. While the differential pair input transistors have two floating gates, one connected to one input signal and the second to a bias voltage, the third FGMOS has three. One is connected to a bias voltage and the other two to the two input signals. These are thus used to compute the average of V_{id}. In [185] source degeneration is combined with FGMOS, achieving an increased linear range at high frequencies (an HD_3 with a 1 V_{pp} at 15 MHz of −86 dB vs. −76 dB when no FGMOS are used). A detailed discussion of floating-gate techniques with a focus on filtering can be found in [189].

Bulk-Driven Input Pair

MOSFETs are symmetric four-terminal devices. Typically, the bulk is biased at a constant voltage. The device effectively has two voltage control nodes, the gate and the substrate, which can be used to modulate the channel current [155]. Thus, instead of applying the input signal to the gate, the bulk input of a transistor can be used as the input terminal of the transconductor. However, the parasitic lateral and vertical BJTs have to remain off [190]. This can be ensured by keeping the current though the bulk small, with appropriate layout [191] and gate biasing [192, 193], in this way keeping the input impedance high. In [182], the authors discuss that the gate bias voltage should be kept low to achieve a maximum DC-input operating range. A bulk-driven transistor needs to be isolated in a separate well. Most CMOS technologies are n-well technologies, where only the pMOS transistors sit in individual n-wells, and thus in such technologies only pMOS transistors can be bulk-driven. The gate-to-source voltage is set to a potential sufficient to form an inversion layer and the bulk potential controls the drain current, as it modulates the channel [190, 191]. The dependence of V_T on the body (i.e. the bulk) potential is known as the body effect and the operation of bulk-driven transistors is based on it. According to (5.41) and (5.73), the drain current is a function of V_T, and the latter is given by [193]

$$|V_{\mathrm{T}}| = |V_{\mathrm{T,0}}| + |\gamma|\left[\sqrt{2|\Phi_{\mathrm{F}}| + V_{\mathrm{BS}}} - \sqrt{2|\Phi_{\mathrm{F}}|}\right], \tag{5.130}$$

where $V_{T,0}$ is the threshold voltage when $V_{BS} = 0$, γ is the body effect coefficient, and Φ_F is the Fermi potential. Thus a transconductance function can be defined due to the dependence of I_D on V_{BS} according to [193]

$$g_{\mathrm{mb}} = \frac{\partial I_{\mathrm{D}}}{\partial V_{\mathrm{SB}}} = \frac{|\gamma| g_{\mathrm{m}}}{2\sqrt{2|\Phi_{\mathrm{F}}| + V_{\mathrm{BS}}}}. \tag{5.131}$$

The transconductance of a bulk-driven transistor (g_{mb}) is typically 0.2–0.4 times the gate transconductance, g_m, leading to a reduction in the DC gain and the gain–bandwidth product of the topology, and it is highly process-dependent [152]. As discussed in [191], the transconductance reduction, together with the increase of the input capacitance, reduces the transition frequency of the transistor (f_T) and thus the maximum possible frequency of operation, while the input-referred noise and offset voltage are also increased. The input common-mode voltage range of such a topology is extended close to rail-to-rail, since part of the input signal does not need to ensure that the device is on [191]. In other words, the dynamic range of the amplifier is increased, as there is no threshold voltage for the bulk terminal. This technique is thus suitable for ultra-low-voltage operation. The effective sub-threshold slope factor for a bulk-driven transistor is equal to $(1 - \kappa)$ [174, 182].

In [152] source degeneration, floating-gate, and bulk-driven techniques, all combined with current division, were compared to each other. The OTA with bulk-driven transistors achieved the widest linear input range (900 mV_{pp}), occupied a small area (0.22 mm^2), but consumed the highest power (4.05 μW) and had the highest input-referred noise (70.3 μV_{rms}). The floating gate OTA occupied the largest area due to the input capacitors (4.65 mm^2) while in terms of noise, power, and linearity its performance was comparable to that of the OTA with source degeneration (1.62 μW vs. 1.35 μW, 26.03 μV_{rms} vs. 17.29 μV_{rms}, and 330 mV_{pp} vs. 242 mV_{pp}). In [182], a bulk-driven input pair was combined with source-degeneration, bump linearization, and gate degeneration. This topology is shown in Fig. 5.27i. Gate degeneration, which was introduced in that paper, is a feedback technique where the current through a transistor is converted into a voltage through a diode and fed back to its gate, in this way decreasing its current [182]. The authors reported a linear range of ± 1.7 V. A bulk-driven input pair, with source and gate degeneration and symmetric bump linearization, as in [170], was presented in [178] achieving a linear region of 1.32 V_{pp}. The output currents were further divided using a transistor R-2R ladder. A bulk-input pair with bump linearization and gate degeneration was used in [175], achieving a linear range of about 1 V.

It should be noted that while the reduced transconductance of the bulk is attractive for low-frequency filters, bulk-driven transistors have been extensively used to realize amplifiers due to the ultra-low-voltage operation possible. As the power-supply is reduced, in order to keep the SNR high, the voltage swings at the various nodes also need to be as high as possible. Thus, rail-to-rail operation becomes increasingly necessary. However, in such cases, the low transconductance is a disadvantage which needs to be compensated for and various techniques have been proposed (e.g. employing partial positive feedback) to achieve this [190, 191].

5.2.4.4 Current Scaling, Transconductance Programming, and Differential Outputs

Current division and scaling are techniques mentioned in previous sections a number of times. In current division, the input pair transistor is split into parallel transistors, to create parallel current paths splitting the tail current, as in Fig. 5.26d, e [152]. The effective transconductance becomes

$$G_{\mathrm{m}} = \frac{g_{\mathrm{m-M}}}{M+1}, \tag{5.132}$$

where $g_{\text{m-M}}$ is the transconductance of the transistor before splitting and M is the ratio of the transistors' dimensions.

An alternative to reducing the transconductance without affecting the linearity of the transconductor is to scale the current in the output branches of e.g. a balanced OTA, using the current mirrors. Instead of using differently sized transistors, series and parallel combinations of transistors of the same unit dimensions can be used to achieve large current scaling. Thus, current scaling is not governed by a ratio of dimensions, but by the number of devices, and it is independent of the current and the transistor inversion region. If a ratio of dimensions is used for e.g. a large division, a short and a long transistor would be necessary, which cannot be matched. The long transistor can be implemented by a series combination of transistors of the same dimensions with that of the short transistor, with their gates connected together. The transistors can then be matched using standard layout techniques, such as common-centroid layout [194]. In this way, better matching and consequently a reduction in random offset is achieved. In addition, such composite transistors demonstrate a much reduced drain–source transconductance ($g_{DS} = 1/r_{DS}$) [195, 196]. Furthermore, series–parallel current division leads to an area efficient current divider. The area efficiency comes because the area is proportional to the square root of the copy factor, M [195]. This was presented in [194, 197], and was applied to achieve large time constants, while in [196] such composite transistors were discussed in terms of their merits with regard to a reduced chip area and a higher cutoff frequency. This was more recently used in [198, 199] to achieve a transconductance of 35 pS for $N = M = 70$. In this way, if P series transistors and S parallel transistors are used to implement a current mirror, then the effective transconductance becomes

$$G_{\mathrm{m}} = \frac{g_{\mathrm{m}}}{S \cdot P} = \frac{g_{\mathrm{m}}}{M}, \tag{5.133}$$

where g_{m} is the transconductance of the input pair transistor. In the more general scenario, M is equal to

$$\frac{I_{\mathrm{in}}}{I_{\mathrm{out}}} = \frac{S \cdot P}{R \cdot Q} = M, \tag{5.134}$$

where P and R are the series and Q and S the parallel unitary transistors, as in Fig. 5.28a. According to [200], for a minimum mismatch offset, P·Q = R·S and the gate area should be large. There is limit on the minimum transconductance achievable with this technique. This is dictated by the sum of the leakage currents (I_{leak}) at the source and drain of each transistor connected in series at the output branch [194, 199]. According to [199], p-doped diffusion leakage is much larger

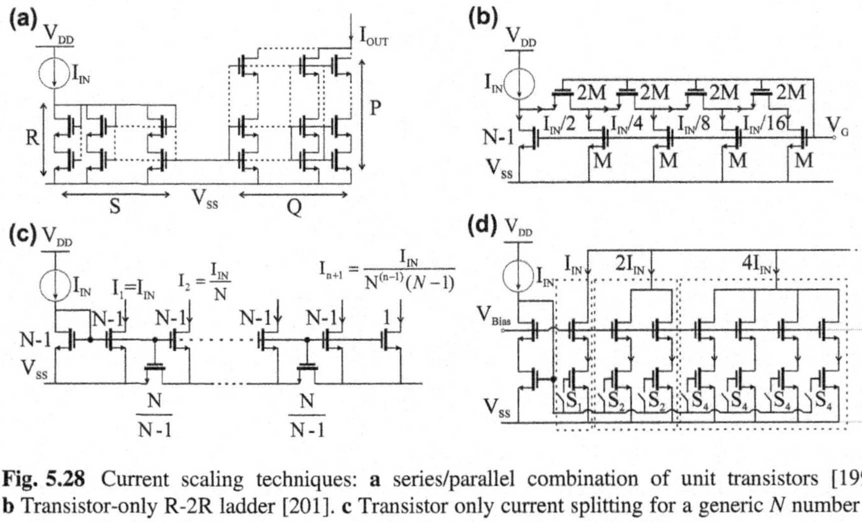

Fig. 5.28 Current scaling techniques: **a** series/parallel combination of unit transistors [199]. **b** Transistor-only R-2R ladder [201]. **c** Transistor only current splitting for a generic N number of stages [112]. **d** The high-swing programmable current mirror used in [202]

than in n-doped diffusions. For a 4 μm × 2 μm n^+ diffusion this can be in the region of 3 fA. The authors proposed that the bias current at the branch should be ten times greater than the sum of all the leakage currents, e.g. using the previous example and if the current being scaled is I_{D1}, then

$$\sum I_{\text{leak}} = \mathrm{P} \cdot I_{\text{leak}} < \frac{I_{D1}}{10S \cdot P}. \tag{5.135}$$

Another approach is to use the current division method presented in [112, 164, 201]. According to [46], a current flowing into a circuit formed by two transistors biased by the same voltage is linearly divided. This division is independent of the current magnitude, the source and drain voltages of the transistors, and whether the devices are in saturation or conduction, irrespective of the inversion region. Current division depends on the transistor dimensions, and is independent of the current. The current produced in the output branches is divided progressively at each branch by the same factor N. Transistors are used to implement R-2R equivalent sections for weighted inherently linear current division. In the circuit of Fig. 5.28b, the even-numbered transistors need to be twice as the others in order to equally split the current. In the alternative realization of Fig. 5.28c, not all transistors are biased in conduction and transistors with even numbers are in saturation [201]. Using switches to control the shorting of the branches can produce any combination of these. The accuracy of the circuit is limited by mismatch of device geometry and oxide thickness, which will affect the division, mismatch of threshold voltages, which affects the linearity of current division in strong inversion, and its accuracy in weak inversion [201]. Drain-induced barrier lowering, velocity saturation and channel length modulation affect only the transistors in saturation. A unit-sized transistor can be used and series–parallel combinations of this can be used to realize the division required.

Using switches to form a weighted summation of currents is an approach used in [202]. This was used to implement a widely tunable, digitally controlled transconductance. The topology was essentially a current mirror with a programmable gain. The output current was the weighted sum of currents flowing through each block of branches. Five groups of one, two, four, eight and 16 parallel branches of equally sized transistors were used as shown in Fig. 5.28d.

A final approach is to use tunable continuously programmable current scaling (as opposed to discrete digital programming) using current mirrors, the gain of which is controlled via currents or voltages. One such example is [203], where a conventional wide-swing current mirror was modified as in Fig. 5.29a. Essentially, the two top transistors are split into two matched transistors each, forming two differential pairs. Two complementary control voltages, V_{G+} and V_{G-}, are required, ensuring that the common source terminals of each differential pair sit at the same potential. This ensures that the potentials across the two bottom transistors, M1 and M3, are equal ($V_{DS1} = V_{DS3}$), and since also $V_{GS1} = V_{GS3}$, $I_{out} = I_{in}$. I_{out} is split into two currents flowing through the two branches of the second differential pair. The splitting ratio depends on the differential control voltage $V_{GC} = V_{G+} - V_{G-}$.

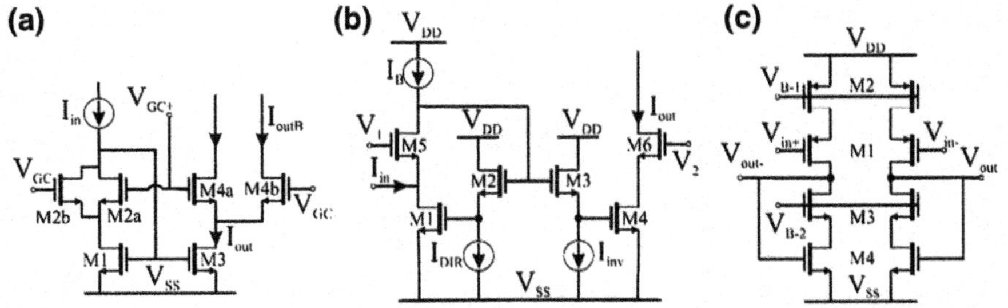

Fig. 5.29 **a** The tunable current mirror of [203]. **b** The translinear loop-based tunable current mirror of [205]. **c** The simple CMFB of [128]

When this is zero, the current splits equally. When it is greater than $\sqrt{2}\ V_{\text{DS-Sat4}}$ all of the current flows through M4a and $I_{\text{outB}} = 0$. When it is less than—$\sqrt{2}\ V_{\text{DS-Sat4}}$ all of the current flows through M4b and $I_{\text{outA}} = 0$. Thus essentially the circuit has a programmable current gain A_I between $0 < A_I < 1$.

Another programmable current scaling technique was presented in [204, 205] and is shown in Fig. 5.29b. This is a balanced arrangement of two translinear loops. All transistors are matched. M2 and M3 are always in saturation as $V_{\text{D2,3}} = V_{\text{DD}}$ and saturation in M1 and M4 can be enforced by using cascode transistors (M5 and M6) to set their drain voltages. To demonstrate the operation of the circuit, the authors use the model of [206], which is a single expression valid in all regions of inversion describing the transistor in saturation. According to this

$$I_{\text{D}} = \text{I}_{\text{spec}}\ln^2\left(1 + \text{e}^{((V_{\text{GS}} - V_{\text{T}})/(2nU_{\text{T}}))}\right). \tag{5.136}$$

Using this it can be shown the input and output current are related by [204, 205]

$$I_{\text{out}} = \text{I}_{\text{spec}}\ln^2\left[1 + \left(\exp\left(\frac{I_{\text{in}}}{I_{\text{spec}}}\right) - 1\right)\xi\right], \tag{5.137}$$

where

$$\xi = \left(\exp\left(\frac{I_{\text{DIR}}}{I_{\text{spec}}}\right) - 1\right) \Big/ \left(\exp\left(\frac{I_{\text{INV}}}{I_{\text{spec}}}\right) - 1\right). \tag{5.138}$$

If all transistors are in weak inversion saturation, then the above reduced to

$$I_{\text{out}} = \frac{I_{\text{DIR}}}{I_{\text{INV}}} I_{\text{in}}. \tag{5.139}$$

In this case the circuit is a linear distortion-free current mirror, the gain of which is controlled by I_{DIR} and I_{INV}. The performance of the topology is in this case limited due to the matching, bandwidth, and adjustment range imposed by weak inversion operation. In strong inversion

$$I_{\text{out}} = I_{\text{in}} + I_0 + 2(I_{\text{in}} I_0), \tag{5.140}$$

where $\sqrt{I_0} = \sqrt{I_{\text{DIR}}} - \sqrt{I_{\text{INV}}}$ and thus the signal will be significantly distorted for gains less than 1. Operation in moderate inversion is a tradeoff and the use of two of these circuits in balanced circuits reduces the distortion. In such an arrangement a THD less than –55 dB was reported for $0.8 < \xi < 1.2$ and $I_B < 10I_{\text{spec}}$. The operation of the circuit is discussed in more detail in [205] and its use in a tunable transconductor in [204]. Nevertheless, as discussed in [207], the stacking of transistors in this topology limits the available swing, in an n-well technology M3 and M4 will suffer from the body effect, as they cannot be placed in a separate well, the methodology consumes additional current and finally, typically voltages applied at high-impedance nodes are preferred instead of currents at low-impedance nodes. The authors thus proposed an alternative based on FGMOS operated in weak or moderate inversion to overcome these limitations.

Finally, it is important to note that for elimination of even-order nonlinearities, fully differential operation is recommended. In such topologies additional circuitry is required to ensure that the output common-mode level is stable and equal in the two output branches. This subject is extensive and there is a wide range of options available depending on the application and the circuit topology used. It should be noted that in complex filter structures, such as the biquad of Fig. 5.24, a number of OTAs may share a single common-mode stabilization circuit if their outputs are shorted as g_{m1} and g_{m2} in Fig. 5.24 and similarly to [77, 208] have a common output node. Here a simple solution will be examined for completeness and the interested reader is referred for more on this subject to [209, 210, 210–216]. The simple topology of Fig. 5.29c has two additional transistors operated in triode and according to [128], the open-loop gain of the CMFB loop is given by

$$A_{\text{CMFB}} = \frac{g_{m3} g_{m4}}{(g_{m3} + g_{d4})} \frac{g_{m3}}{(g_{d3} g_{d4})} \tag{5.141}$$

and if C_p is the capacitance seen at the output node of the transconductor, then the unity gain bandwidth of the CMFB is

$$f_{\text{u-CMFB}} = \frac{g_{m3} g_{m4}}{(g_{m3} + g_{d4})} 2\pi C_p \tag{5.142}$$

According to [128], for reasonable accuracy, A_{CMFB} should be at least 20 dB and $f_{\text{u-CMFB}}$ should be higher than the filter bandwidth. This circuit will ensure that the output DC potential is stabilized, but it requires the stacking of additional transistors, and in contrast to other topologies it is not referenced to a required potential.

5.2.4.5 Capacitive Scaling

Apart from adjusting the transconductance to achieve the cutoff frequency of interest, the capacitance can also be adjusted. For low cutoff frequencies, either the transconductance can be made very small, to emulate a large resistance, or the capacitance can be made large. However, large capacitors are area hungry and typically on-chip capacitors are kept in the pF range. An advantage of using a large capacitance to keep the transconductance large is the reduction of the noise level [149]. According to Ohm's law, the impedance of a device is inversely proportional to the current through it. If less current flows through it for the same voltage across it, then the impedance is increased [149, 217, 218]. This is applicable irrespective of the type of impedance, i.e. whether resistive or capacitive. In the topologies of Fig. 5.30, the input current through the capacitor is sampled, amplified and fed back into the input, scaling the equivalent capacitance [149]. The capacitance is scaled by the current amplification factor controlled in Fig. 5.30a by the transistor dimensions of MSN to MS1 and MSPN to MSP. According to [149] the small signal admittance of the circuit is equal to

$$\frac{i_{\text{in}}}{v_{\text{in}}} = g_{\text{oMSN}} + g_{\text{oMSP}} + sC_{\text{P2}} + s(N+1)C_{\text{i}}\left[\left(1+\frac{C_{\text{i}}}{(N+1)g_{\text{mMS1}}}\right)\Big/\left(1+s\frac{C_{\text{i}}+C_{\text{P1}}}{g_{\text{mMS1}}}\right)\right], \tag{5.143}$$

where g_{oMSN} and g_{oMSP} are the output conductances of the respective transistors, C_{i} is the input capacitance, C_{P1} and C_{P2} are parasitic capacitances. The effect of the two conductances is dominant at low frequencies and it limits the Q-factor of the equivalent capacitor and the accuracy of its value is limited by transistor mismatch and the C_{P2} parasitic. The parasitic pole at $g_{mMS1}/(C_{\text{i}} + C_{\text{P1}})$ should be placed at frequencies higher than the passband of the implemented filter, e.g. 10 times higher [149]. The zero is located at higher frequencies and can typically be ignored. The admittance can thus be approximated by

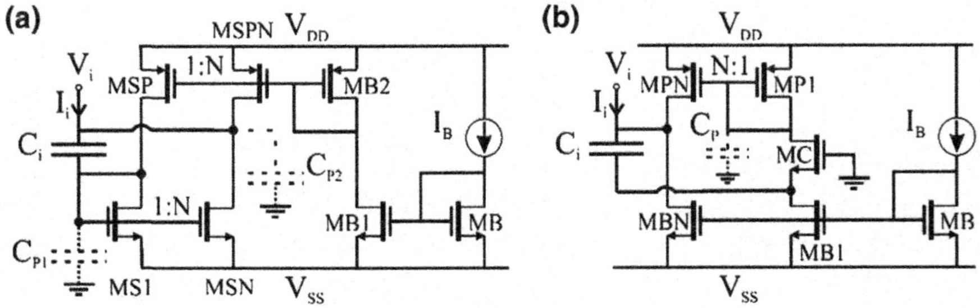

Fig. 5.30 The capacitive scaling circuits proposed in [149]: **a** basic circuit and **b** circuit with improved performance using a cascade transistor (MC)

$$\frac{i_{\text{in}}}{v_{\text{in}}} = g_{\text{oMSN}} + g_{\text{oMSP}} + s[C_{\text{P2}} + (N+1)C_{\text{i}}], \tag{5.144}$$

If the two conductances are decreased and C_{P2} is ignored, then this reduces to [219]

$$C_{\text{L}} = (N+1)C_{\text{i}}, \tag{5.145}$$

To push the parasitic pole to high frequencies, g_{mMS1} must be increased. However, to keep noise levels low, g_{mMS1} must be reduced [149]. This tradeoff can be overcome with the topology of Fig. 5.30b, where C_{i} is connected to a cascade transistor rather than the current mirror. Noise contributions from cascade transistors are negligible and this transistor can be optimized for the frequency response. The scaling factor is controlled by the ratio of the transistor dimensions of MPN to MP1.

5.2.5 Electrical Bioimpedance Instrumentation

5.2.5.1 Overview of Bioimpedance Instrumentation

The basics of bioimpedance and common system level topologies for signal processing and excitation were discussed in Chap. 3. As discussed, a galvanostatic tissue impedance measurement system requires a voltage signal excitation, which is then converted into a current by a VCCS and a front-end voltage recording and signal processing system. Here we will give a general overview of the state of the art in bioimpedance ASICs and then discuss in more detail the various fundamental blocks of a bioimpedance system.

In recent years, there has been an increased interest in bioimpedance in the ASIC community and a number of different systems have been proposed for various applications, which often lead to different design specifications and thus system implementation. This makes a comparison between systems difficult. Examples of complete bioimpedance systems include the following: a system for tissue impedance measurement for ischemia detection was proposed in [60]; an active electrode system for EIT, ECG, and contact impedance measurement was presented in [67] (360 μW); the system in [68] was developed for bioimpedance body composition analysis (52 μW) and its use in a multi-parametric implementation for the detection of glucose, photoplethysmography, ECG and EMG was presented in [69]; cardiovascular parameters (pulse wave velocity (PWV) and respiration) were measured in [220] (1.28 mW); an ASIC combining multi-wavelength near-infrared spectroscopy and bioimpedance for blood glucose estimation was presented in [61] (38 mW); a system for bioimpedance and iontophoresis for transdermal drug delivery was presented in [63]; an inductively powered cardiac monitoring system for wearable applications was described in [29] with a power consumption of 3.9 mW for a 25 electrode system; an EIT system for breast cancer imaging with a 53.4 power consumption was described in [30]; and a 10.4 mW EIT lung ventilation monitoring system was presented in [31]. An implantable ASIC for heart rhythm analysis with a three-lead ECG system with an on-chip QRS feature extractor combined with bioimpedance and an accelerometer readout channel was presented in [221] with a power consumption of 13 μA from a 2.2 V supply. Finally, an implantable bioimpedance ASIC which was inductively powered was presented in [222] consuming 165 μA from a 1.8 V supply.

5.2.5.2 Voltage Signal Excitation

The voltage excitation system is crucial for the operation of both potentiostatic and galvanostatic bioimpedance systems. It determines the frequencies injected into the tissue and to a first approximation the amplitude of the signals. What is meant by the latter statement is that in a galvanostatic system the VCCS will further determine the current injected through its transconductance. In contrast to a current source, making a voltage source is relatively easy, as one of its main characteristics is the need for a low, ideally zero, output impedance such that it can source the current that needs to be delivered through the load, which can have any value. As discussed in Chap. 3, various types of excitation waveforms can be used, especially in multifrequency applications. The type of excitation signal will also determine the front-end and back-end recording electronics needed to decode the required information. Using random noise or binary sequences containing a wide number of harmonics of various amplitudes will, for example, need an FFT algorithm, which will be power-hungry and consume large chip area. As discussed in Chap. 3, one solution is to inject only the frequency components of interest, all with the same constant amplitude. An advantage is that there is no need to measure or store the amplitude of each excitation frequency and if there is an error in the amplitude this is likely to be common for all frequencies. A disadvantage is that the SNR will decrease with increasing frequency, since the impedance decreases and thus the recorded voltage amplitude also decreases in galvanostatic systems. In potentiostatic systems, the opposite situation occurs, where noise will be an issue at low frequencies. In terms of CMOS realizations, as will be seen in the following sections, all proposed systems employ a swept frequency approach, apart from one that considers the multifrequency scenario. Generating square waves, triangle waves, sawtooth waves, and staircase approximations of sine waves is relatively easy. The difficulty is in generating a pure tone sine wave, either through these waveforms or directly. The systems presented in the following sections focus on this aspect. Generally speaking, as discussed in [223] there are two other issues related to generating sine waves on-chip for biomedical applications. For various biomedical applications and in particular in potentiostatic measurements for electrochemical impedance spectroscopy, the measurement frequencies of interest are very low, i.e. from mHz to a few kHz. As discussed in the G_m-C filter section in this chapter, this implies very large time constants, requiring large resistors and/or capacitors.

The second challenge is related to achieving the frequency range of interest, which in galvanostatic applications for interrogating tissues can span from a few kHz to a few MHz.

Filtering-Based Methods

As the name suggests, these methods generate a non-sinusoidal wave of varying fundamental frequency and use banks of filters or programmable filters to reject unwanted harmonics. The generated wave is typically a square wave, triangle wave, or sawtooth wave which can be generated trivially. The main challenge in this system is thus in the design of the filter. It should have a high Q factor and be of high order, and in swept frequency realizations its dominant poles should be programmable over a wide frequency range. One such approach was demonstrated in [223] generating frequencies between 40 MHz to 40 kHz occupying an area of 0.64 mm^2 and achieving a THD less than 0.1%. Since the magnitude of the fundamental frequency of a square wave is much larger from its harmonics than in the sawtooth wave case, and since it contains no even-order harmonics, square waves were preferred. In addition, square waves are much simpler to generate, while their frequency can easily be changed by using dividers. Due to the need for high and accurate pass band tunability, the authors chose a SC filtering approach where the time constant can be accurately controlled by clocks. Since a single frequency component is of interest the flatness of the pass band was not of interest, while a sharp cutoff is required. Also, the attenuation at high frequencies should increase with increasing frequency to attenuate the high-frequency switching noise arising from the fact that an SC topology is used. A fifth-order Chebychev type-I filter with a 0.5 dB pass band ripple was implemented. A similar approach was implemented in [60], where a G_m-C-based ramp integrator was used to generate a triangular wave signal and a subsequent tunable second-order G_m-C BPF filtered this signal to generate a sine wave.

Sine-Shaping Methods

A family of sine generation methods is based on nonlinear circuits that convert a triangular wave signal into a sine wave. These are referred to as triangle-to-sine converters (TSC). The original circuits developed for these methods were based on bipolar transistors. Thus, their CMOS equivalents are based on weak inversion operation. The simplest such circuit is based on a differential pair with or without resistive source degeneration and resistive loads, as in Fig. 5.31a, b. As discussed earlier, the differential output current, I_0, of a differential pair in weak inversion exhibits a tanh nonlinearity. This has a narrow linear region of ~15 mV close to the midpoint of its IV characteristics where the input signal is amplified linearly. On the other hand, close to peaks of the transfer curve, the input signal is amplified logarithmically. This is illustrated in Fig. 5.31c and leads to the rounding of the

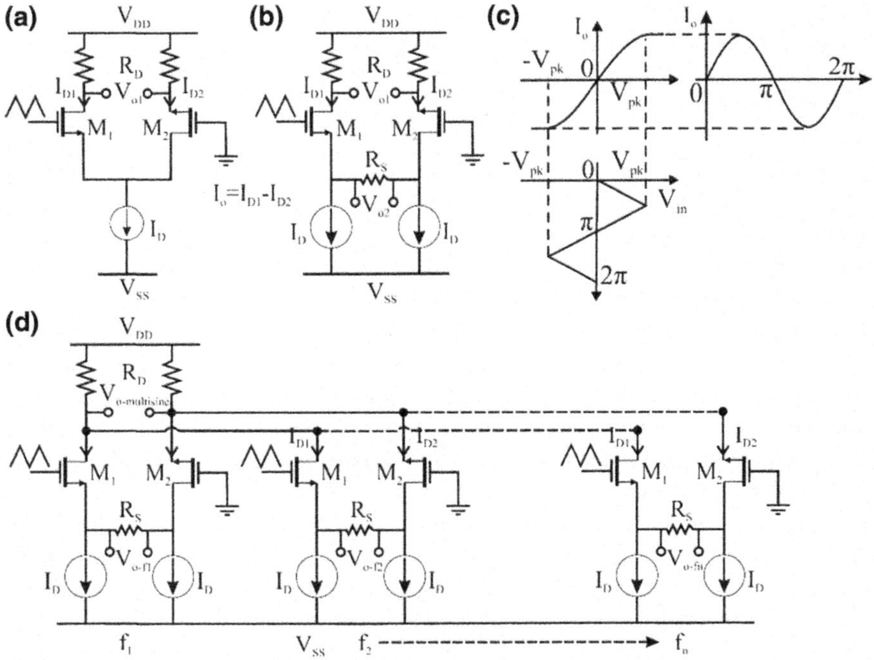

Fig. 5.31 Weak inversion TSCs: **a** without and **b** with resistive source degeneration. **c** IV transfer characteristics of the circuit allowing TSC [158, 224, 226]. **d** Extension of the circuit for multi-sine signal generation, by adding individually-converted sine waves in the current domain [158]

peaks of the triangle wave. The source degeneration resistance allows optimization of the circuit's IV transfer characteristics for an optimum output waveform, thus providing an extra degree of freedom. The output can be taken differentially either across the degeneration resistor or from between the two transistor drains. If the input triangular wave is of an appropriate amplitude, such that the input pair transistors are overdriven to the verge of saturation, then the curvature of the circuit's IV characteristic will smooth the waveform's peaks to approach a sine wave. Thus distortion in the output signal depends on the accurate control of the input triangle wave peak-to-peak voltage amplitude, $V_{\rm pk}$, and the quality of the triangle wave. This operation is also controlled by the current, $I_{\rm D}$, and the degeneration resistor, $R_{\rm S}$. The values of these two parameters are set to fit the first two terms of the Taylor series expansion of the sine. It is important to note that the quality of the nonlinear transfer function is affected by process and temperature variations. Parameters $V_{\rm pk}$, $I_{\rm d}$, and $R_{\rm S}$ are related by

$$\frac{V_{pk}}{nU_T} = \frac{\pi}{2}\frac{I_D R_S}{nU_T} + \pi. \tag{5.146}$$

There is a theoretical range of values for $I_{\rm D}R_{\rm S}/nU_{\rm T}$ given by

$$2 \leq \frac{I_D R_S}{n U_T} \leq 3.3, \quad (5.147)$$

which translates into a theoretical peak input amplitude range of

$$2\pi \leq \frac{V_{pk}}{n U_T} \leq 2.65\pi, \quad (5.148)$$

which has a midrange value of $V_{pk} = \sim 236$ mV. Since R_S, I_D, and g_m (the latter being relatively constant in weak inversion for a constant I_D), the gain or loss of the circuit can only be set via the load resistors R_D. This circuit was examined in its bipolar form in [224]. It was applied in [225] to a 6 GHz DDS to perform the phase-to-amplitude conversion. The MOS form of the circuit was examined [226] using discrete off-the-shelf transistors, achieving a THD of 0.2%. It was then applied to generate sine wave input signals for built-in self-test (BIST) applications for fault diagnosis in analog circuits in [227]. It was also used in [228] in a 16 channel sine wave generator for generating stimulus waveforms for creating electrokinetic forces for particle manipulation. The operation of the circuit was expanded by Kassanos et al. in [158], where multiple differential pairs shared the same R_Ds, such that the output currents of each pair are summed at the same node exploiting Kirchhoff's current law (KCL). With each pair being driven by a triangle wave of the same amplitude but different frequency, this allowed the generation of a multi-sine signal as the differential voltage across the R_Ds. The individual sine waves may be obtained across the R_S of each pair. THDs between 0.255% at 1 kHz and 0.41% at 1 MHz were reported.

As discussed in [229–231] the single differential pair TSC circuit exhibits a finite slope at π/2, whereas a sine has a slope of zero. An improvement to this circuit was presented in [230], where resistive feedback was used between triangular wave input and sine wave output and THDs below 0.07% were reported. Another similar approach also originating from bipolar technology uses multiple (N) differential pairs operated in weak inversion and combined in alternating antiphase, as in Fig. 5.32 [229, 231–235]. The same tail bias current is used for each pair and the same triangular input wave, V_{tri}, while the right-hand side transistor of the pair is

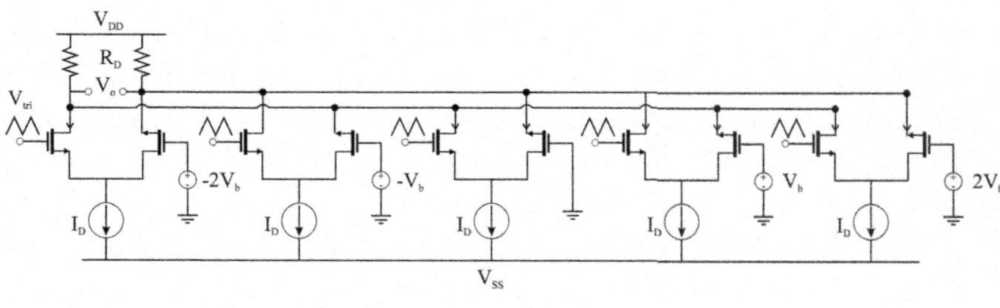

Fig. 5.32 The sine wave generator examined in [231, 234] with $N = 5$

biased by a voltage mV_b, with $-M < m < M$ and $M = (N - 1)/2$. The number of stages, N, defines the number of zero crossings of the generated sine wave within each slope of the triangle wave, and the input and output frequencies are different, as shown in [231]. The correct choice of offset voltages for a certain N allows conversion to the sine. If we define $A = V_{tri}/2nU_T$ and $B = V_b/2nU_T$ and C is the amplitude of the generated sine wave, then if m is large

$$C\sin\left(\frac{A\pi}{B}\right) \approx \sum_{m=-M}^{M} (-1)^n \tanh(A + mB). \tag{5.149}$$

These $m \cdot V_b$ bias voltages, define the voltage range within which each differential pair is in the active region. As discussed in [229], the voltage offset V_b does not need to be generated and applied with additional circuitry, but similarly to the multi-tanh transconductance linearization technique described earlier, it can be generated using differently sized transistors in the differential pair. According to the same paper, if B is small ($B < 2$), then

$$C = \frac{4\pi}{B}\exp\left(-\frac{\pi^2}{2B}\right). \tag{5.150}$$

Another reported TSC is the translinear TSC proposed in [236].

G_m-C Oscillators

In a similar manner to G_m-C filters discussed earlier, OTAs can be used to implement oscillators [237–244]. In this way the oscillating frequency will depend on capacitors and the transconductance of the OTAs. A disadvantage with this method is that it is typically hard to tune such topologies over a wide frequency range, while mismatch will hinder the accuracy of the oscillating frequency and the oscillation condition may not be met. These can be addressed by implementing tuning loops or locking to reference signals, complicating the system significantly. Nevertheless, G_m-C oscillators with frequency tuning between 1 and 25 MHz were reported in [241], 50–130 MHz in [242], 1 kHz–10 MHz in [244] and 74–1 MHz in [240]. The quadrature oscillator topology of Fig. 5.33a was proposed for high-frequency soil impedance measurements and is characterized by the following second-order equation [242]

$$s^2 - bs + \omega_0^2 = 0, \tag{5.151}$$

where

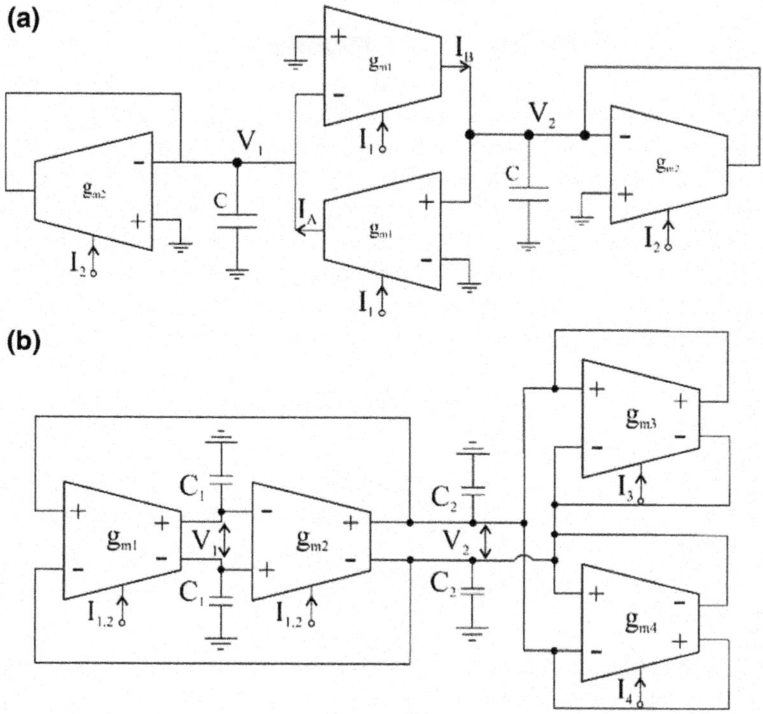

Fig. 5.33 **a** The quadrature G_m-C oscillator proposed in [242]. **b** The fully differential G_m-C oscillator proposed in [241]

$$b = -2\frac{g_{m2}(C - C_2) - g_{m1}C_1}{(C - C_2)^2 + C_1^2}, \tag{5.152}$$

$$\omega_0^2 = \frac{g_{m1}^2 + g_{m2}^2}{(C - C_2)^2 + C_1^2}, \tag{5.153}$$

$$C_1 = \frac{g_{m1}}{\omega_1}, \tag{5.154}$$

$$C_2 = \frac{g_{m2}}{\omega_2} \tag{5.155}$$

and ω_1 and ω_2 are the high-frequency zeros of the respective OTAs. To sustain stable oscillation $b = 0$ must be satisfied and hence the following must be true

$$g_{m2} = g_{m1}\frac{C_1}{C - C_2} \tag{5.156}$$

$$\omega_0 = \frac{g_{m1}}{C - C_2} \tag{5.157}$$

A fully differential quadrature G_m-C oscillator topology is shown in Fig. 5.33b [241], which is based on a non-inverting (g_{m1}-C_1) and an inverting integrator loop (g_{m2}-C_2) [245]. Transconductors g_{m1} and g_{m2} and C_1 simulate an inductor. This, together with C_2, forms an LC resonant circuit. Transconductors g_{m3} and g_{m4} simulate negative and positive resistances that can be set to position the oscillator poles at the necessary location in the unit circle. In this topology different transconductances are responsible for setting parameter b and ω_0 of (5.151), where when not considering OTA non-idealities, these are given by

$$b = \frac{g_{m3} - g_{m4}}{C_2}, \tag{5.158}$$

$$\omega_0^2 = \frac{g_{m1} g_{m2}}{C_1 C_2} \tag{5.159}$$

$$Q = \frac{1}{g_{m4}} \sqrt{\frac{g_{m1} g_{m2} C_2}{C_1}}, \tag{5.160}$$

where Q is the Q-factor of the topology [237]. This implementation occupied a chip area of 0.63 mm^2, had a power consumption of 1.05–1.58 mW, and achieved a THD of 0.66%. Due to non-idealities, the poles may be shifted to the right or the left of the complex frequency plane. Consequently, to ensure oscillation, the poles should be placed initially on the right-half of the complex plane, such that $b > 0$ to ensure a self-starting oscillation. Another difficulty lies in the accurate amplitude

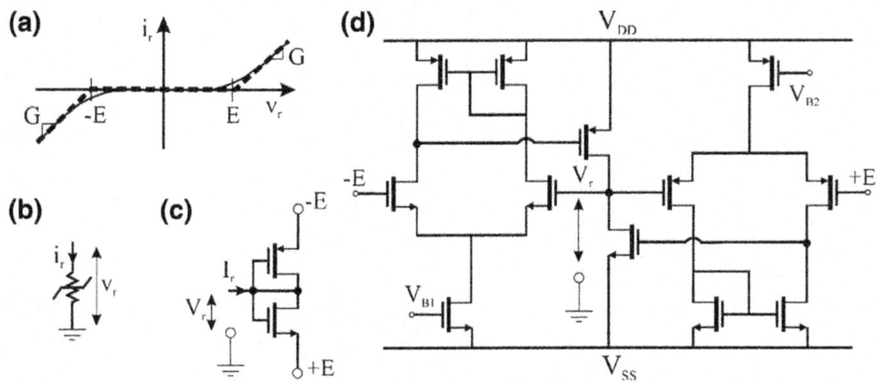

Fig. 5.34 **a** IV characteristic curve of the nonlinear element used for amplitude control. **b** A simple implementation using two diode-connected transistors with the oscillation amplitude controlled by E. **c** An implementation of the nonlinear resistor by two comparators and current sources

control of the generated sine wave. Typically, an external limiter, which is a non-linear element, is used to limit the amplitude and a high Q-factor is used to reduce the harmonic content. Such nonlinear elements can be soft or more often hard nonlinearities, such as comparators. Such topologies are presented in [239, 245] and were applied to the single-ended version of the topology of Fig. 5.33b by connecting them in parallel with C_2. The IV characteristic curve of such an element is shown in Fig. 5.34a, where G is the conductance of the element defined as the slope in the IV curve once the imposed limits E and $-E$ have been exceeded, v_r is the potential across the element and i_r is the current through it; the symbol of such an element is shown in Fig. 5.34b. In this case b and ω_0 are modified to [239, 245]

$$b = \frac{(g_{m3} - g_{m4}) - G \cdot N(A)}{C_2}, \tag{5.161}$$

$$\omega_0^2 = \frac{g_{m1} g_{m2}}{C_1 C_2} + G \cdot N(A), \tag{5.162}$$

where $G \cdot N(A)$ describes the equivalent conductance of the nonlinear element using the describing function. For $v_r(t) = v_{C2}(t) = A \cdot \cos(\omega t)$ and the current through the element being a function of this voltage $i_r = f(v_{C2})$, then the describing function describes the relationship by the first term of the Fourier series expansion [239, 245]

$$N(A) = \frac{1}{AT} \int_0^T \frac{f(A\cos(\omega t)) \cos \omega t}{G} dt, \tag{5.163}$$

where $T = 2\pi/\omega$ and $N(A) = 0$ for $A \ll E$ and $N(A) = 1$ for $A \gg E$. For intermediate values, i.e. $A \geq E$, $N(A)$ is approximated by

$$N(A) = 1 - \frac{2}{\pi} \left[\sin^{-1}\left(\frac{E}{A}\right) + \frac{E}{A} \sqrt{\left(1 - \left(\frac{E}{A}\right)^2\right)} \right]. \tag{5.164}$$

Hence, when $A = 0$ and $N(A) = 0$ if $g_{m3} > g_{m4}$ the oscillation is self-starting, since the poles of the equation will be on the right half-plane. With increasing A, $G \cdot N(A)$ will also increase and consequently the roots will be pulled towards the imaginary axis. When $G \cdot N(A) = g_{m3} - g_{m4}$, this is guaranteed for that specific value of A:

$$A = N^{-1} \left[\frac{g_{m3} - g_{m4}}{G} \right]. \tag{5.165}$$

Two possible realizations for the implementation of such a nonlinear element are shown in Fig. 5.34c, d, where (c) is composed of two diode-connected transistors and (d) comprises two comparators each driving a current source. For the right-hand

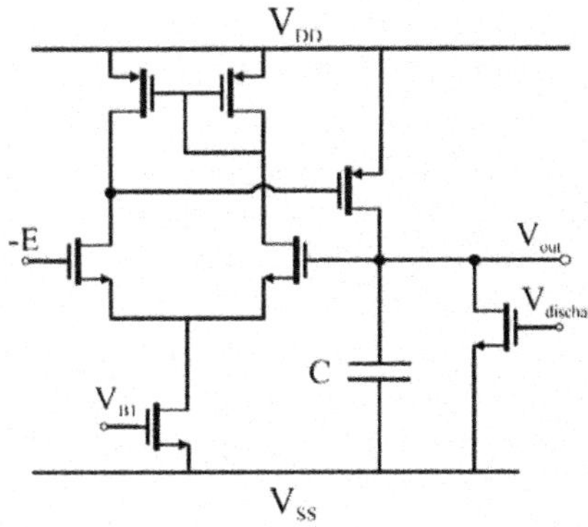

Fig. 5.35 A simple peak detector

part of the circuit, when $V_{C2} < E$, $V_G = V_{SS}$ and the transistor is off and thus the current though it is zero; when $V_{C2} > E$, the transistor in on and a current flows through it, satisfying the transfer characteristics of Fig. 5.34a.

As discussed in [241, 243], this nonlinear element does not need to be an additional circuit element, but one of the OTAs of the oscillator can be exploited to perform this function. In this way, as the oscillating wave is increasing in amplitude at the beginning of its operation (when turned on), one of the transconductors can leave its linear operating region sooner than the rest. In [244], a circuit to compute the squares of the two quadrature output signals of the oscillator and then add them together provides a DC component proportional to the amplitude of oscillation exploiting the $\sin^2 a + \cos^2 a = 1$ trigonometric identity. This is then used to control the transconductance on an OTA. A more complicated but elegant approach is to use automatic gain control (AGC) [237–239]. In this approach, the peak of the oscillator output is calculated using a peak detector. This is then compared with a reference voltage, generating a signal which controls the transconductance of an OTA. The peak detector topology employed in [238] is shown in Fig. 5.35.

At this point the oscillator developed in [29–31, 61] should be mentioned. These are not Gm-C oscillators, but op-amp Wien bridge oscillators, where the frequency and amplitude of oscillation depend on passive components.

Digital-Based Methods

Digital-based methods have a number of advantages which include simple control, good temperature and aging stability and flexibility [246, 247]. However, these are related to increased harmonic distortion as a result of different numerical errors. Digital-based methods can be separated into digital signal processing (DSP)-based synthesizers and resistor-chain digital-to-analog converter-based approaches. As discussed in [246, 248–251], the former may take various forms. These include the use of lookup tables (LUT), interpolation, infinite impulse response (IIR) filters, and polynomials.

Lookup Table (LUT) Method

The traditional approach is the LUT method, where a sine wave is synthesized by successively going through an LUT stored in a read-only memory (ROM). The LUT contains an integer number of equally spaced sampled values of a complete sine wave cycle which are read at each clock pulse of a clock with frequency f_s. LUT values can be generated either by sampling a sine wave or by computation. The sine samples are converted through a DAC into an analog waveform. A continuous-time sinusoid $x(t) = A\sin(2\pi Ft + \theta)$ can be expressed as a discrete-time signal as $x(n) = A\sin(2\pi fn + \theta)$, where $f = F/F_S$ is the relative frequency of the sinusoid, $F_S = 1/T$ is the sampling rate, T is the sampling period or sample interval, and n is related to time, t, by $t = nT = n/F_S$. A sine wave table of length N, containing a single period of the waveform, can be obtained by computing [252]

$$x(n) = \sin\left(2\pi\frac{n}{N}\right), \quad n = 0, 1, \ldots, N-1. \tag{5.166}$$

The table entries are in binary form. Hence, accuracy is determined by word length. The finite word length (e.g. 16 bits) will introduce amplitude quantization errors and hence harmonic distortion. The required sine wave is then generated by reading the LUT at a constant, Δ, the table address increment. The data pointer wraps around at the end of the table when the angle exceeds 2π, which can be implemented using a circular buffer (e.g. modulo-N addressing). The generated frequency is then $f = \Delta F_S/N$ [252]. To avoid aliasing $\Delta \leq N/2$. The circular pointer used can be of the form $k = (m + n\Delta)_{\text{mod } N}$, where m defines the initial phase of the system [252]. The accuracy of the generated frequency depends on N, F_S, and the precision of Δ, which is a function of the decimal places, k, assigned to it.

$$\Delta f = \frac{2^N N}{F_S}. \tag{5.167}$$

To reduce the memory space required by the LUT, only a quarter of the data can be used, since the absolute values are repeated every quarter of a period. In this case, the algorithm will need to consider which wave quadrant is generated and will thus be more complicated. The harmonic distortion in the generated wave can be reduced by reducing time-quantization errors by increasing the table size. Both the frequency and phase resolution are determined by word length. The limitations of the LUT method are the limited set of output frequencies, the dependence of the spectral purity of the generated signal, on the word length, and the large memory required for acceptable frequency and phase resolution [246].

For a given table size, harmonic distortion can be decreased by using interpolation to evaluate values between table entries. A simple method is to use linear interpolation, where sample points between table entries are estimated according to a linear function. This approximation of a sine wave is given by

$$\begin{aligned} x(n+d) &= \sin\left(\frac{2\pi(n+d)}{N}\right) \\ &= \sin\left(\frac{2\pi n}{N}\right) + d\left(\sin\left(\frac{2\pi(n+1)}{N}\right) - \sin\left(\frac{2\pi n}{N}\right)\right) \\ x(n+d) &= x(n) + d[x(n+1) - x(n)], \end{aligned} \tag{5.168}$$

where d is a decimal number between 0 and 1. $[x(n+1) - x(n)]$ is the slope of the line segment between two consecutive sample points from the table. The slopes are calculated by the difference between two consecutive values and also need to be stored in a second table, called the slope table [253]. The LUT method with interpolation is discussed in more detail in [247].

Recursive Method

This method relies on the fact that the impulse response of a filter poles on the unit circle is a sine wave. The frequency of this sine wave depends on the location of these poles. Thus a second-order IIR filter is employed with conjugate complex poles on the unit circle of the z-plane. At time $n = 0$ a digital impulse is required to initialize the system; subsequently the oscillation is self-sustaining, as the system has no damping since $r = 1$ [254]. A system with a transfer function of

$$H(z) = \frac{b_0}{1 + a_1 z^{-1} + a_2 z^{-2}} \tag{5.169}$$

and $a_1 = -2r \cdot \cos(\omega_0)$, $a_2 = r^2$, $r = 1$, and $b_0 = A \cdot \sin(\omega_0)$ has a pair of conjugate complex poles at $\mathrm{p} = \mathrm{re}^{\pm j\omega_0}$, where ω_0 is proportional to the ratio of the desired frequency, F, and the sampling frequency, F_S, according to $\omega_0 = 2\pi F/F_S$. This has an impulse response that can be written in recursive form as

$$h(n) = b_0\delta(n) - a_1 h(n-1) - h(n-2), \tag{5.170}$$

with initial condition $y(-1) = (y - 2) = 0$, where $\delta(n)$ is the Dirac delta function with $\delta(1) = 1$ and for $n \neq 0$, $\delta(n) = 0$ [254]. Alternatively, the system input can be set to zero, in which case the first term in the above equation disappears and the initial conditions become $y(-1) = 0$ and $y(-2) = A \cdot \sin(\omega_0)$. The implementation of this method requires only a few operations and its signal flow graph is shown in Fig. 5.36a [254]. The resonator algorithm must control the propagation of numerical errors, ensure oscillation conditions are satisfied, and keep a low spur level. Due to the limited word length and the related quantization errors, this method can become unstable. If the poles lie outside the unit circle, the output will diverge towards infinity, while if they are within the unit circle the output will decay over time towards zero. Methods to improve the performance of this method are discussed in [249–251, 255].

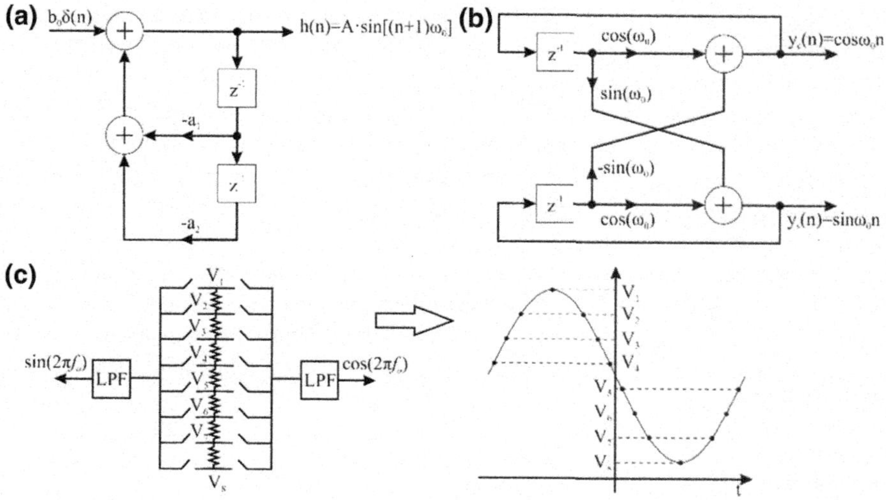

Fig. 5.36 **a** Recursive biquad IIR filter-based digital sinusoidal oscillator. **b** Coupled-form quadrature digital oscillator. **c** Operation of a R-DAC with eight taps

In the SD algorithm for the estimation of the real and imaginary parts of the impedance at the front end, quadrature signals are required. These can be generated by the coupled-form oscillator of Fig. 5.36b, which is obtained by the following trigonometric equations [254]: $\cos(\alpha+\beta) = \cos(\alpha)\cdot\cos(\beta) - \sin(\alpha)\cdot\sin(\beta)$ and $\sin(\alpha+\beta) = \sin(\alpha)\cdot\cos(\beta) + \cos(\alpha)\cdot\sin(\beta)$, where $\alpha = n\omega_0$ and $\beta = \omega_0$. With $y_c(n) = \cos(n\omega_0)u(n)$ and $y_s(n) = \sin(n\omega_0)u(n)$, where $u(n)$ is the unit step function. From these, the two coupled difference equations are obtained

$$y_c(n) = \cos(\omega_0)y_c(n-1) - \sin(\omega_0)y_s(n-1), \tag{5.171}$$

$$y_s(n) = \sin(\omega_0)y_c(n-1) + \cos(\omega_0)y_s(n-1). \tag{5.172}$$

As shown in Fig. 5.36b, there is no system input. The required initial conditions for the system are $y_c(-1) = A\cdot\cos(\omega_0)$ and $y_s(-1) = -A\cdot\sin(\omega_0)$.

The LUT and recursive methods were compared in [253], leading to the following conclusions. The direct lookup method is the fastest to generate a sine wave as it only requires time to access the samples stored in memory. A longer execution time is required if interpolation is used due to the additional computation required. The use of recursive algorithms is the slowest method, but it requires considerably less memory. As discussed earlier, exploiting the symmetry of sinusoids will lead to reduced memory requirements; however, the extra processing required to track the quadrants increases execution time. The direct lookup method is the least accurate and it improves significantly when linear interpolation is used. The recursive method leads to the least distortion. The main source of error in this method is quantization noise, the power of which increases linearly with the iteration number.

Resetting the variable at the end of each sine wave period or after a certain number of iterations can reduce the accumulation of quantization noise. This error can also be reduced by performing the calculations as accurately as possible. With regard to frequency resolution, for the LUT-based methods this is determined by Δf. In the recursive method, with the right choice of initial conditions the frequency can be accurately set with $\Delta f = 0$. The recursive algorithm is much more complex and is very sensitive to the initial conditions, requiring high accuracy calculation in terms of decimal points.

A coordinate rotation digital computer (CORDIC)-based approach implemented both with an ASIC and a field-programmable gated array (FPGA) was presented in [256]. The CORDIC architecture uses only adders and shift registers and is an iterative fixed-point technique. The developed system generated signals from 100 Hz to 1 MHz.

Polynomial Method

Another approach is to use polynomial approximations of trigonometric functions. The sine and cosine functions can be expressed as infinite power series expansions, the accuracy of which depends on the number of terms used. Their Taylor series expansions are

$$\cos(\alpha) = 1 - \frac{\alpha^2}{2!} + \frac{\alpha^4}{4!} - \frac{\alpha^6}{6!} + \cdots, \tag{5.173}$$

$$\sin(\alpha) = \alpha - \frac{\alpha^3}{3!} + \frac{\alpha^5}{5!} - \frac{\alpha^7}{7!} + \cdots. \tag{5.174}$$

This technique uses more mathematical operations, but requires significantly less memory space than the LUT method.

Resistor Chain Method

Instead of storing sine wave samples in an LUT, a resistor chain can be used which is tapped unevenly and according to a sinusoidal waveform as in [257, 258]. In [257] such an R-DAC is used to generate quadrature sine waves of frequencies between 100 Hz and 10 kHz. The output of this is then used by a sub-sampling signal generator, which as the name suggests samples the generated signal below the Nyquist frequency to generate signals from 1 MHz to 100 Hz. As discussed by the authors, if one period of a sine wave is sampled at 14 equidistant intervals, these correspond to eight unique values, which can be set along a resistor chain by using appropriate resistor values. Sequentially, these can be switched according to 1-2-3-4-5-6-7-8-7-6-5-4-3-2 using switches. To generate quadrature signals simultaneously a second set of switches is used, which is switched out of phase by four switches in this example to establish the 90° phase shift between the two signals. LPFs reduce the harmonic content of the signals. This is illustrated in

Fig. 5.36c. Thus in this topology, the LUT has been replaced by the resistor chain. Similarly to the LUT approach, this method cannot generate very low frequency signals efficiently. In the LUT approach, a larger memory is required, while in this case more resistors and switches are required, consuming more chip real estate and complicating routing. A larger number of resistor taps also leads to improved linearity. The number of output taps in the DAC determines the oversampling ratio (OSR). This cannot be too low, as it determines how close the fundamental frequency and its replicas are, in which case the output filter complexity would need to be increased. The authors used an OSR of 64, which required only 33 taps, leading to a spurious-free dynamic range (SFDR) of 33 dB with additional replica suppression provided by the LPF to bring the SFDR in the range of 50 dB. The reduced OSR and use of an output filter, instead of directly using a large OSR and no filter, lead to a chip area reduction of 87%. The output frequency is controlled by the updating clock, obtained by dividing the master clock of the system with a frequency f_M by M. The latter (M) could be set between 16 and 1562, and $f_M = 10$ MHz. The OSR in this example was set to 64 and the system output frequency, f_0, was given by $f_0 = f_M/(64\ M)$. Matching of the resistors is essential for high linearity. However, since the resistors are not equal, as the DAC is sinusoidally tapped, 1,000 equal and matched resistors where used and tapped accordingly. For the subsampling section a single track-and-hold circuit consisting of minimum sized nMOS switches and a capacitor was used followed by a LPF. The LPF was a G_m-C biquad topology with the capacitors implemented by nMOS gate capacitances to save chip area, while the OTA used employed current division and source degeneration similar to what was discussed in Sect. 5.2.4. The complete system achieved a THD less than 0.6%, a phase and amplitude mismatch of less than 0.8° and 3% respectively, and a power of 180 μW in a die area of 1 mm^2. The authors reduced the area of the design by 62% in [258] and managed to generate signals between 4.8 Hz and 39 kHz using a single R-DAC. The main chip area reduction came from the design of the token ring circuit, which tracks the output sample used to construct the sine wave and accordingly changes the next one with each clock pulse. It does so by storing the position with a quarter of the wave and determines the quarter by using the slope and sign states. The system developed in [222] is another example where a quadrature sinusoidal signal generator was implemented using a resistor ladder.

One disadvantage of this topology is the need for a tunable filter that needs to be tuned according to the output frequency generated. The method proposed in [259] further relaxes the specification for the filter, where, similar to the LUT linear interpolation method discussed earlier, the coarse points generated by the R-DAC (Fig. 5.36c) are connected together by a fine staircase signal approximating a linear interpolation. These intermediate points are generated by a linearly weighted R-DAC. The design consumed 330 μW and a 0.32 mm^2 die area generating frequencies between 1 MHz and 100 kHz and a SFDR of 64 dB at low frequencies and 42 dB at high frequencies. In [260] a R-DAC-based system was presented capable of generating sine waves between 1 kHz and 2.046 MHz with a minimum 65.7 dB SFDR and maximum 51.6% THD with a power consumption less than

15.3 mW. The authors used a differential R-DAC followed by charge-sampler discrete-time FIR (finite impulse response)/IIR and subsequent continuous-time second-order G_m-C filters.

In addition to all the above are phased-locked loop-based methods for signal generation and traditional DDS that use various of the above mentioned techniques, such as [225, 236, 261–266]. These are, however, typically implemented for high-frequency applications (in the MHz–GHz range).

5.2.5.3 Voltage-Controlled Current Sources (VCCS)

As discussed in Chap. 3, the design of a good quality VCCS is a challenging task. A key parameter is the output impedance of the circuit, which ideally should be infinite throughout the bandwidth of interest, while the output compliance should accommodate the voltage swings at the output nodes for the load impedance being investigated with a particular output current. This output current should be within safety limits (IEC60601-1) and also capable of producing voltage differences across the sensor, which are resolvable by the voltage recording front end. In addition, no DC current should be injected into the body, once again for safety purposes, and the harmonic distortion should be minimal, such that only frequency components of interest are injected into the tissue. Finally, the VCCS should be a differential floating VCCS, such that there is no direct path to ground and to minimize the common-mode voltage at the load and thus voltage measurement errors at the front end due to limited CMRR at the IA.

A number of different approaches have been implemented over the years in ASIC implementation, but many fail to adequately report many of the above parameters, which would indicate the performance of the VCCS. In any case, here we will provide a general overview of the various approaches implemented in silicon. A simple differential G_m-based VCCS was presented in [60]. The topology had a high-pass output impedance, being low near DC and constant and high up to 20 kHz to minimize offsets from the previous stage driving the VCCS. The topology essentially buffered the output of a G_m-C-based BPF. This is shown in Fig. 5.37a. In the bioimpedance system of [222] a simple transconductor was used to implement the VCCS and the output current was controlled by changing the current copy factor at the output branches of the circuit using parallel transistors and switches. Another G_m-based approach employing an output current sensor resistor and feedback to regulate the current output and hence the output impedance of the circuit was proposed in [267, 268]. The linearity of the output impedance and accuracy have been thoroughly characterized; nevertheless, these are large, complex, and power hungry specialized systems for high-frequency measurements.

A group from Korea has developed a number of bioimpedance systems [29–31, 61], which use a Wien-bridge differential sinusoidal oscillator topology to produce a voltage output, which is then converted into a current by a simple VCCS based on two transistors driven by the differential output of the oscillator, four current sources, and a programmable resistor (as in Fig. 5.37b), which sets the

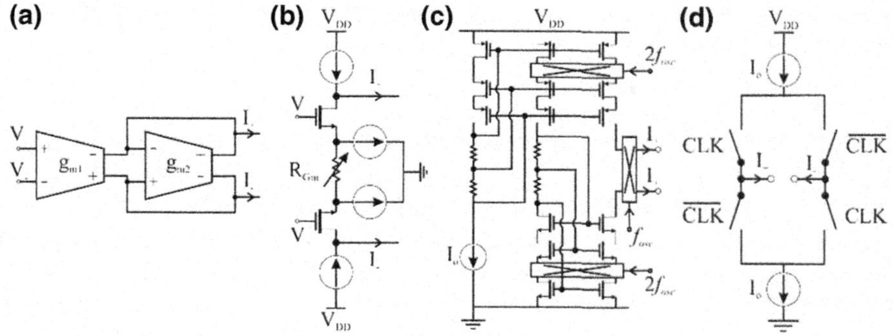

Fig. 5.37 **a** The simple Gm output buffer proposed in [60]. **b** The V–I converter used in [29–31, 61]. **c** The self-biased triple-cascode high-swing current source with dynamic element matching and square wave output current implemented in [68, 69]. **d** The H-bridge-based current excitation circuit implemented in [220]

transconductance of the topology ($G_{\rm m} = 1/R_{\rm Gm}$) to produce currents between 10 μA to 1 mA, while an output impedance greater than 120 Ω was reported in [61].

Various proposed systems use a high output impedance DC current source, which is easy to implement. The output DC current is then modulated at the output with a chopper to create a square wave excitation current. This approach was used in [68, 69] and is shown in Fig. 5.37c; self-biased triple cascode high-swing current sources were used and the DC current was controlled with an eight bit current DAC between 0 and 169 μA in 256 steps. Up to 1.24 MHz the output current demonstrated amplitude variation of 0.3%, up to a load of 1 kΩ. Chopping at twice the impedance measurement frequency, internal to the current source structure, was used for dynamic element matching to average mismatches. A similar approach was used in [32, 63]. Current DACs were used in [221, 269] to implement a pseudo-sine wave generator. As was argued, while square waves are very power-efficient, the measurement accuracy is hampered when using a synchronous demodulation front-end algorithm, due to the folding of the odd harmonics at baseband. The pseudo-sine offers a good power/accuracy tradeoff. By limiting the bandwidth far below the 63rd harmonic, harmonic residuals are attenuated and the accuracy is not hampered. The pseudo-sine generation is based on an LUT to control the DAC.

A number of different systems use an H-bridge topology as in neural stimulation applications (see Sect. 5.2.9 and Fig. 5.44). In [270] a ΔΣ bit stream pattern generated off-chip with an FPGA drives a simple H-bridge, the outputs of which are filtered off-chip with a passive *RC* LPF to create a sine wave. An H-bridge was also used in [220] as shown in Fig. 5.37d. Cascode current source are used to increase their output impedance and the switches control the direction of the current. The authors argue that this topology consumes significantly less power. In the design of [67], an H-bridge topology was used to inject currents controlled digitally, produced by DC current sources. 127 positive and 127 negative current sources were used separated each in 7 groups to inject a current between ±127 μA controlled in steps of 1 μA. An off-chip 100 nF capacitor for AC-coupling was used at the output of the current source.

5.2.5.4 Voltage Recording and Signal Processing

The various analog signal processing methods for the calculation of the electrical impedance from voltage measurements were discussed in Chap. 3. In this section we will discuss ASIC implementations of these methods. By far the most commonly used analog signal processing method for front-end voltage recording systems, due to its simplicity, low power, ability to dispose of unwanted frequency components, and multi-frequency operation capability, is SD (mentioned earlier), which allows the calculation of the real and imaginary parts of the tissue impedance, as discussed in Chap. 3. Apart from tissue impedance measurement applications, it is also used in high-frequency dielectric spectroscopy (MHz–GHz) [5] and electrochemical impedance applications for the detection of biomolecules (DC to a few kHz). An early ASIC implementation of SD was presented in [60], where the recorded signal through a current feedback IA is demodulated simultaneously by two channels—by a 0° signal in one and a 90° signal in the other—to obtain the real and imaginary components simultaneously. The quadrature demodulation signals were obtained by the low-pass and high-pass outputs of a biquad filter used to create a sinusoid from a triangular wave input signal. A passive HPF was used to eliminate voltage offsets from the biquad filter. This HPF was composed of a capacitor and two MOS transistors in weak inversion biased with a 1 nA current to implement the resistors. Following demodulation, the signal was filtered by a cascade of two biquad LPFs for fourth-order filtering. A first-order 12 bit incremental ADC was used with a 50 kHz clock. A similar approach was implemented in [31], where the quadrature demodulation signals were obtained by passive HPF and LPF. A capacitive feedback amplifier was used to record the voltage with bandpass characteristics to eliminate electrode offsets and an 18 dB gain. Further amplification was provided with an amplifier of the same topology and with an AGC with a gain that can vary between 0 and 42 dB in steps of 6 dB. The AGC was implemented in [30] with a digital approach rather than an analog one, where a comparator compared the input voltage to a number of references and with the additional necessary logic; this controlled the gain between 0 dB or 24 dB and in a second stage between 0, 6, 12, and 18 dB. The same approach was followed in [30]. Demodulation was followed by an output LPF implemented using a passive topology implemented with a capacitor and two pMOS transistor pseudo-resistors, the resistance of which is controlled by their gate voltage, as in Fig. 5.38f. This gate voltage is set to 0 V for the first few cycles of the signal and then switched on to allow fast settling of the filter. A 10 bit SAR ADC was used for low power consumption at a sampling rate of 1–200 kS s^{-1}. In [222] a transconductance amplifier was used to record the voltage signals, followed by *I*/*Q* demodulation and a subsequent TI stage to convert the signal into a voltage. This was followed by a 50 Hz biquad filter with buffered outputs.

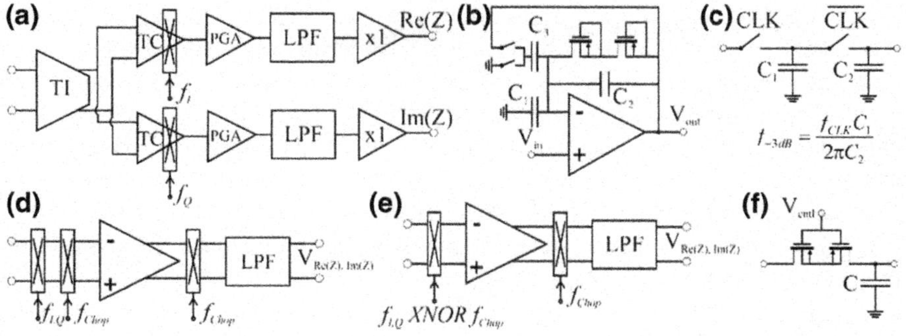

Fig. 5.38 SD front-end topologies. **a** The IA transconductance stage sharing approach implemented in [32, 221, 269]. In this figure TC stands for transconductance stage, not temperature coefficient. **b** The PGA used in (**a**). **c** A simple passive SC LPF. **d** The method proposed in the multi-frequency system of [59]. **e** Implementation of [59] using a single input chopper with a clock combining chopping and SD clocks proposed in [68]. **f** A simple passive LPF using pMOS pseudo-resistors implemented in [30]

A current feedback IA was also used in [32, 221, 269]. In order to save power and chip area the two SD channels shared a common TI, but separate transconductance stages. This is shown in Fig. 5.38a. In [32], the signal was demodulated in the transconductance stage and subsequently amplified (20, 41, and 83 V/V) by the PGA DC-coupled amplifier of Fig. 5.38b and finally filtered at 50 Hz and buffered. The same approach was followed in [221], only that the passive SC LPF used had a corner frequency of 10 Hz. A simple passive SC filter is shown in Fig. 5.38c. An intermediate and a subsequent buffer were used for isolation. An interesting approach that combines detection, filtering, and digitization, developed for potentiostatic electrochemical measurements, was presented in [271], where a dual-slope multiplying ADC was developed.

A system with the capability of measuring the real and imaginary components of up to two frequencies was presented in [59]. In contrast to the other approaches, demodulation is performed before amplification, as in Fig. 5.38d, in order to relax the frequency performance of all subsequent blocks, since the signal is immediately demodulated to DC before amplification. In addition, at high frequencies, when demodulation is performed following amplification, phase errors due to the addition of the delay of the recording amplifier will hinder measurement accuracy. This was combined with chopping in order to reduce the effect of voltage offsets and noise to the measurement. In order to eliminate errors in the chopping at high-frequencies, the chopping clock used in the second chopping stage was first processed by an additional copy of the IA set at the same gain as the signal-recording IA to ensure that the signal is delayed by the same amount as the recorded signal. The system demonstrated significant reduction of offset voltages, which can saturate subsequent signal processing stages, demonstrating accurate operation up to 1 MHz. A current feedback IA was used as in previous examples, and a fourth-order filter implemented as a cascade of two G_m-C biquads provided the filtering. In [68, 69], the approach proposed in [59] was used, except that the SD clock and the chopping clock were combined using an XNOR gate in order to use a single chopper at the input, as shown in Fig. 5.38e. A non-inverting DDA with resistive feedback was used as the PGA, and a fourth-order LPF and a 12 bit SAR ADC were used. As highlighted in [59], when using SD and the demodulation signals are generated though the voltage excitation system, as is typically the case, the phase delay of the VCCS at high frequencies will introduce errors in the calculation of Re(*Z*) and Im (*Z*). This is because the I and Q demodulating signals will no longer be in phase and in quadrature with the injected current.

Apart from SD, another approach, as discussed in Chap. 3, is to perform magnitude/phase detection. This was presented in [58] and is shown in Fig. 5.39a. A measure of the magnitude was obtained by multiplying the recorded voltage signal with a square wave version of itself. This was obtained through a comparator, and thus at low frequencies the square wave had the same phase as the recorded sine wave. Thus the signal was essentially synchronously detected. An LPF removed all frequency components to provide a DC signal proportional to the magnitude of the impedance. The output of the same comparator was used together with a square wave version (through a comparator) of the potential across a reference resistor in series with the unknown impedance to obtain a measure of the phase of the unknown impedance. The series sense resistor generated a signal in phase to the injected signal to the tissue, i.e. a 0° I signal. An XOR gate was used to compare the phase difference of these two square wave signals, generating a pulse

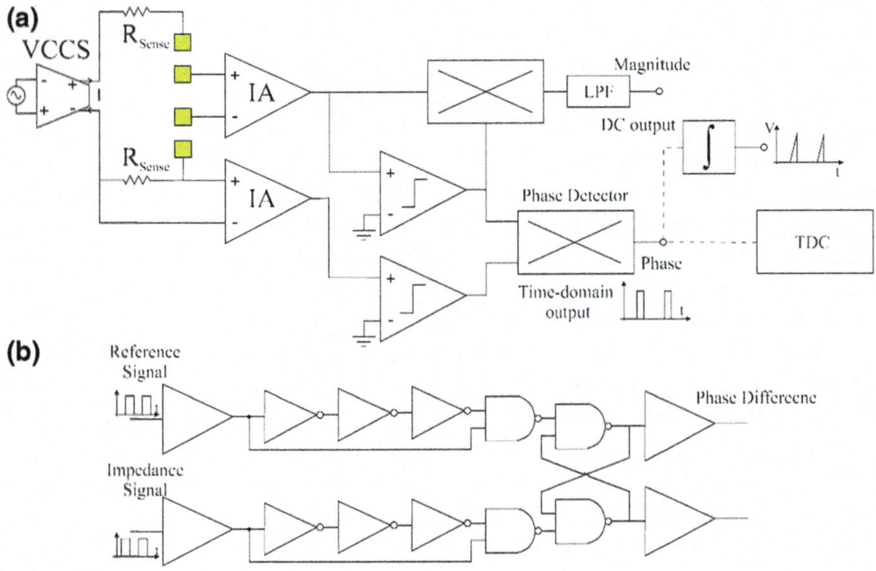

Fig. 5.39 **a** The magnitude phase system proposed in [58]. The phase signal following the phase detector (which can be e.g. a XOR gate or a chopper) can be integrated as in [58] or processed by a TDC circuit as in [272, 273]. **b** The phase detector proposed in [273]

with a pulse width proportional to the phase difference of the two square waves. This was then integrated with a G_m-C integrator to generate a sawtooth waveform, the slope of which was proportional to the phase difference of the two signals. This varying voltage signal can then be digitized and processed further in the digital domain. Since charge and discharge times are different, because the square wave input signal does not have a 50% duty cycle, the integrating capacitor had to be discharged at each cycle. This was achieved with a switch across the capacitor, where its control signal was one of the inputs. Instead of an XOR gate a chopper could have equivalently been used instead to provide the same operation. A disadvantage of the magnitude/phase measurement system is that while the magnitude is a DC voltage signal, as in SD, the phase is a time-domain signal and the pulse width proportional to the phase will vary significantly with changing stimulus frequency. As highlighted in [272], frequency sweeps between 1 kHz and 2 MHz will require time-domain digitization from sub-nanoseconds to microseconds. The signal-processing steps in the phase detection channel were developed further in [272, 273]. The method proposed in [273] used a set-reset latch preceded by rising-edge pulse generators. The phase detector is shown in Fig. 5.39b. The authors argue that this approach is superior to the XOR approach, as it has a monotonic transfer characteristic and is insensitive to the input duty cycle. A time-to-digital converter (TDC) was used to digitize the phase information following the phase detector. The TDC consisted of a time accumulator, digitizing the time the phase detector is high, and a pulse counter, counting the number of phase detector pulses. The 10 bit TDC solution proposed in [272] achieved a phase error less than 1° (0.52°) at 2.048 MHz. An intermediate resolution is initially achieved through a counter-based TDC, and additional bits are obtained by cascading time interpolation stages. Another disadvantage of the magnitude/phase approach is that the two channels are different, in contrast to the SD method. Thus, each channel will be governed by different error sources. In addition, the method is not inherently suitable for multi-frequency operation, as is SD, and additional BPFs are needed to isolate each frequency component. It is thus more suitable for spectroscopy applications. Instead of the multiplying magnitude detector, the circuit of Fig. 5.35 can also be used instead. However, as discussed in [274], the upper frequency limit is limited by the bandwidth and slew rate of the comparator and the lower is limited by the leakage at V_{out}. A different approach, based on the same principles for magnitude/phase detection, was proposed in [274], where both measurands are directly digitized using counters. In the magnitude channel, the input signal is compared with a voltage stored in a capacitor and the result of the comparison controls whether the capacitor will be further charged. The counter counts the number of charging pulses sent to charge the capacitor further. In the phase channel, the sine waves are converted into square waves as before, and these are then input into SR latches, which are reset when both signals go to zero. These are used to remove the effects of high-frequency low-magnitude noise, which may affect the outputs of the comparators. This mechanism, however, creates a tradeoff between maximum phase detection range and noise-inhibiting capability. An XOR gate calculates their phase difference. This is then AND-gated with the clock signal used

(5 MHz internal or larger externally) so that the counter can count only when the signal is '1'. The resolution increases with increasing clock frequency. A lag-or-lead detector extends the detectable phase range to ±180°.

Another important parameter for bioimpedance measurements related to both the front-end recording and the excitation electronics, is the effect of common-mode errors, for example due to electrode impedance mismatch. To reduce their amplitude and thus improve measurement accuracy, CMFB techniques are employed using a frequency selective network. The interested reader is referred to [275–277], where the subject is discussed in detail.

5.2.6 *Amperometric and Voltammetric Sensor Front Ends*

Amperometric and voltammetric sensors and instrumentation were discussed in detail in Chap. 2, focusing more on discrete off-the-shelf component realizations. Nevertheless, the majority of CMOS-based implementations do not differ significantly, as they are largely based on the simple feedback topology of Fig. 2.21, where the cell potential is kept constant by regulating it and the current through the cell is recorded through the working electrode. The most common approach to record the current through the cell is to use a transimpedance amplifier (TIA) connected to the working electrode. A resistor is often used as a feedback element to convert the redox current into an output voltage. However, depending on the application, this may not be a suitable solution, due to the chip area required by the large on-chip resistor necessary to record currents at the pico and nano scales, while thermal noise can be injected back into the sensor. An alternative approach is to use SCs to implement the resistive feedback, as in Fig. 5.40a. This improves the area efficiency, and the LPF characteristics eliminate high frequency noise. However, switching noise will be injected into the sensor in this case, affecting the recorded current. As discussed in [278], charge leakage from the switch can be minimized below 1 fA using the T-switch topology discussed earlier in Sect. 5.2.4

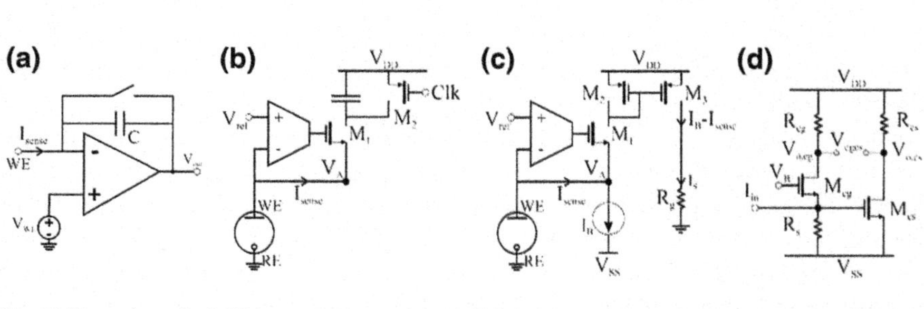

Fig. 5.40 **a** A typical TIA topology used in ASIC implementation. **b** The simple potentiostat topology proposed in [284] and **c** the current mirror-based solution proposed in [286], demonstrated here in a two-electrode amperometric system. **d** The common-gate based amperometric front end proposed in [287]

(Fig. 5.20e). The current is integrated through the feedback capacitor, which is periodically discharged through a switch, and the gain is given by

$$\frac{V_o}{I_{\text{sense}}} = \frac{1}{Cf_s}, \tag{5.175}$$

where f_s is the frequency of the clock driving the switch. This approach was used in [279] with a subsequent PGA, achieving a wide measurement range (from 6 pA to 10 μA) to accommodate sensors with different characteristics. CDS was used to reduce noise and remove offset, while the switch was implemented with a simple minimum sized transmission gate to minimize charge injection and dummy switches were used to minimize clock feedthrough. The amplifier was implemented with a folded cascode topology. Since CDS suppressed $1/f$ noise, the authors used nMOS input transistors, as their larger mobility means that they suffer less from thermal noise. In [280] the same approach was followed, only that the input pair in the folded cascode amplifier was sized differently. In addition, all the analog circuits were surrounded by capacitors connected between V_{DD} and ground to minimize power supply noise and the analog and digital routings were carefully laid out. As a result, an increased sensitivity by a factor of six was achieved (±500 fA sensitivity up to ±47 μA). Chopping can be used to improve the noise performance and reduce offsets, as in [278]. On-chip amperometric recordings using a TIA were also presented in [281], using a single TIA. In [282] a dual slope ADC was used to digitize the signal directly from the working electrode. AZ was used to remove the integrator offset and $1/f$ noise and decrease the mismatch between different array pixels.

An alternative is to use a current conveyer topology [80, 283]. As discussed in [80], in this approach the current is not directly converted into an output voltage, but the recorded current is conveyed to a high-impedance output node, where it is converted to a voltage. A novel implementation of such a circuit was presented in [80], where a comparison with a SC TIA demonstrated that the current conveyer topology was more suitable for small current measurements due to switching noise in the TIA. Chopping was used to reduce $1/f$ noise. This approach was also used in [81] for on-chip DNA analysis. Detailed discussions and comparisons between resistive and capacitive TIAs and current conveyors, especially with regard to noise, can be found in [80, 283]. In [283], measured results showed that the current conveyor topology examined demonstrated the worst noise performance, with capacitive feedback being better than resistive feedback current-recording topologies. The authors employed CDS to reduce $1/f$ noise.

In [284], the issue of on-chip amperometric recording arrays of transient oxidation currents with sub-millisecond resolution and picoampere sensitivity was examined. The system comprised a 2D array of on-chip electrodes, with dedicated front-end electronics to each electrode. As the application was to record chemical neural activity at a cellular level, integration of each sensor and readout was limited to 10–20 μm on each side. In order to achieve this, the authors proposed a topology where all the amplifiers in a given row of the 2D sensor array shared the same half circuit and consequently half the layout array and power were required.

The amplifier was based on a folded-cascode topology and was then used to implement the regulated cascode circuit of Fig. 5.40b, where the topology maintains the electrode at a contact potential of V_{ref} through feedback. The induced current due to the electrochemical reactions taking place at the sensor is buffered by the cascade transistor and integrated by a 50 fF capacitor. The potential across the capacitor is reset at constant intervals before saturation. Consequently, this will lead to linearly decreasing output voltages across the capacitor of different slopes as a function of time due to different input currents. The input impedance of the topology needs to be as low as possible in comparison to the impedance of the sensor interface, and it was estimated to be in the tens of kΩ for input currents in the hundreds of picoamperes, which is much smaller than the interface impedance. The input-referred noise performance of the topology was approximately equal to that of a regular folded cascode amplifier. CDS was used to reduce 1/*f* noise, as due to the area restriction for each cell, the input pMOS transistors cannot be made too big. The same group used a similar approach to [285] and Fig. 5.42a in [286], with the addition of an additional current source with a current I_B, as in Fig. 5.40c. The additional bias current allowed the measurement of bidirectional redox currents. The TI gain was set by the resistor R_g (1 MΩ) and the voltage was recorded with a unity gain buffer. Another system employing a similar topology to [285] is [69]. A common-gate amplifier-based TIA was presented in [287] and is shown in Fig. 5.40d. A common-gate input stage is combined with a common-source stage for noise cancellation, which is further enhanced with chopping. The input current is input to the source of the common-gate stage, which has low input impedance. The common-source stage cancels the non-idealities (noise and distortion) of the common-gate stage by proper common-source transistor sizing and resistor ratios according to

$$\frac{R_{cg}}{R_s} = g_{m,cs}R_{cs}. \tag{5.176}$$

The differential implementation of the topology with chopping achieved an input referred noise of 1.6 pA/$\sqrt{\text{Hz}}$.

A few amperometric front-ends that have been developed are based on quasi-digital converters such as that of [288], shown earlier in Fig. 2.13. Such quasi-digital approaches are discussed in Sect. 5.2.7. A 16-channel CMOS potentiostat with an input current recording sensitivity of 100 fA and a power consumption of 3.4 μW was presented in [289], while a telemetric system capable of recording simultaneously (through two separate channels) both neuro-potential and neuro-chemical activity (with a pA sensitivity measuring up to μA) from the brain with a power consumption of 35 μW was presented in [290]. Both systems used ΔΣ modulators to digitize the recorded signals. A current input ΔΣ modulator was also used in [291, 292] for amperometry and fast-scan cyclic voltammetry (FSCV). Oversampling ΔΣ modulators offer a tradeoff between conversion speed and resolution, and as reported in [292], a third-order ΔΣ modulator provides a good tradeoff between clock frequency, power consumption, and system complexity.

These are important for low-power operation, and for the applications that these systems are intended for, where monitoring of extracellular dopamine levels in the brain vary on sub-second to second timescales. In [292, 293] a stimulation system was also developed to generate neuro-chemical patterns, which were then analyzed with the FSCV front-end system. Appropriate timing ensured that FSCV recording was not performed simultaneously with stimulation. The FSCV system also incorporated on-chip FSCV waveform generation with a sweep rate of 400 V/s. These approaches will not be discussed further at this point, as ΔΣ modulators are discussed in detail in Sect. 5.3.3.

5.2.7 Quasi-digital Sensor Readout Circuits

Quasi-digital converter circuits, such as VFCs, are common in sensing applications. In a VFC, as the name suggests, the acquired voltage signal containing the necessary information is encoded into the frequency of a signal. This can be done using various methods. Typically, square wave oscillators are used, due to their simplicity and because only the frequency is of interest and the amplitude carries no information. Hence, such topologies combine the simplicity of analog circuits and the noise immunity of digital signals. Another advantage is that the square wave signal does not need synchronization and can be read by a microcontroller as with any serial data stream. In addition, VFCs typically require less chip area and power consumption than traditional ADCs with an equivalent resolution and accuracy, and they are not fast converters, being slower than successive approximation ADCs, but comparable to integrating ADCs [294]. As discussed in more detail in [294], VFCs are characterized by a number of parameters, which include:

(a) Input range: The range of input voltages, which correspond to a linear change in output frequency, f_0. There will be a lower and upper limit.
(b) Output range: The range of output frequencies varying linearly for a given input voltage range.
(c) Frequency span (*FS*): The output signal frequency range possible within the linear range.
(d) Slope (*S*), i.e. gain: The variation in frequency over the corresponding variation of input voltage, i.e. the derivative of the transfer function of the VFC, given in Hz V^{-1}. This is computed as $S = f_0/V_{in}$.
(e) Slope error: Normalized percentage change of the measured against the theoretical sensitivity.
(f) Relative error: The normalized percentage change of the measured to the theoretical response at each point within the linear range.
(g) Linearity error: The maximum deviation of the response from a straight line within the range of interest
(h) Offset: A constant frequency added to the expected theoretical output frequency.

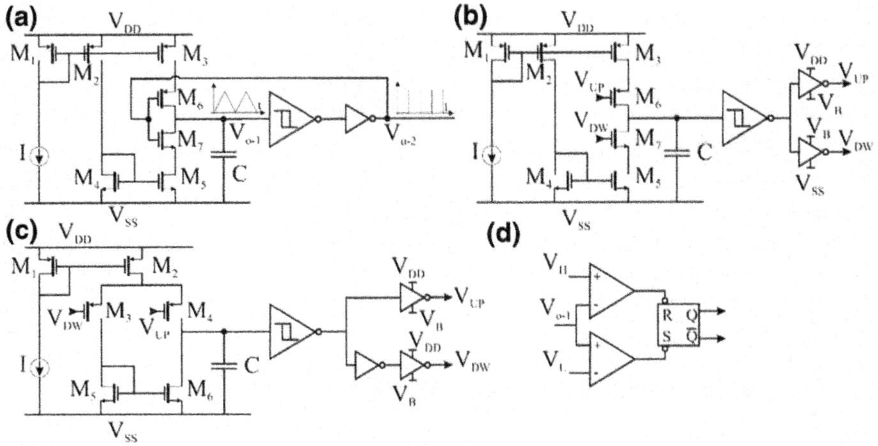

Fig. 5.41 **a** Conventional oscillator topology. **b** Use of separate control signals with different voltage swings. **c** Low-power oscillator topology using a single current source. **d** A standard voltage window comparator

(i) Power supply rejection ratio (PSRR): This quantifies the dependence of the VFC to variations in the power supply. It is expressed in dB and computed as $S^{-1}f_0/V_{DD}$.

Other measures include the temperature dependence of the above, leading to gain drift, linearity error drift, and offset drift [294].

5.2.7.1 Common Oscillator Topologies

As discussed, quasi-digital converters are based on oscillators, the output frequency of which is controlled by the measured signal. The most common topology used is shown in Fig. 5.41a. The circuit is comprised of a current-starved inverter, an inverter, a Schmitt trigger, a current source, a capacitor, and two current mirrors (M_1-M_2 and M_4-M_5), one used to charge and one to discharge the capacitor. Instead of a Schmitt trigger [295] it is common to use a windowed comparator circuit [296], as shown in Fig. 5.41d. In any case, this circuit compares the triangular wave generated across the capacitor with two threshold voltages. When these are reached, the circuit reverses the charging phase. The period (T) of the oscillating frequency is a function of the current (I) charging and discharging the capacitor and the current mirror scaling factor (K), the thresholds of the Schmitt trigger (V_H and V_L) or window comparator, and the capacitor (C) used, according to [158]

$$T = 2CK\frac{V_{\rm H} - V_{\rm L}}{I}. \tag{5.177}$$

All these parameters can be controlled by sensing signals to vary the frequency accordingly. If the sensor is capacitive, it can replace the capacitor in the circuit, and thus the frequency will depend on the capacitance only. If the sensor output is a current, e.g. in amperometric sensors, this can be used to control the current charging and discharging the capacitor. If the sensor is resistive, this can be used to control a voltage setting the integrating current or the bias current generation directly. Finally, if the sensor output is a voltage, it can be used to control the integrating current via a current to voltage converter, e.g. in potentiometric sensors. The control signal can be generated alternatively as in Fig. 5.41b, where the square wave control signal voltage swings can be controlled by the supply voltages used for the two inverters [294]. An alternative topology is shown in Fig. 5.41c. The advantage of this is that only a single current source is used (M_2) and thus this topology consumes less power. The threshold voltages of the standard Schmitt trigger circuit [295] depend on device dimensions, temperature, and supply voltage variations. The voltage-window comparator of Fig. 5.41d is thus preferred to overcome these limitations. A disadvantage is that the threshold voltages V_H and $V_{\rm L}$ need to be generated by additional circuitry, while comparator offsets will further hinder the accuracy of the topology. In addition, the comparator delay must not be comparable to the oscillator output frequency. To increase comparator speed, higher power is required. Another disadvantage is that two comparators are required. One way to circumvent this, as discussed in [294], is to switch off the bottom comparator when the capacitor is charging and the top one when discharging. An alternative is to use a single comparator topology, as discussed in [297].

5.2.7.2 Voltage-to-Current Converter Topologies

Before converting voltage to frequency, typically the voltage is first converted into a current through a voltage-to-current converter. The most common topology to

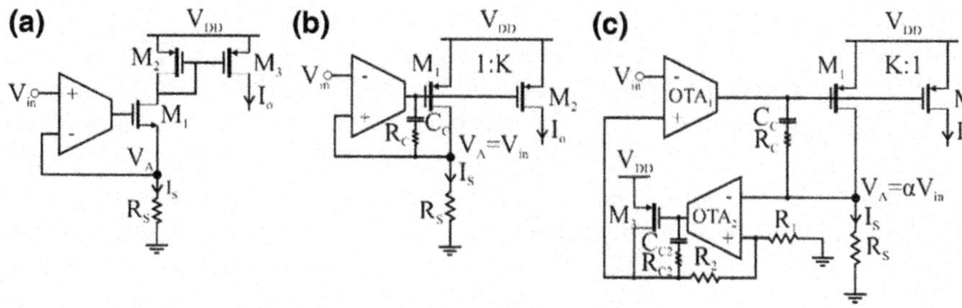

Fig. 5.42 **a** Convectional V-I converter. **b** OTA/common-source amplifier V-I converter. **c** The feedback voltage attenuation V-I converter (FBVA) proposed in [294, 302]

achieve this with high linearity and high input impedance is shown in Fig. 5.42a [298]. Since $V_{in} \approx V_A$, the input voltage appears across R_S, thus defining the current I_S. This is copied to the output node through the pMOS current mirror comprised of M_2 and M_3, which can scale the current depending on the ratio of their dimensions. Thus, ignoring channel length modulation, the transconductance of the circuit, G_M, is described by

$$G_M \approx \frac{1}{R_S} \frac{(W/L)_3}{(W/L)_2} \tag{5.178}$$

The simple current mirror can be replaced by a high-swing current mirror to enhance the output impedance and thus reduce the effects of channel length modulation for more precise copying of the current. However, both V-I topologies require a minimum input voltage to operate of $V_{in.min} = V_{DS1,Sat} + |V_T|$, which significantly reduces the operating range of the V-I converter to $V_{in} < V_{DD} - V_{in,min}$. According to [294], this makes the use of these current mirrors unsuitable for low-voltage operation. Alternative V-I topologies are presented in [299–301]; however, these have a low impedance, which depends on the input resistance and which is kept low to reduce the topology's footprint. To increase the input impedance a buffer is required, increasing the complexity, chip area, power consumption, and noise [294].

The OTA/common-source amplifier topology of Fig. 5.42b is essential a two-stage op-amp and extends the voltage operating range close to the supplies, as V_A can swing up to V_{DD}. Nevertheless, the input range is still limited to $V_{in} < V_{DD} - V_{DS1,Sat}$ to ensure that M_1 is in saturation. To improve the current copying, the transistors can be cascaded to improve the output resistance, in which case an additional $V_{DS,Sat}$ will be added to ensure that all transistors are in saturation. The amplifier formed by M_1 is an inverting amplifier. Thus, negative feedback is achieved by connecting V_A to the positive input of the OTA. The transconductance of the circuit is also given by (5.178). Peaks in the closed-loop frequency response and under-damped oscillations are compensated by the compensation network $R_C C_C$ [294]. There is tradeoff regarding the dimensions of M_1. The larger it is, the smaller its overdrive voltage and thus the larger the input linear range and the more valid the buffering approximation is. However, the bandwidth of the topology decreases with increasing M_1 dimensions.

In order to ensure a rail-to-rail input linear range, i.e. always keeping transistor M_1 in saturation, one approach proposed in [294, 302] is to scale down V_A from V_{in} to αV_{in}. The authors propose three methods to achieve this: feedforward voltage attenuation (FFVA), feedforward current attenuation (FFCA), and feedback voltage attenuation (FBVA). FFVA can achieve significantly smaller bandwidths [294, 302]. Here we will consider only the FBVA topology of Fig. 5.42c. In this topology the scaling factor is achieved by the non-inverting amplifier comprised of OTA_2, M_3, R_1, R_2, R_{C2} and C_{C2}, where $\alpha = R_2/(R_1 + R_2)$. The scaled voltage V_A is converted into a current via R_S and copied to the output node with a scaling factor K:1 via M_2, resulting to an output current I_0 of

$$I_0 = \frac{\alpha v_{in}}{KR_S}. \tag{5.179}$$

The transconductance is given by

$$G_M \approx \frac{\alpha}{R_S} \frac{(W/L)_2}{(W/L)_1}. \tag{5.180}$$

To achieve rail-to-rail operation, OTA_1 should also be a rail-to-rail topology. Since V_A is a scaled version of V_{in}, OTA_2 does not need to be rail-to-rail and a simple topology is sufficient. The transfer function of the topology is given by

$$G_M(s) = \frac{\alpha K}{R_S} \frac{GBW_1(\alpha GBW_2 + s)}{s^2 + \alpha GBW_2 s + GBW_1 GBW_2}, \tag{5.181}$$

where $GBW_1 = g_{m1}G_1R_S/C_{T1}$ and $GBW_2 = g_{m3}G_2(R_1 + R_2)/C_{T2}$, g_{mi} is the transconductances of M_1 and M_3, G_i is the corresponding OTA transconductances, C_{Ti} is the total capacitances at the output of each OTA, given by $C_{T1} = C_1 + C_{gs1} + C_{gs2} + C_C$ and $C_{T2} = C_2 + C_{gs3} + C_{C2}$, and C_i is the output capacitance of the OTA. This transfer function exhibits a zero and two poles, which can be cancelled out to extend the bandwidth if $GBW_1 \ll \alpha^2 GBW_2$. The above topologies can be used to create differential topologies with sensors having differential outputs [303], e.g. Wheatstone bridge-based sensors.

It should be highlighted that resistor R_S in all the above topologies needs to be temperature-independent. To achieve this, two resistors, one with a positive (TC_P) and one with a negative (TC_N) temperature coefficient (*TC*) need to be connected in series. In [294], the former was implemented by a P^+ non-salicide diffusion (PND) layer and the latter by a high-resistivity polysilicon layer. The temperature dependence of the resistors is given by [302]

$$R(T) = R_0\left(1 + TC_1(T - 25) + TC_2(T - 25)^2\right). \tag{5.182}$$

The overall *TC*s are given by

$$TC_i = TC_{iN}\frac{\beta}{\beta + 1} + TC_{iP}\frac{1}{\beta + 1}, \tag{5.183}$$

where TC_i, for $i = 1,2$ corresponds to the first- and second-order thermal coefficients, R_0 is the resistance at room temperature, and β is the ratio of the two resistances at room temperature ($\beta = R_{ON}/R_{OP}$).

5.2.7.3 Examples of Quasi-digital Converters in Sensing

In [304] a VFC was used for pH sensing. The working electrode was DC level shifted by a pMOS source follower, the output of which was used to drive an nMOS current source generating the control current for a topology similar to Fig. 5.41a. Within the normal physiological pH range of 7.38–7.42 the response of the system was linear. In [288], the current flowing between working and counter electrodes in a three-electrode amperometric cell for glucose monitoring was copied and used as the charging current of the topology of Fig. 5.41a using the voltage-windowed comparator of Fig. 5.41d. The potentiostat circuit used is essentially the V-I converter of Fig. 5.42b. A similar approach was used in [305]; however, the potentiostat used was essentially based on the topology of Fig. 5.42a. In [306], a capacitive humidity sensor was used, the capacitance of which controlled the output frequency of the circuit of Fig. 5.41a, essentially realizing a capacitance-to-frequency converter. A similar approach was used for the on-chip detection of the neurotransmitter dopamine [307]. In [308] a resistance-to-frequency converter was presented, where the oscillating frequency is a function of the resistance of a nanogap sensor. Similarly, a resistance-to-frequency converter for resistive gas sensors was presented in [309]. The V-I converter of Fig. 5.42a was used. A resettable Miller integrator was used instead. Temperature-to-frequency converters were presented in [296, 309]. In [296] two approaches were examined. A positive-to-absolute-temperature (PTAT) circuit was used to generate a temperature-dependent current and temperature-independent V_H and V_L in the first, and a temperature-independent current and complementary-to-absolute-temperature (CTAT) voltages in the second.

5.2.8 *Capacitive Sensor Read-Outs*

For completeness, we will briefly discuss here capacitive sensor readout circuits. A large range of sensors perform detection according to the change of capacitance. As discussed in the previous section, one simple approach to detect these changes is through the change of the oscillating frequency of a quasi-digital converter, since the oscillation frequency is determined by the capacitor among other parameters. Nevertheless, for high-accuracy, high-sensitivity measurements in applications such as capacitive accelerometers, other approaches are often preferred. As discussed in [101], important in the design is the high-impedance read-out node and the effects of parasitics and electromagnetic interference (EMI), while the readout packaging also needs to be considered as it will affect the measurement accuracy. Read-outs for this class of sensors are usually based on an AC bridge and a voltage amplifier with subsequent SD as in bioimpedance readouts [310, 311, 70], a TIA [312] or SC circuits [313]. We shall considerer a simple scenario where there are only two capacitors and a reference equal to a sense capacitor, which are driven by two 180° out-of-phase signals. In the AC bridge approach, the bridge is typically driven with square waves and the output of the bridge is proportional to the capacitance

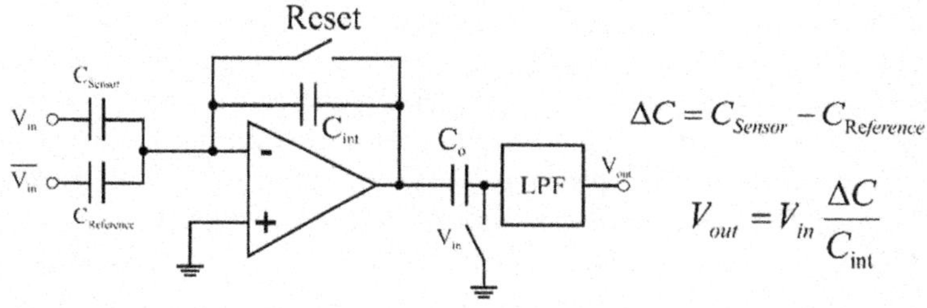

Fig. 5.43 The SC charge integrator of [101, 313]

difference. Shielding the interconnects with the shields driven by a unity gain buffer reduces parasitics, noise, and EMI. Thermal noise and parasitics will limit the accuracy of this approach [101]. Nevertheless continuous-time voltage measurement has a better noise performance than the TIA and SC approaches according to [70]. When the parasitics are small, e.g. when the read-out system and sensor are fabricated on the same substrate, this approach provides the best performance [101]. In the TIA approach, the AC bridge is driven by sinusoids to avoid distortion errors and the output current of the bridge is measured through the feedback resistor. As discussed in [101], there is an optimal drive frequency determined by resonance. An SC approach is preferred in [101]. This is shown in Fig. 5.43. Charge proportional to the capacitive difference is integrated in the feedback capacitor. This approach is associated with a large noise floor due to *kT/C* noise, the thermal noise due to the resistance of the switches and noise folding of sampled systems. The sampled noise of the reset switch is stored on C_0 after reset. The sampled voltage across C_0 is in series with the amplifier output when $\overline{V_{in}}$ is high. This is subtracted from the output and leads to cancellation of kT/C_{int} noise. CDS is used to reduce the effects of *kT/C* noise, but switch noise and noise folding will lead to a larger noise floor than the continuous-time approach for the same power consumption [70]. As discussed earlier, amplifier offset and 1/*f* noise are cancelled using CDS. The SC is ideal when there are large input parasitics, when e.g. sensor and electronics are not on the same substrate. The virtual ground at the inputs of the TIA and SC reduces the effect of the input parasitic capacitance. The SC circuits presented in [101] demonstrated a sensitivity of 75 aF with a sampling rate of 200 kHz and 20 aF with a sampling frequency of 1 MHz, with a 120 pF parasitic capacitance.

In [70], a differential AC bridge approach with fully differential signaling was presented to reduce the effects of common-mode and power supply noise. Chopping was used to reduce 1/*f* noise and offsets and to avoid noise folding. A first stage amplifier with small input transistors for capacitance matching was optimized for noise, providing a small gain. Thus, the second stage provided most of the gain. A DDA was used with three input pairs. One was used with a DC feedback loop and an external capacitor for offset elimination. One of the other pairs was used to eliminate the AC offset. The amplifier inputs were periodically reset to a DC voltage to suppress undesirable charging and eliminate bias voltage drift caused by it. Another approach is to directly interrogate a capacitive sensor through a ΣΔ modulator [314]. As discussed in [102], this leads to high power consumption, since the capacitors need to be charged and discharged at the high rate of the oversampling clock used in the modulator. The authors proposed the use of two C-V converters to sense the capacitances and a subsequent SC amplifier to amplify their difference. For lower power operation a low clock frequency is used for the SC (8 kHz). CDS was used to reduce the effect of the parasitic shunt conductance and chopping was used to eliminate mismatches between the two C-V converters. The low-pass operation of the subsequent ΣΔ modulator (operating at 128 kHz) filters the unwanted modulated frequency components. A CDS C-V converter for an implantable blood pressure sensor with capacitance offset cancellation was presented in [99]. A fully differential topology of a C-V converter using SC circuits and CDS was presented in [99]. The nominal capacitance of the sensor was cancelled out using an 8-bit reference capacitor bank such that a single-ended sensor with wide nominal capacitance range can be used, thus relaxing sensor development specifications. The capacitor bank is cycled through until a capacitor closely matching the capacitance of the sensor is found. A second stage provided a 2.5 V/V gain. A C-V converter for a capacitive pressure sensor was also presented in [315]. A two-stage approach was also implemented here. A pseudo-differential sensor bridge was used, consisting of two reference capacitors and two MEMS pressure sensors. Chopping was used to drive the bridge to minimize 1/*f* noise and offsets. At nominal conditions the bridge is balanced and the output should be zero. The amplifiers were based on the approach presented in [20]. Fabrication tolerances may lead to mismatching and unbalancing of the bridge, leading to an offset voltage that may saturate the recording amplifier. The DAC approach of [99] only provides tens of fF resolution. Consequently, the authors implemented a resistive DAC for fine-tuning the offset with a resolution of 0.78 fF. An SC approach was also used in [7] to realize a reconfigurable sensor interface for the interrogation of resistive, current, voltage, and capacitive sensors. For capacitive measurements, as in the other cases examined earlier, the same voltage signal is applied to the two input capacitors. The difference in the capacitances will lead to a differential voltage, which is then processed by the circuit. The circuit had a conversion gain of 43.7 mV pF^{-1}. A 3 bit capacitor array was used to calibrate the capacitor to accommodate manufacturing tolerances and match the two capacitors. This is important, since the parasitic capacitance can be, in bulk micromachined sensors, for example, three to five times larger than the sensing capacitance [316]. A time multiplexed AZ C-V converter was presented in [316] to provide an improved SNR.

5.2.9 Neural Stimulator Circuits

Although neural stimulator circuits are outside the context of sensor readouts and signal processing, here we will give a very brief overview of this family of circuits, as many implantable neural systems require one and they are necessary in closed-loop neural stimulation systems where, for example, neural stimulation parameters can be adjusted using information collected through a neural recording amplifier or neurochemical sensors at a distant location. Hence, neural stimulators are often combined with front-end amplifiers such as those discussed in Sect. 5.2.1. Neural stimulators are useful for a plethora of biomedical applications, both in the central (CNS) and peripheral nervous systems (PNS), including DBS for tremor and epileptic seizure suppression, vestibular prosthesis, epidural stimulation for paraplegics, retinal stimulation, and implantable hearing aids (cochlear implants). The stimulus signal shape can be optimized for selectively stimulating smaller nerve fibers over larger nerve fibers in the same nerve trunk, for example [317], while the stimulus pulse duration can be used as a parameter for spatial selectivity [318]. Monopolar stimulation electrode systems use a single working electrode in the vicinity of the stimulation target and a distant secondary electrode. Bipolar electrode systems use both electrodes in the vicinity of the target. Tripolar cuff electrodes are often used in the PNS, and multipolar systems have been proposed for a wide range of applications. Spatial selectivity can thus also be achieved by steering the stimulating electric field between a number of electrode pairs and by varying the intensity of the applied field to deliver stimulus to different areas within a volume conductor.

Many different types of stimulus waveforms have also been proposed. The monophasic stimulus is a repeating cathodic pulse. A symmetric biphasic stimulus consists of a repeating pattern of an initial cathodic pulse followed by an anodic pulse of the same amplitude. The cathodic pulse depolarizes the nerve and lowers the local extracellular potential with regard to the membrane potential, evoking in this way action potentials. The anodic pulse hyperpolarizes the nerve and raises the sensory threshold [319]. Nevertheless, the anodic phase reverses potentially damaging electrochemical processes that take place at the electrode/tissue interface and removes accumulated charge. It is thus important to match cathodic and anodic phases and to ensure that there is no residual charge left at the interface, as this will result to a DC current flowing through the interface, leading to electrode and tissue

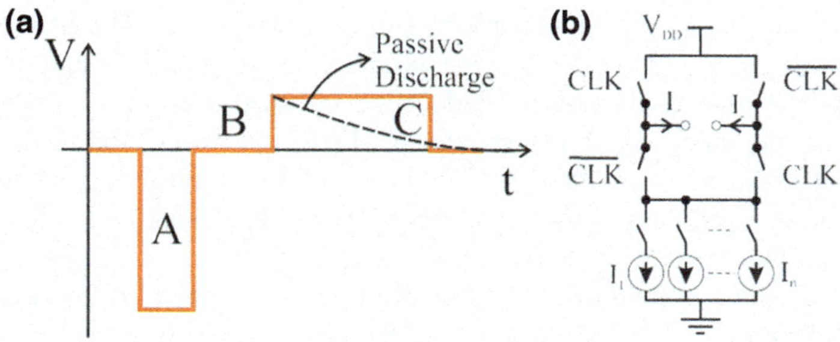

Fig. 5.44 **a** A biphasic stimulus waveform. *A* cathodic pulse, *B* interphase delay, *C* anodic pulse. Exponentially decaying passive discharge is also shown. **b** An H-bridge stimulator with multiple current sources to implement a biphasic stimulus as in [320, 321]

damage. Such a waveform, however, may hinder action potential generation and may lead over time to the need for high stimulus amplitudes for action potential generation. The proposed solution is the use of an asymmetric waveform, where the current amplitudes are different between phases, but the total charges are matched by different pulse durations, and to introduce an interphase delay. The former allows for charge redistribution and reduces fatigue in the cells [320, 321]. The latter allows for the generated action potential to propagate away from the stimulation point before applying the anodic phase. It also reduces crossbar conduction that will waste power [319]. Such a biphasic waveform is shown in Fig. 5.44a.

Imbalance with regard to the total charge between anodic and cathodic phases may be introduced by leakage currents, system failure, crosstalk between stimulation channels in multi-channel systems, and manufacturing tolerances. A large DC blocking capacitor (>100 nF) is often used for safety purposes in series with each electrode with a shunt impedance greater than 5 GΩ to block DC current flow, which also allows for passive (exponential) discharging of the interface. The anodic phase can be exponentially decaying instead of square. However, the DC blocking capacitors are often implemented off-chip due to their large size (up to a few μF), a solution which is restrictive in large multi-channel systems. In addition, the added capacitance at the output of the current source will increase the voltage compliance needed [319]. Many different solutions have been proposed, as reviewed in [33]. The electrode can be discharged passively by shorting it through switches periodically between pulses and during the interphase delay, or actively through a buffer to limit the current flow. In either case, this will be a slow process due to the large electrode *RC* constant and will depend on the charge imbalance and the time available to perform such methods [319]. The large DC blocking capacitor can be used as a feedback element in an integrator to measure the charge with an SC integrator, ensuring that switching transients must be minimized for successful operation. Feedback techniques can be used to monitor the residual charge and compensate it dynamically in order to avoid the use of these capacitors [33]. Matching of the current sources and DACs controlling the currents and the current sources and sinks [319] is another method.

As in bioimpedance applications, either current or voltage can be used to stimulate, but charge-based stimulators have also been proposed [33]. Current stimulators are the most common method used, where active current sources and sinks provide the stimulation current waveform. Once again, high output impedance relative to the load is essential to deliver the intended waveform. Voltage-based stimulators are much more power-efficient [320]. In voltage-based stimulation systems, in order to ensure that the current is limited, the stimulus voltage waveform must approach the voltage waveform that would form across the interface if a current stimulus were to be used [33].

The stimulation system of [322] provides biphasic stimulation and matches the anodic and cathodic phases by calculating the difference between the two stimulus currents and compensating for it, while it also calibrates the DAC used in the current sources. Regulated current sources were used along with methods to increase the output impedance. An H-bridge circuit was used in [320, 321]. The H-bridge allows the same current source to be used at each stimulus phase, guaranteeing in this way that the charges are matched. This is only valid when a balanced symmetric biphasic stimulus is used. An H-bridge for asymmetric biphasic stimulus waveforms that ensures charge balance over time was proposed in [320] and a simplified schematic is shown in Fig. 5.44b. The high cathodic pulse is formed by N unit current sinks connected in parallel to apply the stimulus current for a time T. In the anodic phase only one of the unit current sinks is used to sink the current for a time of NT. In a single cycle the two phases will not be balanced, but if the current sink is switched N times between the N different sinks, over time the residual charge will be removed. A charge mismatch of 0.45% was reported. In the biphasic stimulation H-bridge circuit implemented in [321] a current path to a dummy load is provided before and after each stimulation phase to ensure a smooth transition by minimizing charge build-up and spiking. A dynamic balancing method for a DC-blocking capacitor free stimulator was proposed in [319], achieving a current imbalance of 0.4%, translating to 6 nA. The solution proposed in [323] implemented a multiphase compensation scheme and achieved a current error of 4.5 nA. Various safety measures should be taken to ensure proper system performance and to address various failure mechanisms. In [324] a temperature sensor is incorporated to ensure that the temperature at the implant surface does not rise more than 2 °C. Electrostatic discharge (ESD) protection circuits are also used at the electrode pads, as they can damage the drivers, which may then permanently short electrodes to the supplies, severely damaging the tissue. In addition, the supply is monitored, as if it drops too much uncontrolled stimulus patterns may be applied to the tissue due to improper operation of the implant. The compliance is also monitored to adjust the stimulus current to ensure charge balance. If the electrode impedance is too large, the compliance limits of the stimulator may be reached, leading to inappropriate stimulation. Finally, to ensure charge balancing a dynamic current copying method is implemented, further enhanced with electrode shorting at various potentials [324].

5.3 Data Converters

Analog signal conversion to robust two-level representation in digital form is crucial for a sensing system, particularly when signal transmission over a data link is required. Table 5.1 surveys the latest developments in low-power data converters. The table of energy-efficient data conversion is dominated by two architectures: SAR ADCs and time-based encoding ADCs, including those of the Δ-Σ type. The SAR ADC architecture was originally proposed for robust conversion at moderate data rates and resolution. The typical resolution achievable by a SAR ADC is up to 14 bits. The main advantage of the SAR ADC is that the charge-sharing architecture with minimum active analog blocks allows a very robust conversion. The Δ-Σ

Table 5.1 Comparison of various ADC implementation

References	Type[a]	Resolution[a] (ENOB)	Data rate/ signal BW	Analog power	Digital power	Power supply	FoM[b] E/step	Tech. (μm)	Area (mm^2)	Design highlights
[325]	SAR	8.02	100 ksps	0.62 μW	0.71 μW	0.9 V	53 fJ	0.18	0.15	Cap array sampling up to 2 V_{dd}
[326]	SAR	8.51	240 Msps	0.29 mW	0.4 mW	1 V	7.8 fJ (3 pJ)	0.028	0.003	Extended binary search DAC
[327]	SAR	9.73	16 ksps	72 nW		0.6 V	3.7 fJ	0.18	0.12	LSB-first search DAC, reduces charge leakage error
[328]	ΔΣ	9.6	40 MHz	2.57 mW		0.9 V	42 fJ	0.04	0.017	VCO time-based quantizer
[329]	TB	4.4	1.2 kHz	60.3 nW	160 nW	0.3 V	4.3 pJ	0.13	0.357	0.3 V supply comparator
[330]	SAR	7.56	27 ksps	400 nW	115 nW	1.2 V	89.49 fJ	0.13	0.0326	Combined DAC with input switched-cap amplifier
[331]	DS	6.05	160 Hz	93.5 nW	16.5 nW	3.6 V	5.3 pJ	0.18	0.105	Coarse/fine resolution comparator
[332]	SAR	10.8	135 kHz	2.6 mW	1.44 mW	1.25 V	42.4 fJ	0.065	0.044	Passive amplification of sampling circuit
[333]	ΔΣ	14.8	250 Hz	10.7 μW		1.2 V	760 fJ	0.065	0.2	Two-step architecture versus high-order architecture
[334]	TB	7.9	3.3 kHz	582 nW		0.8 V	565 fJ	0.18	0.045	Closed-loop level-crossing time-based architecture

(continued)

Table 5.1 (continued)

References	Type[a]	Resolution[a] (ENOB)	Data rate/ signal BW	Analog power	Digital power	Power supply	FoM[b] E/step	Tech. (μm)	Area (mm^2)	Design highlights
[211]	SAR	9.2	100 ksps	74.1 nW	315.9 nW	0.6 V	6.7 fJ	0.18	0.103	Hybrid DAC structure
[335]	ΔΣ	16.0	24 kHz	180 μW	100 μW	1.8 V	88 fJ	0.18	1.25	CT-ΔΣ with FIR DAC feedback
[336]	SAR	9.34	200 ksps	180 nW	860 nW	0.6 V	3.8 fJ	0.18	0.082	Adaptive DAC switching sequence
[337]	SAR	7.5	1.1 Msps	1.2 μW		0.5 V	6.3 fJ	0.04	0.0112	Digitally modified tri-level comparator
[338]	ΔΣ	12.8	20 kHz	22 μW	6.6 μW	0.6 V	97 fJ	0.13	0.105	Fourth-order feedforward ΔΣ loop
[339]	TB	5.28	200 Msps	3.6 mW		1.2 V	463 fJ	0.065	0.02	Gate-based time "amplifier"
[340]	TB	7.9	300 kHz	7 μW	7 μW	1/0.4 V	98 fJ	0.09	0.06	Input-time conversion compared with reference ramp time
[341]	SAR	8.75	1 Msps	1.9 μW		1 V	4.4 fJ	0.065	0.026	Stepwise charging DAC cycling
[342]	SAR	9.3	100 ksps	1.3 μW		0.6 V	21 fJ	0.18	0.125	Time-based comparator based on voltage-controlled delays
[343]	SAR	9.1	1 ksps	2 nW	51 nW	1/0.4 V	94.5 fJ	0.13	0.191	Dual supply reducing digital power

(continued)

Table 5.1 (continued)

References	Type[a]	Resolution[a] (ENOB)	Data rate/ signal BW	Analog power	Digital power	Power supply	FoM[b] E/step	Tech. (μm)	Area (mm^2)	Design highlights
[344]	PL	10.4	50 Msps	3.5 mW		1.3 V	52 fJ	0.065	0.16	SAR-pipeline hybrid architecture
[345]	SAR	9.83	768 ksps	21.5 μW	36.5 μW	3.3/ 1.8 V	74 fJ	0.18	0.0106	Digital calibration for capacitor mismatch
[222]	ΔΣ	13.7	100 kHz	139 μW	1.55 μW	1.5 V	54 fJ	0.18	0.49	Time-interleaved single op-amp for third-order modulator
[346]	SAR	7.44	200 ksps	1.49 μW	0.98 μW	0.9 V	65 fJ	0.18	0.7	Rail-to-rail comparator with p/ nMOS input stages
[347]	SAR	7.77	10 Msps	26.3 μW		1 V	12 fJ	0.09	0.055	Asynchronous logic and small unit DAC capacitor

[a]*TB* time-based; *PL* pipeline; *DS* dual slope; *SAR* successive approximation register; *ENOB* effective number of bits
[b]Figure of merit calculated according to FoM = Power/(2 *BW* 2^{ENOB})

converter is invented for high-resolution and low data rate conversions. High resolution is achieved by oversampling, which compensates for the relatively low performance of analog components. Time-based conversion ADCs convert analog signals into a robust time-coded digital representation, much like the frequency modulation schemes used in telecommunication. The advantage of time-encoding ADCs is the minimalistic analog block and digital control circuit requirements. This leads to robustness as well as simplified design efforts and lower power consumption. For the rest of this section, we survey the latest developments of these efficient ADC architectures and provide quantitative as well as transistor-level insights into their designs.

5.3.1 SAR Architecture and the Capacitor Array

The SAR ADC operates by successively approximating the input voltage with a discrete approximation. This is commonly carried out by using a DAC to output the corresponding approximation value (Fig. 5.45a). This approximation value is compared with the input voltage and a better approximation value is derived in the next cycle by the comparator result. In practice, the DAC is commonly implemented with a capacitor array. The selection of different DAC output voltages is accomplished by connecting the capacitors to appropriate reference voltages for each conversion cycle.

To illustrate the operation of an SAR ADC in a simple configuration, consider Fig. 5.45b, where the entire capacitor array with a total capacitance C is connected between the input voltage v_{in} and a reference voltage v_{ref}. The charge sampled on the capacitor array is given by:

$$Q = C(v_{\text{in}} - v_{\text{ref}}). \tag{5.184}$$

In the next cycle, the switches connecting the array to the input and reference voltages are switched off. The capacitor array is instead switched with some capacitors (totaling aC, $a \leq 1$) switched to connect to v_{ref} whereas a total of

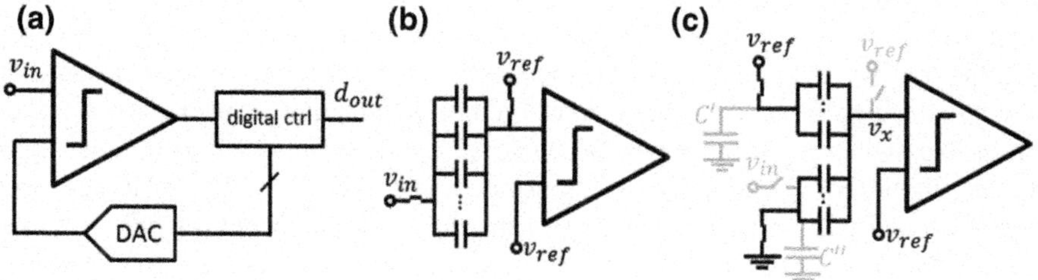

Fig. 5.45 **a** SAR ADC architecture with general DAC. **b** Typical input sampling using SAR ADC. **c** Conversion stage with capacitor array DAC

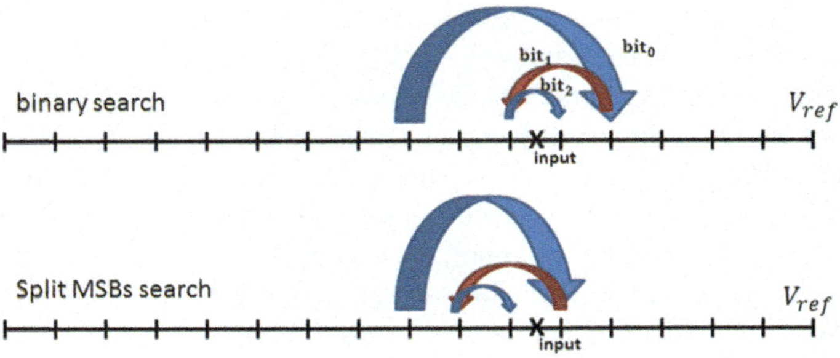

Fig. 5.46 Binary search DAC versus searching with split MSB capacitors

$(1-a)C$ capacitance is switched to connect to ground. Since the sampling switch at the input of the capacitor is off, the total charge held on the capacitors is unchanged (ignoring switch leakage). Connecting one terminal of the capacitor to v_{ref} or ground simply changes the voltage v_x (Fig. 5.45c) applied to the comparator:

$$v_x = v_{\mathrm{ref}}(1+a) - v_{\mathrm{in}}. \tag{5.185}$$

The comparator thus compares av_{ref} and v_{in}. For a binary weighted capacitor array with $1/2C, 1/4C, \ldots, 1/2^N C$, the theoretical (assuming linear array and ignoring noise contributions) finest resolution is given by $1/2^N v_{\mathrm{ref}} = 1\mathrm{LSB}$ (least significant bit). A typical successive approximation algorithm is to carry out a binary search approximating v_{in} by $a_N v_{\mathrm{ref}}/2 + a_{N-1}\ v_{\mathrm{ref}}/4 + \cdots + (a_0\ v_{\mathrm{ref}})/2^N$. This binary search process is illustrated in Fig. 5.46. If the supply voltage V_{dd} is used as the reference voltage, the switches that selectively connect the capacitors to V_{dd} or ground can be replaced by inverters. The switching of the single capacitor terminals between v_{ref} (which often equates to V_{dd}) and ground does not require the shifting of charge equivalent to charging/discharging the capacitors being switched. This is because the overall charge on the capacitor array is constant due to the floating node v_x during the comparison cycle. The capacitors charged/discharged during the SA process are given by the parasitic capacitance at the nodes (C′ and C″ in Fig. 5.46b). These parasitic capacitances scale with the SAR capacitor connected to it and can have significant values. Charging/discharging them constitutes the dynamic power consumption of an SAR ADC. One way of minimizing the power required when switching the voltage on one side of the capacitor array between ground and v_{ref} is to use a stepwise switching scheme, where the charge switched out when we go from v_{ref} to ground is stored on intermediate capacitors and reused when we switch back from ground to v_{ref} [341]. This approach introduces additional digital control circuits that consume extra power. In [217], it has been demonstrated that stepwise switching of the three largest capacitors only achieves significant power reduction over a 1 V supply, where digital power consumption is moderate. For the binary search scheme, the initial MSB (most significant bit) search requires the charging of a parasitic capacitance associated with half of the

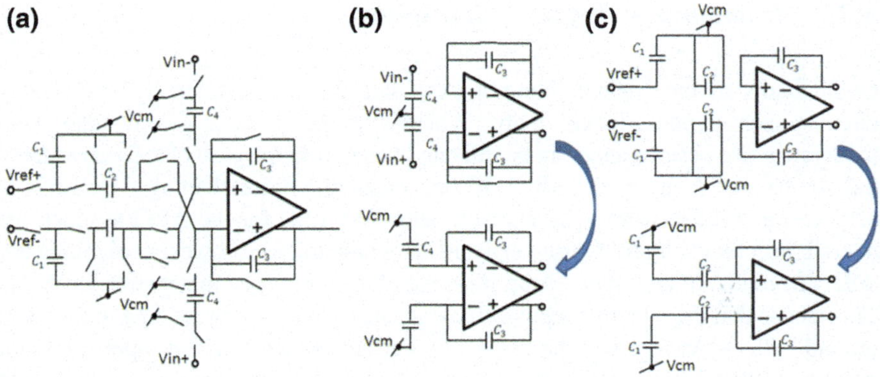

Fig. 5.47 SC SAR ADC. **a** Overall architecture. **b** Input voltage sampling. **c** Conversion steps

capacitors in the array. As well as the fact that charging a large parasitic capacitor in this initial conversion step requires a significant amount of dynamic power, the charging of larger capacitors also requires the driving amplifier of the reference voltage (if V_{dd} is not used as the reference) to be able of driving the maximum load. Depending on the approach, this will either result in an increased bias current consumption of the driving amplifier, or an increase in the design complexity of incorporating class AB output stages. This places a practical lower bound on the supply voltage, resulting in limited power savings. In [326], the MSB capacitors are split into multiple sub-capacitors. The search algorithm is modified (see Fig. 5.46) to accommodate for this. The maximum "leap" thus requires a smaller voltage change on the capacitor, reducing the requirements of the driving amplifier with the switching of the split capacitor interleaved into later steps.

In single-source AD conversion, input signals are usually band-limited to avoid aliasing in signal reconstruction and limit noise folding. The practical implication is that the input signal will not vary much between conversions. In [327], this input property is exploited by initializing the binary search at the previous conversion output. The search then starts at the immediate neighbors of the starting level (i.e. one LSB step) in the hope that the new sample is close to the previously converted sample. This works well in practice, resulting in improved conversion efficiency. Redundant bits are sometimes introduced at the end of the conversion cycles to mitigate comparator offset. This works by cycling backwards assuming the last decision was incorrect for cross-checking.

Instead of utilizing a capacitor array as the DAC in the SAR ADC, an alternative DAC structure based on an SC amplifier circuit has been demonstrated in [206]. Referring to Fig. 5.47a, b, the input sampling is carried out by C_4. Since series capacitors carry the same charge, the charge stored by applying the input voltage gets distributed over C_3 when C_4 is connected to the op-amp, creating an initial voltage gain of C_4/C_3. The conversion of this is then carried out by successively adding or subtracting the binary scaled reference voltage $V_{ref+} - V_{ref-}$ as in Fig. 5.47c.

5.3.2 Comparator Noise Optimization

Comparators are the staple building block of data converters and many other analog circuits, such as phase-locked loops. Whereas in certain cases a comparator with moderate noise performance can be tolerated (such as in the case of ΔΣ converters), the comparator used in an SAR ADC and other non-noise shaping ADCs has to achieve input referred noise levels well below 0.5 LSB for the ADC to reach the overall accuracy target. Noise analysis on comparators, however, significantly differs from that of amplifiers. Amplifiers are often configured in negative feedback. This means that the effective input differential signal is small, producing the "virtual ground" effect. As a result, the respective currents flowing in the input transistor pair of a differential amplifier have little variation during operation (except when the amplifier is output current limited—slewing). The input transistors, which dominate the noise contribution, can be modeled as a constant transconductance g_m device operating in the small signal regime—leading to a much simplified noise analysis. The comparator, in contrast, is mostly used in open loop with large signal swings at the input. This means that the small signal noise model used to analyze and optimize amplifier performance cannot directly be used for comparators. One way of analyzing the noise performance of comparators is to model the comparator as a time-varying piecewise-linear device, with critical transitions during its operation dividing up the piecewise-linear regions. Latch comparators are commonly used in Nyquist rate ADCs such as the SAR ADC due to their discrete-time nature. A typical latch comparator consists of a "degenerative" circuit in positive feedback to amplify the differential input and reset switches for the degenerative circuit to prepare for each comparison cycle. Figure 5.48 shows an example of a latch comparator.

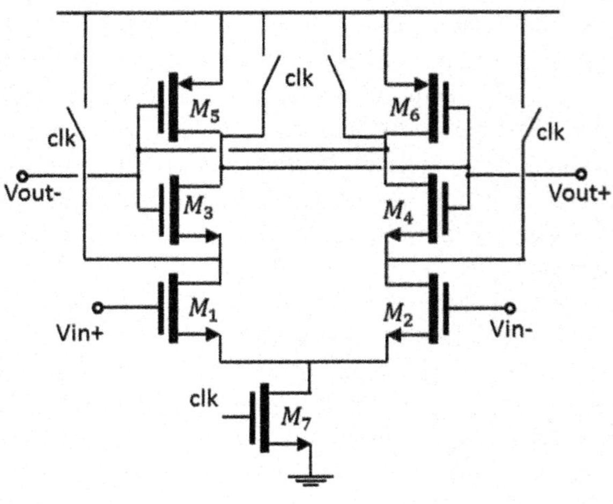

Fig. 5.48 A latch comparator with vertically stacked amplifier and degenerative circuit

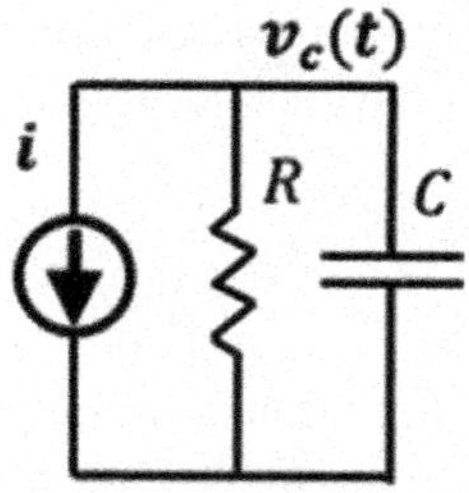

Fig. 5.49 Simple model for each of the piecewise-linearized latch comparator operation phases

To describe the large signal noise behavior of the comparator, one can carry out analysis in the time domain by means of solving a set of differential equations expressing the output voltages at each piecewise linear phase as a function of the inputs. Subsequent analysis is then carried out to express the output statistics (such as variance, which determines the noise power) in terms of the input statistics. A simple model for each of the linearized operation phases of a comparator is shown in Fig. 5.49, consisting of a current source driving an *RC* load. By solving the KCL defined differential equation in $v_c(t)$, we obtain the output voltage given an arbitrary input current $i(t)$:

$$v_c(t) = \exp\left(-\frac{1}{RC}t\right) v_c(0) - \frac{1}{c}\exp\left(-\frac{1}{RC}t\right)\int_0^t \exp\left(\frac{1}{RC}\tau\right) i(\tau)\mathrm{d}\tau. \tag{5.186}$$

For noise analysis of white noise, $i(t)$ is the drain noise current of a transistor with spectral density $4kT\gamma g_m$ with $k = 1.38 \times 10^{-23}$J K^{-1} and γ being a constant between 0.75 and 3 for most transistors. We further write $i(t) = 2k\sqrt{T\gamma g_m} i'(t)$ with $i'(t)$ drawn from a zero mean, unit variance Gaussian distribution. This arrangement allows one to take advantage of the properties of stochastic integrals such that, given that $i(\tau)$ is white noise, we have $i(\tau)d\tau = \mathrm{d}W$. For a Wiener process W, we also have:

$$E\left[\int_0^t G\mathrm{d}W\right] = 0. \tag{5.187}$$

$$E\left[\left(\int_0^t G\mathrm{d}W\right)^2\right] = E\left[\int_0^t G^2\mathrm{d}W\right]. \tag{5.188}$$

The aim of the statistical analysis is to derive the mean $E[v_c(t)]$ and variance $E\left[v_c^2(t)\right]$ of $v_c(t)$ given $E[i(t)]$ and $E[i^2(t)]$. By taking the expectation on both sides of (5.186) and using the properties of integrals of a Wiener processes, one can derive the mean as:

$$E[v_c(t)] = \exp\left(-\frac{1}{RC}t\right)E[v_c(0)]. \tag{5.189}$$

The mean of the output voltage thus decays with time and is dependent on the mean of the initial voltage across the capacitor. In an SC circuit, this initial voltage is also random and depends on the noise of the circuits that previously charged up the capacitor. For the variance, we have:

$$E\left[v_c^2(t)\right] = \mathrm{e}^{-\frac{2}{RC}t}E\left[v_c^2(0)\right] + \frac{2R\gamma kTg_m}{C}\left(1-\mathrm{e}^{-\frac{2}{RC}t}\right). \tag{5.190}$$

For a large resistor R such that the charge from the current source is delivered entirely to the capacitor, one can obtain the expression below by applying L'Hôpital's rule to (5.190):

$$E\left[v_c^2(t)\right]_{R\to\infty} = \frac{4kTg_m}{C^2}t. \tag{5.191}$$

This reveals that the output noise power grows with time for large R. The approximation holds for practical large resistances in the circuit. In practice, the observation time is limited by clock switching to a different operation phase in SC circuits. Based on the analysis of this simple RC circuit, [325] analyzed the

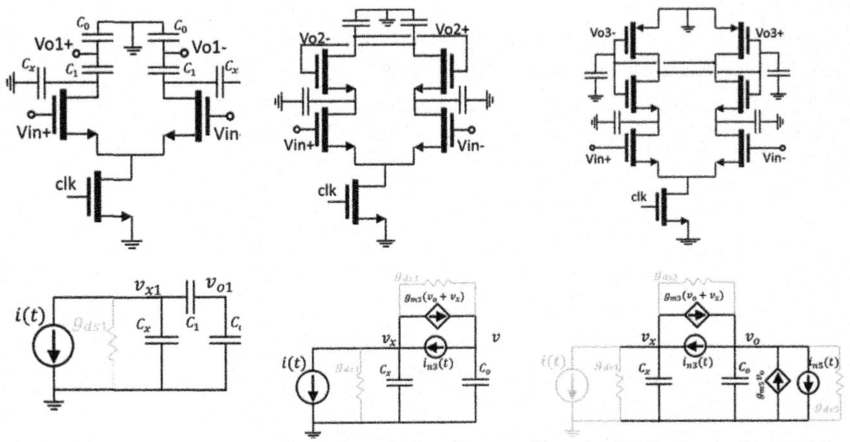

Phase 1: only M_7, M_{1-2} are on. Switches off. Up to v_x reaching $V_{dd} - V_{Tn3,4}$ turning M_{3-4}on.

Phase 2: interval where M_{1-4} are in saturation. Up to $g_{ds1} \geq g_{m1}$ taking M_{1-2} out of saturation

Phase 3: On the degenerative transistors M_{3-6} are active. End of phase 3 happens when transistors in the latch enters the triode region.

Fig. 5.50 Piecewise linear time domain analysis of a latch comparator

comparator in Fig. 5.48 by dividing its operation into three phases according to sequence of transistors switching on. In each of the phases, the linear models resemble noise current charging of *RC* networks. The operation phase and linear approximations are shown in Fig. 5.50.

In Phase 1, if we assume a large output resistance of the input transistors (which is reasonable given they're both in saturation at the beginning of Phase 1), we have the following noise power at the output and node X respectively:

$$E\left[v_o^2(t_1)\right] = \frac{2kT}{C_o}. \tag{5.192}$$

$$E\left[v_x^2(t_1)\right] = \frac{2kT}{C_x} + \frac{4kT\gamma g_{m1}|_1 t_1}{C_x^2}. \tag{5.193}$$

In Phase 2, M_3 and M_4 (Figs. 5.48 and 5.50) start to turn on and are included in the half circuit. We again ignore the output resistance of M_3 (or M_4) and introduce an extra drain current noise. By solving simultaneous equations in v_o and v_x, one arrives at variance expressions as exponential functions of transconductance, capacitor ratios, and the duration of Phase 2. By approximating these exponential functions up to their second-order Taylor series term only, the output noise power at the end of Phase 2 is given by:

$$\begin{aligned} E\left[v_o^2(t_2)\right] = & \frac{C_x^2}{(C_0 - C_x)^2}\frac{(t_2 - t_1)^2}{\tau_2^2}E\left[v_x^2(t_1)\right] \\ & + \frac{\left(C_o - C_x + C_x\frac{t_2 - t_1}{\tau_2}\right)^2}{(C_o - C_x)^2}E\left[v_o^2(t_1)\right] \\ & + \frac{4kT\gamma g_{m3}|_2(t_2 - t_1)}{C_o^2}, \end{aligned} \tag{5.194}$$

with $\tau_2 = C_x C_o / g_{m3}|_2 (C_o - C_x)$. The condition for the engagement of the latch operation is also revealed during the analysis process. This is given as $C_o < C_x$ and is mostly satisfied by an output load capacitor. For Phase 3, transistors M_5 and M_6 are turned on with input transistors having minimum contributions while they are fully turned on. Following a similar process as for deriving the output variance in Phase 2, and assuming the duration of Phase 3 is greater than the time constant dictated by the output node $\tau_3 = C_o/\left(g_{m3}|_3 + g_{m5}|_3\right)$, we have:

$$E\left[v_o^2(t_3)\right] = \left(\frac{2kT\gamma}{C_o} + E\left[v_o^2(t_2)\right]\right)e^{\frac{2(t_3 - t_2)}{\tau_3}}. \tag{5.195}$$

For each phase, one can derive an input–output voltage expression. Combining the three phases, the overall transition function over the piecewise linear model is given by:

$$v_{\mathrm{o}}(t_3) = -v_{\mathrm{in}} \frac{g_{\mathrm{m1}}|_1 t_1 g_{\mathrm{m3}}|_2 (t_2 - t_1)}{C_x C_{\mathrm{o}}} \mathrm{e}^{\frac{t_3 - t_2}{\tau_3}} = v_{\mathrm{in}} G_{\mathrm{eq}}. \tag{5.196}$$

The "input-referred noise" is simply the ratio of $E\left[v_{\mathrm{o}}^2(t_3)\right]$ and G_{eq} combining results from all the phases $\sigma_n^2 = \sigma_1^2 + \sigma_{\mathrm{s1}}^2 + \sigma_{3-6}^2 + \sigma_{\mathrm{s3}}^2$, with contributions $\sigma_1^2, \sigma_{\mathrm{s1}}^2, \sigma_{3-6}^2$, and σ_{s3}^2 given by:

σ_1^2	σ_{s1}^2	σ_{3-6}^2	σ_{s3}^2
M_1, M_2 noise	Noise sampled on C_{o}	Noise from M_3–M_6	Noise on C_x
$\frac{4kT\gamma}{g_{\mathrm{m1}}\|_1 t_1}$	$\frac{2kTC_x^2}{g_{\mathrm{m1}}^2\|_1 t_1^2 C_o}\left[1 + \frac{2}{g_{\mathrm{m3}}\|_2(t_2 - t_1)} + \frac{C_o^2}{g_{\mathrm{m3}}^2\|_2(t_2 - t_1)^2}\right]$	$\frac{2kT\gamma C_x^2}{g_{\mathrm{m1}}^2\|_1 t_1^2 g_{\mathrm{m3}}\|_2(t_2 - t_1)}\left[2 + \frac{C_o}{g_{\mathrm{m3}}\|_2(t_2 - t_1)}\right]$	$\frac{2kTC_x}{g_{\mathrm{m1}}^2\|_1 t_1^2}$

The result of the time domain noise analysis on the piecewise linear comparator model reveals that noise scales inversely with transconductance, as in the case of small signal analysis. While keeping the capacitance C_x at the drain of the input transistors small, one can size the output capacitor with C_x to maximize the time transition t_2. The noise analysis presented in [348] applies to general latch comparators and sheds light on noise optimization in the time domain, introducing operation phase transition time as a key parameter.

Latch comparators operate in synchronization with the clock and are commonly found in clocked ADC circuits. For asynchronous ADCs such as the time encoders detailed below, threshold detection comparators comprising a preamplifier and threshold comparator—essentially open-loop amplifiers—are used. The low-noise design and analysis of threshold comparators based on a simple case is discussed below. We can also begin this from the single transconductance-*RC* model (Fig. 5.51). For white noise drain current $i_n(t)$ with two-sided PSD $S_{xx}(f) = 2kT\gamma G_{\mathrm{m}}$, the output voltage variance due to this noise current is given by:

Fig. 5.51 Comparator pre-amp model

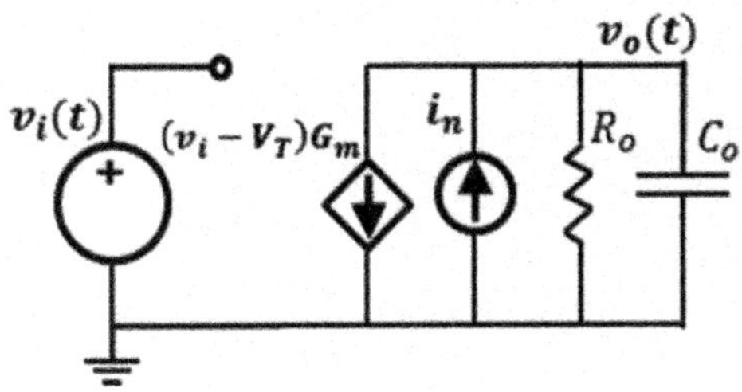

$$\sigma_o^2 = S_{xx}(0) \int_{-\infty}^{t} |h(\tau)|^2 d\tau = 2kT\gamma G_m \int_{0}^{t} \frac{1}{C^2} e^{-\frac{2\tau}{RC}} d\tau = \frac{kT}{C}\gamma G_m R\left[1 - e^{-\frac{2t}{RC}}\right] u(t), \tag{5.197}$$

with $h(\tau)$ being the impulse response from noise current to output voltage and $u(t)$ is the unit step function. Thus the output noise variance is time-varying. Approximating the exponential by a first-order Taylor series term, the output voltage noise resembles a Wiener process with variance scaling linearly with time: $\sigma_o^2(t) = u(t)\frac{2kT\gamma G_m}{C^2}t$. This approximation holds in practice as we consider the initial transient response of the system where $t \ll RC$. For longer elapsed time, the system settles to a steady state and we have a steady state output noise variance of $kT\gamma G_m R G_m$.

From the output noise of the preamp, one can derive the input referred noise of the comparator. Whereas for amplifiers we are interested in the corruption of noise on the output voltage, for comparators the important parameter is the output timing jitter induced by noise. The input-referred noise is thus defined as the equivalent noise source at the input that will result in the same output jitter as the intrinsic noise produced by the transistors (Fig. 5.52). In practice, simulations can be carried out to determine this jitter by measuring the variance of output switching times, while an input ramp is applied to the comparator. The variance is estimated by the ensemble average of a number of transient simulations.

Assume a structure with pre-amplifier G_m followed by an ideal threshold-crossing detector as in Fig. 5.52. The pre-amplifier output noise variance is given by $\sigma_o^2(t)$. With the threshold of the crossing detector set at V_M, we have the timing jitter at the output characterized by:

$$\sigma_t^2 = \frac{\sigma_o^2}{\left(\frac{dv_o}{dt}\right)^2\Big|_{v_o = V_M}}. \tag{5.198}$$

For an ideal comparator, the output jitter should correspond to an input jitter. The input jitter expressed in terms of the input-referred noise v_n^2 of the pre-amplifier is given by:

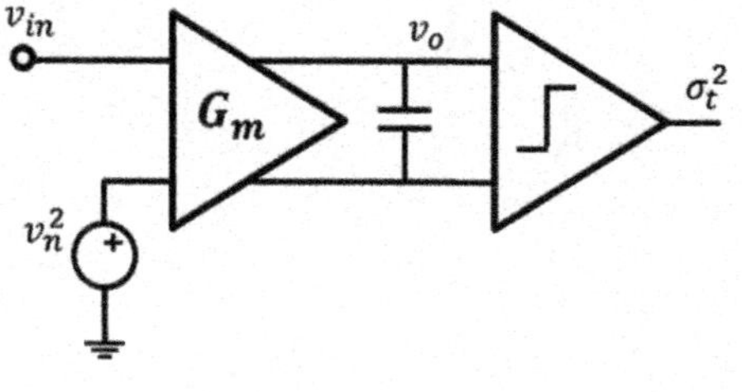

Fig. 5.52 Pre-amp and threshold-crossing comparator

$$\sigma_{\text{tin}}^2 = \frac{v_{\text{n}}^2}{\left(\frac{\text{d}v_{\text{in}}}{\text{d}t}\right)^2\Big|_{v_{\text{in}}=V_{\text{Mi}}}}, \tag{5.199}$$

with V_{Mi} being the amplifier input threshold. Equating the timing jitters reveals the relationship between output and input-referred noise:

$$v_{\text{n}}^2\left[\frac{\left(\frac{\text{d}v_{\text{o}}}{\text{d}t}\right)^2\Big|_{v_{\text{o}}=V_{\text{M}}}}{\left(\frac{\text{d}v_{\text{in}}}{\text{d}t}\right)^2\Big|_{v_{\text{in}}=V_{\text{Mi}}}}\right] = v_{\text{n}}^2A_{\text{N}}^2 = \sigma_o^2. \tag{5.200}$$

The term A_{N}^2 is the noise gain of the comparator, which differs significantly from that of an equivalent amplifier. To evaluate the noise performance of a threshold detection comparator, we thus have to evaluate the ratio of the respective input/output rate of threshold crossing of the pre-amp. Consider Fig. 5.51 again and set $V_{\text{T}} = 0$ for convenience; we can find the ratio of threshold crossing by assuming an input ramp with rate $\text{d}v_{\text{in}}/\text{d}t = \delta$, given by $v_{\text{in}} = t\delta$. The corresponding output voltage is given by:

$$v_{\text{o}}(t) = G_{\text{m}}R_{\text{o}}\delta\left(t - R_{\text{o}}C_{\text{o}}\left(1 - \exp\left(-\frac{1}{R_{\text{o}}C_{\text{o}}}t\right)\right)\right) \quad \text{for } t \geq 0. \tag{5.201}$$

From this we can derive the output crossing rate. Combining this with the input crossing rate γ and the expression for the output white noise in terms of transistor characteristics, the input referred noise becomes [348]:

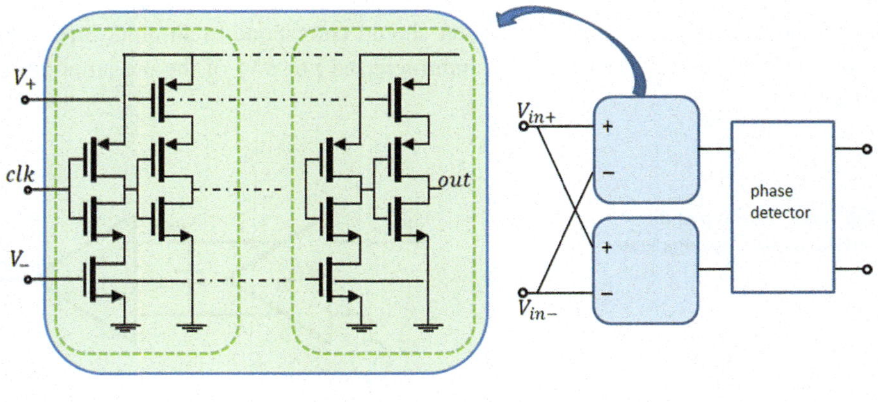

Fig. 5.53 Time-based comparator based on voltage-controlled delay cells

$$v_{\mathrm{n}}^2 = \gamma\left(\frac{kT}{C_0}\right)\left(\frac{1}{G_{\mathrm{m}}R_{\mathrm{o}}}\right)\coth\left[\frac{t_i}{2R_{\mathrm{o}}C_{\mathrm{o}}}\right], \tag{5.202}$$

with $t_i \geq 0$ as the time at which the output crosses its threshold V_{M}. The input referred white noise is independent of the arbitrary input crossing rate δ. A low-noise design can thus begin with increasing the DC gain $G_{\mathrm{m}}R_{\mathrm{o}}$ of the pre-amplifier and the loading capacitor C_{o}. The noise expression also agrees with the noise–bandwidth tradeoff, indicating that the longer it takes to cross the threshold with t_i approaching the time constant $R_{\mathrm{o}}C_{\mathrm{o}}$ of the output stage, the noise reduction due to the contribution from the coth term approaches unity.

Both the dynamic (latch) and threshold comparators utilize an amplifier pre-amp stage. The signal swing and performance of these amplifiers can be severely limited when implemented on deep sub-micron technologies. Leveraging the improved time resolution of the thin-oxide devices, a time-based comparator was proposed in [342]. This is shown in Fig. 5.53 and consists of voltage-dependent delay blocks. The input differential voltage is connected to allow delays controlled by nMOS or pMOS current sources, and the imbalance of the n/pMOS transistors is taken into account by input cross-coupling two parallel voltage-controlled delay blocks.

5.3.3 Delta-Sigma Converters

Delta-Sigma ($\Delta\Sigma$) converters are well known for their robustness and high accuracy at the expense of increased sampling rate. A typical $\Delta\Sigma$-ADC is constructed by feedback loops processing the discrete-time analog signals followed by a digital FIR filter. The output of the feedback loop resembles a low-accuracy (typically 1 bit) digital signal, allowing us to transmit this modulation output directly away from the implant and implement the digital filter on an external device. As a result, the $\Delta\Sigma$-modulator is one of the simplest and most energy-efficient converters typically

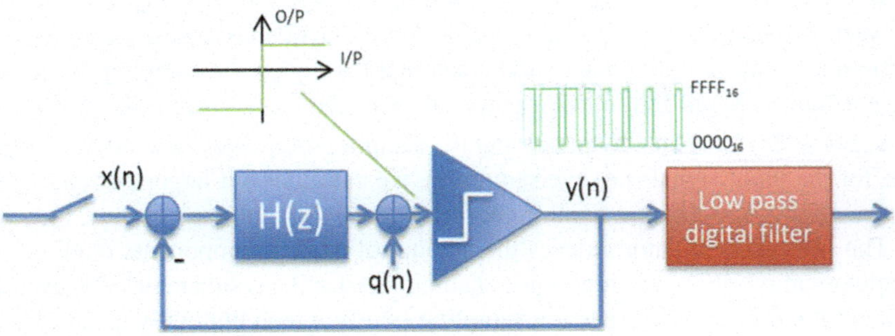

Fig. 5.54 First-order discrete-time $\Delta\Sigma$-modulator

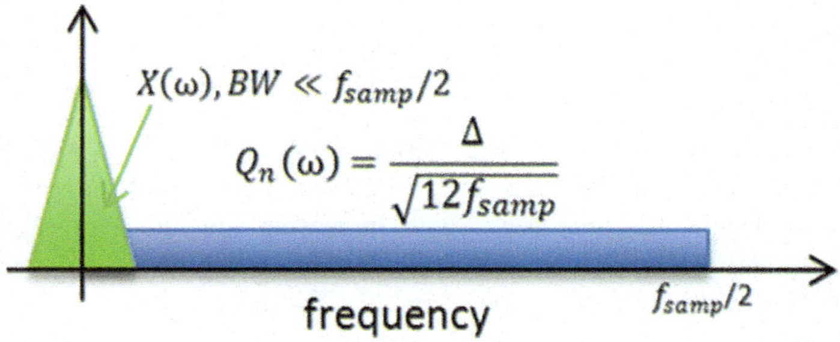

Fig. 5.55 Quantization noise and signal spectrum for oversampling ADCs

used in low-power embedded systems and implants. The operation of a first-order discrete-time ΔΣ-modulator (DTΔΣ) is best illustrated by referring to Fig. 5.54.

The ΔΣ-modulator employs a simple 1-bit quantizer, but achieves high resolution via oversampling and digital filtering. The feedback loop in Fig. 5.54 relates the modulator input $x(n)$ and output $y(n)$ with quantisation noise $q(n)$ by the following transfer domain equation:

$$Y(z) = \frac{H(z)}{1+H(z)}X(z) + \frac{1}{1+H(z)}Q(z) = S(z)X(z) + N(z)Q(z). \quad (5.203)$$

It is obvious that the input signal and the quantization noise are subject to a different transfer function owing to the feedback loop. By selecting an appropriate $H(z)$, the quantization noise can be pre-suppressed. This is known as "noise-shaping". Theoretically, the input-referred noise of the comparator is also subjected to the same in-loop transfer function as the quantization noise, resulting in a relaxation of the quantizer (comparator) noise specifications. A simple example for $H(z)$ is the discrete-time integrator $1/(z-1)$ implemented by a single op-amp SC circuit. This results in the input samples being delayed by a sampling period whereas the quantization noise is subjected to high-pass filtering (i.e. $S(z) = z^{-1}$ and $N(z) = 1 - z^{-1}$). To understand the noise-shaping process, we further articulate that the quantization noise is modeled as white additive Gaussian with noise spectrum given by $\Delta/\sqrt{12 f_{samp}}$ across all frequencies, with Δ being the difference between two quantization levels, as in Fig. 5.55. The quantisation noise is "spread" across the entire sampling bandwidth such that the higher the sampling frequency, the smaller the quantization noise power at each frequency, as the total noise must be equal to $\Delta^2/12$. Thus, by increasing the sampling frequency beyond the Nyquist rate for the input signal (i.e. oversampling), we can filter out larger portions of the quantization noise given a fixed input bandwidth.

The DT ΔΣ-modulator reduces the amount of analog components required to a single op-amp (for a first-order noise-shaping loop) and a comparator. However, the input of the DT ΔΣ-modulator is a sampling switch, which implicitly means that we require an anti-aliasing filter before we apply the signal to a DT ΔΣ-modulator.

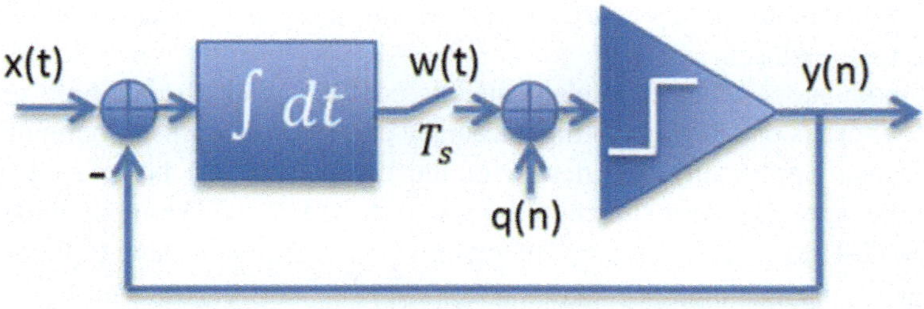

Fig. 5.56 First-order continuous-time ΔΣ-modulator

From a power consumption perspective, the oversampling operation results in increased power requirements typically of the order of 2^N compared to a Nyquist rate ADC. ΔΣ-modulator modulators, however, do not require additional digital control circuits if the reconstruction filter is implemented externally.

Consider the first-order continuous-time ΔΣ-modulator (CT ΔΣ) shown in Fig. 5.56. One immediate observation on the CT ΔΣ modulator compared to the DT ΔΣ is that the continuous-time signal is being sampled in-loop following the integrator. This results in the inherent anti-aliasing properties of the CT ΔΣ modulator and saving of extra band-limiting circuits. The input and quantization noise shaping functions are given by:

$$Y(z) = \frac{k}{z-(1-k)}X(z) + \frac{z-1}{z-(1-k)}E(z), \quad \mathrm{k} = \mathrm{T_s}b/C. \tag{5.204}$$

The stability of the feedback loop thus depends on the sampling frequency $1/T_s$, comparator output value b, and integrator time capacitor C: $2f_s > b/C$. For current measurement units, a Miller integrator consisting of an op-amp and a single capacitor in the feedback loop will be a good choice for a CT ΔΣ implementation, in which case the comparator output will have to be converted to a current. Typical low-power implementations can set $b = 1\mathrm{nA}$ and $C = 1\mathrm{pF}$, resulting in a minimum sampling frequency of 500 Hz. This is very reasonable for low-power implementations. Intuitively, the sampling rate needs to be high enough to capture changes in the integrator output and correct via negative feedback. With a sampling rate below

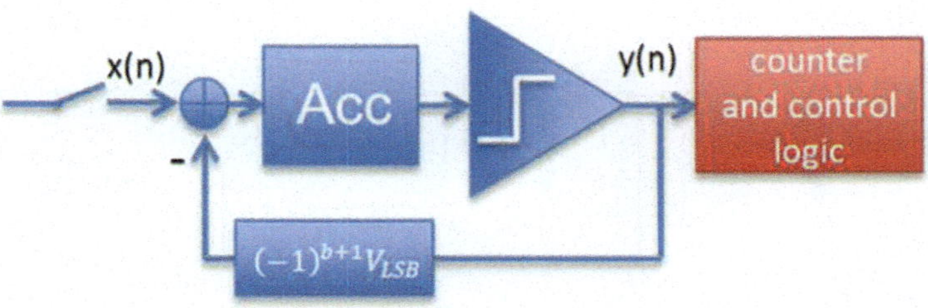

Fig. 5.57 Original incremental ADC with 2^N cycles per conversion

a certain threshold, the integrator output will reach levels that cannot be corrected by negative feedback.

ΔΣ-modulators, whether implemented as continuous or discrete-time versions, provide a means of carrying out high accuracy conversions with sub-optimal analog components. Specifically, the design of the comparator can be relaxed tremendously owing to the noise-shaping property of the ΔΣ loop. The input anti-aliasing filter (for DT ΔΣ), or the in-loop integrator (for CT ΔΣ) still have to be designed with care to achieve high accuracy, i.e. $\sqrt{\langle V_{\mathrm{n}}^2(f)\rangle_{\mathrm{BW}}} < 0.5\,\mathrm{LSB}$, with $V_{\mathrm{n}}^2(f)$ being the input referred noise spectrum and BW the signal bandwidth. In many practical scenarios, however, a single converter often has to handle a number of signal sources. These signal sources are multiplexed at the input of the data converter. ΔΣ converters used specifically for multiplexed inputs are termed incremental ΔΣ ADCs (I-ΔΣ). The first I-ΔΣ converter utilizes a counter structure for conversion (see Fig. 5.57). Subsequent work on I-ΔΣ converters elaborated on the ΔΣ-like properties of the counter-based converter and generalized it for use with multiplexed inputs with an optimal reconstruction filter.

The original I-ΔΣ consists of a loop similar to the time encoder that will be discussed later in this chapter. The main difference between the I-ΔΣ and a regular CT/DT-ΔΣ is that the conversion is set at multiples of 2^N cycles, with N being the accuracy in number of bits. This is significantly slower than most oversampling ΔΣ converters. However, since the ADC carries out conversion for a fixed 2^N cycles before the system is placed in reset and the input is sampled again for the next 2^N cycles of conversion, inputs can be multiplexed for each 2^N cycles. The operation of the I-ΔΣ modulator in Fig. 5.57 effectively counts the number of times the LSB voltage can be subtracted from V_{in}. For each cycle, 1 LSB is added to V_{in} until the input of the comparator reaches zero, in which case 1 LSB is subtracted from V_{in} and the output counter is increased by 1. The effect of multiplexing the input of an I-ΔΣ ADC can be modeled as a regular ΔΣ modulator with a sampled input and windowed digital reconstruction filter. This concept is illustrated in Fig. 5.58.

To see the effect of windowing on the reconstruction function, we express the input $d(n)$ to the reconstruction filter as sampled time domain operations [344]:

$$d(n) = stf(n) \oplus_{\mathrm{M}} [x(n) + u_{\mathrm{n}}(n)] + ntf(n) \oplus_{\mathrm{M}} q(n), \tag{5.205}$$

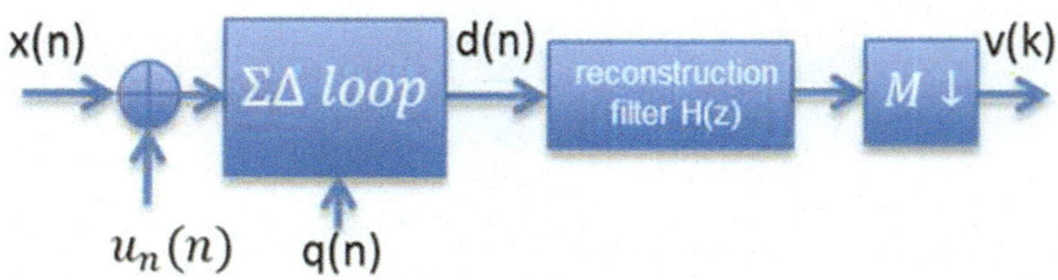

Fig. 5.58 Model of multiplexed ΔΣ-ADC with windowed reconstruction filter

where $u_n(n)$ and $q(n)$ are the respective inputs, referred to as the noise and quantization noise of the ΣΔ-modulator, and $stf(n)$ and $ntf(n)$ are the impulse responses of the signal and noise-shaping transfer functions respectively in the ΣΔ-loop. The symbol $\oplus_{\mathrm{M}}$ denotes a convolution operator. Each output $v(k)$ sample is generated by convolving the M-sample truncated impulse response $h(n) = Z^{-1}[H(z)]$ of the reconstruction filter with the loop output $d(n)$ and N multiplexed inputs:

$$\begin{aligned} v(k) &= \sum_{i=kNM}^{kNM+(M-1)} h([kNM]+[M-1]-i)d(i) \\ &= h(M-1)d(kNM)+h(M-2)d(kNM+1)+\cdots+h(0)d(kNM+[M-1]). \end{aligned} \tag{5.206}$$

Thus, the original impulse response $h(n)$ truncated to M samples from $h(0)$ to $h(M-1)$ is used in the filtering. M samples from the ΣΔ-loop (samples $d(n = kNM)$ to $d(n = kNM + [M-1])$) are used to construct a single output $v(k)$. With the reset operation for every one of these M samples, the output of the ΣΔ-loop can be written as the partial convolution and the input signal contributes to the loop output $d(n)$ by:

$$\begin{aligned} d_x(n = kNM+m) &= stf(n) \oplus_M x(n) = \sum_{i=0}^{m} stf(i)x(kNM+m-j) \\ &= stf(0)x(kNM+m)+stf(1)x(kNM+m-1)+\cdots+stf(m)x(kNM). \end{aligned} \tag{5.207}$$

The output $v(k)$ is generated from the input sequence $x(n)$ as:

$$\begin{aligned} v(k) = &\sum_{n=kNM}^{kNM+(M-1)} \{h([kNM]+[M-1]-n) \\ &\cdot [stf(n) \oplus_{\mathrm{M}} [x(n)+u_{\mathrm{n}}(n)] + ntf(n) \oplus_{\mathrm{M}} q(n)]\}. \end{aligned} \tag{5.208}$$

Overall this becomes:

$$\begin{aligned} v(k) = &\sum_{n=kNM}^{kNM+(M-1)} \{h([kNM]+[M-1]-n) \\ &\cdot \left[\sum_{i=0}^{m} stf(i)x(n-j) + \sum_{i=0}^{m} stf(i)u_{\mathrm{n}}(n-j) + \sum_{i=0}^{m} ntf(i)q(n-j)\right]\bigg\}. \end{aligned} \tag{5.209}$$

To articulate the above equation as a transformation on the input signal and noise contributions, we expand the contribution towards the output with $v(k) = v_{I/P}(k) + v_{\mathrm{n}}(k) + v_{\mathrm{q}}(k)$ and:

$$\begin{aligned} v_{I/P}(k) = h(m-1)[sft(0)x(kNM)] \\ + h(m-2)[stf(0)x(kNM+1) + stf(1)x(kNM)] + \cdots \\ + h(0)[stf(0)x(kNM+M-1) + sft(1)x(kNM+M-2) + \cdots + stf(M-1)x(kNM)]. \end{aligned} \tag{5.210}$$

Note the effect of resetting the loop as well as the reconstruction filter every M samples. Writing this in matrix form:

$$v_{I/P}(k) = [h(0)\ldots h(M-1)] \cdot \begin{bmatrix} stf(M-1) & 0 & 0 & 0 & 0 \\ stf(M-2) & stf(M-2) & \vdots & \vdots & \vdots \\ \vdots & \vdots & \vdots & 0 & \vdots \\ stf(1) & stf(1) & \vdots & stf(1) & 0 \\ stf(0) & stf(0) & \cdots & stf(0) & stf(0) \end{bmatrix} \begin{bmatrix} x(kNM) \\ x(kNM+1) \\ \vdots \\ \vdots \\ x(kNM+M-1) \end{bmatrix} \tag{5.211}$$

If we define the vectors:

$$\underline{x}_k = [x(kNM)\ldots x(kNM+M-1)]^{\mathrm{T}}, \tag{5.212}$$

$$\underline{u}_k = [u_n(kNM)\ldots u_n(kNM+M-1)]^{\mathrm{T}}, \tag{5.213}$$

$$\underline{q}_k = [q(kNM)\ldots q(kNM+M-1)]^{\mathrm{T}}, \tag{5.214}$$

$$\underline{h} = [h(0)\ldots h(M-1)]^{\mathrm{T}}, \tag{5.215}$$

$$\underline{\underline{S}} = \begin{bmatrix} stf(M-1) & 0 & 0 & 0 & 0 \\ stf(M-2) & stf(M-2) & \vdots & \vdots & \vdots \\ \vdots & \vdots & \vdots & 0 & \vdots \\ stf(1) & stf(1) & \vdots & stf(1) & 0 \\ stf(0) & stf(0) & stf(0) & stf(0) & stf(0) \end{bmatrix}, \tag{5.216}$$

and similarly define:

$$\underline{\underline{N}} = \begin{bmatrix} ntf(M-1) & 0 & 0 & 0 & 0 \\ ntf(M-2) & ntf(M-2) & \vdots & \vdots & \vdots \\ \vdots & \vdots & \vdots & 0 & \vdots \\ ntf(1) & ntf(1) & \vdots & ntf(1) & 0 \\ ntf(0) & ntf(0) & ntf(0) & ntf(0) & ntf(0) \end{bmatrix}. \tag{5.217}$$

Now the operation of obtaining a single output sample can be written as:

$$v(k) = \underline{h}^T \underline{\underline{S}}\{\underline{x}_k + \underline{u}_k\} + \underline{h}^{\mathrm{T}} \underline{\underline{N}} \underline{q}_k. \tag{5.218}$$

With a formulated I/O relationship of the I-ΣΔ converter, we can derive the optimal reconstruction filter to maximize the SNR at the output. The signal power assuming a maximum sinusoidal input voltage with amplitude V_{pp} is given by:

$$\begin{aligned} \left\langle v_{I/P}^2(k) \right\rangle &= \left\langle \left[\underline{h}^{\mathrm{T}} \underline{\underline{S}} \underline{x}_k\right] \left[\underline{h}^{\mathrm{T}} \underline{\underline{S}} \underline{x}_k\right]^{\mathrm{T}} \right\rangle = \left\langle \underline{h}^{\mathrm{T}} \underline{\underline{S}} \left[\underline{x}_k \underline{x}_k^{\mathrm{T}}\right] \underline{\underline{S}}^{\mathrm{T}} \underline{h} \right\rangle \\ &= \left\langle \underline{x}_k \underline{x}_k^{\mathrm{T}} \right\rangle \left[\underline{h}^{\mathrm{T}} \underline{\underline{S}}\,\underline{\underline{S}}^{\mathrm{T}} \underline{h}\right] = \frac{V_{\mathrm{pp}}}{2} \left[\underline{h}^{\mathrm{T}} \underline{\underline{S}}\,\underline{\underline{S}}^{\mathrm{T}} \underline{h}\right]. \end{aligned} \tag{5.219}$$

Similarly, the noise power can be derived as:

$$\begin{aligned} \left\langle v_{\mathrm{n}}^2(k) \right\rangle &= \left\langle \left[\underline{h}^{\mathrm{T}} \underline{\underline{S}} \underline{u}_k\right] \left[\underline{h}^{\mathrm{T}} \underline{\underline{S}} \underline{u}_k\right]^{\mathrm{T}} \right\rangle = \left\langle \underline{h}^{\mathrm{T}} \underline{\underline{S}} \left[\underline{u}_k \underline{u}_k^{\mathrm{T}}\right] \underline{\underline{S}}^{\mathrm{T}} \underline{h} \right\rangle \\ &= \left\langle \underline{u}_k \underline{u}_k^{\mathrm{T}} \right\rangle \left[\underline{h}^{\mathrm{T}} \underline{\underline{S}}\,\underline{\underline{S}}^{\mathrm{T}} \underline{h}\right] = \frac{\gamma kT}{C_{\mathrm{in}}} \left[\underline{h}^{\mathrm{T}} \underline{\underline{S}}\,\underline{\underline{S}}^{\mathrm{T}} \underline{h}\right], \end{aligned} \tag{5.220}$$

where we have approximated the input-referred noise power as $\gamma kT/C_{\mathrm{in}}$, where k is the Boltzmann constant as before ($k = 1.38 \times 10^{-23}\,\mathrm{J\,K^{-1}}$) and the γ coefficient is typically equal to 2/3 for long channel input FETs. C_{in} is the input capacitance seen at the input of the ΣΔ loop. For quantization noise power, we have:

$$\begin{aligned} \left\langle v_{\mathrm{q}}^2(k) \right\rangle &= \left\langle \left[\underline{h}^{\mathrm{T}} \underline{\underline{N}} \underline{q}_k\right] \left[\underline{h}^{\mathrm{T}} \underline{\underline{N}} \underline{q}_k\right]^{\mathrm{T}} \right\rangle = \left\langle \underline{h}^{\mathrm{T}} \underline{\underline{N}} \left[\underline{q}_k \underline{q}_k^{\mathrm{T}}\right] \underline{\underline{N}}^{\mathrm{T}} \underline{h} \right\rangle \\ &= \left\langle \underline{q}_k \underline{q}_k^{\mathrm{T}} \right\rangle \left[\underline{h}^{\mathrm{T}} \underline{\underline{N}}\,\underline{\underline{N}}^{\mathrm{T}} \underline{h}\right] = \frac{\Delta^2}{6} \left[\underline{h}^{\mathrm{T}} \underline{\underline{N}}\,\underline{\underline{N}}^{\mathrm{T}} \underline{h}\right], \end{aligned} \tag{5.221}$$

where Δ is the quantization step in the ΣΔ loop. The output SNR is now given by:

$$\begin{aligned} \mathrm{SNR}_{\mathrm{out}} &= 10\log\left(\frac{\frac{V_{\mathrm{pp}}^2}{2}\left[\underline{h}^{\mathrm{T}}\underline{\underline{S}}\,\underline{\underline{S}}^{\mathrm{T}}\underline{h}\right]}{\frac{\gamma kT}{C_{\mathrm{in}}}\left[\underline{h}^{\mathrm{T}}\underline{\underline{S}}\,\underline{\underline{S}}^{\mathrm{T}}\underline{h}\right]+\frac{\Delta^2}{6}\left[\underline{h}^{\mathrm{T}}\underline{\underline{N}}\,\underline{\underline{N}}^{\mathrm{T}}\underline{h}\right]}\right) \\ &= 10\log\left(\frac{\frac{V_{\mathrm{pp}}^2}{2}\left[\underline{h}^{\mathrm{T}}\underline{\underline{S}}\,\underline{\underline{S}}^{\mathrm{T}}\underline{h}\right]}{\underline{h}^{\mathrm{T}}\left[\frac{\gamma kT}{C_{\mathrm{in}}}\left[\underline{\underline{S}}\,\underline{\underline{S}}^{\mathrm{T}}\right]+\frac{\Delta^2}{6}\left[\underline{\underline{N}}\,\underline{\underline{N}}^{\mathrm{T}}\right]\right]\underline{h}}\right). \end{aligned} \tag{5.222}$$

Normally $\underline{\underline{S}}$ and $\underline{\underline{N}}$ are fixed by the ΣΔ loop whereas $\underline{h}$ is to be designed to maximize $\mathrm{SNR}_{\mathrm{out}}$. The cost function for minimizing the quantisation noise is thus given by:

$$\Theta(\underline{h}) = \underline{h}^{\mathrm{T}}\underline{\underline{N}}\,\underline{\underline{N}}^{\mathrm{T}}\underline{h} + \lambda([1\ldots1]\underline{h} - 1), \tag{5.223}$$

where λ is the Lagrange multiplier. The regularization term $[1\ldots1]\underline{h} - 1$ is to ensure that the reconstruction filter taps $\underline{h}$ produce unity gain at low frequencies. The solution to the optimization problem can be derived analytically as:

$$\underline{h}_{\mathrm{opt}} = \frac{1}{[1\ldots1]\left[\underline{\underline{N}}\,\underline{\underline{N}}^{\mathrm{T}}\right]^{-1}[1\ldots1]^{\mathrm{T}}}\left[\underline{\underline{N}}\,\underline{\underline{N}}^{\mathrm{T}}\right]^{-1}[1\ldots1]^{\mathrm{T}}. \tag{5.224}$$

Intuitively, since the ΣΔ loop only runs for M cycles, the noise-shaping filter $\underline{\underline{N}}$ will be less effective than a continuously running ΣΔ loop. The optimal reconstruction filter thus has to take this into account to restore some of the noise-shaping properties. Whereas this technique of designing an optimal reconstruction filter for multiplexed I-ΣΔ ADCs is promising, the reported performance of optimized I-ΣΔ ADCs is still short of continuously running non-multiplexed ADC with typical second-order cascaded integrator filters.

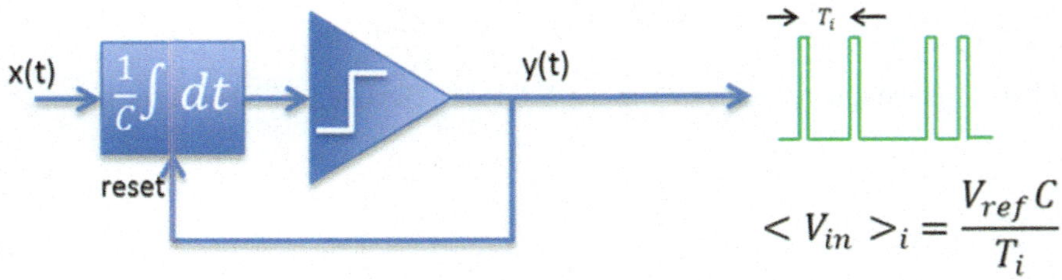

Fig. 5.59 Integrate and fire ADC

5.3.4 *"Integrate and Fire" Converters*

Integrate and fire "neurons" [346] are amongst the simplest form of asynchronous ADCs, where the sample period is non-uniform and often signal-dependent. These exploit alternative robust signal representations other than binary voltage levels. The integrate and fire ADC (IF-ADC) consists of a reset continuous-time integrator (Fig. 5.59), which can be implemented as a Miller integrator for input currents—an op-amp with capacitor feedback. The output of the integrator is fed into a comparator, which feeds to the reset of the integrator. The integrator output ramps with the input current until the comparator threshold V_T is reached. This triggers a reset of the integrator and subsequently triggers another transition at the output of the comparator when the lower switching threshold V_T' is reached. In practice, a comparator without hysteresis can be used to provide some delay in the reset loop, allowing sufficient time for the output of the integrator to reach an acceptable lower level before the reset is terminated. The output waveform of this scheme resembles a frequency modulated signal where the time between two output pulses is inversely proportional to the average input signal within that duration. The output signals can then be applied directly to a telemetry link. A simple decoding method is to drive an edge-reset counter with the output pulse to recover the time (hence average input amplitude) information. The recovered sampling rate is irregular due to the signal-dependent timing.

5.3.5 *Time-Encoding Machine and Related Architectures*

IF-ADCs are among the simplest form of asynchronous ADC, where the sample period is non-uniform. The fact that the frequency-coded output has no real lower frequency limits (e.g. very small inputs) creates ambiguity in the decoding process when such ADCs are practically used. The time-encoding machine (TEM) [349]

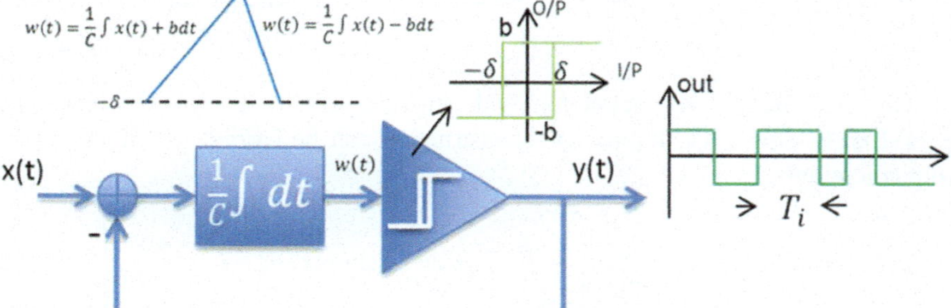

Fig. 5.60 Time-encoding machine (TEM) architecture

mitigates this shortcoming of the integrating ADC by closing the feedback loop, as shown in Fig. 5.60.

In a TEM, the reset action of the integrator is replaced by a comparator with hysteresis. The instantaneous frequency (period) of the output depends on the input signal $|x(t)| < b$. For zero input, the output toggles at twice the natural frequency of the system. In this case, the rise and fall times of the sawtooth waveform $w_T(t)$ are the same and equals to half the natural period T_{k0} as $w_{T0}(t)$ ramps from $-\delta$ to δ [350]:

$$2\delta = \frac{1}{C}\int_{t}^{t+T_{k0}/2} b dt = \frac{bT_{k0}}{2C} \Rightarrow f_{k0} = \frac{1}{T_{k0}} = \frac{b}{4\delta C}. \quad (5.225)$$

Again using the assumption of a current input and Miller integrator implementation with V-I conversion at the comparator output such that $b = 1$ nA and $C = 1$ pF, f_{k0} becomes 250 Hz. This is very reasonable for ultra-low power conversion of slow signals. For general input $x(t)$, the output $w(t)$ of the integrator in the TEM is given in Fig. 5.60 resembling a sawtooth waveform for slowly varying input $x(t)$. The instantaneous period of the output pulse is bounded according to:

$$\frac{2\delta C}{b + x(t)|_{\max}} \le t_{k+1} - t_k \le \frac{2\delta C}{b - x(t)|_{\max}}. \quad (5.226)$$

The integrated input at any period $t_{k+1} - t_k$ can be recovered according to the following:

$$\int_{t_k}^{t_{k+1}} x(t)\mathrm{d}t = (-1)^k[-b(t_{k+1} - t_k) + 2C\delta] \quad (5.227)$$

The original input can be recovered iteratively by fitting a sinc function in the middle of all the sampling periods:

$$\tilde{x}_0(t) = \sum_k \left\{ \left[\int_{t_k}^{t_{k+1}} x(u)\mathrm{d}u \right] \cdot \sin c\left[t - \left(\frac{t_{k+1} + t_k}{2}\right)\right] \right\} \approx x(t) \quad (5.228)$$

This is similar to reconstruction with an ideal LPF. For TEM, however, the period $t_{k+1} - t_k$ varies. Improved reconstruction can be carried out iteratively by implementing:

$$\tilde{x}_{l+1}(t) = \tilde{x}_0(t) + \tilde{x}_l(t) + \sum_k \left\{ \left[\int_{t_k}^{t_{k+1}} \tilde{x}_l(u)\mathrm{d}u \right] \cdot \sin c\left[t - \left(\frac{t_{k+1} + t_k}{2}\right)\right] \right\} \quad (5.229)$$

5.4 Conclusions

It is evident from the introductory discussion that the majority of integrated circuit developments in the field of implantable electronics have been in the field of neural interfacing, either in the CNS or the PNS. Driven by these applications in low-power implantable ASICs and developments in the sensing field, in recent years there have been significant advances in the field of implantable sensing.

From the above, it is evident that there are many solutions and circuit techniques that the modern ASIC designer can employ. The choice of these and the design, according to the various tradeoffs, need to be performed carefully to suit the required application. As always, a good understanding of the fundamentals of the target application and the design specifications is essential. The designer needs to work with a combination of pen and paper and simulations to obtain a deep understanding of the particular designs in the deep submicron technologies of today. Basic building block design, such as op-amps, folded-cascode amplifiers, high-performance current mirrors, etc. have not been discussed, as these can be found in a wide variety of textbooks. Also, a number of fundamental building blocks, such as current and voltage references, have not been discussed, as the focus was more on the front-end signal processing electronics.

Due to the need to lower power consumption, electrochemical sensors are the driving force in the development of implantable sensing. Amperometric systems require sensitivities in the fA-pA range and special circuit techniques must be applied to facilitate this. This involves either careful transistor level design, to ensure low noise, or the use of circuit techniques such as CHS and AZ, and in particular CDS. Similar circuits and techniques are useful for the interrogation of capacitive sensors and electrochemical impedance sensors.

Significantly more challenging is the development of galvanostatic bioimpedance systems with only a handful of integrated systems demonstrated in the literature and even fewer for implantable applications. The complexity of such systems and the different subunits they consist of lead to a wide range of options for the various building blocks. SD methods for obtaining the real and imaginary parts of the impedance allow multi-frequency measurements due to the band-selective properties of the algorithm, which is quite similar to CHS. Magnitude-phase measurement appears at first simpler to implement, but there are a few bottlenecks that still need to be addressed, particularly with the phase measurement channel. A few possible solutions were discussed. In terms of the VCCS, this needs to be a fully custom system designed for the particular application and should be properly characterized within the frequency bands of interest and loading conditions. High-quality VCCS are challenging to design and are typically large power-hungry systems. Analog and digital methods for generating the excitation signal necessary for bioimpedance measurements were also examined. Digital systems result in large complex systems, while analog implementations can be more compact and low-power, but sensitive to circuit imperfections. For completeness, an introductory discussion of stimulators and stimulus waveforms was presented, highlighting the need for asymmetric biphasic stimulus waveforms and the need to ensure that there is no residual charge.

With regard to voltage measurement front-end amplifiers, the AC-coupled capacitive feedback topology was discussed. Although this is a popular design, it has limited input impedance, while the pseudo-resistors used suffer from nonlinearity for large voltage swings. Alternative pseudo-resistive structures to the classic topology were discussed. The DDA is a useful circuit building block that can be used at the front end and also at the back end, while the additional input it provides allows for it to be exploited in AZ amplifiers. The current feedback IA demonstrates a high CMRR at high frequencies and a high input impedance. These amplifiers can be combined with offset/noise compensation methods to enhance their performance, and methods have been presented to increase their input impedance, to eliminate in CHS the modulated residual offset/noise and the spiking transients due to switch imperfections, as well as AC-coupling methods to reject electrode DC offset voltages. Methods to improve the performance of switches were discussed, such as the AT-switch, which has been a popular choice for analog circuit designers in recent years. An in-depth discussion of linearization techniques was presented with a focus on weak inversion transconductors. Source degeneration, bump linearization, and multi-tanh linearization were presented. Depending on the application, one or more of these methods may be applied to achieve a required linear input range. The chapter concluded with a discussion of popular ADC approaches in implantable application, such as SAR ADCs and ΔΣ ADCs.

References

1. C.M. Lopez et al., An implantable 455-active-Electrode 52-channel CMOS neural probe. IEEE J. Solid-State Circuits **49**(1), 248–261 (2014)
2. M. Yin, D.A. Borton, J. Aceros, W.R. Patterson, A.V. Nurmikko, A 100-channel hermetically sealed implantable device for chronic wireless neurosensing applications. IEEE Trans. Biomed. Circuits Syst. **7**(2), 115–128 (2013)
3. P.T. Huang et al., 2.5D heterogeneously integrated microsystem for high-density neural sensing applications. IEEE Trans. Biomed. Circuits Syst. **8**(6), 810–823 (2014)
4. E.E. Aktakka, K. Najafi, A micro inertial energy harvesting platform with self-supplied power management circuit for autonomous wireless sensor nodes. IEEE J. Solid-State Circuits **49**(9), 2017–2029 (2014)
5. M. Bakhshiani, M.A. Suster, P. Mohseni, A broadband sensor interface IC for miniaturized dielectric spectroscopy from MHz to GHz. IEEE J. Solid-State Circuits **49**(8), 1669–1681 (2014)
6. S.S. Ghoreishizadeh, C. Baj-Rossi, A. Cavallini, S. Carrara, G.D. Micheli, An integrated control and readout circuit for implantable multi-target electrochemical biosensing. IEEE Trans. Biomed. Circuits Syst. **8**(6), 891–898 (2014)
7. Y.J. Huang et al., A self-powered CMOS reconfigurable multi-sensor SoC for biomedical applications. IEEE J. Solid-State Circuits **49**(4), 851–866 (2014)

8. M. Shoaran, M.H. Kamal, C. Pollo, P. Vandergheynst, A. Schmid, Compact low-power cortical recording architecture for compressive multichannel data acquisition. IEEE Trans. Biomed. Circuits Syst. **8**(6), 857–870 (2014)
9. Y. Suo, J. Zhang, T. Xiong, P.S. Chin, R. Etienne-Cummings, T.D. Tran, Energy-efficient multi-mode compressed sensing system for implantable neural recordings. IEEE Trans. Biomed. Circuits Syst. **8**(5), 648–659 (2014)
10. H.G. Rhew, J. Jeong, J.A. Fredenburg, S. Dodani, P.G. Patil, M.P. Flynn, a fully self-contained logarithmic closed-loop deep brain stimulation SoC with wireless telemetry and wireless power management. IEEE J. Solid-State Circuits **49**(10), 2213–2227 (2014)
11. Y.J. Huang et al., A CMOS cantilever-based label-free DNA SoC with improved sensitivity for hepatitis B virus detection. IEEE Trans. Biomed. Circuits Syst. **7**(6), 820–831 (2013)
12. Y. Huang, Y. Liu, B.L. Hassler, R.M. Worden, A.J. Mason, A protein-based electrochemical biosensor array platform for integrated microsystems. IEEE Trans. Biomed. Circuits Syst. **7** (1), 43–51 (2013)
13. S.W. Chiu et al., A fully integrated nose-on-a-chip for rapid diagnosis of ventilator-associated pneumonia. IEEE Trans. Biomed. Circuits Syst. **8**(6), 765–778 (2014)
14. M. Haruta et al., An implantable CMOS device for blood-flow imaging during experiments on freely moving rats. Jpn. J. Appl. Phys. **53**(4S), 04EL05 (2014)
15. T.H. Tsai, H.C. Tsai, T.K. Wu, A CMOS micromachined capacitive tactile sensor with integrated readout circuits and compensation of process variations. IEEE Trans. Biomed. Circuits Syst. **8**(5), 608–616 (2014)
16. M. Yip, R. Jin, H.H. Nakajima, K.M. Stankovic, A.P. Chandrakasan, A fully-implantable cochlear implant SoC with piezoelectric middle-ear sensor and arbitrary waveform neural stimulation. IEEE J. Solid-State Circuits **50**(1), 214–229 (2015)
17. Y.K. Lo, K. Chen, P. Gad, W. Liu, A fully-integrated high-compliance voltage SoC for epi-retinal and neural prostheses. IEEE Trans. Biomed. Circuits Syst. **7**(6), 761–772 (2013)
18. M. Monge et al., A fully intraocular high-density self-calibrating epiretinal prosthesis. IEEE Trans. Biomed. Circuits Syst. **7**(6), 747–760 (2013)
19. S. Oh et al., Light-controlled biphasic current stimulator IC using CMOS image sensors for high-resolution retinal prosthesis and in vitro experimental results with rd1 mouse. IEEE Trans. Biomed. Eng. **62**(1), 70–79 (2015)
20. R.R. Harrison, C. Charles, A low-power low-noise CMOS amplifier for neural recording applications. IEEE J. Solid-State Circuits **38**(6), 958–965 (2003)
21. B. Gosselin, M. Sawan, C.A. Chapman, A low-power integrated bioamplifier with active low-frequency suppression. IEEE Trans. Biomed. Circuits Syst. **1**(3), 184–192 (2007)
22. A. Bagheri, M.T. Salam, J.L.P. Velazquez, R. Genov, Low-frequency noise and offset rejection in DC-coupled neural amplifiers: a review and digitally-assisted design tutorial. IEEE Trans. Biomed. Circuits Syst. **11**(1), 161–176 (2017)
23. X. Liu, A. Demosthenous, N. Donaldson, Platinum electrode noise in the ENG spectrum. Med. Biol. Eng. Comput. **46**(10), 997–1003 (2008)
24. H. Kassiri, K. Abdelhalim, R. Genov, Low-distortion super-GOhm subthreshold-MOS resistors for CMOS neural amplifiers, in *2013 IEEE Biomedical Circuits and Systems Conference (BioCAS)* (2013), pp. 270–273
25. R.F. Yazicioglu, P. Merken, R. Puers, C.V. Hoof, A 60 μW 60 nV/$\sqrt{\text{Hz}}$ readout front-end for portable biopotential acquisition systems. IEEE J. Solid-State Circuits **42**(5), 1100–1110 (2007)
26. X. Zou, X. Xu, L. Yao, Y. Lian, A 1-V 450-nW fully integrated programmable biomedical sensor interface chip. IEEE J. Solid-State Circuits **44**(4), 1067–1077 (2009)
27. Y. Tseng, Y. Ho, S. Kao, C. Su, A 0.09 μW low power front-end biopotential amplifier for biosignal recording. IEEE Trans. Biomed. Circuits Syst. **6**(5), 508–516 (2012)
28. A. Demosthenous, I. Pachnis, D. Jiang, N. Donaldson, An integrated amplifier with passive neutralization of myoelectric interference from neural recording tripoles. IEEE Sens. J. **13** (9), 3236–3248 (2013)

29. L. Yan, J. Bae, S. Lee, T. Roh, K. Song, H.J. Yoo, A 3.9 mW 25-electrode reconfigured sensor for wearable cardiac monitoring system. IEEE J. Solid-State Circuits **46**(1), 353–364 (2011)
30. S. Hong et al., A 4.9 mOhm-sensitivity mobile electrical impedance tomography IC for early breast-cancer detection system. IEEE J. Solid-State Circuits **50**(1), 245–257 (2015)
31. S. Hong, J. Lee, J. Bae, H.J. Yoo, A 10.4 mW electrical impedance tomography SoC for portable real-time lung ventilation monitoring system. IEEE J. Solid-State Circuits **50**(11), 2501–2512 (2015)
32. R.F. Yazicioglu, S. Kim, T. Torfs, H. Kim, C.V. Hoof, A 30 μW analog signal processor asic for portable biopotential signal monitoring. IEEE J. Solid-State Circuits **46**(1), 209–223 (2011)
33. A. Demosthenous, Advances in microelectronics for implantable medical devices. Adv. Electron. 981295 (2014)
34. H. Chandrakumar D. Marković, A high dynamic-range neural recording chopper amplifier for simultaneous neural recording and stimulation. IEEE J. Solid-State Circuits **52**(3), 645–656 (2017)
35. N.V. Helleputte, S. Kim, H. Kim, J.P. Kim, C.V. Hoof, R.F. Yazicioglu, A 160μA biopotential acquisition ic with fully integrated IA and motion artifact suppression. IEEE Trans. Biomed. Circuits Syst. **6**(6), 552–561 (2012)
36. M.S.J. Steyaert, W.M.C. Sansen, A micropower low-noise monolithic instrumentation amplifier for medical purposes. IEEE J. Solid-State Circuits **22**(6), 1163–1168 (1987)
37. F. Zhang, J. Holleman, B.P. Otis, Design of ultra-low power biopotential amplifiers for biosignal acquisition applications. IEEE Trans. Biomed. Circuits Syst. **6**(4), 344–355 (2012)
38. K.A. Ng, E. Greenwald, Y.P. Xu, N.V. Thakor, Implantable neurotechnologies: a review of integrated circuit neural amplifiers. Med. Biol. Eng. Comput. **54**(1), 45–62 (2016)
39. W. Wattanapanitch, M. Fee, R. Sarpeshkar, An energy-efficient micropower neural recording amplifier. IEEE Trans. Biomed. Circuits Syst. **1**(2), 136–147 (2007)
40. T. Delbruck and C. A. Mead, Adaptive photoreceptor with wide dynamic range, in *Proceedings of IEEE International Symposium on Circuits and Systems (ISCAS '94)* (1994), pp. 339–342
41. Z. Kárász, R. Fiáth, P. Földesy, Á.R. Vázquez, Tunable low noise amplifier implementation with low distortion pseudo-resistance for *in vivo* brain activity measurement. IEEE Sens. J. **14**(5), 1357–1363 (2014)
42. R.H. Olsson, D.L. Buhl, A.M. Sirota, G. Buzsaki, K.D. Wise, Band-tunable and multiplexed integrated circuits for simultaneous recording and stimulation with microelectrode arrays. IEEE Trans. Biomed. Eng. **52**(7), 1303–1311 (2005)
43. M. Yin, M. Ghovanloo, A low-noise preamplifier with adjustable gain and bandwidth for biopotential recording applications, in *2007 IEEE International Symposium on Circuits and Systems* (2007), pp. 321–324
44. H. Rezaee-Dehsorkh, N. Ravanshad, R. Lotfi, K. Mafinezhad, A.M. Sodagar, Analysis and design of tunable amplifiers for implantable neural recording applications. IEEE. J. Emerg. Sel. Top. Circuits Syst. **1**(4), 546–556 (2011)
45. R.H. Olsson, K.D. Wise, A three-dimensional neural recording microsystem with implantable data compression circuitry. IEEE J. Solid-State Circuits **40**(12), 2796–2804 (2005)
46. K. Abdelhalim, L. Kokarovtseva, J.L.P. Velazquez, R. Genov, 915-MHz FSK/OOK wireless neural recording SoC With 64 mixed-signal FIR filters. IEEE J. Solid-State Circuits **48**(10), 2478–2493 (2013)
47. N. Verma, A. Shoeb, J. Bohorquez, J. Dawson, J. Guttag, A.P. Chandrakasan, A micro-power EEG acquisition SoC with integrated feature extraction processor for a chronic seizure detection system. IEEE J. Solid-State Circuits **45**(4), 804–816 (2010)
48. K.A. Ng, Y.P. Xu, A compact, low input capacitance neural recording amplifier. IEEE Trans. Biomed. Circuits Syst. **7**(5), 610–620 (2013)

49. V. Majidzadeh, A. Schmid, Y. Leblebici, Energy efficient low-noise neural recording amplifier with enhanced noise efficiency factor. IEEE Trans. Biomed. Circuits Syst. **5**(3), 262–271 (2011)
50. Z. Zhou, P.A. Warr, A high input impedance low noise integrated front-end amplifier for neural monitoring. IEEE Trans. Biomed. Circuits Syst. **10**(6), 1079–1086 (2016)
51. K.A. Ng, Y.P. Xu, A low-power, high CMRR neural amplifier system employing CMOS inverter-based OTAs With CMFB through supply rails. IEEE J. Solid-State Circuits **51**(3), 724–737 (2016)
52. R. Muller, S. Gambini, J.M. Rabaey, A 0.013 mm^2, 5 μW, DC-coupled neural signal acquisition IC with 0.5 V supply. IEEE J. Solid-State Circuits **47**(1), 232–243 (2012)
53. W. Biederman et al., A fully-integrated, miniaturized (0.125 mm^2) 10.5 μW wireless neural sensor. IEEE J. Solid-State Circuits **48**(4), 960–970 (2013)
54. A.P. Brokaw, M.P. Timko, An improved monolithic instrumentation amplifier. IEEE J. Solid-State Circuits **10**(6), 417–423 (1975)
55. R. Martins, S. Selberherr, F.A. Vaz, A CMOS IC for portable EEG acquisition systems. IEEE Trans. Instrum. Meas. **47**(5), 1191–1196 (1998)
56. A. Worapishet, A. Demosthenous, X. Liu, A CMOS instrumentation amplifier with 90-dB CMRR at 2-MHz using capacitive neutralization: analysis, design considerations, and implementation. IEEE Trans. Circuits Syst. Regul. Pap. **58**(4), 699–710 (2011)
57. Y.-Q. Zhao, A. Demosthenous, R.H. Bayford, A CMOS instrumentation amplifier for wideband bioimpedance spectroscopy systems, in *2006 IEEE International Symposium on Circuits and Systems* (2006), p. 5082
58. P. Kassanos, I.F. Triantis, A. Demosthenous, A CMOS magnitude/phase measurement chip for impedance spectroscopy. IEEE Sens. J. **13**(6), 2229–2236 (2013)
59. P. Kassanos, L. Constantinou, I.F. Triantis, A. Demosthenous, An integrated analog readout for multi-frequency bioimpedance measurements. IEEE Sens. J. **14**(8), 2792–2800 (2014)
60. A. Yufera, A. Rueda, J.M. Munoz, R. Doldan, G. Leger, E.O. Rodriguez-Villegas, A tissue impedance measurement chip for myocardial ischemia detection. IEEE Trans. Circuits Syst. Regul. Pap. **52**(12), 2620–2628 (2005)
61. K. Song, U. Ha, S. Park, J. Bae, H.J. Yoo, An impedance and multi-wavelength near-infrared spectroscopy IC for non-invasive blood glucose estimation. IEEE J. Solid-State Circuits **50**(4), 1025–1037 (2015)
62. R.F. Yazicioglu, P. Merken, R. Puers, C.V. Hoof, A 200 μW eight-channel EEG acquisition ASIC for ambulatory EEG systems. IEEE J. Solid-State Circuits **43**(12), 3025–3038 (2008)
63. K. Song, U. Ha, J. Lee, K. Bong, H.J. Yoo, An 87-mA.min iontophoresis controller IC with dual-mode impedance sensor for patch-type transdermal drug delivery system. IEEE J. Solid-State Circuits **49**(1), 167–178 (2014)
64. J. Xu, B. Büsze, C.V. Hoof, K.A.A. Makinwa, R.F. Yazicioglu, A 15-channel digital active electrode system for multi-parameter biopotential measurement. IEEE J. Solid-State Circuits **50**(9), 2090–2100 (2015)
65. E. Sackinger, W. Guggenbuhl, A versatile building block: the CMOS differential difference amplifier. IEEE J. Solid-State Circuits **22**(2), 287–294 (1987)
66. K.A. Ng, P.K. Chan, A CMOS analog front-end IC for portable EEG/ECG monitoring applications. IEEE Trans. Circuits Syst. Regul. Pap. **52**(11), 2335–2347 (2005)
67. M. Guermandi, R. Cardu, E.F. Scarselli, R. Guerrieri, Active electrode IC for EEG and electrical impedance tomography with continuous monitoring of contact impedance. IEEE Trans. Biomed. Circuits Syst. **9**(1), 21–33 (2015)
68. H. Ko, T. Lee, J. H. Kim, J. Park, J.P. Kim, Ultralow-power bioimpedance IC with intermediate frequency shifting chopper. IEEE Trans. Circuits Syst. II Express Briefs **63**(3), 259–263 (2016)
69. J. Kim, H. Ko, Reconfigurable multiparameter biosignal acquisition SoC for low power wearable platform. Sensors **16**(12), 2002 (2016)

70. J. Wu, G.K. Fedder, L.R. Carley, A low-noise low-offset capacitive sensing amplifier for a 50-μg/√Hz monolithic CMOS MEMS accelerometer. IEEE J. Solid-State Circuits **39**(5), 722–730 (2004)
71. H. Alzaher, M. Ismail, A CMOS fully balanced differential difference amplifier and its applications. IEEE Trans. Circuits Syst. II Analog Digit. Signal Process. **48**(6), 614–620 (2001)
72. P.K. Chan, J. Cui, Design of chopper-stabilized amplifiers with reduced offset for sensor applications. IEEE Sens. J. **8**(12), 1968–1980 (2008)
73. J.F. Duque-Carrillo, G. Torelli, R. Perez-Aloe, J.M. Valverde, F. Maloberti, Fully differential basic building blocks based on fully differential difference amplifiers with unity-gain difference feedback. IEEE Trans. Circuits Syst. Fundam. Theory Appl. **42**(3), 190–192 (1995)
74. F. Witte, K.A.A. Makinwa, J. Huijsing, *Dynamic Offset Compensated CMOS Amplifiers* (Springer Science & Business Media, Berlin, 2009)
75. M.J.M. Pelgrom, A.C.J. Duinmaijer, A.P.G. Welbers, Matching properties of MOS transistors. IEEE J. Solid-State Circuits **24**(5), 1433–1439 (1989)
76. P.R. Kinget, Device mismatch and tradeoffs in the design of analog circuits. IEEE J. Solid-State Circuits **40**(6), 1212–1224 (2005)
77. C.C. Enz, G.C. Temes, Circuit techniques for reducing the effects of op-amp imperfections: autozeroing, correlated double sampling, and chopper stabilization. Proc. IEEE **84**(11), 1584–1614 (1996)
78. A. Agnes, A. Cabrini, F. Maloberti, G. Martini, Cancellation of amplifier offset and 1/f noise: an improved chopper stabilized technique. IEEE Trans. Circuits Syst. II Express Briefs **54**(6), 469–473 (2007)
79. M.A.T. Sanduleanu, A.J.M.V. Tuijl, R.F. Wassenaar, M.C. Lammers, H. Wallinga, A low noise, low residual offset, chopped amplifier for mixed level applications, in *1998 IEEE International Conference on Electronics, Circuits and Systems. Surfing the Waves of Science and Technology (Cat. No.98EX196)* (1998), pp. 333–336
80. H.M. Jafari, R. Genov, Chopper-stabilized bidirectional current acquisition circuits for electrochemical amperometric biosensors. IEEE Trans. Circuits Syst. Regul. Pap. **60**(5), 1149–1157 (2013)
81. H.M. Jafari, K. Abdelhalim, L. Soleymani, E.H. Sargent, S.O. Kelley, R. Genov, Nanostructured CMOS wireless ultra-wideband label-free PCR-free DNA analysis SoC. IEEE J. Solid-State Circuits **49**(5), 1223–1241 (2014)
82. J. Xu, R.F. Yazicioglu, B. Grundlehner, P. Harpe, K.A.A. Makinwa, C.V. Hoof, A 160 μW 8-channel active electrode system for EEG monitoring. IEEE Trans. Biomed. Circuits Syst. **5** (6), 555–567 (2011)
83. Y. Hu, M. Sawan, CMOS front-end amplifier dedicated to monitor very low amplitude signal from implantable sensors. Analog Integr. Circuits Signal Process. **33**(1), 29–41 (2002)
84. R. Chebli, M. Sawan, Low noise and high CMRR front-end amplifier dedicated to portable EEG acquisition system, in *2013 35th Annual International Conference of the IEEE Engineering in Medicine and Biology Society (EMBC)* (2013), pp. 2523–2526
85. C. Enz, A. Boukhayma, Recent trends in low-frequency noise reduction techniques for integrated circuits, in *2015 International Conference on Noise and Fluctuations (ICNF)* (2015), pp. 1–6
86. C. Menolfi, Q. Huang, A low-noise CMOS instrumentation amplifier for thermoelectric infrared detectors. IEEE J. Solid-State Circuits **32**(7), 968–976 (1997)
87. C.C. Enz, E.A. Vittoz, F. Krummenacher, A CMOS chopper amplifier. IEEE J. Solid-State Circuits **22**(3), 335–342 (1987)
88. C. Menolfi, Q. Huang, A fully integrated, untrimmed CMOS instrumentation amplifier with submicrovolt offset. IEEE J. Solid-State Circuits **34**(3), 415–420 (1999)
89. A. Bakker, K. Thiele, J.H. Huijsing, A CMOS nested-chopper instrumentation amplifier with 100-nV offset. IEEE J. Solid-State Circuits **35**(12), 1877–1883 (2000)

90. J. Xu, Q. Fan, J.H. Huijsing, C.V. Hoof, R.F. Yazicioglu, K.A.A. Makinwa, Measurement and analysis of current noise in chopper amplifiers. IEEE J. Solid-State Circuits **48**(7), 1575–1584 (2013)
91. M.A.P. Pertijs, A. Niederkorn, X. Ma, B. McKillop, A. Bakker, J.H. Huijsing, A CMOS smart temperature sensor with a 3σ inaccuracy of ±0.5 °C from –50 °C to 120 °C. IEEE J. Solid-State Circuits **40**(2), 454–461 (2005)
92. R. Wu, K.A.A. Makinwa, J.H. Huijsing, A Chopper Current-Feedback Instrumentation Amplifier With a 1 mHz Noise Corner and an AC-Coupled Ripple Reduction Loop. IEEE J. Solid-State Circuits **44**(12), 3232–3243 (2009)
93. A. Bakker, J.H. Huijsing, A CMOS chopper opamp with integrated low-pass filter, in *Proceedings of the 23rd European Solid-State Circuits Conference* (1997), pp. 200–203
94. R. Burt, J. Zhang, A Micropower Chopper-Stabilized Operational Amplifier Using a SC Notch Filter With Synchronous Integration Inside the Continuous-Time Signal Path. IEEE J. Solid-State Circuits **41**(12), 2729–2736 (2006)
95. Q. Fan, F. Sebastiano, J.H. Huijsing, K.A.A. Makinwa, A 1.8 μW 60 nV/√Hz capacitively-coupled chopper instrumentation amplifier in 65 nm CMOS for wireless sensor nodes. IEEE J. Solid-State Circuits **46**(7), 1534–1543 (2011)
96. Y. Kusuda, Auto correction feedback for ripple suppression in a chopper amplifier. IEEE J. Solid-State Circuits **45**(8), 1436–1445 (2010)
97. H. Chandrakumar, D. Marković, A simple area-efficient ripple-rejection technique for chopped biosignal amplifiers. IEEE Trans. Circuits Syst. II Express Briefs **62**(2), 189–193 (2015)
98. T. Denison, K. Consoer, W. Santa, A.T. Avestruz, J. Cooley, A. Kelly, A 2 μW 100 nV/√Hz chopper-stabilized instrumentation amplifier for chronic measurement of neural field potentials. IEEE J. Solid-State Circuits **42**(12), 2934–2945 (2007)
99. P. Cong, N. Chaimanonart, W.H. Ko, D.J. Young, A wireless and batteryless 10-Bit implantable blood pressure sensing microsystem with adaptive RF powering for real-time laboratory mice monitoring. IEEE J. Solid-State Circuits **44**(12), 3631–3644 (2009)
100. C.-G. Yu, R.L. Geiger, An automatic offset compensation scheme with ping-pong control for CMOS operational amplifiers. IEEE J. Solid-State Circuits **29**(5), 601–610 (1994)
101. N. Yazdi, H. Kulah, K. Najafi, Precision readout circuits for capacitive microaccelerometers, in *Proceedings of IEEE Sensors* (2004), pp. 28–31
102. W. Bracke, P. Merken, R. Puers, C.V. Hoof, Ultra-low-power interface chip for autonomous capacitive sensor systems. IEEE Trans. Circuits Syst. Regul. Pap. **54**(1), 130–140 (2007)
103. Y.-C. Hung, B.-D. Liu, A 1 V CMOS analog comparator using auto-zero and complementary differential-input technique, in *Proceedings. IEEE Asia-Pacific Conference on ASIC* (2002), pp. 181–184
104. L.H.C. Ferreira, T.C. Pimenta, R.L. Moreno, An auto-zero sample-and-hold circuit in 0.8 μm CMOS, in *5th International Conference on ASIC* (2003), pp. 510–513
105. R. Assaad, J. Silva-Martinez, Optimisation of direct auto-zeroing offset cancellation in low voltage applications using dual level CMFB. Electron. Lett. **45**(16), 809–811 (2009)
106. V. Srinivasan, G.J. Serrano, J. Gray, P. Hasler, A precision CMOS amplifier using floating-gate transistors for offset cancellation. IEEE J. Solid-State Circuits **42**(2), 280–291 (2007)
107. H.P. Forghani-Zadeh, G.A. Rincon-Mora, A programmable 210-μV offset rail-to-rail Gm-C filter. IEEE Trans. Circuits Syst. Regul. Pap. **54**(8), 1636–1646 (2007)
108. M.A.P. Pertijs, W.J. Kindt, A 140 dB-CMRR current-feedback instrumentation amplifier employing ping-pong auto-zeroing and chopping. IEEE J. Solid-State Circuits **45**(10), 2044–2056 (2010)
109. I.E. Opris, G.T.A. Kovacs, A rail-to-rail ping-pong op-amp. IEEE J. Solid-State Circuits **31**(9), 1320–1324 (1996)
110. M.C.W. Coln, Chopper stabilization of MOS operational amplifiers using feed-forward techniques. IEEE J. Solid-State Circuits **16**(6), 745–748 (1981)

111. I.G. Finvers, J.W. Haslett, F.N. Trofimenkoff, A high temperature precision amplifier. IEEE J. Solid-State Circuits **30**(2), 120–128 (1995)
112. B. Linares-Barranco, T. Serrano-Gotarredona, On the design and characterization of femtoampere current-mode circuits. IEEE J. Solid-State Circuits **38**(8), 1353–1363 (2003)
113. M. O'Halloran, R. Sarpeshkar, A 10-nW 12-bit accurate analog storage cell with 10-aA leakage. IEEE J. Solid-State Circuits **39**(11), 1985–1996 (2004)
114. K.B. Mirza, S. Luan, A. Eftekhar, T.G. Constandinou, Towards a fully-integrated solution for capacitor-based neural stimulation, in *2012 IEEE International Symposium on Circuits and Systems* (2012), pp. 2243–2246
115. K. Ishida, K. Kanda, A. Tamtrakarn, H. Kawaguchi, T. Sakurai, Managing subthreshold leakage in charge-based analog circuits with low-VTH transistors by analog T- switch (AT-switch) and super cut-off CMOS (SCCMOS). IEEE J. Solid-State Circuits **41**(4), 859–867 (2006)
116. C. Mayr et al., A biological-realtime neuromorphic system in 28 nm cmos using low-leakage switched capacitor circuits. IEEE Trans. Biomed. Circuits Syst. **10**(1), 243–254 (2016)
117. M. Waltari, K. Halonen, Bootstrapped switch without bulk effect in standard CMOS technology. Electron. Lett. **38**(12), 555–557 (2002)
118. J. Steensgaard, Bootstrapped low-voltage analog switches, in *Proceedings of the 1999 IEEE International Symposium on Circuits and Systems 1999 (ISCAS '99)* (1999), pp. 29–32
119. C. Lillebrekke, C. Wulff, T. Ytterdal, Bootstrapped switch in low-voltage digital 90 nm CMOS technology, in *2005 NORCHIP* (2005), pp. 234–236
120. A.M. Abo, P.R. Gray, A 1.5-V, 10-bit, 14.3-MS/s CMOS pipeline analog-to-digital converter. IEEE J. Solid-State Circuits **34**(5), 599–606 (1999)
121. D. Aksin, M. Al-Shyoukh, F. Maloberti, Switch bootstrapping for precise sampling beyond supply voltage. IEEE J. Solid-State Circuits **41**(8), 1938–1943 (2006)
122. C.J.B. Fayomi, G.W. Roberts, M. Sawan, Low-voltage CMOS analog bootstrapped switch for sample-and-hold circuit: design and chip characterization, in *2005 IEEE International Symposium on Circuits and Systems* (2005), pp. 2200–2203
123. T. Deliyannis, Y. Sun, J.K. Fidler, *Continuous-Time Active Filter Design* (CRC Press, Boca Raton, Fla, 1998)
124. Y. Sun (ed.), *Design of High Frequency Integrated Analogue Filters* (First Edition edition. London, Institution of Engineering and Technology, 2002)
125. R.L. Geiger, E. Sánchez-Sinencio, Active filter design using operational transconductance amplifiers: a tutorial. IEEE Circuits Devices Mag. **1**(2), 20–32 (1985)
126. J.C. Voghell, M. Sawan, Current tuneable CMOS transconductor for filtering applications, in *2000 IEEE International Symposium on Circuits and Systems 2000*. ISCAS 2000 Geneva (2000), pp. 165–168
127. H. Khorramabadi, P.R. Gray, High-frequency CMOS continuous-time filters. IEEE J. Solid-State Circuits **19**(6), 939–948 (1984)
128. A. Kaiser, A micropower CMOS continuous-time low-pass filter. IEEE J. Solid-State Circuits **24**(3), 736–743 (1989)
129. P. Wu, R. Schaumann, Design of a 1-GHz self-resonance integrated GaAs gyrator-C inductor, in *Proceedings of the 33rd Midwest Symposium on Circuits and Systems* (1990), pp. 362–365
130. R. Sarpeshkar, *Ultra Low Power Bioelectronics: Fundamentals, Biomedical Applications, and Bio-Inspired Systems* (Cambridge University Press, Cambridge, 2010
131. P. Crombez, J. Craninckx, M. Steyaert, A linearity and power efficient design strategy for architecture optimization of gm-C biquadratic filters, in *Research in Microelectronics and Electronics Conference. PRIME 2007* (2007), pp. 229–232
132. E. Lebel, A. Assi, M. Sawan, Programmable monolithic Gm-C band-pass filter: design and experimental results. Analog Integr. Circuits Signal Process. **54**(1), 21–29 (2007)
133. J. Silva-Martinez, J. Adut, J.M. Rocha-Perez, M. Robinson, S. Rokhsaz, A 60-mW 200-MHz continuous-time seventh-order linear phase filter with on-chip automatic tuning system. IEEE J. Solid-State Circuits **38**(2), 216–225 (2003)

134. Y. Li, H. Yu, L. Jiang, Z. Ji, Design of low power CMOS band-pass Gm-C filter for 5.8 GHz RF transceiver of ETC system, in *2010 12th IEEE International Conference on Communication Technology (ICCT)* (2010), pp. 1410–1413
135. P. Crombez, J. Craninckx, P. Wambacq, M. Steyaert, A 100-kHz to 20-MHz reconfigurable power-linearity optimized Gm-C biquad in 0.13-μm CMOS. IEEE Trans. Circuits Syst. II Express Briefs **55**(3), 224–228 (2008)
136. A.S. Sedra, K.C. Smith, *Microelectronic Circuits: International edition*, 6th edn. (OUP USA, New York, 2010)
137. C. Toumazou, F. J. Lidgey, D.G. Haigh, *Analogue IC Design: the Current-Mode Approach*, New Ed edition. The Institution of Engineering and Technology (1990)
138. B. Razavi, *Design of Analog Cmos Integrated Circuit*, International edition (McGraw-Hill Higher Education, New York, 2003)
139. E. Sanchez-Sinencio, J. Silva-Martinez, CMOS transconductance amplifiers, architectures and active filters: a tutorial. IEE Proc. Circuits Devices Syst. **147**(1), 3–12 (2000)
140. Y. Tsividis, Z. Czarnul, S.C. Fang, MOS transconductors and integrators with high linearity. Electron. Lett. **22**(5), 245–246 (1986)
141. Z. Czarnul, Y. Tsividis, Independent tuning of quality factor and unity-gain frequency in a transconductance-capacitance integrator. Electron. Lett. **22**(19), 1026–1027 (1986)
142. F. Krummenacher, N. Joehl, A 4-MHz CMOS continuous-time filter with on-chip automatic tuning. IEEE J. Solid-State Circuits **23**(3), 750–758 (1988)
143. D.A. Johns, K. Martin, *Analog Integrated Circuit Design* (Wiley, New York, 1996)
144. A. Lewinski, J. Silva-Martinez, OTA linearity enhancement technique for high frequency applications with IM3 below −65 dB. IEEE Trans. Circuits Syst. II Express Briefs **51**(10), 542–548 (2004)
145. A. Nedungadi, T. Viswanathan, Design of linear CMOS transconductance elements. IEEE Trans. Circuits Syst. **31**(10), 891–894 (1984)
146. R. van Langevelde, F.M. Klaassen, Effect of gate-field dependent mobility degradation on distortion analysis in MOSFETs. IEEE Trans. Electron Devices **44**(11), 2044–2052 (1997)
147. P.E. Allen, D.R. Holberg, *CMOS Analog Circuit Design* (Oxford University Press, Oxford, 2002
148. K.-C. Kuo, A. Leuciuc, A linear MOS transconductor using source degeneration and adaptive biasing. IEEE Trans. Circuits Syst. II Analog Digit. Signal Process. **48**(10), 937–943 (2001)
149. S. Solis-Bustos, J. Silva-Martinez, F. Maloberti, E. Sanchez-Sinencio, A 60-dB dynamic-range CMOS sixth-order 2.4-Hz low-pass filter for medical applications. IEEE Trans. Circuits Syst. II Analog Digit. Signal Process. **47**(12), 1391–1398 (2000)
150. J. Silva-Martinez, S. Solis-Bustos, Design considerations for high performance very low frequency filters, in *Proceedings of the 1999 IEEE International Symposium on Circuits and Systems, 1999 (ISCAS '99)* (1999), pp. 648–651
151. J. Silva-Martinez, J. Salcedo-Suñer, IC voltage to current transducers with very small transconductance. Analog Integr. Circuits Signal Process. **13**(3), 285–293 (1997)
152. A. Veeravalli, E. Sanchez-Sinencio, J. Silva-Martinez, Transconductance amplifier structures with very small transconductances: a comparative design approach. IEEE J. Solid-State Circuits **37**(6), 770–775 (2002)
153. S.-C. Liu, J. Kramer, G. Indiveri, T. Delbruck, R. Douglas, *Analog VLSI: Circuits and Principles* (MIT Press, Cambridge, Mass, 2002)
154. A. Wang, B.H. Calhoun, A.P. Chandrakasan, *Sub-threshold Design for Ultra Low-Power Systems*, Softcover reprint of hardcover 1st ed. 2006 edition (Springer, New York, 2009
155. P.M. Furth, *On the design of optimal continuous-time filter banks in subthreshold CMOS*, in PhD Thesis, John Hopkins, Baltimore, MA (1996)
156. C. Mead, *Analogue and Neural Very Large Scale Integration Systems* (Addison Wesley, Reading, Mass, 1989)
157. P.M. Furth, A.G. Andreou, Linearised differential transconductors in subthreshold CMOS. Electron. Lett. **31**(7), 545–547 (1995)

158. P. Kassanos, I.F. Triantis, A CMOS multi-sine signal generator for multi-frequency bioimpedance measurements, in *2014 IEEE International Symposium on Circuits and Systems (ISCAS)* (2014), pp. 249–252
159. P. Georgiou, C. Toumazou, ISFET characteristics in CMOS and their application to weak inversion operation. Sens. Actuators B Chem. **143**(1), 211–217 (2009)
160. D. Binkley, *Tradeoffs and Optimization in Analog CMOS Design* (Wiley, Chichester, West Sussex, UK, 2008)
161. A.J. Casson, E. Rodriguez-Villegas, A 60 pW gmC continuous wavelet transform circuit for portable EEG systems. IEEE J. Solid-State Circuits **46**(6), 1406–1415 (2011)
162. L. Watts, D.A. Kerns, R.F. Lyon, C.A. Mead, Improved implementation of the silicon cochlea. IEEE J. Solid-State Circuits **27**(5), 692–700 (1992)
163. M. Ismail, T. Fiez, *Analog VLSI: Signal and Information Processing* (McGraw-Hill Education, 1994)
164. K. Bult, G.J.G.M. Geelen, An inherently linear and compact MOST-only current division technique. IEEE J. Solid-State Circuits **27**(12), 1730–1735 (1992)
165. H. Tanimoto, M. Koyama, Y. Yoshida, Realization of a 1-V active filter using a linearization technique employing plurality of emitter-coupled pairs. IEEE J. Solid-State Circuits **26**(7), 937–945 (1991)
166. B. Gilbert, The multi-tanh principle: a tutorial overview. IEEE J. Solid-State Circuits **33**(1), 2–17 (1998)
167. K. Kimura, The ultra-multi-tanh technique for bipolar linear transconductance amplifiers. IEEE Trans Circuits Syst. Fundam. Theory Appl. **44**(4), 288–302 (1997)
168. E. Rodriguez-Villegas, A.J. Casson, P. Corbishley, A subhertz nanopower low-pass filter. IEEE Trans. Circuits Syst. II Express Briefs **58**(6), 351–355 (2011)
169. P. Corbishley, E. Rodriguez-Villegas, A nanopower bandpass filter for detection of an acoustic signal in a wearable breathing detector. IEEE Trans. Biomed. Circuits Syst. **1**(3), 163–171 (2007)
170. P.M. Furth, H.A. Ommani, Low-voltage highly-linear transconductor design in subthreshold CMOS, in *Proceedings of the 40th Midwest Symposium on Circuits and Systems 1997* (1997), pp. 156–159
171. A.G. Katsiamis, E.M. Drakakis, R.F. Lyon, A biomimetic, 4.5 μW, 120 + dB, log-domain cochlea channel with AGC. IEEE J. Solid-State Circuits **44**(3), 1006–1022 (2009)
172. T. Delbruck, 'Bump' circuits for computing similarity and dissimilarity of analog voltages, in *IJCNN-91-Seattle International Joint Conference on Neural Networks* (1991), pp. 475–479
173. M.H. Cohen, A.C. Andreou, MOS circuit for nonlinear Hebbian learning. Electron. Lett. **28** (6), 591–593 (1992)
174. A. Pesavento and C. Koch, A wide linear range four quadrant multiplier in subthreshold CMOS, in *Proceedings of the 1999 IEEE International Symposium on Circuits and Systems (ISCAS '99)* (1999), pp. 240–243
175. M.W. Baker, R. Sarpeshkar, Low-power single-loop and dual-loop AGCs for bionic ears. IEEE J. Solid-State Circuits **41**(9), 1983–1996 (2006)
176. P. Georgiou, C. Toumazou, Towards an ultra low power chemically inspired electronic beta cell for diabetes, in *2006 IEEE International Symposium on Circuits and Systems* (2006), pp. 173–176
177. K.H. Wee, R. Sarpeshkar, An electronically tunable linear or nonlinear MOS resistor. IEEE Trans. Circuits Syst. Regul. Pap. **55**(9), 2573–2583 (2008)
178. Y. Hu, Y. Liu, T.G. Constandinou, C. Toumazou, A 5 s-time-constant temperature-stable integrator for a tuneable PID controller in LOC applications, in *2011 IEEE International Symposium of Circuits and Systems (ISCAS)* (2011), pp. 1387–1390
179. B. Rumberg, D.W. Graham, A low-power and high-precision programmable analog filter bank. IEEE Trans. Circuits Syst. II Express Briefs **59**(4), 234–238 (2012)
180. B. Gosselin, M. Sawan, An ultra low-power CMOS automatic action potential detector. IEEE Trans. Neural Syst. Rehabil. Eng. **17**(4), 346–353 (2009)

181. B. Gosselin, M. Sawan, E. Kerherve, Linear-phase delay filters for ultra-low-power signal processing in neural recording implants. IEEE Trans. Biomed. Circuits Syst. **4**(3), 171–180 (2010)
182. R. Sarpeshkar, R.F. Lyon, C. Mead, A low-power wide-linear-range transconductance amplifier. Analog Integr. Circuits Signal Process. **13**(1–2), 123–151 (1997)
183. T. Delbruck, Bump circuits for comparing similarity and dissimilarity of analog voltages. California Institute Of Technology, Pasadena, California, Memo 26, May 1993
184. A.E. Mourabit, G.-N. Lu, P. Pittet, Wide-linear-range subthreshold OTA for low-power, low-voltage, and low-frequency applications. IEEE Trans. Circuits Syst. Regul. Pap. **52**(8), 1481–1488 (2005)
185. J. Chen, E. Sanchez-Sinencio, J. Silva-Martinez, Frequency-dependent harmonic-distortion analysis of a linearized cross-coupled CMOS OTA and its application to OTA-C filters. IEEE Trans. Circuits Syst. Regul. Pap. **53**(3), 499–510 (2006)
186. R. Chawla, F. Adil, G. Serrano, P.E. Hasler, Programmable Gm–C filters using floating-gate operational transconductance amplifiers. IEEE Trans. Circuits Syst. Regul. Pap. **54**(3), 481–491 (2007)
187. E. Rodriguez-Villegas, H. Barnes, Solution to trapped charge in FGMOS transistors. Electron. Lett. **39**(19), 1416–1417 (2003)
188. E. Rodriguez-Villegas, M. Jimenez, R.G. Carvajal, On dealing with the charge trapped in floating-gate MOS (FGMOS) transistors. IEEE Trans. Circuits Syst. II Express Briefs **54**(2), 156–160 (2007)
189. E. Rodriguez-Villegas, *Low Power and Low Voltage Circuit Design with the FGMOS Transistor* (Institution of Engineering and Technology, Stevenage, 2006)
190. L. Zuo, S.K. Islam, Low-voltage bulk-driven operational amplifier with improved transconductance. IEEE Trans. Circuits Syst. Regul. Pap. **60**(8), 2084–2091 (2013)
191. J.M. Carrillo, G. Torelli, R.P.-A. Valverde, J.F. Duque-Carrillo, 1-V rail-to-rail CMOS OpAmp with improved bulk-driven input stage. IEEE J. Solid-State Circuits **42**(3), 508–517 (2007)
192. J.M. Carrillo, G. Torelli, J.F. Duque-Carrillo, Transconductance enhancement in bulk-driven input stages and its applications. Analog Integr. Circuits Signal Process. **68**(2), 207–217 (2011)
193. J.M. Carrillo, G. Torelli, M.A. Domínguez, J.F. Duque-Carrillo, On the input common-mode voltage range of CMOS bulk-driven input stages. Int. J. Circuit Theory Appl. **39**(6), 649–664 (2011)
194. P. Kinget, M. Steyaert, J. van der Spiegel, Full analog CMOS integration of very large time constants for synaptic transfer in neural networks. Analog Integr. Circuits Signal Process. **2** (4), 281–295 (1992)
195. A. Arnaud, R. Fiorelli, and C. Galup-Montoro, On the design of very small transconductance otas with reduced input offset, in *2005 18th Symposium on Integrated Circuits and Systems Design* (2005), pp. 15–20
196. C. Galup-Montoro, M.C. Schneider, I.J.B. Loss, Series-parallel association of FET's for high gain and high frequency applications. IEEE J. Solid-State Circuits **29**(9), 1094–1101 (1994)
197. M. Steyaert, P. Kinget, W. Sansen, Full integration of extremely large time constants in CMOS. Electron. Lett. **27**(10), 790–791 (1991)
198. A. Arnaud, C. Galup-Montoro, Pico-A/V range CMOS transconductors using series-parallel current division. Electron. Lett. **39**(18), 1295–1296 (2003)
199. A. Arnaud, R. Fiorelli, C. Galup-Montoro, Nanowatt, Sub-nS OTAs, with Sub-10-mV input offset, using series-parallel current mirrors. IEEE J. Solid-State Circuits **41**(9), 2009–2018 (2006)
200. R. Fiorelli, A. Arnaud, C. Galup-Montoro, Series-parallel association of transistors for the reduction of random offset in non-unity gain current mirrors, in *Proceedings of the 2004 International Symposium on Circuits and Systems, 2004 (ISCAS '04)* (2004), pp. 881-884
201. C.C. Enz, E.A. Vittoz, CMOS low-power analog circuit design. Design. Low Power Digital Syst. Emerg. Technol. **1996**, 79–133 (1996)

202. B. Pankiewicz, M. Wojcikowski, S. Szczepanski, Y. Sun, A field programmable analog array for CMOS continuous-time OTA-C filter applications. IEEE J. Solid-State Circuits **37** (2), 125–136 (2002)
203. J. Ramirez-Angulo, S.R.S. Garimella, A.J. Lopez-Martin, R.G. Carvajal, Gain programmable current mirrors based on current steering. Electron. Lett. **42**(10), 559–560 (2006)
204. A.J. Lopez-Martin, J. Ramirez-Angulo, C. Durbha, R.G. Carvajal, A CMOS transconductor with multidecade tuning using balanced current scaling in moderate inversion. IEEE J. Solid-State Circuits **40**(5), 1078–1083 (2005)
205. A.J. Lopez-Martin, J. Ramirez-Angulo, C. Durbha, R.G. Carvajal, Highly linear programmable balanced current scaling technique in moderate inversion. IEEE Trans. Circuits Syst. II Express Briefs **53**(4), 283–285 (2006)
206. Y. Tsividis, K. Suyama, K. Vavelidis, Simple 'reconciliation' MOSFET model valid in all regions. Electron. Lett. **31**(6), 506–508 (1995)
207. A.J. Lopez-Martin, J. Ramirez-Angulo, R.G. Carvajal, L. Acosta, CMOS transconductors with continuous tuning using FGMOS balanced output current scaling. IEEE J. Solid-State Circuits **43**(5), 1313–1323 (2008)
208. J. Silva-Martinez, M.S.J. Steyaert, W. Sansen, Design techniques for high-performance full-CMOS OTA-RC continuous-time filters. IEEE J. Solid-State Circuits **27**(7), 993–1001 (1992)
209. M. Banu, J.M. Khoury, Y. Tsividis, Fully differential operational amplifiers with accurate output balancing. IEEE J. Solid-State Circuits **23**(6), 1410–1414 (1988)
210. P. Wu, R. Schaumann, P. Latham, Design considerations for common-mode feedback circuits in fully-differential operational transconductance amplifiers with tuning, in *IEEE International Symposium on Circuits and Systems 1991* (1991), pp. 1363–136
211. J.F. Duque-Carrillo, Control of the common-mode component in CMOS continuous-time fully differential signal processing. Analog Integr. Circuits Signal Process. **4**(2), 131–140 (1993)
212. J.N. Babanezhad, A low-output-impedance fully differential op amp with large output swing and continuous-time common-mode feedback. IEEE J. Solid-State Circuits **26**(12), 1825–1833 (1991)
213. P.M. VanPeteghem, J.F. Duque-Carrillo, A general description of common-mode feedback in fully-differential amplifiers, in *IEEE International Symposium on Circuits and Systems* (1990), pp. 320–312
214. L. Lah, J. Choma, J. Draper, A continuous-time common-mode feedback circuit (CMFB) for high-impedance current-mode applications. IEEE Trans. Circuits Syst. II Analog Digit. Signal Process. **47**(4), 363–369 (2000)
215. W. Yan, H. Zimmermann, Continuous-time common-mode feedback circuit for applications with large output swing and high output impedance, in *11th IEEE Workshop on Design and Diagnostics of Electronic Circuits and Systems, 2008 (DDECS 2008)* (2008), pp. 1–5
216. M. Maymandi-Nejad, M. Sachdev, Continuous time common mode feedback technique for sub 1 V analogue circuits. Electron. Lett. **38**(23), 1408–1409 (2002)
217. J. Silva-Martinez, A. Vazquez-Gonzalez, Impedance scalers for IC active filters, in *Proceedings of the 1998 IEEE International Symposium on Circuits and Systems, 1998 (ISCAS '98)* (1998), pp. 151–154
218. G. Doménech-Asensi et al., Low-frequency CMOS bandpass filter for PIR sensors in wireless sensor nodes. IEEE Sens. J. **14**(11), 4085–4094 (2014)
219. Y. Li, C.C.Y. Poon, Y.T. Zhang, Analog integrated circuits design for processing physiological signals. IEEE Rev. Biomed. Eng. **3**, 93–105 (2010)
220. W. Lee, S. Cho, Integrated all electrical pulse wave velocity and respiration sensors using bio-impedance. IEEE J. Solid-State Circuits **50**(3), 776–785 (2015)
221. L. Yan et al., A 13 μW analog signal processing IC for accurate recognition of multiple intra-cardiac signals. IEEE Trans. Biomed. Circuits Syst. **7**(6), 785–795 (2013)
222. S. Rodriguez, S. Ollmar, M. Waqar, A. Rusu, A batteryless sensor ASIC for implantable bio-impedance applications. IEEE Trans. Biomed. Circuits Syst. **10**(3), 533–544 (2016)

223. C.L. Wei, Y.W. Wang, B.D. Liu, Wide-range filter-based sinusoidal wave synthesizer for electrochemical impedance spectroscopy measurements. IEEE Trans. Biomed. Circuits Syst. **8**(3), 442–450 (2014)
224. R.G. Meyer, W.M.C. Sansen, S. Peeters, The differential pair as a triangle-sine wave converter. IEEE J. Solid-State Circuits **11**(3), 418–420 (1976)
225. S. Thuries, É. Tournier, A. Cathelin, S. Godet, J. Graffeuil, A 6-GHz low-power BiCMOS SiGe: C 0.25 μm direct digital synthesizer. IEEE Microw. Wirel. Compon. Lett. **18**(1), 46–48 (2008)
226. Z. Tang, O. Ishizuka, H. Matsumoto, MOS triangle-to-sine wave convertor based on subthreshold operation. Electron. Lett. **26**(23), 1983–1985 (1990)
227. A.A. Hatzopoulos, S. Siskos, J.M. Kontoleon, A complete scheme of built-in self-tests (BIST) structure for fault diagnosis in analog circuits and systems. IEEE Trans. Instrum. Measur. **42**(3), 689–694 (1993)
228. J. Gomez-Quiñones, et al., An application specific multi-channel stimulator for electrokinetically-driven microfluidic devices, in *2011 IEEE 9th International New Circuits and Systems Conference* (2011), pp. 350–353
229. B. Gilbert, Circuits for the precise synthesis of the sine function. Electron. Lett. **13**(17), 506–508 (1977)
230. W.A. Evans, V. Schiffer, A low distortion tri-wave to sine converter. Radio Electron. Eng. **47**(5), 217–224 (1977)
231. O. Ishizuka, Z. Tang, H. Matsumoto, MOS sine function generator using exponential-law technique. Electron. Lett. **27**(21), 1937–1939 (1991)
232. B. Gilbert, A monolithic microsystem for analog synthesis of trigonometric functions and their inverses. IEEE J. Solid-State Circuits **17**(6), 1179–1191 (1982)
233. J.W. Fattaruso, R.G. Meyer, Triangle-to-sine wave conversion with MOS technology. IEEE J. Solid-State Circuits **20**(2), 623–631 (1985)
234. Y.G. Li, A.K.M. Arifuzzman, A.E. Pollard, M.R. Haider, An inductorless low-power tunable sinusoidal oscillator for micro-impedance spectroscopy, in *IEEE 15th Annual Wireless and Microwave Technology Conference (WAMICON)* (2014), pp. 1–4
235. T.H. Lee, P.A. Abshire, Accurate, wide range analog sine shaping circuits, in *2014 IEEE 12th International New Circuits and Systems Conference (NEWCAS)* (2014), pp. 285–288
236. C.Y. Yang, J.H. Weng, H.Y. Chang, A 5-GHz direct digital frequency synthesizer using an analog-sine-mapping technique in 0.35-μm SiGe BiCMOS. IEEE J. Solid-State Circuits **46** (9), 2064–2072 (2011)
237. A. Rodriguez-Vazquez, B. Linares-Barranco, J.L. Huertas, E. Sanchez-Sinencio, On the design of voltage-controlled sinusoidal oscillators using OTAs. IEEE Trans. Circuits Syst. **37**(2), 198–211 (1990)
238. B. Linares-Barranco, A. Rodriguez-Vazquez, E. Sanchez-Sinencio, J.L. Huertas, CMOS OTA-C high-frequency sinusoidal oscillators. IEEE J. Solid-State Circuits **26**(2), 160–165 (1991)
239. B. Linares-Barranco, A. Rodriguez-Vazquez, E. Sanchez-Sinencio, J.L. Huertas, Generation, design and tuning of OTA-C high frequency sinusoidal oscillators. IEE Proc. G - Circuits Devices Syst. **139**(5), 557–568 (1992)
240. T. Serrano-Gotarredona, B. Linares-Barranco, 7-decade tuning range CMOS OTA-C sinusoidal VCO. Electron. Lett. **34**(17), 1621–1622 (1998)
241. J. Galan, R.G. Carvajal, A. Torralba, F. Munoz, J. Ramirez-Angulo, A low-power low-voltage OTA-C sinusoidal oscillator with a large tuning range. IEEE Trans. Circuits Syst. Regul. Pap. **52**(2), 283–291 (2005)
242. B. Linares-Barranco, T. Serrano-Gotarredona, J. Ramos-Martos, J. Ceballos-Caceres, J.M. Mora, A. Linares-Barranco, A precise 90° quadrature OTA-C oscillator tunable in the 50–130-MHz range. IEEE Trans. Circuits Syst. Regul. Pap. **51**(4), 649–663 (2004)
243. K.M. Odame, P. Hasler, Theory and design of OTA-C oscillators with native amplitude limiting. IEEE Trans. Circuits Syst. Regul. Pap. **56**(1), 40–50 (2009)

244. N. Revanna, T. R. Viswanathan, Low frequency CMOS sinusoidal oscillator for impedance spectroscopy, in *2014 IEEE Dallas Circuits and Systems Conference (DCAS)* (2014), pp. 1–4
245. B. Linares-Barranco, A. Rodriguez-Vazquez, E. Sanchez-Sinencio, J.L. Huertas, 10 MHz CMOS OTA-C voltage-controlled quadrature oscillator. Electron. Lett. **25**(12), 765–767 (1989)
246. L.L. Presti, G. Cardamone, A direct digital frequency synthesizer using an IIR filter implemented with a DSP microprocessor, in *1994 IEEE International Conference on Acoustics, Speech, and Signal Processing 1994 (ICASSP-94)* (1994), p. III/201-III/204
247. M. Schanerberger, S.S. Awad, The implementation of a digital sine wave oscillator using the TMS320C25: distortion reduction and applications. IEEE Trans. Instrum. Meas. **39**(6), 870–873 (1990)
248. F. Kua, in *Generation of a sine wave using a TMS320C54x digital signal processor*. Texas Instruments, Application Report SPRA819, July 2004
249. A.I. Abu-el-haija, M.M. Al-ibrahim, Digital oscillator having low sensitivity and roundoff errors. IEEE Trans. Aerosp. Electron. Syst. **AES-22**(1), 23–33 (1986)
250. A. Abu-El-Haija, M. Al-Ibrahim, Improving performance of digital sinusoidal oscillators by means of error feedback circuits. IEEE Trans. Circuits Syst. **33**(4), 373–380 (1986)
251. A.I. Abu-El-Haija, M.M. Al-Ibrahim, M.A. Al-Jarrah, Recursive digital sine wave oscillators using the TMS32010 DSP, in *Conference Proceedings. 10th Anniversary. IMTC/94. Advanced Technologies in I M. 1994 IEEE Instrumentation and Measurement Technolgy Conference* (1994), pp. 792–796
252. S.M. Kuo, B.H. Lee, W. Tian, *Real-Time Digital Signal Processing: Fundamentals, Implementations and Applications* (Wiley, New York, 2013)
253. V. Matic, V. Marinkovic-Nedelicki, V. Tadic, Comparison of digital signal processing methods for sine wave signal generation, in *Telecommunications Symposium. ITS '98 Proceedings. SBT/IEEE International* (1998), pp. 290–295
254. J.G. Proakis, D.K. Manolakis, *Digital Signal Processing* (Pearson Higher Ed, 2013)
255. L. Lo Presti, G. Cardamone, A. De Marchi, E. Rubiola, A new architecture for a sinewave output DDS with high spectral purity, in *Direct Digital Frequency Synthesizers* (Wiley-IEEE Press, New York, 1999), pp. 352–357
256. P.R.B. de Carvalho, J.A.A. Palacio, W.V. Noije, Area optimized CORDIC-based numerically controlled oscillator for electrical bio-impedance spectroscopy, in *2016 IEEE International Frequency Control Symposium (IFCS)* (2016), pp. 1–6
257. C. Yang, A.J. Mason, Fully integrated seven-order frequency-range quadrature sinusoid signal generator. IEEE Trans. Instrum. Meas. **58**(10), 3481–3489 (2009)
258. D. Rairigh, X. Liu, C. Yang, A.J. Mason, Sinusoid signal generator for on-chip impedance spectroscopy, in *2009 IEEE International Symposium on Circuits and Systems* (2009), pp. 1961–1964
259. M. Nawito, H. Richter, J.N. Burghartz, Compact wide-range sinusoidal signal generator for in vivo impedance spectroscopy, in *2015 Conference on Design of Circuits and Integrated Systems (DCIS)* (2015), pp. 1–5
260. S.J. Kweon, S.H. Jo, J.H. Park, and H.J. Yoo, A CMOS sinusoidal signal generator based on mixed-time processing for electrical bioimpedance spectroscopy supporting beta dispersion range, in *2016 IEEE Asia Pacific Conference on Circuits and Systems (APCCAS)* (2016), pp. 91–94
261. J. Vankka, M. Waltari, M. Kosunen, K.A.I. Halonen, A direct digital synthesizer with an on-chip D/A-converter. IEEE J. Solid-State Circuits **33**(2), 218–227 (1998)
262. J.M.P. Langlois, D. Al-Khalili, Novel approach to the design of direct digital frequency synthesizers based on linear interpolation. IEEE Trans. Circuits Syst. II Analog Digit. Signal Process. **50**(9), 567–578 (2003)
263. B.-D. Yang, J.H. Choi, S.-H. Han, L.-S. Kim, H.-K. Yu, An 800-MHz low-power direct digital frequency synthesizer with an on-chip D/A converter. IEEE J. Solid-State Circuits **39** (5), 761–774 (2004)

264. A.L. McEwan, S. Collins, Efficient, ROM-less DDFS using non-linear interpolation and non-linear DAC. Analog Integr. Circuits Signal Process. **48**(3), 231–237 (2006)
265. X. Yu, F.F. Dai, J.D. Irwin, R.C. Jaeger, A 12 GHz 1.9 W direct digital synthesizer MMIC implemented in 0.18 m SiGe BiCMOS technology. IEEE J. Solid-State Circuits **43**(6), 1384–1393 (2008)
266. C. Shi, E. Sánchez-Sinencio, 150-850 MHz high-linearity sine-wave synthesizer architecture based on FIR filter approach and SFDR optimization. IEEE Trans. Circuits Syst. Regul. Pap. **62**(9), 2227–2237 (2015)
267. L. Constantinou, R. Bayford, A. Demosthenous, A wideband low-distortion cmos current driver for tissue impedance analysis. IEEE Trans. Circuits Syst. II Express Briefs **62**(2), 154–158 (2015)
268. L. Constantinou, I.F. Triantis, R. Bayford, A. Demosthenous, High-power CMOS current driver with accurate transconductance for electrical impedance tomography. IEEE Trans. Biomed. Circuits Syst. **8**(4), 575–583 (2014)
269. N.V. Helleputte et al., A 345 μW multi-sensor biomedical SoC with bio-impedance, 3-channel ECG, motion artifact reduction, and integrated DSP. IEEE J. Solid-State Circuits **50**(1), 230–244 (2015)
270. M. Crescentini, M. Bennati, M. Tartagni, A high resolution interface for Kelvin impedance sensing. IEEE J. Solid-State Circuits **49**(10), 2199–2212 (2014)
271. R. Genov, M. Stanacevic, M. Naware, G. Cauwenberghs, N.V. Thakor, 16-channel integrated potentiostat for distributed neurochemical sensing. IEEE Trans. Circuits Syst. Regul. Pap. **53**(11), 2371–2376 (2006)
272. S. Shin, S. J. Kweon, J.H. Park, Y.C. Choi, H.J. Yoo, An efficient, wide range time-to-digital converter using cascaded time-interpolation stages for electrical impedance spectroscopy, in *2016 IEEE Asia Pacific Conference on Circuits and Systems (APCCAS)* (2016), pp. 425–428
273. H. Huang, S. Palermo, A TDC-based front-end for rapid impedance spectroscopy, in *2013 IEEE 56th International Midwest Symposium on Circuits and Systems (MWSCAS)* (2013), pp. 169–172
274. T.A. Chen, W.J. Wu, C.L. Wei, R.B. Darling, B.D. Liu, Novel 10-bit impedance-to-digital converter for electrochemical impedance spectroscopy measurements. IEEE Trans. Biomed. Circuits Syst. **11**(2), 370–379 (2017)
275. J. Rosell, P. Riu, Common-mode feedback in electrical impedance tomography. Clin. Phys. Physiol. Measur. **13**(A), 11 (1992)
276. M. Rahal, A. Demosthenous, R. Bayford, An integrated common-mode feedback topology for multi-frequency bioimpedance imaging, in *2009 Proceedings of ESSCIRC* (2009), pp. 416–419
277. P.J. Langlois, Y. Wu, R.H. Bayford, A. Demosthenous, On the application of frequency selective common mode feedback for multifrequency EIT. Physiol. Measur. **36**(6), 1337 (2015)
278. X. Tan et al., 4.3 μW 10 fA-sensitivity dual-mode current converter for implantable glucose monitoring. Electron. Lett. **51**(19), 1484–1486 (2015)
279. C. Yang, Y. Huang, B.L. Hassler, R.M. Worden, A.J. Mason, Amperometric electrochemical microsystem for a miniaturized protein biosensor array. IEEE Trans. Biomed. Circuits Syst. **3**(3), 160–168 (2009)
280. L. Li, X. Liu, W.A. Qureshi, A.J. Mason, CMOS amperometric instrumentation and packaging for biosensor array applications. IEEE Trans. Biomed. Circuits Syst. **5**(5), 439–448 (2011)
281. S. Hwang, C.N. LaFratta, V. Agarwal, X. Yu, D.R. Walt, S. Sonkusale, CMOS microelectrode array for electrochemical lab-on-a-chip applications. IEEE Sens. J. **9**(6), 609–615 (2009)
282. P.M. Levine, P. Gong, R. Levicky, K.L. Shepard, Active CMOS sensor array for electrochemical biomolecular detection. IEEE J. Solid-State Circuits **43**(8), 1859–1871 (2008)

283. D. Kim, B. Goldstein, W. Tang, F.J. Sigworth, E. Culurciello, Noise analysis and performance comparison of low current measurement systems for biomedical applications. IEEE Trans. Biomed. Circuits Syst. **7**(1), 52–62 (2013)
284. S. Ayers, K.D. Gillis, M. Lindau, B.A. Minch, Design of a CMOS potentiostat circuit for electrochemical detector arrays. IEEE Trans. Circuits Syst. Regul. Pap. **54**(4), 736–744 (2007)
285. M.M. Ahmadi, G.A. Jullien, Current-mirror-based potentiostats for three-electrode amperometric electrochemical sensors. IEEE Trans. Circuits Syst. Regul. Pap. **56**(7), 1339–1348 (2009)
286. C.I. Dorta-Quiñones, X.Y. Wang, R.K. Dokania, A. Gailey, M. Lindau, A.B. Apsel, A wireless FSCV monitoring IC with analog background subtraction and UWB telemetry. IEEE Trans. Biomed. Circuits Syst. **10**(2), 289–299 (2016)
287. V. Balasubramanian, P.F. Ruedi, Y. Temiz, A. Ferretti, C. Guiducci, C.C. Enz, A 0.18 μm biosensor front-end based on 1/f noise, distortion cancelation and chopper stabilization techniques. IEEE Trans. Biomed. Circuits Syst. **7**(5), 660–673 (2013)
288. M.M. Ahmadi, G.A. Jullien, A wireless-implantable microsystem for continuous blood glucose monitoring. IEEE Trans. Biomed. Circuits Syst. **3**(3), 169–180 (2009)
289. M. Stanacevic, K. Murari, A. Rege, G. Cauwenberghs, N.V. Thakor, VLSI potentiostat array with oversampling gain modulation for wide-range neurotransmitter sensing. IEEE Trans. Biomed. Circuits Syst. **1**(1), 63–72 (2007)
290. M. Mollazadeh, K. Murari, G. Cauwenberghs, N. Thakor, Wireless micropower instrumentation for multimodal acquisition of electrical and chemical neural activity. IEEE Trans. Biomed. Circuits Syst. **3**(6), 388–397 (2009)
291. M. Roham et al., A wireless IC for wide-range neurochemical monitoring using amperometry and fast-scan cyclic voltammetry. IEEE Trans. Biomed. Circuits Syst. **2**(1), 3–9 (2008)
292. B. Bozorgzadeh, D.P. Covey, C.D. Howard, P.A. Garris, P. Mohseni, A Neurochemical Pattern generator SoC with switched-electrode management for single-chip electrical stimulation and 9.3 μW, 78 pA rms, 400 V/s FSCV sensing. IEEE J. Solid-State Circuits **49** (4), 881–895 (2014)
293. B. Bozorgzadeh, D.R. Schuweiler, M.J. Bobak, P.A. Garris, P. Mohseni, Neurochemostat: a neural interface SoC with integrated chemometrics for closed-loop regulation of brain dopamine. IEEE Trans. Biomed. Circuits Syst. **10**(3), 654–667 (2016)
294. C.A. Murillo, B. C. Lopez, S.C. Pueyo, *Voltage-to-Frequency Converters: CMOS Design and Implementation* (Springer Science & Business Media, Berlin, 2013)
295. I.M. Filanovsky, H. Baltes, CMOS Schmitt trigger design. IEEE Trans. Circuits Syst. Fundam. Theory Appl. **41**(1), 46–49 (1994)
296. C. Azcona, B. Calvo, N. Medrano, S. Celma, 1.2 V-0.18-μm CMOS temperature sensors with quasi-digital output for portable systems. IEEE Trans. Instrum. Meas. **64**(9), 2565–2573 (2015)
297. A.B. Grebene, *Bipolar and MOS Analog Integrated Circuit Design* (Wiley-Interscience, New York, 2002)
298. C.C. Wang, T.J. Lee, C.C. Li, R. Hu, An All-MOS high-linearity voltage-to-frequency converter chip with 520-kHz/V sensitivity. IEEE Trans. Circuits Syst. II Express Briefs **53** (8), 744–747 (2006)
299. V. Srinivasan, R. Chawla, P. Hasler, Linear current-to-voltage and voltage-to-current converters, in *48th Midwest Symposium on Circuits and Systems* (2005), pp. 675–678
300. R. Shukla, J. Ramirez-Angulo, A. Lopez-Martin, R.G. Carvajal, A low voltage rail to rail V-I conversion scheme for applications in current mode A/D converters, in *2004 IEEE International Symposium on Circuits and Systems* (2004), I-916-19
301. A.J. Lopez-Martin, J. Ramirez-Angulo, R.G. Carvajal, ±1.5V 3mW CMOS V–I converter with 75 dB SFDR for 6 V_{pp} input swings. Electron. Lett. **43**(6), 31–32 (2007)

302. C. Azcona, B. Calvo, S. Celma, N. Medrano, P.A. Martínez, Low-voltage low-power CMOS rail-to-rail voltage-to-current converters. IEEE Trans. Circuits Syst. Regul. Pap. **60**(9), 2333–2342 (2013)
303. C. Azcona, B. Calvo, S. Celma, N. Medrano, P.A. Martínez, Ratiometric voltage-to-frequency converter for long-life autonomous portable equipment. IEEE Sens. J. **13**(6), 2382–2390 (2013)
304. R.A. Croce et al., A highly miniaturized low-power CMOS-Based pH monitoring platform. IEEE Sens. J. **15**(2), 895–901 (2015)
305. K. Zhu, S.K. Islam, M.R. Haider, M. Roknsharifi, J. Holleman, Simple oscillators-based readout circuit for low-power biomedical implant system. Analog Integr. Circuits Signal Process. **72**(2), 383–393 (2012)
306. D. Cirmirakis, A. Demosthenous, N. Saeidi, N. Donaldson, Humidity-to-frequency sensor in CMOS technology with wireless readout. IEEE Sens. J. **13**(3), 900–908 (2013)
307. M.S.-C. Lu, Y.-C. Chen, P.-C. Huang, 5 × 5 CMOS capacitive sensor array for detection of the neurotransmitter dopamine. Biosens. Bioelectron. **26**(3), 1093–1097 (2010)
308. A. Bonanno, M. Crepaldi, I. Rattalino, P. Motto, D. Demarchi, P. Civera, A CMOS operational Schmitt trigger R-to-F converter for nanogap-based nanosensors read-out. IEEE Trans. Circuits Syst. Regul. Pap. **60**(4), 975–988 (2013)
309. M. Grassi, P. Malcovati, A. Baschirotto, A 141-dB dynamic range CMOS gas-sensor interface circuit without calibration with 16-bit digital output word. IEEE J. Solid-State Circuits **42**(7), 1543–1554 (2007)
310. W. Yun, R.T. Howe, P.R. Gray, Surface micromachined, digitally force-balanced accelerometer with integrated CMOS detection circuitry, in *Technical Digest IEEE Solid-State Sensor and Actuator Workshop* (1992), pp. 126–131
311. P. Holmberg, Automatic balancing of linear AC bridge circuits for capacitive sensor elements. IEEE Trans. Instrum. Meas. **44**(3), 803–805 (1995)
312. J.A. Geen, S.J. Sherman, J.F. Chang, S.R. Lewis, Single-chip surface micromachined integrated gyroscope with 50°/h Allan deviation. IEEE J. Solid-State Circuits **37**(12), 1860–1866 (2002)
313. N. Yazdi, A. Mason, K. Najafi, K.D. Wise, A generic interface chip for capacitive sensors in low-power multi-parameter microsystems. Sens. Actuators Phys. **84**(3), 351–361 (2000)
314. H. Kulah, J. Chae, N. Yazdi, K. Najafi, Noise analysis and characterization of a sigma-delta capacitive microaccelerometer. IEEE J. Solid-State Circuits **41**(2), 352–361 (2006)
315. W.P. Chan et al., A monolithically integrated pressure/oxygen/temperature sensing SoC for multimodality intracranial neuromonitoring. IEEE J. Solid-State Circuits **49**(11), 2449–2461 (2014)
316. Y.M. Wang, P.K. Chan, H.K.H. Li, S.E. Ong, A low-power highly sensitive capacitive accelerometer IC using auto-zero time-multiplexed differential technique. IEEE Sens. J. **15** (11), 6179–6191 (2015)
317. Z.P. Fang, J.T. Mortimer, Selective activation of small motor axons by quasitrapezoidal current pulses. IEEE Trans. Biomed. Eng. **38**(2), 168–174 (1991)
318. W.M. Grill, J.T. Mortimer, The effect of stimulus pulse duration on selectivity of neural stimulation. IEEE Trans. Biomed. Eng. **43**(2), 161–166 (1996)
319. J.J. Sit, R. Sarpeshkar, A low-power blocking-capacitor-free charge-balanced electrode-stimulator chip with less than 6 nA DC error for 1-mA full-scale stimulation. IEEE Trans. Biomed. Circuits Syst. **1**(3), 172–183 (2007)
320. I. Williams, T.G. Constandinou, An energy-efficient, dynamic voltage scaling neural stimulator for a proprioceptive prosthesis. IEEE Trans. Biomed. Circuits Syst. **7**(2), 129–139 (2013)
321. T.G. Constandinou, J. Georgiou, C. Toumazou, A partial-current-steering biphasic stimulation driver for vestibular prostheses. IEEE Trans. Biomed. Circuits Syst. **2**(2), 106–113 (2008)
322. E. Greenwald, et al., A CMOS current steering neurostimulation array with integrated DAC calibration and charge balancing. IEEE Trans. Biomed. Circuits Syst. **11**(2), 324–335 (2017)

323. D. Jiang, A. Demosthenous, T. Perkins, X. Liu, N. Donaldson, A stimulator ASIC featuring versatile management for vestibular prostheses. IEEE Trans. Biomed. Circuits Syst. **5**(2), 147–159 (2011)
324. N. Tran et al., A complete 256-electrode retinal prosthesis chip. IEEE J. Solid-State Circuits **49**(3), 751–765 (2014)
325. H. Tang, Z.C. Sun, K.W.R. Chew, L. Siek, A 1.33 μW 8.02-ENOB 100 kS/s successive approximation ADC with supply reduction technique for implantable retinal prosthesis. IEEE Trans. Biomed. Circuits Syst. **8**(6), 844–856 (2014)
326. J.H. Tsai et al., A 0.003 mm^2 10 b 240 MS/s 0.7 mW SAR ADC in 28 nm CMOS with digital error correction and correlated-reversed switching. IEEE J. Solid-State Circuits **50**(6), 1382–1398 (2015)
327. F.M. Yaul, A.P. Chandrakasan, A 10 bit SAR ADC with data-dependent energy reduction using LSB-first successive approximation. IEEE J. Solid-State Circuits **49**(12), 2825–2834 (2014)
328. X. Xing, G.G.E. Gielen, A 42 fJ/Step-FoM two-step VCO-based delta-sigma ADC in 40 nm CMOS. IEEE J. Solid-State Circuits **50**(3), 714–723 (2015)
329. X. Zhang, Y. Lian, A 300-mV 220-nW event-driven ADC with real-time QRS detection for wearable ECG sensors. IEEE Trans. Biomed. Circuits Syst. **8**(6), 834–843 (2014)
330. A. Rodríguez-Pérez, M. Delgado-Restituto, F. Medeiro, A 515 nW, 0-18 dB programmable gain analog-to-digital converter for in-channel neural recording interfaces. IEEE Trans. Biomed. Circuits Syst. **8**(3), 358–370 (2014)
331. S. Oh et al., A dual-slope capacitance-to-digital converter integrated in an implantable pressure-sensing system. IEEE J. Solid-State Circuits **50**(7), 1581–1591 (2015)
332. H. Nakane et al., A fully integrated SAR ADC using digital correction technique for triple-mode mobile transceiver. IEEE J. Solid-State Circuits **49**(11), 2503–2514 (2014)
333. C.H. Chen, Y. Zhang, T. He, P.Y. Chiang, G.C. Temes, A micro-power two-step incremental analog-to-digital converter. IEEE J. Solid-State Circuits **50**(8), 1796–1808 (2015)
334. Y. Li, D. Zhao, W.A. Serdijn, A sub-microwatt asynchronous level-crossing ADC for biomedical applications. IEEE Trans. Biomed. Circuits Syst. **7**(2), 149–157 (2013)
335. A. Sukumaran, S. Pavan, Low power design techniques for single-bit audio continuous-time delta sigma ADCs using FIR feedback. IEEE J. Solid-State Circuits **49**(11), 2515–2525 (2014)
336. G.Y. Huang, S.J. Chang, C.C. Liu, Y.Z. Lin, A 1-μW 10-bit 200-kS/s SAR ADC with a bypass window for biomedical applications. IEEE J. Solid-State Circuits **47**(11), 2783–2795 (2012)
337. A. Shikata, R. Sekimoto, T. Kuroda, H. Ishikuro, A 0.5 V 1.1MS/sec 6.3fJ/conversion-step SAR-ADC with tri-level comparator in 40 nm CMOS, in *2011 Symposium on VLSI Circuits —Digest of Technical Papers* (2011), pp. 262–263
338. J. Zhang, Y. Lian, L. Yao, B. Shi, A 0.6-V 82-dB 28.6-μW continuous-time audio delta-sigma modulator. IEEE J. Solid-State Circuits **46**(10), 2326–2335 (2011)
339. K. Kim, Y. Kim, W. Yu, S. Cho, A 7b, 3.75 ps resolution two-step time-to-digital converter in 65 nm CMOS using pulse-train time amplifier, in *2012 Symposium on VLSI Circuits (VLSIC)* (2012), pp. 192–193
340. S. Naraghi, M. Courcy, M.P. Flynn, A 9-bit, 14 μW and 0.06 mm^2 Pulse position modulation ADC in 90 nm digital CMOS. IEEE J. Solid-State Circuits **45**(9), 1870–1880 (2010)
341. M. van Elzakker, E. van Tuijl, P. Geraedts, D. Schinkel, E.A.M. Klumperink, B. Nauta, A 10-bit charge-redistribution ADC consuming 1.9 μW at 1 MS/s. IEEE J. Solid-State Circuits **45**(5), 1007–1015 (2010)
342. S.K. Lee, S.J. Park, H.J. Park, J.Y. Sim, A 21 fJ/conversion-step 100 kS/s 10-bit ADC with a low-noise time-domain comparator for low-power sensor interface. IEEE J. Solid-State Circuits **46**(3), 651–659 (2011)

343. D. Zhang, A. Bhide, A. Alvandpour, A 53-nW 9.1-ENOB 1-kS/s SAR ADC in 0.13-μm CMOS for medical implant devices. IEEE J. Solid-State Circuits **47**(7), 1585–1593 (2012)
344. C.C. Lee, M.P. Flynn, A SAR-assisted two-stage pipeline ADC. IEEE J. Solid-State Circuits **46**(4), 859–869 (2011)
345. R. Xu, B. Liu, J. Yuan, Digitally calibrated 768-kS/s 10-b minimum-size SAR ADC array with dithering. IEEE J. Solid-State Circuits **47**(9), 2129–2140 (2012)
346. H.C. Hong, G.M. Lee, A 65-fJ/conversion-step 0.9-V 200-kS/s rail-to-rail 8-bit successive approximation ADC. IEEE J. Solid-State Circuits **42**(10), 2161–2168 (2007)
347. P.J.A. Harpe et al., A 26 μW 8 bit 10 MS/s asynchronous SAR ADC for low energy radios. IEEE J. Solid-State Circuits **46**(7), 1585–1595 (2011)
348. T.C. Sepke, in *Comparator design and analysis for comparator-based switched-capacitor circuits*. Thesis, Massachusetts Institute of Technology (2006)
349. E. Roza, Analog-to-digital conversion via duty-cycle modulation. IEEE Trans. Circuits Syst. II Analog Digit. Signal Process. **44**(11), 907–914 (1997)
350. P.R. Kinget, A. A. Lazar, L.T. Toth, On the robustness of an analog VLSI implementation of a time encoding machine, in *2005 IEEE International Symposium on Circuits and Systems* (2005), pp. 4221–4224

Chapter 6
Tethered and Implantable Optical Sensors

A. J. Thompson and Guang-Zhong Yang

List of Acronyms

AFI	Autofluorescence imaging
AMD	Age-related macular degeneration
Arch	Archaerhodopsin
ASIC	Application-specific integrated circuit
BE	Barrett's Esophagus
CCD	Charge-coupled device
ChR2	Channelrhodopsin-2
CW	Continuous wave
ECG	Electrocardiogram
GI	Gastro-intestinal
IBD	Inflammatory bowel disease
IOL	Intraocular lens
IR	Infrared
LDA	Linear discriminant analysis
LED	Light-emitting diode
LSI	Laser speckle imaging
LSR	Laser speckle rheology
NpHR	Halorhodopsin
OCT	Optical coherence tomography
OFDI	Optical frequency domain imaging
PCA	Principal components analysis
PPG	Photoplethysmography
PpIX	Protoporphyrin IX
PTT	Pulse transit time
PWV	Pulse wave velocity
RF	Radio-frequency
RGC	Retinal ganglion cell

A. J. Thompson (✉) · G.-Z. Yang
The Hamlyn Centre, Imperial College London, London, UK
e-mail: alex.thompson08@imperial.ac.uk

G.-Z. Yang (ed.), *Implantable Sensors and Systems*
https://doi.org/10.1007/978-3-319-69748-2_6

RP	Retinitis pigmentosa
SECM	Spectrally encoded confocal microscopy
SEM	Scanning electron microscopy
SERS	Surface-enhanced Raman spectroscopy
SLM	Spatial light modulator
SNR	Signal-to-noise ratio
SO_2	Oxygen saturation
SSI	Surgical site infection
TCE	Tethered capsule endomicroscopy
UTI	Urinary tract infection
VEP	Visual-evoked potential
VPU	Visual processing unit
1D	1-dimensional
2D	2-dimensional
3D	3-dimensional
5-ALA	5-aminolevulinic acid
μLED	Micro-scale light-emitting diode

6.1 Introduction

Optical imaging and sensing modalities have been used in medical diagnosis for many years. An obvious example is endoscopy, which allows remote wide-field imaging of internal tissues using optical fibers and/or miniature charge-coupled device (CCD) cameras. While techniques such as endoscopy provide useful tools for clinicians, they do not typically allow a complete diagnosis to be made. Instead, physical biopsies may be required to confirm or refute the presence of disease. Furthermore, endoscopic procedures are both invasive and time-consuming. As such, much research is currently directed toward the development of devices that can provide a complete in vivo diagnosis without the requirement for a physical biopsy. Ideally, such devices should also be minimally or non-invasive, and they should provide immediate identification of disease at the point of care. Additionally, there is significant interest in the development of implantable diagnostic devices that can be left within patients' bodies for extended periods of time (for several days or longer). Such systems could be used for automated disease diagnosis, and example applications include the detection of post-surgical infections as well as monitoring of the health status of patients undergoing chemotherapy. This chapter focuses on the development of optical instruments that can provide in situ diagnosis at the point of care, with an emphasis on progress towards miniature devices that may function as implants in the future.

Most current optical techniques used clinically have tethered formats and use optical fibers for light delivery and collection. Some smaller, untethered biophotonic devices have been reported for research purposes (including early clinical trials) and these may open the door to truly implantable optical sensors. Both tethered and fully implantable systems will have significant clinical utility, and the aim of this chapter is to provide an overview of optical technologies with clinical applications that are currently in use or under development. In keeping with the theme of the book, the scope is limited to addressing only those approaches that are aimed at minimally invasive tethered devices or fully implantable systems. In the following sections a collection of relevant topics are reviewed, and in each case the current state of development is summarized and potential future research directions are discussed.

6.2 Principles of Optical Sensing

Optical sensing for medical diagnostics first involves the illumination of a region of interest—typically a small volume of tissue in vivo or a clinical sample of some kind (e.g. urine or blood) ex vivo—with either broadband (i.e. spanning a wide wavelength range) or narrowband light. A chosen signal is then detected in response to the illumination and, in almost all cases, this is also optical in nature (a notable exception to this is the field of photo-acoustics, where ultrasonic pressure waves are detected in response to pulsed optical excitation). The measured signal can be the light that is reflected, transmitted, or scattered by the sample. Alternatively, it can consist of the fluorescence or phosphorescence that is emitted some time after absorption of the illumination. Generally, the detected light is quantified in some way, and in the simplest case this entails measurement of the light intensity. The detected signal can be also quantified in terms of its wavelength (i.e. spectroscopy), its polarization, or its lifetime (in the case of fluorescence or phosphorescence measurements) among other factors. Importantly, the quantification of different parameters provides different opportunities for contrast (for example, between healthy and diseased tissue).

The observed signals can be based on either endogenous components (i.e. scattering, absorbing, or fluorescent molecules within the sample itself) or on exogenous contrast agents, such as fluorescent dyes. The former provides a direct measurement of the behavior of the sample, while the latter is dependent on the interaction of the sample with the contrast agent; both are useful clinically. Endogenous measurements have the advantage of being label-free and, hence, intrinsically less invasive, while exogenous contrast agents tend to offer higher signal-to-noise ratios and can be tailored to give specific readouts. For example, fluorescent dyes can be designed to be bright in the presence of a chosen biomarker and dim in its absence. Additionally, environment-sensitive dyes are available that change color—i.e. they exhibit changes in their absorbance, reflectance and/or fluorescence profiles—in response to, for example, changes in the pH or oxygen concentration in their immediate vicinity. Optical sensing has multiple prospective clinical applications and many devices have been and are being developed to exploit these opportunities.

Of course, for diagnostic applications it is often desirable to obtain images, and optical sensing techniques can be readily extended to imaging formats. This is typically achieved through the use of scanning mechanisms (e.g. galvanometric scanning mirrors) or pixelated detectors (e.g. CCD cameras) and acts to provide spatial resolution to the measurement, allowing the collection of morphological information. Such an approach is clearly useful in medicine where it is necessary to visualize lesions and to differentiate healthy from diseased tissue, often with tight constraints on the margins between the two. When developing implantable systems, however, it is also important to consider power consumption, as untethered devices will need to operate for a set period of time using either a battery or an external wireless powering technology. Imaging devices will naturally draw more power than systems employing single point detection. This represents an important tradeoff in the fabrication of optical implants. Nonetheless, both approaches can provide useful clinical information.

Indeed, numerous optical sensing and imaging techniques have been deployed clinically, and obvious examples include endoscopes and pulse oximeters. In addition, there has been significant progress in recent years towards the development of implantable devices that could be used, for example, to monitor for signs of infection after surgery. In particular, a number of miniature pulse and blood oxygen sensors have been developed (e.g. [1]) that are based on measurement of the diffuse reflectance (see Sect. 6.4.2). Devices of this sort can also be extended to provide blood pressure monitoring through the addition of electrodes that make electrocardiogram (ECG) measurements [2, 3]. The above are chip-based systems that can be made fully implantable (i.e. untethered) using ASIC (application-specific integrated circuit) technology, and many further examples of optical systems like this can be found in the field of optogenetics (discussed in Sect. 6.6), where a number of miniaturized, implantable devices have been developed for use in freely moving rodents. Tethered implants are also under development for medical applications, and these will still permit minimally invasive measurements to be made. Examples include optical fiber-based glucose and pH sensors (as well as other ionic sensors) that rely on the immobilization of a sensing layer on the tip of the optical fiber (e.g. [4]). The most common method of immobilization is the sol-gel approach, and sensors of this type are discussed in Sect. 6.4.5. Finally, the development of miniature imaging devices is also under way, and this includes the production of imaging capsules (e.g. wireless capsule endoscopy [5] and tethered capsule optical coherence tomography (OCT) [6]), as well as probes designed for insertion through catheters or needles (such as the combined OCT/fluorescence probe reported in [7]). All of the above sensors, along with a variety of other devices, are described in more detail in the following sections.

6.3 Optical Imaging Techniques

6.3.1 Endoscopy

Medical endoscopy involves the remote imaging of internal tissues, such as the gastro-intestinal (GI) tract, the lungs, or the bladder. Traditionally, an incoherent (i.e. non-imaging) optical fiber bundle is used to deliver white light illumination to the tissue of interest, and a second coherent fiber bundle collects the reflected light and relays the image to a detector or eyepiece [8]. In modern systems, miniature distal CCD cameras are usually used instead of imaging fiber bundles due to the higher resolution they provide. Medical endoscopes permit wide-field, reflected light imaging with white light illumination and, as such, provide the clinician with morphological and limited (i.e. three-color) spectroscopic information. For this reason, endoscopy is used as a tool to aid clinicians, but does not provide complete diagnoses. As an example, in a typical colonoscopy, endoscopic imaging is used to locate suspicious regions of tissue, which are then excised using biopsy forceps and sent for histological examination in order to confirm the clinical diagnosis. Not only is this process invasive, it also often leads to the unnecessary removal of benign/healthy tissue. Hence there is significant interest in the development of techniques that can provide a so-called optical biopsy, where a diagnosis is made at the point of care (i.e. during endoscopy) and tissue is only removed when necessary. These techniques include more advanced imaging modalities such as endomicroscopy and OCT, as well as single-point spectroscopic approaches. Devices based on these methods can be incorporated not only into endoscopes, but also into other medical devices, such as catheters and needles. Thus, when deployed using optical fibers, instruments using these methods have the potential to be applied as tethered optical diagnostic implants in the future.

6.3.2 Endomicroscopy

One technique that aims to provide an optical biopsy during endoscopy is endomicroscopy, which involves the microscopic investigation of tissue using optical fibers [9]. The concepts behind endomicroscopy are the same as those in a standard scanning (e.g. confocal or multiphoton) microscope, with optical fibers also used to provide light delivery and collection, permitting remote imaging. Endomicroscopy is currently in clinical use and in some systems the microscopic imaging device is incorporated into the medical endoscope itself (e.g. EG-3870 and EC-3870, Pentax, Japan), while in others a separate probe-based endomicroscope can be inserted into the working channel of the endoscope (e.g. Cellvizio GI, Mauna Kea Technologies, France). In both cases, endomicroscopy is typically used to microscopically examine regions of tissue that have first been identified as suspicious by the clinician during the standard endoscopic procedure. Endomicroscopy provides sub-cellular resolution (albeit within a much smaller field of view than that provided by standard wide-field endoscopy), and hence presents the clinician with significantly more information with which to make a diagnosis. In the future, endomicroscopic interrogation of this sort may replace (or more likely augment) physical biopsy, and at present much research is directed toward this end.

Endomicroscopes record images by sequentially scanning an illumination spot across a sample and detecting the resultant fluorescent or reflected light. They work in much the same way as a confocal or multiphoton microscope [10, 11], and use optical fibers to deliver light to and collect light from a remote sample. Endomicroscopes can also be designed to provide optical sectioning (i.e. confocal imaging), where light is only collected from a thin focal plane. This is important in the context of medical applications, as optical sectioning can provide information analogous to that obtained from the physical sectioning of excised tissue that is performed during histological examination. Typically, the resolution obtained during endomicroscopy is of the order of a micron and the field of view can extend up to approximately 600 μm^2. Additionally, it is worth noting that contrast agents such as methylene blue, fluorescein, or toluidine blue are often used to provide optimum image quality.

At the distal end of the endomicroscope, scanning of the illumination spot is usually achieved using one of two techniques: either distal or proximal scanning [9]. These approaches derive their names from the location of their scanning mechanism and more details on both methods can be found in [12] (proximal scanning) and [13–15] (distal scanning). Both proximal and distal scanning systems are in clinical use (proximal scanning—Cellvizio GI, Mauna Kea Technologies, France; distal scanning—EG-3870 and EC-3870, Pentax, Japan) and these systems have been applied widely to medical diagnostic studies.

6.3.2.1 Advanced Endomicroscopy Systems

Beyond the standard approaches to endomicroscopy, there is also significant research being directed toward the development of more advanced systems. These include the incorporation of multiphoton excitation, adaptations to the standard design in order to allow measurement of additional optical parameters, and the development of novel scanning techniques or imaging modalities to provide improvements over the standard approaches (e.g. in terms of minimum size, imaging speed or sampling rate).

Multiphoton excitation can be achieved simply through the use of a pulsed infrared (IR) laser in place of the standard excitation source—which is typically a continuous wave (CW) laser or a light emitting diode (LED)—and an alternative optical fiber with low dispersion characteristics (such as a dual clad photonic crystal fiber) to account for pulse dispersion (e.g. [16–19]). The adaptation of endomicroscopes to allow measurement of additional optical parameters is also relatively straightforward and examples include multi-spectral detection [20] as well as fluorescence lifetime imaging endomicroscopy [21, 22]. There have also been multiple attempts to improve the sampling rate (and image quality in general) provided by proximal scanning systems while maintaining the small size of the distal tip. Several similar approaches involve the use of a spatial light modulator (SLM) to control the phase of light propagating through the optical fiber that is used for light delivery and collection (either in each individual core of a fiber bundle or in a single multimode optical fiber) in order to synthesize a focused scanning beam at the distal end [23–26]. Alternative approaches to address undersampling use additional optical or mechanical components to "fill the gaps" between the fiber cores in a proximal scanning endomicroscope. This can involve spectrally dispersing the distal light such that it reaches the otherwise unsampled regions of the object [27] or collecting multiple images as the object is shifted relative to the endomicroscope [28].

In practice, faster imaging is always desirable and significant work has been directed toward this end. On the one hand this has involved the development of a number of proximal line-scanning endomicroscopes, which permit faster imaging at the expense of optical sectioning strength [29–31]. Additionally, endocytoscopy [32, 33]—which provides fast, wide-field imaging without optical sectioning—has been widely used clinically and a number of commercial systems are available (e.g. XEC-300, XEC-120, XGIF-Q260EC1 and XCF-Q260EC1, Olympus, Japan). Another approach that provides fast, wide-field microscopic imaging is high-resolution micro-endoscopy [34, 35], which has also recently been modified to permit optically sectioned imaging [36]. One interesting alternative technique to improve imaging speed is spectrally encoded confocal microscopy (SECM), which involves spectrally dispersing white light illumination at the sample such that a line of illumination is obtained with each position in the line corresponding to a different wavelength [37]. By using a spectral detector at the proximal end rather than a simple single-channel detector, it is possible to measure the reflected light intensity at all points along the illumination line simultaneously. This allows the imaging speed to be doubled without compromising image quality, and much research is now being dedicated to optimizing such systems and minimizing the size of the required distal components (e.g. [38–40]). This technique is particularly relevant to this book, as it is has been used in the fabrication of a miniature capsule endomicroscope (see Sect. 6.3.3) that can be swallowed by the patient.

Overall, significant research is still in progress into the development of novel or improved endomicroscopes. Nonetheless, current commercial systems are already in clinical use and many studies have been published discussing the potential for endomicroscopy to be used as an in vivo diagnostic tool.

6.3.2.2 Clinical Applications of Endomicroscopy

Both distal and proximal scanning endomicroscopy have been applied clinically and their use is becoming more widespread. Many research articles have discussed the application of endomicroscopy to a variety of diseases and the possibility of its use as a tool for optical biopsy. This section highlights and summarizes some important studies. This discussion is not exhaustive, however, and for more information readers are directed towards review articles that describe the use of endomicroscopy in specific diseases or tissues, for example [41–47].

Most commonly, endomicroscopy has been used to study the GI tract and diseases that have been investigated include neoplasia [48–50], colonic polyps [48, 51–53], Barrett's Esophagus (BE) [50, 54–57], inflammatory bowel disease (IBD) [49, 58–60], Crohn's Disease [59], and ulcerative colitis [49, 58, 60]. There have been multiple studies evaluating the utility of endomicroscopy for the characterization of colonic polyps [48, 51–53, 61]. While some of these have suggested that endomicroscopy does not improve upon the sensitivity and specificity provided by alternative endoscopic imaging modalities (e.g. narrowband imaging [52]), others have presented significantly improved diagnostic accuracies [48, 51]. One particularly promising study showed that neoplastic and non-neoplastic polyps could be differentiated in vivo with a sensitivity of 97% and a specificity of 99% [48]. These results suggest that endomicroscopy has the potential to become widely used in endoscopy for the examination of colorectal polyps. However, the current conflict in the literature over diagnostic accuracy means that further studies will be required before this occurs.

IBD—including Crohn's disease and ulcerative colitis—is another gastro-intestinal condition that has been widely studied with endomicroscopy [49, 58–60]. Kiesslich et al. [49] showed that dysplastic lesions could be diagnosed with an accuracy of 97% using endomicroscopy, while van den Broek et al. [58] were able to correctly predict the presence of neoplasia in 81% of cases when combining confocal endomicroscopy with high-definition narrowband imaging endoscopy. As a brief aside, it is worth noting here that one of the important features of endomicroscopy is the fact that the images obtained contain similar details to those observed in histology, which means that clinicians can make diagnoses without the need for significant additional training. This is demonstrated for ulcerative colitis in Fig. 6.1, where three regions of tissue have been imaged using both endomicroscopy (top row) and standard histology (bottom row) [49].

Endomicroscopy has also been applied extensively to the examination of suspicious lesions in patients suffering from BE [42, 50, 54–57, 62, 63]. High sensitivity and/or specificity of diagnosis have been reported in a number of studies, for example in the articles by Kiesslich et al. (92% sensitivity, 98% specificity) [50], Sharma et al. (94% sensitivity, 69% specificity) [56], and Bajbouj et al. (60% sensitivity, 95% specificity) [55]. As well as producing good diagnostic accuracy, it has also been shown that endomicroscopy can provide improvements in sensitivity relative to standard white light imaging and narrowband imaging [56], and that it can reduce the number of biopsies required when screening BE patients for neoplastic regions of tissue [54, 56]. Taken together, this implies that endomicroscopy will become increasingly widespread in the inspection of patients with BE.

While it has been mainly applied to the study and diagnosis of GI conditions, endomicroscopy has also been used to investigate other areas and ailments such as the bile duct [64], head and neck cancers [41], and brain tumors [47, 65, 66]. In these cases, most studies are at much more preliminary stages than the GI

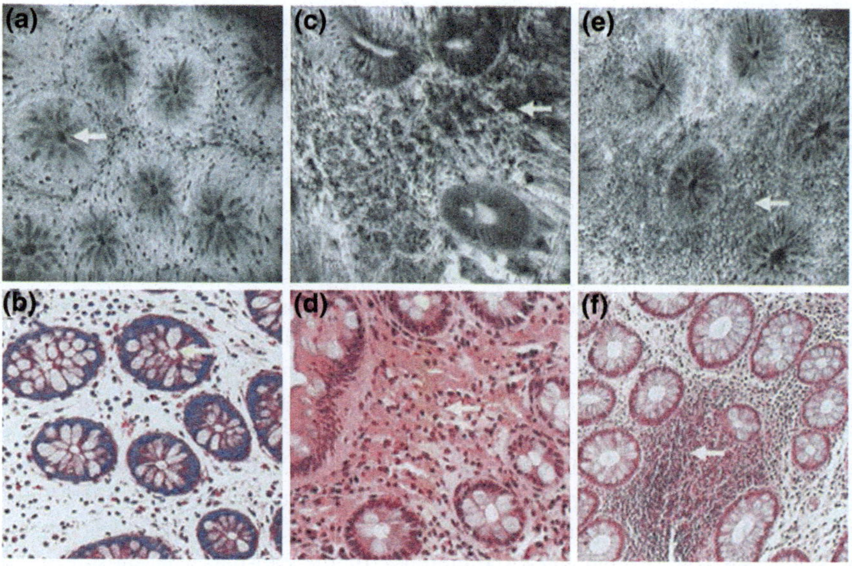

Fig. 6.1 Comparison of images of colon tissue obtained using endomicroscopy (top row) and standard histopathology (bottom row). Note the similar contrast obtained with the two techniques. Images **a** and **b** show normal crypt architecture with regular lumen indicated by the white arrows. Tissue affected by ulcerative colitis is shown in (**c**) and (**d**), where capillaries are distorted and the crypt architecture is altered. In particular, the spacing between crypts is larger than in healthy tissue. **e** and **f** show mixed cellular infiltration of disease with the crypts remaining unaffected. Cellular infiltration is indicated by the white arrows. Figure reproduced from [49], © 2007, with permission from Elsevier

investigations discussed above, and as such, the current focus is on presenting exemplar images rather than carrying out extensive experiments and reporting diagnostic accuracies. Nonetheless, the fact that such studies are under way demonstrates the growing clinical applications of endomicroscopy.

6.3.3 Capsule Endoscopy and Endomicroscopy

An important advancement of endoscopy and endomicroscopy in terms of implantable or ingestible medical devices is the development of capsule-based systems that are swallowed by the patient and provide imaging as they pass through the GI tract. Wireless capsule endoscopy [5] provides wide-field imaging using a small, untethered capsule (approximately 10 mm × 30 mm) that is ingested by the patient. Video images are recorded and transmitted for live viewing as the capsule passes through the patient's body and the image data obtained is similar to that provided by standard endoscopy (i.e. the images are not high resolution and high magnification as provided by endomicroscopy). Importantly, unlike endomicroscopes, capsule endoscopes can be considered as (transiently) implantable. They have also become very useful in the investigation of the small intestine —which is difficult to access using standard endoscopes for the upper or lower GI tract—and are now widely used clinically (e.g. PillCam SB, Given Imaging Ltd., USA; MiroCam, Medivators Inc., USA; Endocapsule 10, Olympus Europa, Germany). In particular, wireless capsule endoscopy has been used to study small intestinal bleeding [67–70]. Photographs of clinically used capsule endoscopes are shown in Fig. 6.2, where the implantable/ingestible nature of this method is evident.

Further research in the area of capsule endoscopy is now mainly focused on the production of systems with additional sensing capabilities or higher imaging resolution. One important technique that was recently reported by Tabatabaei et al. [71] uses a tethered capsule (which is considerably less invasive than a standard endoscope) to provide confocal endomicroscopy. The capsule contains all the necessary optics and scanning mechanics to achieve SECM. A diffraction grating disperses broadband IR light into its constituent wavelengths to provide a line of illumination. A motor then drives rotation of the optics in order to allow an image to be collected from around the circumference of the capsule. The capsule is connected to a thin tether that houses an optical fiber that is used for light delivery and collection, and images covering a field of view of 22 mm × 223 μm can be recorded at a frame rate of 6 Hz. The lateral and axial resolutions provided by this system are respectively 2.1 and 14 μm, and its use has been demonstrated in swine esophagus in vivo as well as in ex vivo human biopsy samples. As such, this system represents a significant step towards the development of truly implantable/ingestible probes for confocal endomicroscopy.

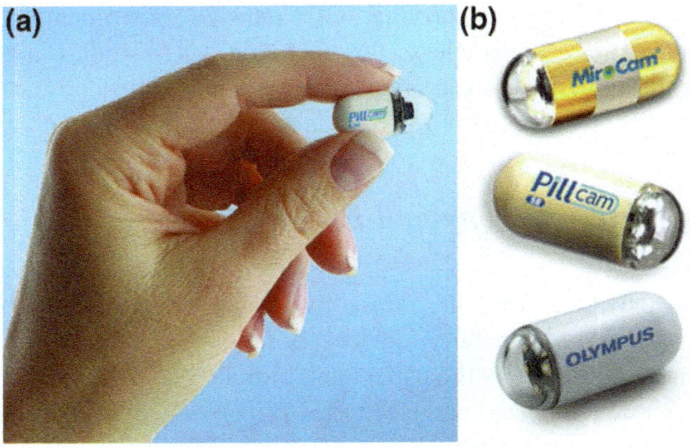

Fig. 6.2 Images of commercially available capsule endoscopy systems. The small size is illustrated in **a** while **b** shows capsules (all of which have very similar features) manufactured by three different companies. **a** PillCam^{TM} SB 2 (Given Imaging, USA). **b** From top: MiroCam^{TM} (IntroMedic Co., Republic of Korea); PillCam^{TM} SB (Given Imaging, USA); EndoCapsule^{TM} (Olympus, Japan)

6.3.4 *Optical Coherence Tomography (OCT)*

Optical coherence tomography (OCT) is an interferometric imaging technique that was first reported by Huang et al. in 1991 [72]. It relies upon the use of low coherence interferometry to provide depth-resolved images. A low-coherence light source (such as a superluminescent diode, an ultrashort pulsed laser, or a supercontinuum source) is used to illuminate the sample. The illumination is split into a reference and a sample arm, which are recombined at the detector to generate an interference pattern. Due to the short coherence length of the illumination source, interference fringes are only observed when the sample and reference arm lengths are very closely matched. This means that interference fringes are only detected at a specific imaging depth. By filtering the recorded signal such that only regions exhibiting interference (i.e. those with high spatial frequencies) are included, it is possible to obtain a measurement of the light intensity at a single specific depth within the sample. By using a CCD camera to record the light intensity across a wide field of view or by using scanning mirrors, it is then possible to obtain an image of the chosen z-plane (i.e. an optical section). Importantly, the optical sectioning strength in OCT is determined by the coherence length of the light source and, with appropriate illumination, axial resolutions as low as 1 μm can be achieved. Traditionally, depth scanning is then realized through a simple adjustment of the length of the reference arm, which acts to alter the imaging depth at which interference is observed. In more modern systems, however, three-dimensional (3D) imaging is often achieved by rapidly scanning the wavelength of the illumination or by using spectrally resolved detection [73–75]. In both cases, the reference arm length remains fixed and depth-resolved imaging is achieved by calculating the Fourier transform of the spectral data in order to reconstruct the depth information. This approach is referred to as frequency domain OCT, Fourier domain OCT, swept source OCT (in the case of rapid illumination wavelength scanning), or optical frequency domain imaging (OFDI).

Using either approach, OCT can provide 3D imaging at very high frame rates, and for this reason it is now widely used in clinical environments, often using optical fiber-based formats [76]. Its clinical uses include retinal imaging [77–83], vascular imaging for the diagnosis of atherosclerosis [84–94] and endoscopic imaging of the GI tract [95–100]. These medical applications of OCT are reviewed in the following section.

6.3.4.1 Clinical Applications of OCT

As discussed briefly above, there have been myriad medical studies using OCT and it is widely used clinically, with multiple commercial systems now available (e.g. NvisionVLE, NinePoint Medical, USA; Cirrus HD-OCT, Zeiss, Germany; Envisu Clinical SDOCT, Bioptigen, USA). The clinical applications of OCT have been wide-ranging and include (but are not limited to): imaging of the skin [101–104];

retinal imaging for studies on glaucoma, diabetic retinopathy, and neurodegeneration [73, 77–83, 105–109]; vascular imaging for the study and diagnosis of atherosclerosis [84–94, 110, 111]; breast cancer detection [112–114]; prostate cancer detection [115]; optical biopsy in the urinary tract [115, 116]; and optical biopsy in the GI tract (for example, in BE) [95–100, 117]. It is beyond the scope of this review to discuss all of these applications. Instead, the text below focuses on those studies that comprise approaches that fall under the broad remit of techniques that could be considered implantable—i.e. catheter-based probes for vascular imaging, endoscopic probes for GI and urinary tract imaging, and needle-based probes. While none of these devices are currently implantable, it is conceivable that they could be deployed as implants in the future, especially in the case of needles and catheters, which are often left within patients' bodies for long periods of time.

As with endomicroscopy, a number of OCT probes have been developed that can be deployed through the working channel of standard medical endoscopes. Such systems (and others) have been used to investigate the potential for OCT in the diagnosis of various conditions in the GI tract. One example is BE, where OCT has been used to assess neoplastic changes in esophageal tissue [100]. Additionally, a number of studies have also reported the use of OCT to investigate specialized intestinal metaplasia [96, 98, 117]. Beyond BE and specialized intestinal metaplasia, OCT has also been applied to the detection of esophageal adenocarcinomas and colonic polyps [95], as well as to the study of cancers of the urinary tract [116].

One of the most common applications of OCT has been in the study and diagnosis of coronary diseases, such as atherosclerosis [86, 89, 118]. In this case, side viewing OCT probes have been incorporated into arterial and venous catheters, and cross-sectional images of the vasculature are typically recorded by rotating the probe (e.g. [119]). 3D imaging of entire blood vessels can then be realized simply through pullback of the probe, allowing it to record cross-sectional images as it moves backward through the vein or artery (e.g. [91, 111, 120]). Cross-sectional images and a 3D rendered volume dataset of vascular tissue are shown in Fig. 6.3 as examples.

Vascular probes like those described above have found multiple applications in coronary medicine. In terms of atherosclerosis, OCT has been used to identify the chemical composition of plaques and to determine their risk of rupture. It has also been used to measure the thickness of the fibrous caps that cover the lipid-rich cores in atherosclerotic plaques, and at present it is the only technique capable of making such measurements. Moreover, it has been used to study acute intracoronary thrombosis, to guide angioplasties, and to assess percutaneous coronary interventions (i.e. the implantation of stents to improve blood flow) [85, 88, 89, 121–123]. Thus, OCT has become widely used in coronary medicine and its applications are likely to grow further in the future as the technology continues to improve.

A further important use of OCT for the purpose of this chapter is its application in needle probes. While less common than the vascular methods described above, several needle-based OCT systems have been reported in recent years. These devices have been developed with the aim of permitting imaging below the tissue surface and have been applied to ex vivo studies of freshly resected tissue from both

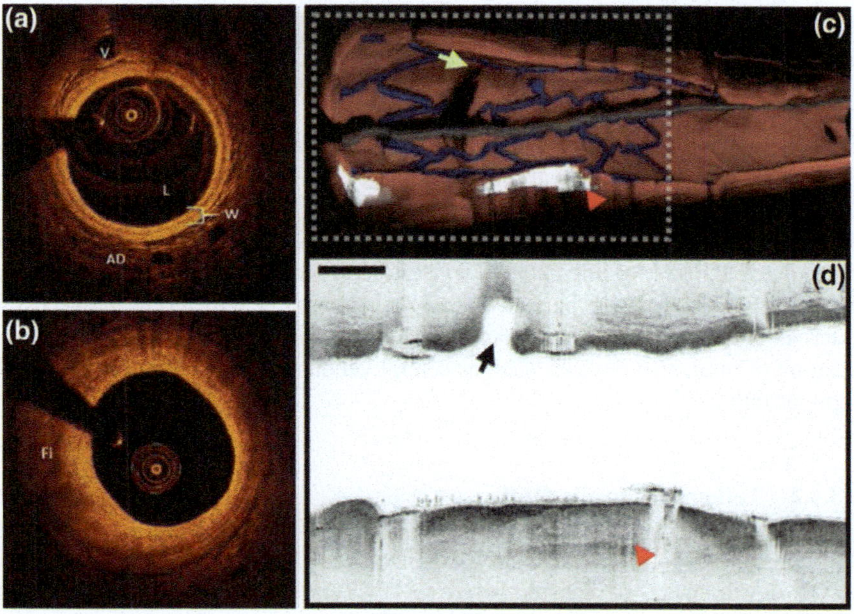

Fig. 6.3 Exemplar intravascular OCT data. **a** Radial cross-sectional image of a healthy blood vessel showing the lumen (L), vessel wall (W) and adventitia (AD) with vasa vasorum (V). **b** Radial cross-section of a blood vessel with a fibrotic atheroma (Fi). **c** 3D reconstruction of a blood vessel containing a stent (blue). A side branch (yellow arrow) and a calcified nodule (white region, highlighted by the red arrowhead) are also indicated. **d** Longitudinal section through the dataset shown in (**c**). Black arrow and red arrowhead indicate the side branch and calcified nodule respectively. **a**, **b** Reproduced with permission from [89], © 2015. **c**, **d** Reproduced from [91], © 2008, with permission from Elsevier

animals and humans. Specifically, imaging of deep skeletal mouse muscle tissue [124] and of the alveoli and other small airways in sheep's lung [125] have been demonstrated using OCT needle probes. In addition, a dual-modality OCT/fluorescence needle probe was recently presented by Scolaro et al. [7], where the fluorescence capability was used to detect fluorescently labeled antibodies in an excised mouse liver. More importantly, freshly resected human tissue samples from patients suffering from breast [112] and prostate cancer [115] have also been successfully imaged using needle-based OCT. Significantly, OCT was able to accurately reproduce the tumor margins obtained from histology in both cases.

OCT needle probes are also minimally invasive: typically, the reported needle diameters are below 1 mm, and the thinnest system presented to date was incorporated into a 310 μm diameter (30-gauge) needle [124, 126]. For illustration, this system is shown in Fig. 6.4 along with a volumetric OCT dataset acquired in ex vivo mouse tissue. Such devices will provide clinicians with the opportunity to image below the tissue surface, hence permitting deeper imaging than is usually available with OCT (or other optical techniques). For these reasons, they are likely to become clinically useful in the coming years.

Overall, OCT is already widely used clinically—particularly in endoscopic and catheter-based applications—and such devices have shown great potential in their capacity to provide in vivo, real-time disease diagnosis (i.e. optical biopsy). Furthermore, the development of needle-based OCT probes also suggests that further minimally invasive clinical systems will become available in the future. In particular, catheter-based and needle-based probes show considerable promise as potential candidates for implantable devices. With probe tips small enough to cause minimal discomfort to the patient, such systems could be tethered to the necessary external optics and electronics. As an example, this could then allow perpetual imaging of implanted devices, such as central venous catheters, to regularly test for signs of failure.

6.3.4.2 Tethered Capsule OCT

A promising advancement of OCT within the scope of implantable optical sensors is the development of capsule-based OCT. Tethered capsule endomicroscopy (TCE) has been developed for OCT in the GI tract and was first reported by Gora et al. in 2013 [6]. In this approach, the patient swallows a small pill that contains all of the distal scanning mechanics necessary to achieve frequency domain OCT. The pill includes a rotary scanning mechanism and is connected to a 1 mm diameter tether that houses an optical fiber that collects the light from the tissue. As such, a

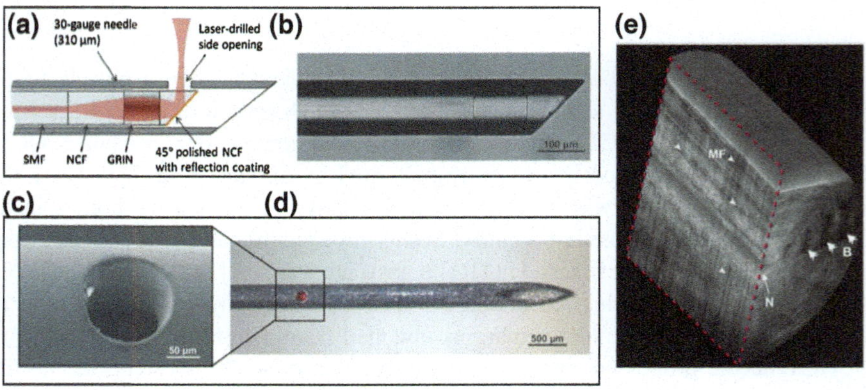

Fig. 6.4 Needle-based OCT. **a** Schematic of 310 μm diameter OCT needle probe. **b** Microscope image of the fiber probe used within the needle. **c** Scanning electron microscopy (SEM) image of the opening in the needle that allows imaging (fabricated using laser drilling). **d** Image of the fully assembled needle probe where red laser light can be seen emerging from the laser-drilled opening. **e** 3D rendered volumetric dataset from normal mouse muscle acquired with the OCT needle probe. Needle tract (N), myofibers (MF) and birefringence artefacts (B) are indicated by the white arrows. Reproduced with permission from [124], © 2014

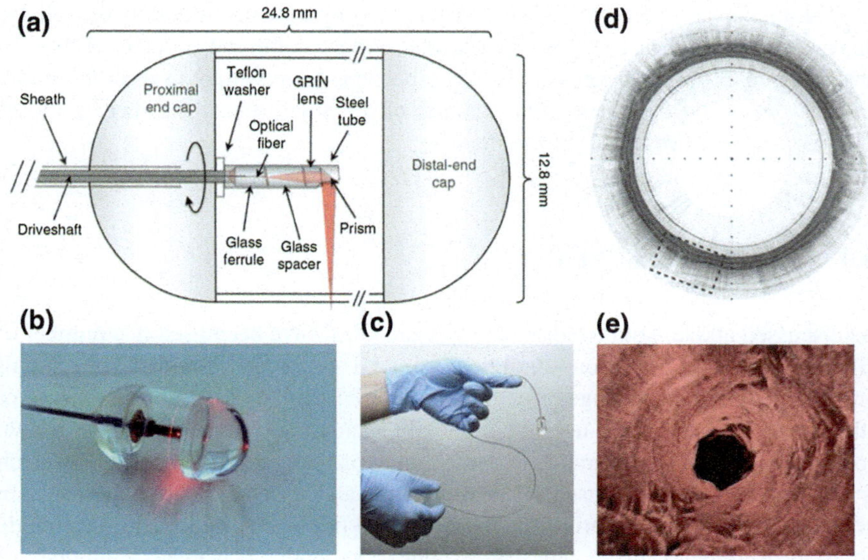

Fig. 6.5 Tethered capsule OCT. **a** Schematic of the tethered capsule OCT probe presented by Gora et al. [6]. Photographs of the tethered capsule system are shown in (**b**) and (**c**), where its small size and minimally invasive nature are evident. **d** Radial cross-sectional OCT image of normal esophageal tissue obtained from a healthy volunteer in vivo. **e** 3D rendered 'flythrough' view of a Barrett's segment of esophagus, also obtained in vivo. Reproduced with permission from Macmillan Publishers Ltd: Nature Medicine [6], © 2013

typical GI imaging procedure is considerably more comfortable for a patient than standard endoscope-based approaches. In general, the pill is swallowed by the patient and allowed to reach the stomach before recording volumetric OCT images while being pulled back through the esophagus. Using this method, imaging of the entire esophagus is possible in 5–6 min and this has been demonstrated in vivo in unsedated patients with BE [6, 127]. A diagram and two photographs of a capsule-based OCT device are shown in Fig. 6.5, along with example data obtained in the esophagi of a healthy patient (d) and a patient suffering from BE (e).

A further tethered capsule OCT device has been reported by Liang et al. [128]. This system has a dual scanning mechanism that allows both high-speed rotary scanning for fast *en face* volumetric imaging (similar to that reported by Gora et al. [6]) as well as slower, accurate 2D scanning to permit high-resolution imaging. The use of this device has been demonstrated in pigs in vivo and its dual scanning mechanisms mean that it may be compatible with both screening applications (which require wide-field imaging) and surveillance applications (which require high-resolution imaging of small, pre-determined regions of tissue) [128]. Interestingly, the 2D distal scanning mechanism means that this capsule could also be utilized for other types of scanning microscopy (e.g. distal scanning endomicroscopy). Its potential for future use as a tool for GI imaging is clear.

Overall, capsule-based OCT is clearly an exciting new methodology that allows in vivo GI imaging in a manner that is much more comfortable for patients than current endoscopic methods. It is likely to become more widely used in the future and may also find applications in imaging of the small intestine, where standard endoscopy is challenging.

6.4 Other Optical Sensors

As discussed above, both endomicroscopy and OCT have been used in a number of different clinical applications—ranging from GI endoscopy for the detection of premalignant tissue changes to intravascular imaging for the diagnosis of atherosclerosis—and their use is continually increasing. While they have shown considerable promise in terms of the potential to provide an optical biopsy, research is still ongoing into the use of other imaging and spectroscopic techniques. In general, this research comprises the use of optical fibers to permit imaging or single point measurements using alternative optical methods. However, a number of capsule- and integrated chip-based devices have also been reported. In both cases, an important aim of the research is to provide the clinician with functional information (e.g. the rate of metabolism at a specific imaging location) rather than simply offering morphological information, as is the case in endomicroscopy or OCT. In some cases, these additional optical approaches have been incorporated into endomicroscopy (see Sect. 6.3.2.1) or OCT probes to provide dual modality imaging systems. This section focuses on research in which the "advanced" optical technique is the central theme of the work (i.e. rather than discussing research involving the incorporation of an additional imaging/sensing modality into an OCT probe or an endomicroscope).

Many optical techniques have been investigated for this purpose and the remainder of this section addresses some of the most common methods. These include laser speckle imaging (LSI), diffuse reflectance spectroscopy, fluorescence spectroscopy (including time resolved fluorescence measurements), and Raman spectroscopy. While LSI (as its name suggests) is typically deployed in an imaging setup, the latter three techniques have mainly been applied in single-point measurement arrangements (although a small number of imaging experiments are also discussed herein). In these single-point approaches, morphological information is dispensed with and light is generally collected from a small (typically less than 1 mm × 1 mm) field of view. This means that data can be collected in shorter acquisition times without the need for expensive, highly sensitive detectors. Such methods are useful in optical modalities where the required acquisition times can be long due to low signal levels or the incorporation of additional dimensionality (e.g. in wavelength-resolved or time-resolved measurements). They often serve as preliminary research to ascertain whether an imaging modality (which can be used in a more intuitive manner by a clinician) would have any merit.

When considering the development of an implantable optical sensor, however, single-point measurements may in fact hold advantages over imaging techniques. The reason for this is that the data collected by the sensor is less complex and is more suitable for automated analysis and diagnosis. While automatic, computerized processing of 2D or 3D image data can be very complicated (often involving image segmentation and recognition techniques), automated analysis of 1D single-point data can be much simpler. Multiple approaches that allow computer-based diagnosis using spectral data have been reported and these typically employ spectral decomposition techniques (such as principal component analysis—PCA) followed by a form of discriminant analysis. The spectral decomposition acts to break the recorded data down into a small number of base components that accurately describe all measurements, and the discriminant analysis then uses the relative contributions of these base components to make a diagnosis for each measurement site based solely on its spectrum. This type of automated data analysis and diagnosis will be vitally important in any implanted sensor, as it will not be feasible to manually assess data that is being collected continuously. Furthermore, the electrical power used in single-point measurements is considerably lower than that required for imaging, which implies that single point approaches will be preferable when developing an untethered implantable device that cannot be wired to an external power source. Thus, single-point measurement techniques are likely to play an important role in the development of optical implants due to their low power requirements and the potential to analyze the data they provide in an automated fashion.

Lastly, it is worth noting that all of the optical techniques discussed below are readily compatible with optical fibers, and hence have the potential to be deployed as minimally invasive, or even tethered, implantable diagnostic devices. In addition, a number of untethered sensors—based on integrated chips and/or capsules—have also been reported, indicating the possible development of fully implantable devices. The following sections provide a description of the various (pre)clinical studies that have been carried out using these approaches.

6.4.1 Laser Speckle Imaging (LSI)

6.4.1.1 Theoretical Background

Laser speckle imaging (LSI) is most often used to produce spatial maps of blood flow in tissue, for example in the retina or brain. A precursor to LSI is laser Doppler flowmetry, which provides measurements of flow rate based on the Doppler shift of backscattered photons [129]. While Doppler approaches to flow measurements provide accurate and reliable assessment of the flow rate, they rarely provide images or spatial maps of flow due to high data processing requirements. For this reason, LSI was developed in order to provide a technique with less intensive data processing that could enable spatial imaging of tissue blood flow.

In LSI, a coherent laser source diffusely illuminates a sample, and a CCD camera is typically used to image the backscattered light. Due to the non-uniformity of the sample surface and the coherent nature of the illumination, a random interference pattern is recorded by the detector. This interference pattern is known as a speckle pattern and, importantly, speckle patterns have been shown to vary due to sub-surface blood flow (as blood flow effects changes in the surface morphology). In LSI, speckle patterns are converted to flow maps (often referred to as speckle contrast images) by assessing the rate of change of the speckle pattern in either time or space. As such, LSI can provide images of vasculature—including relative or absolute flow rates—with mm to cm fields of view [130].

In a typical measurement procedure, a (series of) speckle pattern(s) is recorded and the speckle contrast, K, at all image pixels is calculated as

$$K = \frac{\sigma}{\langle I \rangle} \tag{6.1}$$

where σ is the standard deviation of the speckle intensity fluctuations and $\langle I \rangle$ is the mean intensity value. The mean and standard deviation of the speckle variations can be measured temporally by recording a series of speckle patterns and calculating the mean and standard deviation of subsequent images. Alternatively, they can be calculated from a single image using a spatial approach. In this case, for every image pixel, $\langle I \rangle$ and σ are obtained in a small window (typically 5×5 or 7×7 pixels) around the central pixel. This method sacrifices spatial resolution (as the window introduces an averaging effect), but improves temporal resolution (as only a single image is required).

The speckle contrast, K, can be related to the image acquisition time, T, and the speckle decorrelation time (τ_c—the characteristic time over which the speckle pattern changes) according to

$$K = \beta^{0.5}\left\{\frac{\tau_c}{T} + \frac{\tau_c^2}{2T^2}\left[\exp\left(\frac{-2T}{\tau_c}\right) - 1\right]\right\}^{0.5}. \tag{6.2}$$

Here, β is a constant of proportionality that accounts for the loss of speckle contrast related to the size ratio of the detector and the individual speckles, and due to polarization effects. A complete derivation of this equation can be found in [130, 131]. Nonetheless, it is clear from (6.2) that the speckle contrast is directly related to the rate of decorrelation of the speckle pattern (τ_c). In the case of blood flow measurements, the decorrelation will occur faster in regions of high flow rate, and, as expected, τ_c is inversely related to the flow rate (v):

$$\tau_c = \frac{1}{ak_0 v}. \tag{6.3}$$

In (6.3), k_0 represents the wavenumber while a is a proportionality factor dependent on the scattering properties of the tissue. It is clear from the above equations then that the speckle contrast can be directly related to the absolute flow rate. However, this often proves difficult in practice, as the number of moving particles and their orientations are unknown (i.e. the parameter a is difficult to calculate). For this reason, relative flow rates are most often reported based on ratios of the decorrelation times (or speckle contrast values) at different times or in different spatial locations. This is not always the case though, and some more complex analysis procedures have been presented that allow calculation of the absolute flow rate (e.g. [132]). Despite the challenges involved in the measurement of absolute flow velocities, LSI has still found multiple biomedical applications, and these are described in detail below.

6.4.1.2 Biomedical Applications of LSI

LSI was first described by Fercher and Briers [131] in 1981, where it was used to generate maps of blood flow in the retina. Since then it has found various biomedical applications—which are reviewed extensively in [130]—and these are discussed below.

As described above, the most common application of LSI has been in blood flow mapping. One example of this is in retinal imaging, where LSI has been applied numerous times (e.g. [131, 133–135]). Most studies have been carried out in animals; however, a small number of investigations have now also been published that report on the in vivo measurement of blood flow rates in human retinas [135, 136]. LSI has also been used to monitor skin perfusion (e.g. [137, 138]). While it is difficult to measure flow in individual vessels in the skin (as the top layer of skin contains very few blood vessels and obscures the signal from below), LSI has instead been used to monitor the overall perfusion [139]. In one interesting application, LSI was used to provide feedback during laser therapy of port wine stains [140]. In this study, in vivo imaging was reported in humans and LSI was used to assess the efficacy of laser therapy of port wine stains in real time. A further application of LSI is in small animal brain imaging, where it has been used to study the relative cerebral blood flow in a number of scenarios. These include the study of the brain's response to external stimuli (for example, after electrical stimulation of the paws of rats [141]) as well as the investigation of cortical spreading depressions (the underlying cause of migraine aura) [142–144] and stroke physiology [145–147]. Example LSI data from the brain of a rat is shown in Fig. 6.6, where a raw speckle pattern and a speckle contrast image are presented. The sensitivity to blood flow is clear in the speckle contrast image, with different vessels showing markedly different speckle contrast values.

One particularly interesting use of LSI within the scope of this book is its application to the measurement of the viscoelastic properties of blood [148]. This approach—known as laser speckle rheology (LSR)—has been used in the characterization of blood coagulation [149, 150] and atherosclerotic plaques [151–153]. Both have obvious clinical uses and the assessment of atherosclerosis ties in well with the catheter-based OCT probes discussed previously (see Sect. 6.3.4.1).

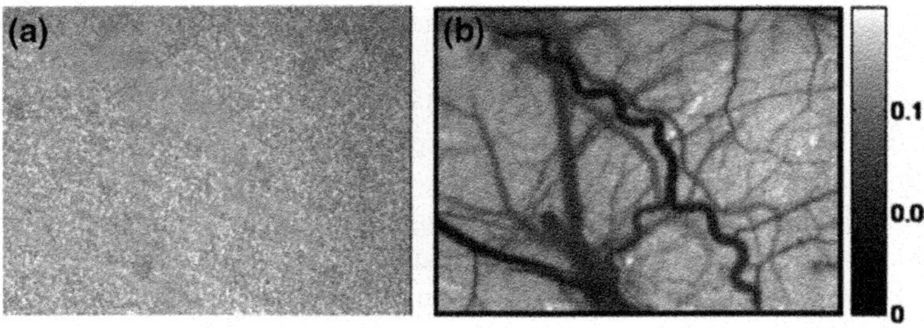

Fig. 6.6 Laser speckle imaging of blood flow in the brain of a rat. **a** Raw speckle image. **b** Speckle contrast image calculated using the spatial speckle approach with a 7 × 7 pixel window. Blood vessels are clearly visualized in the speckle contrast image, as are differences in flow rate. Dark regions with low speckle contrast correspond to high flow rates, while light regions (with higher speckle contrast values) indicate lower flow rates. Figure reproduced with permission from [130], © 2010

Indeed, Wang et al. [154] have reported on the development of fiber bundle-based LSI, and this has the potential to permit speckle measurements in endoscopic or catheter-based systems. Combined with the simplicity and low cost of most LSI setups, the incorporation of optical fiber-based measurements provides a route towards the development of clinically viable (tethered) implantable devices.

6.4.2 Diffuse Reflectance Spectroscopy

When light impinges on biological tissue, reflection can take place in either a specular or a diffuse fashion. Specular reflections return directly from the tissue surface, while the diffusely reflected light undergoes several elastic scattering events within the tissue before re-emerging from the tissue surface. Diffuse reflectance spectroscopy involves the study of this elastically scattered light, and hence provides information about the absorptive and scattering properties of a sample.

In a typical diffuse reflectance spectrometer, a white light source is employed for illumination and spectrally resolved detection is provided by the combination of a dispersive optical element (i.e. a prism or a diffraction grating) and a large area detector (such as a CCD array). Optical fibers are also often used for light delivery and collection, and the positioning of the delivery and collection fibers can be chosen in order to selectively reject specular reflections (see Fig. 6.7a, b).

Optical fiber-based diffuse reflectance spectrometers have been applied to a multitude of clinical studies investigating a variety of different diseases and pathologies. These range from cancers of the skin [156–158], cervix [159–161], bladder [162], and colon [163, 164] to stroke and epilepsy [165–167]. In some cases, diffuse reflectance spectra are used to directly assess disease state. However, in the majority of studies, the measured spectra are used to calculate the tissue or

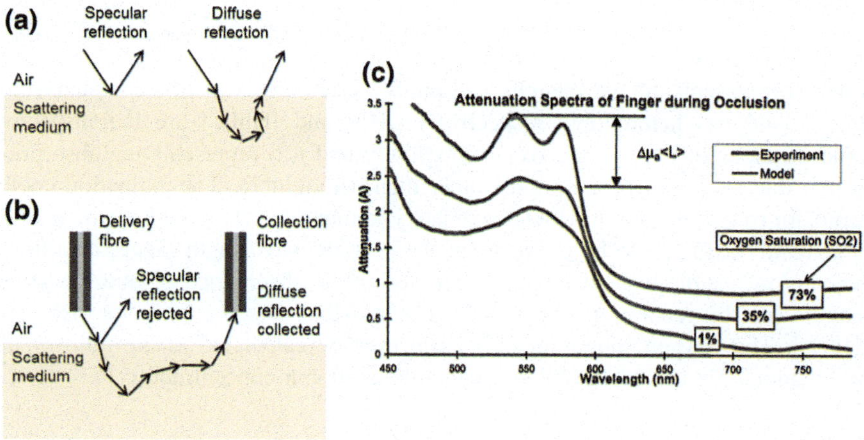

Fig. 6.7 Diffuse reflectance spectroscopy. **a** Diagram illustrating the difference between specular and diffuse reflection. Specular reflection involves direct backscattering of light from the surface of a medium, while diffuse reflection involves multiple scattering events within the medium. Hence, diffusely reflected light carries information about the absorptive and scattering properties of the sample. **b** In fiber-optic diffuse reflectance spectroscopy, the optical fibers used for light delivery and collection can be positioned such that the diffusely reflected light is collected while specular reflections are rejected. **c** Example diffuse reflectance spectral data (in this case, attenuation data is shown) illustrating how the spectra depend upon the tissue oxygen saturation. Graph shown in (**c**) has been reproduced with permission from [155], © 2001

blood oxygen saturation, which can be correlated with the state of health of the tissue (or patient). Three example diffuse reflectance spectra are shown in Fig. 6.7c, where the effect of oxygen saturation on the spectra is clear [155].

In vivo evaluation of the oxygen saturation from diffuse reflectance spectra can be achieved using a number of approaches (e.g. [155, 168–173]) and a detailed description of the relevant mathematics can be found in [174]. Briefly, it is first necessary to calculate the absorption coefficient (μ_a) of the sample according to

$$\mu_a = \varepsilon[C] = -\frac{1}{L}\log\frac{I}{I_0} \tag{6.4}$$

where ε is the extinction coefficient, $[C]$ is the concentration of the absorber, I and I_0 are respectively the detected and incident light intensity, and L is the propagation distance within the sample. If the exact path length (L) is not known, it can be calculated by measuring spectra with multiple different source–detector separations or by employing time- or frequency-resolved techniques (which use pulsed illumination and time-resolved detection or frequency-modulated illumination and detection respectively) [174]. The measured absorption coefficient can then be related to the concentrations and extinction coefficients of oxy- and deoxy-hemoglobin as

$$\mu_a(\lambda) = \varepsilon_{Hb}(\lambda)[Hb] + \varepsilon_{HbO_2}(\lambda)[HbO_2] + f(\lambda) \tag{6.5}$$

where λ represents the wavelength, ε_{Hb} and ε_{HbO2} are the extinction coefficients for deoxy- and oxy-hemoglobin respectively, [Hb] and [HbO_2] are the respective concentrations of deoxy- and oxy-hemoglobin, and $f(\lambda)$ represents the absorption profile due to tissue chromophores other than hemoglobin. The extinction coefficients for oxy-hemoglobin (HbO_2) and deoxy-hemoglobin (Hb) are known, and $f(\lambda)$ is generally assumed to be constant (in the near IR wavelength range) or a linear function (at visible wavelengths). By measuring the absorbance at multiple wavelengths (i.e. by recording a spectrum) it is possible to solve (6.5) for [Hb] and [HbO_2]. The oxygen saturation (SO_2) can then be calculated as the ratio of the oxy-hemoglobin concentration to the total hemoglobin concentration:

$$SO_2 = \frac{[HbO_2]}{[Hb] + [HbO_2]}. \tag{6.6}$$

Blood and tissue oxygen saturation measurements have been used in many clinical studies and it is beyond the scope of this chapter to review all of these. However, examples of the use of SO_2 measurements include the study of stroke physiology [165], monitoring photodynamic therapy [155, 175] and cerebral ischemia [166], as well as the investigation of various cancers [159–163, 176].

A number of commercial devices exist that permit clinical oxygen saturation measurements in vivo. These are typically deployed during or after surgery to ascertain whether the operated tissue is receiving a sufficient blood supply, or to monitor the resuscitation process. Examples of these clinical devices include the O2C (LEA Medizintechnik GmbH, Germany) and a number of finger clip systems (for both clinical and home use) such as the GO_2 (Nonin Medical Inc., USA). Such systems typically provide measurements of the pulse as well as the oxygen saturation, and some devices (e.g. the O2C) also incorporate blood flow monitoring using the laser Doppler approach.

As discussed above, SO_2 measurements are often carried out using optical fiber-based devices (e.g. the O2C is in clinical use) and, as such, their use in a tethered implantable format is not difficult to envisage. Furthermore, the relatively simple device requirements (i.e. a minimum of two illumination wavelengths and a single-point detector) mean that miniaturization for the development of fully implantable devices is also possible. Indeed, work to this end is under way and Chen et al. [1] recently presented a tissue oxygen saturation monitor in which all of the optical and electronic components required for illumination and detection were contained within a single 1.5 × 2 cm integrated chip. Devices of this sort can also be used to make pulse measurements, and in that case only a single illumination wavelength is required. These systems allow the pulse rate to be obtained by monitoring the variations in the reflected or transmitted light intensity that occur as the volume of the vessel or the oxygen content of the blood change in a pulsatile nature. This approach to pulse measurement is commonly referred to as photoplethysmography (PPG).

Optical pulse and SO_2 sensors have also been extended to provide systolic blood pressure measurements through combination with ECG probes, and a number of implantable systems have recently been reported [2, 3]. In these devices the pulse transit time (PTT) is first calculated as the difference between the arrival times of the peak of the ECG signal (the R-wave) and the rising edge (the systolic slope) of the PPG signal. PTT can then be converted to pulse wave velocity (PWV) through knowledge of the distance that the pulse has traveled (i.e. the distance from the heart to the measurement location), which is directly related to blood pressure. The relationship between the PWV and the blood pressure is linear in the physiologically relevant pressure range, but is also dependent on a number of unknown parameters, such as the blood density, the thickness of the vessel wall, and the inner diameter of the vessel [3]. Thus, PTT-based blood pressure measurements require a calibration procedure to be carried out for each individual patient (as the blood density, vessel wall thickness, etc. will vary from patient to patient) and this represents a significant drawback to techniques of this sort. In addition, research in this area is in its early stages, and as such there is considerable variation across the different reported devices due to (among other factors) the choice of the point in the PPG signal from which the transit time is measured. Nonetheless, these devices provide an opportunity to measure a highly important physiological variable (blood pressure) in a continuous and minimally invasive manner. For this reason, they are likely to be developed further in the near future.

An additional format that is of interest when considering implants is the ingestible capsule, and a device of this sort (HemoPill, Ovesco Endoscopy AG, Germany) has recently been launched that permits detection of blood in the GI tract. This untethered pill is swallowed by the patient and blood is detected based on the relative absorbance of light from blue and red LEDs. This system permits rapid diagnosis of internal bleeding at a much lower cost than with standard or capsule-based endoscopes due to the fact that an image sensor is not required. This demonstrates that the development of untethered, ingestible/implantable probes that can detect blood, measure SO_2, or make other optical measurements is entirely feasible.

Due to its ease of miniaturization and compatibility with optical fibers, the development of implantable SO_2 monitors (and other diffuse reflectance-based probes) is likely to continue. Furthermore, as SO_2 measurements have already been shown to have important clinical applications and can be extended to allow detection of additional physiologically useful parameters, such as pulse and blood pressure, the development and use of such devices is almost certain to be successful. For this reason, an increasing number of implantable or minimally invasive SO_2 devices are likely to be presented in the coming years, and these will offer new opportunities to accurately monitor resuscitation, ischemia, tissue viability during/after surgery, and many other important clinical targets.

6.4.3 Fluorescence Spectroscopy

In a fluorescent molecule, the absorption of an incident photon acts to promote an electron within the molecule to a higher energy state (i.e. a higher energy molecular orbital). When the electron later relaxes back into its (lower energy) ground state, energy can be released in the form of light. This emissive behavior is known as fluorescence and has been widely studied in the context of medicine.

A fluorescence spectrometer has a very similar layout to a diffuse reflectance system. However, narrowband excitation is typically employed (using a laser source or a lamp with a narrow excitation filter), as is an emission filter, which ensures that no elastically scattered (reflected) excitation light reaches the detector. Most often the collected light is recorded as a function of wavelength (i.e. as a spectrum), but alternative parameters can also be measured, such as the polarization of the emitted light or the fluorescence lifetime [177].

Fluorescence measurements have been applied to many biomedical studies and some fluorescence-based systems are already in clinical use. One such approach is autofluorescence imaging (AFI) [178–181], which is used during endoscopy to help guide GI biopsies. Furthermore, as described previously, endomicroscopy has been used in conjunction with contrast agents to allow microscopic fluorescence imaging for in vivo assessment of GI disorders. As with many of the techniques described above, optical fibers are often used for light delivery and collection in clinical investigations, as these provide a minimally invasive measurement modality. In the future, such systems could also become useful as tethered implanted devices. As fluorescence measurements are clinically compatible, many studies have reported the investigation of fluorescence as a diagnostic parameter in a variety of diseases. One important example is in the study and diagnosis of cancer, where both spectrally [182–188] and temporally [158, 189–198] resolved fluorescence measurements have been investigated extensively. This has involved single-point measurements [158, 185–190] as well as imaging modalities [193, 194, 199–202] and has entailed the study of the signals from intrinsic tissue fluorophores (referred to as autofluorescence) as well as the use of fluorescent contrast agents [203]. In terms of autofluorescence measurements, observed changes in fluorescence between healthy and diseased tissues have been correlated to metabolic state or to changes in the constituent fluorophores that are present. In other cases, the fluorescence spectrum or lifetime is simply used as a diagnostic parameter with no attempt made to make a link to the physiological changes occurring. In studies where contrast agents are used, a fluorescent dye or stain is chosen to optimize the image contrast. In some cases the chosen stain preferentially accumulates in cancerous cells and one important example is 5-aminolevulinic acid (5-ALA)—a precursor to protoporphyrin IX (PpIX), which accumulates in tumors [203].

The final goal of research into fluorescence as a clinical tool for cancer diagnosis is usually aimed at producing screening systems that can be used to test high-risk patients. Another option, however, is to develop implantable devices that can continually monitor tumors during treatment. Such a system could entail a fiber-optic tethered device, and these are likely to become more prevalent once the size of the required optical apparatus has been sufficiently miniaturized.

Another important use of implantable medical devices is in the diagnosis of infection (for example, during recovery from surgical procedures). As such, optical systems capable of detecting the presence of infection or of specific bacteria would be very useful clinically. Limited studies into the use of fluorescence for this purpose have been reported in the literature and these have typically used contrast agents. One example study was presented by Yamashita et al. [204], where the authors used a chymotrypsin-activated fluorescent probe to optically monitor for signs of pancreatic fistula (a common complication after digestive surgery that can lead to infection) after resection of the pancreas. The results of this study were positive, with high diagnostic sensitivity and specificity reported as well as real-time visualization of pancreatic leakage during surgery. However, while an approach that requires the use of a contrast agent may be feasible during surgery, it is much more difficult to envisage the use of a contrast agent as part of a long-term implantable sensing modality. One option here would be to chemically immobilize the contrast agent on the tip of an optical fiber. However, for some fluorescent dyes this would entail complex chemical synthesis and such a technique would also require a foreign chemical to be implanted within the patient. Although this would involve a very small amount of fluorescent dye implanted for a relatively short amount of time (days, weeks, or months), it is still a sub-optimal arrangement. Research into the use of alternative optical techniques that can report on the presence of infection or bacteria without the need for contrast agents is currently under way. One technique under investigation for this purpose is Raman spectroscopy, which is described in the following section.

6.4.4 Raman Spectroscopy

Beyond absorption, reflection (elastic scattering) and fluorescence, another transition that can occur when light is incident upon an optically active molecule is Raman (inelastic) scattering. As discussed above, in the case of absorption, photons incident on a molecule act to promote electrons within that molecule to electronically excited states. When those electrons decay back to their ground energy state, fluorescence photons are emitted. In the case of elastic scattering (which is also referred to as Rayleigh scattering or reflection), the incident photon perturbs the scattering molecule but does not have sufficient energy to promote an electron to its first (or higher) excited state. Instead, an electron is instantaneously promoted to a so-called 'virtual state' before immediately decaying back to the ground state. No energy transfer occurs between the photon and the molecule; hence the wavelength of the scattered photon remains unchanged. Inelastic scattering—also referred to as Raman scattering—also involves the instantaneous promotion of an electron to a virtual state. However, in this case, the excited electron decays back into a different vibrational energy level (vibronic mode) to the one that it started in. Due to this change in the vibronic mode of the molecule there is a concomitant transfer of energy between the molecule and the scattered photon, which can be detected as a change in the wavelength of the photon. Most often the energy is transferred from the photon to the scattering molecule: the molecule gains vibrational energy and the photon loses energy, leading to an increase in its wavelength (known as a Stokes shift). Alternatively, the opposite can also occur when a photon impinges on a molecule that is already in an excited vibrational state. In this case, the perturbation of the molecule caused by the scattering event can lead to the molecule falling into a lower energy vibronic mode. Energy is then transferred from the molecule to the photon and a reduction in the wavelength of the photon—known as an anti-Stokes shift—is observed. The relevant electronic and vibrational transitions that can occur in an optically active (i.e. scattering/absorbing/fluorescent) molecule are shown in the Jablonski diagram in Fig. 6.8.

In theory, the instrumentation required for Raman spectroscopy is identical to that used in fluorescence spectroscopy. Typically, however, Raman signal levels are orders of magnitude lower than those observed in fluorescence measurements. For this reason, high throughput diffraction gratings as well as cooled, high quantum efficiency CCD detectors are often used for Raman measurements, even though these are not necessarily required for fluorescence experiments. They provide improved detection efficiency and allow data to be collected in reasonable acquisition times and with acceptable excitation laser powers. Additionally, a notch (or band-stop) filter (rather than a long pass filter) is often used to block the excitation

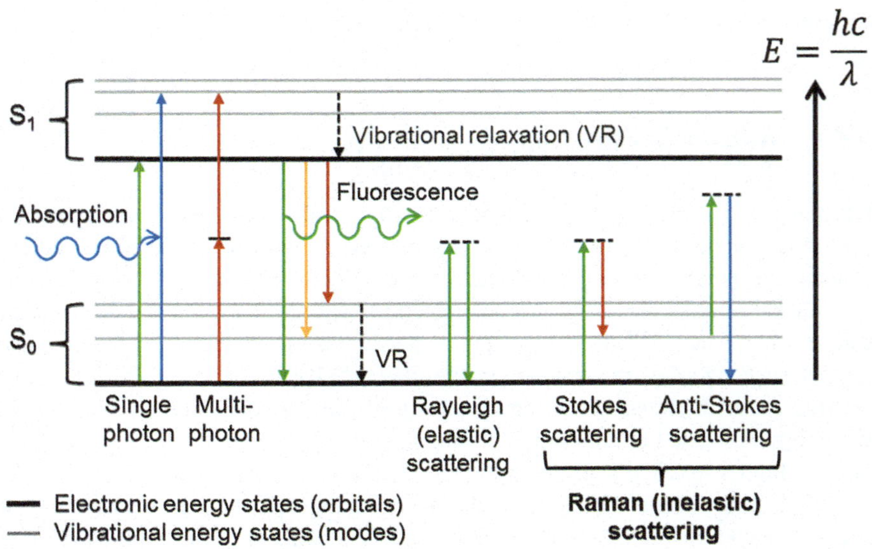

Fig. 6.8 Jablonski energy level diagram showing some electronic and vibrational transitions that can occur in an optically active molecule, including absorption (both single photon and multi-photon), fluorescence, elastic scattering, and Raman (inelastic) scattering. E energy; h Planck's constant; c speed of light; λ wavelength

light. This then allows wavelengths both above and below the excitation wavelength to reach the detector meaning that both the Stokes and anti-Stokes signals can be collected.

Raman spectroscopy is becoming widely used in a variety of different fields and the main reason for this is that it provides very specific signals. In fact, the observed signals are so specific that they can even be used to characterize chemical compounds in terms of their constituent molecular bonds (due to the different bond vibrations that can be excited). Thus, Raman spectroscopy is becoming increasingly common in chemistry, where it is often used for chemical identification [205]. The specificity of the spectra obtained has also meant that Raman spectroscopy has been investigated as a potential clinical tool and this has involved its application to the study of—among many other things—retinal inflammation [206], gastric [207] and oral neoplasia [208], colonic polyps [209], lung cancer [210, 211], brain tumors [212], and dengue virus [213].

As with the techniques described above, Raman spectroscopy is directly compatible with optical fibers, and this compatibility is vital in terms of developing a clinical modality. While the use of optical fibers does present significant hurdles—such as reduced signal collection efficiency and the introduction of background signal (in the form of fluorescence or Raman scattering from the optical fiber itself)—optical fiber-based Raman spectroscopy is now becoming more widespread, with a number of commercial spectrometers and delivery/collection optical fibers now available. Typically, the optical fibers used have optical filters incorporated into them in order to reject both elastically scattered excitation light and any Raman or fluorescence signal from the fiber itself (e.g. RPB Fiber Optic Raman Probe, InPhotonics Inc., USA). Despite the potential limitations imposed by optical fibers, a number of research groups have developed and used optical fiber-based Raman spectrometers (e.g. [214]) for medical applications. Additionally, Raman spectroscopy has been deployed using both endoscopes and catheters [209–211, 215, 216]. Combined with promising results involving the use of Raman spectroscopy for disease diagnosis (e.g. [206–213]), this implies that Raman measurements may well become useful in the clinic in the future.

One particularly promising application of Raman spectroscopy for implantable sensing is in the detection of bacteria for the diagnosis of infection. Raman spectroscopy has been applied to the investigation of bacteria many times, with an important early study being reported by Howard et al. in 1980 [217]. Since then, many bacterial Raman studies have been conducted, showing that Raman spectra can be used to detect the presence of bacteria and to differentiate between different bacterial species (e.g. [218–228]). This has been presented in the context of the food industry where the detection of bacteria is vitally important in terms of food preservation (e.g. [229]). Furthermore, a number of (pre)clinical studies have been reported discussing the use of Raman spectroscopy in the detection of infection. Teh et al. [207] reported the detection of *Helicobacter pylori* infection (a precursor to stomach cancer) in freshly excised gastric tissue samples using a fiber-optic Raman probe, while detection of the bacterium *Shewanella oneidensis* MR-1 using a fiber-optic surface enhanced Raman spectroscopy (SERS) probe was reported by

Yang et al. [226]. In addition, de Siqueira e Oliveira et al. [230] recorded Raman spectra from multiple bacterial species responsible for urinary tract infections (UTIs) using a fiber-based system and demonstrated discrimination of those bacteria by applying PCA to the measured spectra (and this approach is discussed in more detail below).

Considerable work in this area has also been undertaken by the research group of Costas Pitris [218–222]. This group has further demonstrated that Raman spectra can be used to discriminate between multiple bacteria that are relevant to UTIs in an in vitro setting. This was achieved using spectral decomposition techniques and is illustrated in Fig. 6.9. Specifically, it involved the use of PCA to break the spectra down into a small number of base spectral components, followed by the use of linear discriminant analysis (LDA) in order to use the relative contributions of those base spectra to provide an 'optical diagnosis' for each spectrum. In this case, the diagnosis entailed an assessment of the species of bacteria present. This is quite a common approach in optical spectroscopy and has also been used in numerous fluorescence and diffuse reflectance studies. Importantly, it provides a method of automated diagnosis (in this case, automated identification of bacteria), which would be vital in any implanted optical sensor. Interestingly, the authors were also able to use Raman spectra to successfully carry out an antibiogram—i.e. they were able to successfully identify the antibiotic that was best suited to treatment of a specific infection [218–221]. While this work was carried out in vitro, it shows considerable promise as an alternative, improved method for UTI diagnosis and antibiogram, where standard techniques currently require up to two days for a full diagnosis and antibiotic selection.

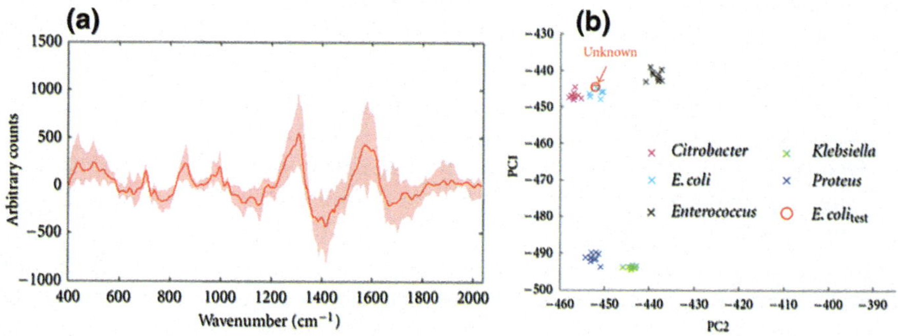

Fig. 6.9 Data illustrating the use of Raman spectroscopy to detect and discriminate between multiple bacteria relevant to urinary tract infections (UTIs). **a** Surface enhanced Raman (SERS) spectra recorded from multiple *Escherichia coli* samples showing the mean spectrum (red line) and the intersample variability (pink). **b** PCA of Raman spectra from multiple bacteria permits accurate classification of each sample. This is illustrated using a plot of the first and second principal components of the spectra calculated using PCA. On this graph, the locations of each of the bacterial species are clearly separated allowing accurate classification. Graphs reproduced from [220] under the terms of the Creative Commons Attribution license CC BY 3.0 (https://creativecommons.org/licenses/by/3.0/), © 2012

An inherent drawback of Raman measurements, discussed briefly above, is the low signal level that is often obtained. This means that long acquisition times and high excitation powers are frequently required in order to obtain an acceptable signal-to-noise ratio (SNR). Neither of these is likely to be compatible with a clinical embodiment; hence much research is currently directed toward the development of methods that provide improved SNRs. One such method is SERS (surface-enhanced Raman spectroscopy, mentioned briefly above) [231], which was first observed (although not initially recognized) by Fleischmann et al. [232]. Typically, SERS involves the use of surface immobilized metal nanostructures to provide an enhancement of the Raman signal. This enhancement is a result of surface plasmons that are excited in the nanostructures and act to amplify the Raman signal, with this amplification being strongest when the plasmon frequency is resonant with the optical radiation (i.e. the excitation light and/or the Raman scattering). A variety of SERS devices have been presented in the literature (e.g. [233–239]). However, enhancement factors (relative to standard Raman scattering measurements) vary greatly between different configurations, and for this reason commercial SERS systems have not yet gained significant market penetration. Nonetheless, and importantly from the perspective of this book, a number of optical fiber-based SERS devices have been reported. These have mostly involved the deposition or immobilization of silver or gold nanoparticles on the tip of an optical fiber, with the fiber then being used to allow fluid measurements via immersion of the fiber tip in the analyte [226, 240–243]. One interesting approach that was recently reported used two-photon lithography to fabricate a 3D structure on an optical fiber tip that provided both surface enhancement and optical focusing, thus also improving the collection efficiency of the probe [244]. Furthermore, optical fiber-based SERS devices have been applied to the measurement and detection of bacteria [226] and it is easy to see how this type of device could become useful clinically in the future. Particularly useful clinical embodiments could include fiber-optic SERS devices incorporated into urinary or central venous catheters. The fiber-optic tip of such a device would be minimally invasive and could feasibly be implanted as a tethered format. SERS could then be used to monitor for signs of infection, and UTIs appear to be a particularly promising target, as these are especially common in patients using urinary catheters and in vitro research has already been published reporting the use of Raman spectra to detect the relevant bacteria (see above and [218–222, 230]). Of course, the external housing of such a tethered implantable device would also need to be miniaturized in order that it could be worn or easily carried by the patient. However, portable handheld Raman spectrometers are available commercially and further miniaturization to produce clinically useful devices should be possible.

Overall, Raman spectroscopy and SERS are promising modalities for clinical applications with many potential diagnostic uses. In particular, the application of Raman measurements to the detection and discrimination of bacteria appears to provide a promising avenue for the development of tethered implantable devices based on optical fibers. Such devices could be used in patients recovering from surgery or in those requiring the use of catheters, for which infections are especially prevalent. In these circumstances, a minimally invasive implanted device that is capable of detecting infections early, before symptoms become severe, would provide considerable clinical benefit. Raman spectroscopy has the potential to be used for this purpose, and research in this area is likely to continue and increase in the coming years.

6.4.5 Other Optical Fiber-Based Sensing Modalities

Beyond the methods described above, there are also many further optical fiber-based sensing techniques—some of which have potential applications in healthcare—and these are described briefly in this section.

Fiber Bragg gratings [245, 246] have been widely used for environmental sensing purposes, for example in the measurement of strain and temperature. However these are typically used in industrial applications and are not particularly relevant here (although a number of fiber Bragg grating sensors have been applied to in vivo monitoring of peristalsis, e.g. [247, 248]). A more promising technique for medical applications is optical fiber-based Fabry–Perot interferometry [249], in which a small Fabry–Perot etalon is deposited on the tip of an optical fiber. The wavelength of the light reflected back from this etalon can then be used to determine a number of parameters, including temperature, pressure, refractive index and even acoustic wave frequency. As with fiber Bragg gratings, Fabry–Perot systems have mainly been developed for industrial applications; however, they are more versatile and could certainly be used clinically due to their forward viewing arrangement. Furthermore, the range of sensing targets is higher, and some—including pressure and acoustic wave frequency—have more obvious clinical applications, for example in minimally invasive ultrasound measurements or blood pressure monitoring. Indeed, a fiber-optic Fabry–Perot blood pressure sensor was recently reported by Wu et al. with in vivo use demonstrated in the aortic arch and right coronary artery of a pig by introducing the probe into the artery through a catheter [250].

Finally, there are several further optical fiber-based sensing modalities that rely upon the immobilization of a sensing layer or membrane on the tip of an optical fiber. This sensing layer exhibits changes in its optical properties (for example, its refractive index, its absorption profile, or its fluorescence emission wavelength) when it comes into contact with a specific chemical compound or biological species. Fiber-optic devices of this sort have been used as chemical sensors (to detect the presence or concentration of particular elements or compounds) [251–255], biosensors (for example, for the detection of glucose [256]), and ion sensors (which allow measurement of ion concentrations and, hence, pH) [251, 254, 257–263]. One approach that has been particularly useful in the development of these types of fiber-optic sensors is the production of so-called sol-gel films [4, 264]. Sol-gel technology allows the fabrication of thin, porous glass layers on the tips of optical fibers using a relatively simple polymerization reaction. The resulting films can have high homogeneity and purity and are also chemically inert. Furthermore, it is possible to encapsulate sensing molecules into the sol-gel glass via simple modifications of the polymerization protocol. This technology has become widely used in the development of optical fiber sensors and one particularly prevalent application has been in pH sensing. For this reason, sol-gel based optical fiber sensing technology has the potential to become useful medically where measurements of pH or other ion concentrations could be very valuable. Additionally, biosensing could also be very useful medically and this has been achieved using modified sol-gel preparation procedures that allow the encapsulation of chosen biomolecules within the sol-gel matrix.

While the sensing modalities addressed in this section are still under development, they certainly have potential merit as medical sensors. Moreover, some can be applied using a single fiber with a (relatively) simple modification (i.e. the production of a Fabry–Perot etalon or a sol-gel film) to the optical fiber tip. For this reason, it is again easy to envisage the use of these techniques as tethered implantable sensors, where the tip of the optical fiber sensor could be incorporated into a clinical catheter or needle in a straightforward manner. Of course, other fiber-optic sensing modalities exist and this review is certainly not exhaustive. However, the techniques reported above should provide a reasonably complete summary of approaches that could be used clinically—perhaps as (tethered) implants—in the future.

6.5 Implantable Visual Prosthetics

In addition to the diagnostic devices described above, implantable optical systems have also been deployed for therapeutic purposes, in particular for the treatment of vision loss. Such implants are used in the treatment of cataracts, retinitis pigmentosa (RP), and age-related macular degeneration (AMD), and are described briefly in the following sections.

6.5.1 Intraocular Lens Implants

Cataracts—clouding of the lens in the eye leading to impaired vision—have long been treated by surgically removing the lens and replacing it with a synthetic implant known as an intraocular lens (IOL). This procedure was first reported in 1949, and found more widespread applications from the late 1970s and early 1980s [265]. Traditionally, the implanted lenses had a single focal length, providing good distance vision but poor near-field sight. Thus, patients with so-called monofocal IOL implants would typically require reading glasses after surgery. To combat this issue, accommodating and multifocal IOLs have been developed over the last 20–30 years, with the latter now widely used clinically [266–268]. Accommodating

IOLs interact with the ciliary muscles within the eye to provide forward and backward movement of the implanted lens, hence allowing the patient to focus on both near and distant objects. While in theory this approach will offer clear vision over all distances, in practice the range of accommodation (i.e. the range of movement of the lens, and hence the range of focal lengths that it provides) often remains insufficient to provide truly spectacle-free post-operative sight [267, 268].

Multifocal IOLs, on the other hand, focus light to multiple points using diffractive or refractive effects (or a combination of both in the case of hybrid multifocal IOLs). Refractive IOLs feature two or more spherical zones with different radii of curvature, which each focus light to a different focal point [267, 268]. As the pupil dilates or constricts, more or less light is focused to each of the discrete focal points. In the majority of patients, pupil diameter changes as they focus on near or distant objects [269] and this effect helps the function of refractive IOLs. When focusing on a distant object the pupil dilates, allowing more light to pass through the outer portion of the IOL, which is optimized for distance vision. Similarly, the pupil constricts when the eye is focused on a near object, causing the light to pass through the central region of the lens, which has a shorter focal length designed for near-field vision. Diffractive IOLs generate multiple points of focus in a different manner. A series of concentric rings of different heights acts as a phase mask that diffracts light to two or more different points simultaneously. In contrast to refractive systems, the foci generated by diffractive IOLs are independent of pupil diameter [268]. Despite this, they can still provide clear vision at multiple focal distances, as patients will primarily perceive only the focused image [268, 270–272].

Both refractive and diffractive multifocal IOLs provide improved near vision relative to their monofocal counterparts [266–268]. However, both also have drawbacks. For example, halos and glare are more common than in monofocal IOLs [268, 273, 274] and the contrast of the retinal image is typically poorer [266, 267]. This is an effect of the defocused images reaching the retina as well as aberrations within the lens itself, and it can be further exacerbated by tilt or displacement of the implanted lens [275]. Several hybrid IOLs have been and are being developed, which combine both refractive and diffractive features into single lenses with the aim of improving image clarity while maintaining good vision over all distances [268]. Despite their limitations, both multifocal and monofocal IOL implants are in widespread clinical use, with surgeons often recommending different lenses for different patients in order to best address the needs of the individual. Indeed, in some cases patients will even have a different type of IOL implanted in each eye in order to obtain the best possible tradeoff between near and distance vision.

6.5.2 Visual Prostheses

Further ocular implants are also used in the treatment of RP and AMD, both of which cause the onset of blindness through degradation of the photoreceptors (the rods and cones) in the retina. Importantly, in both diseases, the inner nerve cells in the retina—the bipolar cells, the retinal ganglion cells (RGCs), and the axons—as well as the tissues responsible for transmission and processing of visual signals in the eye and brain (such as the optic nerve and the visual cortex) remain functional [276–278] (although there is some secondary degeneration and remodeling of the retinal neurons [277, 279–282]). In the healthy retina, the photoreceptive rods and cones absorb light and then send electrical signals to the bipolar cells that are transmitted to the brain via the RGCs, the axons, and the optic nerve. Thus, as only the photoreceptors are damaged in RP and AMD, it should theoretically be possible to induce visual signals (phosphenes) in patients with these conditions by electrically stimulating the bipolar cells, RGCs, axons, or the optic nerve [276–278]. Alternatively, phosphenes can also be generated by stimulating the regions of the brain that are responsible for processing and interpreting visual information, for example, the visual cortex [283, 284]. Indeed, several studies have reported the development of devices that can be implanted in the eye and instigate a degree of vision in blind patients by electrically stimulating the retina (detailed reviews can be found in [276–278]). Furthermore, a small number of systems have also been commercialized and are now available clinically (e.g. Argus II (Second Sight Medical Products, USA) [285], Alpha-IMS (Retina Implant AG, Germany) [286–288], Iris II (Pixium Vision, France) [289]).

These devices are known as visual or retinal prostheses and they can be broadly categorized according to the technology on which they are based or the region of the retina in which they are implanted. The majority of systems require implantation in or on the retina, although stimulation of the brain and optic nerve have also been reported [276–278]. Of the more common retinal implants, the smallest systems consist of an array of photodiodes that is placed close to or within the retina. The eye focuses light onto the photodiodes, which then convert that light into electrical impulses that stimulate (depolarize) the bipolar cells and/or RGCs, hence inducing a visual signal [286]. Photodiodes have the advantage of containing all the necessary electronics to provide synthetic vision within the implant itself, and thus do not theoretically require wires, batteries, or external components. A significant drawback, however, is that photodiodes alone are often incapable of providing sufficient current to efficiently activate the retinal neurons. To circumvent this problem, an example device has been reported that incorporates an amplification circuit within each pixel along with an external head-mounted power source that is used to inductively drive the implant (Fig. 6.10a). This system delivers larger currents to the retinal neurons and more efficiently stimulates the visual pathway [286, 287]. Photodiode-based systems are typically implanted within the retina at the location of the degenerated photoreceptors such that the photodiodes directly replace the photosensitive cells and stimulate the bipolar cells. This approach has the advantage of requiring little or no processing of the stimulation signal as the targeted neurons are those directly proximal to the photoreceptors. Due to the implantation location, these devices are referred to as subretinal implants.

In the most common alternative, the implant contains only the stimulation electronics, and the sensing elements are dispensed with. Instead, a camera mounted

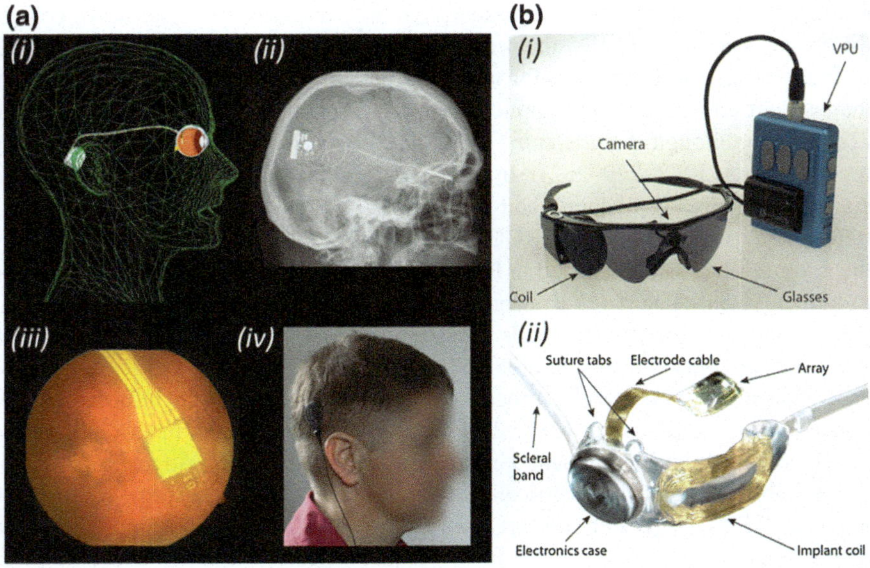

Fig. 6.10 Implantable retinal prostheses: **a** the Alpha-IMS subretinal implant and **b** the Argus II epiretinal implant. **a** Diagrams and images of the Alpha-IMS system showing: (i) 3D diagram and (ii) X-ray image of the retinal prosthesis and the coil implanted behind the ear to allow external control and powering; (iii) photograph of the retina after implantation of the 3 mm × 3 mm chip beneath the fovea; (iv) photograph of the primary external coil, which is magnetically mounted above the implanted coil shown in (i) and (ii), and provides power and signals to the implant via electrical induction. **b** Images of the Argus II visual prosthesis: (i) external components including glasses (with mounted camera and external powering coil) and visual processing unit (VPU); (ii) internal components, including the electrode array, the internal coil for wireless power delivery, the electronics casing, and the suture tabs and scleral band used to fix the implant in place on the retinal surface. **a** Reproduced from [287] under the terms of the Creative Commons Attribution license CC BY 4.0 (https://creativecommons.org/licenses/by/4.0/), © 2013. **b** Reproduced from [290], © 2012, with permission from Elsevier

on a pair of glasses is used to capture the image and an external processing unit converts the visual information into an electrical signal to be delivered by the electrodes in the implant (Fig. 6.10b). This signal is transmitted to the implant by an external power source and a transcutaneous telemetry link very similar to that used in the wirelessly powered photodiode-based system described above [290–292]. In this case, the implants are typically fixed onto the outer surface of the retina (i.e. on the vitreal side, the side closer to the front of the eye) and are thus known as epiretinal implants. This means that the stimulation signal is delivered to more distal parts of the retinal neural processing network—either the RGCs or the axons. In turn, this acts as both an advantage and a disadvantage. On the one hand, stimulating more distal neurons means that lower threshold currents are required to elicit visual signals. On the other hand, addressing the RGCs or axons means that more advanced processing of the visual signal is required (as the bypassed photoreceptors and bipolar cells provide a degree of processing in the healthy eye) and this explains the use of an external visual processing unit (VPU).

Of course, both approaches have pros and cons. The subretinal photodiode-based implants represent the system with the smallest impact on the patient, with the only necessary external component being a wireless power source, which consists of a small telemetry link worn on the head and a battery pack [287]. The external constituents of epiretinal prostheses, on the other hand, typically also entail a pair of glasses to mount a camera as well as a VPU. While this increases the external footprint of the device, it also acts to simplify the electronics of the implanted chip (which entails only stimulation electrodes and not photodiodes or amplification circuits). Furthermore, as epiretinal implants are positioned on—rather than below—the surface of the retina, the implantation procedure is simpler than for subretinal systems and the constraints on the size of the chip are less stringent. This represents a clear advantage of epiretinal implants, but also brings an important drawback: using a head-mounted camera to capture the image rather than an implanted array of photodiodes means that the phosphenes (or the image) perceived by the patient will change only in response to head movements and not as a result of changes in the direction of the patient's gaze (i.e. the direction in which their eyes are looking).

Despite the current limitations of the two methods, commercial systems have been released based on both [276], and devices based on other concepts and/or implant locations have also been reported (and in some cases commercialized) [278]. The Argus II device was recently brought to market by Second Sight Medical Products [285] and has been approved for use in the treatment of late-stage RP in both Europe and the USA. It is an epiretinal system consisting of an implantable array of 60 electrodes, a pair of glasses with a mounted camera and wireless powering unit, and a VPU. A similar system is also available from Pixium Vision—the Iris II Bionic Vision Restoration System—which is CE marked (i.e. it has European regulatory approval) for use in RP and is currently undergoing clinical trials [289]. In addition, the Alpha-IMS is a subretinal device that entails an implantable array of 1500 photodiodes/electrodes and an external power source to drive amplification circuits within the chip [287]. This system has been commercialized by Retina Implant AG [288] and has recently received regulatory approval for the treatment of late-stage RP in Europe.

Naturally, further systems are also under investigation that do not fall perfectly into either of the two categories discussed above. For example, Pixium Vision are developing a subretinal implant [293, 294] that uses an external camera for image detection in combination with an implanted photovoltaic array and a near IR image projection system that is integrated into the goggles that house the camera (PRIMA Bionic Vision Restoration System [295]). This system is effectively a hybrid of the photodiode and camera-based approaches discussed above and has the advantage that the implant does not require a power supply, as the near IR image projection system provides a high enough light intensity to generate sufficient current in the implanted photodiodes to efficiently activate the retinal neurons. As a result, the necessary surgery is much simpler than for implants with extra-ocular wireless powering links. Furthermore, the connection between eye movement and image perception is maintained, as this device uses the eye's optical system to transmit information from the image projector to the retinal implant. However, despite the potential advances offered by this system, it is at a considerably earlier stage of development than the three commercially available devices described above.

In addition to the PRIMA system from Pixium, several other visual prostheses have been reported that generate synthetic vision with alternate approaches to the epiretinal and subretinal implants discussed above. Suprachoroidal implants are typically positioned between the choroid and the sclera. The greater distance from the retina renders the surgery simpler and safer; however, this also acts to increase the stimulation thresholds. Nonetheless, there have been reports of safe stimulation in animals using this approach [296, 297] and recent clinical studies have demonstrated that suprachoroidal systems can stimulate visual percepts in blind patients [298, 299]. Further alternative approaches involve placing the electrodes further from the retina, with the aim of stimulating deeper layers of the visual pathway. Example devices of this kind have been reported for stimulation of the optic nerve [300–302] and the visual cortex [303–305]. While these latter methods offer opportunities to treat blindness caused by degeneration of more than just the photoreceptive cells in the retina, clinical research is at a much earlier stage than for retinal prostheses [276].

For the more common epiretinal and subretinal implants, clinical trials have reached a more advanced stage and several studies have demonstrated improvements in the vision of blind patients. With the Argus II and Alpha-IMS devices (as well as with several other systems including earlier embodiments such as the Argus I), improvements have been reported in the ability of patients to perform simple tasks such as object identification and localization. In addition, patients' visual acuities (a common measure of the clarity of vision) have been shown to improve when using the prostheses [276–278, 286, 287, 290]. While these early results are of course very promising, it is worth noting that the visual acuities reported to date are still well below the level at which patients are classified as legally blind. For this reason, significant work in this field is still required, and this will involve further device development alongside physiological studies aimed at better elucidating the relationship between the electrical stimulation protocol and the phosphenes that are invoked in the patient. Indeed, some particularly interesting questions remain in this area, with one important example being how to increase the number of image pixels perceived by the patient (which is not currently believed to be a simple function of the number of stimulating electrodes [277]). Nonetheless, despite the many outstanding challenges, the use of visual prostheses can be expected to increase significantly in the future as more advanced devices are developed that provide more profound enhancements to the vision of blind patients.

6.6 Optogenetics

As a further example of optical techniques used as implants, this section presents a discussion of the current progress in optogenetics. Optogenetics involves the genetic modification of nerve cells—most often neurons in the brain—in order to render them sensitive to light. Subsequently, the sensitized neurons can be addressed optically (by focusing light onto chosen cells) allowing control of particular functions or events [306, 307]. This is analogous to the electrical activation of neurons used in the visual prostheses described in Sect. 6.5.2. In recent years, optogenetics has been widely applied in the field of neuroscience, where many in vitro cell studies and in vivo animal experiments have been conducted aimed at elucidating and understanding signaling pathways in the brain. Experiments such as these have shown great promise, not only in the study of neurological function, but also in the investigation of the causes and effects of a multitude of brain disorders (including Parkinson's disease [308]). Importantly, this has provided many valuable new insights into possible modes of treatment for such disorders. Furthermore, the growing use of optogenetic techniques has triggered much discussion about the potential development of neural implants based on optogenetics that would allow treatment of a wide range of ailments, including brain injuries, depression, and even drug/alcohol dependency. While this field does not have particular significance in the area of optical sensing, it is certainly relevant in terms of the development of optical implants. For this reason, optogenetics is discussed in this chapter and the following sections describe the basic principles, the development of miniaturized devices suitable for implantation in animals, and progress towards medical optogenetic systems for use in humans. The latter section has a focus on optogenetic retinal prostheses—which aim to restore vision to the blind—but also discusses the possible future development of more ambitious fully implantable systems for use in the brain.

6.6.1 *Theoretical Background*

Optogenetics involves the combination of the genetic modification of neurons (or other nerve cells) to render them sensitive to light with the subsequent use of optical techniques to address the sensitized neurons. It relies heavily on the discovery of light-sensitive ion channels and ion pumps such as Channelrhodopsin-2 (ChR2) [309], Halorhodopsin (NpHR) [310], and Archaerhodopsin (Arch) [311]. If a neuron expresses one (or more) of these light-sensitive opsin proteins then its membrane will be depolarized (for neural excitation) or hyperpolarized (for neural inhibition) when illuminated with light of the appropriate wavelength and intensity [312, 313]. Depolarization of neuronal membranes involves the transfer of positively charged ions (Na^+ or Ca^{2+}) from the extracellular space into the cell cytoplasm. The charge within the cells then becomes more positive, and importantly this induces the firing of action potentials, which are central to cell–cell communication. Conversely, hyperpolarization involves a cell's membrane potential becoming more negative and occurs either through an efflux of potassium (K^+) ions or an influx of chloride (Cl^-) ions. This inhibits action potentials by effectively increasing the threshold for depolarization (by making the cell's resting potential more negative). Thus, by modulating ion channels/pumps it is possible to control certain signaling pathways in the nervous system and hence influence the behavior of tissues or even live, freely moving animals. Using light-sensitive ion channels/pumps (such as ChR2, NpHR, and Arch) it is possible to achieve this neuronal control using optics and this concept is illustrated in Fig. 6.11 [306, 307, 313, 314].

In order to carry out any optogenetic study it is first necessary to modify specific cells in order that they express one (or more) of the light-sensitive opsin proteins discussed above. These proteins allow control of ion flows as they can reversibly bind to the co-factor retinal, with this binding reaction reliant upon the incident light intensity (and wavelength). Depending on their binding state, the proteins will then provide either an "open" or "closed" configuration for the ion channel/pump. Significantly, vertebrate tissues contain naturally occurring retinal, which means that optogenetic control can be permitted simply through the expression of an appropriate opsin protein and without the need for any additional chemicals or components (i.e. without the separate and additional introduction of retinal, for which it would be considerably more difficult to obtain cellular specificity) [306].

The expression of opsins in mammalian cells can be achieved in a number of ways. One option is to use transgenic animals that have been genetically engineered to express the appropriate opsins in specific cells, and a number of transgenic opsin-expressing mouse lines have now been reported (e.g. [315, 316]). An alternative approach is to use a viral expression system. In this case a virus is injected into the host animal, which causes rapid production of a particular opsin. With viral expression, cellular specificity can be obtained by using particular promoters (viral receptors) that act to target the virus to specific types of cells (although this

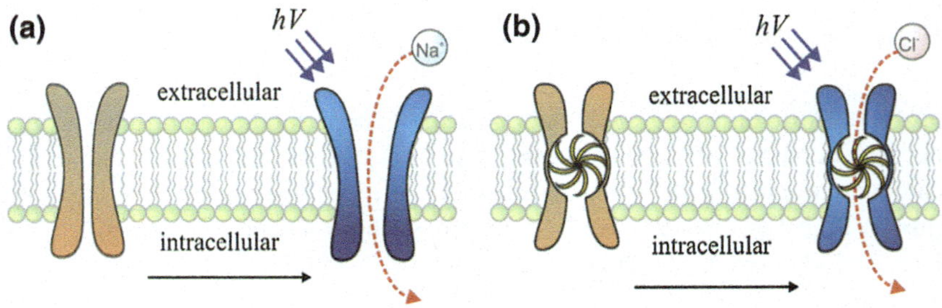

Fig. 6.11 Schematics explaining the function of light-sensitive ion channels (**a**) and ion pumps (**b**). **a** The light-sensitive ion channel ChR2 opens upon irradiation allowing positively charged ions (Na^+) to diffuse into the cell leading to depolarization. **b** The light-sensitive ion pump NpHR is activated when irradiated with light of the appropriate wavelength and induces an influx of Cl^- ions. This leads to hyperpolarization, which inhibits the neuron. Diagrams reproduced with permission from [314], © 2014

approach is actually more efficient in transgenic animals than it is in viral expression in wild type animals). Alternatively, the virus can simply be spatially targeted by injecting a small quantity into a chosen region of tissue, and in practice both techniques are often used in conjunction. Viral expression systems have now been used in multiple live mammals (including mice, rats, and primates) and in all cases the introduced viral vectors have been well tolerated, the proteins have been highly expressed over long periods of time, and no adverse effects have been reported [317]. Interestingly, this reveals the possibility of the use of optogenetics in human subjects, where it could be applied in retinal prostheses or for the treatment of conditions such as depression (see Sect. 6.6.3). Further discussion of the genetic expression of opsin proteins is beyond the scope of this chapter; however, the interested reader can find more details in [307].

With opsin expression successfully achieved, optogenetics then relies solely on optics for the illumination of specific neurons at chosen times in order to elicit a particular response in the subject. This has been realized with a number of different optical systems using filtered lamps, lasers, or LEDs as light sources and with optical fibers often used for light delivery [307, 313]. Using such systems, many researchers have now reported behavioral control in freely moving animals as well as the study of neuronal function and the causes/effects of multiple neurological conditions (e.g. [308, 317–321]). As such, optogenetics is an exciting field of study that has the potential to provide medical advances as well as vital tools for the study of numerous brain disorders. For this reason, considerable research is now directed towards the production of smaller systems that can be used as implants so that truly freely moving animals can be studied. Such implantable systems may also be useful medically in the future and progress towards these miniature systems is discussed in the following section.

6.6.2 Towards Miniaturized Optogenetic Implants

As mentioned above, optogenetics is a promising tool for the study of many neurological disorders and may also hold potential for the treatment of such conditions in the form of medical implants. In order to carry out optogenetic studies in animals it is necessary to mount or implant an optical system on (or inside) the skull of the animal to allow control of the photosensitized neurons. To monitor the effects of the optogenetic excitation/inhibition of neurons in the most physiologically relevant manner it is desirable to have an optical illumination system that does not impair the movement of the animal in any way. To this end, much work has been undertaken aimed at miniaturizing the optical systems used in optogenetics. Ideally, fully implantable wireless systems should be used where the animal is not tethered to any nearby device and can be studied during normal behavior while moving around freely without any restrictions. Such systems will allow high-fidelity optogenetic studies to be carried out and will also provide a potential route for the production of future optogenetic implants for medical use in humans. Progress towards the development of miniaturized optogenetic implants is discussed below and is also reviewed in detail in [313].

The first report of optogenetic control of intact animal brains was published by Aravanis et al. in 2007 [321]. In this study, motor control in living rodents was stimulated using an intracranial multimode optical fiber coupled to a solid state diode laser. Several similar systems using lasers or LEDs coupled to optical fibers for light delivery were reported subsequently (e.g. [327–329]); however, the drawback of this approach is that it is non-specific due to the diffuse illumination, which activates many surrounding neurons. To improve upon this, many further devices have since been developed that allow selective activation of individual neurons. As with less advanced systems, these use either lasers or LEDs as illumination sources, but have more complex methods of light delivery. One such approach involves the use of optical fibers with sharpened tips (produced through chemical etching techniques), which was reported, for example, in [330]. The thinner tips of these sharpened fibers provide improved spatial resolution and also reduce the risk of tissue damage as the probe is inserted into the brain. Additionally, they permit stimulation of deep-lying neurons and provide flexibility of the chosen illumination depth as the etched optical fiber can be inserted into areas below the tissue surface. The drawback, however, is that this method only allows single-site stimulation. An illustration of an example sharpened optical fiber-based optogenetic probe is shown in Fig. 6.12a, where measurement of the neuronal activity is also provided by a second hollow core fiber filled with NaCl, which acts as an electrode [322].

An illumination modality that can provide multi-site stimulation is the use of micro-waveguide arrays. These have a variety of forms, but usually allow laser light to be coupled into a small waveguide that is inserted at a fixed depth in the brain. In some cases numerous waveguides are incorporated into a single device such that simultaneous (or rapid sequential) stimulation of multiple spatially distinct sites is possible. Examples of micro-waveguide array-based optogenetic devices are presented in [323, 331] and one such system is shown in Fig. 6.12b.

The above approaches—based on the use of sharpened optical fibers or waveguide arrays—have typically used lasers as illumination sources due to their high light intensity and narrow spectral bandwidth. Beyond these, a number of devices have also been reported that use LEDs. An obvious advantage is the small size of LEDs (relative to lasers), which means that they can be incorporated into the implantable device itself, precluding the requirement for optical fibers to deliver light from the illumination source to the optogenetic implant. Combined with the low power consumption of LEDs, this means that they are suitable for integration into fully wireless, untethered devices for applications in freely behaving animals. A number of LED-based implantable optogenetic devices have been presented in the literature, with some early systems using standard commercial LEDs [332–334]. In order to reduce size and improve spatial resolution, micro-scale LEDs (μLEDs) have since been used and the formats of these systems have generally been designed to mimic the features of either Utah-style (e.g. [324–326]) or Michigan-style (e.g. [335, 336]) electrode arrays. Both surface-mounted μLED

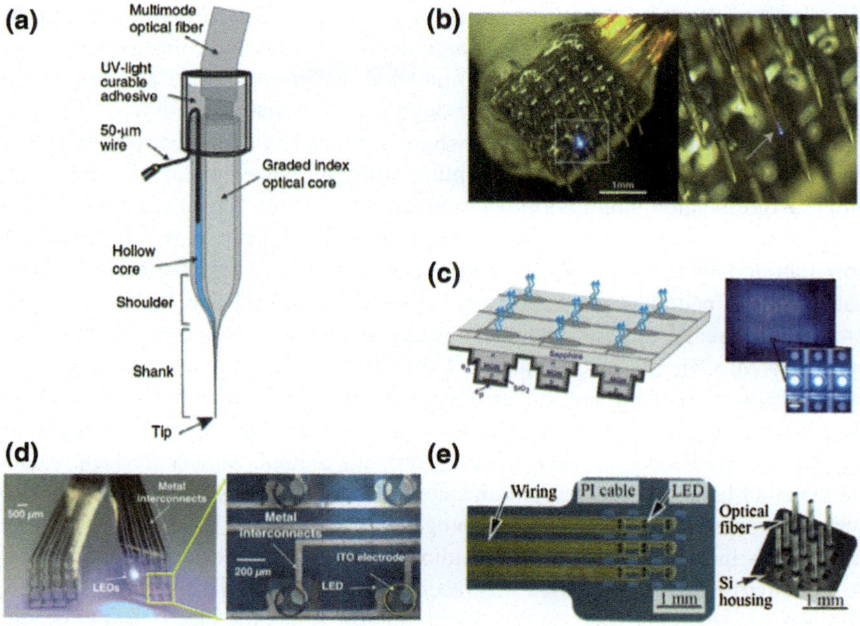

Fig. 6.12 Example implantable stimulation and measurement devices for optogenetics. **a** A dual core optical fiber system with a sharpened tip allowing deep implantation. One fiber core provides optical stimulation while the second is a hollow core fiber that is filled with NaCl. This second fiber core acts as an electrode for measurement of neural activity. Reproduced with permission from Macmillan Publishers Ltd: Nature Methods [322], © 2011. **b** An electrode measurement array in which one electrode has been replaced with a multimode optical fiber to allow optical stimulation. Reproduced with permission from [323]. © IOP Publishing 2012. All rights reserved. **c** A high-density μLED array that permits multi-site surface stimulation. Reproduced with permission from [324]. © IOP Publishing 2010. All rights reserved. **d** A combined μLED and micro-electrode array on a transparent platform allowing both optical stimulation and electrical recording of cortical activity. Reproduced with permission from [325]. © IEEE 2013. **e** A μLED-coupled optical fiber array permitting sub-surface multi-site optical stimulation. Reproduced with permission from [326]. © IEEE 2015

arrays [324, 325] and optical fiber- or waveguide-coupled μLED arrays [326, 337, 338] have been reported, with the latter utilizing short sections of optical fiber or other waveguiding material to permit neural stimulation below the surface of the brain. Importantly, both approaches allow the generation of optical stimulation patterns with high temporal and spatial resolution (on the order of milliseconds and micrometers respectively). Some example implantable μLED-based optogentic devices are shown in Fig. 6.12c–e.

While the techniques described above allow for carefully controlled, multi-site optical stimulation of neurons with high temporal and spatial resolution, there is also a need to read out the effects of the stimulation. For this reason, many of the systems discussed above have also incorporated neural readout electrodes. On the one hand this has involved the addition of measurement electrodes to optogenetic stimulation systems based on sharpened optical fibers [322, 329] (for example, see Fig. 6.12a). Alternatively, micro-waveguides or μLEDs have been incorporated into electrode arrays to allow both stimulation and measurement at multiple sites simultaneously (e.g., [323, 325]). An example system that uses a micro-waveguide for excitation is shown in Fig. 6.12b, where one measurement electrode in an array has been replaced by a multimode optical fiber allowing optical stimulation at a single fixed location along with measurement of multiple surrounding areas [323].

Finally, it is also worth noting that a number of truly wireless optogenetic stimulation devices have been developed, and these can be divided into two distinct categories of neural implants: battery-powered [339–341] and battery-free [342, 343]. As the name suggests, battery-powered neural implants dispense with external powering through the incorporation of a battery in the implant. This provides a straightforward method for the development of entirely untethered devices, but drawbacks include the limited operation time of the implant and the additional weight that the battery imparts to the head of the animal. Battery-free approaches, on the other hand, do not incorporate batteries in the implants and instead rely upon wireless transmission of power. For optogenetic applications this has been achieved using both inductive coupling and radio-frequency (RF) scavenging techniques. A particularly interesting device was recently reported by Park et al. [344] in which wireless RF powering was combined with the use of flexible electronics to produce a soft, stretchable implant that was used for stimulation of the spinal cord or the sciatic nerve in mice. Importantly, these devices remained functional for up to six months after implantation (with at least one use per month). More information on both inductive coupling and RF scavenging can be found in Chap. 7, which discusses progress towards battery-less medical implants and wireless powering telemetries in detail.

Overall, marked progress has been made in recent years on the development of miniature implantable devices for optogenetic stimulation. Alongside this research, much work has also been directed towards the production of more advanced light-sensitive proteins that provide higher transfection efficiencies, improved specificity, or lower activation thresholds. Further discussion of this biological work is beyond the scope of this chapter, but more details can be found in the review published by Fenno et al. [307]. Taken together, these developments have driven significant advances in the field of optogenetics, permitting studies to be carried out in less and less invasive fashions. This research is likely to continue and will provide many useful tools that may be used in biomedical studies of neural function and disease. In the future, this technology may even be used to develop medical devices and/or implants for use in humans. Indeed, work towards this goal is already under way and this is addressed in the following section.

6.6.3 Clinical Optogenetics

The development of optogenetics—particularly miniaturized optogenetic devices suitable for implantation in animals—has raised many interesting questions about the potential use of optogenetics in humans. In terms of neural implants this could entail the treatment of a vast range of disorders, including depression and even substance abuse. However, the development of such medical implants is still only in the conceptual stage. A medical application of optogenetics that has reached a much later stage of development is its use as a form of visual prosthesis, which is aimed at restoring sight in those who suffer from blindness. Importantly, as the eye is transparent, photo-stimulation can be achieved externally, without the need for an implant. Additionally (as discussed in Sect. 6.5.2), in many cases of blindness the photoreceptors in the retina degenerate, causing a complete loss of vision; however, a functional neural link from the retina to the brain is still maintained. In such scenarios, it should be possible to (optically) stimulate those remaining neural links and hence restore the patient's vision.

The basic concept behind the use of optogenetics for visual prostheses—or more specifically, retinal prostheses—would first involve photosensitizing a remaining neural layer in the retina using one of the opsins discussed above. In an ideal case, this alone would lead to complete visual restoration. In reality, however, this may prove very difficult for a number of reasons. Firstly, in a typical case, the patient may have suffered a loss of function in the light-sensitive rods and cones in their retina, meaning that neurons in secondary layers would need to be photosensitized. During normal vision in a healthy subject, a significant degree of retinal processing occurs as signals are transmitted through several layers of nerve cells. Due to this processing, the stimulation that is required in secondary (or higher) retinal nerve layers to produce a specific visual effect will be significantly different to that required in the primary layers (i.e. in the rods and cones). Secondly, the optical powers required for stimulation of most opsin proteins can be quite high (up to several mW mm^{-2}) meaning that ambient light would not be sufficient for photo-stimulation. Thus, an optogenetic retinal prosthesis would most likely also involve the use of a headset or pair of glasses that projects light into the eye in order to stimulate the photosensitized cells.

Considerable research has been undertaken aimed at the development of retinal prostheses based on optogenetics. On the one hand, this has involved studies of the effects of photosensitizing the retinas of animals (typically mice and rats) with degenerative visual conditions. Promising results have been obtained in a number of investigations where the stimulation of action potentials and visual-evoked potentials (VEPs) has been achieved in live, blind animals after transfection with a chosen opsin. The first such study was reported by Bi et al. in 2006 [345], and since then a number of articles have been published discussing similar results in different animals or using different combinations of opsins, promoters, and expression methods [346–349].

Complementary to this research, much work has also been directed towards the development of appropriate display technology for use in optogenetic headsets or glasses. Primarily this entails the production of displays with appropriate resolutions to provide a reasonably complete restoration of vision and with optical powers high enough to permit efficient photo-stimulation of the sensitized retinal cells. The display techniques investigated have included: spatial light modulation (for example, using micro-mirror or liquid crystal arrays) with compact lasers or high-luminance LEDs as light sources; mechanical laser beam scanning; holography, also using laser excitation [350]; and high-power LED arrays [324, 351, 352]. Each has benefits and drawbacks, but the main requirement is that the display can be small enough to be comfortably mounted on a headset or pair of glasses while still providing sufficient resolution and excitation power. Beyond this, an optogenetic headset would also need to incorporate a small camera with which to obtain the visual information to be projected and it would need to provide processing of the recorded image such that the necessary optical stimulation signal can be calculated (i.e. the external components of the prosthesis would be similar to those used in the camera-based visual prostheses described in Sect. 6.5.2).

Of course, research into the development of display systems and transfection techniques for optogenetic retinal prostheses is still under way, and a detailed review of the work carried out to date can be found in [314]. Overall, the use of optogenetics for visual restoration via retinal prosthesis is an exciting field of study with the potential for considerable impact in medicine. Much work is still required before this can be realized though, and this must first involve the evaluation of the safety of virally transfecting human retinas with light-sensitive opsins. Once this work is complete, first-in-human retinal prosthesis trials can begin and these may act as precursors to later investigations into the use of more complex implantable optogenetic systems for neural stimulation in the brain. In the nearer term, however, the true impact of optogenetics is likely to be found not in medical implants and prostheses, but rather in the fundamental biological and neurological studies that it permits. Such investigations have already provided considerable advances in our understanding of a number of diseases (including Parkinson's) and will continue to do so in the future as the optogenetic toolbox is expanded further.

6.7 Conclusions

In this chapter, a variety of optical techniques have been discussed that are being developed with potential future applications as medical implants. The bulk of the chapter is focused on the development of implantable optical sensing modalities that may become useful in disease diagnosis and patient health monitoring. A particularly promising application of miniaturized implantable optical sensors would be in the detection of surgical site infection (SSI), which is a common post-surgical complication that can cause increased patient recovery times and even increased patient mortality [353, 354]. Thus, an implantable or minimally invasive optical device that can provide early detection of SSI would be highly desirable.

Following the discussion of optical diagnostic tools in Sects. 6.3 and 6.4, the fields of both implantable visual prosthetics (6.5) and optogenetics (6.6) were covered. Visual prostheses in current clinical use include IOL implants for the treatment of cataracts and retinal (visual) prostheses for the restoration of sight in patients suffering with RP or AMD. The latter entail electronic devices that are implanted within the eye and provide electrical stimulation of neurons in the visual pathway. While these devices do not provide diagnostic functions, they have been presented in this chapter as they represent an interesting example of an implantable optical technology that is commercially available.

Optogenetics involves the combination of the genetic modification of neurons (such that they express light-sensitive proteins) with the use of photonics in order to permit photo-stimulation of the modified neurons in a minimally invasive manner. Optogenetics is a rapidly growing field that has provided a novel means by which to study neurological and nervous function. For this reason it is becoming widely used within the neuroscience community and the work carried out to date indicates many potential future uses of optogenetics in medicine. These include further optical prostheses for the restoration of vision in the blind—for which considerable work is already underway—as well as the development of more futuristic neural implants for the treatment of illnesses including depression and drug dependency.

When considering the development of diagnostic/therapeutic implants, optogenetics may provide exciting opportunities in the future. However, progress towards this goal can only be expected to take place over a long period of time, as the translation of optogenetics into medical use in humans is likely to face numerous technological and ethical hurdles along the way. The electrical visual prostheses described in Sect. 6.5.2 hold considerable potential for the treatment of blindness, but have not been designed for diagnostic use. On the other hand, the optical imaging and sensing modalities discussed in Sects. 6.3 and 6.4 have a variety of diagnostic applications and are considerably closer to medically deployable formats than optogenetic systems. Indeed, almost all of these methods have been employed for in vivo clinical measurements using optical fibers and some are already available commercially. The commercially available techniques include confocal endomicroscopy and OCT, both of which are finding increasing application owing to their capability to provide in situ diagnoses of a number of medical disorders. Moreover, research continues in these areas and is driven towards the development of smaller, less-invasive devices, including tethered capsule systems, which offer a more comfortable experience for the patient as they undergo endoscopy. Despite this, no long-term (i.e. more than one day) implantable optical systems have yet been reported for use in humans that can provide continuous imaging or sensing capabilities that would permit perpetual monitoring of a patient's state of health.

Any fiber-optic sensing modality has the potential to be deployed as a tethered implant, and many of the techniques discussed in this chapter could be used in this fashion. OCT and endomicroscopy have been incorporated into medical catheters and this could be extended to allow long-term imaging over a number of days (as catheters often remain implanted for days, weeks, or even months). However, such systems provide image data, which is complex and (at least at the present time) typically requires expert analysis before a diagnosis can be made. For this reason, single-point spectroscopic systems—which allow measurement of, for example, diffuse reflectance, fluorescence or Raman spectra—may provide a more feasible route towards the production of tethered, implantable optical sensors. The data acquired with these systems consist of wavelength-resolved intensity measurements that can often provide diagnoses through combination with automated analysis techniques (such as PCA followed by LDA). For this reason they are immediately more compatible with an implant-based format, where it will be impossible for the perpetually recorded data to be analyzed manually. Instead, changes in the spectra corresponding to poor health outcomes could be detected automatically prior to an alert being sent to a clinician.

Many of the spectroscopic techniques described in this chapter have potential applications in this area and all have shown promise in medical diagnostics. For example, diffuse reflectance spectra offer reliable measurements of blood oxygen saturation, which could in turn be used as an indicator of tissue health or viability after surgery. Research has also shown that fluorescence can report on tissue disease state, for example in the detection of pre-cancerous lesions or to monitor infection (albeit with the use of a contrast agent). Raman spectroscopy has also been applied to the detection of various maladies and, importantly, clinical Raman spectra are typically recorded in a label-free manner. In particular, Raman spectra have been shown to effectively detect and discriminate multiple bacteria that are relevant to UTIs, and this capability may be useful in the detection of SSI, which is caused by many of the same bacteria. The development of optical spectroscopic techniques for the detection of bacteria is likely to continue in the future and current research implies that Raman scattering may provide a useful contrast mechanism for this purpose. Once available, such technology could be deployed as a tethered implant and would have the potential to significantly improve patient outcomes. This will be true for monitoring and detecting a number of diseases. SSI, however, will be especially relevant, where an implantable sensor capable of providing early detection would vastly improve patient outcomes by allowing successful treatment of infection using antibiotics rather than requiring repeated additional surgical interventions. Such tethered devices could terminate at a bedside device kept near to the patient, although this would impair mobility. A more desirable format could incorporate miniaturized detection optics into a patch that could be worn by the patient, thus minimizing the impact on everyday activity.

While tethered optical implants like those described above have the potential to provide significant improvements in healthcare, they clearly do not offer an optimum solution, simply due to their tethered format. In an ideal case, a patient should not be attached to any external device, whether portable or not. To this end, the development of fully implantable medical sensors is under way, and this involves optical systems as well as a variety of alternative sensing modalities (some of which are discussed in other chapters of this book). Within optogenetics much research has been undertaken into the development of optical implants, with the main aim being to develop systems suitable for implantation in small animals to allow free behavior during neurological research. This progress is also relevant to the development of fully implantable, untethered optical sensors for healthcare and a small number of devices of this sort have indeed been reported. For example, integrated chips that allow measurements of tissue oxygenation and blood pressure have been developed, and an ingestible capsule for the detection of GI bleeding was also launched recently (see Sect. 6.4.2 for further details). Several technical challenges remain, however, and these include the questions of biocompatibility and the production of battery-free wireless sensors that can be powered externally (also discussed elsewhere in this book). Nonetheless, the development of implantable optical systems for applications in medicine—for both sensing and treatment—will certainly continue. Importantly, this is likely to lead to many exciting medical advances in the future, providing improvements in disease detection rates and reducing patient morbidity and mortality.

Overall, optical sensing modalities offer minimally invasive diagnosis of a number of diseases. Combined with the miniaturization of optical systems, this provides an opportunity for the development of implantable devices that have the potential to revolutionize certain areas of healthcare. In particular, the treatment of SSI would benefit significantly from the development of novel and improved detection modalities that allow continuous patient monitoring via an implant. Early devices are likely to involve tethered formats, where optical fibers will be used for light delivery and collection, with the external optics contained within a wearable patch. Beyond this, the development of fully implantable wireless systems will also continue, with important current research challenges including the questions of biocompatibility, biofouling, and, of course, sufficient miniaturization. Both tethered devices and fully implantable systems offer the prospect of radical advances in medicine and their development will almost certainly bring about significant changes. In conclusion, implantable optical sensing is an exciting field of study, which will attract considerable attention in the coming years. It is likely to witness further advances and to undergo substantial expansion in the near future, and this will doubtless lead to the development of a range of novel devices that can provide vast improvements in the field of healthcare.

References

1. C.M. Chen, R. Kwasnicki, B. Lo, G.Z. Yang, Wearable tissue oxygenation monitoring sensor and a forearm vascular phantom design for data validation, in *2014 11th International Conference on Wearable and Implantable Body Sensor Networks* (*BSN*) (2014), pp. 64–68
2. J. Fiala, P. Bingger, D. Ruh, K. Foerster, C. Heilmann, F. Beyersdorf, H. Zappe, A. Seifert, An implantable optical blood pressure sensor based on pulse transit time. Biomed. Microdev. **15**(1), 73–81 (2013)
3. M. Theodor, D. Ruh, J. Fiala, K. Forster, C. Heilmann, Y. Manoli, F. Beyersdorf, H. Zappe, A. Seifert, Subcutaneous blood pressure monitoring with an implantable optical sensor. Biomed. Microdev. **15**(5), 811–820 (2013)

4. P.C.A. Jeronimo, A.N. Araujo, M. Montenegro, Optical sensors and biosensors based on sol-gel films. Talanta **72**(1), 13–27 (2007)
5. G. Iddan, G. Meron, A. Glukhovsky, P. Swain, Wireless capsule endoscopy. Nature **405** (6785), 417 (2000)
6. M.J. Gora, J.S. Sauk, R.W. Carruth, K.A. Gallagher, M.J. Suter, N.S. Nishioka, L.E. Kava, M. Rosenberg, B.E. Bouma, G.J. Tearney, Tethered capsule endomicroscopy enables less invasive imaging of gastrointestinal tract microstructure. Nat. Med. **19**(2), 238–240 (2013)
7. L. Scolaro, D. Lorenser, W.-J. Madore, R.W. Kirk, A.S. Kramer, G.C. Yeoh, N. Godbout, D. D. Sampson, C. Boudoux, R.A. McLaughlin, Molecular imaging needles: dual-modality optical coherence tomography and fluorescence imaging of labeled antibodies deep in tissue. Biomed. Opt. Express **6**(5), 1767–1781 (2015)
8. G. Berci, K.A. Forde, History of endoscopy: what lessons have we learned from the past? Surg. Endosc. **14**(1), 5–15 (2000)
9. F. Helmchen, Miniaturization of fluorescence microscopes using fibre optics. Exp. Physiol. **87**(6), 737–745 (2002)
10. T. Wilson, Confocal microscopy, in *Confocal Microscopy*, ed. by T. Wilson (Academic Press Limited, London, 1990), pp. 1–60
11. T. Wilson, Resolution and optical sectioning in the confocal microscope. J. Microsc. **244**(2), 113–121 (2011)
12. A.F. Gmitro, D. Aziz, Confocal microscopy through a fiberoptic imaging bundle. Opt. Lett. **18**(8), 565–567 (1993)
13. D.L. Dickensheets, G.S. Kino, Micromachined scanning confocal optical microscope. Opt. Lett. **21**(10), 764–766 (1996)
14. F. Helmchen, M.S. Fee, D.W. Tank, W. Denk, A miniature head-mounted two-photon microscope. High-resolution brain imaging in freely moving animals. Neuron **31**(6), 903–912 (2001)
15. A. Polglase, W. McLaren, S. Skinner, R. Kiesslich, M. Neurath, P. Delaney, A fluorescence confocal endomicroscope for in vivo microscopy of the upper- and the lower-GI tract. Gastrointest. Endosc. **62**(5), 686–695 (2005)
16. W. Gobel, J.N. Kerr, A. Nimmerjahn, F. Helmchen, Miniaturized two-photon microscope based on a flexible coherent fiber bundle and a gradient-index lens objective. Opt. Lett. **29** (21), 2521–2523 (2004)
17. M.T. Myaing, D.J. MacDonald, X. Li, Fiber-optic scanning two-photon fluorescence endoscope. Opt. Lett. **31**(8), 1076–1078 (2006)
18. S. Tang, W. Jung, D. McCormick, T. Xie, J. Su, Y.C. Ahn, B.J. Tromberg, Z. Chen, Design and implementation of fiber-based multiphoton endoscopy with microelectromechanical systems scanning. J. Biomed. Opt. **14**(3), 034005 (2009)
19. H. Bao, M. Gu, Reduction of self-phase modulation in double-clad photonic crystal fiber for nonlinear optical endoscopy. Opt. Lett. **34**(2), 148–150 (2009)
20. F. Jean, G. Bourg-Heckly, B. Viellerobe, Fibered confocal spectroscopy and multicolor imaging system for in vivo fluorescence analysis. Opt. Express **15**(7), 4008–4017 (2007)
21. G.T. Kennedy, H.B. Manning, D.S. Elson, M.A.A. Neil, G.W. Stamp, B. Viellerobe, F. Lacombe, C. Dunsby, P.M.W. French, A fluorescence lifetime imaging scanning confocal endomicroscope. J. Biophoton. **3**(1–2), 103–107 (2010)
22. G.O. Fruhwirth, S. Ameer-Beg, R. Cook, T. Watson, T. Ng, F. Festy, Fluorescence lifetime endoscopy using TCSPC for the measurement of fret in live cells. Opt. Express **18**(11), 11148–11158 (2010)
23. A.J. Thompson, C. Paterson, M.A. Neil, C. Dunsby, P.M. French, Adaptive phase compensation for ultracompact laser scanning endomicroscopy. Opt. Lett. **36**(9), 1707–1709 (2011)
24. T. Cizmar, M. Mazilu, K. Dholakia, In situ wavefront correction and its application to micromanipulation. Nat. Photonics **4**(6), 388–394 (2010)
25. M. Ploschner, T. Tyc, T. Cizmar, Seeing through chaos in multimode fibres. Nat. Photonics **9**(8), 529 (2015)

26. Y. Kim, S.C. Warren, J.M. Stone, J.C. Knight, M.A.A. Neil, C. Paterson, C.W. Dunsby, P. M.W. French, Adaptive multiphoton endomicroscope incorporating a polarization-maintaining multicore optical fibre. IEEE J. Sel. Top. Quantum Electron. **11** (99), 1 (2015)
27. N. Bedard, T.S. Tkaczyk, Snapshot spectrally encoded fluorescence imaging through a fiber bundle. J. Biomed. Opt. **17**(8) (2012)
28. M. Kyrish, R. Kester, R. Richards-Kortum, T. Tkaczyk, Improving spatial resolution of a fiber bundle optical biopsy system. Endoscopic Microsc. V, **7558** (2010)
29. Y.S. Sabharwal, A.R. Rouse, L. Donaldson, M.F. Hopkins, A.F. Gmitro, Slit-scanning confocal microendoscope for high-resolution in vivo imaging. Appl. Opt. **38**(34), 7133–7144 (1999)
30. A. Rouse, A. Gmitro, Multispectral imaging with a confocal microendoscope. Opt. Lett. **25** (23), 1708–1710 (2000)
31. M. Hughes, G.Z. Yang, High speed, line-scanning, fiber bundle fluorescence confocal endomicroscopy for improved mosaicking. Biomed. Opt. Express **6**(4), 1241–1252 (2015)
32. H. Neumann, F.S. Fuchs, M. Vieth, R. Atreya, J. Siebler, R. Kiesslich, M.F. Neurath, Review article: in vivo imaging by endocytoscopy. Aliment. Pharmacol. Ther. **33**(11), 1183–1193 (2011)
33. M. Hughes, T.P. Chang, G.-Z. Yang, Fiber bundle endocytoscopy. Biomed. Opt. Express **4** (12), 2781–2794 (2013)
34. J.S. Louie, R. Richards-Kortum, S. Anandasabapathy, Applications and advancements in the use of high-resolution microendoscopy for detection of gastrointestinal neoplasia. Clin. Gastroenterol. Hepatol. **12**(11), 1789–1792 (2014)
35. M.-A. Protano, H. Xu, G. Wang, A.D. Polydorides, S.M. Dawsey, J. Cui, L. Xue, F. Zhang, T. Quang, M.C. Pierce, D. Shin, R.A. Schwarz, M.S. Bhutani, M. Lee, N. Parikh, C. Hur, W. Xu, E. Moshier, J. Godbold, J. Mitcham, C. Hudson, R.R. Richards-Kortum, S. Anandasabapathy, Low-cost high-resolution microendoscopy for the detection of esophageal squamous cell neoplasia: an international trial. Gastroenterology **149**(2), 321–329 (2015)
36. P.A. Keahey, T.S. Tkaczyk, K.M. Schmeler, R.R. Richards-Kortum, Optimizing modulation frequency for structured illumination in a fiber-optic microendoscope to image nuclear morphometry in columnar epithelium. Biomed. Opt. Express **6**(3), 870–880 (2015)
37. G.J. Tearney, R.H. Webb, B.E. Bouma, Spectrally encoded confocal microscopy. Opt. Lett. **23**(15), 1152–1154 (1998)
38. C. Boudoux, S.H. Yun, W.Y. Oh, W.M. White, N.V. Iftimia, M. Shishkov, B.E. Bouma, G. J. Tearney, Rapid wavelength-swept spectrally encoded confocal microscopy. Opt. Express **13**(20), 8214–8221 (2005)
39. D. Kang, R.W. Carruth, M. Kim, S.C. Schlachter, M. Shishkov, K. Woods, N. Tabatabaei, T. Wu, G.J. Tearney, Endoscopic probe optics for spectrally encoded confocal microscopy. Biomed. Opt. Express **4**(10), 1925–1936 (2013)
40. A. Zeidan, D. Yelin, Miniature forward-viewing spectrally encoded endoscopic probe. Opt. Lett. **39**(16), 4871–4874 (2014)
41. M. Abbaci, I. Breuskin, O. Casiraghi, F. De Leeu, M. Ferchiou, S. Temam, C. Laplace-Builhe, Confocal laser endomicroscopy for non-invasive head and neck cancer imaging: a comprehensive review. Oral Oncol. **50**(8), 711–716 (2014)
42. A. Gupta, B.M. Attar, P. Koduru, A.R. Murali, B.T. Go, R. Agarwal, Utility of confocal laser endomicroscopy in identifying high-grade dysplasia and adenocarcinoma in Barrett's esophagus: a systematic review and meta-analysis. Eur. J. Gastroenterol. Hepatol. **26**(4), 369–377 (2014)
43. F. Salvatori, S. Siciliano, F. Maione, D. Esposito, S. Masone, M. Persico, G.D. De Palma, Confocal laser endomicroscopy in the study of colonic mucosa in IBD patients: a review. Gastroenterol. Res. Pract. (2012)
44. P. Su, Y. Liu, S. Lin, K. Xiao, P. Chen, S. An, J. He, Y. Bai, Efficacy of confocal laser endomicroscopy for discriminating colorectal neoplasms from non-neoplasms: a systematic review and meta-analysis. Colorectal Dis. **15**(1), E1–E12 (2013)

45. V. Subramanian, K. Ragunath, Advanced endoscopic imaging: a review of commercially available technologies. Clin. Gastroenterol. Hepatol. **12**(3), 368 (2014)
46. P.S.-P. Thong, M. Olivo, S.S. Tandjung, M.M. Movania, F. Lin, K. Qian, H.-S. Seah, K.-C. Soo, Review of confocal fluorescence endomicroscopy for cancer detection. IEEE J. Sel. Top. Quantum Electron. **18**(4), 1355–1366 (2012)
47. A.H. Zehri, W. Ramey, J.F. Georges, M.A. Mooney, N.L. Martirosyan, M.C. Preul, P. Nakaji, Neurosurgical confocal endomicroscopy: a review of contrast agents, confocal systems, and future imaging modalities. Surg. Neurol. Int. **5**, 60 (2014)
48. R. Kiesslich, J. Burg, M. Vieth, J. Gnaendiger, M. Enders, P. Delaney, A. Polglase, W. McLaren, D. Janell, S. Thomas, B. Nafe, P.R. Galle, M.F. Neurath, Confocal laser endoscopy for diagnosing intraepithelial neoplasias and colorectal cancer in vivo. Gastroenterology **127**(3), 706–713 (2004)
49. R. Kiesslich, M. Goetz, K. Lammersdorf, C. Schneider, J. Burg, M. Stolte, M. Vieth, B. Nafe, P.R. Galle, M.F. Neurath, Chromoscopy-guided endomicroscopy increases the diagnostic yield of intraepithelial neoplasia in ulcerative colitis. Gastroenterology **132**(3), 874–882 (2007)
50. R. Kiesslich, L. Gossner, M. Goetz, A. Dahlmann, M. Vieth, M. Stolte, A. Hoffman, M. Jung, B. Nafe, P.R. Galle, M.F. Neurath, In vivo histology of Barrett's esophagus and associated neoplasia by confocal laser endomicroscopy. Clin. Gastroenterol. Hepatol. **4**(8), 979–987 (2006)
51. A.M. Buchner, M.W. Shahid, M.G. Heckman, M. Krishna, M. Ghabril, M. Hasan, J.E. Crook, V. Gomez, M. Raimondo, T. Woodward, H.C. Wolfsen, M.B. Wallace, Comparison of probe-based confocal laser endomicroscopy with virtual chromoendoscopy for classification of colon polyps. Gastroenterology **138**(3), 834–842 (2010)
52. T. Kuiper, F.J.C. van den Broek, S. van Eeden, P. Fockens, E. Dekker, Feasibility and accuracy of confocal endomicroscopy in comparison with narrow-band imaging and chromoendoscopy for the differentiation of colorectal lesions. Am. J. Gastroenterol. **107**(4), 543–550 (2012)
53. V. Gomez, A.M. Buchner, E. Dekker, F.J.C. van den Broek, A. Meining, M.W. Shahid, M. S. Ghabril, P. Fockens, M.G. Heckman, M.B. Wallace, Interobserver agreement and accuracy among international experts with probe-based confocal laser endomicroscopy in predicting colorectal neoplasia. Endoscopy **42**(4), 286–291 (2010)
54. K.B. Dunbar, P. Okolo III, E. Montgomery, M.I. Canto, Confocal laser endomicroscopy in Barrett's esophagus and endoscopically inapparent Barrett's neoplasia: a prospective, randomized, double-blind, controlled, crossover trial. Gastrointest. Endosc. **70**(4), 645–654 (2009)
55. M. Bajbouj, M. Vieth, T. Roesch, S. Miehlke, V. Becker, M. Anders, H. Pohl, A. Madisch, T. Schuster, R.M. Schmid, A. Meining, Probe-based confocal laser endomicroscopy compared with standard four-quadrant biopsy for evaluation of neoplasia in Barrett's esophagus. Endoscopy **42**(6), 435–440 (2010)
56. P. Sharma, A.R. Meining, E. Coron, C.J. Lightdale, H.C. Wolfsen, A. Bansal, M. Bajbouj, J.-P. Galiniche, J.A. Abrams, A. Rastogi, N. Gupta, J.E. Michalek, G.Y. Lauwers, M.B. Wallace, Real-time increased detection of neoplastic tissue in Barrett's esophagus with probe-based confocal laser endomicroscopy: final results of an international multicenter, prospective, randomized, controlled trial. Gastrointest. Endosc. **74**(3), 465–472 (2011)
57. M.B. Wallace, P. Sharma, C. Lightdale, H. Wolfsen, E. Coron, A. Buchner, M. Bajbouj, A. Bansal, A. Rastogi, J. Abrams, J.E. Crook, A. Meining, Preliminary accuracy and interobserver agreement for the detection of intraepithelial neoplasia in Barrett's esophagus with probe-based confocal laser endomicroscopy. Gastrointest. Endosc. **72**(1), 19–24 (2010)
58. F.J.C. van den Broek, J.A. van Es, S. van Eeden, P.C.F. Stokkers, C.Y. Ponsioen, J.B. Reitsma, P. Fockens, E. Dekker, Pilot study of probe-based confocal laser endomicroscopy during colonoscopic surveillance of patients with longstanding ulcerative colitis. Endoscopy **43**(2), 116–122 (2011)

59. H. Neumann, M. Vieth, R. Atreya, M. Grauer, J. Siebler, T. Bernatik, M.F. Neurath, J. Mudter, Assessment of Crohn's disease activity by confocal laser endomicroscopy. Inflammat. Bowel Dis. **18**(12), 2261–2269 (2012)
60. C.-Q. Li, X.-J. Xie, T. Yu, X.-M. Gu, X.-L. Zuo, C.-J. Zhou, W.-Q. Huang, H. Chen, Y.-Q. Li, Classification of inflammation activity in ulcerative colitis by confocal laser endomicroscopy. Am. J. Gastroenterol. **105**(6), 1391–1396 (2010)
61. B. Andre, T. Vercauteren, A.M. Buchner, M. Krishna, N. Ayache, M.B. Wallace, Software for automated classification of probe-based confocal laser endomicroscopy videos of colorectal polyps. World J. Gastroenterol. **18**(39), 5560–5569 (2012)
62. S. Gaddam, S.C. Mathur, M. Singh, J. Arora, S.B. Wani, N. Gupta, A. Overhiser, A. Rastogi, V. Singh, N. Desai, S.B. Hall, A. Bansal, P. Sharma, Novel probe-based confocal laser endomicroscopy criteria and interobserver agreement for the detection of dysplasia in Barrett's esophagus. Am. J. Gastroenterol. **106**(11), 1961–1969 (2011)
63. H. Neumann, C. Langner, M.F. Neurath, M. Vieth, Confocal laser endomicroscopy for diagnosis of Barrett's esophagus. Front. Oncol. **2**, 42 (2012)
64. I. Smith, P.E. Kline, M. Gaidhane, M. Kahaleh, A review on the use of confocal laser endomicroscopy in the bile duct. Gastroenterol. Res. Pract. (2012)
65. M. Behbahaninia, N.L. Martirosyan, J. Georges, J.A. Udovich, M.Y.S. Kalani, B.G. Feuerstein, P. Nakaji, R.F. Spetzler, M.C. Preul, Intraoperative fluorescent imaging of intracranial tumors: a review. Clin. Neurol. Neurosurg. **115**(5), 517–528 (2013)
66. F. Acerbi, C. Cavallo, M. Broggi, R. Cordella, E. Anghileri, M. Eoli, M. Schiariti, G. Broggi, P. Ferroli, Fluorescein-guided surgery for malignant gliomas: a review. Neurosurg. Rev. **37**(4), 547–555 (2014)
67. C. Ell, S. Remke, A. May, L. Helou, R. Henrich, G. Mayer, The first prospective controlled trial comparing wireless capsule endoscopy with push enteroscopy in chronic gastrointestinal bleeding. Endoscopy **34**(9), 685–689 (2002)
68. B.S. Lewis, P. Swain, Capsule endoscopy in the evaluation of patients with suspected small intestinal bleeding: results of a pilot study. Gastrointest. Endosc. **56**(3), 349–353 (2002)
69. M. Mylonaki, A. Fritscher-Ravens, P. Swain, Wireless capsule endoscopy: a comparison with push enteroscopy in patients with gastroscopy and colonoscopy negative gastrointestinal bleeding. Gut **52**(8), 1122–1126 (2003)
70. M. Pennazio, R. Santucci, E. Rondonotti, C. Abbiati, G. Beccari, F.P. Rossini, R. de Franchis, Outcome of patients with obscure gastrointestinal bleeding after capsule endoscopy: report of 100 consecutive cases. Gastroenterology **126**(3), 643–653 (2004)
71. N. Tabatabaei, D. Kang, T. Wu, M. Kim, R.W. Carruth, J. Leung, J.S. Sauk, W. Shreffler, Q. Yuan, A. Katz, N.S. Nishioka, G.J. Tearney, Tethered confocal endomicroscopy capsule for diagnosis and monitoring of eosinophilic esophagitis. Biomed. Opt. Express **5**(1), 197–207 (2014)
72. D. Huang, E.A. Swanson, C.P. Lin, J.S. Schuman, W.G. Stinson, W. Chang, M.R. Hee, T. Flotte, K. Gregory, C.A. Puliafito, J.G. Fujimoto, Optical coherence tomography. Science **254**(5035), 1178–1181 (1991)
73. B. Cense, N.A. Nassif, T. Chen, M. Pierce, S.H. Yun, B.H. Park, B.E. Bouma, G.J. Tearney, J.F. de Boer, Ultrahigh-resolution high-speed retinal imaging using spectral-domain optical coherence tomography. Opt. Express **12**(11), 2435–2447 (2004)
74. M.A. Choma, M.V. Sarunic, C.H. Yang, J.A. Izatt, Sensitivity advantage of swept source and Fourier domain optical coherence tomography. Opt. Express **11**(18), 2183–2189 (2003)
75. S.H. Yun, G.J. Tearney, B.E. Bouma, B.H. Park, J.F. de Boer, High-speed spectral-domain optical coherence tomography at 1.3 μm wavelength. Opt. Express **11**(26), 3598–3604 (2003)
76. P.H. Tomlins, R.K. Wang, Theory, developments and applications of optical coherence tomography. J. Phys. D Appl. Phys. **38**(15), 2519–2535 (2005)
77. T. Alasil, P.A. Keane, D.A. Sim, A. Tufail, M.E. Rauser, Optical coherence tomography in pediatric ophthalmology: current roles and future directions. Ophthalm. Surg. Lasers Imaging Retina **44**(6), S19–S29 (2013)

78. J.P. Ehlers, Y.K. Tao, S.K. Srivastava, The value of intraoperative optical coherence tomography imaging in vitreoretinal surgery. Curr. Opin. Ophthalmol. **25**(3), 221–227 (2014)
79. T. Garcia, A. Tourbah, E. Setrouk, A. Ducasse, C. Arndt, Optical coherence tomography in neuro-ophthalmology. J. Francais D: Ophtalmol. **35**(6), 454–466 (2012)
80. B.E. Goldhagen, M.T. Bhatti, P.P. Srinivasan, S.J. Chiu, S. Farsiu, M.A. El-Dairi, Retinal atrophy in eyes with resolved papilledema detected by optical coherence tomography. J. Neuro-Ophthalmol. **35**(2), 122–126 (2015)
81. G.J. Jaffe, J. Caprioli, Optical coherence tomography to detect and manage retinal disease and glaucoma. Am. J. Ophthalmol. **137**(1), 156–169 (2004)
82. N.A. Nassif, B. Cense, B.H. Park, M.C. Pierce, S.H. Yun, B.E. Bouma, G.J. Tearney, T.C. Chen, J.F. de Boer, In vivo high-resolution video-rate spectral-domain optical coherence tomography of the human retina and optic nerve. Opt. Express **12**(3), 367–376 (2004)
83. G. Virgili, F. Menchini, G. Casazza, R. Hogg, R. R. Das, X. Wang, M. Michelessi, Optical coherence tomography (OCT) for detection of macular oedema in patients with diabetic retinopathy. Cochrane Database Syst. Rev. **1** (2015)
84. M.E. Brezinski, G.J. Tearney, N.J. Weissman, S.A. Boppart, B.E. Bouma, M.R. Hee, A.E. Weyman, E.A. Swanson, J.F. Southern, J.G. Fujimoto, Assessing atherosclerotic plaque morphology: comparison of optical coherence tomography and high frequency intravascular ultrasound. Heart **77**(5), 397–403 (1997)
85. A.H. Chau, R.C. Chan, M. Shishkov, B. MacNeill, N. Iftimiia, G.J. Tearney, R.D. Kamm, B. E. Bouma, M.R. Kaazempur-Mofrad, Mechanical analysis of atherosclerotic plaques based on optical coherence tomography. Ann. Biomed. Eng. **32**(11), 1494–1503 (2004)
86. G. Ferrante, P. Presbitero, R. Whitbourn, P. Barlis, Current applications of optical coherence tomography for coronary intervention. Int. J. Cardiol. **165**(1), 7–16 (2013)
87. I.K. Jang, B.E. Bouma, D.H. Kang, S.J. Park, S.W. Park, K.B. Seung, K.B. Choi, M. Shishkov, K. Schlendorf, E. Pomerantsev, S.L. Houser, H.T. Aretz, G.J. Tearney, Visualization of coronary atherosclerotic plaques in patients using optical coherence tomography: comparison with intravascular ultrasound. J. Am. College Cardiol. **39**(4), 604–609 (2002)
88. O.C. Raffel, G.J. Tearney, D.D. Gauthier, E.F. Halpern, B.E. Bouma, I.-K. Jang, Relationship between a systemic inflammatory marker, plaque inflammation, and plaque characteristics determined by intravascular optical coherence tomography. Arterioscler. Thromb. Vasc. Biol. **27**(8), 1820–1827 (2007)
89. T. Roleder, J. Jakala, G.L. Kaluza, L. Partyka, K. Proniewska, E. Pociask, W. Zasada, W. Wojakowski, Z. Gasior, D. Dudek, The basics of intravascular optical coherence tomography. Postepy W Kardiologii Interwencyjnej **11**(2), 74–83 (2015)
90. A. Tanaka, G. J. Tearney, B.E. Bouma, Challenges on the frontier of intracoronary imaging: atherosclerotic plaque macrophage measurement by optical coherence tomography. J. Biomed. Opt. **15**(1) (2010)
91. G.J. Tearney, S. Waxman, M. Shishkov, B.J. Vakoc, M.J. Suter, M.I. Freilich, A.E. Desjardins, W.-Y. Oh, L.A. Bartlett, M. Rosenberg, B.E. Bouma, Three-dimensional coronary artery microscopy by intracoronary optical frequency domain imaging. JACC-Cardiovasc. Imag. **1**(6), 752–761 (2008)
92. G.J. Tearney, H. Yabushita, S.L. Houser, H.T. Aretz, I.K. Jang, K.H. Schlendorf, C.R. Kauffman, M. Shishkov, E.F. Halpern, B.E. Bouma, Quantification of macrophage content in atherosclerotic plaques by optical coherence tomography. Circulation **107**(1), 113–119 (2003)
93. L. Vignali, E. Solinas, E. Emanuele, Research and clinical applications of optical coherence tomography in invasive cardiology: a review. Curr. Cardiol. Rev. **10**(4), 369–376 (2014)
94. H. Yabushita, B.E. Bourna, S.L. Houser, T. Aretz, I.K. Jang, K.H. Schlendorf, C.R. Kauffman, M. Shishkov, D.H. Kang, E.F. Halpern, G.J. Tearney, Characterization of human atherosclerosis by optical coherence tomography. Circulation **106**(13), 1640–1645 (2002)

95. S. Brand, J.M. Poneros, B.E. Bouma, G.J. Tearney, C.C. Compton, N.S. Nishioka, Optical coherence tomography in the gastrointestinal tract. Endoscopy **32**(10), 796–803 (2000)
96. J.A. Evans, B.E. Bouma, J. Bressner, M. Shishkov, G.Y. Lauwers, M. Mino-Kenudson, N.S. Nishioka, G.J. Tearney, Identifying intestinal metaplasia at the squamocolumnar junction by using optical coherence tomography. Gastrointest. Endosc. **65**(1), 50–56 (2007)
97. C.-P. Liang, C.-W. Chen, J. Wierwille, J. Desai, R. Gullapalli, R. Mezrich, C.-M. Tang, Y. Chen, Endoscopic microscopy using optical coherence tomography. Curr. Med. Imag. Rev. **8**(3), 174–193 (2012)
98. M.J. Suter, B.J. Vakoc, P.S. Yachimski, M. Shishkov, G.Y. Lauwers, M. Mino-Kenudson, B.E. Bouma, N.S. Nishioka, G.J. Tearney, Comprehensive microscopy of the esophagus in human patients with optical frequency domain imaging. Gastrointest. Endosc. **68**(4), 745–753 (2008)
99. G.J. Tearney, M.E. Brezinski, J.F. Southern, B.E. Bouma, S.A. Boppart, J.G. Fujimoto, Optical biopsy in human gastrointestinal tissue using optical coherence tomography. Am. J. Gastroenterol. **92**(10), 1800–1804 (1997)
100. B.J. Vakoc, M. Shishko, S.H. Yun, W.-Y. Oh, M.J. Suter, A.E. Desjardins, J.A. Evans, N.S. Nishioka, G.J. Tearney, B.E. Bouma, Comprehensive esophageal microscopy by using optical frequency-domain imaging (with video). Gastrointest. Endosc. **65**(6), 898–905 (2007)
101. S.A. Coulman, J.C. Birchall, A. Alex, M. Pearton, B. Hofer, C. O'Mahony, W. Drexler, B. Povazay, In vivo, in situ imaging of microneedle insertion into the skin of human volunteers using optical coherence tomography. Pharma. Res. **28**(1), 66–81 (2011)
102. T. Gambichler, V. Jaedicke, S. Terras, Optical coherence tomography in dermatology: technical and clinical aspects. Arch. Dermatol. Res. **303**(7), 457–473 (2011)
103. T. Gambichler, A. Pljakic, L. Schmitz, Recent advances in clinical application of optical coherence tomography of human skin. Clin. Cosmet. Investig. Dermatol. **8**, 345–354 (2015)
104. J. Welzel, Optical coherence tomography in dermatology: a review. Skin Res. Tech. **7**(1), 1–9 (2001)
105. A.S. Fjeldstad, N.G. Carlson, J.W. Rose, Optical coherence tomography as a biomarker in multiple sclerosis. Expert Opinion Med. Diagnost. **6**(6), 593–604 (2012)
106. L.M. Simao, The contribution of optical coherence tomography in neurodegenerative diseases. Curr. Opinion Ophthalmol. **24**(6), 521–527 (2013)
107. A. Vidal-Jordana, J. Sastre-Garriga, X. Montalban, Optical coherence tomography in multiple sclerosis. Rev. Neurol. **54**(9), 556–563 (2012)
108. A. Vinekar, S. Mangalesh, C. Jayadev, R.S. Maldonado, N. Bauer, C.A. Toth, Retinal imaging of infants on spectral domain optical coherence tomography. Biomed. Res. Int. (2015)
109. K.A. Townsend, G. Wollstein, J.S. Schuman, Imaging of the retinal nerve fibre layer for glaucoma. Br. J. Ophthalmol. **93**(2), 139–143 (2009)
110. G.J. Tearney, OCT imaging of macrophages a bright spot in the study of inflammation in human atherosclerosis. JACC Cardiovasc. Imag. **8**(1), 73–75 (2015)
111. K. Toutouzas, Y.S. Chatzizisis, M. Riga, A. Giannopoulos, A.P. Antoniadis, S. Tu, Y. Fujino, D. Mitsouras, C. Doulaverakis, I. Tsampoulatidis, V.G. Koutkias, K. Bouki, Y. Li, I. Chouvarda, G. Cheimariotis, N. Maglaveras, I. Kompatsiaris, S. Nakamura, J.H.C. Reiber, F. Rybicki, H. Karvounis, C. Stefanadis, D. Tousoulis, G.D. Giannoglou, Accurate and reproducible reconstruction of coronary arteries and endothelial shear stress calculation using 3D OCT: comparative study to 3D IVUS and 3D QCA. Atherosclerosis **240**(2), 510–519 (2015)
112. R.A. McLaughlin, B.C. Quirk, A. Curatolo, R.W. Kirk, L. Scolaro, D. Lorenser, P.D. Robbins, B.A. Wood, C.M. Saunders, D.D. Sampson, Imaging of breast cancer with optical coherence tomography needle probes: feasibility and initial results. IEEE J. Sel. Top. Quantum Electron. **18**(3), 1184–1191 (2012)
113. C. Zhou, D.W. Cohen, Y. Wang, H.-C. Lee, A.E. Mondelblatt, T.-H. Tsai, A.D. Aguirre, J. G. Fujimoto, J.L. Connolly, Integrated optical coherence tomography and microscopy for

ex vivo multiscale evaluation of human breast tissues. Cancer Res. **70**(24), 10071–10079 (2010)

114. A.M. Zysk, S.A. Boppart, Computational methods for analysis of human breast tumor tissue in optical coherence tomography images. J. Biomed. Opt. **11**(5) (2006)
115. B.G. Muller, D.M. de Bruin, W. van den Bos, M.J. Brandt, J.F. Velu, M.T.J. Bus, D. J. Faber, D. Savci, P.J. Zondervan, T.M. de Reijke, P.L. Pes, J. de la Rosette, T.G. van Leeuwen, Prostate cancer diagnosis: the feasibility of needle-based optical coherence tomography. J. Med Imag. (Bellingham, Wash.) **2**(3), 037501 (2015)
116. M.T.J. Bus, D.M. de Bruin, D.J. Faber, G.M. Kamphuis, P.J. Zondervan, M.P.L. Pes, T.M. de Reijke, O. Traxer, T.G. van Leeuwen, J.J.M.C.H. de la Rosette, Optical diagnostics for upper urinary tract urothelial cancer: technology, thresholds, and clinical applications. J. Endourol. **29**(2), 113–123 (2015)
117. J.M. Poneros, S. Brand, B.E. Bouma, G.J. Tearney, C.C. Compton, N.S. Nishioka, Diagnosis of specialized intestinal metaplasia by optical coherence tomography. Gastroenterology **120**(1), 7–12 (2001)
118. H.G. Bezerra, M.A. Costa, G. Guagliumi, A.M. Rollins, D.I. Simon, Intracoronary optical coherence tomography: a comprehensive review clinical and research applications. JACC Cardiovasc. Interven. **2**(11), 1035–1046 (2009)
119. G.J. Tearney, M.E. Brezinski, B.E. Bouma, S.A. Boppart, C. Pitris, J.F. Southern, J.G. Fujimoto, In vivo endoscopic optical biopsy with optical coherence tomography. Science **276**(5321), 2037–2039 (1997)
120. J. Ha, H. Yoo, G.J. Tearney, B.E. Bouma, Compensation of motion artifacts in intracoronary optical frequency domain imaging and optical coherence tomography. Int. J. Cardiovasc. Imag. **28**(6), 1299–1304 (2012)
121. I. Ben-Dor, M. Mahmoudi, A.D. Pichard, L.F. Satler, R. Waksman, Optical coherence tomography: a new imaging modality for plaque characterization and stent implantation. J. Interven. Cardiol. **24**(2), 184–192 (2011)
122. B.E. Bouma, G.J. Tearney, H. Yabushita, M. Shishkov, C.R. Kauffman, D.D. Gauthier, B.D. MacNeill, S.L. Houser, H.T. Aretz, E.F. Halpern, I.K. Jang, Evaluation of intracoronary stenting by intravascular optical coherence tomography. Heart **89**(3), 317–320 (2003)
123. R.C. Chan, A.H. Chau, W.C. Karl, S. Nadkarni, A.S. Khalil, N. Iftimia, M. Shishkov, G. J. Tearney, M.R. Kaazempur-Mofrad, B.E. Bouma, OCT-based arterial elastography: robust estimation exploiting tissue biomechanics. Opt. Express **12**(19), 4558–4572 (2004)
124. X. Yang, D. Lorenser, R.A. McLaughlin, R.W. Kirk, M. Edmond, M.C. Simpson, M.D. Grounds, D.D. Sampson, Imaging deep skeletal muscle structure using a high-sensitivity ultrathin side-viewing optical coherence tomography needle probe. Biomed. Opt. Express **5** (1), 136–148 (2014)
125. B.C. Quirk, R.A. McLaughlin, A. Curatolo, R.W. Kirk, P.B. Noble, D.D. Sampson, In situ imaging of lung alveoli with an optical coherence tomography needle probe. J. Biomed. Opt. **16**(3) (2011)
126. D. Lorenser, X. Yang, R.W. Kirk, B.C. Quirk, R.A. McLaughlin, D.D. Sampson, Ultrathin side-viewing needle probe for optical coherence tomography. Opt. Lett. **36**(19), 3894–3896 (2011)
127. M.J. Gora, J.S. Sauk, R.W. Carruth, W. Lu, D.T. Carlton, A. Soomro, M. Rosenberg, N.S. Nishioka, G.J. Tearney, Imaging the upper gastrointestinal tract in unsedated patients using tethered capsule endomicroscopy. Gastroenterology **145**(4), 723–725 (2013)
128. K. Liang, G. Traverso, H.-C. Lee, O.O. Ahsen, Z. Wang, B. Potsaid, M. Giacomelli, V. Jayaraman, R. Barman, A. Cable, H. Mashimo, R. Langer, J.G. Fujimoto, Ultrahigh speed *en face* OCT capsule for endoscopic imaging. Biomed. Opt. Express **6**(4), 1146–1163 (2015)
129. M.D. Stern, In vivo evaluation of microcirculation by coherent light scattering. Nature **254** (5495), 56–58 (1975)
130. D.A. Boas, A.K. Dunn, Laser speckle contrast imaging in biomedical optics. J. Biomed. Opt. **15**(1), 011109 (2010)

131. A.F. Fercher, J.D. Briers, Flow visualization by means of single-exposure speckle photography. Opt. Commun. **37**(5), 326–330 (1981)
132. H. Li, Q. Liu, H.Y. Lu, Y. Li, H.F. Zhang, S.B. Tong, Directly measuring absolute flow speed by frequency-domain laser speckle imaging. Opt. Express **22**(17), 21079–21087 (2014)
133. Y. Tamaki, M. Araie, E. Kawamoto, S. Eguchi, H. Fujii, Noncontact, two-dimensional measurement of retinal microcirculation using laser speckle phenomenon. Invest. Ophthalmol. Vis. Sci. **35**(11), 3825–3834 (1994)
134. Y. Tamaki, M. Araie, E. Kawamoto, S. Eguchi, H. Fujii, Non-contact, two-dimensional measurement of tissue circulation in choroid and optic nerve head using laser speckle phenomenon. Exp. Eye Res. **60**(4), 373–383 (1995)
135. M. Nagahara, Y. Tamaki, A. Tomidokoro, M. Araie, In vivo measurement of blood velocity in human major retinal vessels using the laser speckle method. Invest. Ophthalmol. Vis. Sci. **52**(1), 87–92 (2011)
136. H. Isono, S. Kishi, Y. Kimura, N. Hagiwara, N. Konishi, H. Fujii, Observation of choroidal circulation using index of erythrocytic velocity. Arch. Ophthalmol. **121**(2), 225–231 (2003)
137. B. Choi, N.M. Kang, J.S. Nelson, Laser speckle imaging for monitoring blood flow dynamics in the in vivo rodent dorsal skin fold model. Microvasc. Res. **68**(2), 143–146 (2004)
138. B. Choi, J.C. Ramirez-San-Juan, J. Lotfi, J. Stuart Nelson, Linear response range characterization and in vivo application of laser speckle imaging of blood flow dynamics. J. Biomed. Opt. **11**(4), 041129 (2006)
139. K.R. Forrester, J. Tulip, C. Leonard, C. Stewart, R.C. Bray, A laser speckle imaging technique for measuring tissue perfusion. IEEE Trans. Biomed. Eng. **51**(11), 2074–2084 (2004)
140. Y.C. Huang, T.L. Ringold, J.S. Nelson, B. Choi, Noninvasive blood flow imaging for real-time feedback during laser therapy of port wine stain birthmarks. Lasers Surg. Med. **40** (3), 167–173 (2008)
141. A.K. Dunn, A. Devor, A.M. Dale, D.A. Boas, Spatial extent of oxygen metabolism and hemodynamic changes during functional activation of the rat somatosensory cortex. Neuroimage **27**(2), 279–290 (2005)
142. A.K. Dunn, H. Bolay, M.A. Moskowitz, D.A. Boas, Dynamic imaging of cerebral blood flow using laser speckle. J. Cereb. Blood Flow Metab. **21**(3), 195–201 (2001)
143. H. Bolay, U. Reuter, A.K. Dunn, Z. Huang, D.A. Boas, M.A. Moskowitz, Intrinsic brain activity triggers trigeminal meningeal afferents in a migraine model. Nat. Med. **8**(2), 136–142 (2002)
144. C. Ayata, H.K. Shin, S. Salomone, Y. Ozdemir-Gursoy, D.A. Boas, A.K. Dunn, M.A. Moskowitz, Pronounced hypoperfusion during spreading depression in mouse cortex. J. Cereb. Blood Flow Metab. **24**(10), 1172–1182 (2004)
145. H.K. Shin, A.K. Dunn, P.B. Jones, D.A. Boas, M.A. Moskowitz, C. Ayata, Vasoconstrictive neurovascular coupling during focal ischemic depolarizations. J. Cereb. Blood Flow Metab. **26**(8), 1018–1030 (2006)
146. H.K. Shin, A.K. Dunn, P.B. Jones, D.A. Boas, E.H. Lo, M.A. Moskowitz, C. Ayata, Normobaric hyperoxia improves cerebral blood flow and oxygenation, and inhibits peri-infarct depolarizations in experimental focal ischaemia. Brain **130**(Pt 6), 1631–1642 (2007)
147. S. Zhang, T.H. Murphy, Imaging the impact of cortical microcirculation on synaptic structure and sensory-evoked hemodynamic responses in vivo. PLoS Biol. **5**(5), e119 (2007)
148. Z. Hajjarian, S.K. Nadkarni, Evaluating the viscoelastic properties of tissue from laser speckle fluctuations. Sci. Rep. **2** (2012)
149. M.M. Tripathi, Z. Hajjarian, E.M. Van Cott, S.K. Nadkarni, Assessing blood coagulation status with laser speckle rheology. Biomed. Opt. Express **5**(3), 817–831 (2014)
150. Z. Hajjarian, M.M. Tripathi, S.K. Nadkarni, Optical thromboelastography to evaluate whole blood coagulation. J. Biophotonics **8**(5), 372–381 (2015)

151. S.K. Nadkarni, B.E. Bouma, T. Helg, R. Chan, E. Halpern, A. Chau, M.S. Minsky, J.T. Motz, S.L. Houser, G.J. Tearney, Characterization of atherosclerotic plaques by laser speckle imaging. Circulation **112**(6), 885–892 (2005)
152. S.K. Nadkarni, A. Bilenca, B.E. Bouma, G.J. Tearney, Measurement of fibrous cap thickness in atherosclerotic plaques by spatiotemporal analysis of laser speckle images. J. Biomed. Opt. **11**(2), 8 (2006)
153. S.K. Nadkarni, Optical measurement of arterial mechanical properties: from atherosclerotic plaque initiation to rupture. J. Biomed. Opt. **18**(12), (2013)
154. J. Wang, S.K. Nadkarni, The influence of optical fiber bundle parameters on the transmission of laser speckle patterns. Opt. Express **22**(8), 8908–8918 (2014)
155. A.A. Stratonnikov, V.B. Loschenov, Evaluation of blood oxygen saturation in vivo from diffuse reflectance spectra. J. Biomed. Opt. **6**(4), 457–467 (2001)
156. R. Marchesini, N. Cascinelli, M. Brambilla, C. Clemente, L. Mascheroni, E. Pignoli, A. Testori, D.R. Venturoli, In vivo spectrophotometric evaluation of neoplastic and non-neoplastic skin pigmented lesions. II: discriminant analysis between nevus and melanoma. Photochem. Photobiol. **55**(4), 515–522 (1992)
157. N. Rajaram, T.J. Aramil, K. Lee, J.S. Reichenberg, T.H. Nguyen, J.W. Tunnell, Design and validation of a clinical instrument for spectral diagnosis of cutaneous malignancy. Appl. Opt. **49**(2), 142–152 (2010)
158. A.J. Thompson, S. Coda, M. Brydegaard Sørensen, G. Kennedy, R. Patalay, U. Waitong-Bramming, P.A.A. De Beule, M.A.A. Neil, S. Andersson-Engels, N. Bendsoe, P.M.W. French, K. Svanberg, C. Dunsby, In vivo measurements of diffuse reflectance and time-resolved autofluorescence emission spectra of basal cell carcinomas. J. Biophotonics **5** (3), 240–254 (2012)
159. I.J. Bigio, T.M. Johnson, J.R. Mourant, B.J. Tromberg, Y. Tadir, M. Fehr, H. Nilsson, V.C. Darrow, Determination of the cervical transformation zone using elastic-scattering spectroscopy, in *Advances in Laser and Light Spectroscopy to Diagnose Cancer and Other Diseases III: Optical Biopsy, Proceedings*, vol. 2679 (1996), pp. 85–91
160. J.R. Mourant, T.J. Bocklage, T.M. Powers, H.M. Greene, K.L. Bullock, L.R. Marr-Lyon, M. H. Dorin, A.G. Waxman, M.M. Zsemlye, H.O. Smith, In vivo light scattering measurements for detection of precancerous conditions of the cervix. Gynecol. Oncol. **105**(2), 439–445 (2007)
161. J.R. Mourant, T.J. Bocklage, T.M. Powers, H.M. Greene, M.H. Dorin, A.G. Waxman, M.M. Zsemlye, H.O. Smith, Detection of cervical intraepithelial neoplasias and cancers in cervical tissue by in vivo light scattering. J. Lower Genital Tract Dis. **13**(4), 216–223 (2009)
162. J.R. Mourant, I.J. Bigio, J. Boyer, R.L. Conn, T. Johnson, T. Shimada, Spectroscopic diagnosis of bladder cancer with elastic light scattering. Lasers Surg. Med. **17**(4), 350–357 (1995)
163. J.R. Mourant, J. Boyer, T.M. Johnson, J. Lacey, I.J. Bigio, A. Bohorfoush, M. Mellow, Detection of gastrointestinal cancer by elastic scattering and absorption spectroscopies with the los-alamos optical biopsy system, in *Advances in Laser and Light Spectroscopy to Diagnose Cancer and Other Diseases II, Proceedings*, vol. 2387 (1995), pp. 210–217
164. H.W. Wang, J.K. Jiang, C.H. Lin, J.K. Lin, G.J. Huang, J.S. Yu, Diffuse reflectance spectroscopy detects increased hemoglobin concentration and decreased oxygenation during colon carcinogenesis from normal to malignant tumors. Opt. Express **17**(4), 2805–2817 (2009)
165. J.P. Culver, T. Durduran, T. Furuya, C. Cheung, J.H. Greenberg, A.G. Yodh, Diffuse optical tomography of cerebral blood flow, oxygenation, and metabolism in rat during focal ischemia. J. Cerebral Blood Flow Metabol. **23**(8), 911–924 (2003)
166. H. Levy, D. Ringuette, O. Levi, Rapid monitoring of cerebral ischemia dynamics using laser-based optical imaging of blood oxygenation and flow. Biomed. Opt. Express **3**(4), 777–791 (2012)
167. Y. Shang, L. Chen, M. Toborek, G.Q. Yu, Diffuse optical monitoring of repeated cerebral ischemia in mice. Opt. Express **19**(21), 20301–20315 (2011)

168. F.F. Jobsis, Noninvasive, infrared monitoring of cerebral and myocardial oxygen sufficiency and circulatory parameters. Science **198**(4323), 1264–1267 (1977)
169. N. Sato, T. Kamada, M. Shichiri, S. Kawano, H. Abe, B. Hagihara, Measurement of hemoperfusion and oxygen sufficiency in gastric mucosa in vivo. Evidence of mucosal hypoxia as the cause of hemorrhagic shock-induced gastric mucosal lesion in rats. Gastroenterology **76**(4), 814–819 (1979)
170. G.A. Mook, A. Buursma, A. Gerding, G. Kwant, W.G. Zijlstra, Spectrophotometric determination of oxygen saturation of blood independent of the presence of indocyanine green. Cardiovasc. Res. **13**(4), 233–237 (1979)
171. F.W. Leung, T. Morishita, E.H. Livingston, T. Reedy, P.H. Guth, Reflectance spectrophotometry for the assessment of gastroduodenal mucosal perfusion. Am. J. Physiol. **252**(6 Pt 1), G797–G804 (1987)
172. J.W. Feather, M. Hajizadeh-Saffar, G. Leslie, J.B. Dawson, A portable scanning reflectance spectrophotometer using visible wavelengths for the rapid measurement of skin pigments. Phys. Med. Biol. **34**(7), 807–820 (1989)
173. K.H. Frank, M. Kessler, K. Appelbaum, W. Dummler, The erlangen micro-lightguide spectrophotometer empho I. Phys. Med. Biol. **34**(12), 1883–1900 (1989)
174. E.M. Sevick, B. Chance, J. Leigh, S. Nioka, M. Maris, Quantitation of time- and frequency-resolved optical spectra for the determination of tissue oxygenation. Anal. Biochem. **195**(2), 330–351 (1991)
175. I.J. Bigio, J.R. Mourant, G. Los, Elastic-scattering spectroscopy for quantitative measurement of chemotherapy and PDT drug concentrations in vivo, in *Optical Biopsies and Microscopic Techniques III, Proceedings*, vol. 3568 (1999), pp. 26–30
176. J.R. Mourant, I.J. Bigio, J. Boyer, J. Lacey, T. Johnson, R.L. Conn, Diagnosis of tissue pathology with elastic scattering spectroscopy, in *Leos'94—Conference Proceedings*, vol. 1 (1994), pp. 17–18
177. J.R. Lakowicz, *Principles of Fluorescence Spectroscopy*, 3rd edn. (Springer, New York, 2006)
178. J. Haringsma, G.N.J. Tytgat, H. Yano, H. Iishi, M. Tatsuta, T. Ogihara, H. Watanabe, N. Sato, N. Marcon, B.C. Wilson, R.W. Cline, Autofluorescence endoscopy: feasibility of detection of GI neoplasms unapparent to white light endoscopy with an evolving technology. Gastrointest. Endosc. **53**(6), 642–650 (2001)
179. F.J.C. van den Broek, P. Fockens, S. van Eeden, J.B. Reitsma, J.C.H. Hardwick, P.C.F. Stokkers, E. Dekker, Endoscopic tri-modal imaging for surveillance in ulcerative colitis: randomised comparison of high-resolution endoscopy and autofluorescence imaging for neoplasia detection; and evaluation of narrow-band imaging for classification of lesions. Gut **57**(8), 1083–1089 (2008)
180. W.L. Curvers, R. Singh, L.M.W.-K. Song, H.C. Wolfsen, K. Ragunath, K. Wang, M.B. Wallace, P. Fockens, J.J.G.H.M. Bergman, Endoscopic tri-modal imaging for detection of early neoplasia in Barrett's oesophagus: a multi-centre feasibility study using high-resolution endoscopy, autofluorescence imaging and narrow band imaging incorporated in one endoscopy system. Gut **57**(2), 167–172 (2008)
181. M.A. Kara, F.P. Peters, P. Fockens, F.J.W. ten Kate, J.J.G.H.M. Bergman, Endoscopic video-autofluorescence imaging followed by narrow band imaging for detecting early neoplasia in Barrett's esophagus. Gastrointest. Endosc. **64**(2), 176–185 (2006)
182. N. Rajaram, J.S. Reichenberg, M.R. Migden, T.H. Nguyen, J.W. Tunnell, Pilot clinical study for quantitative spectral diagnosis of non-melanoma skin cancer. Lasers Surg. Med. **42**(10), 716–727 (2010)
183. L. Brancaleon, A.J. Durkin, J.H. Tu, G. Menaker, J.D. Fallon, N. Kollias, In vivo fluorescence spectroscopy of nonmelanoma skin cancer. Photochem. Photobiol. **73**(2), 178–183 (2001)
184. M.A. D'Hallewin, L. Bezdetnaya, F. Guillemin, Fluorescence detection of bladder cancer: a review. Eur. Urol. **42**(5), 417–425 (2002)

185. L. Silveira Jr., J.A. Betiol Filho, F.L. Silveira, R.A. Zangaro, M.T. Pacheco, Laser-induced fluorescence at 488 nm excitation for detecting benign and malignant lesions in stomach mucosa. J. Fluoresc. **18**(1), 35–40 (2008)
186. C.R. Kapadia, F.W. Cutruzzola, K.M. O'Brien, M.L. Stetz, R. Enriquez, L.I. Deckelbaum, Laser-induced fluorescence spectroscopy of human colonic mucosa. Detection of adenomatous transformation. Gastroenterology **99**(1), 150–157 (1990)
187. Y. Yang, G.C. Tang, M. Bessler, R.R. Alfano, Fluorescence spectroscopy as a photonic pathology method for detecting colon cancer. Lasers Life Sci. **6**(4), 259–276 (1995)
188. D.C. De Veld, M. Skurichina, M.J. Witjes, R.P. Duin, H.J. Sterenborg, J.L. Roodenburg, Clinical study for classification of benign, dysplastic, and malignant oral lesions using autofluorescence spectroscopy. J. Biomed. Opt. **9**(5), 940–950 (2004)
189. P.A.A. De Beule, C. Dunsby, N.P. Galletly, G.W. Stamp, A.C. Chu, U. Anand, P. Anand, C. D. Benham, A. Naylor, P.M.W. French, A hyperspectral fluorescence lifetime probe for skin cancer diagnosis. Rev. Sci. Instrum. **78**(12), 123101 (2007)
190. S. Coda, A.J. Thompson, G.T. Kennedy, K.L. Roche, L. Ayaru, D.S. Bansi, G.W. Stamp, A. V. Thillainayagam, P.M. French, C. Dunsby, Fluorescence lifetime spectroscopy of tissue autofluorescence in normal and diseased colon measured ex vivo using a fiber-optic probe. Biomed. Opt. Express **5**(2), 515–538 (2014)
191. P.V. Butte, B.K. Pikul, A. Hever, W.H. Yong, K.L. Black, L. Marcu, Diagnosis of meningioma by time-resolved fluorescence spectroscopy. J. Biomed. Opt. **10**(6), 064026 (2005)
192. L. Marcu, J.A. Jo, P.V. Butte, W.H. Yong, B.K. Pikul, K.L. Black, R.C. Thompson, Fluorescence lifetime spectroscopy of glioblastoma multiforme. Photochem. Photobiol. **80** (1), 98–103 (2004)
193. N.P. Galletly, J. McGinty, C. Dunsby, F. Teixeira, J. Requejo-Isidro, I. Munro, D.S. Elson, M.A.A. Neil, A.C. Chu, P.M.W. French, G.W. Stamp, Fluorescence lifetime imaging distinguishes basal cell carcinoma from surrounding uninvolved skin. Br. J. Dermatol. **159** (1), 152–161 (2008)
194. J. McGinty, N.P. Galletly, C. Dunsby, I. Munro, D.S. Elson, J. Requejo-Isidro, P. Cohen, R. Ahmad, A. Forsyth, A.V. Thillainayagam, M.A.A. Neil, P.M.W. French, G.W. Stamp, Wide-field fluorescence lifetime imaging of cancer. Biomed. Opt. Express **1**(2), 627–640 (2010)
195. H.M. Chen, C.P. Chiang, C. You, T.C. Hsiao, C.Y. Wang, Time-resolved autofluorescence spectroscopy for classifying normal and premalignant oral tissues. Lasers Surg. Med. **37**(1), 37–45 (2005)
196. M. Mycek, K. Schomacker, N. Nishioka, Colonic polyp differentiation using time-resolved autofluorescence spectroscopy. Gastrointest. Endosc. **48**(4), 390–394 (1998)
197. T.J. Pfefer, D.Y. Paithankar, J.M. Poneros, K.T. Schomacker, N.S. Nishioka, Temporally and spectrally resolved fluorescence spectroscopy for the detection of high grade dysplasia in Barrett's esophagus. Lasers Surg. Med. **32**(1), 10–16 (2003)
198. V. De Giorgi, D. Massi, S. Sestini, R. Cicchi, F.S. Pavone, T. Lotti, Combined non-linear laser imaging (two-photon excitation fluorescence microscopy, fluorescence lifetime imaging microscopy, multispectral multiphoton microscopy) in cutaneous tumours: first experiences. J. Eur. Acad. Dermatology Venereol. **23**(3), 314–316 (2009)
199. B.W. Chwirot, S. Chwirot, W. Jedrzejczyk, M. Jackowski, A.M. Raczynska, J. Winczakiewicz, J. Dobber, Ultraviolet laser-induced fluorescence of human stomach tissues: detection of cancer tissues by imaging techniques. Lasers Surg. Med. **21**(2), 149–158 (1997)
200. B.W. Chwirot, Z. Michniewicz, M. Kowalska, J. Nussbeutel, Detection of colonic malignant lesions by digital imaging of UV laser-induced autofluorescence. Photochem. Photobiol. **69** (3), 336–340 (1999)
201. S.D. Xiao, L. Zhong, H.Y. Luo, X.Y. Chen, Y. Shi, Autofluorescence imaging analysis of gastric cancer. Chin. J. Digest. Dis. **3**, 95–98 (2002)

202. J. Mizeret, G. Wagnieres, T. Stepinac, H. VandenBergh, Endoscopic tissue characterization by frequency-domain fluorescence lifetime imaging (FD-FLIM). Lasers Med. Sci. **12**(3), 209–217 (1997)
203. J. Haringsma, G.N.J. Tytgat, Fluorescence and autofluorescence. Best Pract. Res. Clin. Gastroenterol. **13**(1), 1–10 (1999)
204. S. Yamashita, M. Sakabe, T. Ishizawa, K. Hasegawa, Y. Urano, N. Kokudo, Visualization of the leakage of pancreatic juice using a chymotrypsin-activated fluorescent probe. Br. J. Surg. **100**(9), 1220–1228 (2013)
205. R. Petry, M. Schmitt, J. Popp, Raman spectroscopy—a prospective tool in the life sciences. ChemPhysChem **4**(1), 14–30 (2003)
206. M. Marro, A. Taubes, A. Abernathy, S. Balint, B. Moreno, B. Sanchez-Dalmau, E.H. Martínez-Lapiscina, I. Amat-Roldan, D. Petrov, P. Villoslada, Dynamic molecular monitoring of retina inflammation by in vivo Raman spectroscopy coupled with multivariate analysis. J. Biophotonics **7**(9), 724–734 (2014)
207. S.K. Teh, W. Zhene, K.Y. Ho, M. Teh, K.G. Yeoh, Z.W. Huang, Near-infrared Raman spectroscopy for optical diagnosis in the stomach: identification of helicobacter-pylori infection and intestinal metaplasia. Int. J. Cancer **126**(8), 1920–1927 (2010)
208. H. Krishna, S.K. Majumder, P. Chaturvedi, M. Sidramesh, P.K. Gupta, In vivo Raman spectroscopy for detection of oral neoplasia: a pilot clinical study. J. Biophotonics **7**(9), 690–702 (2014)
209. M.S. Bergholt, K. Lin, J. Wang, W. Zheng, H. Xu, Q. Huang, J.L. Ren, K.Y. Ho, M. Teh, S. Srivastava, B. Wong, K.G. Yeoh, Z. Huang, Simultaneous fingerprint and high-wavenumber fiber-optic Raman spectroscopy enhances real-time in vivo diagnosis of adenomatous polyps during colonoscopy. J. Biophotonics **9**(4), 333–342 (2016)
210. M.A. Short, S. Lam, A. McWilliams, J.H. Zhao, H. Lui, H.S. Zeng, Development and preliminary results of an endoscopic Raman probe for potential in vivo diagnosis of lung cancers. Opt. Lett. **33**(7), 711–713 (2008)
211. M.A. Short, S. Lam, A.M. McWilliams, D.N. Ionescu, H.S. Zeng, Using laser Raman spectroscopy to reduce false positives of autofluorescence bronchoscopies a pilot study. J. Thoracic Oncol. **6**(7), 1206–1214 (2011)
212. O. Aydin, M. Altas, M. Kahraman, O.F. Bayrak, M. Culha, Differentiation of healthy brain tissue and tumors using surface-enhanced Raman scattering. Appl. Spectrosc. **63**(10), 1095–1100 (2009)
213. M. Saleem, M. Bilal, S. Anwar, A. Rehman, M. Ahmed, Optical diagnosis of dengue virus infection in human blood serum using Raman spectroscopy. Laser Phys. Lett. **10**(3), 5 (2013)
214. Z. Huang, H. Zeng, I. Hamzavi, D.I. McLean, H. Lui, Rapid near-infrared Raman spectroscopy system for real-time in vivo skin measurements. Opt. Lett. **26**(22), 1782–1784 (2001)
215. Z. Huang, S.K. Teh, W. Zheng, J. Mo, K. Lin, X. Shao, K.Y. Ho, M. Teh, K.G. Yeoh, Integrated Raman spectroscopy and trimodal wide-field imaging techniques for real-time in vivo tissue Raman measurements at endoscopy. Opt. Lett. **34**(6), 758–760 (2009)
216. Y. Komachi, H. Sato, H. Tashiro, Intravascular Raman spectroscopic catheter for molecular diagnosis of atherosclerotic coronary disease. Appl. Opt. **45**(30), 7938–7943 (2006)
217. W.F. Howard, W.H. Nelson, J.F. Sperry, Resonance Raman method for the rapid detection and identification of bacteria in water. Appl. Spectrosc. **34**(1), 72–75 (1980)
218. K. Hadjigeorgiou, E. Kastanos, A. Kyriakides, C. Pitris, Complete urinary tract infection (UTI) diagnosis and antibiogram using surface enhanced Raman spectroscopy (SERS), in *Optical Diagnostics and Sensing XII: Toward Point-of-Care Diagnostics and Design and Performance Validation of Phantoms Used in Conjunction with Optical Measurement of Tissue IV*, vol. 8229 (2012)
219. E. Kastanos, K. Hadjigeorgiou, C. Pitris, Rapid identification of bacterial resistance to ciprofloxacin using surface enhanced Raman spectroscopy, in *Optical Diagnostics and*

Sensing XIV: Toward Point-of-Care Diagnostics, SPIE Photonics West (SPIE, San Francisco, CA, 2014)
220. E. Kastanos, A. Kyriakides, K. Hadjigeorgiou, C. Pitris, A novel method for bacterial UTI diagnosis using Raman spectroscopy. Int. J. Spectrosc. (2012)
221. E.K. Kastanos, A. Kyriakides, K. Hadjigeorgiou, C. Pitris, A novel method for urinary tract infection diagnosis and antibiogram using Raman spectroscopy. J. Raman Spectrosc. **41**(9), 958–963 (2010)
222. A. Kyriakides, E. Kastanos, K. Hadjigeorgiou, C. Pitris, Classification of Raman spectra of bacteria using rank order kernels, in *Advanced Biomedical and Clinical Diagnostic Systems XI*, vol. 8572 (2013)
223. A. Sengupta, M.L. Laucks, E.J. Davis, Surface-enhanced Raman spectroscopy of bacteria and pollen. Appl. Spectrosc. **59**(8), 1016–1023 (2005)
224. A. Sengupta, M. Mujacic, E.J. Davis, Detection of bacteria by surface-enhanced Raman spectroscopy. Anal. Bioanal. Chem. **386**(5), 1379–1386 (2006)
225. W.H. Nelson, J.F. Sperry, *Direct Detection of Bacteria-Antibody Complexes Via UV Resonance Raman Spectroscopy* (The Board of Governors for Higher Education, State of Rhode Island and Providence Plantations, 2005)
226. X. Yang, C. Gu, F. Qian, Y. Li, J.Z. Zhang, Highly sensitive detection of proteins and bacteria in aqueous solution using surface-enhanced Raman scattering and optical fibers. Anal. Chem. **83**(15), 5888–5894 (2011)
227. A. Sivanesan, E. Witkowska, W. Adamkiewicz, L. Dziewit, A. Kaminska, J. Waluk, Nanostructured silver-gold bimetallic SERS substrates for selective identification of bacteria in human blood. Analyst **139**(5), 1037–1043 (2014)
228. A.K. Boardman, W.S. Wong, W.R. Premasiri, L.D. Ziegler, J.C. Lee, M. Miljkovic, C.M. Klapperich, A. Sharon, A.F. Sauer-Budge, Rapid detection of bacteria from blood with surface-enhanced Raman spectroscopy. Anal. Chem. **88**(16), 8026–8035 (2016)
229. M. Harz, P. Rosch, J. Popp, Vibrational spectroscopy—a powerful tool for the rapid identification of microbial cells at the single-cell level. Cytometry. Part A: J. Int. Soc. Anal. Cytol. **75**(2), 104–113 (2009)
230. F.S. de Siqueira e Oliveira, H.E. Giana, L. Silveira, Discrimination of selected species of pathogenic bacteria using near-infrared Raman spectroscopy and principal components analysis. J. Biomed. Opt. **17**(10), 107004 (2012)
231. A. Campion, P. Kambhampati, Surface-enhanced Raman scattering. Chem. Soc. Rev. **27**(4), 241–250 (1998)
232. M. Fleischmann, P.J. Hendra, A.J. McQuillan, Raman-spectra of pyridine adsorbed at a silver electrode. Chem. Phys. Lett. **26**(2), 163–166 (1974)
233. Z.A. Nima, A. Biswas, I.S. Bayer, F.D. Hardcastle, D. Perry, A. Ghosh, E. Dervishi, A.S. Biris, Applications of surface-enhanced Raman scattering in advanced bio-medical technologies and diagnostics. Drug Metabol. Rev. **46**(2), 155–175 (2014)
234. A.K. Singh, S.A. Khan, Z. Fan, T. Demeritte, D. Senapati, R. Kanchanapally, P.C. Ray, Development of a long-range surface-enhanced Raman spectroscopy ruler. J. Am. Chem. Soc. **134**(20), 8662–8669 (2012)
235. B. Yan, A. Thubagere, W.R. Premasiri, L.D. Ziegler, L. Dal Negro, B.M. Reinhard, Engineered SERS substrates with multiscale signal enhancement: nanoparticle cluster arrays. ACS Nano **3**(5), 1190–1202 (2009)
236. E. Temur, I.H. Boyaci, U. Tamer, H. Unsal, N. Aydogan, A highly sensitive detection platform based on surface-enhanced Raman scattering for *Escherichia coli* enumeration. Anal. Bioanal. Chem. **397**(4), 1595–1604 (2010)
237. H.Y. Lin, C.H. Huang, W.H. Hsieh, L.H. Liu, Y.C. Lin, C.C. Chu, S.T. Wang, I.T. Kuo, L. K. Chau, C.Y. Yang, On-line SERS detection of single bacterium using novel SERS nanoprobes and a microfluidic dielectrophoresis device. Small **10**(22), 4700–4710 (2014)
238. C. Yuen, Q. Liu, Towards in vivo intradermal surface enhanced Raman scattering (SERS) measurements: silver coated microneedle based SERS probe. J. Biophotonics **7**(9), 683–689 (2014)

239. Z. Fan, R. Kanchanapally, P.C. Ray, Hybrid graphene oxide based ultrasensitive SERS probe for label-free biosensing. J. Phys. Chem. Lett. **4**(21), 3813–3818 (2013)
240. P.R. Stoddart, D.J. White, Optical fibre SERS sensors. Anal. Bioanal. Chem. **394**(7), 1761–1774 (2009)
241. G.F. Andrade, M. Fan, A.G. Brolo, Multilayer silver nanoparticles-modified optical fiber tip for high performance SERS remote sensing. Biosens. Bioelectron. **25**(10), 2270–2275 (2010)
242. G.F.S. Andrade, J.G. Hayashi, M.M. Rahman, W.J. Salcedo, C.M.B. Cordeiro, A.G. Brolo, Surface-enhanced resonance Raman scattering (SERS) using Au nanohole arrays on optical fiber tips. Plasmonics **8**(2), 1113–1121 (2013)
243. F. Deiss, N. Sojic, D.J. White, P.R. Stoddart, Nanostructured optical fibre arrays for high-density biochemical sensing and remote imaging. Anal. Bioanal. Chem. **396**(1), 53–71 (2010)
244. Z. Xie, S. Feng, P. Wang, L. Zhang, X. Ren, L. Cui, T. Zhai, J. Chen, Y. Wang, X. Wang, W. Sun, J. Ye, P. Han, P.J. Klar, Y. Zhang, Demonstration of a 3D radar-like SERS sensor micro- and nanofabricated on an optical fiber. Adv. Opt. Mater. **3**(9), 1232–1239 (2015)
245. J.L. Kou, M. Ding, J. Feng, Y.Q. Lu, F. Xu, G. Brambilla, Microfiber-based bragg gratings for sensing applications: a review. Sensors **12**(7), 8861–8876 (2012)
246. A.D. Kersey, M.A. Davis, H.J. Patrick, M. LeBlanc, K.P. Koo, C.G. Askins, M.A. Putnam, E.J. Friebele, Fiber grating sensors. J. Lightwave Technol. **15**(8), 1442–1463 (1997)
247. J.W. Arkwright, N.G. Blenman, I.D. Underhill, S.A. Maunder, N.J. Spencer, M. Costa, S. J. Brookes, M.M. Szczesniak, P.G. Dinning, A fibre optic catheter for simultaneous measurement of longitudinal and circumferential muscular activity in the gastrointestinal tract. J. Biophotonics **4**(4), 244–251 (2011)
248. J.W. Arkwright, N.G. Blenman, I.D. Underhill, S.A. Maunder, N.J. Spencer, M. Costa, S. J. Brookes, M.M. Szczesniak, P.G. Dinning, Measurement of muscular activity associated with peristalsis in the human gut using fiber Bragg grating arrays. IEEE Sensors J. **12**(1), 113–117 (2012)
249. M.R. Islam, M.M. Ali, M.H. Lai, K.S. Lim, H. Ahmad, Chronology of Fabry-Perot interferometer fiber-optic sensors and their applications: a review. Sensors (Basel) **14**(4), 7451–7488 (2014)
250. N. Wu, Y. Tian, X.T. Zou, Y. Zhai, K. Barringhaus, X.W. Wang, A miniature fiber optic blood pressure sensor and its application in in vivo blood pressure measurements of a swine model. Sens. Actuators B: Chem. **181**, 172–178 (2013)
251. O.S. Wolfbeis, Fiber-optic chemical sensors and biosensors. Anal. Chem. **76**(12), 3269–3283 (2004)
252. M.T. Carter, R.C. Thomas, M.E. Berton, Fiber waveguide chemical sensors for detection of explosive related vapors, in *Detection and Remediation Technologies for Mines and Minelike Targets VII, Pts 1 and 2*, vol. 4742 (2002), pp. 509–519
253. M. Janotta, A. Katzir, B. Mizaikoff, Sol-gel-coated mid-infrared fiber-optic sensors. Appl. Spectrosc. **57**(7), 823–828 (2003)
254. C.A. Browne, D.H. Tarrant, M.S. Olteanu, J.W. Mullens, E.L. Chronister, Intrinsic sol-gel clad fiber-optic sensors with time-resolved detection. Anal. Chem. **68**(14), 2289–2295 (1996)
255. G. Okeeffe, B.D. Maccraith, A.K. Mcevoy, C.M. Mcdonagh, J.F. Mcgilp, Development of a led-based phase fluorometric oxygen sensor using evanescent-wave excitation of a sol-gel immobilized dye. Sens. Actuators B: Chem. **29**(1–3), 226–230 (1995)
256. O.S. Wolfbeis, M. Schaferling, A. Durkop, Reversible optical sensor membrane for hydrogen peroxide using an immobilized fluorescent probe, and its application to a glucose biosensor. Microchim. Acta **143**(4), 221–227 (2003)
257. Y. Qin, S. Peper, A. Radu, A. Ceresa, E. Bakker, Plasticizer-free polymer containing a covalently immobilized Ca^{2+}-selective ionophore for potentiometric and optical sensors. Anal. Chem. **75**(13), 3038–3045 (2003)

258. A.S. Jeevarajan, S. Vani, T.D. Taylor, M.M. Anderson, Continuous pH monitoring in a perfused bioreactor system using an optical pH sensor. Biotechnol. Bioeng. **78**(4), 467–472 (2002)
259. N.A. Yusof, M. Ahmad, A flow cell optosensor for lead based on immobilized gallocynin in chitosan membrane. Talanta **58**(3), 459–466 (2002)
260. A. Lobnik, I. Oehme, I. Murkovic, O.S. Wolfbeis, Ph optical sensors based on sol-gels: chemical doping versus covalent immobilization. Anal. Chim. Acta **367**(1–3), 159–165 (1998)
261. T.M. Butler, B.D. Maccraith, C.M. Mcdonagh, Development of an extended range fibre optic ph sensor using evanescent wave absorption of sol-gel entrapped pH indicators. Chem. Biochem. Environ. Fiber Sens. VII **2508**, 168–178 (1995)
262. B.D. Gupta, D.K. Sharma, Evanescent wave absorption based fiber optic ph sensor prepared by dye doped sol-gel immobilization technique. Opt. Commun. **140**(1–3), 32–35 (1997)
263. E. Alvarado-Mendez, R. Rojas-Laguna, J.A. Andrade-Lucio, D. Hernandez-Cruz, R.A. Lessard, J.G. Avina-Cervantes, Design and characterization of pH sensor based on sol-gel silica layer on plastic optical fiber. Sens. Actuators B: Chem. **106**(2), 518–522 (2005)
264. B.D. Maccraith, C.M. Mcdonagh, G. Okeeffe, A.K. Mcevoy, T. Butler, F.R. Sheridan, Sol-gel coatings for optical chemical sensors and biosensors. Sens. Actuators B: Chem. **29** (1–3), 51–57 (1995)
265. H.P. Williams, Sir Harold Ridley's vision. Br. J. Ophthalmol. **85**(9), 1022–1023 (2001)
266. M. Leyland, E. Zinicola, Multifocal versus monofocal intraocular lenses in cataract surgery: a systematic review. Ophthalmology **110**(9), 1789–1798 (2003)
267. S.R. de Silva, J.R. Evans, V. Kirthi, M. Ziaei, M. Leyland, Multifocal versus monofocal intraocular lenses after cataract extraction. Cochrane Database Syst. Rev. **12** (2016)
268. D. Madrid-Costa, A. Cervino, T. Ferrer-Blasco, S. Garcia-Lazaro, R. Montes-Mico, Visual and optical performance with hybrid multifocal intraocular lenses. Clin. Exp. Optom. **93**(6), 426–440 (2010)
269. M. Chen, Accommodation in pseudophakic eyes. Taiwan J. Ophthalmol. **2**(4), 117–121 (2012)
270. R.F. Steinert, B.L. Aker, D.J. Trentacost, P.J. Smith, N. Tarantino, A prospective comparative study of the amo array zonal-progressive multifocal silicone intraocular lens and a monofocal intraocular lens. Ophthalmology **106**(7), 1243–1255 (1999)
271. M.C. Knorz, Results of a European multicenter study of the true vista bifocal intraocular lens. J. Cataract Refract. Surg. **19**(5), 626–634 (1993)
272. H.V. Gimbel, D.R. Sanders, M.G. Raanan, Visual and refractive results of multifocal intraocular lenses. Ophthalmology **98**(6), 881–888 (1991)
273. J.D. Hunkeler, T.M. Coffman, J. Paugh, A. Lang, P. Smith, N. Tarantino, Characterization of visual phenomena with the array multifocal intraocular lens. J. Cataract Refract. Surg. **28**(7), 1195–1204 (2002)
274. J.S. Chang, J.C. Ng, S.Y. Lau, Visual outcomes and patient satisfaction after presbyopic lens exchange with a diffractive multifocal intraocular lens. J. Refract. Surg. **28**(7), 468–474 (2012)
275. J.L. Alió, D.P. Piñero, A.B. Plaza-Puche, M.J.R. Chan, Visual outcomes and optical performance of a monofocal intraocular lens and a new-generation multifocal intraocular lens. J. Cataract Refract. Surg. **37**(2), 241–250 (2011)
276. R.K. Shepherd, M.N. Shivdasani, D.A. Nayagam, C.E. Williams, P.J. Blamey, Visual prostheses for the blind. Trends Biotechnol. **31**(10), 562–571 (2013)
277. S. Picaud, J.A. Sahel, Retinal prostheses: clinical results and future challenges. C.R. Biol. **337**(3), 214–222 (2014)
278. A.E. Hadjinicolaou, H. Meffin, M.I. Maturana, S.L. Cloherty, M.R. Ibbotson, Prosthetic vision: devices, patient outcomes and retinal research. Clin. Exper. Optometry **98**(5), 395–410 (2015)
279. A.H. Milam, Z.Y. Li, R.N. Fariss, Histopathology of the human retina in retinitis pigmentosa. Progr. Retinal Eye Res. **17**(2), 175–205 (1998)

280. M.S. Humayun, M. Prince, E. de Juan, Y. Barron, M. Moskowitz, I.B. Klock, A.H. Milam, Morphometric analysis of the extramacular retina from postmortem eyes with retinitis pigmentosa. Invest. Ophthalmol. Vis. Sci. **40**(1), 143–148 (1999)
281. M.P. Villegas-Perez, J.M. Lawrence, M. Vidal-Sanz, M.M. Lavail, R.D. Lund, Ganglion cell loss in RCS rat retina: a result of compression of axons by contracting intraretinal vessels linked to the pigment epithelium. J. Comp. Neurol. **392**(1), 58–77 (1998)
282. R.E. Marc, B.W. Jones, C.B. Watt, E. Strettoi, Neural remodeling in retinal degeneration. Progr. Retinal Eye Res. **22**(5), 607–655 (2003)
283. G.S. Brindley, W.S. Lewin, The sensations produced by electrical stimulation of the visual cortex. J. Physiol. **196**(2), 479–493 (1968)
284. W.H. Dobelle, M.G. Mladejovsky, J.P. Girvin, Artifical vision for the blind: electrical stimulation of visual cortex offers hope for a functional prosthesis. Science **183**(4123), 440–444 (1974)
285. Second Sight Medical Products, The Argus® II retinal prosthesis system. http://www.2-sight.com/argus-ii-rps-pr-en.html (2016)
286. E. Zrenner, K.U. Bartz-Schmidt, H. Benav, D. Besch, A. Bruckmann, V.P. Gabel, F. Gekeler, U. Greppmaier, A. Harscher, S. Kibbel, J. Koch, A. Kusnyerik, T. Peters, K. Stingl, H. Sachs, A. Stett, P. Szurman, B. Wilhelm, R. Wilke, Subretinal electronic chips allow blind patients to read letters and combine them to words. Proc. Roy. Soc. B: Biol. Sci. **278** (1711), 1489–1497 (2011)
287. K. Stingl, K.U. Bartz-Schmidt, D. Besch, A. Braun, A. Bruckmann, F. Gekeler, U. Greppmaier, S. Hipp, G. Hortdorfer, C. Kernstock, A. Koitschev, A. Kusnyerik, H. Sachs, A. Schatz, K.T. Stingl, T. Peters, B. Wilhelm, E. Zrenner, Artificial vision with wirelessly powered subretinal electronic implant alpha-IMS. Proc. Roy. Soc. B **280**(1757), 20130077 (2013)
288. Retina Implant AG, Subretinal implant technology. http://www.retina-implant.de/en/patients/technology/default.aspx (2016)
289. Pixium Vision, Iris® bionic vision restoration system. http://www.pixium-vision.com/en/technology-1/iris-vision-restoration-system (2017)
290. M.S. Humayun, J.D. Dorn, L. da Cruz, G. Dagnelie, J.A. Sahel, P.E. Stanga, A.V. Cideciyan, J.L. Duncan, D. Eliott, E. Filley, A.C. Ho, A. Santos, A.B. Safran, A. Arditi, L.V. Del Priore, R.J. Greenberg, Interim results from the international trial of second sight's visual prosthesis. Ophthalmology **119**(4), 779–788 (2012)
291. S. Klauke, M. Goertz, S. Rein, D. Hoehl, U. Thomas, R. Eckhorn, F. Bremmer, T. Wachtler, Stimulation with a wireless intraocular epiretinal implant elicits visual percepts in blind humans. Invest. Ophthalmol. Vis. Sci. **52**(1), 449–455 (2011)
292. A.E. Hadjinicolaou, R.T. Leung, D.J. Garrett, K. Ganesan, K. Fox, D.A.X. Nayagam, M.N. Shivdasani, H. Meffin, M.R. Ibbotson, S. Prawer, B.J. O'Brien, Electrical stimulation of retinal ganglion cells with diamond and the development of an all diamond retinal prosthesis. Biomaterials **33**(24), 5812–5820 (2012)
293. H. Lorach, G. Goetz, R. Smith, X. Lei, Y. Mandel, T. Kamins, K. Mathieson, P. Huie, J. Harris, A. Sher, D. Palanker, Photovoltaic restoration of sight with high visual acuity. Nature Med. **21**(5), 476–482 (2015)
294. K. Mathieson, J. Loudin, G. Goetz, P. Huie, L.L. Wang, T.I. Kamins, L. Galambos, R. Smith, J.S. Harris, A. Sher, D. Palanker, Photovoltaic retinal prosthesis with high pixel density. Nat. Photonics **6**(6), 391–397 (2012)
295. Pixium Vision, Prima bionic vision restoration system. http://www.pixium-vision.com/en/technology-1/prima-vision-restoration-system (2017)
296. M.N. Shivdasani, C.D. Luu, R. Cicione, J.B. Fallon, P.J. Allen, J. Leuenberger, G. J. Suaning, N.H. Lovell, R.K. Shepherd, C.E. Williams, Evaluation of stimulus parameters and electrode geometry for an effective suprachoroidal retinal prosthesis. J. Neural Eng. **7**(3) (2010)

297. R. Cicione, M.N. Shivdasani, J.B. Fallon, C.D. Luu, P.J. Allen, G.D. Rathbone, R.K. Shepherd, C.E. Williams, Visual cortex responses to suprachoroidal electrical stimulation of the retina: effects of electrode return configuration. J. Neural Eng. **9**(3) (2012)
298. T. Fujikado, M. Kamei, H. Sakaguchi, H. Kanda, T. Morimoto, Y. Ikuno, K. Nishida, H. Kishima, T. Maruo, K. Konoma, M. Ozawa, K. Nishida, Testing of semichronically implanted retinal prosthesis by suprachoroidal-transretinal stimulation in patients with retinitis pigmentosa. Invest. Ophthalmol. Vis. Sci. **52**(7), 4726–4733 (2011)
299. L.N. Ayton, P.J. Blamey, R.H. Guymer, C.D. Luu, D.A. Nayagam, N.C. Sinclair, M.N. Shivdasani, J. Yeoh, M.F. McCombe, R.J. Briggs, N.L. Opie, J. Villalobos, P.N. Dimitrov, M. Varsamidis, M.A. Petoe, C.D. McCarthy, J.G. Walker, N. Barnes, A.N. Burkitt, C.E. Williams, R.K. Shepherd, P.J. Allen, C. Bionic Vision Australia Research, First-in-human trial of a novel suprachoroidal retinal prosthesis. PLoS One **9**(12), e115239 (2014)
300. C. Veraart, F. Duret, M. Brelen, J. Delbeke, Vision rehabilitation with the optic nerve visual prosthesis, in *Proceedings of the 26th Annual International Conference of the IEEE Engineering in Medicine and Biology Society*, vols. 1–7, 26 (2004), pp. 4163–4164
301. M.E. Brelen, V. Vince, B. Gerard, C. Veraart, J. Delbeke, Measurement of evoked potentials after electrical stimulation of the human optic nerve. Invest. Ophthalmol. Vis. Sci. **51**(10), 5351–5355 (2010)
302. J.J. Sun, Y.L. Lu, P.J. Cao, X.L. Li, C.S. Cai, X.Y. Chai, Q.S. Ren, L.M. Li, Spatiotemporal properties of multipeaked electrically evoked potentials elicited by penetrative optic nerve stimulation in rabbits. Invest. Ophthalmol. Vis. Sci. **52**(1), 146–154 (2011)
303. R.A. Normann, B.A. Greger, P. House, S.F. Romero, F. Pelayo, E. Fernandez, Toward the development of a cortically based visual neuroprosthesis. J. Neural Eng. **6**(3) (2009)
304. E.M. Schmidt, M.J. Bak, F.T. Hambrecht, C.V. Kufta, D.K. O'Rourke, P. Vallabhanath, Feasibility of a visual prosthesis for the blind based on intracortical microstimulation of the visual cortex. Brain **119**(Pt 2), 507–522 (1996)
305. P.H. Schiller, W.M. Slocum, M.C. Kwak, G.L. Kendall, E.J. Tehovnik, New methods devised specify the size and color of the spots monkeys see when striate cortex (area v1) is electrically stimulated. Proc. Natl. Acad. Sci. U.S.A. **108**(43), 17809–17814 (2011)
306. K. Deisseroth, Optogenetics. Nature Meth. **8**(1), 26–29 (2011)
307. L. Fenno, O. Yizhar, K. Deisseroth, The development and application of optogenetics, in *Annual Review of Neuroscience*, vol. 34, ed. by S.E. Hyman et al. (2011), pp. 389–412
308. V. Gradinaru, M. Mogri, K.R. Thompson, J.M. Henderson, K. Deisseroth, Optical deconstruction of Parkinsonian neural circuitry. Science **324**(5925), 354–359 (2009)
309. G. Nagel, T. Szellas, W. Huhn, S. Kateriya, N. Adeishvili, P. Berthold, D. Ollig, P. Hegemann, E. Bamberg, Channelrhodopsin-2, a directly light-gated cation-selective membrane channel. Proc. Natl. Acad. Sci. U.S.A. **100**(24), 13940–13945 (2003)
310. B. Schobert, J.K. Lanyi, Halorhodopsin is a light-driven chloride pump. J. Biol. Chem. **257** (17), 306–313 (1982)
311. B.Y. Chow, X. Han, A.S. Dobry, X.F. Qian, A.S. Chuong, M.J. Li, M.A. Henninger, G.M. Belfort, Y.X. Lin, P.E. Monahan, E.S. Boyden, High-performance genetically targetable optical neural silencing by light-driven proton pumps. Nature **463**(7277), 98–102 (2010)
312. E.S. Boyden, F. Zhang, E. Bamberg, G. Nagel, K. Deisseroth, Millisecond-timescale, genetically targeted optical control of neural activity. Nature Neurosci. **8**(9), 1263–1268 (2005)
313. B. Fan, W. Li, Miniaturized optogenetic neural implants: a review. Lab Chip **15**(19), 3838–3855 (2015)
314. J.M. Barrett, R. Berlinguer-Palmini, P. Degenaar, Optogenetic approaches to retinal prosthesis. Vis. Neurosci. **31**(4–5), 345–354 (2014)
315. B.R. Arenkiel, J. Peca, I.G. Davison, C. Feliciano, K. Deisseroth, G.J. Augustine, M.D. Ehlers, G.P. Feng, In vivo light-induced activation of neural circuitry in transgenic mice expressing channelrhodopsin-2. Neuron **54**(2), 205–218 (2007)

316. S.L. Zhao, C. Cunha, F. Zhang, Q. Liu, B. Gloss, K. Deisseroth, G.J. Augustine, G.P. Feng, Improved expression of halorhodopsin for light-induced silencing of neuronal activity. Brain Cell Biol. **36**(1–4), 141–154 (2008)
317. F. Zhang, V. Gradinaru, A.R. Adamantidis, R. Durand, R.D. Airan, L. de Lecea, K. Deisseroth, Optogenetic interrogation of neural circuits: technology for probing mammalian brain structures. Nat. Protoc. **5**(3), 439–456 (2010)
318. K.M. Tye, J.J. Mirzabekov, M.R. Warden, E.A. Ferenczi, H.-C. Tsai, J. Finkelstein, S.-Y. Kim, A. Adhikari, K.R. Thompson, A.S. Andalman, L.A. Gunaydin, I.B. Witten, K. Deisseroth, Dopamine neurons modulate neural encoding and expression of depression-related behaviour. Nature **493**(7433), 537 (2013)
319. K.M. Tye, R. Prakash, S.-Y. Kim, L.E. Fenno, L. Grosenick, H. Zarabi, K.R. Thompson, V. Gradinaru, C. Ramakrishnan, K. Deisseroth, Amygdala circuitry mediating reversible and bidirectional control of anxiety. Nature **471**(7338), 358–362 (2011)
320. D. Chaudhury, J.J. Walsh, A.K. Friedman, B. Juarez, S.M. Ku, J.W. Koo, D. Ferguson, H.-C. Tsai, L. Pomeranz, D.J. Christoffel, A.R. Nectow, M. Ekstrand, A. Domingos, M.S. Mazei-Robison, E. Mouzon, M.K. Lobo, R.L. Neve, J.M. Friedman, S.J. Russo, K. Deisseroth, E.J. Nestler, M.-H. Han, Rapid regulation of depression-related behaviours by control of midbrain dopamine neurons. Nature **493**(7433), 532 (2013)
321. A.M. Aravanis, L.P. Wang, F. Zhang, L.A. Meltzer, M.Z. Mogri, M.B. Schneider, K. Deisseroth, An optical neural interface: in vivo control of rodent motor cortex with integrated fiberoptic and optogenetic technology. J. Neural Eng. **4**(3), S143–S156 (2007)
322. Y. LeChasseur, S. Dufour, G. Lavertu, C. Bories, M. Deschenes, R. Vallee, Y. De Koninck, A microprobe for parallel optical and electrical recordings from single neurons in vivo. Nature Meth. **8**(4), 319–325 (2011)
323. J. Wang, F. Wagner, D.A. Borton, J.Y. Zhang, I. Ozden, R.D. Burwell, A.V. Nurmikko, R. van Wagenen, I. Diester, K. Deisseroth, Integrated device for combined optical neuromodulation and electrical recording for chronic in vivo applications. J. Neural Eng. **9**(1) (2012)
324. N. Grossman, V. Poher, M.S. Grubb, G.T. Kennedy, K. Nikolic, B. McGovern, R.B. Palmini, Z. Gong, E.M. Drakakis, M.A.A. Neil, M.D. Dawson, J. Burrone, P. Degenaar, Multi-site optical excitation using chr2 and micro-led array. J. Neural Eng. **7**(1) (2010)
325. K.Y. Kwon, B. Sirowatka, A. Weber, W. Li, Opto-mu ecog array: a hybrid neural interface with transparent mu ecog electrode array and integrated leds for optogenetics. IEEE Trans. Biomed. Circuits Syst. **7**(5), 593–600 (2013)
326. M. Schwaerzle, P. Elmlinger, O. Paul, P. Ruther, Miniaturized 3 × 3 optical fiber array for optogenetics with integrated 460 nm light sources and flexible electrical interconnection, in *2015 28th IEEE International Conference on Micro Electro Mechanical Systems* (*MEMS*) (2015)
327. R.D. Airan, K.R. Thompson, L.E. Fenno, H. Bernstein, K. Deisseroth, Temporally precise in vivo control of intracellular signalling. Nature **458**(7241), 1025–1029 (2009)
328. J.A. Cardin, M. Carlen, K. Meletis, U. Knoblich, F. Zhang, K. Deisseroth, L.H. Tsai, C.I. Moore, Targeted optogenetic stimulation and recording of neurons in vivo using cell-type-specific expression of channelrhodopsin-2. Nat. Protoc. **5**(2), 247–254 (2010)
329. P. Anikeeva, A.S. Andalman, I. Witten, M. Warden, I. Goshen, L. Grosenick, L.A. Gunaydin, L.M. Frank, K. Deisseroth, Optetrode: a multichannel readout for optogenetic control in freely moving mice. Nature Neurosci. **15**(1), 163–170 (2012)
330. E. Stark, T. Koos, G. Buzsaki, Diode probes for spatiotemporal optical control of multiple neurons in freely moving animals. J. Neurophysiol. **108**(1), 349–363 (2012)
331. T.V.F. Abaya, M. Diwekar, S. Blair, P. Tathireddy, L. Rieth, G.A. Clark, F. Solzbacher, Characterization of a 3D optrode array for infrared neural stimulation. Biomed. Opt. Express **3**(9), 2200–2219 (2012)
332. Y. Iwai, S. Honda, H. Ozeki, M. Hashimoto, H. Hirase, A simple head-mountable led device for chronic stimulation of optogenetic molecules in freely moving mice. Neurosci. Res. **70** (1), 124–127 (2011)

333. L. Campagnola, H. Wang, M.J. Zyka, Fiber-coupled light-emitting diode for localized photo stimulation of neurons expressing channelrhodopsin-2. J. Neurosci. Meth. **169**(1), 27–33 (2008)
334. T. Ishizuka, M. Kakuda, R. Araki, H. Yawo, Kinetic evaluation of photosensitivity in genetically engineered neurons expressing green algae light-gated channels. Neurosci. Res. **54**(2), 85–94 (2006)
335. H. Cao, L. Gu, S.K. Mohanty, J.C. Chiao, An integrated μLED optrode for optogenetic stimulation and electrical recording. IEEE Trans. Biomed. Eng. **60**(1), 225–229 (2013)
336. N. McAlinden, D. Massoubre, E. Richardson, E. Gu, S. Sakata, M.D. Dawson, K. Mathieson, Thermal and optical characterization of micro-LED probes for in vivo optogenetic neural stimulation. Opt. Lett. **38**(6), 992–994 (2013)
337. K.Y. Kwon, H.M. Lee, M. Ghovanloo, A. Weber, W. Li, Design, fabrication, and packaging of an integrated, wirelessly-powered optrode array for optogenetics application. Front. Syst. Neurosci. **9**, 69 (2015)
338. H.M. Lee, K.Y. Kwon, W. Li, M. Ghovanloo, A power-efficient switched-capacitor stimulating system for electrical/optical deep brain stimulation. IEEE J. Solid-State Circ. **50** (1), 360–374 (2015)
339. M.A. Rossi, V. Go, T. Murphy, Q. Fu, J. Morizio, H.H. Yin, A wirelessly controlled implantable LED system for deep brain optogenetic stimulation. Front. Integr. Neurosci. **9**, 8 (2015)
340. M. Hashimoto, A. Hata, T. Miyata, H. Hirase, Programmable wireless light-emitting diode stimulator for chronic stimulation of optogenetic molecules in freely moving mice. Neurophotonics **1**(1), 011002 (2014)
341. S.T. Lee, P.A. Williams, C.E. Braine, D.T. Lin, S.W. John, P.P. Irazoqui, A miniature, fiber-coupled, wireless, deep-brain optogenetic stimulator. IEEE Trans. Neural Syst. Rehabil. Eng. **23**(4), 655–664 (2015)
342. C.T. Wentz, J.G. Bernstein, P. Monahan, A. Guerra, A. Rodriguez, E.S. Boyden, A wirelessly powered and controlled device for optical neural control of freely-behaving animals. *J. Neural Eng.* **8**(4) (2011)
343. T.-I. Kim, J.G. McCall, Y.H. Jung, X. Huang, E.R. Siuda, Y. Li, J. Song, Y.M. Song, H.A. Pao, R.-H. Kim, C. Lu, S.D. Lee, I.-S. Song, G. Shin, R. Al-Hasani, S. Kim, M.P. Tan, Y. Huang, F.G. Omenetto, J.A. Rogers, M.R. Bruchas, Injectable, cellular-scale optoelectronics with applications for wireless optogenetics. Science **340**(6129), 211–216 (2013)
344. S.I. Park, D.S. Brenner, G. Shin, C.D. Morgan, B.A. Copits, H.U. Chung, M.Y. Pullen, K.N. Noh, S. Davidson, S.J. Oh, J. Yoon, K.-I. Jang, V.K. Samineni, M. Norman, J.G. Grajales-Reyes, S.K. Vogt, S.S. Sundaram, K.M. Wilson, J.S. Ha, R. Xu, T. Pan, T.-I. Kim, Y. Huang, M.C. Montana, J.P. Golden, M.R. Bruchas, R.W. Gereau Iv, J.A. Rogers, Soft, stretchable, fully implantable miniaturized optoelectronic systems for wireless optogenetics. Nat. Biotech. (Advance online publication, 2015)
345. A.D. Bi, J.J. Cui, Y.P. Ma, E. Olshevskaya, M.L. Pu, A.M. Dizhoor, Z.H. Pan, Ectopic expression of a microbial-type rhodopsin restores visual responses in mice with photoreceptor degeneration. Neuron **50**(1), 23–33 (2006)
346. H. Tomita, E. Sugano, H. Isago, T. Hiroi, Z. Wang, E. Ohta, M. Tamai, Channelrhodopsin-2 gene transduced into retinal ganglion cells restores functional vision in genetically blind rats. Exper. Eye Res. **90**(3), 429–436 (2010)
347. H. Tomita, E. Sugano, H. Yawo, T. Ishizuka, H. Isago, S. Narikawa, S. Kugler, M. Tamai, Restoration of visual response in aged dystrophic RCS rats using AAV-mediated channelopsin-2 gene transfer. Invest. Ophthalmol. Vis. Sci. **48**(8), 3821–3826 (2007)
348. P.S. Lagali, D. Balya, G.B. Awatramani, T.A. Munch, D.S. Kim, V. Busskamp, C.L. Cepko, B. Roska, Light-activated channels targeted to on bipolar cells restore visual function in retinal degeneration. Nature Neurosci. **11**(6), 667–675 (2008)
349. B. Lin, A. Koizumi, N. Tanaka, S. Panda, R.H. Masland, Restoration of visual function in retinal degeneration mice by ectopic expression of melanopsin. Proc. Natl. Acad. Sci. U.S.A. **105**(41), 16009–16014 (2008)

350. I. Reutsky-Gefen, L. Golan, N. Farah, A. Schejter, L. Tsur, I. Brosh, S. Shoham, Holographic optogenetic stimulation of patterned neuronal activity for vision restoration. Nat. Commun. **4** (2013)
351. B. McGovern, R.B. Palmini, N. Grossman, E.M. Drakakis, V. Poher, M.A.A. Neil, P. Degenaar, A new individually addressable micro-LED array for photogenetic neural stimulation. IEEE Trans. Biomed. Circ. Syst. **4**(6), 469–476 (2010)
352. L. Chaudet, M. Neil, P. Degenaar, K. Mehran, R. Berlinguer-Palmini, B. Corbet, P. Maaskant, D. Rogerson, P. Lanigan, E. Bamberg, B. Roska, Development of optics with micro-LED arrays for improved opto-electronic neural stimulation. Optogenetics: Opt. Meth. Cell. Control **8586** (2013)
353. J.T. DiPiro, R.G. Martindale, A. Bakst, P.F. Vacani, P. Watson, M.T. Miller, Infection in surgical patients: effects on mortality, hospitalization, and postdischarge care. Am. J. Health-System Pharmacy **55**(8), 777–781 (1998)
354. C.D. Owens, K. Stoessel, Surgical site infections: epidemiology, microbiology and prevention. J. Hosp. Infect. **70**(Suppl 2), 3–10 (2008)

Chapter 7
Power Harvesting and Data Exchange Links

B. Gil, H. Ip and Guang-Zhong Yang

List of Acronyms

AC	Alternate current
ADC	Analog-to-digital converter
AM	Amplitude modulation
ASIC	Application specific integrated circuit
ASK	Amplitude shift keying
BPSK	Binary phase shift keying
BSN	Body sensor network
BVD	Butterworth–Van Dyke Model
CMOS	Complementary metal-oxide-semiconductor
CW	Continuous wave
DAC	Digital-to-analog converter
DBB	Digital baseband
DC	Direct current
DPSK	Differential phase shift keying
DRAM	Dynamic random access memory
DSSH	Double synchronized switch harvesting
ECG	Electrocardiogram
EM	Electromagnetic
FBAR	Film bulk acoustic resonator
FCC	Federal Communications Commission
FDA	Food and Drug Administration
FM	Frequency modulation
FSK	Frequency shift keying
IC	Integrated circuit
IDT	Interdigital transducer
IoT	Internet of things
ISM	Industrial, scientific and medical band

B. Gil (✉) · H. Ip · G.-Z. Yang
The Hamlyn Centre, Imperial College London, London, UK
e-mail: b.gil-rosa@imperial.ac.uk

G.-Z. Yang (ed.), *Implantable Sensors and Systems*
https://doi.org/10.1007/978-3-319-69748-2_7

KLM	Krimholtz–Leedom–Matthaei model
LC	Inductor–capacitor circuit
LDO	Low drop output
LED	Light-emitting diode
LPF	Low-pass filter
LSK	Load shift modulation
MEMS	Microelectromechanical system
MES	Miller encoding scheme
MFC	Micro-fibre composites
MICS	Medical implant communication service band
MOSFET	Metal oxide semiconductor field effect transistor
MPE	Maximum permissible exposure
MRI	Magnetic resonance imaging
OOK	On-off keying
PA	Power amplifier
PDMS	Polydimethylsiloxane
PCL	Polyprolactone
PLL	Phase-locked loop
PLLA	Polyactide
PM	Phase modulation
PMOS	p-channel MOSFET
PMU	Power management unit
POR	Power-on-reset
PPy	Polypyrrole
PSK	Phase shift keying
PTE	Power transfer efficiency
PUT	Programmable unijunction transistor
PV	Photovoltaic array
PVDF	Polyvinylidene difluoride
PWM	Pulse-width modulation
RC	Resistor–capacitor circuit
RF	Radiofrequency
RFID	Radiofrequency identifier
RLC	Resistor–inductor–capacitor circuit
RX	Receiver
SAR	Specific absorption rate
SAW	Surface acoustic wave resonator
SCR	Silicon controlled rectifier
SDRAM	Synchronous dynamic random access memory
SECE	Synchronous electric charge extractor
SSHI	Synchronized switch harvesting on inductor
TEG	Thermoelectric generator
TX	Transmitter
USB	Universal serial bus

UTET	Ultrasonic transcutaneous energy transfer
UWB	Ultra-wide band
VCO	Voltage-controlled oscillator
WMTS	Wireless medical telemetry service band
WPC	Wireless power consortium

7.1 Introduction

In the previous chapters, we have discussed the basic principle and the current state of the art in implantable sensor design, embodiment, and ASIC-based on-node processing. For data exchange of implantable devices, wireless links are unavoidable except for the case when an indwelling catheter or probe is allowed to establish either a direct or close contact with the implantable sensor. Light transmission via optical fibers can offer a solution to accomplish data exchange. However, without a conductive path to the outside world, the environment found inside the human body for the propagation of electromagnetic radiation poses new challenges. The problem of data exchange in implantable sensors only encounters a contender of the same level when sensor powering comes to play, at least for *active* sensing systems. It is therefore possible to retrieve data from *passive* sensors with no need for DC powering, as will be discussed later in this chapter. Nevertheless, the vast majority of implantable sensors are still actively powered and the subject of power consumption cannot be overlooked [1]. Low power consumption is of paramount importance in implantables to ensure long-term function of the sensor and patient safety.

In practice, the operational longevity of implantable sensors is typically dictated by the time between battery replacements. As the dimensions of implantable devices decrease, so do energy consumption and battery size. However, battery technology still does not follow the tendency towards miniaturization of other electronic components, which compromises the operation of devices in the long term [2]. Electrochemical capacitors can eliminate this gap by providing higher energy and power densities than conventional lithium batteries, which can be further enhanced by micro-supercapacitors through faster charging and discharging cycles [3]. In addition, the development of flexible and stretchable batteries and supercapacitors can revolutionize the future of implantable sensors and packaging [4].

In this chapter, we will describe the current and future approaches to wireless powering and data exchange for implantable sensors. Strategies devoted to reducing power demand must include low-power transducing techniques (capacitive and inductive), integrated electronics, and smart compression algorithms for measurement and transmission data sets.

7.2 Power Sources for Implantables

Alternative power sources for sensors deployed remotely from base stations have been extensively studied in the past for environmental monitoring, military surveillance, agricultural applications, weather forecast, animal movement tracking, and biomedical applications. The types of energy that can be harvested from the surrounding environment include ultrasounds, electromagnetic, magnetic, mechanical, thermoelectric, photovoltaic, and electrochemical. In addition, power harvesting can also be obtained from the physiological environment itself where the implantable sensor is deployed within the body. Reducing the size of the electronic components can lower the power requirements down to the range of mW, as illustrated in Fig. 7.1. Some practical examples of energy consumption for biomedical devices are listed in Table 7.1, whereas the maximum powers harvested from implantable solutions found in the literature are presented in Table 7.2, some of which are further described.

Electrochemical reactions that involve substances other than the natural ones produced by body fluids pose a potential threat to patient health. With this approach, glucose fuel cells have been proposed in different configurations depending on the catalyst used to enable the electrochemical reaction: enzymatic, microbial, and abiotic. For the first type of glucose cells, the examples reported in literature show a maximum value of power density not exceeding a level of 430 μWcm^{-2}, which decreases drastically within hours of continuous operation [5]. Kerzenmacher et al. performed an extended literature review of the performance of implantable abiotically catalysed glucose fuel cells and established a range of 50 and 2.2 μWcm^{-2} for in vitro and in vivo applications, respectively. In either of the configurations, factors such as the expected lifetime of the enzyme and the covalent bond established between enzymes and/or mediators with the electrodes have shown to decrease energy extraction [6]. To overcome these limitations, Cinquin et al. proposed a

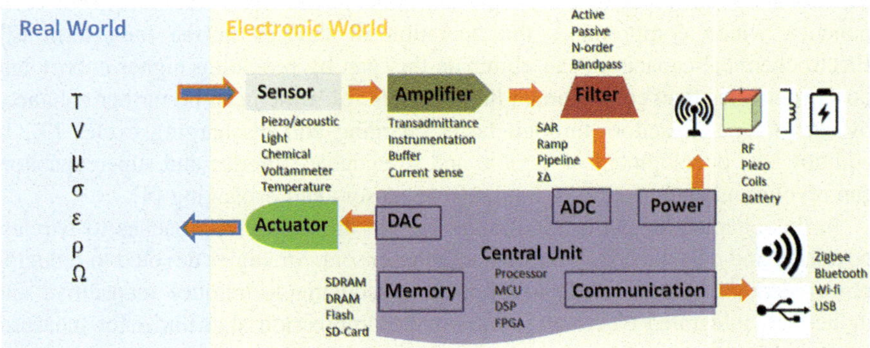

Fig. 7.1 Basic electronic modules for monitoring physiological variables with real-world physical characteristics: T temperature, V electric voltage, μ magnetic permeability, σ electrical conductivity, ε dielectric permittivity, ρ mass density, Ω electrical resistance

Table 7.1 Energy consumption for typical implantable devices and relative power consumed internally by the electronic modules

Implanted device	Power requirement	Electronic module	Portion of power consumption (%)
Cardiac pacemaker	30–100 μW	Sensor and readout	7
Cardiac defibrillator	30–100 μW	Analog-to-digital converter	2
Neurologic stimulator	30 μW to several mW	Microcontroller	2
Drug pump	100 μW to 2 mW	Wireless transceiver	50
Cochlear implants	Several μW to 10 mV	Power management	39

Reprinted with permission from [54], © 2015 IEEE

glucose biofuel cell based on mechanical confinement of redox mediators and enzymes inside the electrodes to extract the energy from glucose and dioxygen in the retroperitoneal space of rats. A stable power of 2 μW under 200 mV was achieved in vivo for a period of 11 days [7]. Nonetheless, the low voltage levels delivered by glucose cells that hardly exceed 0.5 V in physiological conditions have led some researchers to tackle the problem by connecting biofuel cells in electrical series or storing the energy in capacitors/charge pumps and releasing it in short pulse bursts [8]. Experiments performed by connecting five cells independently and in series have proved successful in powering a pacemaker, although impractical for deployment into a single person. The bio-catalytic electrodes are also found to be impractical to implement inside blood vessels unless they are attached to the surface of tissues in close proximity to the capillary bed and in contact with a constant supply of glucose, nutrients, and oxygen. Wearable "tattoos" are also considered as natural candidates for biofuel cells to convert the energy from saliva or sweat. Even the electrochemical gradient that occurs naturally in the inner ear has been exploited to power a cochlear implant, as shown in Fig. 7.2a. Mercier et al. were able to extract an output level of just 1.12 nW from the inner ear of a guinea pig, enough to power a 2.4 GHz RF radio every 40–360 s [9].

Photovoltaic energy seemed impossible for implantable sensors since in-body tissues are not exposed directly to sunlight and this type of energy source should be limited for powering exposed wearable electronics in BSN. However, some recent research in photovoltaic harvesting has proved the contrary. The authors of [10] proposed a photovoltaic array (PV) located beneath the skin to charge implantable cardiac biosensors by using near infrared radiation directly on the skin surface by laser. For this type of irradiation the safety limit imposed by healthcare regulators is 2 mW mm^{-2}. With that in mind, the computational simulations have shown an achievable power of 10–25 mW harvested from this type of radiation. In a different study, Ayazian et al. designed a 2.5 × 2.5 mm^2 CMOS implantable sensor with PV cells to harvest power in the μW range directly from the ambient light that penetrates body tissues. The power level is enough to perform real-time sensing through a set of in vivo polarized electrodes whose signals could be measured directly on the surface of the skin as biopotential recordings [11] (Fig. 7.2b). The CMOS technology employed worked in the subthreshold regime (<500 mV) since a current

Table 7.2 Power harvested by implantable solutions found in literature

Source	References	Output power	Voltage harvested	Frequency	Data link	Distance	Target site
Electrochemical	[5, 6]	50 μW cm^{-2}		–	–	–	In vitro
Electrochemical	[5, 6]	2.2 μW cm^{-2}		–	–	–	In vivo
Electrochemical	[5]	430 μW cm^{-2}		–	–	–	–
Electrochemical	[7]	2 μW	200 mV	–	–	–	Rat
Electrochemical	[9]	1.1–6.3 nW	30–55 mV	2.4 GHz	–	–	Guinea pig
Photovoltaic	[10]	10–25 mW		–	–	–	Simulation
Photovoltaic	[11]	few μW	450 mV	0.1–20 kHz	FSK[a]	3 mm	In vitro
Photovoltaic	[12]	27 μW	450 mV	–	Optical LED (16 kbps)	4.7 mm	Pig
Thermoelectric	[13]	–	11 mV 25 mV	–	–	–	In vitro rabbit
Thermoelectric	[14]	60 μW	100 mV	–	OOK[b] (38.4 kbps)	0 mm	Human chest
Thermoelectric	[15]	130 μW	40 mV	–	–	–	Simulation
Thermoelectric	[18]	645 μW	170 mV	–	–	–	Rat
Mechanical	[19]	30 nW	1.5 V	–	–	–	Pig
Mechanical	[20]	60 nW	3.8 V	–	–	–	Simulation
Mechanical	[22]	3 μW cm^{-3}	–	–	–	–	Simulation
Mechanical	[21]	3 μW	–	–	–	–	Simulation
Inductive	[30]	1–10 mW	1.8 V	6.785 MHz	–	1–10 mm	Air
Inductive	[34]	80 mW	–	8.75 MHz	LSK[c] (36 kbps)	3.5 cm	Air
Inductive	[35]	0.1 mW	–	25 MHz	OOK (2.8 Mbps)	2 cm	Air
Inductive	[37]	42 mW	1.8 V	13.56 MHz (power)	PDM[d] (13.56 Mbps)	1 cm	Air

(continued)

Table 7.2 (continued)

Source	References	Output power	Voltage harvested	Frequency	Data link	Distance	Target site
Inductive	[47]	11 mW	3.3 V	4 MHz	MES[e] (10 kbps)	2.5 cm	Water bearing colloid samples
Inductive	[48]	150 mW	3.6 V	1.056 MHz	–	–	Gastric track
Inductive	[49]	10 mW	2.2 V	1 MHz	–	–	Air
Inductive	[50]	0.6 mW	1.8 V	10 MHz	BPSK[f] (1.12 Mbps)	1.5 cm	Air
Inductive	[51]	50 mW	–	10 MHz	BPSK (1 Mbps)	1 cm	Air
Inductive	[52]	0.2 mW	–	133 kHz	OOK (4 kbps)	1 m	Air
Radio-frequency	[54]	3.5 mW	1 V	–	DPSK[g] (0.01–4.5 Mbps)	–	Fluid body model
Radio-frequency	[72]	–	1.5 V	2.4 GHz	BFSK (42.2 kbps)	10 cm	Pig
Radio-frequency	[73]	513 μW	> 1 V	2.4 GHz (RX) 5.2 GHz (TX)	OOK	3 cm	Saline solution
Radio-frequency	[76]	175 μW	0.75 V	900 MHz	OOK (5 Mbps)	–	Air
Ultrasounds	[88]	20 mW	–	765 kHz	ASK[h] (1200 bps)	–	Water tank
Ultrasounds	[88]	24.7 nW 19.8 nW 14.5 nW	–	24.30 kHz 24.08 kHz 26.23 kHz	–	50 mm	Experimental setup
Ultrasounds	[89]	49 μW	1.46 V	40.43 kHz	–	22 mm	In vitro
Ultrasounds	[91]	100 mW	3.9 V	673 kHz	–	30 mm	Water tank
Ultrasounds	[108]	84 μW	0.8 V	400 kHz	OOK (4800 bps)	30 mm	Lung phantom
Ultrasounds/ inductive	[102]	29 μW	2.4 V	500 kHz	Ultrasonic pulse (220 bps)	70 mm	Experimental setup

[a]Frequency Shift Keying, [b]On/Off Keying, [c]Load Shift Keying, [d]Pulse Delay Modulation, [e]Miller Encoding Scheme, [f]Binary Phase Shift Keying, [g]Differential Phase Shift Keying, [h]Amplitude Shift Keying

of only μA is achieved in subcutaneous tissues from direct sunlight (1.36 mW mm^{-2}). The study shows that the optical loss for subcutaneous tissue is around 40 dB, which is higher than the losses recorded for radiofrequency power delivering systems (20 dB/cm in the 2.4 GHz ISM band). Another subcutaneous sensing device was presented by Sankaragomathi et al. for optical power and data links, which harvests 27 μW of power from an 850 nm infrared source across a 4.75 mm thick pig skin. The efficiency achieved was 4.9% which is still higher when compared to inductive coupling for that transmission distance [12]. In fact, for a 2 × 2 mm^2 implant area, the transmitted optical power from sunlight is 16 mW when accounting for all losses in the path. This means that with 4.9% efficiency up to 0.8 mW of power could be achieved with 100% exposure time. Nonetheless, with exposures under 10%, sensors of the same size can operate at 80 μW at best.

Thermoelectric energy generation involves the presence of a temperature gradient in the vicinity of the implant that limits its location to the skin surface, subcutaneous tissue, or body cavities in contact with air, like the nose or ear, since internal organs intend to keep the same temperature in the physiological condition. Even so, recent advances in the design of thermoelectric materials, such as bismuth telluride, have made it possible to harvest power from just a temperature gradient of 5 K, a difference commonly found between body core temperature and skin. The authors of [13] have shown some theoretical and simulation results regarding the transfer of heat within the human body in order to find the optimal location for a thermoelectric generator (TEG): a higher heat flux is reported as the distance from the body core increases, as well as with the intentional cooling of the skin surface. The authors also reported voltage levels of 11 and 25 mV derived from in vitro and in vivo animal experiments using a commercial TEG with ΔT around 5 K. Wu et al. designed a battery-less wearable ECG monitoring system-in-a-patch capable of delivering an output power of 60 μW from a 5 × 5 cm^2 thermoelectric generator attached to the chest ($\Delta T = 6$ K), depicted in Fig. 7.2c [14]. Similarly, Ashraf and Masoumi designed a fully embodied thermal harvesting system that could operate from input voltages as low as 40 mV to produce voltage levels up to 3 V with $\Delta T = 2$ K by the deployment of a voltage boost converter [15]. The power achieved was enough to power an implantable pacemaker (50 μW and 2.5 V), although a minimum start-up voltage of 60 mV was necessary for charging the storage element of the boost converter. In a different direction, Jo et al. have focused their research on the design of flexible TEGs using polydimethylsiloxane (PDMS) combined with Bi_2Te_3 to achieve an output power of 2.1 μW ($\Delta T = 19$ K) [16], whereas Watkins et al. had long been pursuing the design of Bi_2Te_3-based superlattices, sufficient of generating 100 μW of power with thermal variations in the range from 0.3 to 1.7 K [17]. Rozgic and Markovic developed a thin-film array-based TEG integrated in 65 nm CMOS technology capable of delivering a regulated output power of 645 μW ($\Delta T = 3.5$ K) when operated in vivo in a rat. The stacking in parallel of the TEGs proved efficient to increase output voltage to a level of 170 mV and decrease the source impedance to 6.3 Ω, yielding the high power density reported in [18].

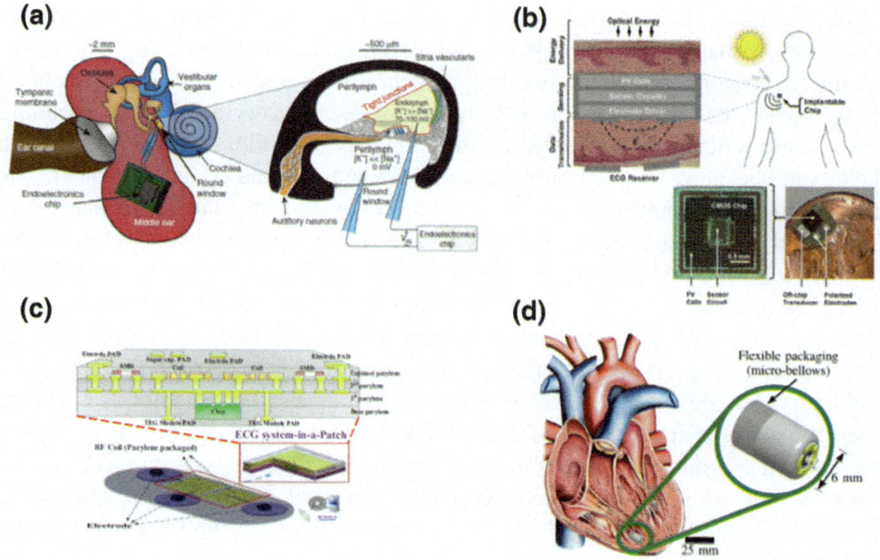

Fig. 7.2 **a** Anatomy and physiology of the inner ear with location of the implantable cochlear device. Reprinted by permission from [9], © 2012 Nature Publishing Group. **b** Photovoltaic-driven CMOS implantable sensor. Reprinted with permission from [11], © 2012 IEEE. **c** Structure of the ECG patch with TEG module and sensing electrodes. Reprinted with permission from [14], © 2015 IEEE. **d** Implantable piezoelectric energy harvester placed within the heart cavity. Reprinted with permission from [22], © 2014 IEEE

Mechanical energy harvesting is mainly limited to the movements of body joints, internal organs, muscles, and circulating fluids to excite (typically) piezoelectric generators that convert mechanical deformation (strain) to electrical voltage passively, i.e. with no use of acoustic pressure waves originated by ultrasounds (to be discussed later). The scavenging of energy by motion of in-body organs is still challenging due to the low frequency content and weak mechanical energy conveyed by these movements, not to mention the undesirable perturbation of the normal functioning of a tissue caused by an energy harvester deployed nearby. For joint movements, the vast majority of piezoelectric generators found in the literature still rely on numerical simulations to attest their designs and performance inside the body. That leaves the cardiovascular system for exploration with some potential power harvesting solutions. Indeed, the heart and circulatory system can provide a stable and inexhaustible mechanical power for the duration of the individual's lifespan by continuous systolic and diastolic motions. It has been shown that the human cardiac output power can reach as high as 1.4 W, a value obtained by the product of the average aortic pressure and cardiac output [19]. Implementation of harvesters directly over the heart muscle can result in the impairment of heart function, and therefore there are significant challenges. The continuous flow of blood in the periphery induces temporary changes in the size and gauge of blood vessels that can be explored as reservoirs of mechanical strain conversion to electrical power. In fact, part of the energy produced by the heart is released via pulsatile expansion of the arteries, which exhibit both high flexibility and adaptability for sensor deployment. In this approach, Zhang et al. developed an electrical power generator for conversion of the pulsating energy provided by the ascending aorta using a flexible polyvinylidene difluoride (PVDF) film coated by aluminum layer and packaged by polyimide membrane to wrap around a latex tube [19]. The maximum voltage and current levels achieved by the implanted harvester were 1.5 V and 300 nA, respectively, with average peak instantaneous output power of 30 nW with 700 ms duration. Fadhil et al. have designed a nanoscale dual layer PVDF film embedded with a biocompatible self-curling silicon artery cuff to produce a power of 60 nW with maximum output voltage level of 3.8 V [20]. Karami and Inman have proposed a bistable piezoelectric harvester capable of keeping an output power above 3 μW for a considerable range of heart rates [21]. More recently, Deterre et al. designed a microspiral-shaped harvester to collect energy from blood pressure changes within the cardiac cycle. The size was compatible with fitting inside an intravenous catheter with cylindrical geometry and capable of producing an output power of 3 μW cm^{-3}, unaffected by heartbeat [22], as depicted in Fig. 7.2d.

Ultrasounds **or** ***electromagnetic power harvesting*** for deeply implanted devices require an efficient source of wave generation in opposition to the natural radiation that can be passively harvested from the environment. Electromagnetic radiation can be further split into two different regimes to exploit different ways of harvesting energy: in the near-field region, the electric and magnetic fields can be decoupled, yielding capacitive and inductive links; in the far-field region, the fields are coupled together and the power is transferred in the form of radio-frequency (RF) waves. The separation of the two regimes is dictated by the frequency of transmission, with the far-field regime emerging in the higher spectrum (>1 GHz). Some aspects regarding electromagnetic radiation are worth mentioning. Although light is also electromagnetic radiation, here we include only low-energy sources of radiation with frequencies up to a few tens of GHz. Secondly, the passive electromagnetic properties of body tissues make it hard to build efficient capacitive links fed by relatively weak electric fields (electrical conductivity) and also contribute to a higher attenuation of RF waves (dielectric permittivity). The magnetic field can pass relative unscathed through the human body since the magnetic permeability is fairly constant among tissues, and because of that inductive links are natural candidates for building wireless links at close distances. More details about wireless powering solutions in the form of induction, RF, and ultrasounds will be discussed in the rest of this chapter. For the three selected links, telemetric capability is also possible through the same carrier wave that powers the device in a sustainable way. Data can be exchanged in both directions: the physiological data sensed by the implant, including its operational condition can be transmitted to the external powering module, whereas programming and actuator commands travel in the opposite direction. The modulation strategy impressed in the carrier signal can increase the quality of the signal in the presence of noise and/or nearby disturbances and improve the security of the link against external attacks aiming to retrieve sensitive patient-related data. Considerations related to power consumption are translated in appropriate designs of modulation and demodulation schemes inside the implantable electronics. Signal amplitude, carrier frequency, bandwidth, number of active channels, and noise interference are all factors that need to be considered to ensure minimally accurate transmission of data. In fact, there exists a communication band expressly designed for implantables that works on the Medical Implants Communication Services (MICS) band, from 402–405 MHz.

For active implants, it is highly recommended to project a Power Management Unit (PMU) inside the electronics to manage the energy harvesting procedure and transfer the energy to a storage element (capacitor or battery) while delivering power to the remaining subsystems [2]. The power-on-reset activates the system that enters an idle state of lower power consumption, interrupted in a timely manner by watchdog circuits that monitor any variation in the working environment worth triggering the active operation of the sensor. When awakened from idle to an active state, the sensor performs any data acquisition and transmission before entering the idle state again for the shortest time interval possible. The power-down sequence is activated when the cycling operation of the device is no longer required (no stimulus arriving from the exterior) or worse when a power failure is verified. The possible integration includes electronic stages for buck/boost converter to increase the harvested voltage amplitude to meet the requirements of the sensor; power routing switches to interface active loads during operation and prevent charge leakage when disconnected (high impedance); non-volatile memory to store critical information about the PMU itself; and pulse generators and oscillator/timer circuits to ensure proper device initialization and operational stability in a myriad of environmental conditions.

Finally, micro-nanotechnology is a new emerging field in power scavenging that offers potential for operating smaller scale sensors with power demands in the range of nanowatts. Every source inside the human body could provide sufficient power to operate such micro-nanodevices from muscle vibrations, human pulse, or even the waste energy from biological processes [23].

7.3 Inductive Link

Wireless powering via inductive means is ubiquitous with wirelessly chargeable smart devices currently widely available in the consumer market. Standards such as Qi and WPC are widely adopted in commercially available ICs, making wireless charging/powering accessible to the vast majority of device designers. Before inductive power transfer became popular amongst wearable gadgets and IoT devices, it had been one of the most successfully deployed wireless powering modalities used in transcutaneous implants such as cochlea prosthetics. The latest research on inductive powering aims to improve efficiency from low to moderate power consumption, facilitating a simultaneous high-speed data link. In this section, we will discuss the latest developments in this area, including power and voltage efficiency, optimal load, coil design and misalignment/deformation, and telemetry schemes. We will conclude this section with a brief discussion of wireless inductive power transfer and safety concerns over Specific Absorption Rates (SARs).

7.3.1 Two Coil Inductive Power Transfer

We start by considering two scenarios of inductive power transfer using two coils, as depicted in Fig. 7.3. Assuming a series configuration in the primary side, where maximum current flows in the inductor coil at resonance frequency $1/2\pi\sqrt{L_p L_c} = 1/2\pi\sqrt{L_s L_s}$. On the secondary side, one can adopt a series or parallel resonance configuration. Some design equations for the series and parallel circuits are summarised in Table 7.3. For wireless power transfer, we are interested in power transfer efficiencies η_s and η_p, with voltage gain also being important as it determines the choice of voltage regulation circuit required. The power transfer efficiency of both parallel and series circuits varies with the load R_L as well as the frequency. Earlier work in inductive power transfer has resulted in guidelines for selecting a fixed frequency or load for optimal power transfer. This is effective only under fixed conditions, such as primary to secondary coil separation and alignment. In the next section, we discuss the latest developments in this field, where adaptive schemes are used to dynamically adjust circuit parameters such that maximum efficiency is always achieved. So far we have summarised both parallel and series configurations on the secondary side of an inductive power link. Power transfer is maximised when the primary and secondary circuit are tuned to the same frequency, with the exception of very tightly coupled systems, where deviation from resonance can have its advantage [24]. Later we will see that for improved data signal integrity a parallel arrangement in the primary side can be advantageous, whereas series configuration can maximise the magnetic field emitted by the primary coil.

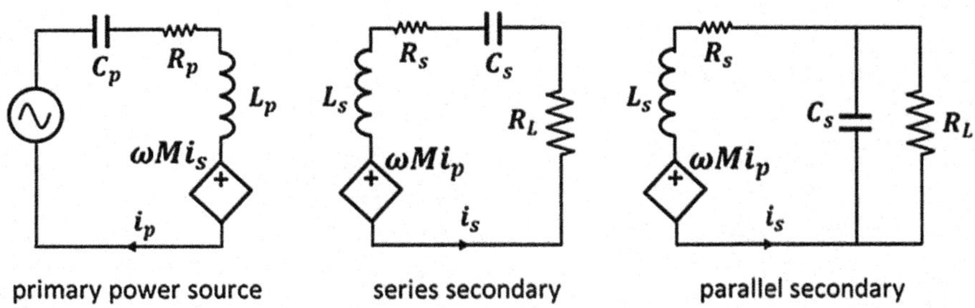

Fig. 7.3 Two-coil inductive power transfer configuration. The primary power source is configured as a series LC resonance circuit. The secondary coil on the implant side can be configured in parallel or series LC resonance. The mutual inductance M is given by $M = k\sqrt{L_1 L_2}$ with geometry dependent coupling coefficient k

Table 7.3 Design equations for two coil inductive power transfer

	Series configuration	Parallel configuration
Voltage gain [25]	$\frac{\omega M}{Z_p Z_{ss} + \omega^2 M^2} R_L$	$\frac{\omega M}{Z_p Z_{sp} + \omega^2 M^2} \left(\frac{1}{j\omega C_s} // R_L \right)$
Frequency at $\boldsymbol{R_L}$ in dependent voltage gain [25]	$\sqrt{\frac{\omega_p^2 + \omega_s^2 \pm \Delta}{2(1-k^2)}}, \Delta = \sqrt{\left(\omega_p^2 + \omega_s^2\right)^2 - 4(1-k^2)\omega_p^2 \omega_s^2}$	$\frac{\omega_p}{\sqrt{1-k^2}}$
Voltage gain at $\boldsymbol{R_L}$ independent frequency [25]	$-j\sqrt{\frac{L_s}{L_p}} \frac{\omega_p^2 + \omega_s^2 \pm \Delta}{(1-2k^2)\omega_p^2 \pm \omega_s^2 + \Delta}$	$-j\frac{1}{k}\sqrt{\frac{L_s}{L_p}}$
Power efficiency [24]	$\eta_s = \left\{ 1 + \frac{R_s}{R_L} + \frac{R_p}{k^2 L_p R_L} \left[L_s + \left(\frac{(R_L + R_s)^2}{L_s} - \frac{2}{C_s} \right) \frac{1}{\omega^2} + \left(\frac{1}{L_s C_s^2 \omega^4} \right) \right] \right\}^{-1}$	$\eta_p = \left\{ 1 + \frac{R_s}{R_L} + \omega^2 R_L R_s C_s^2 + \frac{R_p}{k^2 L_p L_s R_L} \left[\left(\frac{(1-\omega^2 L_s C_s) R_L + R_s}{\omega} \right)^2 + (L_s + R_L C_s R_s)^2 \right] \right\}^2$
Efficiency at resonance [24]	$\eta_s = 1 + \frac{R_s}{R_L} + \frac{R_p (R_L + R_s)^2}{k^2 L_p L_s \omega^2 R_L}$	$\eta_p = \left\{ 1 + \frac{R_s}{R_L} + \frac{R_L R_s}{L_s^2 \omega^2} + \frac{R_p}{k^2 L_p L_s R_L} \left[\frac{R_s^2}{\omega^2} + \left(L_s + \frac{R_L R_s}{L_s \omega^2} \right)^2 \right] \right\}^{-1}$

$\boldsymbol{Z_p, Z_{ss}, Z_{sp}}$ are primary, secondary series, and secondary parallel impedance respectively given by: $\boldsymbol{Z_{ss} = j\omega L_s + \frac{1}{j\omega C_s} + R_s + R_L}$, $\boldsymbol{Z_{sp} = j\omega L_s + R_s + \frac{1}{j\omega C_s} // R_L}$ and $\boldsymbol{Z_p = j\omega L_p + \frac{1}{j\omega C_p} + R_p}$. Primary and secondary resonant frequencies are defined as $\boldsymbol{\omega_P = 1/\sqrt{L_p C_p}}$ and $\boldsymbol{\omega_s = 1/\sqrt{L_s C_s}}$

For the secondary implant side, it can be shown that there exists a "crossover frequency" that, when exceeded, renders secondary series resonance more efficient than secondary parallel resonance [24]. This crossover frequency is given as

$$f_c = \frac{R_L}{2\pi L_s}\sqrt{1 - \frac{k^2 L_p r_s}{L_s r_p}} \tag{7.1}$$

where r_s and r_p are the respective lossy secondary and primary resistors. For small k in the case of implants, the crossover frequency becomes $R_L/2\pi L_s$. For series resonance to dominate and assuming that the tuned frequency is $1/2\pi\sqrt{L_s C_s}$, we have a condition on the load resistor given by (7.2).

$$\sqrt{L_s/C_s} \geq R_L \tag{7.2}$$

Thus, for a fixed load R_L and resonant frequency, one can determine whether series or parallel resonance can be used for derived values of C_s and L_s.

Typically, an inductive power link is constructed with the LC resonance circuits along with a driver amplifier on the primary side as a power source and a rectifier circuit following the LC resonance components on the secondary side. The rectifier, together with voltage regulation circuits, converts the RF magnetic field into a usable DC voltage for subsequent electronics in the implant. Figure 7.4 describes two commonly used rectifier circuits. These provide a low voltage drop compared to standard diode connected transistors and are advantageous for implementation using commercially available CMOS processes where Schottky diodes are not commonly available.

The rectifier shown in Fig. 7.4a has the top diode connected PMOS transistor's bulk tied to the drain. This effectively connects the parasitic PN diode in parallel with the PMOS. The diode drawn is implemented the same way as the top PMOS. With this configuration, V_{sg} of M_x is always two diode threshold voltage $2V_{TD}$ drops, forcing it to switch on and charge up V_{cap} until V_{in} is smaller than the threshold of M_x. Under this condition, $V_{out} = V_{in} - (|V_{TP}| - V_{TD})$, with V_{TD} being the threshold voltage of the bottom PMOS transistor. Thus such a configuration permits a minimal voltage drop across the rectifier. For the rectifier shown in

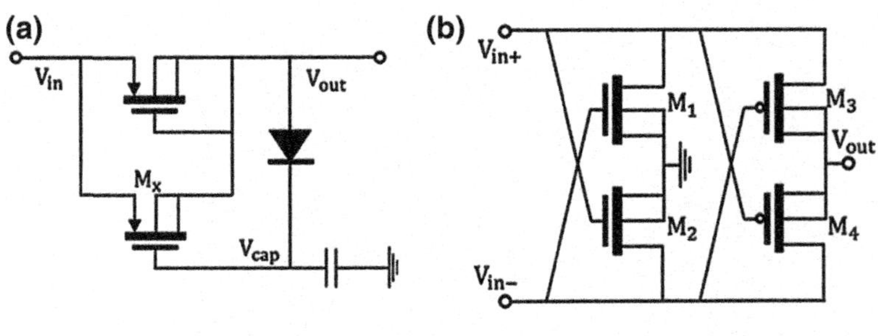

Fig. 7.4 **a** Low voltage drop rectifier diode [26]. **b** Self-driven synchronized rectifier [27]

Fig. 7.4b, the transistor pairs (M_1, M_2) and (M_2, M_3) work in complementary mode, with one pair conducting at a time. The PMOS transistors have the bulk connected such that the parasitic PN junction diode works in parallel with the transistors. Unlike simple bridge rectifiers with diodes, the rectifier shown here utilises the source and drain terminals as input/output. The voltage drop is not equal to the transistor threshold voltages as in the case of diodes, but instead depends on the channel resistance.

7.3.2 *Maximizing Power Transfer*

High power efficiency of the link is important to save energy, minimize tissue RF exposure, and minimize any interference with other electronics. One way of dynamically maintaining efficient power transfer with the varying load and coupling conditions of an inductive link is to note that, for each condition, an optimal frequency of power transfer exists for maximizing the power delivered to the secondary side. Common power transmitters for inductive link powering comprise fixed capacitors and a coil inductor. To vary the transmission frequency, one can switch a capacitor array in and out of the resonance tank. The drawback of direct capacitor switching is the discrete frequency values achievable as well as switch losses. In [28], a soft switching scheme is proposed for continuous frequency variations. Figure 7.5 shows a current-fed resonance converter where switches S_x and S_y switch on alternately. This creates sinusoidal voltage variations across the terminals of the tank capacitor. Whenever one of the terminals (V_x or V_y) reaches beyond a reference voltage (V_{ref}), the auxiliary capacitor at that terminal (c_x or c_y) is switched out of the circuit. These capacitors have an effective value given by:

$$1/2\left[2c_x\left[1-\cos\left(\sin^{-1}\frac{V_{\mathrm{ref}}}{V_x}\right)\right]+c_x\left(\pi-2\sin^{-1}\frac{V_{\mathrm{ref}}}{V_x}\right)\frac{V_{\mathrm{ref}}}{V_x}\right] \tag{7.3}$$

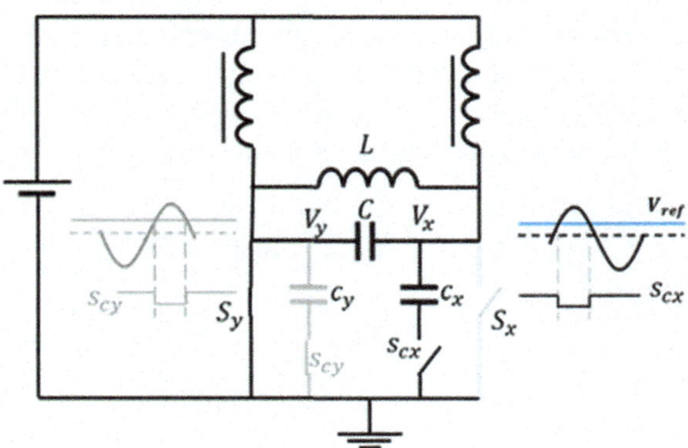

Fig. 7.5 A soft switching circuit emulates the effect of a dynamically varying capacitor for adaptive frequency wireless power transfer, maximizing the efficiency of an inductive power link

By detecting the power delivered to the load by monitoring the load-induced voltage on the primary side, one can vary the frequency accordingly by shifting V_{ref} in the transmitting circuit, which results in a closed-loop wireless power regulator. An alternative dynamic power adjustment scheme takes advantage of the fact that power efficiency is load-dependent. One can maintain maximum efficiency under varying coil distance and alignment by dynamically changing the load of the secondary setup [29]. This is in contrast with the frequency adaptation scheme, which is implemented at the power source and relies on load detection. The energy transfer efficiency of an inductive link can be expressed as the "effective Q" in the system [30] as:

$$\eta = \frac{k^2 Q_1 Q_2'}{1 + k^2 Q_1 Q_2'} \frac{Q_2}{Q_2 + Q_L} \tag{7.4}$$

where Q_1 and Q_2 are the respective quality factors of the primary and secondary coils given by $\omega L_p / R_p$ and $\omega L_s / R_s$. Q_2' is the loaded Q-factor of the secondary side taking into account the effective contribution of the load in series with R_s and given by $Q_2' = \omega L_s / (R_s + R_{L,eq})$. Q_L is the quality factor of the load network given by $Q_L = \omega R_L C_s$ for a parallel RC load. By differentiating the above equation, we can express the optimum load for maximum power by

$$R_{L,opt} = \frac{1}{k \omega C_s} \sqrt{\frac{Q_s}{Q_p}} \tag{7.5}$$

For which, the maximum efficiency by tuning the optimal load is obtained by:

$$\eta_{max} = \frac{k^2 Q_1 Q_2}{(k Q_1 + 1)(k Q_2 + 1)} \tag{7.6}$$

Thus one design strategy to maximize Power Transfer Efficiency (PTE) is building an optimal load (DC equivalent) resistance. This will work for a fixed coupling medium and distance and coil alignment, with k being constant. To carry out load matching over varying coupling coefficients, the authors of [29] have proposed a way of dynamically adapting the load to accommodate a change in coupling factor k, while keeping the effective load at the optimum value. Both the secondary inductor resistance R_s and the load resistance R_L are transformed by periodically switching in/out the load resistor as

$$\begin{aligned} R_{L,eq} &= R_L \left[1 - D + \frac{1}{2\pi} \sin(2\pi D) \right], \\ R_{s,eq} &= R_s + R_{sw} \left[D - \frac{1}{2\pi} \sin(2\pi D) \right] \end{aligned} \tag{7.7}$$

with D being the duty cycle when the load is shorted. When the load is switched out by shorting with a switch, energy is stored in the L_sC_s tank, whereas when the load is switched back in while the switch is open, energy is transferred to the load. The optimum duty cycle D for maximising power efficiency is obtained by the solution of:

$$1-D+\frac{1}{2\pi}\sin(2\pi D)=\frac{R_\mathrm{s}}{R_\mathrm{L}}\sqrt{(1+k^2Q_\mathrm{p}Q_\mathrm{s})} \tag{7.8}$$

The duty cycle of the Q-modulated load can be controlled by a monostable pulse generator [29], with the duty cycle controlled by the threshold voltage of comparators. This voltage is provided by a charge-pump circuit driven by a finite state machine with inputs derived from an envelope detector to measure the result of duty cycle change and adapt accordingly. The maximum power efficiency load adaptation reported in [29] is similar to the load modulation used for back telemetry.

The key to maximising power transfer efficiency is to match the reflected load at the primary side to half of the source resistance. As seen from above, this can be achieved by electronically modifying the effective load closer to its optimum value. An alternative approach is to utilise multiple resonance coils to *transform* the load to a maximum value. Let us consider a general M stage configuration, depicted in Fig. 7.6. For an all-series configuration, the load reflected back from stage $J+1$ to stage j is given by [31]:

$$Z_{j,j+1}=k_{j,j+1}^2\omega_0 L_j Q_{(j+1)L} \tag{7.9}$$

with ω_0 being the identical tuned frequency for all stages and $Q_{(j+1)L}$ computed iteratively as well, in accordance with (7.10), with unloaded quality factors $Q_{j+1}=\omega_0 L_{j+1}/r_{j+1}$ and $Q_\mathrm{mL}=\omega_0 L_\mathrm{m}/(r_\mathrm{m}+R_\mathrm{L})$:

$$Q_{(j+1)L}=\frac{Q_{j+1}}{1+k_{j+1,j+2}^2 Q_{j+1}Q_{(j+2)L}} \tag{7.10}$$

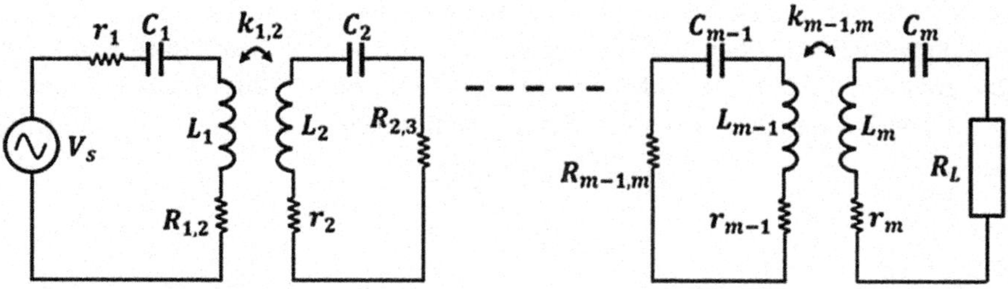

Fig. 7.6 A four-coil inductive power transfer schematic. Considering coupling between adjacent coils only, loading effects are transferred from R_L to $z_{3,4}$ and from $z_{3,4}$ to $z_{2,3}$ and eventually reach $z_{1,2}$

The power transfer efficiency at each stage is defined in (7.11), as the ratio between the reflected load from the next stage and the inductor series resistance.

$$\eta_{j,j+1} = \frac{z_{j,j+1}}{r_j + z_{j,j+1}} \tag{7.11}$$

By iteratively propagating the final load R_L back to the first stage where the power source V_s is present, one can derive the overall power transfer efficiency. The advantage of a multi-coil scheme (also known as *magnetically coupled resonators* [32]), when compared to the two-coil scheme, is the optimal power transfer given a fixed load in the proceeding case, which can be optimised by varying parameters such as coupling coefficient and quality factors across the stages. However, for the two-coil scheme, the PTE is fixed for a given load and Q-factors of the coils. A three-coil scheme was reported in [31] with a single coil on the primary and two coils on the secondary implant side. The overall efficiency of this three-coil scheme is given by:

$$\eta_{3\text{coil}} = \frac{\left(k_{1,2}^2 Q_1 Q_2\right)\left(k_{2,3}^2 Q_2 Q_{3L}\right)}{\left(1 + k_{1,2}^2 Q_1 Q_2 + k_{2,3}^2 Q_2 Q_{3L}\right)\left(1 + k_{2,3}^2 Q_2 Q_{3L}\right)} \frac{Q_{3L}}{Q_L} \tag{7.12}$$

which achieves optimal PTE for the condition given by

$$k_{2,3} = \left(\frac{\left(1 + k_{1,2}^2 Q_1 Q_2\right)\left(1 + \frac{R_L}{R_3}\right)^2}{Q_2^2 Q_{3L}^2}\right)^{1/4} \tag{7.13}$$

One can thus maximise the PTE (given the fixed distance $d_{1,2}$ and R_L) by first tuning the geometries of L_1 and L_2 in order to improve $k_{1,2}^2 Q_1 Q_2$ and then, subsequently, tune L_3 to satisfy the last equation.

While using a single coil for power and telemetry can lead to performance constraints, multi-coils can be used for respective power and data transmission of inductive links. A single coil tuned to two resonant frequencies for simultaneous power and data linking has recently been reported in [33]. The implant side consists of double frequency-tuned LC resonators, which comprise two inductors (one receiver coil and one lumped element) and two capacitors. The resonant frequencies of the dual circuit in terms of the series resonant frequency $f_{LC1} = \frac{1}{2\pi\sqrt{L_1 C_1}}$ and ratios $r_L = L_1/L_2$ and $r_C = C_1/C_2$ are shown in Fig. 7.7.

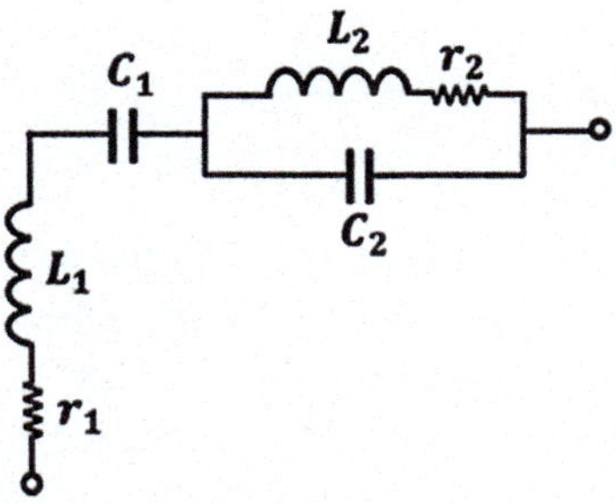

$$f_1 = f_{LC1}\sqrt{\frac{1}{2}+\frac{r_C}{2}\left[1+r_L-\sqrt{\left(\frac{r_L r_C-1}{r_C}\right)^2+2\left(\frac{r_L r_C+1}{r_C}\right)+1}\right]}$$

$$f_2 = f_{LC1}\sqrt{\frac{1}{2}+\frac{r_C}{2}\left[1+r_L+\sqrt{\left(\frac{r_L r_C-1}{r_C}\right)^2+2\left(\frac{r_L r_C+1}{r_C}\right)+1}\right]}$$

Fig. 7.7 Tuning circuit with coil and lumped inductor tuned to two frequencies

7.3.3 *Load-Modulated Data with Power Link and Inductive Telemetry*

In the context of inductive coupling, the power consumption of the secondary implant load is transferred to the primary side in the form of an impedance element in series with the primary coil. This provides the opportunity for the implant to relay information back to the external unit by modulating the implant power consumption. This is known as Load Shift Keying (LSK), widely reported in the literature and commonly implemented in RFID tags. Consider the inductive link shown in Fig. 7.8, where a general load Z_2 is present. The mutual inductive coupling configuration means that series voltages are induced on the primary (secondary) side by the secondary (primary) current flow [34]. The impedance seen at the primary side due to the secondary load current i_s is given by:

$$Z_r = \frac{(\omega M)^2}{Z_2} \tag{7.14}$$

Clearly the load impedance Z_2 at the secondary is reflected back to the primary side. With Z_2 representing capacitor C_2 in parallel with a resistive (AC equivalent) load R_L, the primary load impedance now becomes.

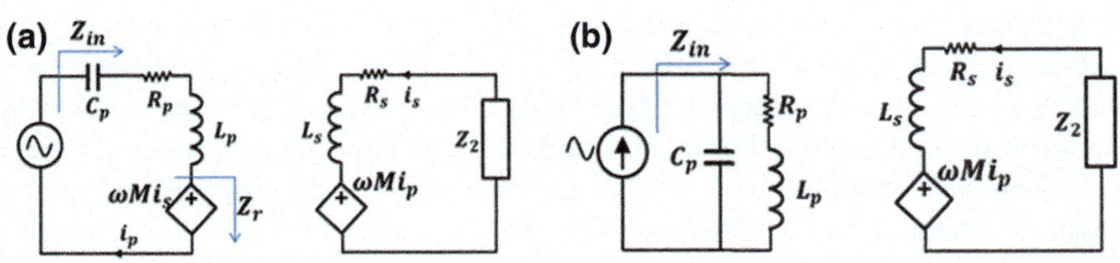

Fig. 7.8 **a** Reflected load in primary circuit at series resonance. **b** Reflected load at parallel resonance

$$Z_{\mathrm{r}} = \frac{(\omega M)^2}{\frac{R_L}{1+\omega^2 C_2^2 R_{\mathrm{L}}^2} + \mathrm{j}\omega\left[L_2 - \frac{C_2 R_{\mathrm{L}}^2}{1+\omega^2 C_2^2 R_{\mathrm{L}}^2}\right]} = k^2 \omega^2 L_{\mathrm{p}} C_2 R_{\mathrm{L}} \tag{7.15}$$

with the assumption that the primary and secondary coils are tuned to the same resonant frequency $1/2\pi\sqrt{L_{\mathrm{p}}C_{\mathrm{p}}}$ and $C_2 R_{\mathrm{L}}^2 \gg L_2$. A common LSK scheme switches R_{L} between two values for binary data transmission. In Fig. 7.8, the primary capacitor and inductor resonate out during resonance. The impedance Z_{in} seen by the power source can then be expressed in terms of the quality factors of the primary and secondary circuit [35] by

$$Z_{\mathrm{in}} = R_{\mathrm{p}}\left(1 + k^2 Q_1 Q_2\right) \tag{7.16}$$

where k is the geometry-dependent coupling coefficient and the quality factors are given by $Q_1 = \sqrt{\frac{L_{\mathrm{p}}}{C_{\mathrm{p}}}}\frac{1}{R_{\mathrm{p}}}$ and $Q_2 = \sqrt{\frac{L_{\mathrm{s}}}{C_2}}\frac{1}{R_{\mathrm{s}}+R_{\mathrm{L}}}$ for Z_2 being a series load with capacitance C_2 and resistance R_{L}. In the case of data transmission utilizing a separate coil for power transmission there is no minimum load R_{L} modeling the current drain of the implant circuit. Maximum modulation can be implemented with binary values represented by setting $R_{\mathrm{L}} = 0$ or open circuit 0. For a parallel primary circuit, the reflected load is given by

$$Z_{\mathrm{in}}\big|_{//} = \frac{L_1}{C_1} \frac{1}{R_{\mathrm{p}}(1 + k^2 Q_1 Q_2)} \tag{7.17}$$

Thus the modulation term now appears in the denominator of the impedance. In contrast, the typically low R_{p} value in a series resonance circuit makes detection of a load-modulated signal difficult. The parallel circuit is also advantageous due to a weaker dependence on the coupling parameter k. The parallel LC configuration can be excited by a transistor acting as a transconductance stage (V-I converter) with its gate driven by a square wave [35]. The gain of the amplifier with the LC load is maximised when the circuit is driven at its resonant frequency $1/2\pi\sqrt{LC}$, and hence the harmonics in the input square wave are filtered out. For this configuration, maximum binary modulation is again achieved by open and closed circuit values for Q_2.

For both parallel and series primary configurations, a simple envelope detector can be used to recover the LSK uplink binary signal, whereas modulation is a simple switch across the secondary terminals. This simplistic approach is sufficient for a low data rate uplink. For more data-demanding applications, such as invasive brain computer interfaces with a large number of recording channels, differential envelope detectors and data recovery synchronised with the carrier by means of a phase-locked loop at the primary receiver can be used for single-coil designs, or a separate data/power coil architecture can be adopted. With LSK, a data only (with modulation circuit separately powered) link in [36] achieved a data rate of 2.8 Mbps over a 25 MHz carrier. Nevertheless, a single-coil LSK scheme limits the data rate to the carrier frequency of power transmission, which is not always sufficient for the application. Furthermore, the inevitable quality factor (power transmitted) to data bandwidth tradeoff for LSK with a single coil could be problematic in some cases. An alternative scheme to LSK for inductive links, known as pulse harmonic modulation, taking advantage of time-domain inter-symbol interference suppression techniques has been proposed in [36]. Efficient data transmission is achieved by taking advantage of the resonating nature of an LC circuit, where applying a short pulse results in sustained ringing of the high Q circuit without using an explicit, often power-hungry, oscillator/amplifier. The long ring time prevents a second bit from being transmitted without introducing inter-symbol interference: a second pulse with the right timing (eliciting an anti-phase ring) is applied to the LC circuit to supress the ringing initiated by the first pulse. The impulse response of the inductively coupled parallel secondary and primary pair is given by

$$\begin{aligned} h(t) = &-\frac{A}{a_2}\mathrm{e}^{-a_1 t}[a_1 \sin(a_2 t) - a_2 \cos(a_2 t)] \\ &-\frac{B}{b_2}\mathrm{e}^{-b_1 t}[b_1 \sin(b_2 t) - b_2 \cos(b_2 t)] \end{aligned} \tag{7.18}$$

where A and B are constants depending on the mutual inductance and the damped oscillations are characterised by the frequency terms $a_2 \approx b_2 \approx 1/\sqrt{L_2 C_2} = 1/\sqrt{L_1 C_1}$ and damping factors $a_1 \approx 1/2r_1 C_1$ and $a_2 \approx 1/2r_2 C_2$. Based on this impulse response model, one can compute the timing (multiples and a half cycle after the first pulse) and the amplitude (coincides with the signal amplitude at the time where the second pulse is applied) of the second pulse for damping out the ringing of the first pulse. The transmission of data within the presence of a power link with a two-coil system is always problematic due to coupling of the power signal into the data line. The authors of [37] have proposed a way of mitigating this interfering effect though the observation that the overall combined waveform of the data and power signals has zero crossings that are correlated with the data bits. With well-timed data pulses synchronised with the power carrier, one can achieve data rates up to the power carrier frequency.

7.3.4 *Effect of Coil Misalignment, Deformation, and Tissue Interface*

So far we have considered an inductive power/telemetry link in a circuit's perspective while paying little attention to the actual coil implementation and assuming perfect coaxial alignment of primary and secondary coils. In practice, coils can be made up of wound wires or implemented on a printed circuit board as well as realised with 3D manufacturing techniques. Less than perfect alignment of the coils results in a power drop. Closed form equations of power transfer have been derived by [38]. For coaxially aligned coils, the power transfer function is given by

$$\eta_0 = \frac{\mu_0^2\pi^2\omega^2}{16R_{tx}R_{rx}}\left[\sum_{i=1}^{n}\frac{a_i^2}{(a_i^2+D^2)^{\frac{3}{2}}}\right]^2\left[\sum_{j=1}^{k}b_j^2\right]^2 = \frac{\mu_0^2\pi^2\omega^2N_{tx}^2N_{rx}^2a^4b^4\omega^2}{16R_{tx}R_{rx}(a^2+d^2)^3} \tag{7.19}$$

where R_{tx} and R_{rx} are the resistive loss in the respective series resonance primary and secondary circuits; n is the number of turns in the primary coil and k is the number of turns in the secondary coil; and a, b, D are the dimensions of the coils as shown in Fig. 7.9.

$$\eta_1 = \frac{\mu_0^2\omega^2}{16R_{rx}R_{tx}\Delta^2}\left[\sum_{i=1}^{n}\left(\frac{m_i}{2a_i\Delta}\right)^{\frac{1}{2}}\left(K(m_i)\Delta + \frac{a_i m-(2-m_i)\Delta}{2-2m_i}E(m_i)\right)\right]^2\left[\sum_{j=1}^{k}b_j^2\right]^2 \tag{7.20}$$

with $m_i = 4a_i\Delta/\left[(a_i+\Delta)^2+d^2\right]$ and $K(m_i)$ and $E(m_i)$ denoting the complete elliptic integrals of the first and second kind, respectively. For short solenoids this can be simplified to

$$\eta_{ls} = \frac{\mu_0^2\omega^2mN_{tx}^2N_{rx}^2b^4}{64aR_{rx}R_{tx}\Delta^3}\left[K(m)\Delta + \frac{am-(2-m)\Delta}{2-2m}E(m)\right]^2 \tag{7.21}$$

and here $m = 4a\Delta/\left[(a+\Delta)^2+d^2\right]$. For angular misalignments, the link efficiency is calculated in accordance with (7.22).

The right-hand side of (7.19) is the well-known result obtained for short solenoid coils. For coils laterally misaligned by an amount Δ, the power transfer equation now becomes

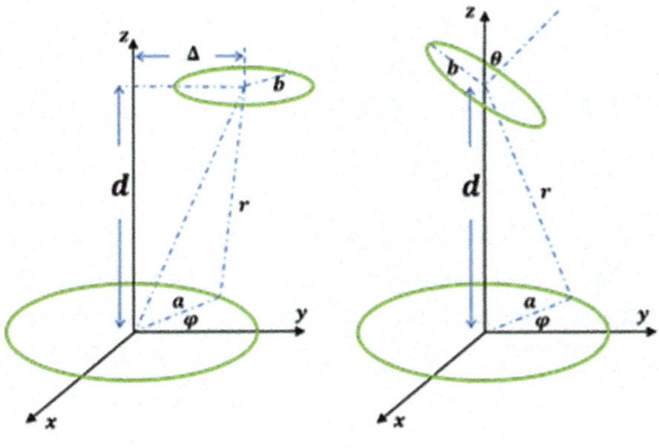

Fig. 7.9 Coil lateral and axial misalignment models

$$\eta_A = \frac{\mu_0^2\omega^2\pi^2}{16R_{rx}R_{tx}}\left[\sum_{i=1}^{n}\frac{a_i\pi\cos\theta^2}{\sqrt{a_i^2+D^2}}\right]^2\left[\sum_{j=1}^{k}b_j^2\right]^2 = \frac{\mu_0^2\omega^2\pi^2N_{rx}^2N_{tx}^2b^4a^4cos\theta^2}{16R_{rx}R_{tx}(a^2+D^2)^3} \tag{7.22}$$

For inductive powering, the increase in frequency results in a higher induced voltage on the secondary side. However, an increase in frequency for a given coil size results in increased radiative component effects and greater tissue absorbance. Furthermore, the near-field magnetic field drops off with distance following a cubic law, rendering a rapid drop in received power with increased distance. Specific studies of energy transfer across tissue medium with mm scale coils have been reported in the literature. While inductive power transfer in the MHz range leads to relatively large coils, the authors in [39] reported that, for small antennas, on the mm scale operating in the body, the mid-field regime is advantageous compared with near-field inductive coupling.

With reduced coil size and source–implant distance, the magnetic component received by a small coil is limited. The mid-field approach carrying energy with a combination of magnetic and radiative components is suitable for efficient power transfer to small implant coils with operating frequency in the low GHz range. When taking into account near- and far-field transmission, the optimal frequency of power transmission in a number of human tissues also coincides with the low GHz range. With the involvement of mid-field radiation, the coil designs are distinct from the coils used for near-field inductive power transfer only. This is due to the fact that mid-field radiation patterns can be concentrated by manipulating the geometry to generate the desired current density of the source antenna.

Apart from misalignment of the transmission and receiver coils, coil deformation also has an impact on inductive power transfer performance. This is not uncommon for implanted coils or RFID coils fabricated on a flexible substrate. Generally, for bent coils the resulting inductance is proportional to the projected coil area with the proportionality constant being geometry-dependent [40, 41]. For a rectangular coil like that shown in Fig. 7.10, a first-order approximation of the inductance is given by

$$L \approx 518 + 420w_{\text{bent}}l_{\text{eff}} + 363l_{\text{eff}} \tag{7.23}$$

The projected width w_{bent} is obtained by resorting to (7.24). A more accurate model utilising second-order terms of w_{eff} and l_{eff} can be found in [41].

$$w_{\text{bent}} = 2r\sin\left(\frac{w_{\text{eff}}}{2r}\right) \tag{7.24}$$

Coils that are inserted in the body as part of an implant system are often hermetically sealed for protection against body fluids. The coil coating and surrounding environment play an important role in shifting the nominal in-air inductance, parasitic capacitance, and resistance of the coil. As a result, implanted coils exhibit different tuning characteristics compared with values calculated

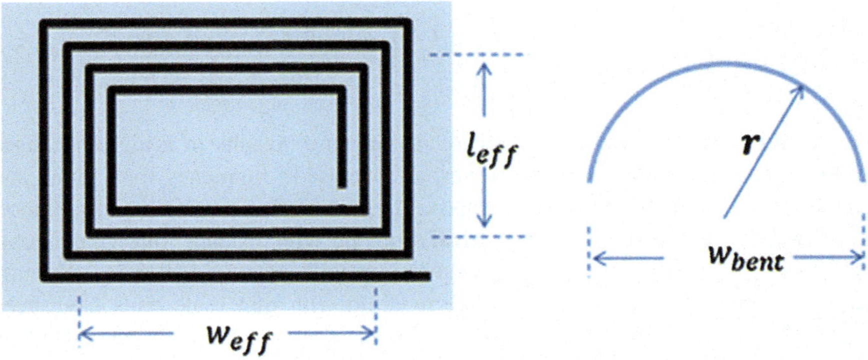

Fig. 7.10 A rectangular coil with average width and length denoted by w_{eff} and l_{eff}. The projected width after bending is denoted by w_{bent}, with the radius of curvature being r

without taking into account the final deployment environment. A model is developed by the authors of [39] such that the electrical parameters of implanted coils can be estimated and the power transmission (e.g. tuned frequency) of a coil pair under implant conditions can be optimised. The study focused on square coils with rounded corners, with the inductance given as

$$L = 1.27\mu \frac{n^2 d_{\text{avg}}}{2}\left[\ln\left(\frac{2.07}{\varphi}\right) + 0.18\varphi + 0.13\varphi^2\right] \tag{7.25}$$

where n is the number of turns, μ is the permeability, and d_{avg} the average width of the coil; φ is the fill factor defined by $(d_{\text{o}} - d_{\text{i}})/(d_{\text{o}} + d_{\text{i}})$ with outer and inner widths d_{o} and d_{i}, respectively. The parallel parasitic capacitance depends on the dielectric constants and thicknesses of the coils' surrounding materials by

$$C_{\text{p}} = C_{\text{e}} l_{\text{c}} + C_{\text{ov}} \tag{7.26}$$

with l_{c} being the overall conductor length of the coil, calculated as $l_{\text{c}} = 4n(d_{\text{o}} - w) - (2n+1)^2(s+w)$, where s and w represent the track spacing and width of the printed coil, respectively; C_{ov} is the overlapping capacitance by the routing of inductor terminals out from the inner part of the coil and obtained by (7.27); and C_{e} is the capacitance value that mimics the effect of various layers surrounding the coil.

$$C_{\text{ov}} = \varepsilon_0 \frac{\varepsilon_{r5}+1}{2} + \frac{\varepsilon_{r5}-1}{2}\left(1+\frac{12}{\frac{w}{t_5}}\right)^{-\frac{1}{2}} - \frac{\varepsilon_r - 1}{4.6}\frac{t_0/t_5}{\sqrt{w/t_5}} \tag{7.27}$$

The explicit formula for C_{e} involves a considerable number of terms and, with reference to Fig. 7.11, it can be expressed as

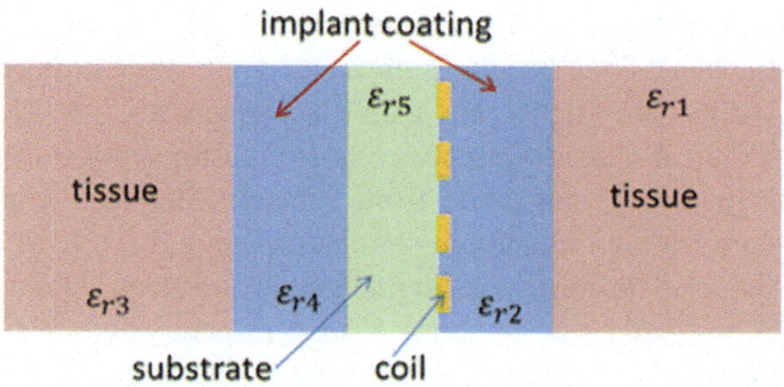

Fig. 7.11 Implant coil modeling taking into account the coating and tissue interface [42]

$$C_e = \varepsilon_0 \frac{K(k_0')}{K(k_0)} \left[1 + \frac{\alpha}{2} \left[(\varepsilon_{r1} - 1) \frac{K(k_1')}{K(k_1)} + (\varepsilon_{r2} - \varepsilon_{r1}) \frac{K(k_2')}{K(k_2)} + (\varepsilon_{r3} - 1) \frac{K(k_3')}{K(k_3)} + + (\varepsilon_{r4} - \varepsilon_{r3}) \frac{K(k_4')}{K(k_4)} + (\varepsilon_{r5} - \varepsilon_{r4}) \frac{K(k_5')}{K(k_5)} \right] \right] \tag{7.28}$$

with $k_0 = 2s/(s + 2w)$, $k_0' = \sqrt{1 - k_0^2}$, $\alpha = K(k_0')/K(k_0)$, $k_i = \tan\mathrm{h}\left(\frac{\pi s}{4t_i}\right)$, and $k_i' = \sqrt{1 - k_i^2}$, and $K(k_i)$ is the complete elliptic integral of the first kind. In turn, the effective series resistance taking into account the skin effect and eddy currents can be modeled by

$$R_s = \rho_c \frac{l_c}{wt_0} \left[\frac{t_0}{\delta(1 - e^{-t_0/\delta})} \frac{1}{1 + t_0/w} + \frac{1}{10} \left(\frac{w}{\omega_{crit}} \right)^2 \right] \tag{7.29}$$

with ρ_c being the resistivity, $\delta = \sqrt{\rho_c/\pi\mu f}$ the skin depth and $\omega_{crit} = 3.1(s + w)R_m/\mu_0 w^2$. R_m is the sheet resistance of the metal used. Given the parameters L, C_p, and R_s, the optimal power transfer efficiency for a fixed load can then be realized by tuning the coil geometries accordingly in a two-coil setting. Maximization of PTE is obtained as described in the previous sections.

7.3.5 Inductive Powering and EM SAR Safety Limits

International guidelines on electromagnetic exposure limits are based on Specific Absorption Rate (SAR). For standards set by the International Commission on Non-ionizing Radiation Protection (ICNIRP) [43], this is limited to 2 W kg^{-1} averaged over 10 g of tissue for local exposure and an average of 0.08 W kg^{-1} for the whole body limit. The IEEE C95.1-1999 standard has a more restricted local exposure limit based on averaging over 1 g of tissue for 1.6 W kg^{-1} local SAR, whereas the subsequently revised IEEE C95.1-2010 has a more relaxed limit of 2 W kg^{-1}. The authors of [44] report simulation results correlating with electromagnetic field measurements of coupled magnetic resonators for power transfer. Their results indicate that the most stringent exposure limits defined in IEEE C95.1-1999 are exceeded given an 0.5 A_{rms} 8 MHz current flowing in identical transmit/receive coils of six-turn spirals with outer diameter of 5.8 cm and a pitch of 1 cm made from 2.54 mm diameter copper wires. The SAR limiting currents are obtained in simulation using a six years old male high-resolution MRI model. The corresponding magnetic and electric field limits are of the order of a few A m^{-1} and few hundred V m^{-1} respectively. In a study related to the Wireless Power Consortium (WPC) commercial standard for small electronic equipment, such as smartphones, the authors of [45] reported the resulting SAR on human models during exposure to corresponding wireless charging equipment. The transmission coils used in this study are indicative of those recommended by the WPC standard and measures an inner diameter of 12 mm with outer diameter of 40 mm over 20 turns of 0.5 mm diameter wires with backing magnetic plates. The study in [45] shows SAR values way below international limits when the charging coil pairs are placed at distances down to 10 mm from the human body, indicating the effectiveness of the magnetic plates in containing the energy field. This however, has limited application in the implant scenario, as inductively powered implants have tissue between the transmitting and receiving coils. For specific electromagnetic field safety for brain implants, the study in [46] investigated the peak SAR exposure averaged over 10 g of tissues over a range of power and data transmission frequencies. The coils used are up to 4 cm outer diameter with 18–20 turns using 0.5 mm tracks and have a transmitter–receiver separation of 1 cm. The authors found peak SAR increasing as a function of frequency, with brain tissue absorbing significantly more power compared to bone and skin. With a fixed power level set to 1 W, 2 W kg^{-1} (10 g tissue average) is reached at 20 MHz for brain tissue. While keeping a constant transmission power at 1 W, the SAR also reduces with coil radius. This is not unexpected, as increased coil radius results in the power spreading out over a larger area, reducing the peak exposure.

7.4 Radio-Frequency Link

Radio-frequency transmissions dominate the wireless communications for many years due to the success of cellular phones, thus opening a new era for personal data exchange and interconnectivity. The need for higher bandwidths to withstand the constant streaming of information has led to the separation of the electromagnetic spectrum into frequency communication bands controlled by international standards for the radio and microwave ranges. The success of communication protocols such as WiFi, Bluetooth, and ZigBee has assured the interconnectivity between smart sensors, mobile/cell phones, computers, and the internet. In the context of implantable medical devices, radio-frequency telemetry is still a preferred way for data exchange for untethered devices and the exclusive allocation of the MICS band (and others) sets the high expectations of the international community for RF-based implantable devices. However, the propagation of low-energy EM waves in the interior of the human body is different from open-air transmission due mainly to the differences in the dielectric permittivity of body tissues recorded in the far-field regime. Antenna design is thus one of the most critical aspects during the projection of any RF-operating implantable devices, not only due to dimension constrains and/or biocompatibility of the casing but also the effective radiative power within safety limits. Power harvesting from external RF sources has long played a secondary role with respect to communication, but recent developments in materials science and low-power electronics have brought the possibility of power extraction from the RF carrier wave to the domain of feasibility, as a valid alternative to inductive powering. In fact, the inaccurate geometrical alignment between an implantable sensor and external RF devices no longer limits the efficiency of power transference, making it more orientation-immune with larger operational distances. These topics and many others involving in RF powering and data exchange will be further explored in the incoming sections.

7.4.1 Biological Models, Sources of RF, and Safety Limits

The human body is a known harsh environment for propagation of RF fields. The electromagnetic response of biological tissues imposes a serious reduction on the power and integrity of RF waves as they travel across tissues and their boundaries. The human body can thus reflect and absorb these waves, inducing attenuation, delay and phase distortion of the RF signals. The heterogeneous distribution of dielectric permittivities among biological tissues is responsible for this phenomenon, which impacts not only the strength of the external RF waves, but also the operability of implantable antennas to receive and transmit RF data without detuning when surrounded by a highly variable physiological environment. The effect is more pronounced as we move towards higher frequencies in the far-field regime, so it would be more logical to employ low-frequency wireless transmissions. However, lower frequencies imply lower communication speeds, limited range of operation, and a larger antenna implant. Therefore, finding a tradeoff between all the physical variables involved in the selection of the operational frequency is crucial for the proper design of RF links. Moreover, modern wireless communications outside the body environment account also for non-negligible sources of interference that further compromise the efficiency of the RF-based communication channels for implantable devices.

Standard regulations for medical device radio communications concerning biological exposure limits and power-level requirements are controlled by the Federal Communications Commission (FCC), which sets a maximum output power of –16 dBm on the transmitter side and a channel bandwidth of 300 kHz with differential phase-shift keying modulation for the MICS band (402–405 MHz). Yet medical devices are also allowed to operate in other radio bands, such as the Industrial, Scientific and Medical band (ISM), centred at 900 MHz, 2.45 GHz, and 5.8 GHz;

the Wireless Medical Telemetry Service band (WMTS) with frequencies allocated to 608–614, 1395–1400, and 1429–1432 MHz; and the emerging Ultra Wide-Band (UWB) that explores the spectrum located between 3.1 and 10.6 GHz. Regulation may be found to be more relaxed or attractive for operation in one particular radio band than the remaining ones; however, the FCC still defines limits to the exposure to uncontrolled sources of RF that are incidental on the body surface or absorbed by a specific volume of tissue, through the criteria for Maximum Permissible Exposure (MPE) and SAR. For instance, the averaged value for the MPE over a 30 min exposure is 10 W m^{-2} at 2.4 and 5.8 GHz, whereas for SAR the limit is set at 0.08 W kg^{-1} averaged over the whole body and 1.6 W kg^{-1} for instantaneous power peaks deposited over a cube weighing 1 g of body tissue [53]. These restrictions function as a guideline for the design of radiating antennas and associated electronics to evaluate how much RF power can be withstood by the biological tissues without causing damage, as shown in Fig. 7.12. The value for power flux density at any point on the radiating path can be obtained as

$$S(x) = \frac{P_{\mathrm{EIRP}}}{4\pi x^2} = \frac{P_{\mathrm{T}} G_{\mathrm{T}} e_{\mathrm{T}}}{4\pi x^2} \tag{7.30}$$

where x is the distance to the radiating source and P_{EIRP} is the equivalent isotropically radiated power from the antenna [52]. The latter can be further expanded as the product of antenna input power P_{T} with antenna gain G_{T} and efficiency e_{T}. The RF power absorbed by a cubic volume of tissue is expressed as

$$P_{\mathrm{SAR}} = l^2(S_0 - S_l) = l^2 S_0\left(1 - e^{-\frac{2l}{\lambda_{\mathrm{d}}}}\right), \quad \lambda_{\mathrm{d}} = \sqrt{\frac{1}{\pi f \mu_0 \sigma}} \tag{7.31}$$

(a)

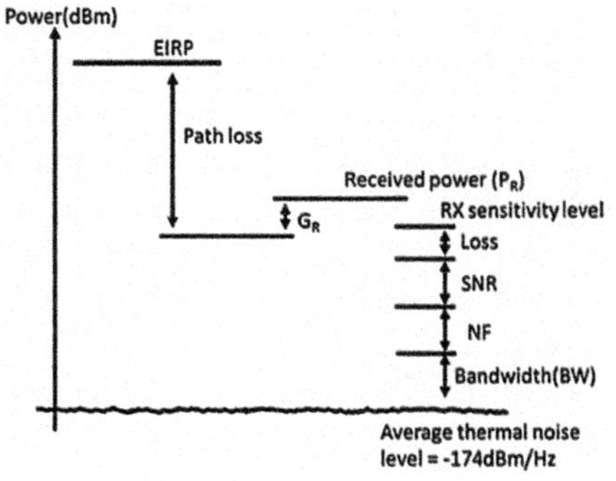

(b)

Powering Frequency	*Implant Location*	*Average Power Flux Density at Implant (W/m^2)*	*Average Usable Power Received by Implant (μW)*
915MHz	Between skin and adipose	4.93	84.4
	Between adipose and muscle	4.64	79.4
	1 cm into muscle	2.37	40.6
	2 cm into muscle	1.21	20.7
2.4GHz	Between skin and adipose	4.38	10.9
	Between adipose and muscle	3.16	7.86
	1 cm into muscle	1.57	3.90
	2 cm into muscle	0.561	1.40
5.8GHz	Between skin and adipose	4.88	2.08
	Between adipose and muscle	2.20	0.936
	1 cm into muscle	0.129	0.0549
	2 cm into muscle	0.00150	0.000638

Fig. 7.12 **a** Link budget analysis for an RF implantable sensor. Reprinted with permission from [53], © IEEE 2015. **b** Power flux densities at different implant depths and corresponding maximum average power at the disposal of the implant. Reprinted with permission from [52], © IEEE 2013

with l being the edge length of the cube, S_0 the power flux density that enters the cube, and S_l the remaining flux that leaves the cube.

The second equality in the last equation reflects the exponential decay of the power flux density as it crosses biological tissue, characterized by a skin depth given by λ_d that depends on the radiating frequency f, the permeability of free space μ_0, and tissue conductivity σ. Translated into safety levels, the power flux density that enters the cube of tissue must be inferior to 1 W m^{-2} for whole-body SAR and 17 W m^{-2} for peak SAR when powering at 2.4 GHz. The calculated values bound the average MPE limit of 10 W m^{-2} for 30 min exposure, which then imposes an antenna–tissue separation larger than 9 cm, in accordance with (7.30), for the case of an isotropic antenna (gain of 0 dB) with a 100% matching network and driven by an input power of 1 W. In practice, the radiated power from the transmitting antenna suffers from path loss in free space, which further lowers antenna–tissue separation. For the purpose of power harvesting, a separation of 1 cm is reasonable to be assumed between the external antenna and body surface, which implies a power reduction close to −13 dBm to meet MPE specifications, a reduction to be shared between the effective power delivered to the antenna, the antenna's characteristics, and path loss.

At the reception side, the maximum received power by the implantable device can be calculated from (7.32), which only accounts for the RF link budget analysis.

$$P_R = P_{EIPR} - P_L + G_R \tag{7.32}$$

The receiver antenna's gain is now represented by G_R, whereas P_L is the path loss and fading margin of RF channels. Maximum power received does not translate necessarily into maximum available power for operation of the deeply implanted device, since one part is lost to the surrounding tissues and embedded electronics. A more realistic approximation to the useable power at the disposal of the implant is given by (7.33), which depends on the average power flux density at the depth of the implant S_d (exponential decay from the flux reported at the surface S_0, as previously mentioned), the wavelength of the RF signals λ, and the efficiencies of conversion electronics e_E and matching network e_R at the receiver.

$$P_R = S_d \frac{e_E e_R G_R \lambda^2}{4\pi} \tag{7.33}$$

The conversion electronics is responsible for converting the incoming RF power to a DC voltage for powering, whereas the matching network tries to adapt the antenna's impedance to the overall input impedance posed by the implantable electronics. Although the efficiency in power conversion increases with the amplitude of the RF signals, reported levels in the literature mention an average reachable level of only 20% [52]. The antenna's gain is also shown to be moderately low (>–20 dB) since it is particularly difficult to design an efficient antenna whose size and mode of operation is constrained by the implant itself. Translated into numbers and expecting maximum matching network efficiency, the average usable power at the disposal of the implantable sensor would be around 2 μW if the sensor were to be deployed 1 cm below the skin and operating at 2.4 GHz.

In the case of RF links that possess telemetric capabilities, the successful demodulation of data with integrity checking requires that the received power must be larger than the sensitivity level of the RF channel employed, as [53]

$$\text{Sensitivity}\, RF = -174 + 10\log_{10}(BW) + SNR + NF + \text{Loss} \quad (7.34)$$

where BW is the data bandwidth, SNR is the minimum signal-to-noise ratio required for the type of data modulation, NF is the noise figure of the receiver front end, and Loss accounts for the remaining implementation losses. A minimum path loss of around −65 dB must be accommodated by the implantable device in order to comply with the sensitivity requirement imposed by the IEEE802.15.6 standard [52].

From the above discussions, it is important to note that the design of RF powering and data exchange systems for implantable devices must take into account several factors, including the radiation pattern of the transmitting antenna, the antenna–tissue surface separation, and RF power levels attainable for transmission and reception, so as to ensure the link's efficiency within biological safety levels. One safety restriction not mentioned regards the localized rise in tissue temperature produced by EM waves, which can lead to thermal damage [54]. Since, in principle, the trusted external RF source will not produce an injury of this kind due to the supervision of a clinician and correct handling, the same is not valid for deeply implanted devices, where thermal damage can occur either from the EM waves produced by the sensor itself or from heat dissipation during circuit operation.

7.4.2 RF Antenna Design

Antenna design is the most critical process in RF link projection, as all the modules in the electronic chain that follow the antenna will be affected by its performance. It plays the same role as coils in near-field inductive links and piezoelectric transducers in ultrasounds. Modeling of antennas to be deployed for operation in implantable devices is particularly challenging due to the high level of heterogeneity of the human body, composed by manifold layers of tissues with different dielectric properties. The major considerations faced by implantable antenna design that does not have counterparts in large-scale telecommunication systems are tissue absorption, tissue reflection, and antenna matching in the implanted environment [55]. The attenuation of RF signals with distance caused by tissue absorption of part of the carried energy follows an exponential decay, governed by the attenuation coefficient. The coefficient is a function of the frequency of transmission and dielectric properties of the medium, being particularly adverse for loss biological tissues. RF signal reflections result from electromagnetic impedance mismatches found at the interfaces of different layers of tissue that decrease the strength of the signal that effectively crosses the boundary in any direction. Finally, the environment where the antenna is located further contributes to a change in the impedance and resonance frequency from free-space performance values that can detune the antenna.

7.4.2.1 Types of Antennas

An antenna is a system designed to radiate or receive electromagnetic waves as defined by the IEEE. There exist a broad range of sizes and shapes available for antennas, and the design ultimately dictates their radiation patterns. Antennas are commonly divided into different types, such as wire, aperture, reflector, microstrip (patch), dielectric, lens, fractal, and array antennas. The main parameters to assess the performance of an antenna (for a given topology) are the radiation pattern, directivity, efficiency, gain, and input impedance.

The *radiation pattern* is a mathematical representation of the radiative properties of the antenna that depends on the spatial coordinate system adopted. Due to the vast number and intricate configurations of antennas, the analysis of the mathematical radiation pattern can be very complicated, it being almost impossible to obtain a closed-form solution for the boundary value problem in RF propagation [56]. The spherical coordinate system shown in Fig. 7.13a is commonly adopted to derive the radiative field of an antenna in all spatial directions. An *isotropic* antenna is defined as having an equal radiation pattern in all directions, whereas a *directional* antenna radiates electromagnetic waves more efficiently in some directions than others and an *omnidirectional* radiator possesses a non-directional pattern in a given plane and a directional radiation pattern in any other orthogonal plane, as shown in Fig. 7.13b.

The space surrounding an antenna can be further divided into three zones depending on the distance R between the antenna and the observation point: the reactive near-field, the radiating near-field (Fresnel zone), and the far-field (Fraunhofer zone). The reactive near-field is the zone in the vicinity of the

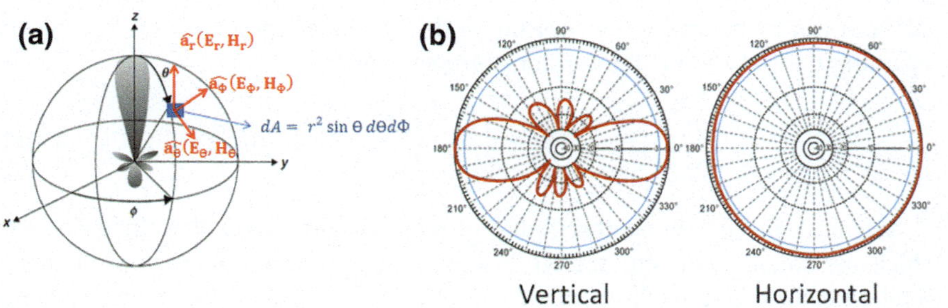

Fig. 7.13 **a** Coordinate system employed in antenna design. **b** Omnidirectional pattern of an antenna

antenna where the reactive field of the EM waves predominates. The transition towards the radiating near-field is given at a distance $= 0.62\sqrt{L^3/\lambda}$, with L the largest dimension of the antenna and λ the wavelength. The radiating near-field is characterized by the predominance of the radiation fields, whose radial components (E_r, H_r) are still appreciable. The border with the far-field region is located at $R = 2L^2/\lambda$ and from this point on, the angular field distribution in both θ and ϕ is essentially independent of the distance from the antenna, and the radial components vanish.

The *average power* radiated by the antenna as electromagnetic energy can be calculated by integrating the normal component of the Poynting vector over a closed spherical surface with radius r as

$$\begin{aligned} P_{\mathrm{rad}} &= \oiint_S W_0 \mathrm{d}S \\ &= \frac{1}{2}\oiint_S \mathrm{Re}(\boldsymbol{E}\times\boldsymbol{H}^*)\mathrm{d}S \\ &= \int_0^{2\pi}\int_0^{2\pi} W_0(r)r^2 \sin\theta \mathrm{d}\theta\mathrm{d}\phi = 4\pi r^2 W_0(r) \end{aligned} \tag{7.35}$$

where W_0 is the radial power density uniformly distributed over the spherical surface (isotropic) and $r^2 \sin\theta\mathrm{d}\theta\mathrm{d}\phi$ is the infinitesimal area $\mathrm{d}A$. The average power can also be calculated using the element of solid angle $\mathrm{d}\Omega$ of a sphere as

$$P_{\mathrm{rad}} = \oiint_\Omega U_0 \mathrm{d}\Omega = \int_0^{2\pi}\int_0^{\pi} U_0 \sin\theta\mathrm{d}\theta\mathrm{d}\phi = 4\pi U_0 \tag{7.36}$$

with U_0 representing the isotropic *radiation intensity* defined over the far-field region. Either W_0 or U_0 are used in the computation of the parameters that determine antenna performance in free space. Hence, the *directivity* of an antenna is defined as the ratio between the radiation intensity produced by the antenna in one direction to the radiation intensity averaged in all directions in space, as given by (7.37). It can be regarded as a measure of the directional properties of the antenna regarding the radiation pattern of an isotropic source.

$$D = \frac{U}{U_0} = \frac{4\pi U}{P_{\mathrm{rad}}} \tag{7.37}$$

Antenna *efficiency* is another figure of merit that measures the efficiency with which the antenna converts the RF power present at its terminals into radiation. Total antenna efficiency is calculated in accordance with (7.38), which includes factors related to losses related to signal reflections arising from the mismatch between the transmission line and the antenna (e_r) and I^2R power waste produced by conduction (e_c) and dielectric (e_d) losses in the antenna's material.

$$e_0 = e_r e_c e_d = e_{cd}\left(1 - |\Gamma|^2\right) \tag{7.38}$$

The conduction and dielectric losses can be combined to yield the radiation efficiency, e_{cd}, separated from the reflection coefficient, which is obtained by itself as

$$\Gamma = (Z_{IN} - Z_0)/(Z_{IN} + Z_0) \tag{7.39}$$

with Z_{IN} and Z_0 being the impedances of the antenna and the transmission line, respectively.

Antenna *gain* relates both antenna efficiency and its directional properties, expressing the ratio of the intensity that would be obtained if all power at the input terminals of the antenna were to radiate isotropically, so gain does take into consideration losses arising from impedance mismatch, in accordance with (7.40):

$$\text{Gain} = \frac{4\pi U}{P_{IN}} = e_{cd}\frac{4\pi U}{P_{rad}} = e_{cd}D \tag{7.40}$$

The *input impedance* is calculated by the voltage-to-current rate registered at its terminals. The value is divided into a real number related to the resistance and an imaginary number that includes the reactance of the antenna:

$$Z_A = R_A + jX_A = (R_r + R_L) + jX_A \tag{7.41}$$

The resistance of an antenna can be further divided into two components: R_r, which represents the radiation resistance of the antenna, and R_L, which accounts for all the resistive losses of the antenna. Lossless antennas are thus ideal, as wave transmission will not be undermined by internal losses.

When connecting the antenna to a wave generator whose impedance is given by Z_g, the transmission line will create a voltage divider at the terminals of the antenna, and the circulating current can be expressed as

$$I_g = \frac{V_g}{(Z_A + Z_g)} = \frac{V_g}{(R_r + R_L + R_g) + j(X_A + X_g)} \tag{7.42}$$

with V_g representing the voltage wave produced by the generator. The power rated at each component of the network is calculated by (7.43) and includes the power delivered to the antenna for radiation, P_r, the power dissipated as heat by antenna's losses, P_L, and also the power dissipated as heat by the internal resistance of the generator, P_g.

$$\begin{cases} P_{\mathrm{r}} = \frac{1}{2}\left|I_{\mathrm{g}}\right|^2 R_{\mathrm{r}} = \frac{\left|V_{\mathrm{g}}\right|^2}{2}\left[\frac{R_{\mathrm{r}}}{\left(R_{\mathrm{r}}+R_{\mathrm{L}}+R_{\mathrm{g}}\right)^2+\left(X_{\mathrm{A}}+X_{\mathrm{g}}\right)^2}\right] \\ P_{\mathrm{L}} = \frac{1}{2}\left|I_{\mathrm{g}}\right|^2 R_{\mathrm{L}} = \frac{\left|V_{\mathrm{g}}\right|^2}{2}\left[\frac{R_{\mathrm{l}}}{\left(R_{\mathrm{r}}+R_{\mathrm{L}}+R_{\mathrm{g}}\right)^2+\left(X_{\mathrm{A}}+X_{\mathrm{g}}\right)^2}\right] \\ P_{\mathrm{g}} = \frac{1}{2}\left|I_{\mathrm{g}}\right|^2 R_{\mathrm{g}} = \frac{\left|V_{\mathrm{g}}\right|^2}{2}\left[\frac{R_{\mathrm{g}}}{\left(R_{\mathrm{r}}+R_{\mathrm{L}}+R_{\mathrm{g}}\right)^2+\left(X_{\mathrm{A}}+X_{\mathrm{g}}\right)^2}\right] \end{cases} \tag{7.43}$$

The maximum power that can be delivered to the antenna occurs in the situation of conjugate matching between the impedances of the generator and the antenna. In the case of a lossless antenna ($R_{\mathrm{L}} = 0$) matched to the transmission line (with $R_{\mathrm{L}} = R_{\mathrm{g}}$ and $X_{\mathrm{A}} = -X_{\mathrm{g}}$), the powers are calculated according to:

$$\begin{cases} P_{\mathrm{r}} = \frac{\left|V_{\mathrm{g}}\right|^2}{8R_{\mathrm{r}}} \\ P_{\mathrm{g}} = \frac{\left|V_{\mathrm{g}}\right|^2}{8R_{\mathrm{g}}} \end{cases} \tag{7.44}$$

Of the total power obtained from the generator on the transmitter side, half will be dissipated as heat by the resistance of the generator and the rest will be delivered to the equivalent radiation resistance of the antenna. On the receiver side, in the condition of maximum efficiency, half of the power captured by the antenna is delivered to the electric load and the remaining half is scattered in the medium.

Regarding all the aforementioned mathematical formulae involved in antenna design for implantable devices, the most common type of antenna is reduced either to wire, patch, or array. A wire antenna is made of a simple conductive wire with an omnidirectional radiation pattern, which can take different configurations, such as dipole, helix, and loop. The loop can adopt any geometrical shape (square, circle, rectangle, triangle, etc.). Two categories are established for loop antennas: electrically small, whose length is less than one-tenth of the wavelength, or electrically large, otherwise. Small-loop antennas for implantable devices are also referred to as magnetic loops due to their inductive behaviour. A helix antenna possesses a configuration similar to a helix. The dipole antenna constitutes two opposing electrical charges (or magnetic poles) with a small separation between them. The length of a dipole antenna is usually set at λ/2 and the feed point for injection of the RF wave is located at the centre of the dipole, where the current experiences maximum amplitude in opposition to the lowest levels registered at the ends of the conductors, or wings. The gain of the dipole antenna also increases with length and resistance decreases from the ends toward the centre of the antenna.

The patch antenna is composed of a metallic radiator conductor bound to a grounded dielectric substrate. This type of antenna is simple, lightweight, inexpensive, and conformal to surfaces [57]. Common shapes for the radiator include squares, circles, ellipses, triangles, and dipoles. The ground plane underneath is effective in redirecting radiation in the forward direction, whereas the dielectric constant of the substrate slows down the propagation of the RF waves: higher dielectric constants imply slower propagation. The disadvantage is a lower overall efficiency of the antenna due to losses in the feed network and substrate material. The antenna can be regarded as an open-ended transmission line with current peaks located at the centre of the patch (minimum voltage levels) and minimum values at the edges (maximum voltage); hence, resistance is higher at the edges and zero at the centre. Feeding the patch antenna can be performed by four different methods: microstrip line, probe, aperture-coupled, and proximity-couple feeding [58].

Finally, array antennas are composed of multi-radiator elements that are disposed on a regular arrangement to obtain a characteristic radiation pattern. The total field of the array is determined by vector addition of each individual field produced by the elements, which can interfere constructively or destructively in the desired direction and pattern. Factors such as the geometrical configuration of the array, spacing between elements, and excitation profile in amplitude and phase of the individual elements are used to shape the overall pattern of the radiation field. Within this approach, the antenna arrays can be divided into two groups: scanning and non-scanning. In the former, the beam can be directed electronically by changing the phase of the elements (phased arrays), whereas for non-scanning arrays changes to the main beam lobe are only possible by moving the antenna direction.

7.4.2.2 Far-Field Electromagnetic Modeling of an Antenna

The analysis of radiation problems involves the resolution of Maxwell's equations for a particular set of temporal and geometrical boundary conditions, which significantly reduce the complexity or high abstraction level encountered in the system of equations originally proposed by Maxwell. The process typically starts by defining the sources of electric and magnetic currents, and then uses the mathematical simplifications posed by the geometry of the problem itself to calculate an approximation to the fields radiated by these sources throughout the domain of simulation. A common approach in the determination of such fields is the introduction of auxiliary functions, which may reveal themselves as just a mathematical curiosity with no physical meaning, to tackle the complexity of the EM system of equations and still yield a valid solution. In this regard, vector potential functions such as the magnetic vector potential (represented by $\boldsymbol{A}$) and electric vector potential (represented by $\boldsymbol{F}$) are used to estimate the magnetic and electric flux densities ($\boldsymbol{B}$ and $\boldsymbol{D}$, respectively), in accordance with:

$$\begin{cases} \boldsymbol{B}_A = \nabla \times \boldsymbol{A} \\ \boldsymbol{D}_F = \nabla \times \boldsymbol{F} \end{cases} \tag{7.45}$$

The auxiliary functions relate the sources of electric and magnetic currents (represented by $\boldsymbol{J}$ and $\boldsymbol{M}$, respectively) to the radiated electric and magnetic field intensities ($\boldsymbol{E}$ and $\boldsymbol{H}$). The only requirement is that the computation of such vectors must be restricted over the bounds of the sources $\boldsymbol{J}$ and $\boldsymbol{M}$, and so they are a function of the observation point coordinates [56].

The introduction of $\boldsymbol{A}$ and the scalar potential ψ in the EM problem at hand causes the field intensity $\boldsymbol{E}$ resulting from the presence of electric current sources only to be written for the harmonic regime as

$$\boldsymbol{E}_A = -\nabla\psi - \frac{\partial \boldsymbol{A}}{\partial t} = -\nabla\psi - \mathrm{j}\omega\boldsymbol{A} \tag{7.46}$$

Equivalently, Ampère's circuit law for the curl of the magnetic field intensity $\boldsymbol{H}$ will look similar to the one given by

$$\nabla\times\boldsymbol{H}_A = \boldsymbol{J} + \mathrm{j}\omega\varepsilon\boldsymbol{E}_A = \frac{1}{\mu}\nabla\times\nabla\times\boldsymbol{A} \tag{7.47}$$

From the algebraic properties of the double curl operator, further simplifications can be made to the last equation:

$$\nabla\times\nabla\times\boldsymbol{A} = \nabla(\nabla\cdot\boldsymbol{A}) - \nabla^2\boldsymbol{A} = \mu\boldsymbol{J} + \mathrm{j}\omega\mu\varepsilon\boldsymbol{E}_A \tag{7.48}$$

Replacing $\boldsymbol{E}_A$ by its new definition (with $k^2 = \omega^2\mu\varepsilon$) leads to:

$$\nabla^2\boldsymbol{A} + k^2\boldsymbol{A} = -\mu\boldsymbol{J} + \nabla(\nabla\cdot\boldsymbol{A} + \mathrm{j}\omega\mu\varepsilon\phi) \tag{7.49}$$

According to Helmholtz's theorem, any vector field is fully characterised if and only if its divergence and curl operators are uniquely defined. The curl of $\boldsymbol{A}$ has been well defined previously, but not its divergence. The degree of freedom available for the selection of the divergence of $\boldsymbol{A}$ is typically used to impose a gauge invariance condition that better fits the electromagnetic phenomena being modelled. For the present electromagnetic problem, the Lorentz condition is the best choice that leads to the definition of the divergence of the magnetic vector potential as

$$\nabla\cdot\boldsymbol{A} = -\mathrm{j}\omega\mu\varepsilon\phi \tag{7.50}$$

which nullifies part of the right-hand side of (7.49) to get the time-varying differential wave equation that models the electromagnetic phenomena in the presence of electrical current sources in the form expressed by

$$\nabla^2\boldsymbol{A} + k^2\boldsymbol{A} = -\mu\boldsymbol{J} \tag{7.51}$$

The closed-form solution for the inhomogeneous vector potential wave equation has been formerly derived in terms of $\boldsymbol{A}$ by multiplying the static solution ($k = 0$) of Poisson's equation with the evanescent radial traveling wave term e^{-jkr} to yield

$$\boldsymbol{A} = \frac{\mu}{4\pi}\iiint_V \boldsymbol{J}\frac{\mathrm{e}^{-jkr}}{r}\mathrm{d}V \tag{7.52}$$

By a similar approach, the electric vector potential $\boldsymbol{F}$ can be obtained in the presence of a magnetic current source when they arise in the context of volume or surface equivalence theorems. The solution for the electric vector potential is

$$\boldsymbol{F} = \frac{\varepsilon}{4\pi} \iiint_V \boldsymbol{M} \frac{\mathrm{e}^{-\mathrm{j}kr}}{r} \mathrm{d}V \tag{7.53}$$

The total field intensities in the domain of simulation are finally computed as

$$\begin{cases} \boldsymbol{E} = \boldsymbol{E}_A + \boldsymbol{E}_F = -\mathrm{j}\omega \boldsymbol{A} - \mathrm{j}\frac{1}{\omega\mu\varepsilon}\nabla(\nabla \cdot \boldsymbol{A}) - \frac{1}{\varepsilon}\nabla\times\boldsymbol{F} \\ \boldsymbol{H} = \boldsymbol{H}_A + \boldsymbol{H}_F = \frac{1}{\mu}\nabla\times\boldsymbol{A} - \mathrm{j}\omega\boldsymbol{F} - \frac{1}{\omega\mu\varepsilon}\nabla(\nabla \cdot \boldsymbol{F}) \end{cases} \tag{7.54}$$

In the far-field region ($r \rightarrow \infty$), the radiated fields are spherical waves and variations due to the radial component are negligible when compared to the θ and ϕ components in spherical coordinates [56]. So they can be separated in the calculation of the integrand found in (7.52) and (7.53), yielding wave equations of the form

$$\begin{cases} \boldsymbol{A} \simeq \left[\hat{a}_r A_r(\theta,\phi) + \hat{a}_\theta A_\theta(\theta,\phi) + \hat{a}_\phi A_\phi(\theta,\phi)\right] \frac{\mathrm{e}^{-\mathrm{j}kr}}{r} \\ \boldsymbol{F} \simeq \left[\hat{f}_r F_r(\theta,\phi) + \hat{f}_\theta F_\theta(\theta,\phi) + \hat{f}_\phi F_\phi(\theta,\phi)\right] \frac{\mathrm{e}^{-\mathrm{j}kr}}{r} \end{cases} \tag{7.55}$$

where $\hat{a}_r$, $\hat{a}_\theta$, and $\hat{a}_\phi$ are unity vectors and A_r, A_θ, and A_ϕ the components of the magnetic vector potential for the r, θ, and ϕ components, respectively; whereas $\hat{f}_r$, $\hat{f}_\theta$, and $\hat{f}_\phi$ are also unity vectors and F_r, F_θ, and F_ϕ the components of the electric vector potential.

Finally, each component of the radiated fields $\boldsymbol{E}$ and $\boldsymbol{H}$ in the far-field region produced by electric and magnetic current sources (like an antenna) can be calculated in accordance with (7.56) with $\eta = \sqrt{\mu/\varepsilon}$ being the impedance of the propagating medium.

$$\begin{cases} E_r \simeq 0 \\ E_\theta \simeq -\mathrm{j}\omega A_\theta - \mathrm{j}\omega\eta F_\phi \\ E_\phi \simeq -\mathrm{j}\omega A_\phi + \mathrm{j}\omega\eta F_\theta \\ H_r \simeq 0 \\ H_\theta \simeq \mathrm{j}\frac{\omega}{\eta} A_\phi - \mathrm{j}\omega F_\theta \\ H_\phi \simeq -\mathrm{j}\frac{\omega}{\eta} A_\theta - \mathrm{j}\omega F_\phi \end{cases} \tag{7.56}$$

7.4.2.3 Reciprocity Between a Pair of Antennas

Any functional RF link involves at least the presence of a pair of antennas that can interchange the operations of transmission and reception. For this scenario, if the mode of operation (or location) of one antenna is interchanged with the other antenna the parameters being measured inside a closed domain should remain equivalent, as imposed by the Lorentz's reciprocity theorem. In the case of two arbitrary systems—denoted by 1 and 2—with two sets of electric current sources—($\boldsymbol{J}_1$, $\boldsymbol{J}_2$)—the resultant electromagnetic field pair—($\boldsymbol{E}_1$, $\boldsymbol{H}_1$) and ($\boldsymbol{E}_2$, $\boldsymbol{H}_2$)—must satisfy the following volume integral:

$$\int_{\infty} \nabla \cdot (\boldsymbol{E}_1 \times \boldsymbol{H}_2 - \boldsymbol{E}_2 \times \boldsymbol{H}_1) \mathrm{d}V = 0 \tag{7.57}$$

Applying the vector identity to the topological operators involving divergence and curl of the form

$$\nabla \cdot (\boldsymbol{E}_1 \times \boldsymbol{H}_2) = \boldsymbol{H}_2 \cdot \nabla \times \boldsymbol{E}_1 - \boldsymbol{E}_1 \cdot \nabla \times \boldsymbol{H}_2 \tag{7.58}$$

and the integrand in the reciprocity theorem can be replaced by

$$\int_{\infty} (\boldsymbol{H}_2 \cdot \nabla \times \boldsymbol{E}_1 - \boldsymbol{E}_1 \cdot \nabla \times \boldsymbol{H}_2 - \boldsymbol{H}_1 \cdot \nabla \times \boldsymbol{E}_2 + \boldsymbol{E}_2 \cdot \nabla \times \boldsymbol{H}_1) \mathrm{d}V = 0 \tag{7.59}$$

Further mathematical simplifications to the equation above are provided by the relations between electromagnetic quantities in the harmonic regime for systems #1 and #2 and given below.

$$\begin{cases} \nabla \times \boldsymbol{H}_1 = \mathrm{j}\omega \boldsymbol{D}_1 + \boldsymbol{J}_1 = \mathrm{j}\omega\varepsilon_1 \boldsymbol{E}_1 + \boldsymbol{J}_1 \\ \nabla \times \boldsymbol{H}_2 = \mathrm{j}\omega\varepsilon_2 \boldsymbol{E}_2 + \boldsymbol{J}_2 \\ \nabla \times \boldsymbol{E}_1 = -\mathrm{j}\omega \boldsymbol{B}_1 = -\mathrm{j}\omega\mu_1 \boldsymbol{H}_1 \\ \nabla \times \boldsymbol{E}_2 = -\mathrm{j}\omega\mu_2 \boldsymbol{H}_2 \end{cases} \tag{7.60}$$

The replacement of the curl operations involving the electromagnetic quantities for both systems by previous relations is performed in (7.61), with de-coupling between the magnetic and electric phenomena.

$$\int_{\infty} [\mathrm{j}\omega(\mu_2 - \mu_1)\boldsymbol{H}_2\boldsymbol{H}_1 + \mathrm{j}\omega(\varepsilon_1 - \varepsilon_2)\boldsymbol{E}_1\boldsymbol{E}_2 + \boldsymbol{J}_1\boldsymbol{E}_2 - \boldsymbol{J}_2\boldsymbol{E}_1] \mathrm{d}V = 0 \tag{7.61}$$

Expanding the integrand into smaller factors and imposing $\mu_1 = \mu_2$ and $\varepsilon_1 = \varepsilon_2$ for a linear, isotropic, and homogeneous medium, we obtain

$$\int_{\infty} \boldsymbol{J}_1 \boldsymbol{E}_2 \mathrm{d}V = \int_{\infty} \boldsymbol{J}_2 \boldsymbol{E}_1 \mathrm{d}V \tag{7.62}$$

For a filamentous configuration of the current-carrying conductors of sources, the current density flowing through a differential volume can be given by the current circulating along a one-dimensional differential path as

$$\begin{cases} \boldsymbol{J}_1 \mathrm{d}V = -|J_1| \mathrm{d}a \mathrm{d}l = -\boldsymbol{I}_1 \mathrm{d}l \\ \boldsymbol{J}_2 \mathrm{d}V = -|J_2| \mathrm{d}a \mathrm{d}l = -\boldsymbol{I}_2 \mathrm{d}l \end{cases} \tag{7.63}$$

which limits the interval of integration over a path Γ_{12} as given by

$$\int_{\Gamma_{12}} \boldsymbol{I}_1 \boldsymbol{E}_2 \mathrm{d}l = \int_{\Gamma_{12}} \boldsymbol{I}_2 \boldsymbol{E}_1 \mathrm{d}l \tag{7.64}$$

The last equation starts to resemble the relation between the current and electric field in the two-port network, which allows us to define two voltages as

$$\begin{cases} V_{12} = \frac{1}{I_1} \int_{\Gamma_{12}} \boldsymbol{I}_1 \boldsymbol{E}_2 \mathrm{d}l \\ V_{21} = \frac{1}{I_2} \int_{\Gamma_{12}} \boldsymbol{I}_2 \boldsymbol{E}_1 \mathrm{d}l \end{cases} \tag{7.65}$$

where V_{21} represents, in the antenna context, the induced open-circuit voltage in antenna #2 referred to the current at its input terminals due to the electrical distribution map produced by #1, whereas V_{12} means the induced open-circuit voltage in antenna #1 due to the field radiated by antenna #2.

The mathematical relations just derived show that antenna performance depends not only on its own current but also on the current distribution of neighbouring elements or, in electronics language, the mutual effects arising from each element in the domain. The input impedance of an antenna is no longer dictated by the self-impedance (in the absence of other elements) but also by the mutual impedance between the driven antenna and the others. For a two-port network, the voltage-current relation can be computed as

$$\begin{cases} V_1 = Z_{11} I_1 + Z_{12} I_2 \\ V_2 = Z_{21} I_1 + Z_{22} I_2 \end{cases} \tag{7.66}$$

with Z_{11} and Z_{22} the input impedances of port #1 port #2 when radiated in an unbounded medium, respectively, whereas Z_{12} is the mutual impedance at port #1 (open-circuited) due to a current in port #2 and Z_{21} the mutual impedance at port #2 (open-circuited) due to a current in port #1. If the network is reciprocal, then $Z_{12} = Z_{21}$, as previously demonstrated.

The reciprocity theorem can also be employed to calculate the power transfer between the two-port systems when completely matched in terms of impedance: the generator impedance Z_g is matched to antenna #1 and the load impedance Z_L is matched to antenna #2. From (7.44), the power delivered by the generator to antenna #1 is

$$P_1 = \frac{|V_g|^2}{8R_1} \tag{7.67}$$

and, if we assume a transfer admittance in the system equal to Y_{21} (or equivalently $1/Z_{21}$), the power delivered to antenna #2 in the end of the transmission line is given by

$$P_2 = \frac{1}{2}\mathrm{Re}\left[V_2 I_2^*\right] = \frac{1}{2}\mathrm{Re}\left[Z_2 (V_g Y_{21})(V_g Y_{21})^*\right] = \frac{1}{2} R_2 |V_g|^2 |Y_{21}|^2 \tag{7.68}$$

The power ratio obtained when #1 is transmitting and #2 is receiving is given by (7.69) or, for the reverse operation ($Y_{12} = Y_{21}$), one gets (7.70).

$$\frac{P_2}{P_1} = 4R_1 R_2 |Y_{21}|^2 \tag{7.69}$$

$$\frac{P_1}{P_2} = 4R_2 R_1 |Y_{12}|^2 \tag{7.70}$$

7.4.2.4 Spherical Mode Theory for Small Size Antennas

Without a transmission line, the antenna is radiating in an unbounded domain and the impedance will depend on many factors, including operational frequency, geometry, method of excitation, and proximity to surrounding objects. Due to the complexity of the problem, the input impedance of an antenna is typically determined experimentally, whereas for special cases it can be determined by numerical tools such as the boundary value method, the transmission line method, or the Poynting vector method. For antenna scale modeling, characteristics such as impedance, pattern, and polarization are found to be invariant to a change in the size of the antenna if and only if a similar change is also produced in the operating frequency. The limitations of electrically small antennas are derived using the spherical mode theory, which sets the Q-factor and bandwidth of antennas, as introduced originally by Chu and followed by Harrington [59]. The spherical mode theory assumes that the entire antenna structure is included within a sphere of radius r and the radiated field outside the sphere is a combination of orthogonal spherical vector waves or modes that radiate in an outward radial direction from the antenna. This model reduces the antenna space problem to a number of equivalent electric circuits for each spherical wave mode and with no coupling between them. Only the propagating modes contribute to the overall radiated power of the antenna, whereas all modes contribute to the reactive power (energy storage), and the Q-factor for each mode translates to the ratio between the stored and radiated energy. For a lossless antenna, the equivalent electrical circuit for every single mode is made of a series capacitor and a shunt inductor, producing a ladder-like network of LC tanks with a terminal shunt resistor in the complete equivalent circuit of the antenna. The Q-factor of the complete circuit can be calculated via [56]:

$$Q = \frac{1 + 2(kr)^2}{(kr)^3\left[1 + (kr)^2\right]} \tag{7.71}$$

The last equation represents the fundamental limit on the antenna's electrical size by relating the theoretically lowest achievable Q to the largest linear dimension of the antenna. The fractional bandwidth of the circuit is also related to the Q-factor as

$$BW = \frac{\Delta f}{f_0} = \frac{1}{Q} \tag{7.72}$$

where f_0 is the centre frequency and Δf is the bandwidth. Bandwidth can only be improved if the geometrical configuration of the antenna utilises the available volume within the sphere efficiently. Fractal antennas are one such type of antenna that makes efficient use of the spherical space. These antennas use fractal geometries that represent structures recursively generated by computers and commonly found in nature in the form of plants, leaves, trees, clouds, etc. Besides providing smaller Q and larger bandwidths, fractal antennas can also be made to resonate at a particular frequency and have input impedances larger than in the classic geometries like dipoles and loops.

7.4.2.5 Bio-Inspired Antenna Design

Antenna size with respect to the wavelength of transmission is the most critical factor that influences an antenna's field of radiation. Small geometrical antennas require higher frequencies for operation where efficiencies are consonant to the electrical size of the antenna. A standard dipole antenna would require a nominal length of 2.6 cm (given by $\lambda/2$) in the 5.8 GHz ISM band, whereas the common area for a patch antenna in MICS is $7 \times 6\ \text{cm}^2$. These lengths are impractical for implantable devices if no other technique for shrinking the antenna's size is employed that takes advantage of processes involved in 3D antenna folding (zigzag, meander), fractal antenna design, and the employment of higher-k dielectric matrices [60]. However, as the size is scaled down, the resonant frequencies of the substrate material increase and the antenna can start operating in a regime near or above the self-resonance of the external readout antenna [61].

Regarding the antenna's bandwidth, higher frequencies allow higher data exchange rates, shortening the transmission time (smaller duty cycles), which saves power, since wireless communications tend to be the most power-hungry module inside the implantable device. Higher data bandwidths in miniature antennas are feasible for higher frequencies at the expense of some transmission loss of the radio channel through biological tissues and lower Q of the antenna. Compensating for such losses by increasing an antenna's gain has proved as challenging as for any other parameter, because most compact antennas have a low gain as a result of their limited aperture.

For implanted antennas inside the human body, the influence on the performance of the antenna of muscles, skin, and other tissue layers is quite different. If the skin thickness is smaller than the wavelength, which happens for frequencies above 50 GHz, reflections from the surface of the body are reduced, as the skin would look transparent to the incoming RF waves. The dielectric properties of the underlying fat tissue are very similar to that of the air, and hence detuning and impedance mismatch are minimized within this tissue, and only muscle and other nearby tissues can effectively deteriorate the antenna's performance, since they are highly dielectric [59].

Sensor devices with RF capabilities that demand smaller form factors and an increase in power and transmission frequency are still regarded as the best way to obtain higher efficiencies for the antenna, accepting the greater power loss through tissue attenuation and reflection. Some strategies found in the literature to achieve RF operability in size-restricted antennas are described below; some have been implemented in tangible structures, while others are still in the domain of computational simulation.

Shakib et al. have proposed an Egret-Beak shaped patch tri-band antenna for communication in the MICS, WMTS, and ISM bands with dimensions $10 \times 12 \times 1.57\ \text{mm}^3$, as shown in Fig. 7.14a. The antenna is composed of the radiator plane printed on the top layer of the substrate and a ground plane on the bottom layer, with a shorting pin connected in-between [62]. The positioning of the pin allows not only a reduction in the overall antenna size, but also modification of the resonant modes of operation, introducing inductive effects on the input impedance of the antenna. Simulation results obtained for an antenna implanted 2 mm under the skin of the arm exhibited a triple bandwidth of 260 MHz (74%) for MICS, 150 MHz (11%) for WMTS, and 680 MHz (27%) for ISM when parameter S11 is set to be at a level lower than −10 dB. However, due to SAR restrictions, the nominal input power of 1 W for wideband antenna operation had to be decreased to 19.2 mV, 23.50 mW, and 118.3 mW in the MICS, WMTS, and ISM bands, respectively, which can place considerable technical issues on the electronic modules that will interface the antenna.

Karnaushenko et al. recently proposed a compact helical antenna for operation in the ISM band with a total size of $5.5 \times 0.5\ \text{mm}^2$ and fabricated by a strain engineering process in polymeric-metal heterostructures, as depicted in Fig. 7.14b. The unique Swiss roll geometry of the antenna allowed the authors to tune antenna parameters such as diameter, number of windings of the helix, and separation between windings to match the required input impedance of 50 Ω with no additional electric components or materials [63]. The antenna design proved successful to communicate with a mobile phone and, when embedded in a tooth model surrounded by biological tissue (head model compliant with SAR levels), the simulations have shown an achievable distance of up to 2 m, within the typical receiver sensitivity of −110 dBm standard for mobile applications.

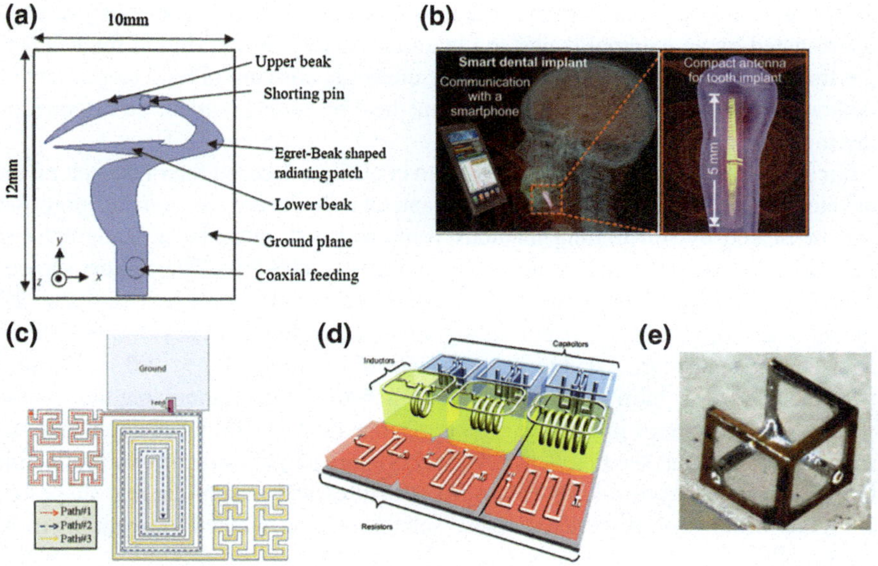

Fig. 7.14 **a** Geometry of the Egret-Beak shaped patch tri-band antenna. Reprinted with permission from [62], © IEEE 2014. **b** Compact helical antenna for the wireless tooth implant. Reprinted with full acknowledgment of [63]. **c** Hilbert-based curve 3D folded antenna with Archimedean spirals. Reprinted with permission from [60], © IEEE 2012. **d** Microelectronic components produced by 3D manufacturing techniques. Reprinted with full acknowledgement of [64]. **e** $500 \times 500\ \mu m^3$ cubic shaped antenna. Reprinted with permission from [59], © IEEE 2013

Yang et al. [60] have also designed an antenna for teeth implants with dimensions of $8 \times 11.5 \times 8\ mm^3$, combining an Archimedes spiral design with the Hilbert fractal antenna, as depicted in Fig. 7.14c. The high dielectric substrate employed in the antenna's fabrication was zirconium dioxide (ZrO_2), which exhibits not only good biocompatibility but is also a ceramic material, thus suitable for use in denture implants (as opposed to metals) [60]. The design-inspired antenna with 3D folding allowed broadening of the bandwidth of operation to reach the MICS band. The results obtained by the authors after implant had shown a shift in the operational frequency whenever the mouth was opened or closed, although the reported bandwidth (57–62 MHz) and gain (−6.78 dBi around 400 MHz) were sufficient to accommodate the variations of the antenna.

Wu et al. demonstrated the feasibility of three-dimensional manufacturing of electrical components for passive wireless sensors [64]. They designed hollow micro-channels on 3D polymer microstructures using multi-nozzle printing, to be filled by liquid metal paste. The fundamental electrical components were fabricated in this way: resistors, from meander-shaped conductive wire; inductors, with the shape of a spiral coil; and capacitors, in the form of two parallel-plates, as shown in Fig. 7.14d. With the basic electrical components at their disposal, the authors successfully built functional electric circuits, namely an LC tank with dimensions

$10 \times 20 \times 6.4\ mm^3$ and resonant frequency of 0.53 GHz that could be passively interrogated by an external reader to obtain the quality of liquid food by a shift in the frequency produced by a change in the dielectric constant of food (capacitance). Although the tank was not applied to biomedical purposes, it could be extended to the monitoring of body fluids.

In a slightly different approach, Anacleto et al. have focused their research on the development of micro-device integrated antennas for energy harvesting purposes and fabricated by self-folding methods combined with multi-layer lithography to pattern on silicon substrates at the micro- and nano-scales [59]. The authors showed that the self-folding properties of the micro-antennas could surpass the 2D restrictions imposed by current lithographic fabrication processes, and hence achieve 3D modeling. A cubic antenna with dimensions $500 \times 500 \times 500\ \mu m^3$ was modeled and fabricated that exhibits a 6 dB improvement on the overall wireless link efficiency in the low gigahertz range (1–10 GHz) when compared to a square loop antenna; see Fig. 7.14e. With this approach, 2D structures with flexible hinges can be made to fold to obtain 3D structures with reconfigurable shapes by applying heating or residual stresses, suitable for untethered microgrippers or implantable drug delivery systems.

Another trend in antenna design targeting the increase of energy harvesting deploys ultra-wideband antennas that can extract power from various available frequencies instead of a single frequency, thus maximizing DC output power. However, the identification of realistic RF ambient sources is difficult because the level of scavengeable power can vary significantly, depending on several factors such as the distance to a nearby RF source, telecommunication traffic density, transmission media, and antenna orientation [65]. Typical mobile phone infrastructures and broadcast sources are included in the roll of energy-scavenging sources with RF power levels from −40 to −10 dBm. Single-frequency conversion requires rectification, and harvesting signals from various sources (multi-tone) typically results in the design of tuneable impedance matching networks that independently convert a single RF carrier to its equivalent DC power. However, this is not always the most efficient way to harvest energy from RF sources. Moreover, it has been demonstrated that the performance of typical diodes in the rectifier degrades for higher frequencies due to a higher junction capacitance presented by diodes, which ultimately narrows the frequency contents of the available environmental sources. With these concerns in mind, Shariati et al. recently proposed an efficient dual resonant rectifier for operation at 490 and 860 MHz that uses two separate single-rectifier circuits: the frequency-dependent nonlinearities imposed on the rectifier's impedance are matched to the 50 Ω output of a single antenna at the operating frequencies, over a power range from −40 to −10 dBm. The equivalent schematic of the circuit is shown in Fig. 7.15, which allowed an effective efficiency of 54.3% (in opposition to an average 20% as previously mentioned) and an output DC voltage level of 712.8 mV for a multi-tone input power of −10 dBm.

When both power and data exchange must coexist on a single implantable device, the strategy followed by many authors is to use two different systems (antennas) tuned at different frequencies for the aforementioned operations. Then,

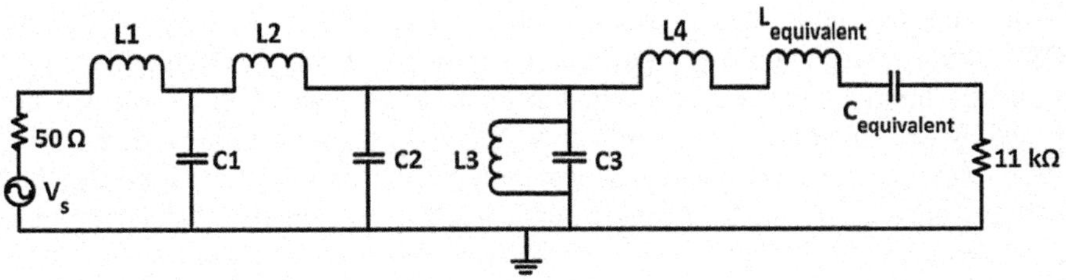

Fig. 7.15 Schematic of the dual resonant rectifier for environmental RF energy scavenging. Reprinted with full acknowledgement of [65]

on one side, the harvesting efficiency is not biased by an impedance variation of the antenna induced by backscattered transmission operations, while on the other side, data exchange does not lower the efficiency of the harvest [66]. Suggestions have been made to perform energy extraction at frequencies lower than 13.56 MHz, that is, in the near-field regime where the EM field is predominantly magnetic (small antenna), and data exchange would occur at the standard MICS, WMTS, and ISM bands, as long as the final size of the implant is not compromised.

7.4.3 *Low-Power RF Electronics*

The modules that follow the antenna in the electronic circuit also need to be carefully planned such that the first module must match the antenna's impedance and the following ones convert RF power into DC and retrieve data from the carrier wave. Since power is carried as an AC wave, conversion to DC involves diode rectification with the lowest forward-bias voltage achievable in order to yield the highest possible DC amplitude after rectification. The full diode rectifier bridge found in typical AC-to-DC powering systems poses a voltage drop close to 1 V, which cannot afford to be wasted when dealing with power scavenging. Solutions that use single-diode or double-diode rectification are thus more convenient. Diode rectification also eliminates any data modulated to the AC carrier, so data extraction must be performed before rectification. An interesting approach that combines data extraction with single-diode rectification at the same time was proposed by Gulick and Towe, based on the nonlinearities encountered in diode operation during rectification [67]. As the diode rectifies the wave signal it also distorts it, which shows up as harmonics at multiples of the drive frequency: the relative ratios of the detectable harmonics to each other can then be used to build a harmonic data decoder that, according to the authors, uniquely defines the operation of diode-type neural devices, returning information about current flow in the implant. However, errors in the AC drive, DC bias, and load of the circuit may lead to inefficiency of the data decoder.

In order to comply with standard data encoding techniques, such as ASK and PSK, while harvesting power from RF signals, a more complex electronic circuit is needed. The modified Cockcroft–Walton multiplier shown in Fig. 7.16a was proposed by Ahmed et al. to function within an RFID tag for smart biological sensing [68]. Maximum power transfer in the circuit depends on impedance matching between the antenna and the Schottky rectifiers. The matching is achieved through a series inductor (with shunt capacitor) that also functions as a voltage booster by providing higher peak input voltages to the double-diode rectifier. The amount of power harvested with this approach reaches a maximum of 42 μW (900 mV) at radiation levels of −3 dBm.

Power harvested by the RF-DC converter is subsequently delivered to a low dropout regulator (LDO) that provides a constant voltage level to the analog and digital modules involved in signal conditioning, modulation and demodulation for wireless communication at 900 MHz. The most critical piece of hardware inside the wireless modules is the design of the Phase-Locked Loop (PLL) that generates the clock for signal recovery from input ASK data and outbound communication. PLL design involves circuitry such as a phase frequency detector, charge pumps, and a Voltage-Controlled Oscillator (VCO) that can be very power demanding in achieving high system performance in order to overcome thermal and phase noise. Even so, the authors were able to design an RFID with power consumptions of 2.86 μW for PLL, 0.2 μW for the modulator/demodulator, and 13.66 μW for LDO and the analog front-end, which are still below the harvested power level by a manageable margin.

A two-stage Cockcroft–Walton multiplier was proposed by Chow et al. to achieve high power conversion efficiency with relative low input power levels [69]. Schottky diodes on the schematic are chosen due to their low forward-bias voltage (150–300 mV) and higher performance for operations that require higher frequencies, as depicted in Fig. 7.16b. The authors also used periodic pulsing of the RF signal instead of continuous wave stimulation, not only to meet the SAR exposure limits with higher output voltages, but also to take some advantage of the nonlinear increase in rectifier performance as large instantaneous power levels are

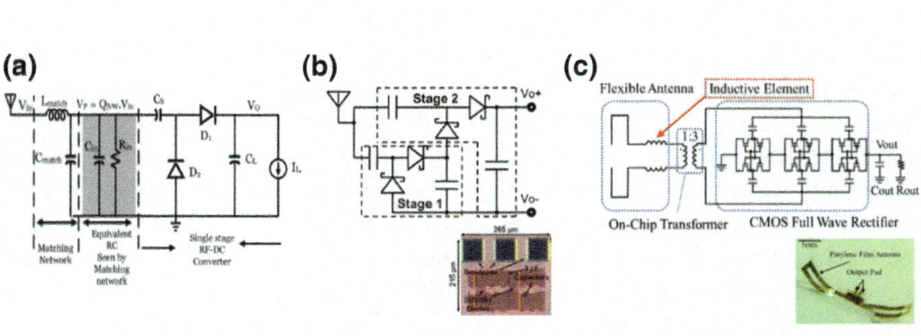

Fig. 7.16 **a** Schematic of a single-stage RF-DC converter with matching network. Reprinted with permission from [68], © IEEE 2012. **b** Two-stage RF rectifier schematic integrated on CMOS technology. Reprinted with permission from [69], © IEEE 2011. **c** WPT device with on-chip transformer and flexible parylene film antenna. Reprinted with permission from [70], © IEEE 2015

used. This approach has achieved a voltage level of 1 V across a 27 kΩ load at a distance of 5 cm with 1.5 W transmitting power, and also charged a 20 μF capacitive powering array to 2.5 V in less than half a second in an in vivo implanted setting with 1 inch separation distance to the transmitter at a relative power of 0.1 W.

A full diode rectifier bridge implemented on CMOS technology has recently been proposed by Okabe et al. for RF-powered implantable neural interfaces. The CMOS rectifier presents input impedances expressed as a series combination of high resistance and capacitance that needs to be matched to a flexible antenna; see Fig. 7.16c. The strategy for impedance matching followed by the authors was to implement on-chip transformers with a turn ratio of 1:3 and distribute some inductance along the flexible antenna to cancel out the capacitance from the rectifier. The device, with integrated antenna, transformer, and rectifier, generates more than 1.5 V with a power transmission efficiency of 0.086% at 825 MHz when the RF power is transmitted 10 cm away from the solution tank [70].

After rectification, the focus of electronic design can be switched to the planning of power-efficient wireless transmitters (or receivers) inside the implantable device. Together with the PLL design, the VCO implementation is also critical, since it requires modulation of the data to be transmitted in a standard binary encoding scheme. The cross-coupled VCO topology with an LC tank depicted in Fig. 7.17a is most attractive for its low phase noise and low power consumption [71]. LC-tank oscillators rely on passive matching and depend on the effective parallel resistance in the tank to limit power consumption, as given by

$$R = Q\omega L \tag{7.73}$$

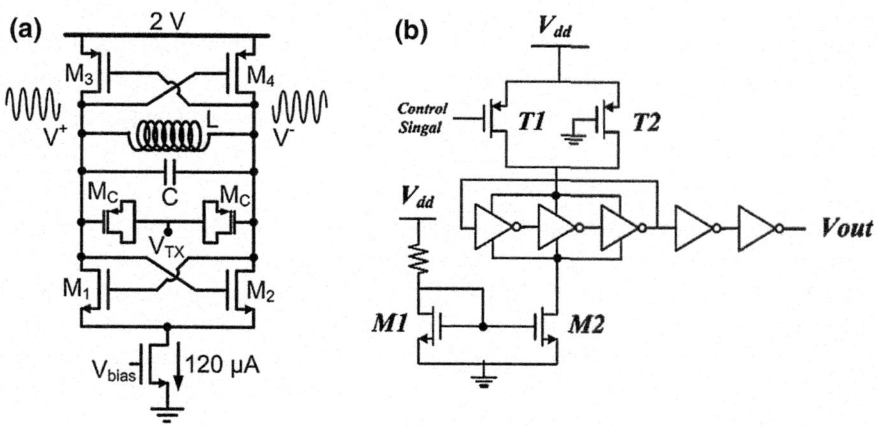

Fig. 7.17 **a** Schematic of a VCO topology with an LC-tank for data telemetry. Reprinted with permission from [72], © IEEE 2009. **b** VCO topology based on a single-ended inverter ring oscillator. Reprinted with permission from [60], © IEEE 2012

Ideal LC tanks have infinite parallel resistance, which is unattainable in real applications due to the limited power quality factor Q of on-chip inductors. By increasing the frequency of operation, the resistance will also increase, which reduces power consumption. However, obtaining high oscillation frequencies with the same inductor is only achieved for some capacitance values of the tank, which in turn, limits the range for frequency tuning (narrowband). The quality factor in CMOS technologies is reduced due to a combination of factors related to size-constrained devices, which include resistive losses and stray capacitances that lead to energy dissipation in the lossy substrate. Some design and layout methodologies to improve the Q of the inductor have been reported, such as octagonal spiral design, patterned ground shielding to isolate the inductor from substrate, and a differential topology that reduces substrate resistance and increases the magnetic flux [71]. The latter method reduces not only the required area for inductor placement but also improves the overall phase noise. VCO performance in biomedical applications can also be seriously compromised due to variations on the supply voltage (harvesting efficiency) and temperature. Hence the design of an oscillator immune to the aforementioned factors is of paramount importance to achieve high accuracy and communication reliability without frequency detuning. Techniques to improve VCO accuracy can employ electronic circuitry to stabilize bias current, monitor temperature variations, and provide symmetric loads with built-in self-calibration schemes.

The authors of [60] have proposed a low-power VCO for operation in the MICS band, based on ring oscillators. As opposed to LC-tank oscillators, they possess the advantage of ease of integration, small area for implementation (no inductors needed), and a wide tuning range. Although phase noise performance has been shown to be higher for LC-tank oscillators, in situations where data exchange rates are not that fast this parameter might not be so critical. Power consumption can also be a more limiting parameter for ring oscillators. The schematic for the ring oscillator is shown in Fig. 7.17b and the frequency of oscillation is found to be inversely proportional to the delay time, t_d, of the inverter and number n of inverters employed in the circuitry, as given by

$$f_{osc} = \frac{1}{2nt_d} \tag{7.74}$$

The delay time is controlled by changing the current flowing through the inverter, which determines the frequency of operation. Due to the small supply of current in harvesting devices, slight variations of current can produce substantial frequency deviations of the oscillator. To compensate for that, the authors have deployed two PMOS transistors located on top of the schematic that fine tune the current flowing through T1 using the voltage level presented at its gate, achieving a coverage of 5 MHz within the MICS frequency band.

In terms of modulation encoding, Amplitude Shift Keying (ASK) is more advantageous for power purposes since it provides a constant power flow to implantable devices with limited on-chip energy storage. Nonetheless, Phase or

Frequency Shift Keying schemes are more immune to signal amplitude attenuation produced by body tissues. These schemes require an on-chip clock and carrier synchronization to receive and transmit data, which significantly increases energy consumption [73]. Conventional PSK and FSK modulation transmitters can be made using mixed-based or PLL-based topologies, depicted in Fig. 7.18a, b. In the former, the waveform is generated in the digital baseband (DBB) and converted to its analog equivalent, then low-pass filtered (LPF) and up-converted to the RF carrier by a quadrature mixer before power amplification (PA). The performance of the mixer-based topology depends on the accuracy of the quadrature oscillator being completely immune to noise, nonlinearities of electronic components, mismatch, and voltage/temperature variations. Hence, complex architectures with higher power consumption levels (and eventual extra calibration circuits) are needed to ensure signal quality.

However, PLL-based topologies have a more attractive ratio of hardware-to-power efficiency. The baseband data is converter to Amplitude (AM) and Phase (PM) Modulation at the same time and, whereas AM is applied directly to the PA, PM undergoes extra modulation through the PLL and Frequency-Modulation Digital-to-Analog Converter (FM-DAC). This topology dismisses the use of analog/RF blocks like filters and mixers, which reduce power consumption while providing different modulation schemes and data rates since modulation is generated in the digital domain [53].

On the receiver side, two types of topologies can be considered: Heterodyne Low-IF and Homodyne Zero-IF. The first involves the generation of quadrature signals at a lower frequency than the RF carrier by means of a divider, reducing power consumption and preventing a dynamic DC offset. The homodyne by itself directly down-converts the RF wave to DC at higher power consumption levels; however, it does not suffer from image interference resulting from down-converting

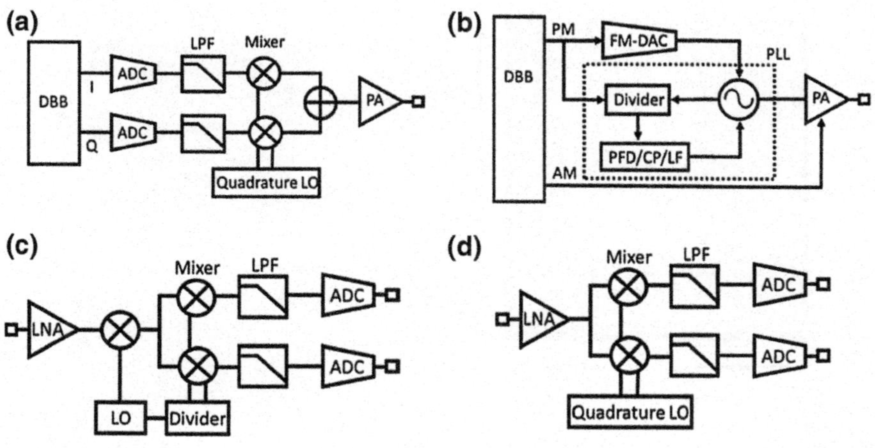

Fig. 7.18 **a** Illustration of a mixer-based module transmitter. **b** PLL-based module transmitter. **c** Low-IF receiver. **d** Zero-IF receiver. Both reprinted with permission from [53], © IEEE 2015

to a frequency other than the RF one. Hence the heterodyne topology may require an external high-Q bandpass or poly-phase filter to deal with image interference, but that comes with an increase in power consumption and area size. Thus homodyne zero-IF architecture is preferable if DC offset can be corrected by using either ADCs with higher dynamic range or automatic offset cancellation techniques.

Ba et al. have recently designed wireless transceivers for operation in the MICS band with PLL-based transmitter and Homodyne Zero-IF receiver topologies for medical implantable applications [53]. Fabricated in 40 nm CMOS technology with 1 V voltage supply, the power consumption levels required for transmission and reception are 1.78 and 1.49 mW, respectively, with data exchange rates from 11 kbps to 4.5 Mbps. The die occupies an area of 1.8 mm × 1.7 mm and is depicted in Fig. 7.19a. In turn, Chow et al. designed an ASIC chip on a Texas Instruments 130 nm CMOS process for wireless monitoring of the cardiovascular pressure at 2.4 GHz and integrated with a medical stent; see Fig. 7.19b. The design of the VCO employed a differential inductor topology to modulate data using binary FSK, located in the middle of the chip [71]. The pressure resolution achieved by the sensor was 0.5 mmHg over a range of pressures between 0 and 50 mmHg, with transmission rates of 42.2 kbps and external power level set at 35 dBm for a separation distance of 10 m, yielding an average rectified voltage of 1.5 V.

Arsalan et al. developed a single-chip implantable wireless sensor for intraocular pressure monitoring to be attached to a MEMS pressure sensor, which uses separate receiver (RX) and transmitter (TX) on-chip antennas at 2.4 and 5.2 GHz, respectively [74]. The purpose of employing separate RF communication channels was to facilitate full duplex operation and avoid intra-band interference. The chip, implemented in a standard six-metal 0.18 μm CMOS process, employs an on-off keying protocol to communicate with the external device while receiving incoming

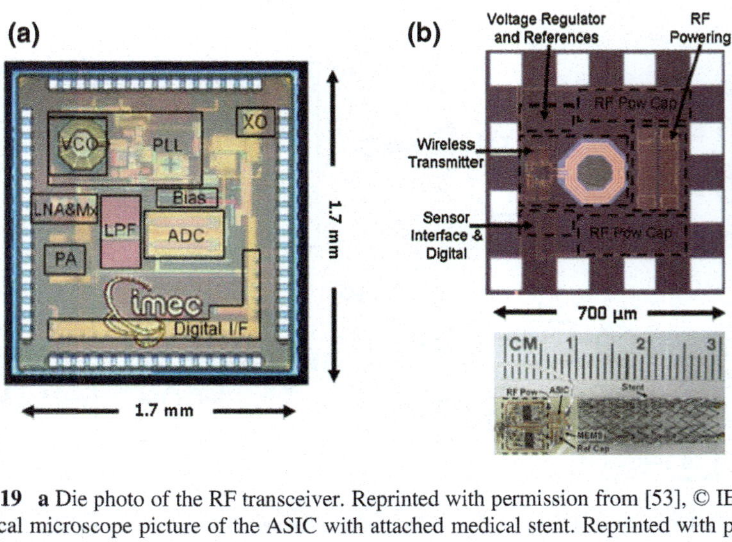

Fig. 7.19 **a** Die photo of the RF transceiver. Reprinted with permission from [53], © IEEE 2015. **b** Optical microscope picture of the ASIC with attached medical stent. Reprinted with permission from [71], © IEEE 2010

RF signals when the total power harvested is sufficient to turn on all the electronic modules (peak consumption of 513 μW). Finally, Mao et al. presented a sub-gigahertz ultrawideband transmitter for RF-powering applications [75]. The sensor exploits subharmonic injection locking to eliminate complex frequency synthesis circuitry or bulky crystal references. The radio clock is locked by the received carrier, which facilitates time synchronization, and hence both energy and clock are obtained from the same signal carrier. The injection-locked ring oscillator makes use of a three-stage ring oscillator with output frequency stabilized at 450 MHz, which corresponds to a subharmonic of the received carrier at 900 MHz, avoiding frequency collision in the form of self-jamming. The energy consumption of the sensor is 175 μW (peak amplitude of 0.75 V) with active area less than 150 μm × 60 μm implemented in a 180 nm CMOS process.

7.4.4 Passive RF Sensing with Unconventional Materials

The discussion so far about RF powering and data exchange link has been based on the assumption that the implantable device requires a minimum of DC power involved in the operation of the analog front ends dedicated to signal sensing and digital blocks for signal processing and wireless transmission. The materials employed in the design of such devices were also based on standard electronic components, despite some exceptions in antenna modeling. However, recent trends in material engineering are widening the limits towards the development of new materials that mimic the typical behaviour of RLC resonators in passive sensing measurements with increased performance occasionally. Passive sensing reduces power consumption to unprecedented levels (if any level at all) not achievable by active electronics, requiring only the modification of a physical property of the sensing material that can be passively coded as a change in the RF reflected impedance by the implantable device. Some examples of these sensors are briefly described below and include the emerging field of biodegradable materials, microelectromechanical (MEMS) technology and metamaterials for RF telemetry operation.

Starting with biodegradable materials, Boutry et al. have proposed an implant made of conductive polymer composites PLLA-PPy and PCL-PPy that mimic an RCL resonator for wireless communications [76]. The polymer composites are fabricated on a biodegradable matrix of polylactide (PLLA) or polyprolactone (PCL) with conductive polymer nanoparticles of polypyrrole (PPy). The concept of the implantable sensor is depicted in Fig. 7.20a and includes a sensing biodegradable layer and the RLC resonator. The stimulus (change of permittivity) causes a variation in the resonant frequency of the device that is detected by an external reader. The conductivities of the polymer composites were fully characterized by the authors in terms of the frequency and adjusted to the Jonscher and Papathanassiou models of conductivity, showing a conductivity increase in the range from 10^5 to 10^7 Hz, which can be used for future electrical circuit design at several frequencies of interest. Chen et al. designed a wireless pressure monitoring system using millimetre-scale sensors that can overcome operational frequency limits commonly found in traditional small measurement systems and remain relatively insensitive to lossy tissue environments [61]. The structure of the sensor includes a pressure-sensitive capacitive element within an inductive antenna to form an LC tank, as depicted in Fig. 7.20b. Size reduction is achieved by stacking a deformable dielectric layer between the two inductive spirals in a sandwich-like structure. The spiral layers are printed on a flexible polyimide structure, whereas the sensitive dielectric layer is implemented by a microstructured styrene-butadiene-styrene elastomer. Under pressure, the distance between the spiral layers decreases, causing an increase in capacitance, which in turn lowers the resonant frequency. The authors have tested several sizes for the sensor as low as $1 \times 1 \times 0.1\ \text{mm}^3$ with varying numbers of spiral turns to achieve peak resonance at frequencies around 1 GHz and a pressure-to-frequency linear response in the physiological range of 0–100 mmHg. The authors went further to completely monitor the waveform produced by the human arterial pulse using a sensor with dimensions $2.5 \times 2.5 \times 0.1\ \text{mm}^3$ and the intracranial pressure in vivo by deploying a two-turn $3 \times 3\ \text{mm}^2$ sensor in the epidural space of mice.

Kim et al. introduced an RF sensor system based on oscillation frequency deviations recorded at 2.4 GHz with application to the detection of biomolecular binding systems like biotin-streptavidin and DNA hybridization [77]. The device is composed of a planar resonator, a surface acoustic wave (SAW) filter, and a power detector. The principle of operation consists in the change of oscillation frequency as the impedance of the RCL resonator varies due to biomolecular immobilization on the substrate. The deviation in frequency is further increased by the SAW filter and the results obtained by the authors have shown a resonance in the vicinity of 2.4 GHz with 3 dB bandwidth of 150 MHz, suitable to reflect the difference in weight-mass of target molecules. In a similar approach, the authors of [78] designed a miniature rectangular meandered line resonator with operational frequency centred at 9.2 GHz to be employed in the detection of the glucose level in human serum. The structure also contains a sensitive air-bridge cap made of Au on the top

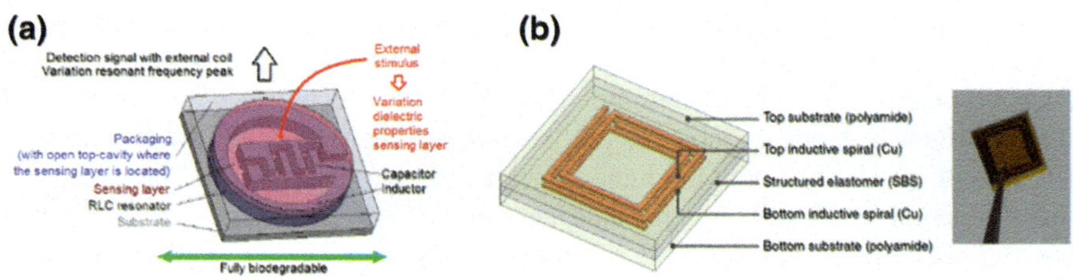

Fig. 7.20 **a** Biodegradable implant based on an RCL resonator. Reprinted with permission from [76], © IEEE 2012. **b** Structure of the sensor device that uses a pressure-sensitive element with an inductive antenna to form a resonant device. Reprinted by permission from [61], © Nature Publishing Group 2013

surface to enhance capacitance variations between the serum and the air-bridge. The measurement principle is grounded in the change of dielectric permittivity produced by different concentrations of glucose, which then induces variations in the electrical behaviour of the line resonator. The results obtained by the structure showed a resonance peak shift downward with increasing glucose levels and sensitivity in detecting 1 MHz changes produced per 1 mg dl^{-1} variation in the glucose level, with minimal concentrations of 25 mg dl^{-1}. The initial resonance response of the sensor was also recovered after every consecutive serum testing followed by washing, which demonstrates the reusability of the device.

Finally, Melik et al. proposed a metamaterial-based RF strain sensor to monitor bone fracture healing by evaluating resonant frequency variations produced by mechanical loading [79]. The employment of metamaterials allowed a higher sensor performance when compared to traditional RF materials by exhibiting a higher *Q*-factor at lower frequencies, the ones that avoid absorption by soft tissue. The higher electric field density found between the gaps in the splits of the metamaterial produces additional capacitance which yields higher frequency deviations and better SNR and linearity of the measurements. The metamaterial was fabricated by vapour deposition of 0.1 μm thick Si_3N_4 onto the silicon substrate, followed by patterning of a 0.1 μm Au layer on the dielectric film to obtain one functional cell. The final size of the sensor had dimensions 1.5×1.5 cm^2 and consisted of a 5×5 array of cells.

7.5 Ultrasonic Link

Ultrasounds has long been used in medical imaging to assess fetal development in pregnant women, in breast cancer scanning, and in blood flow measurements. Only recently has ultrasound technology gained interest for power harvesting purposes, especially to deliver energy to implanted devices, due to its long transmission depths in tissues and electromagnetic immunity. Ultrasounds uses the piezoelectricity property found in some materials to convert acoustic waves into electrical energy. As such, piezoelectric transducers replace the coils employed in the inductive link: at the external device, the transducer converts electrical power into a pressure wave that is transmitted transcutaneously and collected by the implant transducer where the inverse process takes place, and the energy is transformed back to electrical. The power transfer efficiency is higher than the inductive and RF links and it does not suffer from signal interference arising from devices at close range, either inside (for instance pacemakers) or outside the human body.

Although promising, there are still some issues hindering the efficiency of power transfer by means of ultrasounds. The small device area limits the size of the implantable transducer, making it more sensitive to the orientation of the surface transducer, to the depth of deployment within the body, and the acoustic reflections from surrounding medium. Nonetheless, in opposition to the recurrent misalignment verified between implantable and surface coils in inductive links (and antennas in RF), some of these geometrical constraints can be attenuated to some extend by apodization of the acoustic beam at the surface transducer, as will be demonstrated later.

7.5.1 Piezoelectricity and Sources of Ultrasound Fields

The piezoelectric effect occurs in certain crystalline structures by converting mechanical strain into an electric voltage. Inside a crystal lattice, the balance between negative and positive charges neutralizes crystal polarization. However, if this distribution is disturbed by mechanical stress, the displacement of charge carriers will produce a current in the crystal. In the opposite direction, an external voltage applied to the crystal will force instability in the neutral charge balance, producing mechanical deformation. This type of transduction has been shown to offer higher power densities when compared to electromagnetic transduction, with the additional advantage that microfabrication of piezoelectric elements by MEMS processes is better adjusted than electromagnetic ones [80]. The coupling between the mechanical and electrical domains is given by the constitutive equations for piezoelectricity using Einstein's summation convention as

$$\begin{cases} T_{\mathrm{p}} = c_{\mathrm{pq}}^{\mathrm{E}} S_{\mathrm{q}} - e_{\mathrm{kp}} E_{\mathrm{k}} \\ D_{\mathrm{i}} = e_{\mathrm{iq}} S_{\mathrm{q}} + \varepsilon_{\mathrm{ik}}^{\mathrm{S}} E_{\mathrm{k}} \end{cases} \tag{7.75}$$

where T, S, E, and D are the mechanical stress, mechanical strain, electric field, and electric induction, respectively; $c_{\mathrm{pq}}^{\mathrm{E}}$ represents Young's modulus for a steady electric field, e_{iq} the piezoelectric coefficient and $\varepsilon_{\mathrm{ik}}^{\mathrm{S}}$ the clamped dielectric permittivity under a constant strain [81]. The electromechanical coupling factor defined as k_{iq} relates the conversion between mechanical and electrical energy:

$$k_{\mathrm{iq}}^2 = \frac{W_{\mathrm{i}}^{\mathrm{electrical}}}{W_{\mathrm{q}}^{\mathrm{mechanical}}} = \frac{e_{\mathrm{iq}}^2}{\varepsilon_{\mathrm{ik}}^{\mathrm{S}} c_{\mathrm{pq}}^{\mathrm{E}} + e_{\mathrm{iq}}^2} \tag{7.76}$$

The efficiency of the harvester at resonance, when the energy conversion is at its maximum, can be given by the coupling factor k^2 and the quality factor of the generator, Q, as expressed by

$$\eta = \frac{\frac{k^2}{2(1-k^2)}}{\frac{1}{Q} + \frac{k^2}{2(1-k^2)}} \tag{7.77}$$

The different coefficients determined by the type of piezoelectric material are fundamental to selecting the best harvesters in the design of the electronic device. The balance between the mechanical characteristics (which define the robustness of the material) and the electrical properties (voltage coefficient) allows high energy density piezoelectric materials to be obtained. The most common materials employed in ultrasounds are mainly PZT (and its derivatives), barium titanate ($BaTiO_3$), aluminium nitride (AlN), PVDF polymers, micro-fiber composites (MFC), and zinc oxide (ZnO). Although PZT generally exhibits a superior figure of merit for coupling, its lead content raises safety issues (toxicity) in the deployment into human tissues without proper encapsulation [81]. That leaves the biocompatible materials $BaTiO_3$, AlN, and ZnO for use in biomedical applications. Among them, the piezoelectric coefficient for $BaTiO_3$ is ten-fold of that registered by AlN and ZnO. However, the latter possesses both semiconductor and piezoelectric properties that can be micro-fabricated on a nanometric scale without losing the piezoelectric field.

In practice, standalone configurations using monolithic piezoelectric materials are rarely used, as they are embedded in mechanical devices (cantilever beams, matrix composite, and cymbal structures), often laminated into layers (like the transducer probes employed for ultrasound imaging) or made into fibers to enhance/reduce voltage peaking or material robustness [82]. The modeling of complex piezoelectric transducers typically uses a quasi-static assumption as an approximation to the aforementioned constitutive equations [83], by ascribing scalar equations (instead of tensors) in the form

$$\begin{cases} l = s^{\mathrm{V}} P + h\Phi \\ Q = hP + C^{\mathrm{F}}\Phi \end{cases} \tag{7.78}$$

where l is the displacement, Q is the charge on the electrodes (not to be confused with the quality factor defined previously), Φ is the corresponding voltage, P is the force exerted on the device, h is the piezoelectric coefficient, s^{V} is the compliance in constant voltage, and C^{F} is the capacitance in constant force. When both the force and voltage are in equilibrium, the change in the free energy of the transducer is equal to the work exerted by the force plus the work done by the voltage (power source) as given by

$$\delta F = P\delta l + \Phi\delta Q \tag{7.79}$$

The state of the transducer can thus be characterized as a function of two independent variables, $F = F(l, Q)$, which, for relative small variations δl and δQ yields the Helmholtz free energy of the transducer at the equilibrium as

$$\left[\frac{\partial F(l,Q)}{\partial l} - P\right]\delta l + \left[\frac{\partial F(l,Q)}{\partial Q} - \Phi\right]\delta Q = 0 \Rightarrow \begin{cases} P = \frac{\partial F(l,Q)}{\partial l} \\ \Phi = \frac{\partial F(l,Q)}{\partial Q} \end{cases} \tag{7.80}$$

Knowledge of $F(l, Q)$ allows calculation of the force and voltage levels required to produce a certain displacement and charge [83].

The equations just derived are in general nonlinear, which reflects the difficulty in modeling the thermodynamic framework for electromechanical coupling. Linearization of the equations is often pursued to reduce mathematical complexity and simplify the operation mode of a piezoelectric transducer. This is the case for dielectric elastomers recently proposed for realistic energy conversion solutions and characterized by a three-dimensional mesh of flexible polymers held together by crosslinks. The enclosed areas in the thermodynamic planes represent the total electrical and mechanical energy processed over one operational cycle, as derived by (7.81). If losses are excluded from the conversion process, the total mechanical energy is rated equal to the output electrical energy from the harvester. This equality in both domains can be further explored, for instance, to adjust the displacement electrically and increase the amount of mechanical energy converted to electricity by providing a bias voltage whenever reasonable.

$$\begin{cases} W^{\text{electrical}} = \oint \Phi \mathrm{d}Q \\ W^{\text{mechanical}} = \oint P \mathrm{d}l \end{cases} \tag{7.81}$$

Finally, the operation mode of the material also plays a crucial role in energy harvesting. In the so-called h_{31} mode an induced electric field appears in the z-direction as a result of a stress along the x-axis (see Fig. 7.21a, b). In turn, in h_{33} the stresses are in the same direction as the induced electric field (Fig. 7.21c, d). Typically, h_{33} mode has a larger voltage coefficient compared with h_{31}, which implies greater output voltage and potentially higher output power: letting the direction of operation coincide with the length of the beam results in a generator

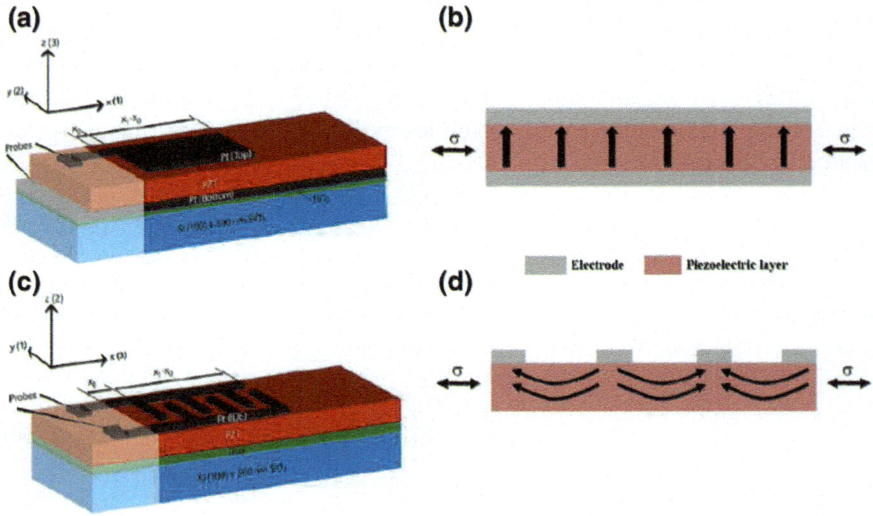

Fig. 7.21 **a** Top and bottom electrode configuration. **b** Polarization in h_{31} mode. **c** Interdigitated electrode configuration. **d** Polarization in h_{33} mode. Both reprinted with permission from the Creative Commons Attribute license, © 2014 by the authors; licensee MDPI, Basel, Switzerland

that is independent of the piezoelectric thickness. Active research continues to progress in the selection of the best mode generation to harvest the maximum power as well as in the fabrication and deposition of electrodes along the surface of the piezoelectric material, so the reader is directed to references [81] for in-depth insights of the technologies that are clearly out of the scope of this chapter.

7.5.2 *Transducer Design*

The transducer is the most critical component in the design of wireless power and data exchange links for ultrasound. The efficiency of energy conversion and bandwidth for data transmission are largely dictated by the characteristics of the selected piezoelectric transducer and can only be increased by a small factor by means of auxiliary electronics designed specifically to boost the performance of the sensor device. In the case of passive sensors, the performance is entirely dependent on the type of transducer employed. Piezoelectric transducers can be made of several shapes and sizes that ultimately alter the corresponding resonant frequency, bandwidth, and both electrical and acoustic impedances. Thus, an equivalent model is essential to describe together the acoustic and electrical characteristics of the transducer.

7.5.2.1 Electromechanical Equivalent Circuit

Typical transducers are flat, with one dimension larger than the other dimensions and two electrodes attached to the top and bottom. Piezoelectric materials are dielectric by nature, which translates into a static or clamped capacitance C_0 originated from the condition of zero strain as stated by

$$C_0 = \varepsilon^S \frac{A}{d} \tag{7.82}$$

The equivalent electrical circuit of the transducer is completed by adding the radiation impedance Z_A with real and imaginary parts given by R_A and X_A, respectively. The value of R_A is found by equating the power delivered by the real electrical source to the transducer and the acoustic power radiated from each side of the transducer into the surrounding medium with impedance Z_C. The radiated reactance is obtained through a Hilbert transform performed to R_A [82], yielding the equation for the transducer impedance as a function of the frequency in the form expressed by

$$Z_T(f) = R_A(f) + \mathrm{j}\left[X_A(f) - \frac{1}{\omega C_0}\right] \tag{7.83}$$

When the equations for Z_A are difficult to obtain, the transducer can be modelled by a simpler electrical lumped element circuit, with its associated limitations. Being a purely electrical model, it does not take into account the acoustic response of the transducer. It can only render the electrical characteristics of the transducer as a first approximation to a more complete electromechanical circuit described by the Krimholtz–Leedom–Matthaei (KLM) or Butterworth–Van Dyke (BVD) models.

7.5.2.2 KLM Model

The KLM model provides a valid representation for transducers whose aspect ratio (width/thickness) is greater than 10 and operate in the pure one-dimensional thickness vibration mode. In this configuration, the transducer is modelled as three-port network, composed of one electrical port and two acoustic ports, which constitute the front and back faces of the transducer, as shown in Fig. 7.22a. The clamped capacitance is also represented by C_0, whereas X_A is obtained by the following formula [82] as

$$X_A = \frac{h_{33}^2 \sin\left(\frac{\omega d_0}{c_0}\right)}{Z_0 A_0 \omega^2} \tag{7.84}$$

where h_{33} is the piezoelectric coefficient, d_0 the thickness of the transducer, c_0 the acoustic velocity, ω the angular velocity, A_0 the radiating surface of the transducer and Z_0 the bulk acoustic impedance. The transduction between mechanical and electrical domains is modeled by the electromechanical transformer placed in the middle of the transmission line with a turn ratio of φ given by

$$\varphi = \frac{2h_{33} \sin\left(\frac{\omega d_0}{2c_0}\right)}{Z_0 A_0 \omega} \tag{7.85}$$

At resonance operation ($d_0 = \lambda_0/2$) the equality $\omega_0 = \pi c_0/d_0$ is verified, which nullifies the series reactance and maximizes the turn ratio of the transformer. The

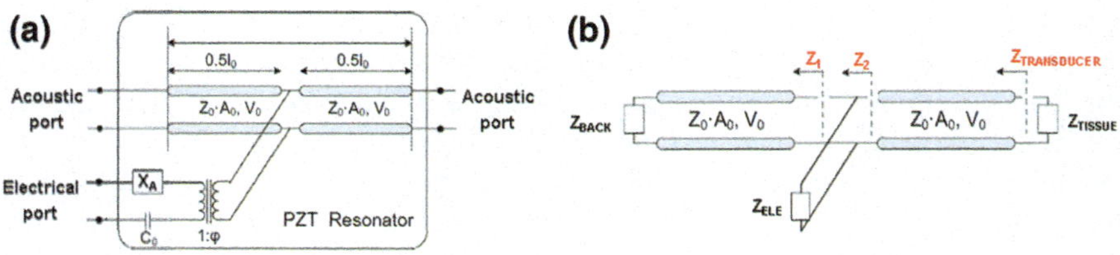

Fig. 7.22 **a** KLM model of a piezoelectric transducer with one electrical port and two acoustic ports. **b** Transmission line model used to evaluate the impedance at different sections of the piezoelectric elements (both taken from [84] with modifications). Reprinted from [84], © 2014, with permission from Elsevier

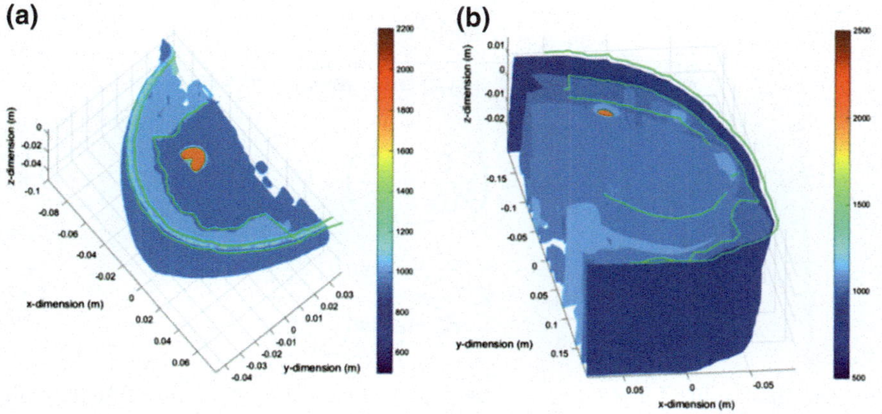

Fig. 7.23 **a** Volume-mass density for the tissues composing the 3D breast phantom with the implantable device shown in orange (higher density material) on a frontal view. **b** Volume-mass density for the tissues composing the 3D abdominal phantom, with the implantable sensor located in the liver (half-domain transversal slices seen from the back of the body)

operation of the transducer close to its fundamental resonance or any odd frequency harmonics $(3\omega_0, 5\omega_0, 7\omega_0, \ldots)$ gives the maximum conversion between mechanical to electrical energy.

The transmission line is complete when the transducer is connected to the sensor electronics in the electrical port (modeled by Z_{ELE}) and to the backing and tissue layers in the acoustic ports (Z_{BACK} and Z_{TISSUE}) as depicted in Fig. 7.22b. Moreover, symbols Z_1, Z_2, and Z_{TRANS} represent the impedance seen at the input terminals of the left transmission line terminated by Z_{BACK}; the parallel impedance equivalent formed by Z_1 and Z_{ELE}; and the total impedance seen at the input of the transducer, respectively. Independently of the complexity involved, the circuit can be described by a passive series network comprising real and imaginary electronic parts. Maximum power transfer at this point is obtained by conjugate matching the clamped capacitance and series reactance of the transducer. The total electrical network can then be expressed in accordance with

$$Z_{\mathrm{ELE}} = \left(R_{\mathrm{ELE}} + \mathrm{j}X_{\mathrm{ELE}} + \frac{1}{\mathrm{j}\omega C_0} + \mathrm{j}X_A\right)\varphi^2 \tag{7.86}$$

For power transfer, the backing layer of the transducer is usually replaced by a short circuit ($Z_{\mathrm{BACK}} = 0$), since the damping of ultrasonic vibrations might not be required, as in imaging applications. With this simplification in mind, one can derive the complete equation for the input impedance of the transducer (as seen by biological tissues as seen in Fig. 7.23a and 7.23b) using transmission line theory over the chain of equations from (7.87) to (7.89).

$$Z_1 = Z_0 \frac{Z_{BACK} + Z_0 \tan h\left(\frac{\omega d_0}{2c_0}\right)}{Z_0 + Z_{BACK} \tan h\left(\frac{\omega d_0}{2c_0}\right)} = Z_0 \tan h\left(\frac{\omega d_0}{2c_0}\right) \quad (7.87)$$

$$Z_2 = \left(\frac{1}{Z_{ELE}} + \frac{1}{Z_1}\right)^{-1} \quad (7.88)$$

$$Z_{TRANS} = Z_0 \frac{Z_2 + Z_0 \tan h\left(\frac{\omega d_0}{2c_0}\right)}{Z_0 + Z_2 \tan h\left(\frac{\omega d_0}{2c_0}\right)} \quad (7.89)$$

The final value for Z_{TRANS} is thus dependent on the operation frequency, the bulk impedance of the piezoelectric material and the electrical load presented by the sensor electronics. Optimization of the electrical termination (with compensation for the internal static capacitance of the transducer) and selection of materials with acoustic impedance closer to that of biological tissues not only ensures higher conversion efficiencies, but also minimizes signal reflections for data exchange between the implant and the external device. For the vast majority of piezoelectric materials, the mismatch in acoustic impedance in comparison to that of tissues in Table 7.4 is still the limiting factor in transducer design. The intercalation of a passive layer with impedance Z_{LAYER} between the transducer and tissue can attenuate the reflections and matching parameters obtained by solving a matrix system for the pressure wave propagation given by

$$\begin{bmatrix} p_{TISSUE} \\ c_{TISSUE} \end{bmatrix} = \begin{bmatrix} \cos\left(\frac{\omega d_{LAYER}}{c_{LAYER}}\right) & jZ_{LAYER} \sin\left(\frac{\omega d_{LAYER}}{c_{LAYER}}\right) \\ \frac{1}{jZ_{LAYER}} \sin\left(\frac{\omega d_{LAYER}}{c_{LAYER}}\right) & \cos\left(\frac{\omega d_{LAYER}}{c_{LAYER}}\right) \end{bmatrix} \cdot \begin{bmatrix} p_{TRANS} \\ c_{TRANS} \end{bmatrix} \quad (7.90)$$

Table 7.4 Properties of human tissues and piezoelectric transducer materials [82]

Tissue	c (m s^{-1})	ρ (kg m^{-3})	Z (MRayls)	Material	c (m s^{-1})	ρ (kg m^{-3})	Z (MRayls)
Blood	1584	1060	1.679	Water 20 °C	1482.3	1	1.482
Bone	3198	1990	6.364	PZT-5A	4350	7750	33.71
Brain	1562	1035	1.617	PZT-5H	4560	7500	34.31
Breast	1510	1020	1.540	$BaTiO_3$	5470	5700	31.18
Fat	1430	928	1.327	$LiNbO_3$	7360	4640	34.2
Heart	1554	1060	1.647	Quartz	5000	2650	13.3
Kidney	1560	1050	1.638	PVDF	2200	1780	3.92
Liver	1578	1050	1.657	PMN-PT	4646	8060	37.45
Muscle	1580	1041	1.645	PZN-PT	4030	8310	33.49
Spleen	1567	1054	1.652	Navy VI	4575	7500	34.31

Or equivalently, in terms of impedance,

$$Z_{\mathrm{TISSUE}} = \frac{p_{\mathrm{TISSUE}}}{c_{\mathrm{TISSUE}}} = \frac{\cos\left(\frac{\omega d_{\mathrm{LAYER}}}{c_{\mathrm{LAYER}}}\right) Z_{\mathrm{TRANS}} + jZ_{\mathrm{LAYER}} \sin\left(\frac{\omega d_{\mathrm{LAYER}}}{c_{\mathrm{LAYER}}}\right)}{\frac{1}{jZ_{\mathrm{LAYER}}} \sin\left(\frac{\omega d_{\mathrm{LAYER}}}{c_{\mathrm{LAYER}}}\right) Z_{\mathrm{TRANS}} + \cos\left(\frac{\omega d_{\mathrm{LAYER}}}{c_{\mathrm{LAYER}}}\right)} \tag{7.91}$$

It has been demonstrated that, for quarter-wave transmission layers, the simplest matching is given by the mean of the impedances at stake as

$$Z_{\mathrm{LAYER}} = \sqrt{Z_{\mathrm{TRANS}} Z_{\mathrm{TISSUE}}} \tag{7.92}$$

The major disadvantage of a single layer is that the resulting impedance might still be restricted to materials that are not biocompatible. Furthermore, the adhesive glue that holds together the transducer and matching layer can degrade the quality of the matching previously achieved [85]. The solution involves employing multilayer impedance matching with the multiplication of a chain of transfer matrices of the same form as above. In fact, the complete ultrasound link comprises the source and receiver transducers plus all the matching layers and tissues in the path. The power transmission efficiency of the link is then determined by the ratio of power delivered to the electrical load in the receiver to the input power at the source as expressed by

$$\eta = \left| \frac{V_{\mathrm{LOAD}} I_{\mathrm{LOAD}}}{\mathrm{Re}(V_{\mathrm{IN}} I_{\mathrm{IN}})} \right| \tag{7.93}$$

The transducer and matching layers can also be fabricated by microelectromechanical (MEMS) technology into a single structure by processes such as etching, deposition of metal films, lithography, and sputtering of piezoelectric layers. Zhu et al. reported the design of a two-degree-of freedom piezoelectric MEMS that converts mechanical energy into electrical in both the *x*- and *y*-directions of displacement, independently, by exciting the structure with two different resonant frequencies [86]. The structure was fabricated in a commercial silicon-on-insulator (SOI) MEMS foundry with a comb-finger configuration of the electrostatic capacitors. Fowler et al. used the same foundry process to design a single 3D MEMS system capable of harvesting electrical energy in both the plane directions and out-of-plane (*z*-direction) [87]. Similarly, He et al. reported a MEMS-based 3D array piezoelectric ultrasonic transducer prepared by sputtering niobium-doped PZT on a pre-prepared curved substrate [88]. The lower value for the piezoelectric coefficient achieved can increase the transmission range of the ultrasonic wave when deployed into implantables. In terms of acoustic matching layers, it has been reported that PEEK plastic or graphite [89] bonded to the front surface of the transducer by cyanoacrylate glue minimizes impedance mismatches between PZT tissues. Regarding biocompatibility issues, the PZT can be sealed and graphite can be replaced by pyrolytic carbon [90].

7.5.2.3 Pressure Field Shape Calculation

The electrical excitation of the piezoelectric transducer will generate a distribution of pressure acoustic waves over the radiated area. The knowledge of this distribution will allow defining the intensity of the beam that reaches the implantable sensor, from which some inferences can be made: power delivered to the implant and operation within safety limits. The shape of the pressure field is controlled by the geometry of the transmitter transducer and type of electrical excitation. The power delivered to the receiver will depend on its location and orientation within the pressure field and the size of the implantable transducer. As opposed to inductive links, ultrasound imaging already implements apodization of the excitation beam that produces a non-uniform distribution of the acoustic field over the radiated area by employing a non-planar wave pattern emerging from the transmitter's surface. Moreover, the use of more than one single piezoelectric element for the transmitter's transducer can contribute to sharpening the acoustic field to meet the requirements of tissue imaging; therefore, most ultrasonic scanning probes are composed of an array of piezoelectric elements.

The acoustic field pattern generated by a transducer can be obtained by using the Huygens Principle, which states that all points on transducer can be seen as an independent source of radiation. Hence the pressure field at any observation point in the radiated area is calculated by the Rayleigh integral as [90]

$$p(x,y,z;t) = \rho_0 \int \frac{\dot{u}_p\left(x',y';t-\frac{r}{c_0}\right)}{2\pi R} \mathrm{d}S \tag{7.94}$$

where ρ_0 is the medium density, $\dot{u}_p$ the vibration velocity over the radiating cross section in the coordinates of the transducer's frame, c_0 the speed of sound, R the distance from the point source to the observation point, and S the transducer area. The closed-form solution for the integral is difficult to obtain for a more general transducer's shape and vibration distribution. Simplifications can be adopted as in the case of a uniformly distributed disk-shaped transducer excited by a sinusoidal continuous wave (CW). The pressure field is thus obtained by

$$p(x,y,z;t) = \frac{jk\rho_0 c_0 u_0}{2\pi} \mathrm{e}^{\mathrm{j}\omega t} \int \frac{\mathrm{e}^{-\mathrm{j}kR}}{R} \mathrm{d}S \tag{7.95}$$

The wavenumber k is typically calculated as the ratio between the angular frequency and phase velocity ($k = \omega/c$) for non-dissipative mediums. In the presence of biological tissues, the absorption coefficient α should be added to the model as the imaginary component for k. The exact solution of the last equation is given in terms of spherical coordinates by taking advantage of coordinate symmetry as

$$p(r,\theta;t) = \frac{\mathrm{j}a\rho_0 c_0 u_0}{r} \mathrm{e}^{\mathrm{j}(\omega t - kr)} \frac{J_1(ka \sin\theta)}{\sin\theta} \tag{7.96}$$

with a the transducer radius, θ the angle between the observation point and acoustic axis, r the distance between the centre of the transducer to the observation point and J_1 the first-order Bessel function. Maximum power transfer is thus obtained at one Rayleigh distance from the transducer's face, in accordance with (7.97), where beam spreading is at a minimum.

$$L_{\text{Rayleigh}} = \frac{(2a)^2 - \lambda^2}{4\lambda} \tag{7.97}$$

The concentration of wave intensity on a narrow strip of the radiated area can have side-effects on the underlying body tissue if the intensity is beyond the safety limit imposed by the FDA (≤ 94 mW cm^{-2}). The average power intensity of the wave can be obtained by integrating the instantaneous pressure multiplied by the velocity, which yields (7.98). Another index to measure acoustic beam quality is the so-called acoustic energy density given by (7.99).

$$I_{\mathrm{P}} = \frac{p^2}{2Z} \tag{7.98}$$

$$E_{\mathrm{D}} = \frac{p^2}{\rho_0 c_0^2} = \frac{I_{\mathrm{P}}}{c_0} \tag{7.99}$$

Reduction of both focus intensity and side lobe transitions has been possible to achieve by the generation of a Gaussian beam at the transmitter instead of uniform excitation [89]. This will spread the energy of the wave more evenly through the radiated area with reduced pressure variations in the near field region and better side lobes suppression while the beam maintains its Gaussian shape along the acoustic axis. Calculation of the pressure field involves the summation of the pressure fields produced individually by each concentric layer of elements in the array, and apodization of the beam is achieved by imposing an excitation voltage pattern that approximates a Gaussian mask.

7.5.2.4 Scattered Field Calculation for Realistic for Realistic Biological Phantom

The calculation of the pressure field in the last section has relied on a pure mathematical description of the model of the excitation transducer when operating in a background medium with no scatterers or attenuators present to modify the shape of the field. The power delivered to a deeply implanted device in the body depends on the acoustic pressure map produced by the excitation piezo and the biological tissues in the propagation medium, as exposed previously. Moreover, from the theory of wave propagation, the potential locations for the implantable device can be obtained with regard to the safe biological limits for exposure to ultrasound. These mathematical relations cannot be computed without simplifications to the 3D model in the presence of multiple acoustic materials. The total acoustic field is then given by a sum of the incident field (background medium) with the scattered field, the latter accounting for all domain inhomogeneities as expressed by

$$p^{\mathrm{tot}}(x) = p^{\mathrm{inc}}(x) + p^{\mathrm{sct}}(x) \tag{7.100}$$

The p^{inc} term can be calculated by any of the mathematical equations derived in the previous section, since it is only dependent on the geometry of the source transducer and regime of excitation. For numerical computations, the incident field can be approximated by a convolution between the source transducer model and the Green's function as [91]

$$p^{\mathrm{inc}}(x) = \mathrm{j}\omega\rho_0 \int\limits_{x' \in \mathbb{D}} \mathfrak{G}(x - x')s(x')\mathrm{d}V \tag{7.101}$$

where $s(x')$ represents the domain points in $\mathbb{D}$ allocated to the source transducer, the angular frequency, ρ_0 the volume-mass density, and $\mathfrak{G}$ the Green's function. In the expression above, p^{sct} is calculated from the wave scattering theory in its integral form as [92]

$$p^{\mathrm{sct}}(x) = \int\limits_{x' \in \mathbb{D}} \mathfrak{G}(x - x')\Delta\gamma p^{\mathrm{tot}}(x')\mathrm{d}V + \int\limits_{x' \in \mathbb{D}} \mathfrak{G}(x - x')\nabla \cdot [\Delta\rho \nabla p^{\mathrm{tot}}(x')]\mathrm{d}V \tag{7.102}$$

with $\Delta\gamma$ and $\Delta\rho$ encompassing differences in the complex propagation coefficient γ and volume-mass density ρ within the model, respectively, as depicted in Fig. 7.23 for the breast and lower abdominal regions; ∇ is the gradient operator while ∇ represents the topological divergence. The linearized total pressure field is computationally implemented in the form of a forward operator, $\mathfrak{l}^{\mathrm{sct}}$, that maps all the geometric relations and acoustic material properties verified in any point $x \in \mathbb{D}$ to the entire domain, as (7.103) attests.

$$p^{\mathrm{inc}}(x) = \mathfrak{l}^{\mathrm{sct}}[p^{\mathrm{tot}}(x)] \tag{7.103}$$

The last equation is solved numerically by the Conjugate Gradient (CG) method in terms of p^{tot} when the incident field remains constant (pre-computed). The method converges to the solution of the total acoustic field with an update iteration given by

$$\begin{cases} \mathbf{p_0} = \mathbf{0} \\ \mathbf{p_{n+1}} = \mathbf{p_n} + \alpha_n \xi_n \\ \mathbf{r_n} = p^{\mathrm{inc}} - \mathfrak{L}^{\mathrm{sct}}[\mathbf{p_n}] \\ \mathrm{error} = \frac{\|\mathbf{r_n}\|_{\mathbb{D}}}{\|p^{\mathrm{inc}}\|_{\mathbb{D}}} \end{cases} \tag{7.104}$$

where $\mathbf{p_0}$ is the initial estimation of the method (e.g. null vector) and $\mathbf{p_{n+1}}$ represents its iterative update, with α_n and ξ_n being the iteration step and update direction, respectively. The scattered field is represented by $\mathbf{r_n}$ with a normalized *error* calculated in each iteration in order to control the convergence of the method. Three-dimensional simulations using realistic biological phantoms of the breast and abdomen are shown in Figs. 7.24 and 7.25 for the incident and scattered pressure fields, respectively. The implantable device located on both phantoms can be made to vary its geometrical shape and composition using the piezos from Table 7.4 and, thus, simulate the pressure field level recorded at the exterior of the human body in order to determine the configuration that yields the maximum detectable signal. On the other end, the pressure levels recorded over the implantable device should give an approximation to the voltage equivalent harvested by the device if a model of the electromechanical response of the transducer is available for the regimes and frequencies of excitation.

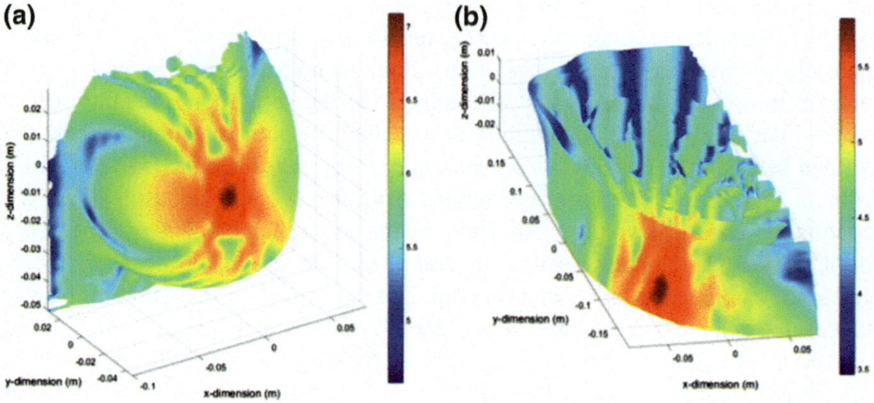

Fig. 7.24 **a** Incident pressure field produced by a transducer located close to the breast nipple, with colour levels projected onto the walls of the biological tissues present in the 3D model (values in logarithmic scale: $\log_{10}[p^{\mathrm{inc}}]$). **b** Incident pressure field produced by a transducer located in the skin at the height of the liver in the lower abdominal region ($\log_{10}[p^{\mathrm{inc}}]$)

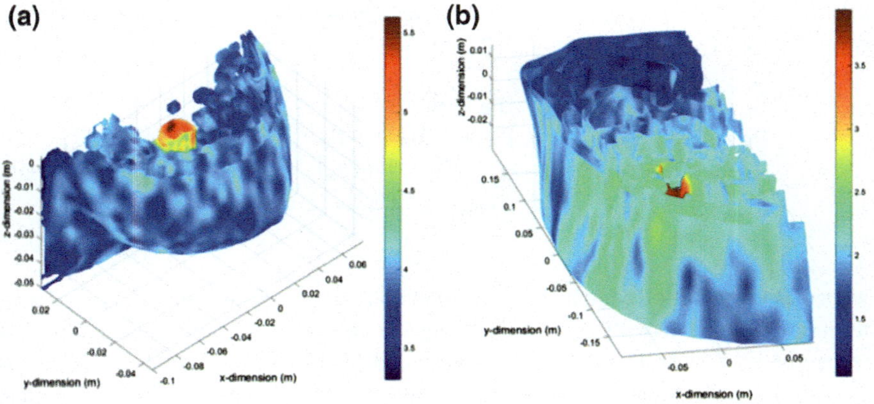

Fig. 7.25 **a** Scattered pressure field in logarithmic scale projected onto the walls of the tissues present in the breast phantom, which is cut off from the topmost transversal slices to highlight the highest levels occurring over the implantable device. **b** Scattered pressure field in logarithmic scale produced by the implant deployed inside the liver

7.5.3 Ultrasonic Power Harvesting

Ultrasonic Transcutaneous Energy Transfer (UTET) has been shown to be a reliable technology that combines both power transfer efficiency, compactness, and electromagnetic immunity. Unlike RF and magnetic links, the coupling coefficient between the in-body sensor and external recorder is higher and the sources of interference are lower when compared to electromagnetic ones [93]. The efficiency of the ultrasonic harvester will depend on the size and type of the implantable piezoelectric materials, depth of location within the body, orientation relative to the on-skin transducer, and acoustic reflections from the surrounding medium (impedance matching layers). Although data exchange in both directions can be performed between the implantable sensor and exterior transducer, this section will be entirely focused on DC power conversion from time-varying acoustic waves to a continuous electrical voltage that will provide active functionality to the sensor device. The main topologies will be described, which can be made of discrete electronic components or system-on-chip. The limitation on the amount of power that can be delivered to the implant in order to avoid tissue damage sets a maximum of 100 mW to power a DC load. This is roughly the value obtained for an implantable device with an effective area of 1 cm^2 and receiving the maximum acoustic intensity allowed by the FDA.

7.5.3.1 Transducer Output and Standard Rectification

The piezoelectric transducer can be described as a sinusoidal current source in parallel with the capacitance C_0. The current I_0 is proportional to the derivative of the applied force, and the time-varying signal depends on the mechanical displacement and frequency of excitation as stated by

$$I(t) = I_0 \sin(\omega_0 t) \qquad (7.105)$$

The alternating current is not directly usable by the load circuitry because of the high impedance of the piezoelectric transducer, especially at low frequencies, when the capacitive effect is predominant: the voltage and current output signals will be 90° out of phase, which decreases the instantaneous power. Hence conditioning electronics must be added to the circuit to extract the maximum power produced by the harvester. Moreover, voltage peaks might be experienced during energy conversion, which can cause the breakdown or depoling of the piezoelectric material [94].

In order to gain the benefits of AC to DC power conversion, the piezoelectric element is directly connected to a (passive) diode rectifier circuit. The topology in Fig. 7.26a is the standard full-bridge rectifier connected directly to the load R_L and a smoothing capacitor C_r with a larger value than C_0. For the time intervals in which all diodes are reverse-biased (off conduction) the piezoelectric current source is charging its internal capacitor C_0; after the bridge is turned on, the same source provides current to the load. However, in half of the conduction intervals, the power transfer is negative, which indicates that energy returns from the electrical domain to the mechanical one. Together with the charging and discharging cycles for C_0, these factors limit the energy conversion efficiency. One solution to attenuate the problem to some extent is to employ a simple voltage doubler topology, as depicted in Fig. 7.26b. When $V(t)$ is positive for half the cycle, D_1 is forward biased and D_2 is off, which then charges C_1. In the negative sequence of the cycle, both diodes exchange status and C_2 begins to charge. Due to the series combination of C_1 and C_2, the load receives twice the voltage level as that delivered by the typical full-bridge configuration [95].

The maximum power produced by the voltage doubler is, nonetheless, the same between topologies, since only half the current flows in the voltage doubler. The advantage is found when the diode forward voltages are included in power calculations, as it only has two voltage drops in the circuit. The control of the power flow associated with the charging and discharging of the piezoelectric capacitance is therefore the major parameter in active research to further improve the standard

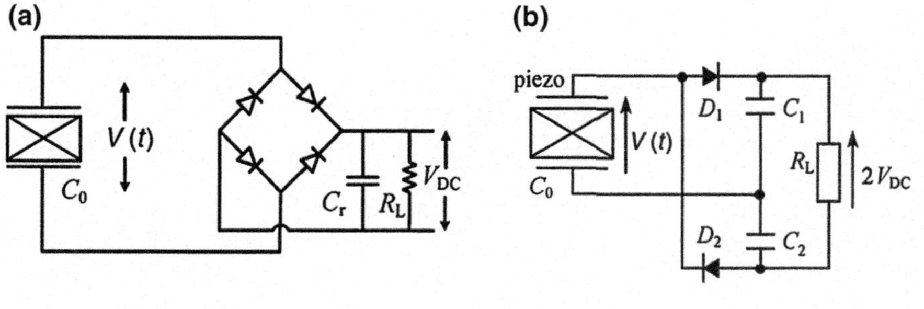

Fig. 7.26 **a** Standard full-bridge rectifier. **b** Voltage doubler interface

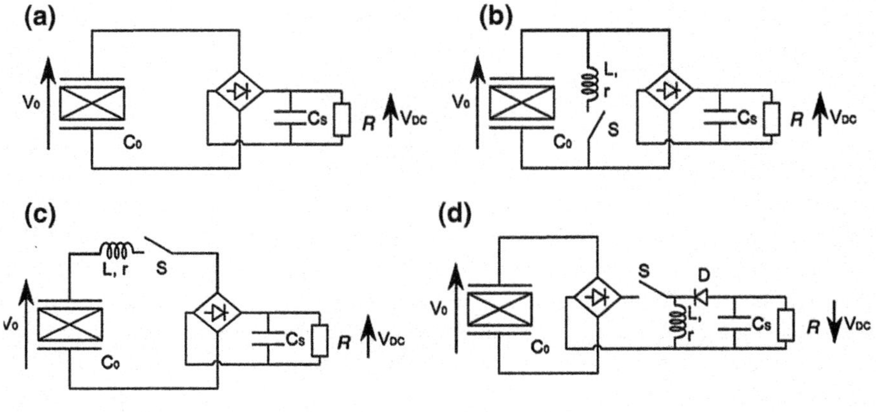

Fig. 7.27 **a** Schematic for the standard energy harvesting topology. **b** Parallel SSHI topology. **c** Series SSHI topology. **d** SECE topology. Reprinted with permission from [96], © SAGE Publications 2009

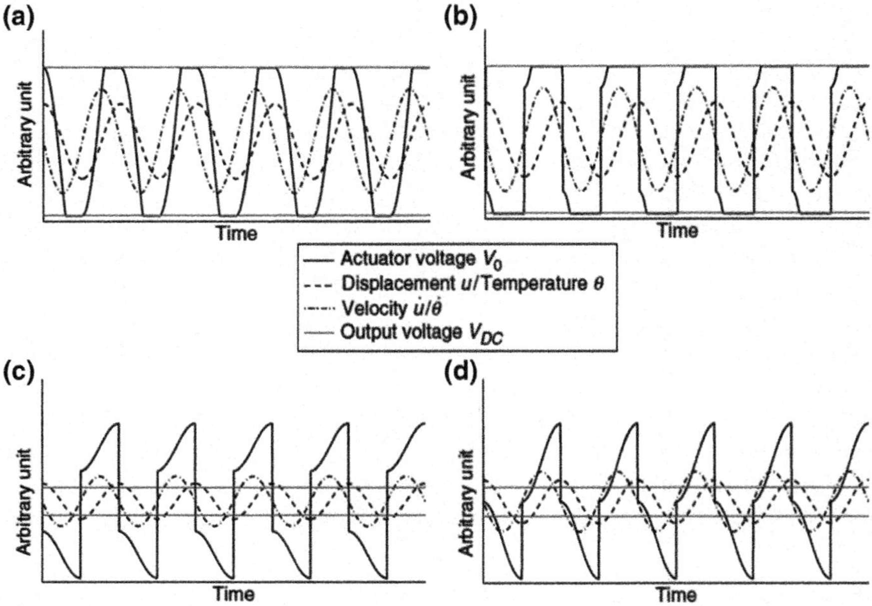

Fig. 7.28 **a** Typical waveforms for the standard energy harvesting topology. **b** Waveforms for the parallel SSHI. **c** Waveforms for the series SSHI topology. **d** Waveforms for the SECE topology. Reprinted with permission from [96], © SAGE Publications 2009

full-bridge rectifier using nonlinear harvesting approaches. Figure 7.27 shows the most common topologies for nonlinear power harvesting that can be kept as references in circuitry design, as well as the corresponding waveforms depicted, in turn, in Fig. 7.28.

7.5.3.2 Synchronized Switch Harvesting on Inductor Interfaces (SSHI)

The SSHI interfaces use the clamped capacitance of the piezoelectric material to explore the synchronization process between charge extraction from C_0 and transference of the corresponding electrical energy to the load or a secondary capacitor. This nonlinear approach has been reported to enhance the total power harvested by operating the rectifier bridge in a switch mode [97]: the combination of digital switches and inductors within the bridge at special locations can result in a significant increase of the output voltage amplitude from the transducer, due to a voltage inversion process. There are two types of SSHI: a parallel-SSHI where the nonlinear processing circuit is placed in parallel with the transducer, and a series-SSHI where the location is within one of the branches of the transducer-rectifier.

In the p-SSHI the current source from the piezoelectric element charges C_0 from V_m to V_{DC} and then delivers power to the load. Just after the same current reaches zero, the electronic switch is activated: this moment corresponds to the maximum energy transfer process when the mechanical displacement reaches an extreme. The closure of the switch creates an oscillating electrical network composed by the inductor L and capacitor C_0 and, as a consequence, the voltage V in the capacitor will be inverted after a time period given by half of the pseudo-period for the LC_0 tank. The period must be selected to be smaller than the mechanical vibration period T (7.106) so that the voltage on C_0 can reach larger amplitudes, since electrical charges are being generated by mechanical strains.

$$t_{SW} = \pi\sqrt{LC_0} \ll T \tag{7.106}$$

During part of the inversion process, the absolute value of V is lower than V_{DC} on the capacitor C_r and the rectifier bridge is blocked. Due to internal losses, the inversion process is characterized by a coefficient γ that depends on the quality factor Q of the network as

$$\gamma = e^{-\pi/(2Q)} \tag{7.107}$$

After the inversion process completes, the switch is turned off and the cycle can be repeated. In conclusion, there are three main phases in the process—open-circuit, harvesting, and inversion phases—which make the generator voltage increasing all the time.

For the s-SSHI the harvesting phase occurs at the same time as inversion. Now, whenever the switch is in open-circuit, the rectifier bridge is disconnected from the piezoelectric transducer. When on, the output power is slightly smaller when compared to that obtained by p-SSHI, since the load is also in the path connecting L and C_0. So, energy harvesting occurs at every switching instant or every voltage peak.

7.5.3.3 Synchronous Electric Charge Extraction (SECE)

In SECE, the power harvesting process is performed in two steps [98]. Firstly, the electrostatic energy on the piezoelectric element is converted to electromagnetic energy by turning the switch on, thus connecting together C_0 to the inductor L when the displacement reaches an extreme (maximum electrostatic charge). Then, the piezoelectric element is disconnected from the circuit when the piezoelectric capacitor voltage is zero again and the energy stored in the inductor will flow to a secondary storage capacitor C_s. The intercalation of a diode in the loop formed by L and C_s blocks the flow of current back to the inductor. The activation time for the switch must be very short, typically only 1/50 of the vibration period T. The advantage of SECE over SSHI is an effective load decoupling interface, since there is no direct connection between the transducer and the load. The disadvantage is that C_0 needs to be charged from the condition of 0 V, which implies extracting energy from the vibrations, damping the piezo movement.

7.5.3.4 Active Harvesting Approach

Active energy harvesting is yet another approach for power conversion that operates on the limits of both the piezoelectric material and the electronic circuitry. Unlike the other topologies that transfer power to the DC load only at certain transient conditions, the power flow in the active topology occurs essentially all the time. This requires faster switching frequencies to be achieved that can control the piezoelectric voltages and currents more efficiently in opposition to the switching frequencies employed in other topologies, which are dictated by the mechanical vibration of the piezoelectric material.

The power electronics usually involve a full-bridge inverter circuit operated by MOSFETs and controlled by pulse-width modulation techniques. Also, an inductor is included to filter out possible pulse-width modulation voltages that can damage both the piezoelectric element and power electronics [100]. The rate of voltage change is now controlled by PWM and maximum energy extraction is therefore a function of the mechanical force and applied voltage ($\pm V_{DC}$). The increase in magnitude of the peak-to-peak voltage can theoretically increase the amount of harvested energy independently of force excitation. The proper operation of the active energy harvester also requires a bidirectional power flow in the sense that the circuit must have initial energy in store to start the harvesting process and drive the MOSFET switches. Passive energy harvesting schemes like the switch trigger circuits presented before can provide the initial energetic impulse and then switch the harvesting operation to an active mode.

Implementation of the different topologies can be made by relying on discrete components, but if the size of the implantable sensor is critical, system-on-chip is preferable. The choice of CMOS technology for power harvesting can be critical due to the expected piezoelectric voltage, which can vary from the millivolt range to an appreciable number of volts. In order to avoid oxide breakdown, HV transistors

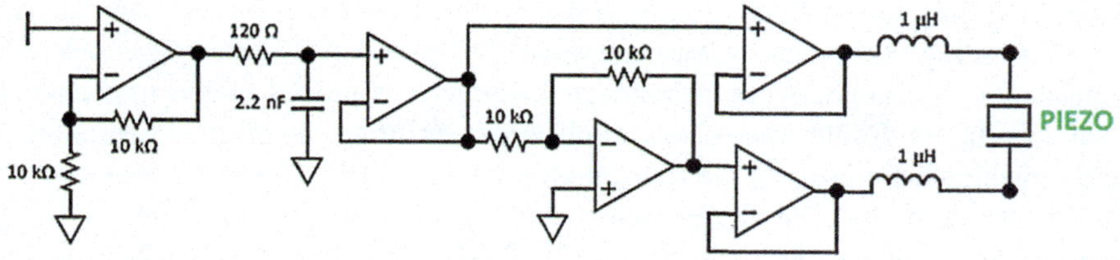

Fig. 7.29 UTET system with two out-of-phase resonance half-bridge branches for excitation of the piezo transducer

must be selected for the front-end architecture [99]. Moreover, since inductors are hard to implement in CMOS technology, due to their chip size and parasitic resistance, some authors opt to place a capacitor in parallel with the clamped capacitance of the piezoelectric element, whose value is greater than at least 10 times the value of C_0. This way, the moments for switching the capacitor are synchronized with the maximum and minimum piezoelectric voltage [99]. On the opposite side of the link, the power is delivered to the implant through a UTET device that consists roughly of a PA with higher efficiency in the conversion from electrical to piezoelectric current. Figure 7.29 shows an example of a battery-powered UTET with topology based on two out-of-phase half-bridge legs that promote the attenuation of harmonics while doubling the transducer's voltage. Many other configurations for the PA are possible to produce higher pressure fields as well as different power sources, since the UTET or readout device does not need to be battery-powered and may even rely on the same probe employed in ultrasound imaging.

7.5.3.5 Combination of Other Energy Sources with Piezoelectricity

The availability of UTET devices to deliver power to implantable devices is not always possible due to technical, physical, or even budgetary constraints. The continuous operation of UTETs also carries health risks for patients in extended exposure to acoustic fields. Thus one must rely on alternative sources for continuous power harvesting that may seem unconventional at first glance but which can be very practical in suppressing the power demands of the implantable device when consumption rates are not that high. In this context, power harvesting solutions that use piezoelectric transduction in combination with other materials that exhibit distinctive energy conversion processes will be discussed.

Magnetic fields can be used along with piezoelectric transduction to improve the performance of harvesters, since they can cross the human body relatively unharmed and have little influence on the normal functioning of body organs. Sanni et al. designed a subcutaneous interface platform with inductive and ultrasound sub-systems operating in conjunction to power a deeply implanted device, allowing a power of 29 µW to be delivered at a distance of 70 mm inside the body [101].

The platform uses a subcutaneous coil that retrieves energy from the external coil, converting the energy into ultrasonic waves by means of a primary piezoelectric transducer. A secondary piezoelectric transducer is deployed in the implant that converts the ultrasonic waves back to electrical, ensuring intermittent operation of the sensing electronics and the transmission of digitalized data every 800 ms.

Cheng et al. focused their research on the development of a new type of nano-generator based on biocompatible BZT-BCT nanowires [102]. The structure of the nano-generator is depicted in Fig. 7.30a and consists of a PDMS layer mixed with BZT-BCT, sandwiched by two glass substrates coated with Ag and covered on the top electrode by a PDMS layer with iron powder. The principle of operation is quite simple: an inhomogeneous external magnetic field is applied, the iron powder squeezes the device, and the BZT-BCT converts the mechanical displacement into electrical energy. The maximum output voltage and current levels obtained by the harvester were 3.9 V and 1.17 μA, respectively, which drop to 1.2 V and 510 nA after being implanted in a rabbit. The decrease in performance can be attributed to the distance between the device and the external electromagnet and the layers of tissue covering the device, which reduce the mechanical deformation.

The combination of piezoelectric and triboelectric energy harvesting systems in a hybrid generator can produce high output power from body movements due to straightforward cooperation between the piezo- and triboelectric mechanisms in a single press-and-release cycle. The generator proposed by Jung et al. is designed by vertically stacking together two layers: an arch-shaped piezoelectric generator with Au/PVDF/Au structure on top and a triboelectric generator with PTFE/Al structure on the bottom, in accordance with Fig. 7.30b. The pre-strained piezoelectric generator is kept in an arched rest position by polyimide substrate and produces electricity via the h_{31} mode when pressed [103]. The application of an external

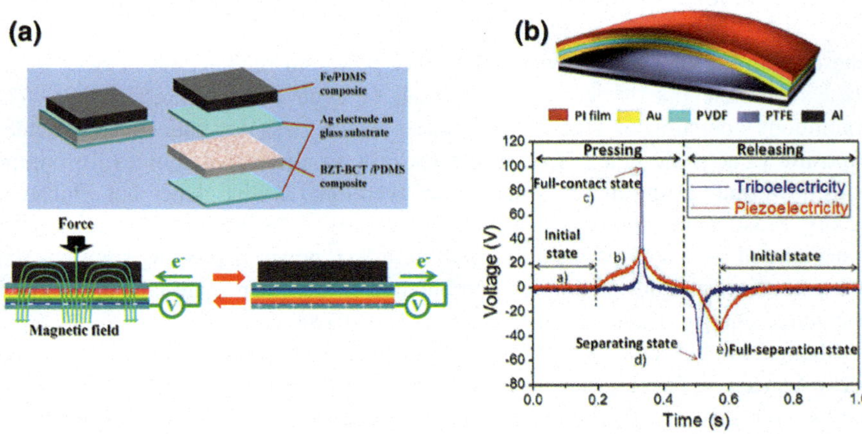

Fig. 7.30 **a** Structure of the wireless nanogenerator based on biocompatible BZT-BCT nanowires. Reprinted from [102], © 2015, with permission from Elsevier. **b** Hybrid piezo/triboelectric generator. Reprinted with full acknowledgment of [103]

force generates a positive piezoelectric potential between the Au electrodes produced by mechanical deformation; then, when the bottom Au electrode comes into contact with PTFE, a transfer of electrostatic charge will occur to the triboelectric element, resulting in maximum piezoelectric and triboelectric voltage outputs, as shown in Fig. 7.30b. After removing the force, the structure starts to bounce back to the rest position (compressive force) which changes the flow of charges, thus producing both negative piezoelectric and triboelectric voltages. The hybrid generator allows current densities of 12 μA cm^{-2} and an average power density of 4.5 mW cm^{-2} to be obtained with 0.2 N of applied force. The high output power verified is mainly due to the triboelectric peak voltage and, if not properly conditioned in the context of implantables, can cause tissue damage.

7.5.4 Data Exchange

Data exchange in an ultrasonic link can be implemented by two different mechanisms: active and passive communication strategies. Active links generate their own transmission wave to broadcast data back to the external readout circuitry (downlink), independently of the interrogator signal (uplink). This is achieved by employing two different frequencies for the downlink and uplink communication channels. However, the generation of a propagating acoustic wave internally to the implant requires DC power, provided either by a storage element—capacitor or rechargeable battery—or by the operation of the UTET link when activated. The active transmission also requires some type of modulation of the data into the carrier wave which also constitutes a burden to power consumption. The alternative to power exhaustion is to transmit data from the implantable sensor by modulating the incoming ultrasonic energy and radiating the energy back to the readout circuit, in a process known as *backscattering*. In the context of ultrasound imaging, an analogy can be established for the backscattering process: it is the same as modifying the echo pulse signals received by the ultrasound machine. Modulation of the ultrasonic energy in the backscattered echoes is possible by modifying the load of the implantable transducer in the electric port, as perceived by the readout circuitry, thereby modifying the coefficient of reflectivity of the propagation medium. The advantages of passive interrogation are manifold, including the removal of any significant energy storage unit and, as a consequence, any hindrance to size scaling of the implantable sensor can be reduced. Nonetheless, the vast majority of transduction mechanisms still require a small amount of power to acquire physiological data from body tissues, so a *semi*-passive strategy combining active transduction with passive data transmission is closer to reality.

The constraints on power consumption and implantable size narrow the number of electronic stages at our disposal in sensor design, and in ultrasonic links the ideal scenario would be to replace the front-end electronics by a piezoelectric element, a set of sensing electrodes, one transistor, and a pair of resistors, as depicted in Fig. 7.31 for a miniaturized brain–machine interface [104].

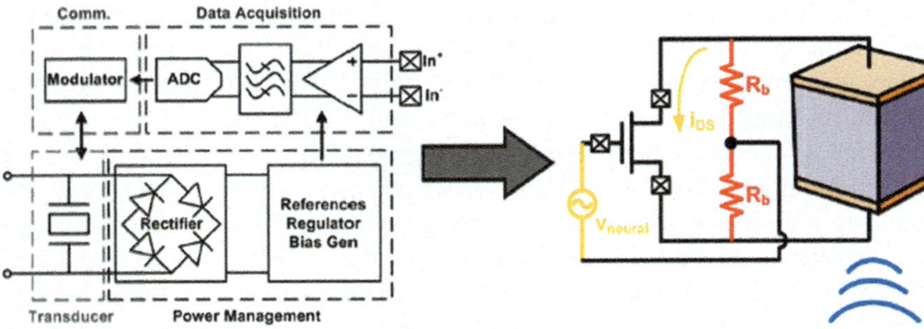

Fig. 7.31 Neural dust: minimal deployment of electronic modules in the sensing device by a process of elimination. Reprinted with full acknowledgment of [104]

Within this topology, the piezoelectric element harvests energy as an AC voltage and transmits it to the FET transistor. The transistor modulates the current I_{DS} as a function of the gate-to-source voltage V_{GS}, which is proportional to the physiological quantity to be measured, V_{neural} in this application. Some special requirements must be met by the AC-operating FET transistor: since the electrical drain and source terminals are swapped in every half cycle of the AC signal, a strategy needs to be deployed to prevent the diodes located on the body-to-source and body-to-drain from becoming forward-biased and limiting the modulation to only one half of the cycle. Resistors R_b remove this restriction by imposing an electrical potential on both the left and right terminals of the FET and on the AC signal from the transducer. Modulation of I_{Ds} produces an impedance variation seen by the piezoelectric element, and in this way the transistor modulates the amplitude of the backscattered signal. In order to achieve the best reflectivity, the values for resistors and transistor width must be selected to take into account the clamped capacitance of the transducer and the bandwidth of operation. The sensitivity of the sensor to detect neural action potentials can be expressed as [104]

$$S = \frac{V_{neural}}{I_{DS} + V_{DS}/2R_b} \frac{\partial I_{DS}}{\partial V_{GS}} = V_{neural} \frac{g_m}{I_{DS} + V_{DS}/2R_b} \tag{7.108}$$

where g_m is the transconductance of the FET. The operation of the transistor in the deep sub-threshold region and its design characteristics to couple with both thermal noise and harvested power determine the achievable level in the differential signal recorded between electrodes that can be effectively received by the readout device.

The exchange of data in binary format instead of a voltage equivalent carried by the backscattered wave can help to increase the discrimination level in the coded parameter retrieved by the readout device by compensating for tissue attenuation in the amplitude of the wave, although the bandwidth for signal transmission is limited. Thus a low-power digital modulator must also be included in the implantable electronics to deal with the digital transmission and, at the readout device, a demodulator to differentiate between only two voltage levels (instead of a full analog range) for a specific baud rate. Typical communication schemes use LSK modulation to switch the electrical load by the piezo [84].

Another possibility for acoustic communication that does not rely on the variation of the amplitude levels is PSK modulation. Considering that ultrasound attenuation in body tissues is stronger than dispersion, this type of coding scheme is more prone to pass through body tissues relatively unscathed and also takes into account the thickness of the communication medium. In PSK, the digital bit pattern is encoded in the carrier wave by phase modulation. The logic state of each bit depends on the previous one: if a 180° shift is verified, the logic levels between adjacent bits are different; if not, the logic levels are the same, as demonstrated by Wild and Hinckley [105].

From the ultrasonic links described above, it is evident that they only possess downlink capabilities. Data has been exclusively flowing from the implantable sensor to the external readout device and not the other way around. This prevents the sensor from receiving, for instance, external update commands due to a change in the experimental conditions or triggering a biological event in the target tissue, such as the opening of ion channels. Moreover, FSK modulation has not yet been implemented in ultrasonic links which just reinforces the first statement: the single-frequency operation of the link to transmit data (and harvest power). The availability of more carrier frequencies would provide a more robust way to broadcast instructions via uplink and power harvesting could even be set to a specific (lower) frequency. In fact, FSK is less susceptible to amplitude-conditioning non-idealities of the transmission path through tissue. However, this separation in the spectrum would imply a broader bandwidth for the transducer, which in turn is inversely proportional to the quality factor of the piezoelectric material. Possible solutions include the deployment of different piezoelectric elements in the implantable sensor (increase in size) or using magnetic or RF transmissions as auxiliary modules to operate the implantable device. Within this approach, the authors of [106] have proposed a mm-sized implantable device with ultrasonic energy transfer shown in Fig. 7.32. The features of the device also include an ultrasonic link employed for low data-rate control signals, along with an

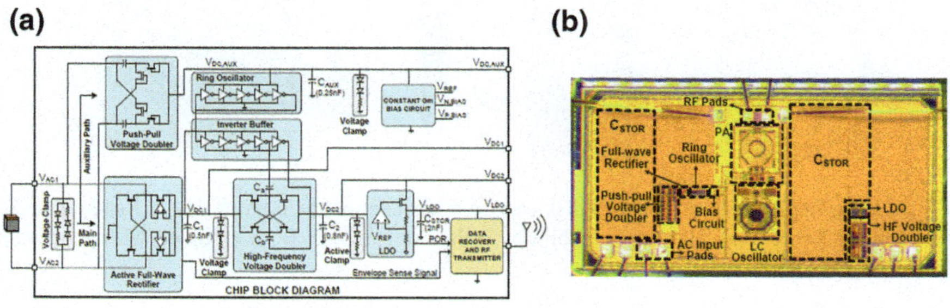

Fig. 7.32 **a** Hybrid two-path block diagram of the implantable device with ultrasonic and RF links. **b** Die photo of the IC designed with TSMC 65 nm general-purpose process technology. Both reprinted with permission from [106], © IEEE 2015

RF link to support higher data rates between the implant and the external device for the purpose of imaging.

7.5.4.1 Active Implantable Sensor for Monitoring of Physiological Parameters Using LSK Modulation

We have recently (2016) developed an implantable sensor to be deployed in soft tissue to measure variables such as electrolyte concentration (lactate/glucose), pH, temperature, and biopotential, using acoustic waves for power harvesting and data transmission [107]. The sensor operates with voltage levels as low as 0.8 V with a current consumption close to 60 μA when excited by a frequency of 400 kHz in the sinusoidal regime. The sensor is depicted in Fig. 7.33 along with the external readout device responsible for producing the incident pressure field around the implant and retrieving the data from the sensor via LSK modulation, which is done by superimposing the binary pattern over the voltage signal present at the input terminals of the implantable transducer.

The embedded electronics on the implant side uses ultra-low-power discrete components available on the market, like the MSP430L092 microcontroller from Texas Instruments® and TS1001 op-amp from Silicon Labs®. The DC level obtained by the harvester when deployed inside a phantom model of the torso region is shown in Fig. 7.34, along with a temporal recording of the temperature, biopotential, pH, and lactate provided by the sensor.

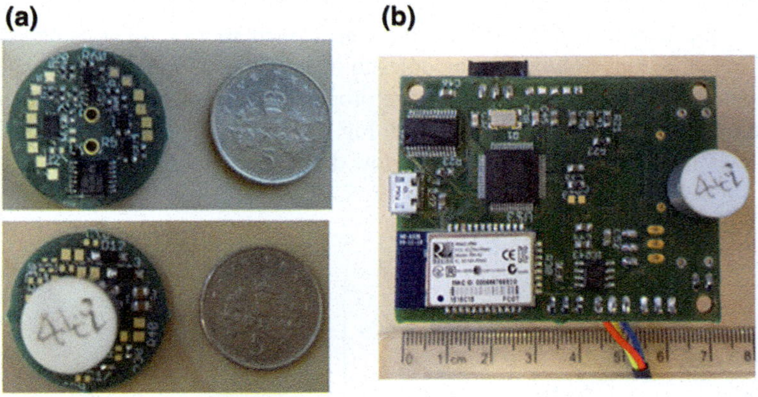

Fig. 7.33 **a** Implantable sensor with disk-shaped transducer on the bottom side. **b** Readout device that delivers power and telemetry capabilities to the implant side. Both reprinted with permission from [107], © IEEE 2016

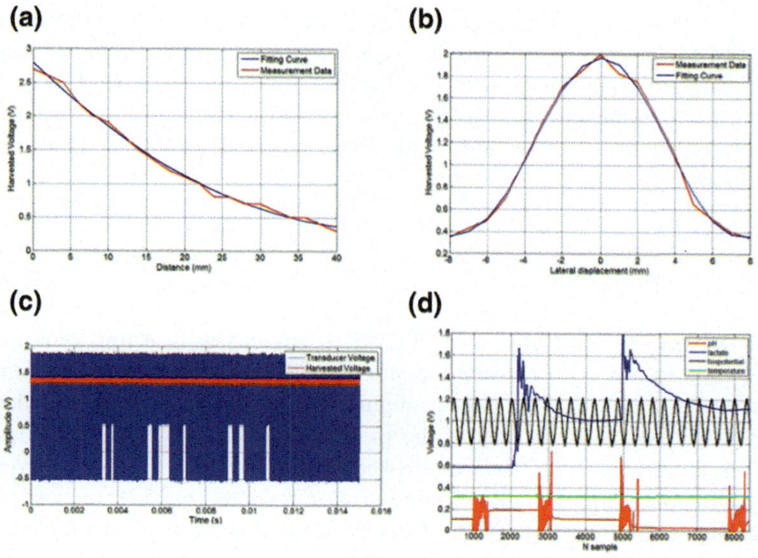

Fig. 7.34 **a** Voltage levels achieved by the implantable device by modifying the geometrical distance towards the readout device. **b** Voltage harvested as a function of the misplacement of the transducers relative to the acoustic axis. **c** LSK modulation imposed over the 400 kHz voltage signal with DC level harvested by the implantable transducer. **d** Analog readings obtained from the acquisitions channels: pH solutions vary between the values of 6, 4, 6 (again), and 10; lactate concentrations change between 0.5, 4, and up to 8 mmol L^{-1}; biopotential is drawn from a sinusoidal external source with amplitude equal to 20 mV; and, temperature remains constant during measurements. Both reprinted with permission from [107], © IEEE 2016

7.5.5 Future Trends in Implantable Electronics: Towards "Zero" Power Sensing

The development of highly integrated and miniaturized acoustic wave biosensors for remote operation and passive interrogation has been growing in the recent years. The sensitivity of the high-frequency waves (MHz–GHz) travelling through the piezoelectric substrate to detect changes in sensor mass loading, viscosity, temperature, pH, etc., makes them ideal to monitor physiological, chemical, and biological variables inside harsh environments like the human body, without active power sources for operation and data exchange. The fabrication process of the sensors naturally follows the latest trends in MEMS technology, ensuring not only compatibility of the piezoelectric layer with the embedded CMOS electronics but also robustness of the biosensor in a single small-sized and low-cost chip, suitable for mass production. In fact, the incorporation of an acoustic wave sensor and signal processing circuits on a single silicon substrate provides a high-resolution system to record any change in the acoustic properties of the piezo film—such as resonant frequency, sound velocity, attenuation or time delay—triggered by some external interaction with the sensing layer. The acoustic wave-based MEMS devices developed so far are divided into two types, depending on the fabrication process and detection mechanisms as film bulk acoustic wave resonators (FBAR) or surface acoustic wave resonators (SAW). The principle of operation of each resonator will be explored in the next sections with some practical examples for biomedical applications.

7.5.5.1 FBAR Technology

FBAR devices are fabricated with MEMS techniques from a thin central piezoelectric layer to which metal electrodes are micropatterned on either side of the film. The acoustic waves will propagate unguided through the volume and detection occurs at the opposite surfaces of the piezoelectric film. The high quality factor Q makes them ideal for interfaces between solid materials and air. However, in biological application, the solid–liquid interface reduces Q and the acoustic waves are no longer confined to the FBAR cavity. To counterbalance this effect, microfluidic channels created at the top of the FBAR need to have heights comparable to the acoustic wavelength to minimize the dissipation of the waves into the liquid and improve the quality factor.

One improvement to the quality factor of the MEMS FBAR is described in [108] by placing ZnO on top of an acoustic mirror and integrated above a silicon substrate. The crystallization of the piezoelectric film with a tilt angle allows operation in the shear mode whereas the Au top electrode is coated with a receptor layer sensitive to the bioanalyte [108]. The attachment of the latter increases the resonator's mass load, which in turn lowers the resonant frequency of the sensor. The insertion of air vacuum gaps between the FBAR sensor and liquid is another technique to protect from liquid damping and Q degradation [108]. Parylene microposts placed on the top electrode support a diaphragm of the same material, which is in contact with the liquid: once the liquid loads the diaphragm it is transferred to the FBAR through the microposts, thus separating the liquid damping effect from the operating frequency of the FBAR and increasing Q with no mass sensitivity loss.

7.5.5.2 SAW Devices

As opposed to FBAR, SAW devices are composed of Interdigital Transducer (IDT) excitation electrodes fabricated on one side of the piezoelectric elements, which then allows the propagation of the acoustic waves along the surface of the piezoelectric substrate. The design of the IDTs dictates the operation of the SAW sensor as a resonator or delay line: for resonators, the IDTs are placed at the centre with reflection gratings on each side of the input and output IDTs to lock the acoustic wave within a cavity; for a delay line, the reflectors are omitted from the design and a larger distance is set between the input and output IDTs. When excited by a sinusoidal voltage at the input IDT, an oscillation mechanical strain is produced in the SAW delay line that originates traveling surface acoustic waves that are detected at the output IDT and converted back to an electrical voltage with characteristic amplitude and phase. To function as a biosensor, a biological layer is created on top of the substrate for specific bioanalyte binding. The changes in mass loading, viscosity, or temperature cause a time delay on the propagation of acoustic waves between IDTs. The incorporation of reflectors beside the input and output IDTs can improve the performance of the sensor by trapping the waves from the central IDT. This reduces wave losses and creates a pronounced resonant peak by a shift in frequency of the traveling wave.

The decay time of the echo signal is proportional to the quality factor of the resonator: longer decay times involve higher accuracy in the detection of the resonant frequency. The Q value is related to the number of IDT finger pairs, number of reflection gratings, manufacturing technology, and size of the acoustic aperture of the resonator [109].

The applications of MEMS SAW sensors in implantable devices include monitoring of seizures by recording temperature shifts within the brain, measurement of inner body temperature [110] and blood pressure monitoring [111]. Figure 7.35 shows one passive sensor based on the 434 MHz SAW filter. In temperature monitoring by means of a SAW delay line sensor, a piezoelectric material like lithium niobate is better suited to detect the changes in the propagation wave parameters.

The delay time t of the signal can be calculated by (7.109) which depends on the wave velocity on the substrate c and the distance l between the input and output IDTs.

$$t = l/c \tag{7.109}$$

An elevation in temperature produces an increase of the delay time by thermal expansion of the substrate and decrease of the sound velocity. These two quantities are related by the temperature coefficient α as

Fig. 7.35 Schematic diagram of the 434 MHz SAW filter for monitoring of brain seizures with micrograph of the sensor. Reprinted with permission from [109], © IEEE 2007

$$\alpha = \frac{1}{t}\frac{\mathrm{d}t}{\mathrm{d}T} = \frac{1}{l}\frac{\mathrm{d}l}{\mathrm{d}T} - \frac{1}{c}\frac{\mathrm{d}c}{\mathrm{d}T} \tag{7.110}$$

where T represents the temperature of the piezoelectric film. For small changes in temperature, the change of delay time Δt is given by

$$\Delta t \approx \alpha t_0 \Delta T \tag{7.111}$$

with t_0 the nominal delay time of the SAW device. For pressure monitoring, a SAW resonator is employed for the determination of the shift in the resonant peak of the device. The frequency f is determined by the velocity of the acoustic waves in the surface of the film c and the period of IDT and reflection gratings λ in accordance with

$$f = \frac{c}{\lambda} \tag{7.112}$$

The application of an external pressure produces the bending of the substrate, which modifies both c and λ and, consequently, a change Δf in the resonant frequency. The resonant frequency can also be expressed by the displacement ε as

$$f_\varepsilon = \frac{c_\varepsilon}{\lambda_\varepsilon} = \frac{c_0(1+k\varepsilon)}{\lambda_0(1+\varepsilon)} = f_0\frac{1+k\varepsilon}{1+\varepsilon} \approx f_0\varepsilon(k-1) + f_0 \tag{7.113}$$

where k represents the material coefficient of the piezoelectric film, f_0, c_0, and λ_0 are the SAW resonant frequency, sound velocity, and IDT period for the condition of absent pressure, respectively, and f_ε, c_ε, and λ_ε are the same parameters under pressure.

Although active research in MEMS SAW technology in the last few years has led to the development of successful biosensors for antibody/antigen recognition systems, there remains the unsolved problem of data transmission via acoustic waves to the external readout device instead of radio-frequency. In fact, the changes in resonant frequency and delay time of the acoustic waves are still coded and transmitted by means of RF embedded circuitry using CMOS chip fabrication processes, which makes the data exchange link not purely ultrasonic as desired. Moreover, the degradation in performance of SAW due to the damping associated with liquid loading still requires further innovations in the technology to couple with the decrease in device's sensitivity in solid–liquid interfaces.

7.6 Conclusion

In this chapter, the main sources of power harvesting for implantable devices have been discussed. These include electrochemical, photovoltaic, thermoelectric, mechanical, inductive, RF, and ultrasounds, with more focus on the EM ones, since they can provide more realistic solutions to fulfil the demand for power scavenging on the implant side. In fact, many of the approaches involving photovoltaic and thermoelectric energies are still in the domain of numerical simulations to evaluate the effective DC power that can be retrieved from sources within or around the human body. Electrochemical reactions are good candidates for micro-sized sensors, with dimensions comparable to the cellular environment as they can use small amounts of glucose to initiate and/or monitor biological processes with a time-limited duration. At a different scale, mechanical harvesters can use body and tissue motions to convert to an electrical equivalent as long as the operation of the harvester does not interfere with the well-being of the patient. The low-frequency content of the movements limit the output power obtained to the μW range, although a stable output can be achieved if using the cardiac pacemaker that permanently produces variations in the entire cardiovascular system that can be converted to electrical energy.

Whenever the output voltage level produced by the harvester is lower to drive circuit electronics, boost converters, and power management units can be used to deliver the required DC value during the active state of the sensor while saving the rest of the energy in capacitors or other reservoirs for active expenditure. When telemetry capabilities are at stake, only inductive, RF, and ultrasounds can effectively assist the implantable device to communicate with the outside world. Besides being the most effective way of carrying energy in the form of electromagnetic and acoustic waves (active links), they can also be used to code the instructions/commands (uplink) and retrieve sensor data (downlink) in different modulation schemes that better fit a particular application. ASK modulation is best suited for applications where the implantable sensor does not rely on digital control and the data can be sent as a true voltage level modulated to the carrier wave, with the full scale range; however, amplitude attenuation caused by tissues and nearby devices help to move to a frequency-based encoding instead if the dispersive effects can be neglected in the transmission path; or even a pure digital LSK encoding scheme by simple changing the impedance of the implanted coil, antenna, or piezo element between two values.

The actual power requirement for commercial implantable devices (pacemaker, defibrillator, drug pump) dictates that only near-field magnetic radiation can reach the mW level required at close distances to the source. For RF and ultrasounds, this level is still one to two orders of magnitude higher than the harvested power obtained. Investigation towards designing higher efficient coils, antennas and/or piezos has led to the discovery of new materials that can be microelectromechanically engineered to meet the requirements of small-sized implants while preserving proper signal bandwidth and DC levels required to power up the embedded electronics in the case of active sensors. For passive ones, DC power might not be required to sustain the functionality of the sensor and the change of reflectivity imposed by the sensor to the external readout device (or source) is enough to code the desired physiological parameter.

In RF applications, antenna design is the most problematic issue to address since it needs to be conformal not only geometrically but also impedance matched to the surrounding biological tissues in order to capture the radiated power more effectively. Parameters such as antenna size, gain, directivity, and impedance are factors to deal with carefully to make the most from the antenna performance while operating inside a harsh dielectric environment like the human body. From all the parameters, size can be the most challenging issue if the device is expected to operate at a licensed frequency inside an FCC approved band. Since the size is related to the wavelength of transmission, the one-dimensional length of the antenna can still be large even for the higher spectrum (10 GHz). Strategies using 3D antenna folding (zig-zag, meander lines), fractal topologies, and employment of high-*k* dielectric matrices can be a practical solution to shrink the size while attaining the correct frequency of transmission. Even the use of biodegradable conductive polymer composites has been reported to model LC resonant tanks correctly as well as SAW filters and RF metamaterials.

Finally, the piezoelectric transducer remains the most critical component in the ultrasound harvester, since its design affects the mode of operation, resonant frequency, signal bandwidth, and both the electrical and acoustic impedances in a three-port system. Technologies to produce high-efficiency piezoelectric devices have been carried out for many years, and these include MEMS technology for substrates embedded with sensitive detection layers by means of FBAR and SAW devices, and also the combination of magnetic or triboelectric substances in order to increase the level of voltage harvested. The small implantable devices limit the size of the transducer, preventing the use of lower frequencies that can cross tissues with lesser attenuation, making them more orientation-sensitive towards the position of the external transducer. Nonetheless, the ultrasonic links can eventually be made more power-efficient than inductive and RF counterparts for deeply implanted devices since they are immune to any source of electromagnetic interference. Moreover, the apodization of the acoustic beam can compensate for some geometrical constraints in the propagation path by moving between the near- and far-field regions of the pressure distribution map.

References

1. K. Bazaka, M.V. Jacob, Implantable devices: issues and challenges. Electronics **2**, 1–34 (2013)
2. A. Dewan et al., Alternative power sources for remote sensors: A review. J. Power Sources **245**, 129–143 (2014)
3. D. Pech et al., Ultrahigh-power micrometre-sized supercapacitors based on onion-like carbon. Nat. Nanotechnol. **5**, 651–654 (2010)

4. S. Xu et al., Stretchable batteries with self-similar serpentine interconnects and integrated wireless recharging systems. Nat. Commun. **4**, 1543 (2013)
5. J. Olivo et al., Biofuel cells and inductive powering as energy harvesting techniques for implantable sensors. Sci. Adv. Mater. **3**, 420–425 (2011)
6. S. Kerzenmacher et al., Energy harvesting by implantable abiotically catalyzed glucose fuel cells. J. Power Sources **182**, 1–17 (2008)
7. P. Cinquin et al., A glucose biofuel cell implanted in rats. PLoS ONE **5**(5), e10476 (2010)
8. E. Katz, K. MacVittie, Implanted biofuel cells operating in vivo—methods, applications and perspectives—feature article. Energy Environm. Sci. **6**, 2791–2803 (2013)
9. P.P. Mercier et al., Energy extraction from the biologic battery in the inner ear. Nat. Biotechnol. **30**(12), 1240–1243 (2012)
10. M. Khan et al., A novel SPICE implementation of MPPT technique for implantable solar powered cardiac biosensors, in *9th International Conference on Industrial and Information Systems (ICIIS)* (2014)
11. S. Ayazian et al., A photovoltaic-driven and energy-autonomous CMOS implantable sensor. IEEE Trans. Biomed. Circuits Syst. **6**(4), 336–343 (2012)
12. K. Sankaragomathi et al., A 27 μW subcutaneous wireless biosensing platform with optical power and data transfer, in *IEEE Proceedings of the Custom Integrated Circuits Conference*, pp. 1–4 (2014)
13. Y. Yang et al., Suitability of a thermoelectric power generator for implantable electronic devices. J. Phys. D Appl. Phys. **40**, 5790–5800 (2007)
14. C. Wu et al., A pliable and batteryless real-time ecg monitoring system-in-a-patch, in *IEEE Transactions on Very Large Integration Systems*, 2015
15. M. Ashraf and N. Masoumi, A thermal energy harvesting power supply with an internal startup circuit for pacemakers, *IEEE Transactions on Very Large Integration Systems* (2015)
16. S.E. Jo et al., Flexible thermoelectric generator for human body heat energy harvesting. Electron. Lett. **48**(16), 1013–1015 (2012)
17. C. Watkins et al., Low-grade-heat energy harvesting using superlattice thermoelectrics for applications in implantable medical devices and sensors, in *International Conference on Thermoelectrics* (2005)
18. D. Rozgic, D. Markovic, A 0.78 mW/cm^2 autonomous thermoelectric energy-harvester for biomedical sensors, in *Symposium on VLSI Circuits Digest of Technical Papers* (2015)
19. H. Zhang et al., A flexible and implantable piezoelectric generator harvesting energy from the pulsation of the ascending aorta: in vitro and in vivo studies. Nano Energy **12**, 296–304 (2015)
20. N. Fadhil et al., Energy harvesting using nano scale dual layers PVDF film for blood artery, in *IEEE Long Island Systems, Applications and Technology Conference*, pp. 1–6 (2013)
21. M.A. Karami, D.J. Inman, Powering pacemakers from heartbeat vibrations using linear and nonlinear energy harvesters. Appl. Phys. Lett. **100**, 042901 (2012)
22. M. Deterre et al., Micro blood pressure energy harvester for intracardiac pacemaker. J. Microelectromech. Syst. **23**(3), 651–660 (2014)
23. Z. Lin Wang, Self-powered nanotech, in *Scientific American*, pp. 82–87 (2008)
24. R. Jagadeesan, Y. Guo, Topology selection and efficiency improvement of inductive power links. IEEE Trans. Antennas Propag. **60**(10), 4846–4854 (2012)
25. W. Zhang et al., Analysis and comparison of secondary series- and parallel-compensated inductive power transfer system operating for optimal efficiency and load-independent voltage-transfer ratio. IEEE Trans. Power Electron. **29**(6), 2979–2990 (2014)
26. T. Le et al., Piezoelectric micro-power generation interface circuits, *IEEE J. Solid State Circuits*, 41(6), 1411–1420 (2006)
27. S. Mandal, R. Sarpeshkar, Low-power CMOS rectifier design for RFID applications. IEEE TCAS-I Regul. Pap. **54**(6), 1177–1188 (2007)
28. P. Si et al., A frequency control method for regulating wireless power to implantable devices. IEEE TBCAS **2**(1), 22–29 (2008)

29. M. Kiani et al., A Q-modulation technique for efficient inductive power transmission, IEEE J. Solid State Circuits (2015) (accepted and available online)
30. M.W. Baker, R. Sarpeshkar, Feedback analysis and design of RF power links for low-power bionic systems. IEEE Trans. Biomed. Circuits Syst. **1**(1), 28–38 (2007)
31. M. Kiani et al., Design and optimization of a 3-coil inductive link for efficiency wireless power transmission. IEEE Trans. Biomed. Circuits Syst. **5**(6), 579–591 (2011)
32. A. Sample et al., Analysis, experimental results, and range adaptation of magnetically coupled resonators for wireless power transfer. IEEE Trans. Ind. Electron. **58**(2), 544–554 (2011)
33. M.L. Kung, K.H. Lin, Enhanced analysis and design method of dual-band coil module for near-field wireless power transfer systems. IEEE Trans. Microw. Theory and Tech. **63**(3), 821–832 (2015)
34. Z. Tang et al., Data transmission from an implantable biotelemeter by load-shift keying using circuit configuration modulator. IEEE Trans. Biomed. Eng. **42**(5), 524–528 (1995)
35. S. Mandal, R. Sarpeshkar, Power-efficient impedance-modulation wireless data links for biomedical implants. IEEE Trans. Biomed. Circuits Syst. **2**(4), 301–315 (2008)
36. F. Inanlou, M. Ghovanloo, Wideband near-field data transmission using pulse harmonic modulation, *IEEE Trans*. Circuits Syst. I Regul. Pap. **58**(1), 186–195 (2011)
37. M. Kiani, M. Ghovanloo, A 13.56 Mbps pulse delay modulation based transceiver for simultaneous near-field data and power transmission. IEEE Trans. Biomed. Circuits Syst. **9** (1), 1–11 (2015)
38. K. Fotopoulou, B. Flynn, Wireless power transfer in loosely coupled links: Coil misalignment model. IEEE Trans. Magnet. **47**(2), 416–430 (2011)
39. J.S. Ho, A.S. Poon, Midfield wireless powering for implantable systems. Proc. IEEE **101**(6), 1369–1378 (2013)
40. S. Leung, D. Lam, Performance of printed polymer-based RFID antenna on curvilinear surface. IEEE Transactions on Electronics Packaging Manufacturing **30**(3), 200–205 (2007)
41. G. Fotheringham et al., Parameterization of bent coils on curbed flexible surface substrates for RFID applications, in *59th Conference on Electronic Components and Technology* (2009)
42. U.M. Jow, M. Ghovanloo, Modeling and optimization of printed spiral coils in air, saline, and muscle tissue environments. IEEE Trans. Biomed. Circuits Syst. **3**(5), 339–347 (2009)
43. ICNIRP: Guidelines for limiting exposure to time-varying electric, magnetic and electromagnetic fields (up to 300 GHz), Health Phys. 74, 494–522 (1998)
44. A. Christ et al., Evaluation of wireless resonant power transfer system with human electromagnetic exposure limits. IEEE Trans. Electromagn. Compat. **55**(2), 265–274 (2013)
45. T. Sunohara et al., Induced field and SAR in human body model due to wireless power transfer system with induction coupling, in *IEEE International Symposium on Electromagnetic Compatibility*, Tokyo (2014)
46. A. Al-Kalbani et al., Electromagnetic interference in brain implants using multiple coils: biosafety and data communication performance. IEEE Trans. Electromagn. Compat. **56**(2), 490–493 (2014)
47. C. Sauer et al., Power harvesting and telemetry in CMOS for implanted devices, *IEEE Trans*. Circuits Syst I Regul Pap. **52**(12), 2605–2613 (2005)
48. B. Lenaerts, R. Puers, An inductive power link for a wireless endoscope. Biosens. Bioelectron. **22**, 1390–1395 (2007)
49. K. M. Silay et al., Load optimization of an inductive power link for remote powering of biomedical implants, in *IEEE International Symposium on Circuits and Systems*, pp. 533–536 (2009)
50. Y. Hu, M. Sawan, A fully integrated low-power BPSK demodulator for implantable medical devices, *IEEE Trans*. Circuits Syst. I Regul. Pap. **52**(12), 2552–2562 (2005)
51. M. Piedade et al., Visual neuroprosthesis: a non invasive system for stimulating the cortex, IEEE Trans. Circuits Syst. I Regul. Pap. **52**(12), 2648–2662 (2005)

52. R.A. Bercich et al., Far-field RF powering of implantable devices: safety considerations. IEEE Trans. Biomed. Eng. **60**(8), 2107–2112 (2013)
53. A. Ba et al., A 0.33 nJ/bit IEEE802.15.6/proprietary MICS/ISM wireless transceiver with scalable data rate for medical implantable applications. IEEE J. Biomed. Health Inf. **19**(3), 920–929 (2015)
54. R.E. Diaz, T. Sebastian, Electromagnetic limits to radiofrequency (RF) neuronal telemetry. Nat. Sci. Rep. **3**, 3535 (2013)
55. E.Y. Chow et al., Implantable RF medical devices. IEEE Microwave Mag. **14**(4), 64–73 (2013)
56. C.A. Balanis, *Antenna Theory: Analysis and Design*, 3rd edn. (Wiley, New Jersey, 2005)
57. I. Singh, V.S. Tripathi, Micro strip patch antenna and its applications: a survey. Int. J. Comput. Appl. Technol. **2**(5), 1595–1599 (2011)
58. A. Kumar et al., Performance analysis of different feeding techniques. Int. J. Emerg. Technol. Adv. Eng. **3**(3), 884–890 (2013)
59. P. Anacleto et al., Micro antennas for implantable medical devices, in *IEEE 3rd Portuguese Meeting in Bioengineering*, pp. 1–4 (2013)
60. C.L. Yang et al., Low-invasive implantable devices of low-power consumption using high-efficiency antennas for cloud health care. IEEE Journal on Emerging and Selected Topics in Circuits and Systems **2**(1), 14–23 (2012)
61. L.Y. Chen et al., Continuous wireless pressure monitoring and mapping with ultra-small passive sensors for health monitoring and critical care. Nat. Commun. **5**, 5028 (2013)
62. M. Shakib et al., Design of a tri-band implantable antenna for wireless telemetry applications, in *IEEE International Microwave Workshop Series on RF and Wireless Technologies for Biomedical and Healthcare Applications*, pp. 1–3 (2014)
63. D.D. Karnaushenko et al., Compact helical antenna for smart implant applications. Nat. Publ. Group Asia Mater. **7**, e188 (2015)
64. S.Y. Wu et al., 3D-printed microelectronics for integrated circuitry and passive wireless sensors. Nat. Microsyst. Nanoeng. **1**, 15013 (2015)
65. N. Shariati et al., Multi-service highly sensitive rectifier for enhanced RF energy scavenging. Nat. Sci. Rep. **5**, 9655 (2015)
66. J. Walk et al., Improvements of wireless communication and energy harvesting aspects for implantable sensor interfaces by using the split frequencies concept, in *IEEE Radio and Wireless Symposium*, pp. 406–409 (2011)
67. D.W. Gulik, B.C. Towe, Characterization of simple wireless neurostimulators and sensors, in *IEEE 36th Annual International Conference of the Engineering in Medicine and Biology Society*, pp. 3130–3133 (2014)
68. R. Ahmed et al., An RFID front end for smart biological sensing, in *IEEE 55th International Midwest Symposium on Circuits and Systems*, pp. 778–781 (2012)
69. E.Y. Chow et al., Wireless powering and the study of RF propagation through ocular tissue for development of implantable sensors. IEEE Trans. Antennas Propag. **59**(6), 2379–2387 (2011)
70. K. Okabe et al., A thin film flexible antenna with CMOS rectifier chip for RF-powered implantable neural interfaces, in *IEEE 18th International Conference on Solid-State Sensors, Actuators and Microsystems*, pp. 1751–1754 (2015)
71. E.Y. Chow et al., Fully wireless implantable cardiovascular pressure monitor integrated with a medical stent. IEEE Trans. Biomed. Eng. **57**(6), 1487–1496 (2010)
72. P. Cong et al., A wireless and batteryless 10-bit implantable blood pressure sensing microsystem with adaptive RF powering for real-time laboratory mice monitoring. IEEE J. Solid State Circuits **44**(12), 3631–3644 (2009)
73. D. Venuto, J. Rabaey, RFID transceiver for wireless powering brain implanted microelectrodes and backscattered neural data collection. Microelectron. J. **45**, 1585–1594 (2014)
74. M. Arsalan et al., A 5.2 GHz, 0.5 mW RF powered wireless sensor with dual on-chip antennas for implantable intraocular pressure monitoring, in *IEEE MTT-S International Microwave Symposium Digest* (2013)

75. J. Mao et al., A subgigahertz UWB transmitter with wireless clock harvesting for RF-powered applications, *IEEE Trans*. Circuits Sys. II Express Briefs **61**(5), 314–318 (2014)
76. C. M. Boutry et al., RF conductivity of biodegradable conductive polymers used for a new generation of partially/fully resorbable wireless implantable sensors, in *IEEE 25th International Conference on Micro Electro Mechanical Systems*, pp. 468–471 (2012)
77. S.G. Kim et al., A highly sensitive and label free biosensing platform for wireless sensor node system. Biosens. Bioelectron. **50**, 362–367 (2013)
78. N.Y. Kim et al., A reusable robust radio frequency biosensor using microwave resonator by integrated passive device technology for quantitative detection of glucose level. Biosens. Bioelectron. **67**, 687–693 (2015)
79. R. Melik et al., Metamaterial-based wireless RF-MEMS strain sensors, in *IEEE Sensors*, pp. 2173–2176 (2010)
80. J.H. Lee et al., High temperature, high power piezoelectric composite transducers. Sensors **14**, 14526–14552 (2014)
81. R. Calio et al., Piezoelectric energy harvesting solutions. Sensors **14**, 4755–4790 (2014)
82. T.L. Szabo, *Diagnostic Ultrasound Imaging: Inside Out* (Elsevier Inc, London, 2004)
83. Z. Suo, Theory of dielectric elastomers. Acta Mechanica Solida **23**(6), 549–578 (2010)
84. S. Ozeri, D. Shmilovitz, Simultaneous backward data transmission and power harvesting in an ultrasonic transcutaneous energy transfer link employing acoustically dependent electric impedance modulation. Ultrasonics **54**, 1929–1937 (2014)
85. A. Denisov, E. Yeatman, Ultrasonic vs. inductive power delivery for miniature biomedical implants, in *IEEE International Conference on Body Sensor Networks*, pp. 84–89 (2010)
86. Y. Zhu et al., Ultrasonic energy transmission and conversion using a 2-D MEMS resonator. IEEE Electron Dev. Lett. **31**(4), 374–376 (2010)
87. A. Fowler et al., An omnidirectional MEMS ultrasonic energy harvester for implanted devices. J. Microelectromech. Sys. **23**(6), 1454–1462 (2014)
88. Q. He et al., MEMS-based ultrasonic transducer as the receiver for wireless power supply of the implantable microdevices. Sens. Actuators A **219**, 65–72 (2014)
89. S. Ozeri et al., Ultrasonic transcutaneous energy transfer using a continuous wave 650 kHz Gaussian shagged transmitter. Ultrasonics **50**, 666–674 (2010)
90. S. Ozeri, D. Shmilovitz, Ultrasonic transcutaneous energy transfer for powering implanted devices. Ultrasonics **50**, 556–566 (2010)
91. M. Xu, L.V. Wang, Photoacoustic imaging in biomedicine. Rev. Sci. Instrum. **77**, 041101 (2006)
92. K.W. Dongen, W.M. Wright, A forward model and conjugate gradient inversion technique for low-frequency ultrasonic imaging. J. Acoust. Soc. Am. **120**(4), 2086–2095 (2006)
93. F. Mazzilli et al., Ultrasound energy harvesting system for deep implanted-medical-devices (IMDs), in *IEEE International Symposium on Circuits and Systems*, pp. 2865–2868 (2012)
94. Y. Liu et al., Active piezoelectric energy harvesting: general principle and experimental demonstration. J. Int. Mater. Sys. Struct. **20**, 575–585 (2009)
95. J. Qiu et al., Comparison between four piezoelectric energy harvesting circuits. Front. Mech. Eng. China **4**(2), 153–159 (2009)
96. D. Guyomar et al., Energy harvesting from ambient vibrations and heat. J. Intell. Mater. Sys. Struct. **20**, 609–623 (2009)
97. J. Park et al., The effect of switch triggering offset and switch on-time duration on harvested power in synchronized switch harvesting on inductor. Int. J. Smart Home **7**(3), 207–218 (2013)
98. A. Nechibvute et al., Piezoelectric energy harvesting using synchronized switching techniques. Int. J. Eng. Technol. **2**(6), 936–946 (2012)
99. M.L. Navaii et al., An ultra-low power RF interface for wireless-implantable microsystems. Microelectron. J. **43**, 848–856 (2012)

100. Y. Ammar, S. Basrour, Non-linear techniques for increasing harvesting energy from piezoelectric and electromagnetic micro-power-generators, in *Dans Symposium on Design, Test, Integration and Packaging of MEMS/MOEMS*, Stresa, Italy (2006)
101. A. Sanni et al., Inductive and ultrasonic multi-tier interface for low-power, deeply implantable medical device. IEEE Trans. Biomed. Circuits Sys. **6**(4), 297–308 (2012)
102. L. Cheng et al., Wireless, power-free and implantable nanosystem for resistance-based biodetection. Nano Energy **15**, 598–606 (2015)
103. W.S. Jung et al., High output piezo/triboelectric hybrid generator. Nat. Sci. Rep. **5**, 9309 (2015)
104. D. Seo et al., *Neural Dust: An Ultrasonic, Low Power Solution for Chronic Brain-Machine Interfaces* (Cornell University Library, Berkeley, 2013)
105. G. Wild, S. Hinckley, Acoustic transmissions for wireless communications and power supply in biomedical devices, in *Proceedings of 20th International Congress on Acoustics* (2010)
106. J. Charthad et al., A mm-sized implantable medical device (IMD) with ultrasonic power transfer and a hybrid bi-directional data link. IEEE J. Solid-State Circuits **50**(8), 1741–1753 (2015)
107. B.M. Rosa, G.Z. Yang, Active implantable sensor powered by ultrasounds with application in the monitoring of physiological parameters for soft tissue, in *IEEE Conference on Body Sensor Networks* (2016)
108. I. Voiculescu, A.N. Nordin, Acoustic wave based MEMS devices for biosensing applications. Biosens. Bioelectron. **33**, 1–9 (2012)
109. N. Gopalsami et al., SAW microsensor brain implant for prediction and monitoring of seizures. IEEE Sens. J. **7**(7), 977–982 (2007)
110. G. Martin et al., Measuring the inner body temperature using a wireless temperature saw-sensor-based system, in *IEEE Conference: Ultrasonics Symposium*, 4 (2005)
111. X. Ye et al., Studies of a high-sensitive surface acoustic wave sensor for passive wireless blood pressure measurement. Sens. Actuators A **169**, 74–82 (2011)

Chapter 8
Wireless Datapaths and Security

B. Gil, H. Ip and Guang-Zhong Yang

List of Acronyms

ACL	Control access list
AOA	Angle of arrival
ASIC	Application specific integrated circuit
BCC	Body-coupled communication
CA	Cellar automata
CMOS	Complementary metal-oxide semiconductor
DTOA	Differential time of arrival
ECC	Error correcting codes
EEG	Electroencephalogram
EMA	European Medicines Agency
ESDS	ECG-based secret data sharing
FDA	Food and Drug Administration
GE	Gate equivalents
IID	Independent-identically distributed
IMD	Implantable device
LFSR	Linear feedback shift register
LPN	Learning parity in the presence of noise
MAC	Message authentication code
MICS	Medical Implant Communication System
NFC	Near-field communication
NFSR	Nonlinear feedback shift register
OTP	One-time pads
PKI	Public-key infrastructure
PRG	Pseudorandom generator
RAM	Random access memory
RF	Radiofrequency
RFID	Radiofrequency identification

B. Gil (✉) · H. Ip · G.-Z. Yang
The Hamlyn Centre, Imperial College London, London, UK
e-mail: b.gil-rosa@imperial.ac.uk

G.-Z. Yang (ed.), *Implantable Sensors and Systems*
https://doi.org/10.1007/978-3-319-69748-2_8

RN	Random number
RSSI	Received signal strength indicator
RV	Random variable
SHA	Secure hash algorithm
TOA	Time-of-arrival
USRP	Universal software radio peripheral
WMTS	Wireless Medical Telemetry Services

8.1 Introduction

For Implantable Medical Devices (IMD), we have discussed in the previous chapters the technical challenges related to biocompatible materials, flexible fabrication processes, system-on-chip design, low-power operation, and packaging. IMDs currently available on the market include pacemakers, implantable cardiac defibrillators, neurostimulators (e.g. for Parkinson's disease, epilepsy, dystonia, or depression), drug delivery systems (pain relievers for cancer or insulin pumps), and biosensors that monitor a myriad physiological variables susceptible to adverse events. The sophisticated and intelligent computing capabilities found in IMDs and the advances in networking technology further broaden the spectrum of future functions of IMDs. In fact, success in the deployment of IMDs in human subjects has sparked a vision of pervasive healthcare systems capable of identifying, monitoring and assisting patients without physical or temporal constraints. This cascade of innovation can only make a real impact on healthcare when a high level of security is incorporated in the IMDs regarding the patient's private data, vital signals, and/or therapeutic dosages which can ultimately threaten a patient's life if incorrectly or maliciously manipulated by unauthorized parties or attackers. The wireless data paths therefore represent the main pathway to launch remote attacks against the IMDs when the security of the communication protocol is unsophisticated, which is the case for most of the IMDs on the market today. Although credited regulatory agencies like the FDA in the United States and the EMA in Europe have already started issuing guidelines to manufacturers to take effective measures regarding cybersecurity, there is yet to be a common standard for integrated security on the implant side without compromising its original (and approved) incumbency.

Thus far, many strategies have been proposed to mitigate the security problem for implants, but it is far from being solved. Many even argue that, possibly, a completely secure system is not feasible given the restricted resources available in the implants, and a trade-off is inevitable. For those who are interested in solving the problem, this chapter may serve as a starting point, by identifying some of the security questions and providing a technical overview of the latest developments in lightweight security systems suitable for implants.

We begin the chapter by recognizing the relationship between the interacting parties in an IMD security system, followed by a discussion of the resource constraints, and then give a qualitative overview of the strategies commonly used to provide a secure implant system. The last part of the chapter will give a technical description of some of the lightweight and non-hardware-intensive algorithms that can be deployed for implants.

8.2 Relationship Between IMD, Patient, and Third Parties

An IMD by definition is an electronic device to be deployed inside the human body to treat a medical condition. Life expectancy of the IMD depends on the type of condition and functionality of the device: a passive IMD or biosensor will transmit vital signals from the patient when requested (e.g. glucose monitor), whereas an active one will be responsible for delivering therapy capabilities to the human tissue, such as neurostimulators. Reduction of power consumption by the implant is important so as to preserve battery life or prevent the constant exposure of the human body to radiated wireless fields. When non-rechargeable batteries are the sole powering option, extending their discharging periods is even more important. A typical lithium battery inside a human body can store about 3 kJ of energy [1]. For neurostimulators, batteries can last for 8 years without replacement; on the other hand, up to 10 years of operation is commonly reported for pacemakers [2].

The most recent implants are designed with wireless telemetric capability that transmits sensitive medical information from the patient to an external device supervised by a trusted party, who may be a clinician, caregiver, or the IMD manufacturer. In this way, vital signals and personal data are sent from the IMD to an external device and therapeutic actions and commands are sent in the opposite direction. In many cases, the data exchange link is not well protected by a security protocol, which exposes the IMD to attack or misuse.

Attackers may remotely access and modify the implant with potentially fatal consequences to the patient. At the moment, there are no reported attacks on IMDs outside the scope of scientific research, but with an annual growth rate of 9% in the number of pacemaker and insulin pump implant operations reported in the United States, it is expected that malicious attacks on IMDs will become a reality [3]. Some speculate that the attack rate will be in line with those registered for device recalls. The percentage of computer-related IMD recalls was between 30 and 40% in the period extending from 2006 to 2011, and from the defects analysed it was found that in 33% of cases there was potential for causing harmful effects to the patient.

The nature of the attacker can be divided into two main categories: passive, if the attacker can only perform eavesdropping over the messages exchanged in the network, compromising the confidentially and disclosure of patient data; and active, which is more severe, since the attacker now possesses extended capabilities to modify the messages exchanged in the network, putting in jeopardy both authorization and authentication processes, or even the integrity and availability of the network by jamming the radio channels. It should also be emphasized that the attacker may possibly be the person wearing the IMD in a conscious or unconscious effort to have access to some operations of the IMD that were originally blocked by the manufacturer.

Moreover, the variety of operating modes of an implant makes the implementation of any security modality even harder. IMDs are expected to operate in both normal and emergency scenarios. While in normal mode, strong control mechanisms can prevent unauthorized access to the implant, whereas in an emergency room the same mechanisms may need to be disabled in order for clinicians to have easy access to the patient's vital data stored inside the implant, or even to allow immediate shutdown of a faulty IMD. Emergency scenarios can give rise to the need to bypass an IMD's security protocol, and hence the notion of "break-the-glass" access. Therefore, secure and safe access mechanisms with due consideration for different scenarios are essential.

8.3 Technical Resources at the Disposal of IMDs and Attackers

Implantable devices are highly specialized and designed with the smallest amount of resources indispensable to their operation in order to minimize physical space and save battery life [4]. Memory and computation capabilities are required to perform any security-oriented operation inside the IMD. Long computation cycles can easily drain batteries, and the same is true for operations involving access to the memory, in particular random-access memory (RAM). Size and resource constraints have led to the design of IMDs with limited computational power and memory space. For instance, the storage capabilities for the most traditional implants are typically 2–8 kB for pacemakers, 128–1024 kB for defibrillators and 8 kB for glucose controllers [2]. To add even more constraints to security designers, about 75% of the storage capability must be used for recording physiological data, which leaves only 25% of the storage to accommodate intermediate and final results derived from the security algorithms.

In fact, some existing research has already started to specify the footprint of an ASIC devoted to perform security checks [5]. Martin et al. proposed counting the area of Gate Equivalents (in the form of GE) to be allocated to the lightweight protection algorithms. They argue that, in the design of security-based algorithms, a balance must be found between operations involving pure arithmetic and those involved in logic and memory control. Table 8.1 illustrates the GEs obtained by performing basic operations with different bit-length variables and for two CMOS logic libraries. The value of GE is computed from the area occupied by each circuit in proportion to the area of a standard NAND gate. As a reference, the authors also state that for tiny RFID tags, the range is from 250 to 4 K GE. It is quite evident that the area cost for registers is higher than for multiplexers, and bitwise operations are preferred over addition and multiplication, even if these operations are computed over a large number of cycles. However, in the estimation of the total area occupied by the algorithm, they found that it depends on numerous factors, like the architecture design, the skills of the programmer, and specific design constraints. In turn, Hosseini-Khayat focused on the design of another robust lightweight algorithm using the minimum possible number of hardware modules inside the ASIC. That required a 64-bit cipher decoder, a 32-bit counter, a 32-bit register, some control logic, and no need for RAM, all suitable for implementation in deep submicron technology, where static power consumption dominates [1]. With the goal of zero power consumption by the security scheme, Yang et al. proposed an on-chip security guard that fetches power from external wireless signals and performs authentication using hash encryption with a consumption of 1 mW [6].

Nonetheless, in the vast majority of implants, the constraints on resources cannot be lifted, raising a series of trade-offs that are unique to this field [2]: IMD security versus patient safety, in the sense that security measures can indeed threaten a patient's life in an emergency scenario; battery life versus IMD capabilities, since the more computational power devoted to security, the more energy will be drained from battery; and finally, battery duration versus IMD response times, to respond to trusted sources or withstand attacks by denial-of-service operations. The latter is largely dependent on the wireless infrastructure deployed for data exchange between the IMD and a trusted device while remaining undetectable to unauthorized ones. It is common sense to assume that longer messages exchanged in networking with larger temporal separation can help save computational and power resources, rather than using continuous "bursts" of data [7]. The radio channel *per se* still remains the main source of energy expenditure and it must operate in non-continuous mode, spending most of its time in "silence". In the event of an

Table 8.1 Number of GEs occupied by common mathematical operations and registers in cryptographic ASIC design using the AMI 0.35 μm and UMC 90 nm libraries. Reprinted with permission from [5], © IEEE 2014

Library	Element	32 bit	64 bit	96 bit	128 bit
UMC 90 nm	AND	39.70	79.39	119.08	158.78
	OR	39.70	79.39	119.08	158.78
	XOR	79.39	158.78	238.17	317.56
	ADD	239.66	477.84	716.01	954.19
	Multiplexer	71.00	143.00	214.00	285.81
	Register	147.00	287.00	441.00	588.24
AMI 0.35 μm	AND	42.66	85.33	127.99	170.64
	OR	53.33	106.65	160.03	213.22
	XOR	74.66	149.34	224.00	298.70
	ADD	203.35	406.00	608.69q	811.32
	Multiplexer	85.33	170.66	256.00	341.55
	Register	214.33	435.66	651.66	869.00

attack, repeatedly issuing denial-of-service can deplete IMD battery, whereas data tampering targeted on longer radio messages requires some sort of data correction, which also increases power demand on the implant side.

The technical requirements for wireless IMDs extend from the selection of frequency band and signal modulation to minimize power consumption (in the range from 10 to 100 mW), to communication range in the air (typically one to five meters), data transfer rate (0.1 bps–10 kbps), operating conditions, and signal level attenuation when crossing biological tissues. The frequency bands and standards defined for IMDs are [2]: the US standard for Wireless Medical Telemetry Services (WMTS) with operating frequency bands in the range 608–614 MHz, 1395–1400 MHz, and 1427–1432 MHz; the Medical Implant Communication System (MICS) with frequency band located in the 402–405 MHz range and characterized by a low-power (25 mW) and low-bandwidth communication channel to operate over a distance range of 2 m, which is suitable for signal propagation through the human body without interfering with other external devices; and the extended MICS standard, which covers the frequency band from 401 to 406 MHz. Due to some relaxation in the legislation controlling security in implants, they can operate in different frequency bands other than the aforementioned ones. The telemetric link may also rely on other types of radiation source apart from radio-frequency (RF). For instance, some pacemakers use the 175 kHz band to communicate with the external programmer and require the presence of a magnetic field to trigger a magnetic switch in the implant before starting to exchange messages [8]. Other IMDs do not even have a reverse link, that is, without any transmission from the implant side to the programmer.

From the discussion above regarding the typical resources available in IMDs, the attacker is left with nothing but a handful of weak points in the network to explore actively, if he or she is effective in disguising their identity over the network and eliminating any trace of the attack. Attacks can also be carried out easily using inexpensive radio equipment. The attacker does not need to be extremely close to the patient (i.e. not even sharing the same physical space) to launch the attack. Radio links can reach from a couple of meters to 10 m depending on the frequency band for communication and high-gain antennas can be deployed to cope with eventual signal attenuation in the path. The equipment will most probably rely on a Universal Software Radio Peripheral (USRP) device with open-source GNU radio libraries commonly found on the Internet to decode the frequency band, demodulate the signal symbols, and re-transmit forged messages, as shown in Fig. 8.1 [9]. In fact, it is rather curious to realize that this situation constitutes the perfect scenario for attackers by combining the open-source functionally of the intruder platform with the "in-the-open" communication of sensitive data performed by most IMDs in circulation today.

The skills, motivation or curiosity of the attacker will dictate how much damage he or she is willing to inflict on the patient's health. Some proposed methods found in the literature to counterattack these attacks will be presented in the next section, along with their limitations.

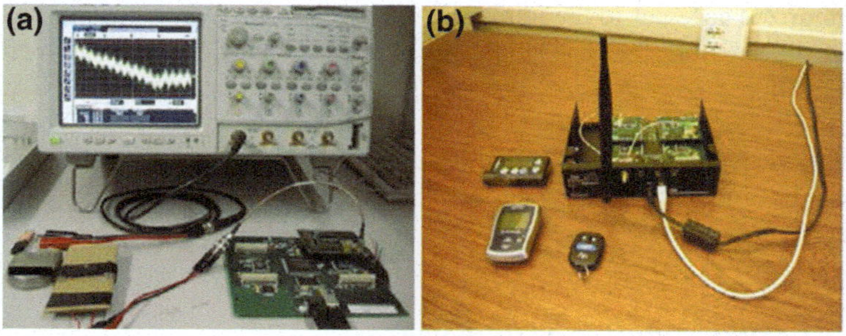

Fig. 8.1 **a** Equipment used to attack an IMD, including an oscilloscope, eavesdropping antenna, and USRP board. Reprinted with permission from [9], © IEEE 2008. **b** Daughter USRP board employed to intercept radio communications from an insulin delivering system. Reprinted with permission from [10], © IEEE 2011

8.4 Security Strategies for IMDs: Merits and Drawbacks

The security mechanisms available to implants are diverse and were originally derived from common defense models deployed in the much broader area of sensor networking. However, the field of implantable technology is unique, and reliable solutions to the security problem must be derived from analysis of the patient's behavior towards the surrounding environment and how this interaction can help promote (or degrade) security. For IMDs, this is particularly relevant when the environment changes from the comfort of private houses to the emergency room, as previously described. Many of the recent security schemes still rely on conceptual attack scenarios that are yet to be proven to be realistic. In a changing world where technologies evolve at unprecedented rates, the behavior of the patient can change, and the same is valid for the attacker towards what might be truly desirable to protect or hack, respectively. Although some mechanisms might prove efficient in theory, in practice the patient (or clinician) will have the ultimate decision to use or reject the IMD based on the level of responsibility involved, the level of knowledge required to correctly operate the device, and the potential health risks involved from incorrect handling by trusted and/or untrusted sources. Thus, the human factor cannot be overlooked when deploying an IMD, a topic that will not be addressed in this chapter.

8.4.1 Simplest Security Mechanisms

For the vast majority of IMDs, especially the older generation of implants, there is no active security mechanism involved and the patient must be aware of potential sources of attack and take pro-active defensive action. For the newest generations of IMDs with wireless telemetry, it is possible to monitor authorized and non-authorized accesses to the implant through a log file. When memory is critical, it is possible to transfer this task to an external device, such as a smartphone. However, this strategy does not prevent attacks and can only provide evidence for subsequent judicial actions if the attacker does not cover the traces left behind. Nevertheless, if the security mechanism detects a recurrent intrusion by looking at the historic log files, it can trigger an alert to the patient.

8.4.2 Cryptographic Mechanisms

Cryptography is by far the most common and most reliable scheme for protecting communication channels and defense against unauthorized access. Cryptographic primitives employed in security algorithms can be divided into two groups: symmetric and asymmetric-key distribution primitives. The first uses a fixed key known only to trusted entities. For IMDs, the key shared by the implant and other authentic devices like the programmer is used to generate authentication and encrypt communications by means of symmetric ciphers, message authentication codes, and identification primitives. Asymmetric cryptography, also known as public-key cryptography, uses two keys; one of them is kept secret, whereas the other is made public, which makes this type of encryption more flexible in terms of key distribution. The TLS protocol that is ubiquitous on the Internet is the best-known form of key distribution to servers [11]. Protocols involving public-key cryptographic mechanism usually require the exchange of multiple messages between trusted entities in order to set up a secure communication channel, which limits their application in IMDs to prevent excessive battery drainage. Some compression techniques have been proposed to apply before the message encryption process to reduce the overall workload involved in wireless transmissions. Nonetheless, symmetric-key approaches are still preferable for IMDs, since they can reduce the footprint in ASIC design. The major concern with this type of security scheme involves externalizing the secret key without suspicious sources getting access to it. Some authors suggest storing the secret key in an external device, like a smart bracelet, or even printing the number on the patient's skin with ultraviolet pigmentation, to be revealed only by medical personnel in the case of emergency. In fact, any type of encryption has the potential to prevent rapid access of medical professionals to sensitive patient data in emergencies when the key is lost. Building a secure worldwide database to store IMD model-dependent universal keys as a "back-up" solution to gain rapid control of the implant by medical request is still far from reality.

Finally, the cryptographic schemes just described can become more attractive when the shared keys are renewed periodically, preferably in every communication session between the implant and programmer. Fixed keys, by their nature, are more prone to attacks due to their longstanding exposure to unwanted entities, which can be avoided by constant refreshing of the keys. Automatic key updates using random processes or metrics is advantageous whenever the patient is unable to configure keys with sufficient strength. Metrics based on the received signal strength indicator (RSSI), chip-die temperature, or any physiological variable can supply a source for key randomness generation.

8.4.3 Close-Range Communication Mechanisms

This security technique establishes the access decision based on the distance between the IMD and programmer. This distance can be computed from the time delay between transmitted and received messages, providing a simple but intuitive way of limiting radio attacks. A distance-bounding protocol usually measures RSSI variations to differentiate trusted devices on the same body (or close by), since these differences remain stable in opposition to the strength of signals sent by a remote attacker. In the context of RF, RSSI protection can be overcome if the attacker, for instance, possesses a strong enough transmitter and high-gain antenna to fool the security at distances within 10 m. Other researchers propose the use of ultrasounds to estimate a distance bound between the programmer and IMD, based on the premise that the attacker cannot send audio signals that propagate faster than the speed of sound. Another solution that is gaining considerable attention is near-field communication (NFC), which allows messages to be exchanged over distances of up to 20 cm, thus more difficult to track by the opponent's RF equipment [4], as shown in Fig. 8.2.

Distance bounds can also be obtained by using body-coupled communication (BCC), which relies on the use of the human body as the transmission medium for signals. The signal will not radiate considerably from the body if the radios operate at lower frequencies, in the range from 10 kHz to 10 MHz. Key distribution schemes can thus be implemented over the human body if electrical signals stay within the physiological range. The main drawbacks of BCCs are the permanent physical contact with the patient's body (which in fact is not a disadvantage in emergencies) and the low data rates achieved by the protocol. Another underestimated attack scenario for BCCs arises from the novel field of capacitive sensing.

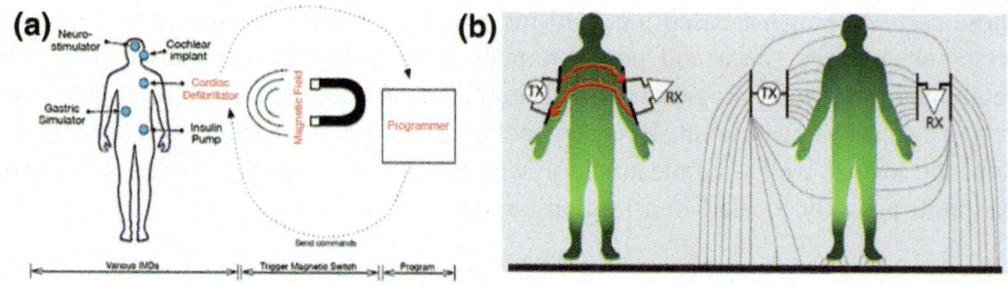

Fig. 8.2 **a** Near-field communication between the programmer and several IMDs [8]. **b** Body-coupled communication for transmission of signals through the body or coupled to nearby electrical fields. Reprinted with permission from [4], © IEEE 2014

Even if BCCs can remain almost invisible to wireless communications, the same is not true for capacitive measurements. The human body in BCCs will act as a capacitor plate that changes its capacitance relative to physical objects in the environment or ground plane. Innovation in this field has made it possible to measure a patient's ECG at distances of more than one meter from the surface of the body, and this type of sensing can be made even more effective if one superimposes external electrical signals over physiological ones due to their generally higher amplitudes and repetition rates.

8.4.4 Out-of-Band Mechanisms

Out-of-band techniques employ auxiliary authentication channels that are outside the data communication channel between trusted entities [8]. Using these channels, key distribution schemes can be negotiated between the implant and programmer if they can be resistant to eavesdropping attacks. There are several types of auxiliary signals that can be deployed, including audio, visual, and tactile, that complement wireless transmission. For instance, with a zero-power RFID device, a random key may be generated and transmitted at the same time as it gives an audible or vibration alert to the patient about the processes taking place, but not to attackers within range.

8.4.5 Access Control Mechanisms

Access control mechanisms are used to guarantee authorized access only to trusted entities before any further action takes place, such as data transmission or programming. The level of privilege of the requested source will be evaluated regarding the action it intends to perform on the IMD side. Control Access Lists (ACL) or Public-Key Infrastructures (PKI) are some of the proposed solutions that differ in the permanent or transitory character of their implementation inside the implant, respectively [2]. Similar to log files used to prevent repudiation attacks, these certificate or list-based mechanisms will compete for the scarce resources in terms of memory space and power requirements inside the implant. In PKI, the relationship between programmer and implant is temporary and requires authentication in each new session. In emergencies, PKIs can be a solution if a patient's keys are lost, as well as in the situation where the patient needs to attend a different hospital, possibly located in a foreign country.

8.4.6 Anomaly Detection Mechanisms

It is the purpose of anomaly detection to automatically identify an IMD's resource depletion and malicious communications, distinguishing between safety and security events. In this regard, anomaly detection completes access control mechanisms. The exchange of messages between the IMD and the external source usually starts by first trying to authenticate the source. In the scenario that authentication fails, communications between the IMD and the external source must be interrupted in order to save implant resources. If this were not the case, the external source in the form of an attacker could repeatedly attempt to communicate with the IMD, leaving the latter in a constant state of issuing denial-of-service. Standard cryptography methods do not prevent this type of attack when an authentication failure occurs [2, 4]. Anomaly detection involves observing changes in behavioral patterns over time, such as physiological changes (e.g. acoustic bowel sounds in the case of insulin pumps) or the IMD's access patterns (sequence of commands, date, and location). For insulin pump devices in the particular case, Hei et al. proposed a supervised learning approach (PIPAC) to get daily insulin infusion patterns from patients by monitoring dosage quantity, rate, and infusion time. These helped them to generate regression models that could be used to dynamically configure a secure infusion on identifying an abnormal infusion command [3]. This study followed a report by the authors of [10] in which they were able to hijack the functions of a normal insulin pump by reverse engineering the cryptographic-absent communication protocol between the pump and glucose monitor.

8.4.7 Biometric Mechanisms

Biometric information has long been a popular source for key generation and distribution based on the physical characteristics of the patient, such as fingerprints, iris patterns, voice, and physiological signals—especially ECG and EEG—although other variables can be used such as blood glucose level, blood pressure, temperature [8], and even gait analysis. Biometric information can be pre-stored inside the implant, which is advantageous in emergency scenarios since it allows medical personnel to simply decode IMD's encryption if they have the correct biometric scanning tool. Hei and Du [12] proposed a two-level secure access control for implants during emergencies. At the first level, basic biometric information of the patient—fingerprint, height, and eye color—is used as a fast authentication mechanism that can defend against attackers who do not possess this information beforehand from the patient. The second level then uses the patient's iris information as a means to achieve effective authentication. Constraints on IMD capabilities have led the authors to propose only partial iris verification instead of entire iris analysis, without compromising the false acceptance and rejection ratios.

If pre-storage of keys is not possible inside the implant, some authors suggest generating keys on-the-fly by simultaneously recording physiological variables by the implant and external device, in order to extract common features that can be used to produce cryptographic keys. With this approach, ECG has been traditionally selected as the leading biometric source due to its high level of randomness (entropy source) and the fact that it can be easily measured anywhere in the body, without altering its common features [13]. Also, the market shared by pacemakers and defibrillators makes ECG a logic choice. Even more, the first recorded attempt to sabotage an implant was indeed an attack on a pacemaker, by the authors of [9]. Equipped with simple USRP radios, Halperin et al. were able to crack the frequency and modulation scheme of a commercial pacemaker, enabling them to intercept patient data and modify the messages. The sole requirement for ECG capture is physical contact with the patient, although capacitive ECG sensors already exist (as previously mentioned), which can help both the patient (galvanically, but not capacitively isolated from the exterior) and attacker (remote capacitive coupling).

Zheng et al. first presented an ECG-based secret data sharing (ESDS) scheme for securing IMDs using fuzzy-vault primitives at both ends of the communication channel. ECG features extracted by the IMD are used to calculate polynomial values, which are sent to the programmer. The programmer then uses its own ECG features to reconstruct the polynomial and reveal the secret [14]. Although the scheme was able to meet the design goal of security, it did not consider either the possibility of an attacker decoding the protocol using historical ECG signals or the level of noise present in a real-world measurement. To overcome these drawbacks, the same authors performed a second study to propose an ECG-based encryption using a combination of classic one-time pads (OTPs) with error-correcting codes (ECCs) [15]. The authors were able to show that OTP keys generated this way could not be probed, duplicated, or guessed due to the randomness of the ECG bit strings, temporal variance (no historical memory), and distinctiveness (when measured at different parts of the body).

8.4.8 Delegation to External Device Mechanisms

The last security mechanism to be mentioned will use an external device other than the programmer to manage access to the IMD, thereby providing confidentiality for wireless communications and protection against untrusted sources. The device is not implanted inside the patient's body but will take an active part in the security scheme between the IMD and the programmer. Since, in principle, the external device will have higher computational power and will be wired to a continuous source of power, the aforementioned constraints in the security mechanisms can now be lifted, leaving IMD with very little control over the entire process. A single external device can perform the required authentication and access control processes while implementing active filtering and remote attack detection. The following examples of security delegation assume that the entity external to the IMD is a trusted source and cannot be replaced by a malicious one.

The first example of a wearable external device is the cloaker, proposed by Denning et al. [16]. The implant constantly performs checks to confirm the presence of the cloaker: if present, the IMD will remain silent to any external source, providing security to the patient. When absent, the IMD will communicate openly to all readers in the range. For emergencies, this approach is ideal, since it only requires removing the cloaker to gain complete access to the implant. However, a malicious programmer could jam the communication channels and convince the IMD of the cloaker's absence to make it exchange data. Another proposed device, called the shield, works by listening for and jamming all IMD transmissions and unauthorized commands [17]. It is designed to receive and jam any incoming message from the implant, so that no-one in the network is able to decode them except the shield. It then encrypts IMD messages and re-transmits them to a legitimate programmer, at the same time jamming any transmission sent directly to the implant. All the encrypted commands sent by the programmer need also to pass through the shield, which relays legitimate instructions to the IMD. This dual functionality of the shield is possible given its proximity to the implant and the capability of cancelling out its own jamming signal to decode IMD transmissions. Thus, the shield ensures the confidentiality of IMD messages and mitigates unauthorized access to sources whose attack distance is covered by the jamming power of the shield, as illustrated in Fig. 8.3a.

The authors of [18] have proposed another solution that combines some features extracted from the cloaker and shield by merging them together in a non-key-based security scheme called BodyDouble, deployed in emergency scenarios (Fig. 8.3b). The system uses an external device as an authentication proxy to control any access to the implant and prevent unauthorized connections. In the event of an external attack, a communication channel is established between BodyDouble and the attacker so as to make the latter believe that is effectively communicating with the

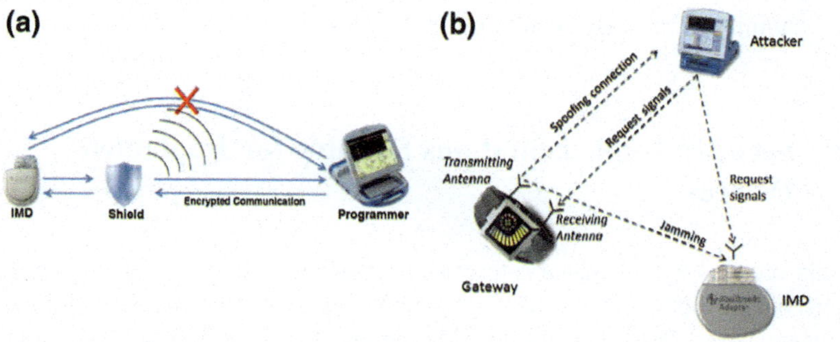

Fig. 8.3 **a** Shield operating mechanism, jamming all IMD transmissions and unwanted commands. Reprinted with permission from [4], © IEEE 2014. **b** Security scheme deployed by BodyDouble to establish a spoofing connection with the attacker while jamming the signal at the IMD antenna. Reprinted with permission from [18], © IEEE 2014

IMD and not with a spoofed connection, thwarting repeated attacks from the same adversary. In emergency mode the gateway is simply turned off and the IMD can be once again accessed directly.

Yet another medical security scheme uses an external device called IMDGuard and ECG-based key authentication between the device and IMD [19]. The guardian controls in which mode the implant must work at any given time: regular or emergency. In regular mode, any request from a programmer must first be authenticated by the guardian, which then issues appropriate keys for the implant and programmer. This key establishment scheme is based on the assumption that both the IMD and the guardian will simultaneously sample the ECG from the patient and extract a symmetric key from common features. The authentication of the programmer by the guardian on behalf of the implant will then save some computational operations by the latter. If the IMD fails to detect the presence of the guardian, it should enter emergency mode immediately. Moreover, to resist attacks from untrusted sources, the guardian can still be calibrated to jam the transmission of the challenge message if an adversary tries to persuade the IMD to enter emergency mode.

Finally, a security monitor called MedMon is proposed by Zhang et al. that detects anomalous transmissions by snooping in all the wireless channels [20]. Anomalies are identified by examining the physical characteristics of the transmitted signal, such as signal strength (RSSI), time of arrival between messages (TOA), differential time of arrival (DTOA), and angle of arrival (AOA). If a suspected wireless transaction is detected by violating the threshold levels assigned for the aforementioned characteristics, MedMon can act into two different modes: passive, by simple notifying the patient, or active, by jamming the transmissions towards the IMD. Although it is effective against battery-draining attacks, as messages that are too frequent in the network may belong to an adversary, it cannot ensure patient confidentiality or privacy against eavesdropping and channel jamming by external attackers. Moreover, in order to become fully operational, MedMon needs to learn the characteristics of normal signal behavior, so it must be placed in a fixed position in the network and remain there after the training stage and beginning of normal operation.

8.5 Security Implementations Suitable for Implanted Devices

In this section, we introduce a few security schemes suitable for implant deployment. We restrict ourselves to "lightweight" methods that require a minimum of hardware logic, implementable on low-power microcontrollers. These should be well suited to remotely powered implants that do not rely on batteries or other long-term energy storage elements and are designed to consume microwatts of power or even less. Some of the techniques mentioned here are taken from the passive wireless smartcard (RFID) literature, where the setting is similar to an implant wirelessly interacting with an external device to read data and/or control the settings on the implant. Here we use the terminology "tag" and implant interchangeably as well as occasionally referring to the external device as a "reader" to be consistent with some of the literature sources.

8.5.1 Digital Signature with Lightweight Hash Function

The digital signature scheme allows authentication of an entity in possession of a private key by a verifier. The private key is protected from the verifier though transformation of the hash functions. The digital signature scheme is a realization of the general zero knowledge proof concept, where generally the verifier has no access to the private key. A typical secure hash algorithm (e.g. SHA-1) provides an output 160-bit long, with 80-bit security (based on a brute force attack). This is used as a one-way encoder, and digital signatures are checked by verifying that the encoded password coincides with the encoded user input. SHA-1 is also "collision-resistant", meaning that inputs that lead to the same hash value are hard to identify. Suppose a hash function is not collision-resistant and inputs that result in the same hash can be easily found by an adversary; then such information can be used to forge signatures and change the signed message without being detected. While the 80-bit security provided by SHA-1 (in practice this can be shorter) falls short of the standards for banking security, such as remote payments, it is considered sufficient for applications in remote sensing and data storage such as low-cost, remotely powered RFIDs, and may fit the model of wireless and remotely powered implants. In [21], an 8-bit implementation of the SHA-1 algorithm is proposed. The original SHA-1 algorithm is based on 32-bit architectures and operates on messages with length $< 2^{64}$ in 512-bit blocks. 80 iterations of hash computation are carried out, resulting in a 160-bit message digest. The SHA-1 protocol does not specify a particular hash function. In [21], the hash computation is carried out by a discrete elliptic curve logarithm. Elliptic curve cryptography is based on recursively computing the intersection of an elliptic curve with its tangent at a given point. The elliptic curve is defined as $y^2 = x^3 + ax + b$ for constants a and b. The recursive computation is a straightforward problem, whereas the reverse problem of solving the number of times the curves intersect given the final intersection point and the starting tangent point is considered hard. To calculate the number of recursion steps, typically the elliptic curve is restricted to a prime number with all results larger than this restriction computed with modulo arithmetic.

The hash function forms the first step of the elliptic curve digital signature algorithm. To understand elliptic curve cryptography and how a digital signature scheme can be implanted with it, we first review some operations on an elliptic curve. These are shown in Table 8.2 and illustrated graphically in Fig. 8.4. Note that in practice, when calculating the point multiplication, the intersection of the tangent and the elliptic curve C can diverge into large numbers, hence we calculate $n \times P$ for large n using modulo arithmetic. This means that the curve will wrap around by modulo m_c, capping the y-axis to m_c. The tangent extension is also calculated in modulo m_c. The resulting coordinates for $n \times P$ are thus limited to $|m_c|$.

Digital security based on elliptic curve cryptography consists of scheme-defining parameters as presented in Table 8.3. Similar to prime factoring-based digital signature schemes, the entity intended to digitally sign a message possesses a private key integer x with $1 \le x \le n - 1$ and a public key in the form of a curve point $P(x_p, y_p) = x \times G$. The signature pair (r, s) is computed with the information $\{H, m, C, G, x\}$ as follows:

$$r = a \bmod n, r \neq 0 \tag{8.1}$$

$$\begin{gathered} s = k^{-1}\left(H(m)_{\text{leftmost bit}} + rx\right) \bmod n, s \neq 0, \text{with} \\ (a, b) = k \times G, \text{with cryptographically secure random integer } k \in \mathbb{Z} \text{ and} \\ 1 \le k \le n - 1 \end{gathered} \tag{8.2}$$

The integer k must be generated for each message. To verify the digital signature, we use the information $\{H, C, G, m, (r, s), P\}$:

$$\begin{aligned} (c, d) = {} & \left\{\left[H(m)_{\text{leftmost bit}}\left(s^{-1} \bmod n\right)\right] \bmod n\right\} \times G \\ & + \left\{\left[r\left(s^{-1} \bmod n\right)\right] \bmod n\right\} \times P \\ & \text{Verify that } r = c \end{aligned} \tag{8.3}$$

Revisiting the signing equations, one first needs to generate a random integer k, followed by the elliptic point multiplication $k \times G$. The result of the point multiplication is used in combination with the hash function of the message to generate

Table 8.2 Elliptic curve operations

Negation $'' - P''$, $P = (x_p, y_p)$	$-P = (x_p, -y_p)$	Elliptic curve symmetrical about x-axis
Addition of two points $P(x_p, y_p) + Q(x_q, y_q) = R(x_r, y_r)$, negation of the intersection of C and line PQ. For $P = Q$, we use the tangent to C at P	$x_r = \left(\frac{y_q - y_p}{x_q - x_p}\right)^2 - x_p - x_q$ $y_r = \left(\frac{y_q - y_p}{x_q - x_p}\right)(x_p - x_r) - y_p$	For a line jointing any two points P, $Q \neq -P$, there line must intersect the curve C on one extra point
Q point multiplication from point $P = nXP$ This is repeating the operation $''P + P''$ n times	Step through the bits of $n = n_0 + 2n_1 + \ldots + 2^m n_m$. Iteratively compute $Q_m = Q_{m-1} + P$ and $P_m = 2P_{m-1}$. This takes $\log_m n$ cycles	Generally very hard to invert $nP = Q$ to find n given Q and P. This forms the hard problem for elliptic curve cryptography

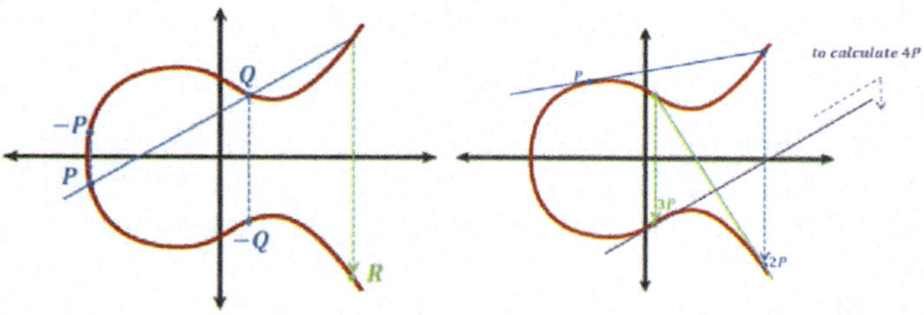

Fig. 8.4 Elliptic curve geometry arithmetic. The graph on the left illustrates the "dot" operation between two points P and Q, which is defined as R: the x-axis reflection of the intersection point between the line PQ and the elliptic curve. Elliptic curve point multiplication is illustrated on the right, where a point is added to itself successively. The addition operation on a single point P is carried out by computing the x-axis reflection of the intersection point between the tangent at P and the elliptic curve. The trajectory of generating $2P$ and $3P$ are shown in the graph

Table 8.3 Elliptic curve cryptography digital signature scheme parameters

H, cryptographic hash function, message m	e.g. SHA-1, SHA-2
C, the elliptic curve used	Typically in the form $y^2 = x^3 + ax + b$
G, base point for the point multiplication, and prime number n	G must have large prime order n. Meaning n is prime and satisfy $n \times G = 0$

the signature (r, s). While the random number k and $k \times G$ can be pre-computed and stored in the tag (known as "coupons"), the generation of the signature parameter $s = k^{-1}\left(H(m)_{\text{leftmost bit}} + rx\right) \bmod n, s \neq 0$ is to be carried out online due to the message m dependence. This operation requires a multiplication of two 160-bit long integers r and the private key x. This can be carried out using shift-and-add multipliers to reduce area consumption.

We now describe the hash function implementation on 8-bit hardware. The SHA-1 algorithm expects the input message to be a multiple of 512 bits + 448 bits (i.e. 448 mod 512). A 64-bit number is subsequently appended to the message, making the overall message dividable into blocks of 512 bits: $M_1, \ldots, M_N$. Each 512-bit block is again considered as a concatenation of 16 × 32-bit words: $M_i = m_{i,0}||\ldots||m_{i,15}$. These 16 × 32-bit words are then expanded into 80 × 32-bit words according to:

$$\begin{aligned} w_j &= m_{i,j} \text{ for } 0 \leq j \leq 15, \text{ and} \\ w_j &= \left[w_{j-3} \oplus w_{j-8} \oplus w_{j-14} \oplus w_{j-16}\right]_{\text{circular left rotate by 1}}, \text{ for } 16 \leq j \leq 79 \end{aligned} \tag{8.4}$$

These 80 × 32-bit words are used in a hashing algorithm to update the 5 × 32-bit variables h_0, h_1, h_2, h_3, h_4 over 80 iterations to produce the final 160 bit "message digest". The hashing algorithm begins by initializing temporary variables

a_i, b_i, c_i, d_i, e_i for the ith iteration to initial values of the hash output h_0, h_1, h_2, h_3, h_4:

$$[a_{-1} = h_0], [b_{-1} = h_1], [c_{-1} = h_2], [d_{-1} = h_3], [e_{-1} = h_4] \tag{8.5}$$

For i from 0 to 79 evaluate the following function and constant:

$$f_i = (b_{i-1} \cdot c_{i-1}) or \left(\overline{b}_{i-1} \cdot d_{i-1}\right), k = \text{0x5A827999 for } 0 \leq i \leq 19 \tag{8.6}$$

$$f_i = b_{i-1} \oplus c_{i-1} \oplus d_{i-1}, k = \text{0x6ED9EBA1 for } 20 \leq i \leq 39 \tag{8.7}$$

$$f_i = (b_{i-1} \cdot c_{i-1}) or (b_{i-1} \cdot d_{i-1}) or (c_{i-1} \cdot d_{i-1}), k = \text{0x8F1BBCDC for } 40 \leq i \leq 59 \tag{8.8}$$

$$f_i = b_{i-1} \oplus c_{i-1} \oplus d_{i-1}, k = \text{0xCA62C1D6 for } 60 \leq i \leq 79 \tag{8.9}$$

and update the temporary variables *a*–*e* according to

$$\begin{aligned} &t = a_{i-1}|_{\text{left rotate } 5} + f_i + e_{i-1} + k + w_i \\ &[e_i = d_{i-1}], [d_i = c_{i-1}], \left[c_i = b_{i-1}|_{\text{left rotate } 30}\right], [b_i = a_{i-1}], \text{ and } [a_i = t] \end{aligned} \tag{8.10}$$

Since the SHA-1 algorithm above is based on hashing with 32-bit data chunks, implementations on 32-bit platforms are most suitable. For resource-constrained scenarios such as when only 8-bit microcontrollers are available (or the equivalent low-power implementation), one can adopt an implementation by breaking the original 512-bit message block into 64 × 8 bit blocks [21], i.e. $M_i = m_{i,0}^{(8)} || \ldots || m_{i,63}^{(8)}$. Instead of 80 iterations of 32-bit logic, 320 iterations of 8-bit operations are to be carried out on the expansion of the original 64 × 8-bit blocks:

$$\begin{aligned} &w_j^{(8)} = m_{i,j}^{(8)} \text{ for } 0 \leq j \leq 63, \text{ and} \\ &w_j^{(8)} = \left[w_{j-12}^{(8)} \oplus w_{j-32}^{(8)} \oplus w_{j-56}^{(8)} \oplus w_{j-64}^{(8)}\right]_{\text{circular left rotate by } 1}, \text{ for } 64 \leq j \leq 319 \end{aligned} \tag{8.11}$$

Here the circular left rotation needs to take into account neighboring $w_j^{(8)}$ values as well as being able to implement the original 32 rotations with 8-bit operands by introducing extra operations for every 4 blocks of 8 bits, as depicted in Fig. 8.5. For the next steps, the logic operations are again identical to those of 32-bit operations. However, the computation of the updating temporary variables defined by $a_i = [a_i(0), a_i(1), a_i(2), a_i(3)], \ldots e_i = [e_i(0), e_i(1), e_i(2), e_i(3)]$ again involves circular rotations over 32-bit blocks. In order to implement these rotations with 8-bit registers, similar delay-and-multiplex logic as in the last figure can be used.

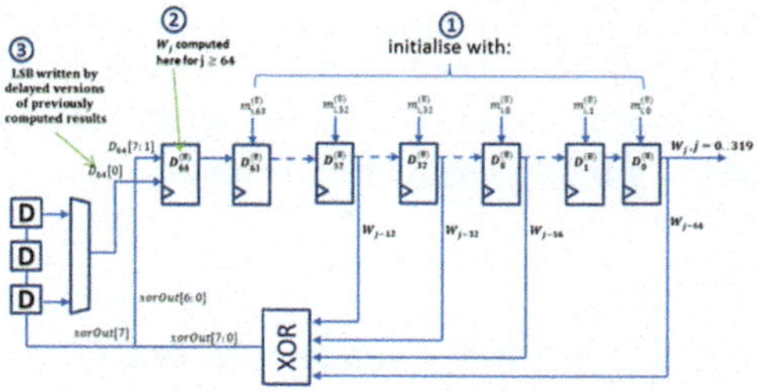

Fig. 8.5 8 bit hardware implementing the 32 bit operations. SHA-1 algorithm

8.5.2 *One-Time Padding with Shared Security Key*

A challenge–response protocol with "pseudonym throttling" is one of the first lightweight cryptographic approaches proposed by Juel [22]. In this scheme, the remote tag stores a list of updatable random identifiers known as pseudonyms, and upon request it transmits sequentially one of the pseudonyms. With added delay to the tag response, this prevents practical phishing of the information from the tag, as this would require continuous access to the tag over an extended period. Juel's proposed security scheme is computationally efficient, at the expense of storage requirements. Consider a tag rotating through a multiple of pseudonyms $\vec{\alpha} = \{\alpha_1, \alpha_2, \ldots, \alpha_n\}$. A two-way authentication is carried out, starting from the tag transmitting α_i and the reader responding with a corresponding key β_i unique to α_i. This authenticates the reader to the tag and is followed by authentication of the tag through transmitting a third authentication key γ_i, again unique to the initial pseudonym α_i. To reduce memory requirements, instead of having a very long set of rotating pseudonyms $\vec{\alpha}$, values of $\alpha_i, \beta_i, \gamma_i$ are updated after successful mutual authentication. By using one-time data padding encryption, the vulnerability of having these updated values eavesdropped can be reduced. The one-time pad cipher consists of XOR operations. Given a shared private key δ, of length l, the ciphertext is constructed by $M \oplus \delta$, with M being the message with length l. The one-time padding is not the most storage-efficient, as the message is restricted to the same length as the random key; however, the computational cost is minimal with logic XOR operations. For the one-time padding to be secure a 1 bit key is required for each bit of the message and these bits should be generated by a pseudorandom generator. Practically there are two approaches, one being pre-generated and stored random keys. As an added level of security, given limited storage, Juel [1] proposed using a key updating scheme that consists of a rotating key for each message M_k: $\Delta_k = \left\{\delta_k^{(1)}, \delta_k^{(2)}, \ldots \delta_k^{(m)}\right\}$. Each time, only $\delta_k^{(1)}$ is used to cipher the message M_k and the key $\delta_k^{(1)}$ subsequently rotated back to the end of Δ_k. The new key is generated by an XOR operation of a transmitted symbol $\tilde{\delta}_k$, and the next key down the list $\delta_k^{(2)}$. This improves security even when the new key is intercepted during key update. Oh et al. [23] proposed a scheme for generating new pseudonyms for one-time padding based on concatenating two arbitrary one-off numbers (nonce) N_T

and N_v through $N_T \oplus N_v$ or $N_T||N_v$. The exact method used depends on the bit length of the shared secret. To reduce storage requirements for a practically large number of shared secrets, two secret values are drawn from the stored secret value set at a time—creating P_n^2 permutations.

8.5.3 Shared Pseudorandom Generator

The main bottleneck of the one-time padding scheme is the requirements of the pseudorandom secret key, which is often pre-generated and stored to reduce computational requirements for generating random bits on-the-fly. The authors of [24] propose a method of sharing a pseudorandom number generator between the remote device and reader. With this the storage requirements are minimal, with some computational effort allocated to the random number generation. A pseudorandom number generator (PRG) is cryptographically secure if there are no probabilistic polynomial-time algorithms that can predict the next bit following some number of bits of observation with probability $\gg$1/2. A PRG requires a seed, which is a truly random binary string, often of length much shorter than the output of the PRG. PRG-based security systems are constructed given the performance predicted by statistical tests of the PRG. For example, [24] cited the EPCGen2 standard specified 16-bit PRG meeting the following bounds:

$$\frac{0.8}{2^{16}} < Prob[\text{number drawn}] < \frac{1.25}{2^{16}} \tag{8.12}$$

and the random number should not be predictable with a probability greater than 0.025%. To further reduce the predictability of the PRG for security implementation, one might periodically refresh the seed of the PRG. A two-way authentication based on a "shared" PRG (identical and with same seed) can be initialized by the external device first sending out a query. The implant then responds with the first random number (RN1), followed by the external device sending out the second (RN2) upon checking the authenticity of RN1. The implant, having received RN2, authenticates the external device and responds with the final number (RN3). If the communication is interrupted, up to five RNs can be used in a single session. The shrinking generator proposed in [25] is one of the power/area-efficient PRGs and uses two random bit sequences to create a third sequence with reduced predictability. This is based on a simple procedure of keeping the bits in the first sequence only when the corresponding bit in the second sequence is "1". In most cases, unless the entire second sequence consists of "1"s, the resulting sequence is of shorter length than the input sequences; hence this is termed the "shrinking generator". The input to the shrinking generator can be random bits generated by simple linear feedback shift registers (LFSR), as shown in Fig. 8.6.

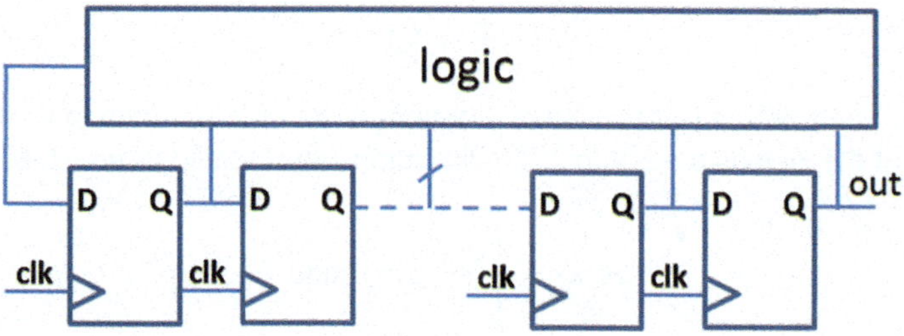

Fig. 8.6 Linear feedback shift registers for pseudorandom bit stream generation

Given two sequences $\vec{A}$ and $\vec{S}$ as input sequences generated by LFSRs, the period of the resulting sequence $\vec{Z}$ is exponential in both $length(\vec{A})$ and $length(\vec{S})$), whereas to generate equivalently the resulting sequence $\vec{Z}$ with a single LFSR one would require an LFSR of length at least exponential in $length(\vec{S})$. This is known as the linear complexity of the sequence, and a low linear complexity indicates higher predictability. The correlation between the two bits is bounded by $O\left(\frac{l}{2^{length(A)}}\right)$ and the output sequence $\vec{Z}$ also has a balanced number of "1" and "0"s. This can be measured by comparing the statistics of $X = sum(z_1, z_2, \ldots z_n)$ and an independent-identically distributed (IID) binary sequence $= sum(y_1, y_2, \ldots y_n)$, with $y_i \sim \{0, 1\}$ at equal probabilities. The expected value of the sum X is bounded by $\frac{n^2}{2^{length(A)}}$, whereas the variance between X and Y is bounded by $(n^3 + n^2)/2^{length(A)}$.

8.5.4 *Learning Parity in the Presence of Noise (LPN) Cryptography*

The LPN-based cryptographic scheme is based on the LPN as a hard problem and is resistant to quantum computation at the time of writing. We begin with an introduction to the LPN problem. Suppose we have a vector of length k with binary values

$$\vec{m} = \{m_1, m_2, m_3, \ldots m_k\}, m_i \in \{0, 1\} \tag{8.13}$$

and the values of $\vec{m}$ are drawn from a uniform distribution. Also consider a Bernoulli distribution with parameter p, Ber_p, denoting the probability of a binary random variable taking the value "1". If we take a vector random vector $\vec{s}$ consisting of independent and identically distributed binary RVs, and a single binary variable b obtained by the inner product of $\vec{m}$ and $\vec{s}$ XORed with a random bit drawn from Ber_p, i.e.

$$b = \langle \vec{m}, \vec{s} \rangle \oplus t, t \sim Ber_p \tag{8.14}$$

where the inner product of two binary vectors is a single bit defined as the sum of products of the elements modulo 2 (i.e. the sum is represented as an XOR function):

$$\langle \vec{m}, \vec{s} \rangle = \left[\sum_i m_i s_i \right] \bmod 2 \tag{8.15}$$

So this system transforms an input vector $\vec{m}$, by drawing a random (uniform) vector $\vec{s}$ and a single random bit t drawn from Ber_p. The combined output is the concatenation of $\vec{m}$ and b of length $k + 1$.

The learning parity with noise problem is that if we are given access to this **oracle** (a black box) as described above, can we be sure that $\vec{m}$ has not been tampered with? You are also given the Bernoulli parameter p. Suppose we also have access to n random outputs from this LPN oracle, i.e. n random vectors $\vec{s}_1, \vec{s}_2, \ldots, \vec{s}_n$ and n random bits generated from the inner product and XOR operations: $\{b_1, b_2, \ldots, b_n\}$, generated by the following matrix operations:

$$\begin{bmatrix} \text{-----} & \vec{s}_1 & \text{-----} \\ \text{-----} & \vec{s}_2 & \text{-----} \\ \text{-----} & \vdots & \text{-----} \\ \text{-----} & \vec{s}_n & \text{-----} \end{bmatrix} \begin{bmatrix} m_1 \\ m_2 \\ \vdots \\ m_n \end{bmatrix} \oplus \begin{bmatrix} t_1 \\ t_2 \\ \vdots \\ t_n \end{bmatrix} = \begin{bmatrix} b_1 \\ b_2 \\ \vdots \\ b_n \end{bmatrix} \tag{8.16}$$

In the above equation, t_i is unknown. However, we know that it is drawn from Ber_p. For $p = 0$, we have deterministic sets of linear equations which we can solve to obtain $\vec{m}$. However, for random $\vec{t}$, this becomes an NP-hard problem, i.e. it cannot be solved within constraints equal to or less than polynomial functions of n, the message length. The security of the LPN methods relies on the system being able to distinguish a uniformly random matrix compared to the LPN oracle generated output. Thus the critical element of implementation is the random binary vector generation, as the entire scheme can potentially be compromised by correlation attacks if biased generators are used. A number of lightweight cryptography schemes based on the LPN problem have been proposed in the literature, initially by Hopper and Blum originally for human authentication [26] and later improved by a number of researchers (e.g. [27]) for use with low-power passive smart cards. This is potentially applicable to an implant setting, with minimum hardware requirements. These schemes are termed "HB" schemes, and notably the scheme presented in [27]—termed $HB^{\#}$—can be described as in Table 8.4. The tag and the external reader share the secret $(\boldsymbol{X}, \boldsymbol{Y})$ matrixes. The first steps of authentication involve the tag and reader exchange challenges in the form of random vectors $\vec{b}$ and $\vec{a}$. This is followed by applying XOR twice (identity operation, since XOR is inverse of itself) and calculating the Hamming weight (H_w = number of "1"s) of the received string. If this is less than string length multiplied by the probability of drawing '1' in the Bernoulli distribution, then the tag is authenticated.

Looking at the scheme above, one might notice there is a possibility that an authentic tag can be rejected. Authentication allows both the implant and the external device to verify that they are communicating with a legitimate entity. Authentication, however, works best with a secure channel. A typical secure channel employs cryptography protocols such as the RSA—the Rivest-Shamir-Adleman, which could be too resource-hungry to be implemented with implant hardware. In [28], the authors proposed a lightweight secure authentication scheme based on the exact implementation of the learning parity with noise problem. As described above, the LPN problem is to distinguish between noisy linear equations from uniformly random signals [29]. The authentication scheme in [28] assumes that each implant device possesses a unique ID T_{id}, which is a shared secret between the implant and an external device. The implant holds two keys: one permanent key S', and a session key S_i''. The use of a session key, refreshed after each successive session i, enhances security for the non tamper-resistant channel. The secure update of the session key relies on authentication. Formally, we can write the exact LPN problem as fixing the Hamming weight of the random vector $\vec{t} = \{t_1, t_2, \ldots, t_n\}$ to be the nearest integer of product of binomial probability p and vector length n, i.e. pn. Public parameters include the binomial probability p, length of the message being encoded m, and number of bits in the random bit stream k. The authentication protocol based on the exact LPN scheme and a challenge–response communication is shown in Table 8.5. To update the session key S_{i+1}'', one follows the operation detailed in Table 8.6.

Additional authentication can be carried out involving a third entity such as a server [28]. While the scheme described above is not totally minimalistic and consisting of only logic operations, the multiplication operator presented in the scheme's cipher operation can be carried out iteratively by a microcontroller.

8.5.5 Rabin Public Key Scheme

The drawback of the non-exact LPN-based authentication schemes is that the randomness of the challenge response inevitably leads to some cases of authentic

Table 8.4 *HB*$^{\#}$ authentication scheme

Tag (e.g. implant)	External device
Choose $\vec{b} \in_R \{0,1\}^{k_y} \xrightarrow{\vec{b}}$	
	$\xleftarrow{\vec{a}}$ choose $\vec{a} \in_R \{0,1\}^{k_x}$
Choose $\vec{v} \in_R \{0,1\}^m$ s.t. $\Pr(v_i = 1) = \eta$	
Compute $z = \vec{a}\boldsymbol{X} \oplus \vec{b}\boldsymbol{Y} \oplus \vec{v} \xrightarrow{\vec{z}}$	Accept iff: $H_w\left(\vec{z} \oplus \vec{a}\boldsymbol{X} \oplus \vec{b}\boldsymbol{Y}\right) \leq \eta m$

Table 8.5 Exact LPN authentication scheme

Tag (e.g. implant)	External device
	← send random v-bit challenge string $\vec{s}$
(a) generate signature σ_i with session key S''_i: $\vec{\sigma_i} = S''_i\vec{s}$ (b) k-bit commitment of message $\vec{m}$ generated by: $\vec{r} = S_i(Id\|\|\vec{s}) \oplus \vec{t}$ S_i created by concatenating the permanent and session keys S'_i and S''_i; the term $(Id\|\|\vec{m})$ is the concatenation of device ID and message (c) generate random v-bit challenge $\vec{s'}$ for external device: $(\vec{r}, \vec{\sigma_i}, \vec{s'}) \rightarrow$	
	(a) verify signature $\vec{\sigma_i}$ is generated by challenge $\vec{s}$ that it sent out by verifying $S''_i\vec{s} = \vec{\sigma_i}$ (b) authenticate tag by verifying: $H_w(S_i(Id\|\|\vec{s}) \oplus \vec{r}) = k(\frac{1}{4} + \frac{\varrho}{2})$, and $H_w(\vec{s'}) = length(\vec{s'})/2$

Table 8.6 Session key updates for exact LPN authentication scheme

Tag (e.g. implant)	External device
	(a) Generate new session key via a non-singular binary matrix Q, with $S''_{i+1} = QS''_i$. Compute: $P''_i = S''_i S''^{+}_i Q$ (b) k bit commitment of message $\vec{s'}$ originally received from tag: $\vec{r'} = S_i(Id\|\|\vec{s'}) \oplus \vec{t}$ $< - (P''_i, \vec{r'})$
(a) authenticate external device by verifying: $H_w(S_i(Id\|\|\vec{s'}) \oplus \vec{r'}) = \lfloor k(\frac{1}{4} + \frac{\varrho}{2}) \rceil$ (b)update session key if external device authenticated: $S''_{i+1} = P''_i S''_i$	

tags being rejected. One can follow the exact LPN scheme as detailed above; alternatively, [30] proposed a message authentication code (MAC) generation scheme based on a square hash that does not require a random number generator. The typical use of this scheme is to allow the implant to have a non-secret identity I together with some secret information S. The scheme should protect S being extracted by an adversary and to allow authentication by means of using S to reply a challenge from the implant reader. These schemes are generally known as zero-knowledge schemes. The digital signature schemes presented earlier in this chapter allow us to do this with moderate computational resources. The digital signature scheme provides a method of authentication with the verifying entity not having access to the private key possessed by the implant. In many cases the verifier can share the private key without compromising security. Unlike the LPN authentication scheme, [30] reports a hash function for a challenge–response protocol based on the Rabin scheme. The Rabin scheme encrypts a message m with key n as a checksum for authentication, with n being the product of at least two unknown integers with bit length >1000.

The message m in a challenge–response protocol can be the concatenation of the original challenge bits and the response bits (such as a secret implant ID), i.e. $m(challenge, secret)$. Encripting this message with a hash function secures the implant ID from adversaries that can obtain many responses by issuing a number of chosen challenges. The challenger will verify the response by computing again the forward hash function with its known secret ID from a database. Using the Rabin scheme, the ciphertext c is generated according to:

$$c = m^2 \bmod n \tag{8.17}$$

This is a one-way function for authentication. However, it is not collision resistant, which means that different messages m will map to the same ciphertext. Trivial cases include $\{m, -m, m+n, -m+n\}$, indicating a 4–1 function. The requirements on the size of m and the squared operation pose considerable problems for resource-strapped implants. The squared operation is thus replaced by a "squashed square hash" function constructed by setting n as a Mersenne number in the form $2^k - 1$ (integer k). The advantage is the efficiency of calculating $mod(2^k - 1)$, being equivalent to shift and carry operations.

8.5.6 Cellular Automata as a Lightweight Cryptography Tool

So far we have described a few cryptographic schemes based on simple binary operations and pseudorandom number generators. These schemes have comparative advantages and complexity. Cellular automata (CA) are a framework for studying the temporal evolution of "cells" defined by simple operations (known as the transition function) with neighboring cells. In the most common arrangement, cells are arranged in a regular 2D grid with "neighbors" defined as those in the immediate spatial neighborhood of each cell. A binary CA constitutes cells with two possible states {0,1} (more in non-binary CAs) and operations between cells are simply binary functions completely definable with a truth table. A typical way of visualizing the temporal evolution of cellular automata is given in Fig. 8.7. A CA is completely defined by giving the neighborhood, local transition function, and possible states of each cell as well as the dimension of the system.

$\{a_{i-1}(t), a_i(t), a_{i+1}(t)\}$	$a_i(t+1)$
000	0
001	1
010	1
011	1
100	1
101	0
110	0
111	0

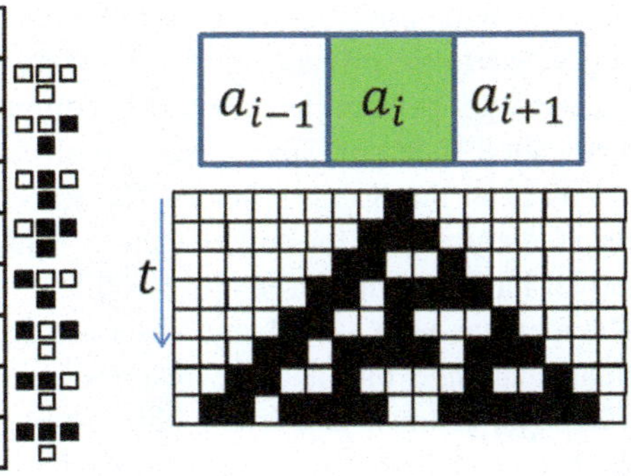

$$e.g. a_i(t+1) = a_{i-1} \oplus (a_i || a_{i+1})$$

Fig. 8.7 Example cellular automata "rule 30" (output column of truth table = $00011110_2 = 30_{10}$) involving two immediate neighbors only with example pattern for a starting state with a single "1". Darkened squares signify a "1" whereas non-darkened squares signify a "0"

For CAs with two adjacent cells, in each time iteration the cells a_i are updated according to:

$$a_i(t+1) = f_1(a_{i-1}(t), a_i(t), a_{i+1}(t)) \tag{8.18}$$

In binary CA, the transition function f_1 can be defined as a truth table such as in the example given in the last figure. If we restrict the neighbor region to be the immediate left and right neighbors of each cell only (known as "first neighborhood"), there will be a total of eight combinations over three cells (including the middle cell itself) of two states {0,1}. Each of these eight possibilities produces one of the two states in the middle cell. This results in $2^8 = 256$ possible scenarios and truth tables, making up the 256 rules summarized by Wolfram [31]. These rules are logic operations and often expressed as a truth table with inputs from the three cells and output as the value of the center cell in the next time iteration. Rules that exclusively use the XOR operation are known as linear rules, since XOR is equivalent to modulo 2 addition. In a form similar to a linear feedback shift-register, Wolfram [32] proposed using *rule* 30 of the CA in a circular register of N cells as a pseudorandom number generator, and the initial states of the registers as the seed. The cycle length of this pseudorandom generator is approximately $2^{0.61N}$. This scheme of using circular registers is also known as "CA with cyclic boundaries". Other ways to handling boundary conditions for a CA include reflective (boundary cells are always a reflection of cells near the boundary) and fixed (boundary cells fixed at specific states). Some CA rules are invertible: namely, given the state of the system, one can follow an "inversion" rule to roll back to earlier states, eventually reaching the initial condition. The roll-back rules do not necessarily equal the forward generation rule of the CA, and some roll-back rules require further temporal neighbors, such as states of the cell at the two previous iterations. As an example, the seed for the *rule* 30 PRNG can be recovered by simply inverting *rule* 30 [33]. Some rules defined on the 1D CA satisfy the linear property [34] such that

$$f(a_{i-r}(t), \ldots, a_i(t), \ldots, a_{i+r}(t)) = \sum_{r}^{\bmod 2} b_r a_{i+r}(t) \tag{8.19}$$

where the modulo 2 summation is equivalent to a bitwise XOR operation and $b_r \in \{1, 0\}$.

These linear rules, together with their complement rules with the rule number adding up to $2^{2^{2r+1}}$, result in affine transformations when applied directly to a message as a cipher operation [35]. Affine ciphers created in this way have severe limitations and the ciphering operation (the key, now defined by the CA rule) in this case is less varied and relatively easy to guess. For example, out of the 256 elementary CA rules there are only 16 affine possibilities, namely rules $\{0, 15, 51, 60, 85, 90, 102, 105, 153, 150, 165, 195, 170, 204, 255\}$. An affine operation on a piece of data can simply be represented by a constant scale and an addition, i.e. definable over just two parameters. This means that by getting hold of two ciphertext–plain text pairs, the two parameters can be recovered via solving linear equations, rendering the ciphering scheme useless. The use of CAs with affine update rules for cryptography should be avoided. The LPN approach addresses the issue of general cryptography with affine functions.

One desirable property of a random number generator used in a cipher is the production of equiprobable "1" and "0"s at the output. This can be linked to CA rule numbers with equal occurrence of "1" and "0"s. For an elementary 1D CA, the total number of balanced rules is given by $C_8^4 = \frac{8!}{4!(8-4)!} = 70$. One can see this in the typical truth table with eight output possibilities, considering the case when half of the output bits are "1". To improve the randomness and irreversibility of the CA-based PRNG, researchers have proposed the use of mixed rules in the CA PRNG. Considering there exist a large number of rules even for the simplest 1D CA (256 rules in this case), a genetic algorithm has been used to optimize for the best rule combinations to be used [36]. The use of PRNG in cryptography can be in the form of stream ciphers, where the rules and initial seeds are shared as private keys between two PRNGs with the encrypted data generated by the XOR function with the random bit stream.

An alternative approach to CA-based symmetric-key cryptography relies on second-order CAs. To define update rules for second-order CAs one requires two rules. The transition function can now be geometrically illustrated as a 3 × 3 binary matrix as depicted in Fig. 8.8. From the figure, the rules used depend on the value at two time steps ago. If $a_i(t-1)$ = "1", rule R_1 will be followed; otherwise, if $a_i(t-1)$ = "0", we follow rule R_2. Often rules R_1 and R_2 are related by

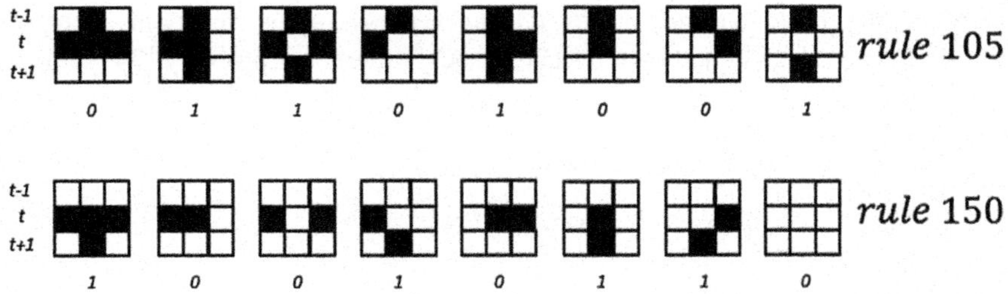

Fig. 8.8 Example of a second-order CA based on two rules that are the reverse of each other

$$R_2 = 2^{2^{r+1}} - R_1 - 1 \tag{8.20}$$

with parameter r being the neighboring radius of the transition function. Rules 105 and 150, shown in the above example, are related by setting r=2. Rules computed in this manner are termed "reversible CA". Now with the second-order CA we require two sets of initial conditions. One of the initial conditions is defined at $t = 0$, whereas the other set of initial conditions is defined at $t = 1$. One way of making use of this second-order dependency in cryptography is to set initial values of the CA $\vec{a}(t = 0)$ as random and the second initial value $\vec{a}(t = 1)$ being the plain text to be encrypted [37]. Given the cipher text and the final data, and with the reversible rule pair and the number of iterations k, one can recover the message as well as the random initialization $\vec{a}(0)$. The process is done by simply placing the final data as $\vec{a}(0)$ and the cipher text as $\vec{a}(1)$ and applying the known rules for k steps (Fig. 8.9). The random initialization is thus not necessary for decoding the message. However, random initial states of the CA facilitate the randomness of the ciphertext and improve security. In practice, the seed to generate this random number can be inferred from environmental aspects, such as the time required executing the encryption. In the case of two-neighbor rules, $r = 2$, there are a total of 256 possible rules to generate the ciphertext. For an increased number of rules (e.g. $r = 3$ gives 65,535 possibilities) and a combination of sequential CA evolution and shifting of ciphertexts, this quickly renders a brute force attack infeasible [37]. The use of second-order CA rules in reversible pairs allows decoding of the cipher text using

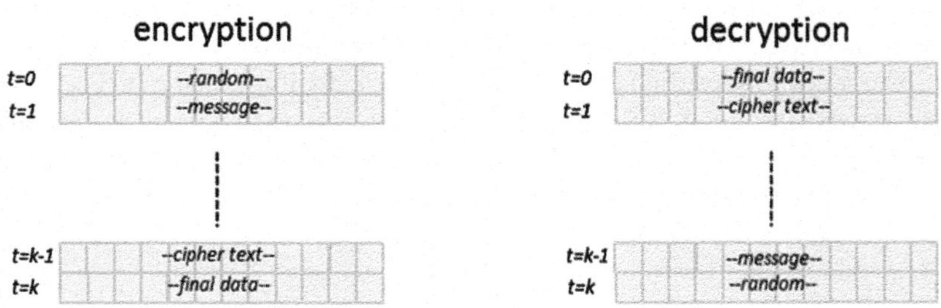

Fig. 8.9 Cryptography based on reversible CA rule pairs

the rule pairs as a key. These reversible rules generate outputs at the next time step by looking at the data from two previous steps. A more general configuration consists of rules that look back by k steps. These are known as linear memory cellular automata.

So far we have looked at 1D CAs with updates computed along a single direction in the 2D grid to illustrate the evolution of the CA with time. 1D CAs have a neighborhood along a single axis. In [38], a 2D CA with the neighborhood defined as the eight immediate neighbors of each cell is used for image encryption. Each CA cell can take a state within a finite set $S = \{0, 1, \ldots, c-1\}$ of pixel intensities. For an nth-order CA with state update equation dependent on the state k time steps back, and transition function varying across the time-recursion index, with update equation given by

$$\begin{aligned} a_{ij}^{t+1} = \sum_{m=1\ldots k} \Big[f_m \Big(& a_{i-1,j-1}^{t-m+1}, a_{i-1,j}^{t-m+1}, a_{i-1,j+1}^{t-m+1}, a_{i,j-1}^{t-m+1}, a_{i,j}^{t-m+1}, a_{i,j+1}^{t-m+1}, \\ & a_{i+1,j-1}^{t-m+1}, a_{i+1,j}^{t-m+1}, a_{i+1,j+1}^{t-m+1} + a_{ij}^{t-k+1} \Big] \bmod c \end{aligned} \tag{8.21}$$

the update equation of the corresponding reverse CA of the above transition function is then

$$\begin{aligned} a_{ij}^{t+1} = \sum_{m=1\ldots k-1} \Big[f_m \Big(& a_{i-1,j-1}^{t-k+m+1}, a_{i-1,j}^{t-m+1}, a_{i-1,j+1}^{t-k+m+1}, a_{i,j-1}^{t-m+1}, a_{i,j}^{t-k+m+1}, a_{i,j+1}^{t-k+m+1}, \\ & a_{i+1,j-1}^{t-m+1}, a_{i+1,j}^{t-k+m+1}, a_{i+1,j+1}^{t-k+m+1} + a_{ij}^{t-k+1} \Big] \bmod c \end{aligned} \tag{8.22}$$

and a general transition function mapping f_m can be defined as:

$$\begin{aligned} & \sum_{\alpha,\beta \in \{-1,0,1\}} \lambda_{\alpha,\beta} a_{i+\alpha,j+\beta}^{t-m+1} \bmod c, \\ & \lambda_{\alpha,\beta} \text{randomly chosen from} \{0, 1, \ldots, c-1\} \end{aligned} \tag{8.23}$$

This is essentially a convolution with a random mask filter repeated m times. At first glance, using the transition rule as a key will result in a brute force attack having to try out all of the $n \cdot c^8$ combinations. Just like in the 1D CA case, for an nth-order reversible CA one initiates the CA operation with n images. The first $n-1$ images are generated randomly and the nth image is the message being encrypted. The number of iterations carried out to encode the image is chosen to be larger than half of its maximum dimension, such that "effect" of the center pixel has reached beyond the edge pixels. The last n encoded images are the result of the encryption algorithm, as the decoding will also require n images. A combination of a number of high-dimensional CAs has been shown to improve security by extending the parameter space of brute force key trials required [39].

8.6 Cryptanalysis and Security Vulnerability

The aim of a cryptanalysis attack is to derive ways of computing the secret information (e.g. key) at significantly lower resource than is required in a blind exhaustive search through the parameter space. Even for batteryless and wirelessly powered implants, where direct attacks often requires the adversary to be close enough to the person wearing the IMD while accessing the external device, the upstream of data from the external device to a database is vulnerable in many other ways due to long range wireless/wired transmissions, legitimate devices that interfaces with the implant will have less power constraints and have most of the cryptographic tools at their disposal to ensure information security. Some of the light weight efficient security schemes introduced in the previous section have been subjected to cryptanalysis that proves the security of such system is yet to be improved. In this section we highlight some of the cryptanalysis techniques focusing in areas that are relevant for the evaluation of light weight, implant ready security schemes.

8.6.1 Background: Boolean Functions

Cryptanalysis concerns the task of recovering secret information given some incomplete parameters or measurements of a cryptographic scheme's output. For most of the lightweight protocols introduced in this chapter, which rely on hardware with feedback shift registers, we consider a few cryptanalysis methods targeted on stream and block ciphers, which also relies on linear/nonlinear feedback shift registers for their operation. We first take a brief look at Boolean functions and introduce some properties that are useful in cryptanalysis [40].

For Boolean functions, we represent an array of binary variables as $\vec{b} = [b_1, b_2, \ldots, b_n]$ belonging to the space $GF(2)^n$, with $GF(2) = \{0, 1\}$ being the binary field of two states "0" and "1". Two operations, multiplication and addition, are defined on $GF(2)$. Multiplication is represented by the standard "AND" binary operation, whereas addition (modulo 2) is the "XOR" binary operation. Any function $f : GF(2)^n \rightarrow GF(2)$ that maps a vector (string of bits) $\vec{b} \in GF(2)^n$ to an output of two possible binary states can be stated as:

$$f(b_0, b_1, \ldots, b_n) = b_0 \oplus [a_1 b_1 \oplus \cdots \oplus a_n b_n] \oplus [a_{1,2} b_1 b_2 \oplus \cdots \oplus a_{n-1,n} b_{n-1} b_n] \oplus \cdots \oplus [a_{1,2\ldots n} b_1 b_2 \ldots b_n] = \sum a_K \prod_k b_k \tag{8.24}$$

Here the summation represents an XOR operation as defined for $FG(2)$ and $a_i \in GF(2)$. This is also known as the algebraic normal form (ANF) of a binary function. The truth table of the binary function is a display of all the possible input/output mappings of f, produced by stepping through all possible input values. The degree of a binary function (polynomial) f is given by the largest number of bits across the product terms in the ANF. Previously we have revealed that Boolean affine transformations can be insecure when used in a cryptographic scheme. An affine Boolean function can be defined, written in the form

$$f_A = \sum_k a_k b_k \tag{8.25}$$

i.e. the maximum degree of all terms in the algebraic normal form is 1. Often we have to investigate if a Boolean function is balanced, that is, does it equally map half of all possible inputs to "1" and the other half to "0"? Note that identically and independently distributed input bits b_i with $P(b_i = 1) = 0.5$ does not imply that the output of f will have $P\left(f\left(\vec{b}\right)\right) = 0.5$. We define cardinality (the number of elements) as the of the set of inputs that yields the result of $f\left(\vec{b}\right) = 1$. This is also the Hamming weight of f, denoted as $wt(f)$, as the number of instances when the output is "1". Balanced functions can thus be defined as f to satisfy (8.26).

$$wt(f) = 2^{n-1} \tag{8.26}$$

The XOR operation of two binary variables is balanced, hence XOR with any balanced input will produce a balanced output. However, XOR of any binary variable with the variable itself results in "0" all the time, and hence is not balanced.

The nonlinearity of a binary function can be thought of as the number of bits it must change in its truth table to get to the closest affine function:

$$nl(f) = \min[d(f, g)] = \min[wt(f + g)] \text{ with } g \text{ being all possible affine functions} \tag{8.27}$$

The distance $d(f, g)$ between two binary functions f and g is given by the Hamming weight $wt(f + g)$ of their sum. The nonlinearity of a Boolean function is bounded as a function of the number of bits n and is given by $2^{n-1} - 2^{n/2-1}$. A Boolean function that achieves this nonlinearity upper bound is known as "bent". One way of constructing a bent function $f_{2b} : GF(2)^{2n} \to GF(2)$ from any Boolean function $f_b : GF(2)^n \to FG(2)$ is by writing:

$$f_{2b}(\vec{x}) = x_1x_2 + x_3x_4 + \cdots + x_{2n-1}x_{2n} + f_b(x_1, x_3, x_5, \ldots, x_{2n-1}) \tag{8.28}$$

Consider a Boolean function g, such that $f\left(\vec{b}\right)g\left(\vec{b}\right) = 0$ for all binary vectors $\vec{b} \in GF(2)^n$, g is known as the annihilator of f. The *algebraic immunity* $\mathcal{AI}(f)$ of f is defined by the minimum degree of an annihilator function g satisfying

$f\left(\vec{b}\right)g\left(\vec{b}\right) = 0$ or the minimum degree of a multiple g in $\left(f\left(\vec{b}\right) \oplus 1\right)g\left(\vec{b}\right) = 0$. To assess the statistical properties of a Boolean function, we define the Walsh transform of the function f:

$$W_f(\vec{u}) = \sum_{\vec{x} \in GF(2)^n} (-1)^{f(\vec{x}) \oplus \vec{u}\vec{x}} W_f(\vec{u}) \in \mathrm{N} \text{ and } \vec{u}\vec{x} = u_1x_1 \oplus u_2x_2 \oplus \cdots \oplus u_nx_n \quad (8.29)$$

Thus the Walsh transform computes an integer from an input binary vector $\vec{u}$. Each of these computations involves the integer summation computation while we step through all the possible vectors in $GF(2)^n$. The Boolean function f is balanced if and only if its Walsh transform evaluated at $\vec{0}$ is zero, i.e. $W_f\left(\vec{0}\right) = 0$. We also define the autocorrelation of f as:

$$C_f(\vec{u}) = \sum_{\vec{x} \in \mathrm{F}_2^n} (-1)^{f(\vec{x}) \oplus f(\vec{x} \oplus \vec{u})} \quad (8.30)$$

A low autocorrelation and statistically balanced Boolean function produces bit streams with lower predictability, and hence more immune to cryptography attacks. The randomness of a Boolean function can be accessed via the "strict avalanche criterion" [41], such that whenever a single input bit is toggled, each of the output bits will change with a 0.5 probability. This can be tested with the autocorrelation function stated in (8.31).

$$C_f(\vec{u}_1) = 0 \text{ for } wt(\vec{u}_1) = 1 \quad (8.31)$$

The *correlation immunity m* of the Boolean function $z = f(b_0, b_1, \ldots, b_{n-1})$, assuming the inputs bits are identically and independently distributed, is defined by having every subset of m binary random variables input being statistically independent of the function output z. An example of input–output correlation can be seen in the function $f(b_0, b_1, b_2) = b_1b_2 \oplus b_1b_3 \oplus b_2b_3$, where we have a probability of the output being equal to any of the inputs as 0.75. For balanced functions, the algebraic degree sets an upper bound on the correlation immunity as the two quantities sum to the number of binary inputs to the function [42]. This is intuitive, as more the variables in the product terms, the more influence it has on the output via the "AND" function when the particular input bit is "0". Hardware-efficient block ciphers consist of one or more linear feedback shift registers and nonlinear Boolean functions taking a number of bits from the linear shift register as the input. The output of the ciphers is the "XOR" between the plaintext to be encrypted and the output of the binary function. The algebraic properties of the Boolean function f mentioned here are thus important for evaluating the immunity of the cipher against attacks [43]. The output of the cipher can be modelled as

$$a_0 = f\left(\vec{k}(t)\right) \tag{8.32}$$

where the function f is public. The cryptanalysis problem is to recover the key $\vec{k} = [k_1, \ldots, k_n]$ from m samples of a_i. For a partially known plain text attack, some bits of the plaintext and the corresponding ciphertext bits are known. One common example of known plaintext is the headers of standard files.

8.6.2 Algebraic Attack

An algebraic attack on LFSR-based ciphers is carried out by reducing the cipher operation to a low-degree polynomial generating a set of over-defined systems of multivariate algebraic equations. Consider a more specific case of Toyocrypt [44], consisting of a 128-bit linear feedback shift register and nonlinear function:

$$\begin{aligned} f(k_0, \ldots k_{127}) &= k_{127} + \sum_{i=0}^{62} k_i k_{\alpha_i} + k_{10}k_{23}k_{32}k_{42} \\ &+ k_1k_2k_9k_{12}k_{18}k_{20}k_{23}k_{25}k_{26}k_{28}k_{33}k_{38}k_{41}k_{42}k_{51}k_{53}k_{59} \\ &+ \prod_{i=0}^{62} k_i \end{aligned} \tag{8.33}$$

with $\{\alpha_0, \ldots \alpha_{62}\}$ a permutation of $\{63, \ldots, 125\}$. The Toyocrypt Boolean function thus consists of a degree 17 monomial and a degree 63 monomial with other low-order terms. The high-order monomial evaluates to zero most of the time given purely random bits as inputs, since any of its constituent bits being zero will result in a zero through the AND-product. We can consider a quadratic approximation of the remaining terms $g(k_0, \ldots, k_{127}) = k_{127} + \sum_{i=0}^{62} k_i k_{\alpha_i} + k_{10}k_{23}k_{32}k_{42}$. Neglecting the probability of the order 63 term being equal to zero. The probability of the 17 degree term being zero is $(1/2)^{17}$. Thus the probability of the quadratic approximation being exact,

$$f\left(\vec{k}\right) = g\left(\vec{k}\right) = k_{127} + \sum_{i=0}^{62} k_i k_{\alpha_i} \tag{8.34}$$

is $1 - (1/2)^{17}$. The input $\vec{k}$ is produced by a linear feedback shift register and can be modeled as a linear polynomial of 128-bit output $\vec{r}$ of the LFSR,

$$k_i = L^t(\vec{r}) \tag{8.35}$$

at each time instance t and the output of the cipher is given by:

$$\begin{aligned} a_{t_1} &= f(L^{t_1}(r_0, r_1, \ldots r_{127})) \\ a_{t_2} &= f(L^{t_2}(r_0, r_1, \ldots r_{127})) \\ &\vdots \\ a_{t_m} &= f(L^{t_m}(r_0, r_1, \ldots r_{127})) \end{aligned} \tag{8.36}$$

So we have m output bits of information as well as known forms of f and the LFSR $L^t(\vec{r})$. The problem is to infer the values of $\vec{r} = [r_0, \ldots, r_{127}]$. With a probability of $1 - (1/2)^{17}$, the following holds:

$$a_{t_m} = g(L^{t_m}(r_0, r_1, \ldots r_{127})) \tag{8.37}$$

When we collect m sample bits and consider the overall approximation to be true in excess of 1/2, we have this probability being $\left[1 - (1/2)^{17}\right]^m = 1/2$. This bring us to $m \approx 2^{16}$, i.e. if we collect 2^{16} output bits a_{t_m}, not necessarily consecutive bits, for the probability of all these bits to be exact we have the low-order approximation $a_{t_m} = g(L^{t_m}(r_0, r_1, \ldots r_{127}))$ is 1/2. For each sample we have an equation of degree 4. Thus there are a total of 2^{16} fourth-degree equations and 128 unknowns in the case of Toyocrypt. Depending on the nonlinear function f and the LFSR definition, one can write the equations in the following general form:

$$\begin{aligned} l_{t_1} &= f(L^{t_1}(r_0, r_1, \ldots r_n)) - a_{t_1} = 0 \\ l_{t_2} &= f(L^{t_2}(r_0, r_1, \ldots r_n)) - a_{t_2} = 0 \\ &\vdots \\ l_{t_i} &= f(L^{t_i}(r_0, r_1, \ldots r_n)) - a_{t_i} = 0 \\ &\vdots \\ l_{t_m} &= f(L^{t_m}(r_0, r_1, \ldots r_n)) - a_{t_m} = 0 \end{aligned} \tag{8.38}$$

One way of solving the above set of over-defined equations is extended linearization [44]. As seen from the process of analyzing the nonlinear function in the Toyocrypt scheme, the form of the equations depends on the nonlinear function f used. An algebraic attack follows a similar approach as above and results in a set of equations with lower order by exploiting the existence (if any) of a lower degree function g (an annihilator of f) that satisfies (8.39).

$$\begin{aligned} f(\vec{x})g(\vec{x}) &= 0 \text{ or } f(\vec{x})g(\vec{x}) \\ &= h(\vec{x}), h(\vec{x}) \text{being a third polynomial of lower degree than } f \end{aligned} \tag{8.39}$$

For equations originated from $a_{t_i} = 1$, we can immediately use the property $f(\vec{x})g(\vec{x}) = 0$ and reduce this to an equation of lower degree $g(\vec{x}) = 0$. For $a_{t_i} = 0$, we use $f(\vec{x})g(\vec{x}) = h(\vec{x})$ to deduce $h(\vec{x}) = 0$, again resulting in an equation of lower degree than the original f. Algebraic immunity of the function f thus sets a lower bound on the degree of the sets of equations and the computational cost of computing the underlying states $\vec{r}$ of the LFSR. One example is the LILI-128 cipher [45], with the binary function reducing from the original degree 6 to only a subset of 10 variables via logic reduction. From this, a suitable approximation $g(\vec{x})$ is found to be x_8x_{10}, a quadratic polynomial. These examples demonstrate the cryptanalysis of block ciphers where an attack on the cipher is implemented by solving a set of over-defined multivariable equations [46]. Cryptanalysis can, on the one hand, reduce the order of these equations and results in a potential reduction in the computation required compared to a blind search across all possibilities, and on the other hand it renders cryptographic systems completely redundant in some cases.

8.6.3 Correlation Attack

A correlation attack exploits the correlation between the output sequence $\vec{y}$ of the stream cipher and the input sequence to the output Boolean function f applied to the result of an LFSR. Registers that produce bits with a high correlation with the input can be initially targeted for an attack. In the extreme case, where a particular input bit is highly correlated with the output, this can be directly analyzed. Having recovered the most correlated input bits, the attack then moves on to other input bits that are less correlated with the input. Generally well-defined f with high algebraic degree and higher nonlinearity will be more immune to correlation attacks. It is best to understand correlation attack in a simple case. Again consider the output function of the form

$$f(b_0, b_1, b_2) = b_1b_2 \oplus b_1b_3 \oplus b_2b_3 \tag{8.40}$$

with 3/4 of the time the output computed by this function being the same as any of the input variables. In the case of a simple stream cipher, the output $f\left(\vec{b}\right)$ is XORed with the text bit p_i to yield the cipher text bit c_i:

$$c_i = f_i(b_0, b_1, b_2) \oplus p_i \tag{8.41}$$

In an encryption system, the information directly available will be the cipher text $\{c_0, c_1, \ldots\}$. In the case of some plain text being known (e.g. common protocol headers), one can recover the output of the keystream generator $f_i\left(\vec{b}\right)$. So, typically a correlation attack starts with a collection of keystream values $\{f_t, f_{t+1}, f_{t+2}, \cdots f_{t+n}\}$ and our task is to find the initial key values $\vec{b}(t=0) = \left[b_0^{t=0}, b_1^{t=0}, b_2^{t=0}\right]$. Since we know that there is a 75% chance that the output is equal to one of the inputs (say, $f_{t+i} = b_0^{t+i}$), we can sweep through the two possibilities of $b_0^{t=0} = \{1, 0\}$ while fixing the other two bits and simulate the LFSR with output function $f\left(\vec{b}\right)$ until we get roughly 75% of matching of our model and the known bit streams $\{f_t, f_{t+1}, f_{t+2}, \ldots f_{t+n}\}$. We can thus recover the initial value of b_0, which is a single bit of the key. Once we have recovered $b_0^{t=0}$, we can repeat the process to recover $b_1^{t=0}$ and $b_2^{t=0}$. The total number of key trials now becomes $2 \cdot 3 = 6$. This vastly simplifies the number of simulations required if we were to simulate all the possibilities of $\vec{b} = [b_0, b_1, b_2]$ of 2^3. In this simple case, all three binary inputs are correlated with the output at 75%. Practical cipher output functions will involve many more Boolean variables and be carefully designed to minimize correlation. Correlation attacks can be carried out directly on LFSR-based ciphers. A good example of correlation attack on a stream cipher with nonlinear output functions is on "Grain" [47]. The Grain stream cipher utilizes nonlinear as well as linear feedback shift registers and an output function of first-order correlation immunity. The structure is illustrated in Fig. 8.10 [48].

The defining equations for the NFSR and LFSR used in the Grain cipher are given respectively by

$$\begin{aligned} x_{80} &= y \oplus g_G(x, x_1, \ldots, x_{79}) \\ y_{80} &= y_{62} \oplus y_{51} \oplus y_{38} \oplus y_{23} \oplus y_{13} \oplus y \end{aligned} \tag{8.42}$$

with x_{80}, y_{80} being the feedback values of the respective NFSR and LFSR, and their outputs of the shift registers are x and y, respectively. The nonlinear function g_G is given by

$$\begin{aligned} g_G(\vec{x}) = {} & x_{63} \oplus x_{60} \oplus x_{52} \oplus x_{45} \oplus x_{37} \oplus x_{33} \oplus x_{28} \oplus x_{21} \oplus x_{15} \\ & \oplus x_9 \oplus x \oplus x_{63}x_{60} \oplus x_{37}x_{33} \oplus x_{15}x_9 \oplus x_{60}x_{52}x_{45} \\ & \oplus x_{33}x_{28}x_{21} \oplus x_{63}x_{45}x_{28}x_9 \oplus x_{60}x_{52}x_{37}x_{33} \\ & \oplus x_{63}x_{60}x_{21}x_{15} \oplus x_{63}x_{60}x_{52}x_{45}x_{37} \\ & \oplus x_{33}x_{28}x_{21}x_{15}x_9 \oplus x_{52}x_{45}x_{37}x_{33}x_{28}x_{21} \end{aligned} \tag{8.43}$$

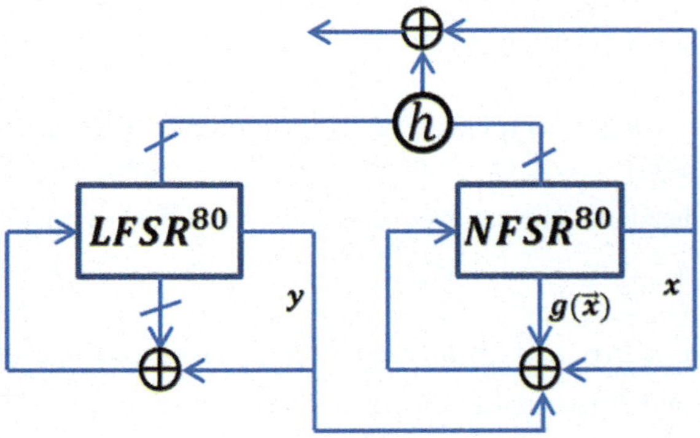

Fig. 8.10 Grain cipher architecture [48] utilizing linear feedback (LFSR) and nonlinear feedback (NFSR) shift registers

The output of the Grain cipher is given by the nonlinear Boolean function

$$z = x \oplus h(y_3, y_{25}, y_{46}, y_{64}, x_{63}) = x \oplus x_{63} p_y \oplus q_y$$
$$p_y = (1 \oplus y_{64} \oplus y_{46}(y_3 \oplus y_{25} \oplus y_{64})), q_y \tag{8.44}$$
$$= (y_{25} \oplus y_3 y_{46}(y_{25} \oplus y_{64}) \oplus y_{64}(y_3 \oplus y_{46}))$$

The Grain stream cipher starts by loading a 80-bit key K into the NFSR, followed by loading the first 64 bits of the LFSR with a 64-bit initialization value (IV) and filling the remaining bits with "1". The cipher is then clocked 160 times with feedbacks to the NFSR and LFSR being the "XOR" of the respective shift register output (x, y) and the overall output z. Generally, an attack aims to recover the unknown key K with known IV. A suitable attack is to derive a sufficient number of linear approximations involving the 80 bits of the initial LFSR, later used to recover the key K. Consider the output function again, consisting of XOR of the outputs $\vec{x}$ of both LFSR and the NFSR $\vec{y}$: $z = x \oplus h(y_3, y_{25}, y_{46}, y_{64}, x_{63})$. In [47], it has been shown that while the function $g_G(\vec{x})$ is balanced, $\vec{x} g_G(\vec{x})$ has a slightly biased probability of 522/1024 of being equal to "1". This has been exploited by considering the XOR function of the output delayed by 80 cycles and itself as:

$$z \oplus z_{80} = x \oplus h(y_3, y_{25}, y_{46}, y_{64}, x_{63}) \oplus x_{80} \oplus h(y_{83}, y_{105}, y_{126}, y_{144}, x_{143}) \tag{8.45}$$

The authors of [47] also observed that the imbalance is conditioned on x_{63}, since $z = x \oplus x_{63} p_y \oplus q_y$. The last expression can then be further simplified by finding linear approximations to $p_y \oplus q_y$, the best linear approximations being $y_3 \oplus y_{25} \oplus y_{64} \oplus 1$ and $y_{25} \oplus y_{46} \oplus y_{64} \oplus 1$. The search for the initial condition key thus reduces to a search given the imbalance probability of $\vec{x} g_G(\vec{x}) = 1$, probability of $x_{63} = 1$, and the probability of the linear approximation being valid. The original nonlinear function of mixed shift register variables is now reduced, depending on the output of the linear shift register only.

8.6.4 Attack on Lightweight Authentication Schemes

In this section, we revisit one of the lightweight authentication protocols proposed for batteryless smart cards, which could also be suitable for implant security deployment. Through the analysis of the shortcomings of the authentication scheme we demonstrate cryptanalysis in this area. We consider the HB$^+$ scheme of [27] in which the security relies on the difficulty of the learning parity with noise problem. The authentication scheme is shown in Table 8.7. Aside from the fact that a legitimate tag can be rejected depending on the randomness of bit stream generation on the remote device, the HB$^+$ scheme is vulnerable to an active attack with the protocol given in Table 8.8.

Table 8.7 HB$^+$ scheme authentication

Remote device (secret x, y)		Base station (secret x, y)
Draw random v with $P(v) = \eta$ Choose random vector k bit binary vector $\vec{b}$	$\xrightarrow{\vec{b}}$	
	$\xleftarrow{\vec{a}}$	Choose random k bit binary vector $\vec{a}$
Using secret $\vec{x}$ and $\vec{y}$, compute $z = \vec{a}\vec{x}^T \oplus \vec{b}\vec{y}^T \oplus \vec{v}$	$\xrightarrow{\vec{z}}$	Check $\vec{z} = \vec{a}\vec{x}^T \oplus \vec{b}\vec{y}^T \oplus \vec{v}$, repeat for N rounds. Accept remote device if this is satisfied with an acceptance threshold

Table 8.8 Attack on the HB$^+$ authentication scheme

Remote device (secret x, y)		Base station (secret x, y)
Draw random v with $P(v) = \eta$ Choose random vector k bit binary vector $\vec{b}$	$\xrightarrow{\vec{b}}$	
	$\xleftarrow{\vec{a}'}$	Choose random k bit binary vector $\vec{a}$ With constant $\vec{\delta}$, generate $\vec{a}' = \vec{a} \oplus \vec{\delta}$
Using secret $\vec{x}$ and $\vec{y}$, compute $z' = \vec{a}'\vec{x}^T \oplus \vec{b}\vec{y}^T \oplus \vec{v}$	$\xrightarrow{\vec{z}'}$	Check $\vec{z}' = \vec{a}\vec{x}^T \oplus \vec{b}\vec{y}^T \oplus \vec{v}$

The active attack on the HB$^+$ scheme assumes full access to the communication between the remote device and the base station. A perturbation on the challenges sent from the reader is carried out by an XOR function of a constant binary vector $\vec{\delta}$ of length k with the challenge $\vec{a}$. With access to the final rejection/acceptance information, one can reveal one bit of information at a time by setting a single bit in $\vec{\delta}$ to be "1" while keeping the rest "0". This is due to the fact that, for a successful authentication, we have high probability of $\vec{\delta}\vec{x}^T = 0$, whereas for rejection we have a high probability that $\vec{\delta}\vec{x}^T = 1$. After retrieving $\vec{x}$, the attacker can also impersonate the remote implant by sending to the station $\vec{b} = 0$. Emulation of a remote device is also possible by initializing a chosen $\vec{b}$ to the reader while returning $\vec{a}\vec{x}^T$ upon recycling a challenge $\vec{a}$ from the base station. In this way, $\vec{y}$ can be recovered by choosing $\vec{b}$ to be a set of linearly independent vectors that covers the entire bit sequence.

8.7 Conclusion

Implantable devices have the potential to monitor physiological parameters from within the human body and actuate in the surroundings at the same level. A high level of security must be provided by IMDs regarding the patient's private data, vital signals, and therapeutic dosages. Due to restrictions on the size of these devices, embedded electronics, and power supply levels, the integrity and validity of the data being broadcasted cannot be completely assured, as it may not withstand attacks from unauthorized sources. Many strategies covered in this chapter have been proposed to mitigate this problem by exploring the relation between the patient, caregivers, and attackers concerning the propagation of the correct IMD signal during the active and emergency operation scenarios. The cryptographic mechanisms ultimately constitute the last refuge for securing the patient's data when all spatial and physical constrains have been broken by attackers. Metrics that rely on the generation and sharing of cypher keys using biometric information have still to prove the level of randomness needed when applied in lightweight security schemes. The cryptanalysis section described in the last section of the chapter helps to understand some of the pitfalls of lightweight algorithms. Whereas it is difficult to map established cryptographic protocols onto the limited implant hardware, these algorithms have survived the test of time. However, the vast majority of lightweight protocols have underlying security issues that can be easily broken to allow unauthorized access to the implant. In some cases this security compromise can lead to patient fatality by unexpected implant shutdown or malfunction.

References

1. S. Hosseini-Khayat, A lightweight security protocol for ultra-low power ASIC implementation for wireless medical devices, in *IEEE 5th International Conference on Medical Information & Communication Technology (ISMICT)*, pp. 6–9 (2011)
2. C. Camara et al., Security and privacy issues in implantable medical devices: a comprehensive survey. J. Biomed. Inform. **55**, 272–289 (2015)
3. X. Hei et al., PIPAC: patient infusion pattern based access control scheme for wireless insulin pump system, in *IEEE Proceedings INFOCOM*, pp. 3030–3038 (2013)
4. M. Zhang et al., Trustworthiness of medical devices and body area networks. Proc. IEEE **102** (8) (2014)
5. H. Martin et al., An estimator for the ASIC footprint area of lightweight cryptographic algorithms. IEEE Trans. Ind. Inf. **10**(2) (2014)
6. Q. Yang et al., An on-chip security guard based on zero-power authentication for implantable medical devices, in *IEEE 57th International Midwest Symposium on Circuits and Systems (MWSCAS)*, pp. 531–534 (2014)
7. K. Daniluk E.N. Szynkiewicz, Energy-efficient security in implantable medical devices, in *IEEE Proceedings of the Federal Conference on Computer Science and Systems (FedCSIS)*, pp. 773–778 (2012)
8. M. Rushanan et al., SoK: security and privacy in implantable medical devices and body area networks, in *IEEE 14th Symposium on Security and Privacy*, pp. 524–539 (2014)

9. D. Halperin et al., Pacemakers and implantable cardiac defibrillators: software radio attacks and zero-power defences, in *IEEE Symposium on Security and Privacy*, pp. 129–142 (2008)
10. C. Li et al., Hijacking an insulin pump: security attacks and defences for a diabetes therapy system, in *IEEE 13th International Conference on e-Health Networking Applications and Software (Healthcom)*, pp. 150–156 (2011)
11. M. Rostami et al., Balancing security and utility in medical devices?, in *ACM/EDAC/IEEE 50th Design Automation Conference (DAC)*, pp. 1–6 (2013)
12. X. Hei, X. Du, Biometric-based two-level secure access control for implantable medical devices during emergencies, in *IEEE Proceedings on INFOCOM*, pp. 346–350 (2011)
13. Z. Ankarali et al., A comparative review on the wireless implantable medical devices privacy and security, in *EAI 4th International Conference on Wireless Mobile Communication and Healthcare (Mobihealth)*, pp. 246–249 (2014)
14. G. Zheng et al., Securing wireless medical implants using an ECG-based secret data sharing scheme, in *IEEE International Symposium on Communications and Information Technologies (ISCIT), 2014*
15. G. Zheng et al., Encryption for implantable medical devices using modified one-time pads. IEEE Open Access J. **3**, 825–836 (2015)
16. T. Denning et al., Absence makes the heart grow fonder: new directions for implantable medical devices security, in *Proceedings 3rd Conference on Hot Topics in Security, no. 5, USENIX Association, 2008*
17. S. Gollakota et al., IMD Shield: Securing Implantable Medical Devices (Poster) (USENIX Association, 2011)
18. G. Zheng et al., A non-key based security scheme supporting emergency treatment of wireless implants, in *IEEE ICC Symposium on Communication and Information Systems Security, 2014*
19. F. Xu et al., IMDGuard: securing implantable medical devices with the external wearable guardian, in *IEEE Proceedings on INFOCOM*, pp. 1862–1870 (2011)
20. M. Zhang et al., MedMon: securing medical devices through wireless monitoring and anomaly detection. IEEE Trans. Biomed. Circuits Syst. **7**(6), 2013
21. M. O'Neill, M.J.B. Robshaw, Low-cost digital signature architecture suitable for radio frequency identification tags. IET Comput. Digit. Tech. **4**(1), 14–26 (2010)
22. A. Juels, Minimalist cryptography for low-cost RFID tags, in *Proceedings International Conference Security in Communication Networks CSN 2004, Amalfi, Italy*
23. J.H. Oh et al., A light-weight security protocol for RFID system, in *7th IEEE International Conference on Computer and Information Technology, 2007*
24. M. Burmester, J. Munilla, Lightweight RFID authentication with forward and backward security, in *ACM Transactions on Information and Systems Security, May 2011*
25. D. Coppersmith et al., The shrinking generator, *Proceedings Advances in Cryptology* (LNCS, Springer, Berlin, 1993) pp. 22–39
26. N.J. Hopper, M. Blum, Secure human identification protocols, in *Proceedings 7th International Conference on the Theory and Application of Cryptology and Information Security: advances in Cryptology*, ed. by C. Boyd, pp. 52–66 (2001)
27. H. Gilbert et al., $HB^{\#}$: increasing the security and efficiency of HB^{+}, *Lecture notes in Computer Science*, vol 4965 (Springer, Berlin, 2008), pp. 361–378
28. M. S. Mamum, A. Miyaji, A fully-secure RFID authentication protocol from exact LPN assumption, in *Proceedings IEEE International Conference on Trust, Security, and Privacy in Computing and Communications, 2013*
29. A. Jain et al., Commitments and efficient zero-knowledge proofs from learning parity with noise. ASIACRYPT 2012 Lect. Notes Comput. Sci. **7658**, 663–680 (2012)
30. A. Shamir, SQUASH—a new MAC with provable security properties for highly constrained devices such as RFID tags, in *Proceedings Fast Software Encryption*, pp. 144–157 (2008)
31. S. Wolfram, *A New Kind of Science* (Wolfram Media Inc., Champaign, IL, 2002)
32. S. Wolfram, Cryptography with cellular automata. *Lecture Notes in Computer Science*, vol 218 (Springer, Berlin, 1986), pp. 429–432

33. C.K. Koc, A.M. Apohan, Inversion of cellular automata iterations. IEEE Proc. Comput. Digital Tech. **144**(5), 279–284 (1997)
34. P. Ping et al., Image encryption based on non-affine and balanced cellular automata. Sig. Process. **105**, 419–429 (2015)
35. S.R. Blackburn et al., Comments on—theory and applications of cellular automata in cryptography. IEEE Trans. Comput. **46**(5), 637–639 (1997)
36. M. Scaban et al., Collective behaviour of rules for cellular automata-based stream ciphers, in *Proceedings Congress Evolutionary Computation, Vancouver, 2006*
37. M. Seredynski et al., Reversible cellular automata based encryption, *Lecture Notes in Computer Science*, vol 3222, pp. 411–418 (2004)
38. G. Alvarex et al., A secure scheme to share secret color images. Comput. Phys. Commun. **173** (1–2) (2005)
39. R.J. Chen, J.L. Lai, Image security system using recursive cellular automata substitution. Pattern Recognit. **40**(5), 1621–1631 (2007)
40. E. Pasalic, C. Carlet, Algebraic attacks and decomposition of Boolean functions, in *Proceedings EUROCRYPT, Lecture Notes in Computer Science*, vol 3027, pp. 474–491, 2204
41. J.C. Castro et al., The strict avalanche criterion randomness test. Math. Comput. Simul. **68**, 1–7 (2005)
42. T. Siegenthaler, Correlation-immunity of nonlinear combining functions for cryptographic applications, in *IEEE Transactions on Information Theory, 31, 1985*
43. N. Courtois, W. Meier, Algebraic attacks on stream ciphers with linear feedback, *Advances in Cryptology, EUROCRYPT 2003, Lecture Notes in Computer Science*, vol 2656 (Springer, Berlin, 2003), pp. 346–359
44. N.T. Courtois, Higher order correlation attacks, XL algorithm and cryptanalysis of toyocrypt. Inf. Secur. Cryptol. Proc. ICISC 2002 Lect. Notes Comput. Sci. **2587**, 182–199 (2002)
45. L.R. Simpson et al., LILI keystream generator. Lect. Notes Comput. Sci. **2012**, 248–261 (2001)
46. N. Courtois, J. Pieprzyk, Cryptanalysis of block ciphers with overdefined systems of equations. ASIACRYPT 2002 Lect. Notes Comput. Sci. **2501**, 267–287 (2002)
47. C. Berbain et al., Cryptanalysis of Grain, in *Proceedings FSE'06, Lecture Notes in Computer Science*, vol 4047 (2006)
48. M. Hell et al., Grain—a stream cipher for constrained environments. ECRYPT 2005 Int. J. Wireless Mobile Comput. **2**(1), 86–93 (2007)

Index

G.-Z. Yang (ed.), *Implantable Sensors and Systems*
https://doi.org/10.1007/978-3-319-69748-2

The manufacturer's authorised representative in the EU is Springer Nature Customer Service Centre GmbH, Europaplatz 3, 69115 Heidelberg, Germany. If you have any concerns regarding our products, please contact ProductSafety@springernature.com

Printed and bound by CPI Group (UK) Ltd, Croydon, CR0 4YY

23/09/2026

02221536-0004